Springer Finance

Lecture Notes

Springer Finance Lecture Notes

Springer Finance is a programme of books addressing students, academics and practitioners working on increasingly technical approaches to the analysis of financial markets. It aims to cover a variety of topics, not only mathematical finance but foreign exchanges, term structure, risk management, portfolio theory, equity derivatives, and financial economics.

This subseries of Springer Finance consists of lecture notes.

More information about this series at http://www.springer.com/series/11356

David Nicolay

Asymptotic Chaos Expansions in Finance

Theory and Practice

Springer

David Nicolay
London
UK

ISSN 1616-0533 ISSN 2195-0687 (electronic)
ISBN 978-1-4471-6505-7 ISBN 978-1-4471-6506-4 (eBook)
DOI 10.1007/978-1-4471-6506-4

Library of Congress Control Number: 2014943936

Mathematics Subject Classification: 35B40, 35C20, 41A60, 60H15, 91G20, 91G30, 91G80, 91G99

Springer London Heidelberg New York Dordrecht

Printed on acid-free paper

Springer is part of Springer Science+Business Media (www.springer.com)

Depth must be hidden.
Where?
On the surface.

—János Arany

This book is dedicated to my parents

Preface

In recent years, derivative trading and research has refocused on liquid instruments, and in particular on European options. Increased market turbulence, unrelenting standardisation and stronger regulatory oversight altogether call for robust and well-calibrated models of the static smile. Furthermore, many agents (e.g. banks, hedge funds) now deploy sophisticated strategies, involving both assets and options, to capture some type of *alpha* or relative value. These algorithms demand accurate, non-arbitrable modelling of the joint dynamics of the underlying and its implied volatility surface.

In principle, Stochastic Volatility (SV) model classes (such as SABR, Heston, LSV or SV term structure frameworks) offer the most potential to fulfil these objectives. Indeed they can reach the statics and represent the dynamics of the smile in a rich, realistic and flexible fashion. In practice however, their lack of tractability makes classical SV models difficult to manage. The primary cause is that the derivation of the smile's exact shape and dynamics from the model's SDE is rarely achievable in closed form, which leaves only numerical methods. This is an issue not only for calibration, but also for computing and hedging the risk of complex derivatives (especially Vega risk) and thus for model design and analysis.

The academic answer to these limitations of stochastic *instantaneous* volatility (SInsV) models has been twofold. The first tack has been to develop numerous *approximation* methods for the static smile of specific SInsV models, mostly using small-time asymptotic techniques up to some low order. These methods exploit *either* an analytic (i.e. PDE) *or* a probabilistic (i.e. SDE) approach, and include for instance heat kernel and WKB expansions, singular perturbations, Malliavin calculus or saddlepoint approximations. Yet none of these approximation methods is flexible enough to provide arbitrary precision across a wide range of SInsV models, and neither do they address the dynamics of the smile. Therefore they cannot adapt easily to rapidly changing and challenging market conditions.

The second academic direction has led to new SV *market model* classes, which take as input *some* representation of the option price surface, such as implied volatility (SImpV models), local volatility (SLocV) or variance swaps (SVarS). Although these frameworks are very informative and theoretically promising, they have not been widely adopted by practitioners, mainly for issues of validity or speed.

This book presents a *third way*: the Asymptotic Chaos Expansion approach (ACE). The ACE algorithm links the SInsV and SImpV classes by *combining* standard PDE and SDE approaches, to provide pure asymptotics of the smile's shape and dynamics. These differentials are computed in closed form, at any order, and are established for a generic SV model.

Its model versatility allows ACE to cover vanilla models such as SABR, Heston or FL-SV, but also the case of fixed—and stochastic-weights baskets, or powerful interest rates/term-structure frameworks such as SV-HJM and SV-LMM.

At low order, the ACE results can be used for rapid model design and analysis. For instance, given some model specification they explain the influence of each parameter on the smile's level, skew and curvature, as well as on their joint evolution (e.g. the backbone). As a corollary, ACE shows easily which systematic bias affects the *most probable path* heuristic for local volatility models. Conversely, the model's parameters can be replaced by these smile-related quantities, which are more meaningful for trading, leading to an *intuitive re-parametrisation*.

Also, since its algorithm is programmable, ACE gives straightforward access to higher orders, which provides fast and arbitrary-precision approximations. The latter can naturally be employed for calibration, since *shape* proxies are traditionally used to mark the model to the static smile. But approximations of the *dynamics* can also be matched to the time series for the underlying and its (implied and/or realised) volatility: this is often called *dynamic calibration*. These approximations also benefit the valuation and hedging of structured trades. For instance, in an American Monte Carlo context, such fast access to the smile associated to the model state variables allows us to significantly refine the exercise boundary.

Importantly, although ACE results can be used on their own, they can also complement other asymptotic approaches, which are usually specialised to a given model. For instance they can increase the precision of the static approximations, and/or provide the missing dynamic information.

This monograph is based on my Ph.D. manuscript [1]. Among other improvements, a large part of the calculus has been simplified and/or extracted from the main body. This has increased clarity and will hopefully help convince the reader that ACE's practical mechanics are actually quite simple. Other enhancements incorporate some updated references, as well as new comments and interpretations pertaining to both results and applications.

The intended audience of this book includes researchers and academics interested in stochastic volatility, market models and/or asymptotics. It also comprises practitioners, especially *quants* supporting vanilla/structured trading desks who wish to improve their calibration or hedging procedures.

London, December 2012 David Nicolay

Reference

1. Nicolay, D.: Volatility dynamics. Ph.D. thesis, Ecole Polytechnique (2011)

Acknowledgments

This book has unsurprisingly turned into a long journey: often exciting, sometimes stressful, never dull and always worthwhile. Looking back on such a rich experience, I feel that it is defined as much by the road and the scenery as it is by the people met along the way. Among these, I believe that several have effectively contributed to this monograph—either directly or in a more peripheral way—and to my broader understanding of mathematical finance. For that reason I am sincerely grateful to them, and would now like to name but a few.

First I wish to thank my Ph.D. supervisor, Nicole El Karoui. In my opinion, and to my benefit, she is one of the few senior academics who combine an encyclopedic awareness (as well as deep specialist knowledge) of financial mathematics, with a deep understanding of the marketplace in general, and of the derivatives business in particular. This means that my inclination to provide an applicative angle to my research has always received an encouraging but circumspect welcome.

It is now difficult to assess which part of my *modus operandi* or of my thought process is nature *vs* nurture, the latter achieved by ways of so many discussions with Nicole. Nevertheless there are *some* practices that she undeniably adheres to, and that I have consciously tried to emulate, albeit at a much more immature and inefficient level. For instance her effort to build intuition during the early stages (often using toy cases) then the seemingly perpetual process of first simplifying and cleaning, then extending and unifying. And finally the deliberate redactional effort, geared towards clarity, honesty and precision. Overall it has been a pleasure and a privilege to work with her.

Many other CMAP[1] members at Polytechnique (students, academics, research engineers or support personnel) have exchanged ideas with me on subjects related to this monograph, while others have guided or helped me throughout the Ph.D. stage. Some have become friends over the years, and all have my gratitude.

[1] Centre de Mathematiques Appliquées.

By the same token I must thank Bernard Lapeyre at ENPC,[2] who has provided me with his complementary and invaluable expertise during the initial phase of the thesis. In all things numerical for example, I have benefitted from a pragmatic input coming either from Bernard, or from several other members of the ENPC, UMLV[3] and INRIA[4] teams.

As will become obvious in the course of the Introduction, this book would simply not exist—at least not in its current form—if it were not for Valdo Durrleman, his own Ph.D. thesis and one of his first submissions. Put simply, Valdo showed us the way towards a new kind of asymptotics, and I merely followed in his footsteps.

Last but not least in the academic realm, I would like to thank the reviewers of my dissertation as well as the members of the jury, who convinced me to publish this work.

I also wish to express my gratitude to the Springer team, for their professionalism and flexibility.

Most students or budding authors rely on a comprehensive support system—whether emotional, logistical or financial—to get through several years of a very exclusive focus. I was no exception, and therefore I thank my wife, my family and my friends for their encouragement and their patience.

Although this thesis is by no means a literature review, I apologise to anyone who thinks that their work should have been cited. Finally, all errors herein are mine, and mine only.

---

[2] Ecole Nationale des Ponts et Chaussées.

[3] Université de Marne-La-Vallée.

[4] Institut National de Recherche en Informatique et Automatique.

Contents

Acronyms

ACE	Asymptotic Chaos Expansion
ATM	At The Money (option, strike)
BFGS	Broyden Fletcher Goldfarb Shanno (algorithm)
BGM	Brace Gatarek Musiela
CEV	Constant Elasticity of Variance (function, model)
CF	Closed Form
CLT	Central Limit Theorem
CMS	Constant Maturity Swap
DCC	Day Count Convention
ESMM	Extended Skew Market Model
FD	Finite Differences
FFT	Fast Fourier Transform
FL-SV	Forward Libor—Stochastic Volatility (model)
FRA	Forward Rate Agreement
HJM	Heath Jarrow Morton (framework)
HKE	Heat Kernel Expansion
IATM	Immediate At The Money
IR	Interest Rates (environment, derivatives)
ITM	In The Money (option, strike)
IV	Implied Volatility
IZDC	Immediate Zero Drift Condition
LDD	Lognormal Displaced Diffusion (function, model)
LMM	Libor Market Model
LNIV	LogNormal Implied Volatility
LV	Local Volatility
MC	Monte Carlo
MMA	Money Market Account (process)
MPP	Most Probable Path
MTM	Mark To Market
NAA	No Arbitrage Assumption

NIV	Normal Implied Volatility
OTC	Over The Counter
OTM	Out of The Money (option, strike)
PCA	Principal Component Analysis
PDE	Partial Differential Equation
SABR	Stochastic Alpha Beta Rho (model)
SDE	Stochastic Differential Equation
SImpV	Stochastic IMPlied Volatility (model framework)
SInsV	Stochastic INStantaneous Volatility (model framework)
SIV	Stochastic Implied Volatility (surface)
SLocV	Stochastic LOCal Volatility (model framework)
SMM	Swap Market Model
SV	Stochastic Volatility
SVarS	Stochastic VARiance Swap (model framework)
TS	Term Structure
WKB	Wentzel Kramers Brillouin (method)
YC	Yield Curve
ZC	Zero Coupon (bond)
ZDC	Zero Drift Condition

Chapter 1
Introduction

1.1 Motivation

This study initially stemmed from my general interest in incomplete markets, which then shifted more specifically to the issue of unobservable and/or unrepresented state variables and dynamics, before settling on the more general notion of *model risk*. Also, underlying the obvious academic potential was a more practical focus on how the calibration and the hedging algorithms should be managed and coordinated, in order to mitigate that model risk.

My first attempts exploited discrete models (essentially trees), which are fine to build up intuition on a few steps but tend to lose their appeal on longer horizons. I then turned to the concepts of dynamic utility and of convex risk measures. The latter appeared promising as it structurally incorporated model uncertainty and also generated bid-ask spreads. But if the static calibration issue could be reasonably tackled, the dynamic hedging seemed elusive at best. At that point it became apparent that a central feature of this financial problem was the existence of a liquid European options market. The latter is represented through an implied volatility surface but is equivalent to the specification of the underlying's marginal distributions (see [1]). Focusing therefore on the joint dynamics of the underlying and of that smile, I decided to consider models incorporating *some* form of stochastic volatility. Indeed the latter is by itself a legitimate source of model ambiguity, allowing to (artificially) distinguish the *endogenous* driver of the underlying from the *exogenous* driver which is specific to the smile. These choices allowed us not only to define a clear mathematical perimeter for market incompleteness, permitting in particular to restrict the scope to continuous processes, but also to provide a natural environment in which to express both the calibration and the hedging procedures. This is because the vanilla options must clearly be included in the calibration set, and ideally should be included in the hedging set as well.

The problem could then be reformulated into three more explicit questions. First, how to fit at least the static shape (and if possible the dynamics) of the implied volatility surface with an instantaneous volatility model. Second, how to use the

D. Nicolay, *Asymptotic Chaos Expansions in Finance*,
Springer Finance Lecture Notes, DOI 10.1007/978-1-4471-6506-4_1

corresponding vanilla options to dynamically hedge a contingent claim, absorbing not only the *Vega* but also the model risk. Third and conversely, how to extract information about the underlying dynamics from the shape and dynamics of the smile, information which can then be used for trading activities other than pure hedging, such as arbitrage strategies. Clearly the problem had to be split into two symmetric issues: the *direct* problem consisting in navigating from the underlying's dynamics to the shape and dynamics of the smile, and the *inverse* problem corresponding to the opposite journey.

It seems pertinent to justify the exclusion *a priori* of discontinuous processes (jump-diffusions, Levy processes) despite the substantial evidence supporting their presence in the market. The first argument is mathematical: the method presented here relies on the assumption that the smile-related processes admit a finite limit *a.s.* for asymptotically short expiries. Although this condition is quite strong, relaxing it would require a workload disproportionate with its payoff, from the chosen angle. The second reason to exclude jumps is financial: they cannot be hedged in the general case. Indeed, usually the *probability* of jump can be replicated, but not the jump itself. Since the present approach is heavily motivated by improving the hedge strategies, it would seem rather inconsistent to select a framework that *guarantees* a tracking error.

1.2 State of the Art

One of the characteristics of the present approach that has the most enabled innovation is simply its positioning. In effect, it finds itself at the intersection of two mathematical domains: on one hand a set of *market model* classes, and on the other hand the *asymptotic* family of analysis and approximation methods. Indeed, on the modelling side, the idea will be to exploit and connect several classes that are capable of describing the afore-mentioned complex joint dynamics. As discussed before, we are restricted to (continuous) Itô processes and therefore to those classes exhibiting *some* form of stochastic volatility. However, the ultimate focus is applicative, hence these joint dynamics or connections must be explicit. In that respect, approximations are acceptable, as long as the error is known and can (ideally) be controlled. Also, a strong emphasis is put on versatility and on a generic approach, both in terms of model coverage and of precision order.

1.2.1 Stochastic Volatility Market Models

In fact, the stochastic volatility (SV) modelling framework regroups several very distinct classes. By far the best-known and most widely used family is of course the stochastic *instantaneous* volatility (SInsV) category. In full generality that framework will include the *local volatility* (a.k.a. Dupire) class, so that the reader is referred to seminal works such as [2–6] or [7], among many others. It is worth noting that most practical SInsV models (e.g. Heston or SABR) are expressed with a pair of state variables driven by a bi-dimensional Wiener process. Indeed some rich dynamics can be generated with only a scalar *endogenous* driver for the underlying, and another scalar *exogenous* driver for its volatility. Since the endogenous driver can be shared by both processes, it is possible to correlate those. More recently a new breed of *market models* has emerged, which takes the marginals of the underlying as model observable input, and then specifies their dynamics. Alternatively, the input process can be taken as some functional of these marginals, or as the equivalent (Dupire) local volatility. The term *market model* denotes here that this multi-dimensional process is represented (or parametrised) directly *via* market observables, such as the implied volatility surface or the term structure of variance swaps. This approach is analogous to the HJM methodology for interest rates (see [8]) which incorporates the yield curve as a state variable, instead of modelling the (instantaneous) short rate. The reader will be assumed familiar with the SInsV framework, and therefore the sequel will be devoted to describing the other market models.

Let us first discuss the Stochastic *Implied* Volatility (SImpV) models. Their shared concept is to take as input the shape of the implied volatility, associated to a single underlying, and then to model the dynamics of that map. Variations on this very general theme include the choice of the driver dimension (which can be infinite), of the Markovian state variables (if any) or of the coordinates (the *moneyness*). However, any SImpV model is subjected to a pair of constraint fields, the first on its static and the second on its dynamic specification.

The static constraint forces all implied marginal distributions, deduced from the price surface by strike-differentiation, to be individually and collectively *valid*. Indeed they must be compatible with *some* martingale diffusion for the underlying, which imposes both intra- and inter-maturity conditions. First each marginal must define an acceptable probability measure by itself, which translates for the implied volatility (IV) into a complex non-linear second-order PDE, rendering difficult the characterisation of non-arbitrable IV surfaces. Then marginals of a diffusion process for different maturities must also be compatible with one another, which in particular places some bounds on the maturity-differentials of the smile.

The dynamic constraint stems from the fact that European options are assets, and that the smile determines uniquely (through Black's formula, for instance) their rebased price process. Hence the latter must be martingale under the numeraire measure. As we will see in Sect. 2.2, this dynamic condition manifests itself as another complex stochastic PDE linking the smile's shape to its dynamic coefficients.

Overall this SImpV class has not yet had much success with practitioners, at least as a legitimate modelling tool, essentially because of the difficulty of satisfying both these arbitrage constraints. It remains, however, a rich and interesting framework with much left to offer, and the present study shows that it can be used with success at least for analysis.

In terms of bibliography, we point the reader first towards [9] and [10] for a statistical analysis of smile deformation modes, which justifies and provides some basis to the risk-neutral modelling concept. Then we must quote [11, 12] and [13], then [14, 15] and [16] as some of the first studies formalising the SImpV framework and the associated dynamic no-arbitrage restrictions. Some more technical issues of existence for such complete models (depending on the choice of measure and filtration) or for the solution of the specified SDE system are tackled in [17–19] and [20].

Note that certain of these papers also introduce the notions of *forward implied volatility* and *local implied volatility* while generalising the framework significantly. Also, on the subject of market completeness issues it is worth consulting [21]. As for implementation, we suggest [22, 23] and [24] for simple and practical considerations on the concept of a dynamic smile model, and then [25, 26] and [27] for more involved instances. We also single out [28] for an interesting way of constructing an arbitrage-free SImpV model. The SImpV class will be discussed again in Sect. 2.1.3.2 [p. 37] under a different angle, leaving us now to present the next market model.

Another new but very promising class is formed by Stochastic *Local* Volatility (SLocV) models. This approach was pioneered in [29] and developed in [30] and [31], but one of the most complete investigations can be found in [32] and [33]. Simply put, these models specify the (possibly exogenous) stochastic evolution of a *local volatility* surface. At any given time, its current snapshot is a bi-dimensional function, and taken as the onward diffusion coefficient for some shadow underlying. In turn, these virtual dynamics uniquely define a vanilla price surface. As befitted by intuition, this framework is in principle equivalent to a SImpV model. In practice, however, it differs in two respects. First and on the upside, the static validity constraints are much simpler to express, to implement and to verify dynamically. Indeed they amount to the positivity of the local variance function. Second, on the downside and in the general case, both the statics and dynamics of the vanilla price surface must a priori now be obtained numerically. The former can typically be achieved by solving a bi-dimensional parabolic PDE, often through a finite differences (FD) scheme.

Note that this class is not to be confused with so-called *Local Stochastic Volatility* models (LSV), which belong to the SInsV class. Indeed their *diffusion coefficients* combine local and stochastic volatility, as they are functions of the underlying and of an extra state variable. Their calibration, however, is specific: typically the SV component is specified exogenously, while the LV coefficient is non-parametric and determined numerically to fine tune the static smile (see [34] for instance).

The last SV market model class of interest specifies *variance swaps* dynamics. Those *stochastic variance swaps* (SVarS) models are covered (among others) in [35], in the series [36–39] and in [40, 41]. In the case of the continuous payoff definition, a (structurally path-dependent) variance swap is equivalent to a *log-contract* European option. Assuming a maturity continuum of such products, still in the fashion of HJM, it is then possible to specify dynamics for that *unidimensional* map. Since each marginal is thereby summarised *via* the log functional, some of the information relative to the underlying process is lost. Thus a modeller could intuitively assume this class to be a poorer framework than its SImpV or SlocV counterparts. After all, it does not provide any *explicit* control on the shape (and even less on the dynamics) of the smile. In fact, in terms of attainable dynamics, the model hierarchy is not necessarily what it seems. Mainly because, providing enough regularity, most market models can be set in bijection with the SInsV class. In this book, only an asymptotic connection between the latter and the SImpV framework will be presented. However, some of my associated work on SLocV models (edited from this version) suggests that the same principles *do* apply throughout the aforementioned market models. This equivalence argument is, however, conceptual rather than practical: the link with the SInsV class is essentially formal. For instance, it suffices to consider the issue of compatibility between model-specific filtrations, or the transfer of Markovian state variables between model classes. These are non-trivial, fundamental modelling aspects for academics and practitioners alike, and to the best of my knowledge these topics have not been covered comprehensively in the literature.

Overall, there is no apparent dominance among these SV market model classes. Each of them focuses on a specific feature of the state variable process, and/or on a given product class. Each of them is adapted to a certain modelling concern, and therefore comes with its pros and cons. For instance SVarS models are the most efficient, economical and sparse way of modelling the term-structure of log-contracts, but would be an *a priori* awkward choice for anything else. Similarly, SLocV models are the easiest (from a validity perspective) but also the most numerically intensive way of modelling the shape and dynamics of the smile.

Although these market model classes do allow complex joint dynamics between the underlying and the European options (or equivalently the marginals) they do not usually provide that information in a fashion which is simultaneously *exact, inherently non-arbitrable, explicit and practical*. However, if such a rich model is to be calibrated (statically and/or dynamically) to price and hedge some contingent claim using the underlying and/or the vanillas, then the joint dynamics *must* be provided with these four properties. For instance, it is well-known that the list of SInsV models that provide the smile in closed or semi-closed form is very limited. Arguably, the richest workable instances in that class belong to the affine processes family, such as Heston's model (see [5]). Conversely, with stochastic *implied* volatility models it is difficult to specify complex dynamics for the underlying while guaranteeing no arbitrage, both statically and dynamically.

Practical usability and lack of arbitrage opportunity being paramount here, it is *a priori* some *approximation* of the link between the underlying's dynamics and the term structure of European options that should be found. However, the precision of that proxy must be known, and if possible controllable. Put another way, some *partial* connection (in either direction) is acceptable. Several published methods do fit part of this mandate, but they usually concern themselves with restricted model classes. More generally, they tend to lack the degree of versatility that is expected here, with regard to the modelling framework and to the attainable precision. It is nonetheless interesting to cover them briefly: not solely as part of the necessary review process, but also because the approach developed herein (ACE) can be used to *complement* those methods, rather than purely compete with them.

As a demonstrative example, let us mention first the popular *Markovian projection technique* (MPT) as described in [42] and [43]. Although efficient within its mandate, this approach is limited to finding equivalent time-homogeneous models *via* efficient parameters. In addition, it provides results not only with a fixed precision, but also according to a criteria which is quite rigid. Hence the MPT does not seem to match our demands, as its stands. However, and as will be discussed in Chap. 4, heavy locality and in particular non-stationarity of the underlying's dynamics tend to degrade the performance of the present approach. Therefore it seems very tempting to *pre-process* such inhomogeneous models through the MPT first, before applying a method, such as the present, that benefits from stationarity.

Among the other available types of approximations susceptible to better fulfil our requirements, the rich arsenal of *asymptotic methods* seems to provide the most promising leads (if not the only realistic alternative). Let us now briefly cover this domain, focusing on financial applications and filtering these methods w.r.t. our concerns for generic features, applicability and smile dynamics.

1.2.2 Asymptotic Methods

In simple terms, within the general context of European option pricing under given SV underlying dynamics, these methods essentially exploit a deviation from some kind of central reference. The latter can be a model, a marginal distribution or a pricing formula, and will be called the *baseline*. It will usually refer to a Normal or a Lognormal process, but it can potentially be more complex (as with Heston for instance).

The majority of these techniques correspond to *small-time* asymptotics and exploit some form of *perturbation method*, invoking concepts such as heat kernel expansions, singular perturbations[1] or WKB expansions. Most of these will follow an analytical approach, usually working with the backward (Feynman-Kac) PDE and the infinitesimal generator of the diffusion, and more occasionally with the forward (Fokker-Planck) formulation and/or the local time. Alternatively, a few approaches are more probabilistic in nature: they can focus, for instance, on the transition probability or on the SDE itself, using techniques related to Wiener chaos expansions or to Malliavin calculus. Another group of methods rely on *saddlepoint approximations* and are therefore related to Fourier/Laplace transform techniques.

First up, *singular perturbation* techniques have proven efficient in mathematics, physics and mechanics. They are especially standard in fluid dynamics, where the parabolic nature of the problems presents a strong analogy with mathematical finance. Applied to solve a given pricing PDE, their general principle is to scale the coefficient of the highest-order differential by an ε factor. Then the solution is identified term-by-term as an ε-expansion, which frequently invokes the WKB method. The *singular* aspect signifies that setting ε to zero fundamentally changes the nature of the equation: as opposed to regular perturbations, the solution to the limit PDE is then a priori *not* the first-order term of the expansion. In a mathematical finance context, the coefficient of the highest order term (i.e. second order in space) is usually some form of volatility, which explains why it will usually degenerate at the boundary.

[1] Regular perturbations can be combined (as in [44]) but are very rarely found on their own in practical SInsV models, whose volatility coefficient is usually degenerate at the boundary.

The *heat kernel expansion* (HKE) approach presents some similarities. It was arguably introduced in finance by [45], but this mathematical technique is well established and its fields of application are numerous. It consists in a geometric (geodesic) method exploiting the natural Riemannian metric, is again strongly linked to WKB expansions and can be applied in either a probabilistic or an analytic setting. The former approach often focuses on the transition probability and tends to be more tolerant of the volatility's boundary behaviour. Whereas the latter (PDE) approach is considered more practical (from a computational perspective) for higher-order extensions.

Most *probabilistic methods* will start by a scaling of each SDE coefficient by some power of ε. Then it is usually the sensitivity of the scaled process' marginals (or of some functional (payoff) of these densities) w.r.t. that parameter which is computed. The output naturally takes the form of an ε power series, centred around the solution provided by the baseline process ($\varepsilon = 0$). But the interpretation and the properties of that series are quite different from the analytical case. The computations themselves can involve a range of sophisticated and related techniques, whose coverage falls outside the scope of this book. These include Malliavin-Watanabe and Watanabe-Yoshida calculus, Kusuoka-Stroock asymptotic expansion theory, or on a different tack the Freidlin-Wentzell theory of large deviations.

Of course the method employed presents a degree of coupling with the class of models on which it can be applied, and both influence the type and quality of the results that can be obtained. Since the intention here is specifically to link various market models, it is worth noting that asymptotic methods in finance focus almost exclusively on the SInsV class. Also, they tend to provide their approximations for the smile *shape* only, ignoring its volatility structure.[2] The literature in that domain is very rich, and difficult to synthesize as each will present a certain combination of method, model and results. Hence the following list does not claim to be exhaustive, but instead (hopefully) representative.

Let us focus first on the simplest *pure local volatility* (LV) model class. In that framework [46] constitutes an essential reference, as it gathers most of the (rare) exact and closed-form results produced by asymptotic methods for that class. By transferring Dupire's formula to the Lognormal implied volatility (LNIV), one obtains a classical PDE which is naturally parabolic, but also degenerate and quasilinear. However, by limiting the argument to the *immediate domain* (i.e. zero expiry) an exact closed form expression for the LNIV is reached, as the *harmonic mean* of the local volatility. Interestingly this result can be linked to the theory of large deviations (see [47] for instance) and to the Riemannian metric (mentioned earlier) corresponding to the inverse of the LV diffusion coefficient. In particular, the well-known zero-expiry convergence of the ATM implied volatility to the current value of the local volatility coefficient can be established this way. Also, using the same starting point, but after more involved analytical work, comes an exact formula for the implied variance at *extreme strikes* (i.e. infinite log-moneyness), as the time-integral of the local variance taken at that same limit.

---

[2] However [6] for instance provides a useful analysis of the first-order smile dynamics *through* the static formulation.

Among other classical results should be cited [48], where ad hoc singular perturbations methods are employed to provide proxies for the Call price and the LNIV. Note that the model is defined with separable arguments for the time-inhomogeneous volatility, and the expansion is provided up to the fourth order. Although this approach misses the expected generic features, it is routinely used by practitioners, which shows that—with reasonably well-behaved local volatilities—this level of precision is often sufficient for applications.

In [49] as well as in [50] (within a more general setting incorporating stochastic rates) it is a more general *probabilistic* approach which is taken. Although the volatility is as usual proportionally scaled by ε, the latter serves there to define the underlying as a *parametric* process. Hence following [51] the corresponding tangent (or first variation) process of the stochastic flow can be defined. Then the ε-differential of the European price is expressed from its (closed-form) Greeks under the simple, first-order baseline model, which is proposed either as Normal or as Lognormal. The expansion is provided up to third order, but although the procedure is described for higher levels *via* Malliavin calculus, it does not seem easily implementable. Interestingly, a rigorous bound on the expansion error is obtained, and the considered European payoff is generic. The method can also potentially be extended to other model classes, seen as deviations of the Lognormal baseline, and has been applied to the time-inhomogeneous Heston model in [52].

In [53], in the case of a time-inhomogeneous LV model and following several HKE approaches, approximation formulae for the implied volatility are provided at the first and second order in time-to-maturity. A general pattern for extending to higher-orders is given, but it imposes limitations on the model, while its implementation does not seem straightforward (it would rely on a full-blown formal calculus engine). Finally, another interest of this study is that it provides a useful panorama of the various HKE methods.

The more recent [54] exploits Yoshida's approach to HKE to deal with both homogeneous and inhomogeneous local volatility models. The expansion is specified for a generic order, but computing the coefficients requires iterative integration, which in the general case is only realistic with a symbolic calculus engine.

Turning now to *stochastic instantaneous volatility* classes, let us first evoke the classic [55], which gives an ambitious and rigorous PDE characterisation of the LNIV, within a very generic SInsV model defined with a finite number of state variables. Similarly to [46] the article establishes links with large deviations theory, which it exploits (invoking notions such as the signed geodesic distance and the effective local volatility) to produce full-domain and asymptotic (short expiry) results. The latter are of particular interest to us, but unfortunately within such a rich framework no generic solution can be given. However, a couple of low-order examples are provided, which suggests some potential as a practical tool, in specific cases.

By contrast, Hagan et al. [6] develops one of the best known and most widely used approximations. That article introduces and analyses the SABR stochastic volatility model *class*, which will be covered in Sects. 4.2 and 4.3. This work is a natural extension of [48], which allows for an insightful comparison of local and stochastic volatility models. It uses a singular perturbation technique which is both involved and customised, providing proxies first for the Call prices, then for the Normal and Lognormal implied volatilities. The latter is the most commonly applied by practitioners to the CEV instance, often well outside of its domain of asymptotic validity, so that Hagan and al's approximation has somehow come to stigmatise the limitations of expanding the implied volatility. However, although that expansion is limited to the second order, its accuracy is remarkable, especially w.r.t. the space dimension and for reasonable expiries. Since this study does not fit our purposes (if only for lack of a generic approach w.r.t model and order) instead it will be used as a benchmark for whole-smile extrapolations (see Sect. 4.5). In the same vein, one should consider closely related work, such as [56], which focuses on the implied distribution, or [57], which brings some exact corrections to Hagan and al's formula. The latter in particular demonstrates the combinatorial approach that I support, as will be discussed in Sect. 4.5.3.

In the same methodological family, an important reference is [44] (and its sequel [58]), as well as some associated works (see [59] or [60] for instance) and some forthcoming extensions. Arguably, the interest of that line of work lies as much with the modelling motivation as with the asymptotic technique itself. Indeed, the framework is only bi- or tri-dimensional and based on a Lognormal baseline, but the volatility is defined as a rich, mean-reverting and potentially multi-scale process. Supported by historical market statistics and capable of mimicking some jump-diffusive features, these dynamics present the mean-reversion coefficients as the natural candidates for the expansion parameters. In other words, the *original* model incorporates already and structurally a strong scaling, which enhances the efficiency of the asymptotic technique. As for the latter, it exploits a perturbation technique on the backward PDE (which can be both singular and regular in nature) and is presented in a concise, inductive manner; furthermore, the authors provide an estimate for the expansion error. It must be noted that the same approach can be followed on a different baseline, such as Heston's dynamics, although the overall methodology seems to be an unlikely target for automation.

On a distinct tack, the approach of [61] and [62][3] presents some similarities with the present methodology. It delivers some powerful results fulfilling several of our criteria, within a simple bi-dimensional SInsV model framework which covers popular models such as Heston, SABR and FL-SV. The starting point is the backward PDE expressed on the Lognormal implied volatility (LNIV). Then an intuitive change of variable followed by a classical perturbation method provide the solution as a power series w.r.t. both time-to-maturity and log-moneyness, whose coefficients can be computed by induction up to any order. Although these closed forms are quite involved, they can be handled by a formal calculus engine. Regarding our demands, the main limitations of this method are first the restricted model framework, and second the exclusive focus on the statics of the smile. Also, it is therein noted that the expansion's convergence is quite slow, exhibiting the usual large-moneyness invalid density issues. With hindsight however, it *is* known that in practice such convergence

[3] See also [63] and [64] in a jump-diffusion context, but at lower orders.

depends strongly on the choice of the space variable. However, the latter has been chosen as the log-moneyness, but a classical result (see [65] for instance) imposes that sub-square-root growth is necessary to prevent arbitrage. It follows that in order for any asymptotic method to provide good extrapolation properties, it must offer the capacity to adapt the expansion variables to the problem at hand. In that perspective, another limitation of the method is that (as noted in [61]) the induction formula is heavily dependent on the PDE structure, which is itself entirely conditioned by the choice of its variables.

With similar ambitions but using original features, Fournie et al. [66] considers a time-homogeneous SInsV model centred on *Normal* dynamics. That study stands out first by its motivation, as it suggests to employ the asymptotic expansion result for (Monte-Carlo) *importance sampling* purposes. Indeed, with most papers in this domain the applicative focus tends to be on calibration, rather than on pricing and hedging. The second distinction comes from the perturbation methodology, which starts classically from the backward PDE, but manages to invoke Feynman-Kac's formula *with right-hand side*. It follows that the coefficients of the expansion are expressed as time integrals, which regularises the solution artificially but explicitly along the maturity axis. Finally, these coefficients come as the result of an induction procedure, which is clearly defined up to any order. Although involved, that algorithm should be programmable within a formal calculus engine. Despite these qualities, it is clear, however, that the study misses several of our criteria. In particular, its generalisation to wider model classes, although theoretically possible, would require significant ad hoc derivations. All things considered, this approach will be used as an important benchmark, essentially for its inductive and integral nature.

Overall, a significant number of the more computationally involved approaches follow the HKE avenue. However, it can be argued that few cover the subject in a way which is as comprehensive, rigorous and pedagogic as [67]. The latter presents in a consistent manner many of the fundamental concepts underlying that approach, in particular the connection between, on one hand, local and stochastic volatility models, and, on the other hand, (hyperbolic) Riemann manifolds. The same author has applied this methodology to general SInsV classes (including the mean-reverting λ-SABR) in [68] and to the LMM framework in [69]. Also, in the same HKE family, one can consult [70], for instance.

In a very different probabilistic vein can be found several approaches that exploit either Wiener chaos expansions or some form of Malliavin calculus, and which for that reason can be associated with [49]. Let us first quote [71], which proposes rich but technical expansions for both the Call prices and the implied volatilities, under a generic multi-dimensional diffusion. A simpler version of these methods, applied to the SABR model, can be found in [72]. Both these presentations are quite attractive from our perspective, and are only limited by the difficulty involved in obtaining higher orders. Indeed, the derivation of an inductive closed-form solution seems unlikely. Let us also cite [73] as a comprehensive review of the authors' application of Malliavin-Watanabe calculus to option pricing. Among the latter we should highlight [74], which covers interest rates products (a rare occurrence) in an HJM framework. Also, for those readers interested in the cross-currency context, let us mention [75] which offers a third-order expansion, in a limited displaced-Lognormal framework and using a simpler presentation.

The notion of *saddlepoint approximation* can be seen as an extension of Laplace's integration method. This is a very mature technique commonly used in various fields, from statistics to financial mathematics (see [76, 77] or [78]). It consists in approximating the density of some distribution from the cumulant generating (CG) function, or alternatively from one of its variations, such as the Laplace, characteristic or moment generating function. The approximation comes as an asymptotic expansion, whose leading term itself is called the saddlepoint approximation. Some advantages of this method are that it gives flexibility over the expansion base (which is usually Gaussian) and that it provides good proxies for the tails of the distribution.

In the current context, this technique is usually employed to approximate the European Call prices (or some type of implied volatility) associated to an underlying process, whose characteristic function (or one of its variations) is either known or easily computable. For that reason, the technique is mainly applied to Levy or affine processes, and in particular to (variants of) the Heston model. In the latter case, references include [79, 80] or [81]. More diverse model classes with the same property are tackled by [82, 83] or [84].

However, our objectives include model genericity: among saddlepoint approaches, this is theoretically provided by [85]. When the Laplace or characteristic function is unknown (the *limiting* case) this parametric expectation is approximated by an arbitrary-order small-time expansion, based on iterated applications of the infinitesimal generator. The method is demonstrated on a generic scalar diffusion, but for our purposes it could in principle be applied to a multi-dimensional process, typically a SInsV model such as SABR. Nonetheless, as no performance test is provided therein for the limiting case, the practical behaviour of this *nested* expansion would have to be investigated. Also, the method's output consists of exclusively static quantities, which does not satisfy our hedge-focused requirements.

The related work [86] develops a generic backward PDE approach for multi-dimensional processes, which is then linked to the saddlepoint approximations of [85]. The solution to the European option pricing problem is expressed as a deviation from a closed form, itself corresponding to an auxiliary (or baseline) model, such as Black-Scholes. The solution to the backward PDE satisfied by this deviation is approximated by a small-time expansion, using an infinitesimal generator power series, in a similar fashion to [85]. Interestingly, this analytic approach easily incorporates price sensitivities, and therefore the conventional (although restricted) vision of hedges.

To conclude this short presentation of asymptotic approaches in finance, let us stress that perturbation methods do not *guarantee* convergence in general. Instead, asymptotic series tend to be the norm rather than the exception, which explains our interest in error control results. Note finally that very strong links exist between most of these approaches, as well as with *time change* techniques, for instance.

There exist other related and interesting veins of research around this subject. For instance, one such thread focuses on the tails of the marginals generated by stochastic instantaneous volatility models. This usually falls under the denomination of asymptotics because it is directly related to the smile at *extreme strikes*. Although it belongs to another methodological domain altogether, these results can be used in conjunction with the short-expiry methods described above, in order to *tame* the expansions and *anchor* the wings of the smile. In that field should be mentioned at least [87–89] and the review [90]. To be used in practice, however, these results often require that the characteristic or moment generating function of the underlying's marginals be known. Hence they mainly concern *affine processes*, which explains why a significant part of that literature is focused on Heston's model and its descendance.

Similarly, there is an interest in studying the extreme maturity or long term (LT) behaviour of implied volatility (IV). Considering an underlying process with mean-reverting stochastic volatility, [59] and [60] characterise in closed-form the LT Lognormal IV. Within a very general setting this time, [91] exploits large deviations principles to proxy the same limit in probabilistic terms, using in particular the cumulant generating function and providing some error control.

A more comprehensive approach can be found in [92], which unifies three types of asymptotics: small-time, long maturities and extreme strikes. The corresponding limit behaviour of implied volatility (and variance) is characterised in terms of log-price, at an arbitrary precision level. Importantly this technique remains model-independent throughout, by identifying specific regimes. Hence its purpose and ability is *not* to address either the direct or inverse[4] problem, as defined above.

1.3 Objectives

Having discussed above the practical interests and constraints of calibrating and hedging within SV frameworks, and having reviewed most existing mathematical solutions, the requirements for a new, more efficient approach can now be listed. The aim is to develop an asymptotic method with highest possible *genericity*, in many respects:

- The method should cover most currently active stochastic volatility models (used in academia and/or in the industry).
- The error of the approximation should not only be quantified but also controlled, in exchange for a higher computational expense. This means that if an expansion method is used, then all differentiation orders should be attainable (provided the convergence radius is sufficient).
- Whichever algorithm is built should be reasonably simple, and programmable if necessary. It should also be stable, both logically and numerically, because sensitivities are used for calibration (optimisation) and for hedging.

[4] The inverse problem has not been extensively covered in the literature. Interesting input and references on this topic can be found in [93], for instance.

Importantly, and as motivated above, this book aims at connecting the underlying's instantaneous dynamics (the model's SDE) not only to the shape but also to the *dynamics* of the smile. Furthermore, the interest lies in both directions of this link, which are the *direct* problem (from model to smile) and the *inverse* one.

1.4 Asymptotic Chaos Expansion

Before outlining the book's structure, let us first succinctly describe the core principles of the methodology that is developed and advocated herein. This approach is called ACE, which stands for *Asymptotic Chaos Expansion*, and stands at the junction of the two domains described above, namely the *market model* family and the field of *asymptotic methods*. The rendition of ACE in this book represents both a variation and a substantial extension of the pioneering work found in [94–96] and, with a slightly different focus,[5] in [95].

In this book, the fundamental idea supporting ACE is to consider two distinct SV market model classes, describing simultaneously the same derivative market, and to establish structural asymptotic links between them. On one hand is assumed a generic version of the stochastic *instantaneous* volatility (SInsV) class. It comes as an SDE system, defined formally as an adapted Wiener chaos, and whose state variables are left unspecified. On the other hand is considered the sliding stochastic *implied* volatility (SImpV) class. This market model is also generic, as it describes explicitly but formally the state and the joint dynamics of the (same) underlying and of its European option surface. The no-arbitrage property of the SImpV model is guaranteed by the Zero Drift Condition (ZDC), a stochastic PDE that links all its functionals. This equation lies at the core of the ACE methodology developed here, and represents an obvious (but not structural) deviation from earlier presentations.

The structural links are established sequentially (by *layer*) between a group of SInsV diffusion coefficients and a set of asymptotic differentials for the SImpV functionals (specifying the IV surface's shape and dynamics). The asymptotic domains in question are first the zero-expiry or *immediate* region, and then the immediate At-The-Money (IATM) point. Hence ACE can be seen as relating the *depth* of the SInsV chaos with the *order* of the SImpV differentiation. This approach exposes the fundamental constraints imposed on both market models, as well as the natural asymmetry between the direct (from SInsV to SImpV) and the inverse problems.

The ACE algorithm can be automated to provide any differentiation order, and programmed without resorting to symbolic calculus. Also, because of its purely asymptotic focus, ACE is defined for a generic model framework, and therefore can be applied to most SV classes. Such versatility allows this book to progress easily from a simple single-underlying/bi-dimensional setup, first to a multi-dimensional configuration, and then to a term-structure context.

---

[5] See also [97] for an extension to jump processes.

1.5 Outline and Main Results

The global structure of this book is articulated in two parts. Part I focuses on a single underlying, first in a scalar and then in a multi-dimensional (basket) context. Part II extends the approach to term-structures and in particular to interest rates derivatives. Let us present both parts sequentially.

Part I

In Chap. 2 are presented the basic principles of the methodology. A market is envisaged, which includes a martingale underlying and an associated continuum of European options. The modelling framework gathers both a stochastic instantaneous volatility (SInsV), being defined as a generic Wiener chaos, and a stochastic Lognormal implied volatility (SImpV) model which is parametrised in log-moneyness y and in time-to-maturity θ. The (stochastic) PDE constraining the SImpV model to be arbitrage-free is established; this Zero-Drift Condition (ZDC) constitutes the central object of the ACE methodology. It is indeed the specific structure of that equation that allows us to reach all precision orders. Differentiating the ZDC and then considering its asymptotics, first in $\theta = 0$ (the *Immediate* region) and then in $y = 0$ (defining the Immediate At-The-Money point), structural links between both models are obtained. The elements being associated are, on one hand, the chaos coefficient of the SInsV model, and, on the other hand, the IATM differentials of the SImpV framework. These two groups are in fact organised in *layers*, the first of which is covered in this chapter. Hereby the *direct* and *inverse* problems are covered, which consist respectively in transferring information from the SInsV to the SImpV setup and vice-versa. Then the possible applications of the methodology are discussed, and the results are illustrated first with the local volatility (LV) class, in comparison to several other exact and approximate approaches. Finally these computations are extended to the *Extended Skew Market Model* (which includes popular SInsV classes such as SABR, Heston or FL-SV) and used to illustrate the strong decoupling between local and stochastic volatility effects.

In Chap. 3 the possible extensions of the method are discussed, and several of these are explored in more detail. The first natural avenue is to increase the IATM differentiation order of the SImpV class, which goes in parallel with deeper chaos dynamics for the SInsV framework. It is shown—in a bidimensional context, for simplicity—that any cross-order can be attained for the direct problem, and the associated procedure is described as part of the proof. This generic methodology is then applied to provide static and dynamic smile differentials with strong financial significance. Most of these are the main *descriptors* of the smile shape, such as the *twist* or the *flattening*. The computations are quite heavy but programmable if necessary, and demonstrate the *ladder effect*: differentials are organised in layers, and increasing the θ-order by one requires to first derive the y-differentials by either one or two further orders. Some of these results are interpreted and discussed, in particular the interaction between, on one hand, the volatility's mean-reversion and volatility, and, on the other hand, the smile flattening.

Next the impact of a higher-dimensional context is investigated, especially the incompleteness of the inverse approach. The direct formulae for the first layer are nonetheless derived, and applied to the notion of a basket with fixed or stochastic weights. These results will be useful later, in chapters dedicated to IR derivatives, since their underlyings can often be seen as baskets of either rebased bonds or rates. Among the other available extensions, the possibility of using an alternative *baseline* to the Lognormal process is explored. This is a simple model which must provide a closed-form option pricing formula and an associated implied parameter. The *Normal baseline* is quickly run through and several other possibilities are discussed, leading in particular to the notion of *baseline transfer*, a generic numerical procedure allowing to convert the IATM differentials from one convention to another.

In Chap. 4 the application of ACE to whole-smile extrapolations is illustrated. First some general principles are discussed, such as which expansion variable(s) to use. Then after selecting the SABR and FL-SV model classes, their relevant IATM differentials are computed,[6] stressing the importance of induction and symmetries. Some numerical examples are presented for the CEV-SABR model, exhibiting a performance comparable or superior to Hagan and al's benchmark. Although it thus appears efficient as a generic method, it is suggested to use ACE to *complement* or *correct* other approaches, rather than to replace them altogether.

Part II

In Chap. 5 the method is extended to term-structures (TS) where the underlyings, numeraires, martingale measures and options are all indexed by maturity. It is found that beyond the multi-dimensional aspect, new terms appear in the ZDC and are carried over to the IATM point. This comprehensive approach, allowing us to envisage the joint dynamics of the whole smile, is contrasted to a collection of term-by-term, single-underlying derivations. The focus then turns to the natural (but not exclusive) field of application of these results, which is the interest rates environment.

In Chap. 6 the results of Chap. 5 are applied to the most liquid interest rates derivatives—caplets, swaptions and bond options—within a generic stochastic volatility Heath-Jarrow-Morton (SV-HJM) framework, where the yield curve dynamics are specified *via* the Zero-Coupons. First the options are cast into the generic TS framework. Then to obtain the smile differentials it suffices to compute the chaos dynamics of the respective underlyings. This is performed for the three option types, up to the first layer, and the results are analysed when pertinent. Finally, the (tempting) possibility of casting the assets rather than the associated rates is discussed.

Chapter 7 proceeds likewise but within a more fashionable and generic stochastic volatility Libor Market Model (SV-LMM). Again, the considered options are caplets, bond options and swaptions. Finally, the latter product occasioning the most complex computations, the impact of the traditional basket approach for the swap rate is measured.

---

[6] The FL-SV computation is presented as an exercise.

1.6 General Spirit and Edited Material

As stated before as part of the objectives, this book aims at providing very *generic* results. The intention is also to build a consistent, self-contained but comprehensive coverage of the topic. An alternative mandate would have been to specialise the study early on, and to dwell further in a few chosen direction(s). For instance it would have been possible to focus on a particular model, some payoff feature or a specific numerical aspect.

One (naively unforeseen) consequence of this methodological choice is the significant volume of this book, courtesy of the calculus-intensive nature of asymptotic expansions. And with such large documents, it is easy for the reader to lose track of the global picture, and for the essence of the message to get lost in the maze of computations. Therefore a very (sometimes overly) structured narrative has been provided, incorporating many cross-references and—when necessary—overt repeats. Particular attention has been given to *interpreting* the results, and to anticipating their practical applications. It is my sincere hope that these efforts have contributed in delivering a comprehensible and useful study.

At a technical level, the choice was made to explore the widest domain of applicative results, at the expense of conservative assumptions. The alternative would have been to secure a smaller set of objectives, but supported by looser conditions.

It is also noteworthy that four chapters have been omitted from the final cut of this book, because they took too much space and/or were not finalised. The first (and most interesting) of these chapters was dedicated to the Stochastic Local Volatility (SLocV) model class, formalised for instance by [31] and [33]. That variant of ACE exploits the strong theoretical links between, on one hand, (Dupire's) local volatility (LV), and, on the other hand, either implied or instantaneous volatility. In practice, the associated asymptotic results are quite powerful, because for many SInsV models, the SDE incorporates some local volatility coefficient.

The second edited chapter dealt with two related subjects. The first topic was the static and dynamic validity of SImpV models, especially the consistency between the respective expansions of the shape and of the dynamic maps. The second subject was the large moneyness asymptotics, i.e. with $K \searrow 0$ and $K \nearrow +\infty$. Indeed, some very strong static and dynamic no-arbitrage results do constrain the specification of these asymptotics, thus complementing the small-time, IATM information.

Finally, two chapters were dedicated to panoramas of stochastic volatility models, first in the single underlying context and then in the interest rates environment, with a particular focus on the LMM framework.

With these four chapters excluded from the book, it is possible that some comments might appear expedited, or lacking in context. I do apologise in advance for any possible shortcoming: all mistakes herein are mine, and mine only.

References

1. Breeden, D., Litzenberger, R.: Prices of state-contingent claims implicit in option prices. J. Bus. **51**, 621–651 (1978)
2. Dupire, B.: Pricing and Hedging with Smiles. Working Paper, Paribas Capital Markets (1993)
3. Stein, E.M., Stein, J.C.: Stock price distributions with stochastic volatility: an analytic approach. Rev. Financ. Stud. **4**(4), 727–752 (1991)
4. Hull, J., White, A.: The pricing of options on assets with stochastic volatilities. J. Finance **42**, 281–300 (1987)
5. Heston, S.L.: A closed-form solution for options with stochastic volatility with applications to bond and currency options. Rev. Financ. Stud. **6**(2), 327–343 (1993)
6. Hagan, P.S., Kumar, D., Lesniewski, A.S., Woodward, D.E.: Managing smile risk. Wilmott Mag. **1**(September), 84–108 (2002)
7. Andersen, L., Brotherton-Ratcliffe, R.: Extended Libor Market Models with Stochastic Volatility. Technical Report, Bank of America (2001)
8. Heath, D., Jarrow, R.A., Morton, A.: Bond pricing and the term structure of interest rates: a new methodology for contingent claims valuation. Econometrica **60**, 77–105 (1992)
9. Cont, R., da Fonseca, J.: Dynamics of implied volatility surfaces. Quant. Finance **2**, 45–60 (2002)
10. Cont, R., da Fonseca, J., Durrleman, V.: Stochastic models of implied volatility surfaces. Econ. Notes **31**(2), 361–377 (2002)
11. Lyons, T.J.: Derivatives as tradeable assets. Seminario de Matematica Financiera **98–99**(2), 213–232 (1997)
12. Schönbucher, P.J.: A market model for stochastic implied volatility. Philos. Trans. R. Soc. **357**(1758), 2071–2092 (1999)
13. Brace, A., Goldys, B., Klebaner, F., Womersley, R.: Market Model of Stochastic Implied Volatility with Application to the BGM Model. Technical Report, University of New South Wales (2001)
14. Babbar, K.A.: Aspects of stochastic implied volatility in financial markets. Ph.D. thesis, Imperial College London (2001)
15. Ledoit, O., Santa-Clara, P., Yan, S.: Relative Pricing of Options with Stochastic Volatility. Technical Report, Anderson Graduate School of Management (UCLA) (2002)
16. Davis, M.H.A.: Complete-market models of stochastic volatility. In: Proceedings of the Royal Society, pp. 11–26 (2004)
17. Wissel, J.: Arbitrage-Free Market Models for Option Prices. Technical Report, NCCR FINRISK No. 428 (2007)
18. Schweiser, M., Wissel, J.: Term structures of implied volatilities: absence of arbitrage and existence results. Math. Finance **18**(1), 77–114 (2008)
19. Schweiser, M., Wissel, J.: Arbitrage-free market models for option prices: the multi-strike case. Finance Stochast. **12**, 469–505 (2008)
20. Jacod, J., Protter, P.: Risk neutral compatibility with option prices. Finance Stochast. **14**(2), 285–315 (2010)
21. Davis, M., Obloj, J.: Market Completion Using Options. Technical Report, Imperial College London (2008)
22. Balland, P.: Deterministic implied volatility models. Quant. Finance **2**, 31–44 (2002)
23. Rousseau, N.: How to keep the smile? Dynamic vega hedges and volatility derivatives. Ph.D. thesis, Universite de Nice Sophia Antipolis (2007)
24. Rogers, L.C.G., Tehranchi, M.R.: Can the implied volatility surface move by parallel shifts? Finance Stochast. **14**(2), 235–248 (2010)
25. Hafner, R.: Stochastic implied volatility: A data based model. In: Lecture Notes in Economics and Mathematical Systems, vol. 545. Springer, Berlin (2004)
26. Hafner, R., Schmid, B.: A Factor-Based Stochastic Implied Volatility Model. Technical Report, Risklab Germany (2005)

27. Fengler, M.R.: Semiparametric modeling of implied volatility. In: Springer Finance Lecture Notes. Springer, Berlin (2005)
28. Zilber, A.: A Market Model for Stochastic Smiles. Working Paper, University of Twente (2006)
29. Derman, E., Kani, I.: Stochastic implied trees: arbitrage pricing with stochastic term and strike structure of volatility. Int. J. Theor. Appl. Finance **1**(1), 61–110 (1998)
30. Alexander, C., Nogueira, L.M.: Hedging with Stochastic and Local Volatility. Working Paper 2004–2011, ISMA Centre, University of Reading (2004)
31. Alexander, C., Nogueira, L. M.: Stochastic local volatility. In: Proceedings of the Second IASTED International Conference, pp. 136–141 (2004)
32. Carmona, R., Nadtochiy, S.: An infinite dimensional stochastic analysis approach to local volatility models. Commun. Stochast. Anal. **2**(1), 109–123 (2008)
33. Carmona, R., Nadtochiy, S.: Local volatility dynamic models. Finance Stochast. **13**(1), 1–48 (2009)
34. Ren, Y., Madan, D., Qian, M.Q.: Calibrating and pricing with embedded local volatility models. Risk Magazine, **120**(9), 138 (2007)
35. Potter, C.W.: Complete Stochastic Volatility Models with Variance Swaps. Technical Report, Oxford University (2004)
36. Bergomi, L.: Smile dynamics. Risk 117–123 (2004)
37. Bergomi, L.: Smile dynamics II. Risk 67–73 (2005)
38. Bergomi, L.: Smile dynamics III. Risk 90–96 (2008)
39. Bergomi, L.: Smile dynamics IV. Risk 94–100 (2009)
40. Buhler, H.: Consistent variance curve models. Finance Stochast. **10**(2), 178–203 (2006)
41. Buhler, H.: Volatility markets: consistent modeling, hedging and practical implementation of variance swap market models. Ph.D. thesis, Technical University Berlin (2006)
42. Piterbarg, V.V.: Stochastic volatility model with time-dependent skew. Appl. Math. Finance **12**, 147–185 (2005)
43. Piterbarg, V.V.: Markovian projection for volatility calibration. Risk Mag. **20**, 84–89 (2007)
44. Fouque, J.-P., Papanicolau, G., Sircar, K.R.: Derivatives in Financial Markets with Stochastic Volatility. Cambridge University Press, Cambridge (2000)
45. Lesniewski, A.: WKB Method for Swaption Smile. Technical Report, BNP Paribas (2002)
46. Berestycki, H., Busca, J., Florent I.: Asymptotics and calibration of local volatility models. Quant. Finance **2**(1), 61–69 (2002)
47. Varadhan, S.R.S.: Large deviations. Ann. Probab. **36**(2), 397–419 (2008)
48. Hagan, P.S., Woodward, D.E.: Equivalent black volatilities. Appl. Math. Finance **6**, 147–157 (1999)
49. Benhamou, E., Gobet, E., Miri, M.: Expansion formulas for European options in a local volatility model. Int. J. Theor. Appl. Finance **13**(4), 603 (2010)
50. Benhamou, E., Gobet, E., Miri, M.: Analytical Formulas for Local Volatility Model with Stochastic Rates. Technical Report, Universite de Grenoble (2009)
51. Kunita, H.: Stochastic Flows and Stochastic Differential Equations. Cambridge University Press, Cambridge (1990)
52. Benhamou, E., Gobet, E., Miri, M.: Time Dependent Heston Model. Technical Report, Universite de Grenoble (2009)
53. Gatheral, J., Hsu, E.P., Laurence, P., Ouyang, C., Wang, T.H.: Asymptotics of Implied Volatility in Local Volatility Models. Technical Report, City University of New York (2010)
54. Gatheral, J., Hsu, E., Laurence, P., Ouyang, C., Wang, T.-H.: Asymptotics of implied volatility in local volatility models. Math. Finance **22**, 591–620 (2012)
55. Beresticki, H., Busca, J., Florent, I.: Computing the implied volatility in stochastic volatility models. Commun. Pure Appl. Math. **LVII**, 1352–1373 (2004)
56. Hagan, P.S., Lesniewski, A.S., Woodward, D.E.: Probability Distribution in the SABR Model of Stochastic Volatility. Report, Bloomberg LP (2005)
57. Obloj, J.: Fine-Tune Your Smile: Correction to Hagan et al. Technical Report, Imperial College London (March) (2008)

58. Fouque, J.-P., Papanicolau, G., Sircar, K.R., Solna, K.: Multiscale Stochastic Volatility for Equity, Interest Rate, and Credit Derivatives. Cambridge University Press, Cambridge (2011)
59. Fouque, J.-P., Papanicolau, G., Sircar, K.R.: Financial modeling in a fast mean-reverting stochastic volatility environment. Asia-Pacific Financ. Markets **6**, 37–48 (1999)
60. Fouque, J.-P., Papanicolau, G., Sircar, K.R.: Mean-reverting stochastic volatility. SIAM J. Control Optim. **31**, 470–493 (2000)
61. Medvedev, A.: Asymptotic Methods for Computing Implied Volatilities Under Stochastic Volatility. Technical Report, National Center of Competence in Research (2004)
62. Medvedev, A.: Implied Volatility at Expiration. Technical Report, Swiss Finance Institute (2008)
63. Medvedev, A., Scaillet, O.: A Simple Calibration Procedure of Stochastic Volatility Models with Jumps by Short Term Asymptotics. Technical Report, University of Geneva (2003)
64. Medvedev, A., Scaillet, O.: Approximation and Calibration of Short-Term Implied Volatilities Under Jump-Diffusion Stochastic Volatility. Technical Report, University of Geneva (2006)
65. Lee, R.W.: Implied volatility: statics, dynamics, and probabilistic interpretation. In: Baeza-Yates R., et al. (eds), Recent Advances in Applied Probability, pp. 241–268. Springer, New York (2005)
66. Fournie, E., Lebuchoux, J., Touzi, N.: Small noise expansion and importance sampling. Asymptot. Anal. **14**(4), 361–376 (1997)
67. Henry-Labordere, P.: In: Analysis, Geometry and Modeling in Finance—Advanced Methods in Option Pricing. CRC Financial Mathematics Series. Chapman & Hall, London (2008)
68. Henry-Labordere, P.: A General Asymptotic Implied Volatility for Stochastic Volatility Models. Technical Report, Barclays Capital (2005)
69. Henry-Labordere, P.: Unifying the BGM and SABR Models: A Short Ride in Hyperbolic Geometry. Technical Report, Societe Generale (2006)
70. Forde, M., Jacquier, A.: Small-Time Asymptotics for Implied Volatility Under a General Local-stochastic Volatility Model. Technical Report, Imperial College London (2009)
71. Osajima, Y.: General Asymptotics of Wiener Functionals and Application to Mathematical Finance. Report, Graduate School of Mathematical Sciences, University of Tokyo (2007)
72. Osajima, Y.: The Asymptotic Expansion Formula of Implied Volatility for Dynamic SABR Model and FX Hybrid Model. Report, Graduate School of Mathematical Sciences, University of Tokyo (2006)
73. Kunitomo, N., Takahashi, A.: Applications of the Asymptotic Expansion Approach Based on Malliavin-Watanabe Calculus in Financial Problems. Report, Graduate School of Mathematical Sciences, University of Tokyo (2003)
74. Kunitomo, N., Takahashi, A.: The asymptotic expansion approach to the valuation of interest rate contingent claims. Math. Finance **11**(1), 117–151 (2001)
75. Jaeckel, P., Kawai, A.: An asymptotic FX option formula in the cross currency Libor market model. Wilmott Mag. 74–84 (2007)
76. Daniels, H.E.: Saddlepoint approximations in statistics. Ann. Math. Stat. **25**, 631–650 (1954)
77. Jensen, J.: Saddlepoint Approximations. Oxford University Press, Oxford (1995)
78. Butler, R.W.: Saddlepoint Approximations with Applications. Cambridge University Press, Cambridge (2007)
79. Forde, M., Jacquier, A.: Small-time asymptotics for implied volatility under the Heston model. Int. J. Theor. Appl. Finance **12**, 861–876 (2009)
80. Forde, M., Jacquier, A., Mijatovic, A.: Asymptotic Formulae for Implied Volatility in the Heston Model. Technical Report, Imperial College London (2010)
81. Forde, M., Jacquier, A., Lee, R.: The Small-Time Smile and Term Structure of Implied Volatility Under the Heston Model. Technical Report, Imperial College London (2012)
82. Rogers, L.C.G., Zane, O.: Saddlepoint approximations to option prices. Ann. Appl. Probab. **9**(2), 493–503 (1999)
83. Carr, P., Madan, D.: Saddlepoint methods for option pricing. J. Comput. Finance **13**(1), 49–61 (2009)

84. Jacquier, A.: Asymptotic Skew Under Stochastic Volatility. Technical Report, Birkbeck University (2007)
85. Ait-Sahalia, Y., Yu, J.: Saddlepoint approximations for continuous-time Markov processes. J. Econ. **134**, 507–551 (2006)
86. Kristensen, D., Mele, A.: Adding and subtracting Black-Scholes: a new approach to approximating derivative prices in continuous-time models. J. Financial Econ. **102**, 390–415 (2011)
87. Lee, R.W.: The moment formula for implied volatility at extreme strikes. Math. Finance **14**(3), 469–480 (2004)
88. Benaim, S., Friz, P.: Regular variation and smile asymptotics. Math. Finance **19**, 1–12 (2009)
89. Benaim, S., Friz, P.: Smile Asymptotics II: Models with Known Moment Generating Function. Technical Report, University of Cambridge (2008)
90. Benaim, S., Friz, P., Lee, R.: On the Black-Scholes Implied Volatility at Extreme Strikes. Technical Report, University of Cambridge, University of Chicago (2008)
91. Tehranchi, M.: Asymptotics of implied volatility far from maturity. J. Appl. Probab. **46**(3), 629–650 (2009)
92. Gao, K., Lee, R.: Asymptotics of Implied Volatility to Arbitrary Order. Technical Report, University of Chicago (2011)
93. Engulatov, A., Gonzalez, R., Scaillet, O.: Recovering Nonlinear Dynamics from Option Prices. Technical Report, University of Geneva (2011)
94. Durrleman, V.: From implied to spot volatilities. Ph.D. thesis, Princeton University (2003)
95. Durrleman, V.: From implied to spot volatilities. Finance Stochast. **14**(2), 157–177 (2010)
96. Durrleman, V., El Karoui, N.: Coupling smiles. Quant. Finance **8**, 573–590 (2008)
97. Durrleman, V.: Convergence of At-the-Money Implied Volatilities to the Spot Volatility. Technical Report, CMAP, Ecole Polytechnique (2007)

Part I
Single Underlying

In this first part we focus on the simplest—but most demonstrative—case of a single (scalar) underlying process. We lay the foundations of the ACE methodology. In particular, we establish the Zero Drift Condition and its variants, and walk through the first order computations (the *first layer*).

We then extend the methodology in several directions. For instance, by establishing the ACE algorithm for arbitrary orders, migrating the results to other implied measures of price than Lognormal volatility, or introducing a multi-dimensional framework.

Throughout, we illustrate the results with examples, such as local volatility models, the case of basket underlyings or the computation of two further layers.

Finally, we demonstrate the practical capabilities of the approach by approximating the implied volatility surface generated by the SABR model.

Chapter 2
Volatility Dynamics for a Single Underlying: Foundations

Abstract In this first and fundamental chapter we lay out the core principles of the Asymptotic Chaos Expansion (ACE) methodology. We investigate the relationship between *stochastic instantaneous volatility* (SInsV) and *stochastic implied volatility* (SImpV) models, in the simple case of a single underlying, and when the endogenous driver is scalar. We discuss both the inverse (or recovery) and the direct problem, initially limiting the asymptotic expansion to its lowest order, which we call the *first layer*. We illustrate these asymptotic results first with the local volatility (LV) class, and then with a comprehensive extension to stochastic volatility (SV) dynamics.

In Sect. 2.1, we define the market environment: the underlying, the numeraire and the liquid European options. We define and justify the re-parametrisation of the option price surface *via* a *sliding* implied volatility map. We can then introduce both stochastic volatility models (SInsV and SImpV) as well as some sufficient regularity assumptions. Finally we state our objectives, which we split into a *direct* and an *inverse* problem.

In Sect. 2.2, we establish the fundamental result of this chapter, and of the ACE methodology. This is the *Zero Drift Condition* (ZDC), a PDE constraining the shape and dynamics of the stochastic *implied* volatility model in the whole strike/expiry domain, in order to respect the no-arbitrage assumption (NAA). We then specialise that result to the immediate or zero-expiry sub-domain, which leads us to a pair of *Immediate ZDCs*. Finally we specialise again these results to the *Immediate At-The-Money* (IATM) point, which is our most limited but fertile asymptotic, and quote the IATM Identity linking the implied and instantaneous volatilities.

In Sect. 2.3, we solve part of the *inverse* problem, which is to recover the *instantaneous* model from the *implied* one. First we establish arbitrage constraints between the coefficients of the SImpV model at the IATM point, which emphasises the structural over-specification of that class. Then we show that at a given level of precision (the *first layer*, which involves a group of low-order IATM differentials of the smile) the implied model injects itself into the instantaneous class.

In Sect. 2.4, we tackle the more popular *direct* problem, which consists in generating the smile, and more generally the *implied* model associated to a given *instantaneous* class. For the first layer, we establish the opposite connection from before, which confirms a full correspondence between the two classes (at that level of

D. Nicolay, *Asymptotic Chaos Expansions in Finance*,
Springer Finance Lecture Notes, DOI 10.1007/978-1-4471-6506-4_2

precision). We comment, illustrate and contrast these results against the available literature, within the simple class of local volatility models, thereby exposing some shortcomings in a popular heuristic.

In Sect. 2.5 we turn to some practical applications of these results, which we classify as either pure asymptotic, whole-smile or sensitivity-oriented. In terms of pure asymptotics, we define a stochastic instantaneous volatility model class, covering most popular SV models, for which we provide the first layer differentials. In particular we analyse the respective merits and properties of the Lognormal displaced diffusion (LDD) and CEV specifications as skew functions. We also use the pure asymptotic results to re-parametrise such instantaneous volatility models into more *intuitive* versions, based on the smile that they generate. Then we briefly discuss the caveats involved in extrapolating the whole smile in the naive way, *via* Taylor expansions.

Eventually, in Sect. 2.6 we conclude this chapter and open onto the more specialised subjects covered thereafter. We also provide a diagram gathering the main proofs of this chapter, and hence capturing the main body of the ACE approach, as a basis and a comparison tool for the more advanced results coming thereafter. Note that a general roadmap for this chapter is provided by Fig. 2.4 [p. 113].

2.1 Framework and Objectives

2.1.1 Market and Underlyings

We consider a market equipped with the usual filtered objective probability space $(\Omega, \mathscr{F}, \mathbb{P}, \mathscr{F}_t)$. Unless explicitly specified otherwise, all processes mentioned thereafter will be assumed to be continuous and $\mathscr{F}_t$-adapted.

Although we impose the No-Arbitrage Assumption (NAA), we do not, however, demand market completeness. This choice is obviously motivated by the stochastic volatility specification, and therefore in the sequel the term "risk-neutral measure" should be understood as "chosen risk-neutral measure" with respect to the *volatility risk premium*. Some considerable literature has been devoted to the economic significance and modelling of this risk premium. We have chosen not to dwell on this interesting subject, since it is of less relevance in our "completed" framework (which includes vanilla options), and also because it presents less interest in the prospect of pure hedging (using dynamically those options). Those readers interested in the subjects of market equilibrium, and market price of volatility risk, can refer to [1] or [2], for instance.

As for the theoretical market, we start by considering a single, scalar asset, with continuous price process $\widehat{S}_t$. We also select a numeraire asset N_t, so that under the risk-neutral measure $\mathbb{Q}$ and using Lognormal conventions, their dynamics come as

$$d\widehat{S}_t/\widehat{S}_t = r_t dt + \sigma_t dW_t^{\mathbb{Q}} \qquad \text{and} \qquad dN_t/N_t = r_t dt + \lambda_t dW_t^{\mathbb{Q}}$$

with $W_t^{\mathbb{Q}}$ a scalar $\mathbb{Q}$-Wiener process, while both σ_t and λ_t are undefined but continuous scalar volatility processes. We then define the *deflated or rebased asset* S_t and its martingale measure $\mathbb{Q}^N$ by writing

$$S_t = \widehat{S}_t / N_t \qquad \text{and} \qquad dW_t^{\mathbb{Q}^N} = dW_t^{\mathbb{Q}} - \lambda_t dt$$

and specifying that $W_t^{\mathbb{Q}^N}$ be a $\mathbb{Q}^N$-Wiener process. But now the process S_t is not *a priori* a tradeable asset any more. For all intents and purposes, it should be seen as an index, a reference defining the payoffs of our soon-to-come vanilla options. For that reason, it will be called an *underlying*, with dynamics coming driftless as

$$\frac{dS_t}{S_t} = \sigma_t dW_t^{\mathbb{Q}^N}. \qquad (2.1)$$

Nevertheless, as a matter of convention S_t will often be called the "spot" process in the sequel. Also, regarding notation, we will forgo the $\mathbb{Q}^N$ identifier for the relevant Wiener process and simply denote it by W_t.

2.1.2 Vanilla Options Market and Sliding Implied Volatility

2.1.2.1 Definitions and Notations

On top of the underlying S_t and of the numeraire N_t, we now assume a *market continuum* of prices $C(t, S_t, T, K)$ for Call options written on S_t. Their payoff is either defined or equivalent (using NAA arguments) to the following cashflow, transferred at time T:

$$C(T, S_T, T, K) = N_T \left(S_T - K\right)^+. \qquad (2.2)$$

The continuum is assumed both in maturity (until a finite horizon $T_{\max}$) and in strike (for all $K \in]-\infty, \infty[$). In fact we could consider Puts or Straddles instead of Calls: a smooth continuum in strike is indeed equivalent to assuming that the full marginal distribution is given.

Providing it is *valid*,[1] this surface of option prices is associated to an implied volatility mapping $\Sigma(t, S_t, K, T)$ via the classical Lognormal re-parametrisation:

$$C(t, S_t, T, K) = N_t C^{BS}\left(S_t, K, \Sigma(t, S_t, K, T).\sqrt{T-t}\right) \qquad (2.3)$$

[1] In other words, that the implied marginal densities satisfy the usual criteria, see Sect. 4.1.

where $C^{BS}(x, k, v)$ is the *time-normalised* Black functional (see [3]), which we now define. Denoting by

$$y(x,k) \triangleq \ln(k/x)$$

the *log-strike* relative to the spot, *a.k.a.* "log-moneyness", we set

$$C^{BS} : \mathbb{R}^{+2} \times \mathbb{R}^{+*} \to \mathbb{R}^{+} \qquad C^{BS}(x,k,v) \triangleq x\mathcal{N}(d_1) - k\mathcal{N}(d_2) \tag{2.4}$$

$$\text{with} \qquad d_{1/2}(x,k,v) = \frac{-y}{v} \pm \frac{1}{2}v \qquad \text{and} \qquad \mathcal{N}(z) = \frac{1}{\sqrt{2\pi}} \int_{-\infty}^{z} e^{-\frac{1}{2}s^2} ds.$$

Note that, since both the Calls and the numeraire have been chosen as traded assets, their ratio as per (2.3) will naturally be martingale under $\mathbb{Q}^N$: it is the numeraire N_t that ensures the necessary link between the underlying and the payoff ((2.1) *vs* (2.2)). The whole construction (asset, numeraire, payoff) will appear arch-classical to any reader familiar with the interest rates environment, and might even look restrictive. In practice, however, it covers most existing vanilla products and market conventions.

Besides, it is possible to extend this simple framework to less classical configurations. For instance, we could theoretically deal with different drifts between the "asset", the numeraire and the Calls. We could also consider different pricing functionals from the Black formula, and/or look at other payoff definitions than Call options.

Most of these possibilities will be discussed in Chap. 3, which deals with extensions of the basic framework. Some of these configurations, in particular drift misalignment, will be used out of necessity in Chap. 5, dealing with the term structure framework. But all in all, the basic setup that we consider here is a good starting point, simply because, by killing the drift, it will enable us to derive shortly a clean, simple Zero-Drift Condition (2.18), which is the foundation of our results.

As will be made clearer in Sect. 4.1, the validity of the *price* mapping itself is reasonably simple to establish. The static aspect for instance can often be checked visually. However, doing so through the Implied Volatility re-parametrisation is quite technical and can prove counterintuitive. Therefore, for the moment, we will put that issue aside and simply assume that the IV surface is statically and dynamically valid.

We now associate to these "absolute" quantities $C(\cdot)$ and $\Sigma(\cdot)$ their "sliding" counterparts, respectively $\widetilde{C}(\cdot)$ and $\widetilde{\Sigma}(\cdot)$. Let us recall that at any given time a sliding quantity can be made to match an absolute one, but that their *dynamics* will thereafter diverge and will therefore provide different insight. These new maps are parametrised w.r.t. a couple of new quantities:

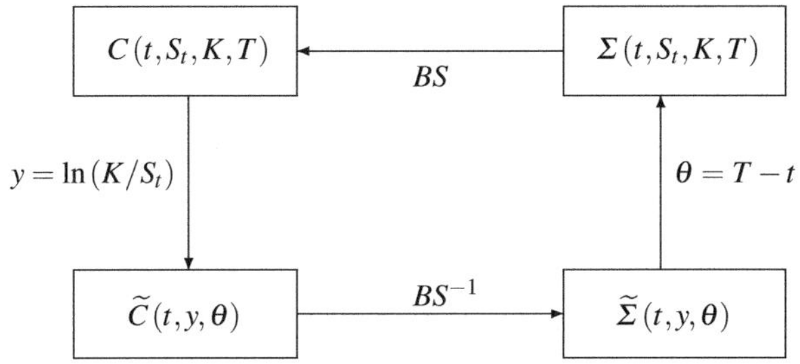

Fig. 2.1 Option price and implied volatility: absolute *vs* sliding

The *time-to-maturity*[2] $\theta \triangleq (T - t)$.

The *log-moneyness* $y \triangleq \ln(K/S_t)$.

Beware that the log-moneyness is defined here as the opposite of $\ln(S_t/K)$, which itself tends to be found in many related papers (since it represents a term of the Black Call pricing formula). Formally we write (see Fig. 2.1)

$$C(t, S_t, K, T) \triangleq \widetilde{C}(t, y, \theta) \quad \text{and} \quad \Sigma(t, S_t, K, T) \triangleq \widetilde{\Sigma}(t, y, \theta).$$

More generally, we will use the superscript $\widetilde{\cdots}$ to identify all sliding quantities, in strike and/or in maturity. However, it must be understood as simultaneously affecting both coordinates, if these are present among the arguments.

Of particular interest in the sliding representation are two regions of the map:

Immediate	will refer to	$\theta \equiv 0$.
At-The-Money (ATM)	corresponds naturally to	$y \equiv 0$.

In our asymptotic context, the intersection of both domains is pivotal, hence we denote by **IATM** (Immediate ATM) the point $(\mathbf{y = 0}, \theta = \mathbf{0})$.

Since a large part of this study will be spent differentiating absolute and sliding functionals with respect to their arguments, it makes sense to gather in a single place all transition formulae between the two configurations: this is the object of Appendix B [p. 431].

[2] "TTM" in shorthand.

2.1.2.2 Motivation and Properties of the Black IV Representation

Since the Black formula assumes Lognormal dynamics for the underlying asset, re-parameterising with the normalised BS implied volatility seems appropriate when S_t is not only martingale (under the measure associated to N_t) but also exhibits "close to Lognormal" dynamics. In some practical instances, other simple dynamics such as the *Normal* framework[3] can prove efficient, as will be discussed in Sect. 3.3.3. But, in most markets, the support of the asset marginal distribution is constrained (or assumed) to be asymmetric, typically bounded on the left. Therefore the (displaced) Black-Scholes implied volatility has proven to be a robust[4] candidate for the re-parametrisation of the price map.

In a more general manner, it is in fact the "implied parameter" approach, which consists in considering prices through a simple "baseline" model, which allows us to compare "raw" prices for different strikes and/or maturities. The Normal and Lognormal dynamics are merely instances of that approach, albeit very common and important ones (more on this in Sect. 3.3.3).

Another advantage of the IV map over the price map is its *regularisation effect*, which is ironically a consequence of its more limited domain of definition.

Indeed, the Black-Scholes formula (2.4) is only specified in the domain $\theta > 0$, since at $T = t$ the option price naturally equals the intrinsic payoff $[S_T - K]^+$. The latter, however, does not provide C^1 regularity at-the-money. This is an issue since our method happens to be of an *asymptotic* nature. It uses expansions intensively and therefore requires/provides differentials of some transform of the price, taken precisely at that same IATM point $(t, y = 0, \theta = 0)$. Alternatively, if we assume that the implied volatility is well behaved for short maturities, typically if it admits a finite limit in $T = t$ along with a sufficient number of its differentials, we can extend the IV map by continuity.[5] The Black-Scholes formula itself becomes valid in the full domain and allows us to fall back effortlessly onto the intrinsic value. This re-parametrisation effectively *contains* the irregularity of the price functional to the Black formula itself, allowing the new functional (the implied volatility) to be infinitely smooth, if required.

In other words, the price always exhibit a singularity (a "kink") at the IATM point, while the implied volatility *can* be infinitely smooth. It then becomes clear that an additional and major attraction of re-parameterising with the Implied Volatility is that it enables, at low cost, the local regularity that our methodology requires.

In the same vein, the expansion method that we use is necessarily less precise for strikes far from the money. Therefore, since the Vega dies out in these regions, using volatility (as opposed to price) expansions artificially limits the resulting *pricing* error, which is most important trading-wise. In other words, in terms of magnitude

[3] And therefore the Bachelier formula.

[4] For other definitions and sources of robustness for Black-Scholes, see [4], for instance.

[5] It is for this reason that, in the sequel, any value of the implied volatility taken with $\theta = 0$, typically in $(t, y, 0)$ or $(t, 0, 0)$, must be viewed as a notation abuse, in fact a limit.

the IV is usually a more uniform, precise albeit dangerous (*c.f.* validity issues) representation than the price itself.

To complement this point, it is also interesting to note that the implied volatility is in practice strongly linked to the Delta, hence to the hedge, and especially so in the FX world.

As a final word of caution, we stress that the argumentation above is valid for vanilla Call and Put options, but might not be so for other products, such as binaries: there will be more on this point in Sect. 3.3.3.

2.1.2.3 Motivation and Properties of the Sliding Representation

We first put this technique into perspective and comment on its relevant mathematical properties. We then discuss the financial attractiveness of this simultaneous "time and strike" slide.

The general concept and the use of relative variables are certainly not new. In the rates environment for instance, it is common practice to denote a Libor rate either with fixed maturity or with fixed accrual[6]: each notation has its specific pros and cons (see [5] or [6], among others). In an option framework, sliding strikes are also frequently used in order to account for "stickiness": certain smiles are "strike sticky" while most are "Delta sticky".

Besides, we emphasise that the nature of the benefit brought by this sliding convention, *in our specific framework*, is more style than substance. Indeed, it does not lead to fundamental or technical results which an absolute setting could not reach. This is a positive feature, since our choice of a strike representation (log-moneyness) is partly subjective and certainly no panacea. It is therefore comforting that our results can effectively be transferred to another convention: the practical aspects of this move are discussed in Sect. 2.1.3.2.

It remains that in principle there are many such ways to define the slide, especially in strike. An obvious candidate is proportional moneyness (K/S_t), but any other adequate function of K and S_t can be considered: such adequacy obviously requires a bijectivity in K and also a sufficient regularity, especially at the money. In [7] one can find a general definition for the strike slide, called simply *moneyness*. But it is stressed therein that the choice should be made on an *ad hoc* basis, an assertion that we support. Indeed, for a given market, a good parametrisation should provide a smile dynamically as stable and stationary as possible. The overall principle consists in conditioning the smile w.r.t. our only observable state variables, i.e. t and S_t. For a complementary discussion on this subject, refer to Sect. 3.3.1.1 [p. 148].

We believe, however, that our specific choice of a sliding convention is justified, for reasons that we expose now. First of all, and on a mathematical level, we elected to use Lognormal dynamics to define an implied parameter: this is in no way mandatory and is simply the most common market practice. However, it leads to Black's formula, which itself clearly makes of the log-moneyness y the *natural* variable to consider.

---

[6] $L(t,T,U)$ *vs* $L^\delta(t,T) \triangleq L(t,T,T+\delta)$.

Also, the results that we present herein are structurally complex, hence any approach that clarifies the interpretation and the role of the various terms is *a priori* welcome. In particular, we find that the sliding representation usually allows a better understanding of stationarity and time-homogeneity issues, especially in the second Part devoted to term structure models.

Furthermore, and as will be covered in the various application sections, the practical efficiency of our methodology depends as much on the pure asymptotic results as on the chosen representation of the variables. This is generally true for most extrapolation methods, but also for numerical reasons as well as for the analysis of model behaviour. For the latter in particular, when the model and/or smile specification themselves are (pseudo-) sliding or time-shift homogeneous, we find that better efficiency is attained by using the sliding versions of our results.

In more formal terms, the main attraction towards sliding versions of the price and implied volatility mappings lies in the dynamic and stochastic properties brought by this change of coordinates. Indeed, reducing the number of arguments from an absolute representation (four arguments: t, S_t, T and K) to a sliding one (three arguments only: t, y and θ) effectively "transfers" the underlying S_t, and therefore the driver W_t *into* the functional (here $\widetilde{\Sigma}$ or $\widetilde{C}$). Let us quickly illustrate this point with two simple examples.

Example 2.1 (Markovian dimension of the IV for a pure local volatility (LV) model) First let us assume that the smile is generated by a pure LV model, as defined by

$$\frac{dS_t}{S_t} = f(t, S_t)dW_t \qquad \text{with} \qquad f(s,x) \geq 0 \,\forall (s,x) \in \mathbb{R}^{+2}. \tag{2.5}$$

Then the Markovian state variables are simply t and S_t, so that both the absolute price surface $C(t, St; K, T)$ and the implied volatility surface $\Sigma(t, St; K, T)$ are entirely deterministic (although *a priori* not explicit) functions of their four arguments. However, the *sliding* implied volatility surface $\widetilde{\Sigma}(t, y, \theta)$ is itself a stochastic function of its three arguments: when the log-moneyness y and time-to-maturity θ are fixed the IV functional becomes *parametrised* by S_t.

Such local volatility models provide an easy understanding of the concept, but they cannot incorporate the notion of *unobservable* state variables. Let us therefore present a more complex illustration, involving a multi-dimensional driver.

Example 2.2 (Markovian dimension of the IV with an independent stochastic volatility) We assume a stochastic instantaneous volatility model with state variables (t, S_t, σ_t) and driven by a bi-dimensional Wiener. Furthermore, we take the dynamics of the volatility σ_t to be purely exogenous as per

$$\begin{cases} \dfrac{dS_t}{S_t} = \sigma_t dW_t \\[2ex] d\sigma_t = f(t, \sigma_t)dt + \vec{h}(t, \sigma_t)^{\perp} d\vec{Z}_t \qquad \text{with} \qquad W_t \perp \vec{Z}_t. \end{cases}$$

Then the absolute Call prices $C(t, S_t, \sigma_t; K, T)$ are deterministic functions of their five arguments. In turn, the same is true of the absolute Lognormal implied volatility $\Sigma(t, S_t; \sigma_t; K, T)$.

However, that map can also be seen as a deterministic function of the four market variables $(t, S_t; K, T)$ which has been parametrised by the hidden (or unobservable) process σ_t. In that case the $\Sigma()$ functional becomes stochastic, is driven purely by the exogenous driver $\overrightarrow{Z}_t$ and is therefore measurable w.r.t. the latter's filtration. As for the sliding IV surface $\widetilde{\Sigma}(t, y, \theta)$, it can be seen as a deterministic function of three variables, parametrised by S_t and σ_t.

In this example, the orthogonality assumption is only there to reinforce the point that the filtration w.r.t. which the sliding "function" $\widetilde{\Sigma}$ is measurable is necessarily much finer that the one sufficient to measure its absolute counterpart Σ. In other words, we have incorporated the driver W_t into the sliding quantity.

Having described the relevant mathematical features of the slide, let us now turn to its financial motivation. It revolves mainly around human factors, in particular our limited ability to comprehend high-dimensional and noisy patterns. We expose the practicality and interest of comparing market and model smile dynamics, and how the sliding representation can help in that process.

In our view, one defining characteristic of a well-chosen model is to *minimise re-calibration*, in other words to exhibit stable (or stationary) calibrated parameters. Ideally, we would like to maintain these constant, and explain all joint movements of the underlying and of the smile "through" the model and its drivers. After all, if the market was kind enough to follow known and stationary laws, as is mostly the case in physics or mechanics, that is exactly what would happen. And there is no argument from practitioners that such a (hypothetical) situation is rather more palatable than having to frequently re-adjust these parameters, a procedure that generates additional Mark-To-Market (MTM) and tracking error noise, and therefore hedging (transaction) costs.

However, in order to even get close to such a stationary behaviour, i.e. to be very "realistic", stochastic instantaneous volatility models generally need to use a significant number of parameters (roughly half a dozen for SABR or for Heston, and even then re-calibration is too frequent). Furthermore, these parameters have very distinct individual impacts on the smile, both in quality and in magnitude. Therefore, attempting to analyse a model's stationarity by observing a collection of historical time series (one for each parameter) might be an interesting academic exercise but *a priori* not a very practical or useful idea.

A much more intuitive approach, in contrast, is to compare the actual market *smile*, at each historical sampling time t_i, to the "prediction" given by the model, itself (statically) calibrated at the previous time t_{i-1}. Obviously the notion of prediction must be made precise, in the sense that this smile's dynamics must be made measurable w.r.t. a given filtration or observable state variables.

In a two-dimensional model such as Heston, SABR, or even some multi-scale extensions,[7] a simple approach consists in using the spot S_t and a single, very short expiry At-The-Money option (C_t^{IATM}) in order to access the full filtration. Indeed, these market instruments are usually very good proxies for the actual state variables. Using then a Euler approximation, it is typically possible to unequivocally associate driver increments $[\Delta W_t, \Delta Z_t]$ to an historical market movement $[\Delta S_t, \Delta C_t^{IATM}]$. Then one can generate the whole "conditional" smile at t_i, which is to be compared to the actual market smile observed. Repeating this process, it is even possible to compute maximum likelihood estimators, and/or to gauge the descriptive/predictive quality of the model.

Such a comparison of smiles, resulting in a *surface of differences*, is easy to interpret, and can also be visualised in motion, in order to assess the dynamic properties of the calibrated model. But in order to facilitate this interpretation, it is important to choose a common representation (i.e. axis coordinates) that tends to stabilise both smiles, and therefore their difference. In other words, we are looking at a change of coordinates that will make both the market-observed and the model-generated smiles as stationary as possible.

It happens, though, that most market smiles demonstrate obvious and simple sliding properties, both in maturity and in strike. The latter is often referred to as "stickiness" (see [10], for instance, for the sticky-strike or sticky-Delta rule). It also happens that this is actually a property that stochastic volatility models are well suited to capture (see [11], for instance, which provides a good comparison with local volatility models).

In summary, the choice of the new coordinates is certainly motivated by the Black formula itself, and in particular the LOG-moneyness $y = \ln(K/S_t)$ that allows scaling and brings some symmetry to the strike axis. But this choice is also brought forward by a healthy desire for stationarity within a realistic modelling framework, and is in no way binding.

2.1.2.4 Illustration and Limitations

Let us demonstrate how to "cast" into our framework a simplistic model/smile combination.

Example 2.3 (SV Normal model with deterministic rates and carry) We place ourselves in the arch-classical case where X_t is the price process of a traded asset, modelled with a Normal[8] volatility (which can be local and/or stochastic),

---

[7] This is not the case, in particular, for "double mean-reverting" models such as "Double Heston", which are tri-dimensional: see [8] and [9].

[8] This is a writing convention, and chosen mainly for demonstration purposes, since it is rather unusual in the Equity world.

and where the short rate $r(t)$ and the dividend/carry rate $d(t)$ are considered deterministic:

$$dX_t = [r(t) - d(t)]\,dt + \gamma_t dW_t,$$

where W_t is a risk-neutral driver, or $\mathbb{Q}$-Wiener process.

We simply select the deterministic function $D(t) = \exp(\int_0^t [r(s) - d(s)]\,ds)$ as our numeraire, which is nothing else than a capitalisation process or money market account. We check that it is a tradeable asset, simply because it is deterministic and therefore can be replicated with liquid assets, the zero-coupons.

Denoting by $S_t \triangleq X_t/D(t)$ the "discounted" asset, and $\sigma_t \triangleq \gamma_t X_t$ the "Lognormal" volatility, we indeed obtain the required dynamics (2.1). As for the option field, the payments are deemed to occur at time T, for an amount of

$$C(K,T) = (X_T - K)^+ = D(T)\left[S_T - \frac{K}{D(T)}\right]^+.$$

It therefore suffices to modify the unit in which we measure the strike, from "cash" K to "discounted" $K/D(t)$, to complete the "cast". Note that the measure has not changed: it is still the risk-neutral measure $\mathbb{Q}$.

Beyond this trivial example, the chosen framework can fit a wide range of underlyings and options. Of course, it also exhibits several limitations, which prevent the coverage of more complex modelling configurations.

A first restriction is that we defined the setup for a *scalar* driver W_t, which excludes the description of full multi-underlying dynamics. Typical cases of multi-dimensional frameworks occur naturally in the FX or equity environments, with baskets or indexes, for instance.

Its second shortfall is that it is not suitable to deal with a *term structure*, and in particular the case of stochastic rates. Indeed, if one wishes to define a whole smile, then the same numeraire N_t must be invoked in the payoff (2.2), whatever the expiry T. In Example 2.3 we bypassed this issue by choosing a deterministic numeraire, but in the general case of stochastic rates the whole setup must be based on a single maturity T, and must be financially meaningful, it can only deal with the associated implied volatility.

This can be seen as both a special case and an extension of the multi-dimensional framework. Indeed, even under a Black-Scholes model, if the short rate is made stochastic then pricing a Call of maturity T_2 usually entices us to use the forward measure $\mathbb{Q}^{T_2}$ and the associated Zero Coupon $B_t(T_2)$ as numeraire. Should we be only interested in that single expiry, then the problem could be treated in the multi-dimensional setup mentioned above. But then for $T_1 < T_2$ the considered payoff would be $B_{T_1}(T_2)\left[X_{T_1}/B_{T_1}(T_2) - K\right]^+$, which is not a liquid product. Therefore the setup would lose its consistency and its financial appeal.

We will therefore extend the current simple setup to cover both these cases, respectively in Sect. 3.4 and in Chap. 5. It will turn out that the multi-dimensional case actually lies in the same conceptual class as the current framework and provides very similar results, and also that its added difficulty is mainly computational. The term structure extension, however, shares the same technical difficulty as the former, but is structurally on a very distinct level.

2.1.3 The Two Stochastic Volatility Model Frameworks

From now on, we will focus on the Sliding Implied Volatility Surface $\widetilde{\Sigma}(t, y, \theta)$ associated to a given model: we are interested in its shape and also in its joint dynamics with the underlying. In our framework, we specifically want this map to exhibit stochastic dynamics, which should be driven by two orthogonal Wiener processes:

- The *endogenous* driver of the underlying, denoted W_t. In full generality the dimension of this driver will be notated as n_W, but initially it will be taken as mono-dimensional since our underlying itself has been defined as a scalar. The consequences of relaxing this assumption will be exposed in Sect. 3.4.
- The *exogenous* driver $\vec{Z}_t$, which enables movements of the implied volatility surface independently of the underlying dynamics. It is taken as multi-dimensional (with finite dimension n_z) to allow for the complex deformation modes observed in practice.

By convention, we will take W_t and $\vec{Z}_t$ to be independent, and all multidimensional Brownian motions (including $\vec{Z}_t$) to be uncorrelated (i.e. to exhibit a *unit* covariance matrix). One might question why we chose to express our dynamics along two uncorrelated Wiener processes. Indeed, other authors have opted for a single, unified driver: this is the case, for instance, in [12]. Clearly this is mathematically insignificant, and purely a matter of presentation. Our view is that it brings two main advantages, for only one drawback.

The first advantage is technical, and is analogous to manipulating independent (as opposed to correlated) Gaussian vectors. The volatility and correlation structures are then combined in (products of) tensorial coefficients, which simplifies the computation of brackets $\langle d\cdot, d\cdot\rangle$.

The second advantage is linked to modelling and interpretation. Indeed, we are attached to the incompleteness, endogenous/exogenous interpretation detailed above. Also, for option pricing purposes and certainly in numerical terms, it makes sense to orthogonalise the drivers.

The single shortfall that we see is also interpretational, in that using our convention we somewhat lose the intuition that drivers' increments are actually "representing" variations in "physical" quantities, such as asset prices, yields or instrumental processes. For instance, the Heston model (see [2]) is traditionally defined as

$$\frac{dX_t}{X_t} = \mu dt + \sqrt{V_t} dW_t \tag{2.6}$$

$$\langle dW_t, dB_t \rangle = \rho dt$$

$$dV_t = \kappa \left[\theta - V_t\right] dt + \varepsilon \sqrt{V_t} dB_t. \tag{2.7}$$

But it might make more financial sense to write the correlation structure as

$$\langle \frac{dX_t}{X_t}, dV_t \rangle = \rho \varepsilon V_t dt \qquad \text{rather than} \qquad dB_t = \rho dW_t + \sqrt{1-\rho^2} dZ_t.$$

Going back to our framework, it is clear that one role of $\overrightarrow{Z}_t$ is to introduce market incompleteness. It can, in particular, embody *model ambiguity*, since it underscores a finer filtration than the σ-field generated by S_t. As mentioned, it certainly makes it possible for the (sliding) implied volatility surface to move independently of the underlying S_t (i.e. not to be purely *local*) or of its driver W_t (i.e. to exhibit an *exogenous* component). Such a rich behaviour of the $\widetilde{\Sigma}(t, y, \theta)$ map can be generated in several ways. In this chapter, we consider only a couple of these distinct model classes: the stochastic *instantaneous* volatility model, and the stochastic *implied* volatility model.[9]

Formally, the specifications of these two model classes (instantaneous SV and implied SV) only share the generic underlying's dynamics (2.1). Indeed, they both describe the shape and the joint dynamics (of the underlying and of the vanilla options) but in different ways; let us now introduce and formalise both these setups.

2.1.3.1 The Generic Stochastic Instantaneous Volatility Model

In this framework, the shape and dynamics of the smile are generated by specifying "in depth" the dynamics of the instantaneous volatility σ_t, using a Wiener chaos representation. Formally, we have to assume a system of SDEs, starting with

$$dS_t/S_t = \sigma_t dW_t \tag{2.8}$$

$$d\sigma_t = a_{1,t}dt + a_{2,t}dW_t + \overrightarrow{a}_{3,t}^{\perp} d\overrightarrow{Z}_t \qquad \text{with} \qquad W_t \perp \overrightarrow{Z}_t. \tag{2.9}$$

The stochastic coefficients $a_{1,t}$, $a_{2,t}$ and $\overrightarrow{a}_{3,t}$ are deemed to be processes, but only imposed to be Markovian and adapted, hence the "generic" denomination for the model. Indeed, most stochastic instantaneous volatility models that are used in practice fall into the *parametric diffusion* category. In that case, the $a_{i,t}$ coefficients are actually parametric functions of a finite collection of state variables, which usually include t, S_t and σ_t. Although the Markovian dimension can get higher than three in this framework, for instance with multi-scale processes,[10] we do not restrict ourselves to such cases. Instead our framework contains those parametric diffusion models, and also gets its "universality" from the fact that the dynamics of $a_{1,t}$, $a_{2,t}$ and $\overrightarrow{a}_{3,t}$

---

[9] It is also possible to employ stochastic *local* volatility models, see [13].

[10] In general, this implies an increase in the dimension of the driver $\overrightarrow{Z}_t$.

are themselves defined by a "chaos" structure. For instance, we specify the (input) dynamics of $a_{2,t}$ in the following way:

$$da_{2,t} = a_{21,t}dt + a_{22,t}dW_t + \vec{a}^{\perp}_{23,t}d\vec{Z}_t. \tag{2.10}$$

However, the dynamics of a multi-dimensional coefficient, such as $\vec{a}_{3,t}$, generate a practical difficulty. Evidently, its dynamics can be similarly symbolised as

$$d\vec{a}_{3,t} = \vec{a}_{31,t}dt + \vec{a}_{32,t}dW_t + \overset{\Rightarrow}{a}_{33,t}d\vec{Z}_t,$$

where $\overset{\Rightarrow}{a}_{33}$ is an $n_z \times n_z$ matrix. If we continued the specification, then $\overset{\Rrightarrow}{a}_{333,t}$ would be a tensor of order 3, and so on. This is clearly not a promising way to conduct computations, at least by hand. Nevertheless, we will explore this avenue in Sect. 3.4, but with a view towards (computerised) *automation*.

Note finally that in order to simplify notations, the time dependency will often be omitted in the sequel (for instance $a_{2,t}$ will become a_2).

In this generic SInsV framework, we define the *depth* of a coefficient simply by the number of digits forming its index. For instance, coefficient $a_{2,t}$ has depth one, while $a_{231,t}$ has depth three, etc. We will also see that the coefficients can be arranged according to another logic, by "layers", which naturally appear in the inductive computation of the smile's asymptotic (IATM) differentials. For example, the first layer will contain σ_t, $a_{1,t}$, $a_{2,t}$, $a_{3,t}$ and $a_{22,t}$ (see Fig. 2.2). Each layer will also be used to designate a group of corresponding smile IATM differentials.

The depth of the *model* itself is defined as the highest depth reached by all the coefficients describing its dynamics. For instance, the model described by (2.8)–(2.10) would have a depth of 2. Since t, S_t, σ_t and the other $a_{i,t}$ coefficients/processes represent the state variables of the model, if the latter's depth is finite then so is its Markovian dimension.

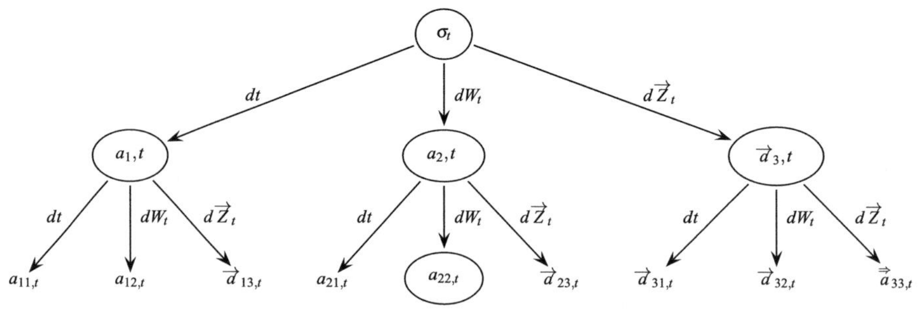

Coefficients of the first layer are circled

Fig. 2.2 Chaos structure of the generic stochastic instantaneous volatility model

The framework described above, however, is obviously *not* an actual model *per se*. Instead, it consists simply in a *cast*, into which we can arrange any real stochastic instantaneous volatility model, such as the aforementioned parametric diffusions. In practice, most real-life models have a finite and fairly small Markovian dimension (it is 3 for Heston and for SABR) whereas their generic cast can exhibit an infinite number of state variables (infinite depth). We will see, however, that this is not an issue in our asymptotic framework, since—schematically—the higher the depth of a coefficient, the higher the degree of precision it brings to the smile description *via* asymptotic smile differentials.

Therefore, if we consider for instance a SABR model (see [11] and Sect. 4.2 for a description of this local-stochastic volatility model) then the Call price will be a deterministic function of the three Markovian state variables (t, $\overrightarrow{S_t}$ and α_t), of the parameter set specifying the diffusion (correlation, vol of vol) and of the option parameters (K and T). Unfortunately, this pricing function is not explicit, as currently only approximations are available (see Sect. 4.2). On the other hand, when casting SABR into our generic framework, we obtain an infinite depth. But the specified dynamics also generate a unique option price surface, by applying no-arbitrage under the chosen measure. This functional cannot *a priori* be expressed either, but we will see that we can instead obtain its asymptotic, potentially infinitely precise description; it simply comes at the cost of an infinite but artificial Markovian dimension.

2.1.3.2 The (Sliding) Stochastic Implied Volatility Model

Modelling the sole dynamics of the implied volatility surface is not a very new idea in itself. A number of empirical studies have been conducted on real data in order to statistically infer the deformation modes of this surface, either in a parametric fashion or not. Usually and rather logically, these empirical investigations have been conducted on very liquid equity indexes, hence minimising the use and influence of interpolation/extrapolation methods.

In [14] the authors analyse the S&P500 and FTSE100 liquid options, with daily frequency. They use a Karhunen-Loève decomposition, which is a generalisation of the Principal Component Analysis to higher dimensional random fields. They uncover, among other interesting features, a typical level/slope/curvature repartition of the leading eigenmodes, as well as characteristic values for the mean-reversion of these modes.

Another interesting presentation can be found in [15], which focuses on semi-parametric modelling, but also covers several inference techniques, as well as practical data processing pitfalls (smoothing in particular).

Modelling the *joint* dynamics of the underlying with the smile, in particular establishing and respecting the structural no-arbitrage constraints, represents a more involved exercise. Apart from stochastic *local* volatility models,[11] the main academic

[11] See [13]: this class is not to be confused with the local stochastic volatility (LSV) models such as those described in [16], for instance.

attraction has been with stochastic *implied* volatility (SImpV) models. A specificity of this class is to define the initial shape of the smile as an input, and then to model the joint underlying/smile dynamics.

In that respect, there is a strong similitude with the approach that [17] introduced with a market model for the whole yield curve, a.k.a. the HJM framework. Indeed, one could summarily consider that the role of the short rate r_t is now taken by the underlying S_t and its volatility σ_t, while the one-dimensional map of Zero-Coupon prices $T \rightarrow B_t(T)$ is replaced by a bi-dimensional mapping $(K, T) \rightarrow \Sigma(K, T)$. We will see that the resemblance is carried even further, in the sense that σ_t is asymptotically included in the Σ dynamics, in the same way that r_t is embedded in the $B_t(T)$ term structure. We shall first cover some of the papers dealing with the general framework and its structural constraints, before moving on to actual models.

The concept of a SImpV model has been introduced in stages and rather independently by several authors, first for a single option and then for whole smiles. Historically, let us first cite [18] and [19], with the latter using a sliding representation. In [20] the authors present four different versions of the no-arbitrage condition for smile dynamics: for the implied volatility or implied variance, and in absolute or sliding coordinates. Then they apply these results in several practical contexts, including a single Caplet within a BGM model.

Modelling-wise [7] proposes a factor-based instance of the class. This article provides a good interpretation for certain equations, makes a deliberate effort to relate the model to other classes, and contains real-market (DAX) applications. Interestingly, it introduces the notion of a generalised *moneyness*, which can be $\ln(K/S_t)$ but is not restricted to that case: in principle it should be chosen so as to render the *stickiness* of the smile considered.

Another seminal collection is [21–24], which focuses not on the smile but on the variance curve. These articles analyse the intrinsic limitations of SInsV models, such as Heston. They propose several dynamic models for the forward variance, based on a Markovian factor representation, using either a continuous or discrete structure.

In [25] one can also find a focus on volatility derivatives, but also the parametrisation of a specific SImpV model. That class (the *Market Model of Implied Volatility*) exploits *local* dynamics, i.e. a diffusion involving only the following state variables: t, S_t and the smile itself. A particular instance is then proposed (the *Skew market Model*) which models the smile as a parabolic function of log-moneyness.

In essence, stochastic implied volatility models represent the next logical step in the natural evolution of modelling practice, within a given derivative market. Indeed, as more liquid derivative products appear, those need to be included in the calibration. The two main avenues are therefore to complexify the existing models and/or to assume that the new set of products represents an input.

In the matter at hand this fits rather well the recent modelling history. Starting from a situation where only ATM options were liquid, Black-Scholes with deterministic volatility was sufficient. As OTM and ITM options became liquid, the model was complexified, upgraded to instantaneous stochastic volatility. First with local volatility (Dupire), then with stochastic volatility (e.g. Heston, hence increasing the Markovian dimension), and currently with a combination of both (SABR, FL-SV).

As the liquidity of the smile increases, so does the need to describe its dynamics, if only because more and more exotic products depend on it, whether in their definition (volatility derivatives) or for their hedge. The next logical step therefore seems to be the incorporation of that smile within the model, along with its dynamics.

The practical difficulty with that model class, however, resides in the parametrisation of an implied volatility surface that starts and stays valid. By valid we mean that the associated option price surface must satisfy the usual non-arbitrage conditions, everywhere in the (K, T) domain, at any time and almost surely. Arguably, this has been the strongest hurdle in the practical introduction of the stochastic implied volatility model class. However, we believe that this model class may ultimately become just as successful as the HJM framework once was, and also as successful as the LMM[12] framework currently is.

As for our version of the SImpV model, it is defined as follows

$$\begin{cases} dS_t/S_t = \sigma_t dW_t & (2.11) \\ d\widetilde{\Sigma}(t, y, \theta) = \widetilde{b}(t, y, \theta)dt + \widetilde{\nu}(t, y, \theta)dW_t + \overrightarrow{n}(t, y, \theta)^{\perp} d\overrightarrow{Z}_t & (2.12) \end{cases}$$

$$\text{with} \qquad y \triangleq \ln(K/S_t) \qquad \theta \triangleq T - t > 0 \qquad W_t \perp \overrightarrow{Z}_t$$

where the drivers W_t and $\overrightarrow{Z}_t$ are independent[13] standard Brownian motions under the (chosen) martingale measure $\mathbb{Q}^N$. All dynamic coefficients (σ_t, $\widetilde{b}$, $\widetilde{\nu}$ and $\overrightarrow{n}$) are taken as generic stochastic processes, so that (2.12) cannot be considered *a priori* as an explicit *diffusion*. However, note that, should these coefficients be specified as deterministic (i.e. local) functions (and irrespective of the drivers' dimensions), then the Markovian dimension of the model would still *a priori* be infinite. Again, this situation is analogous to a bi-dimensional version of the HJM framework, where every point on the yield curve represents a state variable.

At first sight, the fact that no specification is given for σ_t might seem surprising, as it appears to make the model under-determined. But in fact that instantaneous volatility is entirely defined through arbitrage by the implied volatility map $\widetilde{\Sigma}(t, y, \theta)$, as we will soon establish.

Similarly, note that neither the nature nor the dynamics of the other coefficients $\widetilde{b}$, $\widetilde{\nu}$ and $\overrightarrow{n}$ are specified. In principle they can be Itô processes also driven by W_t and/or $\overrightarrow{Z}_t$, or even by entirely new (and orthogonal) Wiener processes. It happens that, for our intents and purposes, this level of definition is in fact irrelevant, and the justifications for such a simplifying feature will become apparent as the book progresses. In particular, these topics will be discussed in Sect. 2.2.4 and in Chap. 3, which is dedicated to more elaborate versions of the methodology. In summary, the SImpV model (2.11)–(2.12) is well-defined and self-contained as it is.

---

[12] Libor Market Model, see Chap. 7.

[13] To lift any remaining ambiguity, this implies that the correlation matrix of $\overrightarrow{Z}_t$ is diagonal.

In spirit, moving from a *stochastic instantaneous volatility* model to a *stochastic implied volatility* model is similar to moving from a short-rate model (such as Vasicek) to an HJM model: it can be seen as a simple matter of number of parameters *vs* number of constraints. Indeed, most short rate models are only capable of generating a certain functional class of yield curves (YC), so that calibration to the bond market is already an issue. In order to recover a given yield curve, one will usually have to add a time-dependent drift, while matching further constraints, such as some marginal distributions of the YC, will require a substantial complexification of the model class: the Hull and White Extended Vasicek [26] comes to mind, which is nothing other than an instance of the HJM class.

On the other hand, opting for an HJM model enables the modeller to calibrate to any yield curve, because this map becomes an integral part of the model. The remaining parameters, or degrees of freedom, are then used to calibrate to the marginal or joint distributions of the YC, to liquid Interest Rates options such as Caps, Swaptions, Bond options, CMS, etc. But as more and more products become liquid, the calibration set tends to increase, so that eventually there are too many constraints and not enough parameters. This is where *stochastic implied volatility* models can come into play: because the liquid Call prices become an integral part of the model, new degrees of freedom become available to calibrate to other (more recent) liquid options.

Although the underlying's instantaneous volatility σ_t is apparently a free parameter of the (sliding) implied volatility model, it is actually to be considered as a formal expression; indeed, we will see that arbitrage constraints impose that σ_t be entirely determined by the stochastic map $\widetilde{\Sigma}(t, y, \theta)$ (see (2.36), p. 50 and Remark 2.4, p. 51). This precision explains why the dynamics of σ_t are not explicitly specified in the *stochastic implied volatility* model, as was the case with the *stochastic instantaneous volatility* model: indeed they are already included.

2.1.3.3 Comparison, Assumptions and Remarks

The two models share the same dynamics for the underlying, and in particular the Lognormal instantaneous volatility. For the SInsV it is the dynamics of σ_t that are defined in a chaos expansion, while the SImpV specifies the smile dynamics: in other words the SInsV model is defined "in depth" while the SImpV is specified spatially. The choice of a single-dimensional driver W_t for the underlying, as specified respectively by (2.8) and (2.11), is actually benign in both cases and for similar reasons. The only interest in employing a multi-dimensional endogenous driver $\vec{W}_t$ and volatility $\vec{\sigma_t}$ is to describe the joint dynamics of the underlying asset S_t along with another process. In the SImpV case, that process is the smile shape $\widetilde{\Sigma}$, while for the SInsV model it will be the coefficients in the Wiener chaos decomposition of σ_t: in both cases, these are *a priori* infinite dimensional processes. But in both instances, any component *not* driving can be formally allocated to the independent, exogenous driver $\vec{Z}_t$, so that in the end only the modulus $\|\vec{\sigma_t}\|$ matters. In essence, this all boils down to the very definition of an exogenous noise, whose instantaneous covariation with S_t must be null.

We will however work with a multi-dimensional $\overrightarrow{W}_t$ later on, in Chap. 3, in order to investigate the basket problem. In that context, we will have to express the joint dynamics of each individual underlying and of the basket itself, which does warrant a vectorial endogenous driver.

In terms of presentation, the fact that we organised the drivers into orthogonal components is clearly artificial, but not binding. In purely economic terms, there is no such thing as clearly identifiable independent factors. It serves several purposes. In mathematical terms, it simplifies the computations and ensures that all correlation-related quantities are represented by linear algebra products. As for interpretation, it clearly divides the picture between an endogenous/observable/complete component, and the exogenous/non-observable/incomplete part. Obviously there is a cost to pay: in particular the bracket between two processes now comes as a scalar product. For instance the notion associated to $\langle \frac{dS_t}{S_t}, d\sigma_t \rangle$ is easy to grasp, even to graphically chart on historical series, whereas $a_{2,t}$ might initially seem a bit abstract. Nevertheless, the logic used throughout this book is to clarify (even artificially) the computation, and re-formulate the output results for interpretation.

Remark 2.1 The choice of the Lognormal convention to write the dynamics of the underlying S_t (see (2.8) and (2.11)), as well as the fact that we are re-parameterising the price surface using a Lognormal (Black) convention, might seem subjective and possibly restrictive. In fact, as will be proven and discussed in Sect. 3.3.3, p. 164, once expansion results are available for a simple model such as the Lognormal dynamics, it is relatively simple to transfer those to most parametrisations, such as the Normal dynamics, or the CEV, etc. In consequence the choice we made is merely practical, and in practice not binding.

In order to facilitate the coming proofs, we add the following technical restrictions:

Assumption 2.1

$$\textit{Almost surely} \qquad S_t > 0 \qquad \forall t \geq 0 \tag{2.13}$$

$$\textit{Almost surely} \qquad \sigma_t > 0 \qquad \forall t \geq 0 \tag{2.14}$$

$$\textit{Almost surely} \qquad \widetilde{\Sigma}(t, y, \theta) > 0 \qquad \forall (t, y, \theta) \in \mathbb{R}^+ \times \mathbb{R} \times \mathbb{R}^+ \tag{2.15}$$

Remark 2.2 In practice, Assumption 2.14 is not as restrictive as it might seem. As will become apparent in Sects. 2.3 and 3.4, the positivity of σ_t is in fact equivalent to the positivity of $\widetilde{\Sigma}(t, 0, 0)$, which in a broader multi-dimensional context, is equivalent to the positivity of the modulus $\| \overrightarrow{\sigma_t} \|$. Therefore, in such a framework, one could see any of the components of $\overrightarrow{\sigma_t}$ go null: as long as one component remains either strictly positive or negative, the following computations are valid.

In fact, because our approach is asymptotic, Assumption 2.13 needs only hold at the current time t. We could still express results should it be breached, but they would be trivial and without financial interest.

2.1.4 The Objectives

The objectives of this chapter are to *establish the links between the stochastic (sliding) implied volatility model and the instantaneous volatility model.*

- The *direct problem* is to derive the shape and dynamics of the implied volatility surface $\widetilde{\Sigma}$ from the value and dynamics of the instantaneous volatility σ_t.
- The *inverse problem* is to derive the value and dynamics of the instantaneous volatility σ_t from the shape and dynamics of the implied volatility surface $\widetilde{\Sigma}$.

The (static) *calibration procedure* is often viewed as an inverse problem. Indeed, when a practitioner exploits one of the many popular stochastic instantaneous volatility models, the implied volatility $\widetilde{\Sigma}$ must be marked to market in some respect, while σ_t is intrinsically model-dependent. In numerical terms, however, this inverse problem is usually solved by an optimisation process. The latter consists in minimising the market error, which itself requires numerous calls to the pricer, and is thus associated to the *direct* problem.

Let us therefore begin by examining the structural constraints of the SImpV model.

2.2 Derivation of the Zero-Drift Conditions

The stochastic implied volatility model, as defined by (2.11)–(2.12), is just an SDE system. It describes the dynamics of an infinite-dimensional state vector (the underlying S_t and the smile $\widetilde{\Sigma}(t, y, \theta)$) with no built-in notion of how they relate to each other, hence it does not intrinsically guarantee no-arbitrage, a condition that must be imposed externally.

We start by establishing the main Zero-Drift Condition, which is valid in the full domain (t, y, θ). Then we specialise it to the Immediate domain, i.e. $\theta = 0$, which provides a pair of Immediate ZDCs. Finally, we restrict ourselves to the Immediate ATM position, which is the starting point of our asymptotics.

2.2.1 The Main Zero-Drift Condition

Let us first transfer the dynamics of the sliding smile back into absolute coordinates.

Lemma 2.1 (Dynamics of the absolute implied volatility surface) *In our framework, the dynamics of the absolute implied volatility are*

$$d\Sigma(t, S_t, K, T) = b(\propto)\, dt + \nu(\propto)\, dW_t + \vec{n}(\propto)^{\perp}\, d\vec{Z}_t \tag{2.16}$$

where the coefficients are given by

$$
\begin{aligned}
b(\propto) &= \widetilde{b}(\circ) - \widetilde{\Sigma}'_\theta(\circ) + \frac{1}{2}\sigma_t^2\left[\widetilde{\Sigma}''_{yy}(\circ) + \widetilde{\Sigma}'_y(\circ)\right] - \sigma_t\,\widetilde{\nu}'_y(\circ) \\
\nu(\propto) &= \widetilde{\nu}(\circ) - \sigma_t\,\widetilde{\Sigma}'_y(\circ) \qquad (2.17)\\
\vec{n}(\propto) &= \vec{\widetilde{n}}(\circ)
\end{aligned}
$$

with absolute and sliding arguments defined as $(\circ) \triangleq (t, y, \theta)$ *and* $(\propto) \triangleq (t, S_t, K, T)$.

Proof Let us invoke the Itô-Kunita formula as in Theorem A.1. Taking $\vec{\alpha}_t \triangleq \binom{y}{\theta}$ leads to

$$dy = -\frac{1}{S_t}dS_t + \frac{1}{2}\frac{1}{S_t^2}\langle dS_t\rangle = \frac{1}{2}\sigma_t^2 dt - \sigma_t dW_t \quad \text{and} \quad \langle dy\rangle = \sigma_t^2 dt.$$

Therefore the dynamics of the absolute $\Sigma(t, S_t, K, T)$ surface are

$$
\begin{aligned}
d\Sigma &= \widetilde{b}(\circ)dt + \widetilde{\nu}(\circ)dW_t + \vec{\widetilde{n}}(\circ)^\perp d\vec{Z}_t + \widetilde{\Sigma}'_y(\circ)dy + \widetilde{\Sigma}'_\theta(\circ)d\theta \\
&\quad + \frac{1}{2}\widetilde{\Sigma}''_{yy}(\circ)\langle dy\rangle - \widetilde{\nu}'_y(\circ)\sigma_t dt \\
&= \widetilde{b}(\circ)dt + \widetilde{\nu}(\circ)dW_t + \vec{\widetilde{n}}(\circ)^\perp d\vec{Z}_t + \widetilde{\Sigma}'_y(\circ)\left[\frac{1}{2}\sigma_t^2 dt - \sigma_t dW_t\right] \\
&\quad - \widetilde{\Sigma}'_\theta(\circ)dt + \frac{1}{2}\widetilde{\Sigma}''_{yy}(\circ)\sigma_t^2 dt - \widetilde{\nu}'_y(\circ)\sigma_t dt.
\end{aligned}
$$

Then grouping the finite and non-finite variation terms provides the desired result. □

Having now moved to an absolute setup enables us to use the martingale property, and therefore to express our main result.

Proposition 2.1 (Zero Drift Condition for a single underlying) *The shape and dynamics functionals of the sliding SImpV model* (2.11)–(2.12) *are constrained by arbitrage to fulfil the following* ***Zero-Drift Condition****:*

In the general domain $(\circ) = (t, y, \theta) \in \mathbb{R}^+ \times \mathbb{R} \times \mathbb{R}^{+*}$ *we have a.s.*

$$\widetilde{\Sigma}^3(\circ)\,\widetilde{b}(\circ) = \theta D(\circ) + E(\circ) + \frac{1}{\theta}F(\circ) \qquad (2.18)$$

with

$$D(\circ) = \frac{1}{8}\widetilde{\Sigma}^4(\circ)\left[\left[\widetilde{\nu}(\circ) - \sigma_t\widetilde{\Sigma}'_y(\circ)\right]^2 + \|\vec{\widetilde{n}}\|^2(\circ)\right] \qquad (2.19)$$

$$E(\circ) = \widetilde{\Sigma}^3(\circ)\left[\widetilde{\Sigma}'_\theta(\circ) - \frac{1}{2}\sigma_t^2\widetilde{\Sigma}''_{yy}(\circ) + \sigma_t\widetilde{\mathrm{v}}'_y(\circ) - \frac{1}{2}\sigma_t\widetilde{\mathrm{v}}(\circ)\right] \tag{2.20}$$

$$F(\circ) = \frac{1}{2}\widetilde{\Sigma}^4(\circ) - \frac{1}{2}\sigma_t^2\widetilde{\Sigma}^2(\circ) - y\sigma_t\widetilde{\Sigma}(\circ)\left[\widetilde{\mathrm{v}}(\circ) - \sigma_t\widetilde{\Sigma}'_y(\circ)\right]$$
$$- \frac{1}{2}y^2\left[\left[\widetilde{\mathrm{v}}(\circ) - \sigma_t\widetilde{\Sigma}'_y(\circ)\right]^2 + \|\vec{\widetilde{n}}\|^2(\circ)\right]. \tag{2.21}$$

Proof In order to obtain the Call price dynamics, we start by applying Itô's Lemma to the normalised Black-Scholes functional (2.3). For the sake of clarity, we omit the multiple dependencies of the implied volatility Σ, of the normalised Greeks Δ ("Delta"), Γ ("Gamma"), $\mathscr{V}$ ("Vega"), Λ ("Volga"), ϑ ("Vanna") and of the absolute diffusion coefficients b, v, n. We obtain simply:

$$\begin{aligned} dC(t, S_t, T, K) = & -\mathscr{V}\Sigma(2\sqrt{\theta})^{-1}\mathbf{dt} + \Delta S_t\sigma_t dW_t \\ & + \mathscr{V}\sqrt{\theta}\left[b\mathbf{dt} + \mathrm{v}dW_t + \vec{n}^{\perp}d\vec{Z}_t\right] \\ & + \frac{1}{2}\Gamma S_t^2\sigma_t^2\mathbf{dt} + \frac{1}{2}\theta\vartheta\left(\mathrm{v}^2 + \|\vec{n}\|^2\right)\mathbf{dt} + \Lambda\sqrt{\theta}S_t\sigma_t\mathrm{v}\mathbf{dt} \end{aligned} \tag{2.22}$$

with

$$\Delta = \frac{\partial C^{BS}}{\partial x} \qquad \mathscr{V} = \frac{\partial C^{BS}}{\partial v} \qquad \Gamma = \frac{\partial^2 C^{BS}}{\partial x^2} \qquad \Lambda = \frac{\partial^2 C^{BS}}{\partial x \partial v} \qquad \vartheta = \frac{\partial^2 C^{BS}}{\partial v^2}.$$

The No-Arbitrage Assumption forces C_t into a martingale under $\mathbb{Q}^N$, so that (2.22) leads classically to the following zero-drift condition:

$$\begin{aligned} 0 = & -\mathscr{V}\Sigma(2\sqrt{\theta})^{-1} + \mathscr{V}\sqrt{\theta}b + \frac{1}{2}\Gamma S_t^2\sigma_t^2 \\ & + \frac{1}{2}\theta\vartheta\left(\mathrm{v}^2 + \|\vec{n}\|^2\right) + \Lambda\sqrt{\theta}S_t\sigma_t\mathrm{v}. \end{aligned} \tag{2.23}$$

Computing the normalised Greeks $\mathscr{V}$, Γ, ϑ and Λ involved in (2.23) presents no difficulty, as detailed in Appendix C. Factorising with the Vega, the resulting expressions are given by

$$\mathscr{V} = S_t\mathcal{N}'(d_1) \qquad \vartheta = \mathscr{V}\left[\frac{y^2}{\left(\Sigma\sqrt{\theta}\right)^3} - \frac{1}{4}\Sigma\sqrt{\theta}\right]$$

$$\Gamma = \frac{\mathscr{V}}{S_t^2\Sigma\sqrt{\theta}} \qquad \Lambda = S_t^{-1}\mathscr{V}\left[\frac{1}{2} + \frac{y}{\Sigma^2\theta}\right]$$

Note that these expressions are well-defined and finite, thanks to the technical assumptions (2.13) and (2.15). Note also that, although the normalised Delta and Gamma correspond to the same expressions as for the classical BS formula, the derivatives in v (namely Vega, Vanna and Volga) differ slightly from their classical expressions. Note finally that here θ is NOT the Greek associated to time t!

Substituting the Greeks, the zero-drift condition becomes

$$
\begin{aligned}
0 = \mathcal{V}\Bigg[& -\frac{\Sigma}{2\sqrt{\theta}} + \sqrt{\theta} b + \frac{1}{2}\frac{1}{\Sigma\sqrt{\theta}}\sigma_t^2 \\
& + \frac{1}{2}\theta\left[\frac{y^2}{\Sigma^3\theta^{\frac{3}{2}}} - \frac{1}{4}\Sigma\sqrt{\theta}\right]\left(\nu^2 + \|\vec{n}\|^2\right) + \left[\frac{1}{2} + \frac{y}{\Sigma^2\theta}\right]\sqrt{\theta}\sigma_t\nu\Bigg].
\end{aligned}
$$

Using the strict positivity of $\mathcal{V}$ and of θ (since we are restricted to $\Theta \in \mathbb{R}^{+*}$) we divide both sides by the product $\mathcal{V}\sqrt{\theta}$, then isolate $b(t, y, \theta)$ and end up with

$$
b = \frac{1}{2}\frac{\Sigma}{\theta} - \frac{1}{2}\frac{1}{\Sigma\theta}\sigma_t^2 - \frac{1}{2}\left[\frac{y^2}{\Sigma^3\theta} - \frac{1}{4}\Sigma\theta\right]\left(\nu^2 + \|\vec{n}\|^2\right) - \left[\frac{1}{2} + \frac{y}{\Sigma^2\theta}\right]\sigma_t\nu.
$$

We can now use Lemma 2.1 and replace the absolute processes with their sliding counterparts. We get the sliding drift as follows, where all processes are evaluated in (t, y, θ) and with terms gathered by power of θ:

$$
\begin{aligned}
\widetilde{b} = \theta & \left[\frac{1}{8}\Sigma\left[\left(\widetilde{\nu} - \sigma_t\widetilde{\Sigma}'_y\right)^2 + \|\vec{n}\|^2\right]\right] \\
& + \left[\widetilde{\Sigma}'_\theta - \frac{1}{2}\sigma_t^2\left(\widetilde{\Sigma}''_{yy} + \widetilde{\Sigma}'_y\right) + \sigma_t\widetilde{\nu}'_y - \frac{1}{2}\sigma_t\left(\widetilde{\nu} - \sigma_t\widetilde{\Sigma}'_y\right)\right] \\
& + \frac{1}{\theta}\left[\frac{1}{2}\Sigma - \frac{1}{2}\frac{\sigma_t^2}{\Sigma} - \frac{1}{2}\frac{y^2}{\Sigma^3}\left[\left(\widetilde{\nu} - \sigma_t\widetilde{\Sigma}'_y\right)^2 + \|\vec{n}\|^2\right] - \frac{y}{\Sigma^2}\sigma_t\left(\widetilde{\nu} - \sigma_t\widetilde{\Sigma}'_y\right)\right].
\end{aligned}
$$

Finally, simplifying the second bracket and multiplying the whole expression by $\widetilde{\Sigma}^3(t, y, \theta)$ provides the desired expression (2.18), which concludes the proof. □

The ZDC (2.18) links the four parametric processes—all bi-dimensional—describing the sliding IV: the shape $\widetilde{\Sigma}$ and the dynamic coefficients $\widetilde{b}, \widetilde{\nu}$ and $\vec{\widetilde{n}}$. We emphasise that this relationship is valid in the full domain (y, θ) as opposed to most of the coming *asymptotic* results, which are either Immediate ($\theta = 0$) or IATM ($y = 0, \theta = 0$). We note the positivity of the highest θ-order term $D(t, y, \theta)$, a property which will prove useful in the sequel. The recurrence of the term

$$
[\widetilde{\nu} - \sigma_t\widetilde{\Sigma}'_y]
$$

is also remarkable, although this term is no stranger to us. Indeed, through (2.17) it is identified as

$$\nu(t, S_t, K, T)$$

which is the endogenous volatility of the *absolute* stochastic IV surface. In other words, the compensation term

$$\sigma_t \widetilde{\Sigma}'_y(\circ)$$

appears to neutralise the space slide associated with movements of the underlying S_t. It would be wrong, however, to assume that it removes *all* dependency on S_t, i.e. that it corresponds to some *unconditional* endogenous volatility of the smile. Indeed, the absolute coefficient ν itself can very well incorporate a local component. We note then that the term

$$\left[\widetilde{\nu} - \sigma_t \widetilde{\Sigma}'_y\right]^2 + \| \vec{n} \|^2$$

which appears in $D(\circ)$ and $F(\circ)$ represents the *quadratic variation* of the stochastic *absolute* IV surface $\Sigma(t, S_t, K, T)$, and is linked to its *sliding* counterpart with

$$\nu^2 + \| \vec{n} \|^2 = \left[\widetilde{\nu} - \sigma_t \widetilde{\Sigma}'_y\right]^2 + \| \vec{n} \|^2 = \widetilde{\nu}^2 + \| \vec{n} \|^2 - \sigma_t \widetilde{\Sigma}'_y \left[2\widetilde{\nu} - \sigma_t \widetilde{\Sigma}'_y\right].$$

In that expression, we will see that IATM (i.e. with $y = 0$ and $\theta = 0$) the last term converges to a *negative* correction (see (2.51) [p. 62]). We cannot stress enough that the stochastic PDE (2.18) constitutes the actual basis for most of the subsequent asymptotic expressions: manipulating the ZDC (i.e. differentiating w.r.t. y and θ) then imposing some regularity assumptions, and finally taking the limits in $(t, 0, 0)$, forms the backbone of the ACE methodology.

An interesting insight into the ZDC structure can be given by exploiting the concept of *local volatility* (LV) as per [27]. Indeed, the LV surface $f_{t,S_t}(K, T)$ can be seen as a re-parametrisation of the absolute IV surface $\Sigma_{t,S_t}(K, T)$ *via* an auxiliary diffusion process. But its square is also interpreted as the expectation of the instantaneous variance σ_t^2 at a given future date T, conditional on the underlying's value $(S_T = K)$. In this context we can switch seamlessly to the sliding coordinates y and θ, defining the equivalent *sliding local volatility* as

$$\widetilde{f}(t, y, \theta) \triangleq f_{t,S_t}(K, T) \qquad \text{so that} \qquad \widetilde{f}^2(t, y, \theta) = \mathbb{E}_t\left[\sigma_{t+\theta}^2 | S_{t+\theta} = S_t e^y\right],$$

which allows us to naturally isolate the *relative* local variance $\xi_t(t, y, \theta)$ with

$$\xi_t(t, y, \theta) \triangleq \widetilde{f}^2(t, y, \theta) - \sigma_t^2 = \mathbb{E}_t\left[\int_t^{t+\theta} d\sigma_u^2 | S_{t+\theta} = S_t e^y\right].$$

We now get slightly ahead of ourselves, by introducing a topic which will be covered in more detail later in this chapter (see p. 66). Indeed, we invoke the classical re-parametrisation of Dupire's formula in terms of Lognormal implied volatility (which can be found in [28]):

$$\widetilde{f}^2(t,y,\theta) = \frac{\widetilde{\Sigma}^4(\circ) + 2\theta\,\widetilde{\Sigma}^3\,\widetilde{\Sigma}'_\theta(\circ)}{\left[\widetilde{\Sigma}(\circ) - y\widetilde{\Sigma}'_y(\circ)\right]^2 - \frac{1}{4}\theta^2\widetilde{\Sigma}^4\,\widetilde{\Sigma}'^{\,2}_y(\circ) + \theta\,\widetilde{\Sigma}^3\,\widetilde{\Sigma}''_{yy}(\circ)},$$

where $(\circ) = (t,y,\theta)$ still denotes the whole domain. First this expression rewrites as

$$\begin{aligned}\widetilde{\Sigma}^3\,\widetilde{\Sigma}'_\theta(\circ) = & -\frac{1}{8}\theta\,\widetilde{f}^2\,\widetilde{\Sigma}^4\,\widetilde{\Sigma}'^{\,2}_y(\circ) + \frac{1}{2}\widetilde{f}^2\,\widetilde{\Sigma}^3\,\widetilde{\Sigma}''_{yy}(\circ) \\ & + \frac{1}{\theta}\frac{1}{2}\widetilde{f}^2\left[\widetilde{\Sigma} - y\widetilde{\Sigma}'_y\right]^2(\circ) - \frac{1}{\theta}\frac{1}{2}\widetilde{\Sigma}^4(\circ).\end{aligned}$$

We identify the l.h.s. as the first component of the term $E(t,y,\theta)$ in (2.26). Hence substituting the r.h.s. into the main ZDC (2.18) leads to

$$\widetilde{\Sigma}^3\widetilde{b}(\circ) = \theta D^*(\circ) + E^*(\circ) + \frac{1}{\theta}F^*(\circ) \tag{2.24}$$

with

$$D^*(\circ) = \frac{1}{8}\widetilde{\Sigma}^4 G^\star(\circ) \tag{2.25}$$

$$E^*(\circ) = \widetilde{\Sigma}^3(\circ)\left[\frac{1}{2}\xi_t\widetilde{\Sigma}''_{yy}(\circ) + \sigma_t\widetilde{\mathrm{v}}'_y(\circ) - \frac{1}{2}\sigma_t\widetilde{\mathrm{v}}(\circ)\right] \tag{2.26}$$

$$F^*(\circ) = \frac{1}{2}\xi_t\widetilde{\Sigma}^2(\circ) - y\widetilde{\Sigma}(\circ)\left[\sigma_t\widetilde{\mathrm{v}}(\circ) + \xi_t\widetilde{\Sigma}'_y(\circ)\right] - \frac{1}{2}y^2G^\star(\circ) \tag{2.27}$$

where

$$G^\star(\circ) \triangleq \widetilde{\mathrm{v}}\left[\widetilde{\mathrm{v}} - 2\sigma_t\widetilde{\Sigma}'_y\right](\circ) - \xi_t\,\widetilde{\Sigma}'^{\,2}_y(\circ) + \|\vec{\widetilde{n}}\|^2(\circ),$$

which emphasises the pivotal role played by the relative local variance $\xi_t(t,y,\theta)$.

2.2.2 The Immediate Zero Drift Conditions

Let us now focus on the *immediate smile*, which is the limit process (assuming its existence) of the implied volatility surface when time-to-maturity θ tends to 0. In other words, or rather in market terms, the limit of $\widetilde{\Sigma}(t,y,\theta)$ when $\theta \searrow 0$ is the implied volatility of a Call option maturing tomorrow, or even in a few hours.

In this context, the assumption of an option continuum (see Sect. 2.1.2.1 [p. 25]) is vital, but it also brings some practical issues. Indeed, *interpolating* between existing Call prices (in order to "fill" the map) is one thing, but *extrapolating* from the liquid Call with the shortest expiry, down to $\theta = 0$, is another modelling and technical problem altogether.

Besides, beyond the continuum and extrapolation hypothesis, we clearly need to assume some additional regularity, in order to ensure the *existence* of a limit. Effectively the ZDC has been established for strictly positive time-to-maturities only, i.e. $(t, y, \theta) \in \mathbb{R}^+ \times \mathbb{R} \times \mathbb{R}^{+*}$. This was due to the use of Black's formula within the proof, as a medium between the price functional (necessarily static in $\theta = 0$, since it identifies with the payoff function) and the implied volatility (which can *a priori* afford some dynamics in the "immediate" area $\theta = 0$). Now we need to somehow extend the ZDC into this asymptotic area, which is why we introduce the following sufficient (strong) conditions:

Assumption 2.2 (*Immediate regularity*) Each process $\widetilde{\Sigma}$, $\widetilde{\Sigma}'_y$, $\widetilde{\Sigma}''_{yy}$, $\widetilde{\Sigma}'_\theta$, $\widetilde{b}$, $\widetilde{\mathrm{v}}$, $\widetilde{\mathrm{v}}'_y$ and $\widetilde{n}$ admits its own finite (stochastic) limit when $\theta \searrow 0$. These limits are (abusively) denoted with argument $(t, y, 0)$.

If this assumption *package* might appear blunt in the mathematical sense, this is not the case in modelling (i.e. financial) terms. It is true that implied and realised volatilities tend to "pick up" shortly before expiry, often due to the *pinning* effect on very liquid maturities/strikes (see [29] for instance). But to our knowledge such behaviour does not warrant any assumption of instability or of explosion at expiry, as is observed with deltas when a barrier option knocks in/out. For all intents and purposes, at vanishing maturities all our smiles can realistically be considered smooth and well-behaved, in their statics and in their dynamics.

Corollary 2.1 (The Immediate Zero-Drift Conditions) *As a consequence of the ZDC* (2.18), *shape and dynamics functionals of the sliding SImpV model* (2.11)–(2.12) *are constrained by arbitrage to respect the following two equivalent Immediate Zero-Drift Conditions (IZDCs):*

In the Immediate domain $(\bullet) = (t, y, 0)$ *we have a.s.*

- *The Primary IZDC:*

$$0 = F(t, y, 0) = \widetilde{\Sigma}^4(\bullet) - \sigma_t^2 \widetilde{\Sigma}^2(\bullet) - y 2\sigma_t \widetilde{\Sigma}(\bullet) \left[\widetilde{\mathrm{v}}(\bullet) - \sigma_t \widetilde{\Sigma}'_y(\bullet)\right]$$
$$- y^2 \left[\| \vec{\widetilde{n}}\,(\bullet) \|^2 + \left[\widetilde{\mathrm{v}}(\bullet) - \sigma_t \widetilde{\Sigma}'_y(\bullet)\right]^2 \right]. \qquad (2.28)$$

- *The Secondary IZDC:*

$$\widetilde{\Sigma}^3 \widetilde{b}(t, y, 0) = E(t, y, 0) + F'_\theta(t, y, 0). \qquad (2.29)$$

Proof Assumption 2.2 applied to the ZDC (2.18) implies that both its l.h.s. and r.h.s. admit a finite limit. Then considering the last term $\theta^{-1}F(t, y, \theta)$ alone leads to

$$\lim_{\theta \searrow 0} F(t, y, \theta) = 0, \tag{2.30}$$

which proves the Primary IZDC (2.28). Let us now compute the limit of $\widetilde{b}(t, y, \theta)$ when $\theta \searrow 0$. Using a small-θ expansion on $F(t, y, \theta)$ we get that

$$F(t, y, \theta) = F(t, y, 0) + \theta F'_\theta(t, y, 0) + O(\theta^2). \tag{2.31}$$

In light of the Primary IZDC (2.28), and since *a priori* $F'_\theta(t, y, 0)$ is non-null, L'Hopital's rule grants us therefore that

$$\lim_{\theta \searrow 0} \quad \frac{1}{\theta} F(t, y, \theta) = F'_\theta(t, y, 0), \tag{2.32}$$

so that, still in the general domain (t, y, θ), the ZDC (2.18) rewrites as:

$$\widetilde{\Sigma}^3 \widetilde{b}(t, y, \theta) = \theta D(t, y, \theta) + E(t, y, \theta) + F'_\theta(t, y, 0) + O(\theta). \tag{2.33}$$

Invoking now the regularity package of Assumption 2.2, we can take the limit of (2.33) in $\theta = 0$, which provides the Secondary IZDC (2.29). □

Remark 2.3 (The "true" IZDC) Note that the Secondary Immediate ZDC (2.29) can be considered the legitimate heir of the main ZDC (2.18) in the immediate domain $(t, y, 0)$, whereas what we call the Primary IZDC (2.28) is simply induced by the regularity assumptions.

Nevertheless, for our purposes the latter is simultaneously more practical and more intuitive, as it involves a smaller group of functionals. Indeed, we can observe that in (2.28) the processes $\widetilde{b}$, $\widetilde{\Sigma}'_\theta$, $\widetilde{\Sigma}''_{yy}$ and $\widetilde{\mathrm{v}}'_y$ play no direct role, by opposition to the general ZDC (2.18) and the Secondary Immediate ZDC (2.29).

Consequently, in the sequel the denomination IZDC will always refer, unless otherwise specified, to the Primary version (2.28).

Interestingly, the limit result (2.30) also provides us with a lower bound for the convergence speed of $F(t, y, \theta)$, which is faster than θ as that time-to-maturity tends to zero. Note, however, that this statement is conditional on the regularity Assumption 2.2, which provides the *existence* of a finite limit for $F(t, y, \theta)$ in the first place.

Looking at the IZDC (2.28) we note that, should the sliding smile $\widetilde{\Sigma}(t, y, \theta)$, in the Immediate domain, either be static (in y and θ) or exhibit uniform volatilities $\widetilde{\mathrm{v}}(\bullet)$ and $\widetilde{n}(\bullet)$, then the IZDC would be consistent with a parabolic expression for the Immediate Smile. Indeed, (2.28) can be rewritten as

$$\widetilde{\Sigma}^4(\bullet) = \underbrace{\left[\sigma_t \widetilde{\Sigma}(\bullet) + y\left[\widetilde{\mathrm{v}}(\bullet) - \sigma_t \widetilde{\Sigma}'_y(\bullet)\right]\right]^2}_{\text{Endogenous}} + \underbrace{y^2 \|\vec{\widetilde{n}}(\bullet)\|^2}_{\text{Exogenous}}. \tag{2.34}$$

In our view, this expression stresses the role of the exogenous volatility $\overrightarrow{n}()$ as a generator of pure smile convexity. Indeed, there is no interaction with a y term, therefore there cannot be any influence on the skew. Recall also from (2.17) that we can simplify the term $[\widetilde{\nu}(\bullet) - \sigma_t \widetilde{\Sigma}'_y(\bullet)]$ into the absolute endogenous coefficient $\nu(t, S_t, K, T = t)$.

Hence (2.34) can be seen as a "Pythagorean" relationship: in the Immediate domain, the square of the implied "variance" $\widetilde{\Sigma}^2$ is the sum of two quadratic and orthogonal terms: one endogenous, the other exogenous. Note also that log-moneyness y acts as "leverage" so that the endogenous term in (2.34) can be interpreted as the endogenous move created at $y = 0$ (i.e. IATM) plus a "torque" proportional to y.

The information provided by the IZDC through (2.34) is actually richer. For simplicity's sake, let us restrict ourselves to purely endogenous models, which include (but are not restricted to) local volatility models. Given the regularity of all process involved, and in particular assuming that (2.15) extends to the immediate domain, this implies in turn that the sign of the bracket above must stay positive:

$$\widetilde{\Sigma}^2(\bullet) = \sigma_t \widetilde{\Sigma}(\bullet) + y\left[\widetilde{\nu}(\bullet) - \sigma_t \widetilde{\Sigma}'_y(\bullet)\right] \geq 0.$$

We can then divide both sides by $\sigma_t \widetilde{\Sigma}^2(t, y, 0)$ to obtain

$$\frac{1}{\sigma_t} = \left[\frac{1}{\widetilde{\Sigma}(\bullet)} - y\frac{\widetilde{\Sigma}'_y}{\widetilde{\Sigma}^2}(\bullet)\right] + y\frac{\widetilde{\nu}(\bullet)}{\sigma_t \widetilde{\Sigma}^2(\bullet)} = \left[\frac{y}{\widetilde{\Sigma}(\bullet)}\right]'_y + y\frac{\widetilde{\nu}(\bullet)}{\sigma_t \widetilde{\Sigma}^2(\bullet)}. \tag{2.35}$$

It is interesting to compare the above expression with similar results in the pure local volatility (LV) case. For instance—and again getting ahead of ourselves—[30] provides the Immediate implied volatility as the *harmonic mean* of the local volatility, over the $[S_t, K]$ segment (see (2.62) [p. 67]). Hence by integrating (2.35) we can interpret the bracket on the r.h.s. as some *local* component.

2.2.3 The IATM Identity

Having expressed the ZDC in the general domain, and then specialised that no-arbitrage constraint to the Immediate domain, it is time to focus further and quote the well-known relationship between instantaneous and implied volatility when taken at the IATM point.

Corollary 2.2 (The IATM Identity) *As a consequence of the IZDC* (2.28) *and hence by arbitrage:*

At the Immediate ATM (IATM) point $(\star) \triangleq (t, y = 0, \theta = 0)$ *we have a.s.*

$$\sigma_t = \widetilde{\Sigma}(t, 0, 0). \tag{2.36}$$

Proof We simply take the Immediate Zero-Drift Condition (2.28) at the origin point $(t, 0, 0)$, which immediately gives us the fundamental identity. □

The fundamental identity (2.36) in itself is not new, although it has usually been stated in the case of the *absolute* implied volatility surface: see [12], for instance. In practice, this feature of any SInsV model is commonly used in order to "complete the market", under the assumption that the exogenous driver Z_t is scalar.

In more general terms, hedging stochastic instantaneous volatility can be achieved by incorporating a well-chosen European option in the replication portfolio (refer to [31], for instance). In that respect, selecting a short-dated ATM option often makes sense, for liquidity reasons. Indeed, the At-The-Money Call maturing the soonest tend to be very liquid, if not the most liquid, within the whole price surface. Trading this Call is the natural proxy for trading $\widetilde{\Sigma}(t, 0, 0)$ and therefore hedging σ_t.

Another way to comprehend the IATM identity (2.36) is to consider the Gamma-Theta trading of that same option: the implied volatility determines the time decay, a.k.a. Theta, while the instantaneous volatility conditions the Gamma. Therefore, should we not have (2.36), a very obvious (and exploitable) arbitrage opportunity would arise.

Remark 2.4 (Minimal specification of the SImpV model) A consequence of the fundamental result (2.36) is that the (sliding) implied volatility model presented in Sect. 2.1 can be rewritten in a sparser way. Indeed, denoting the generic point by $(\circ) = (t, y, \theta)$ we have

$$\begin{cases} \dfrac{dS_t}{S_t} = \widetilde{\Sigma}(\mathbf{t}, \mathbf{0}, \mathbf{0}) dW_t \\[2ex] d\widetilde{\Sigma}(\circ) = \widetilde{\Sigma}^{-3} \left[\theta D + E + \dfrac{1}{\theta} F \right] (\circ) dt + \widetilde{\mathrm{v}}(\circ) dW_t + \vec{n}(\circ)^{\perp} d\vec{Z}_t. \end{cases}$$

In other words, the inclusion of the stochastic instantaneous volatility σ_t within the definition of the sliding implied volatility model is redundant. Indeed, the specification of the sole stochastic map $\widetilde{\Sigma}(t, y, \theta)$ formally includes its asymptotics, provided that we assume finite limits in $\theta = 0$. Therefore the statics and dynamics of the V surface entirely determine the SImpV model.

This configuration shows (again) strong similarities with the HJM framework, where the dynamics of the Zero-Coupon are

$$\frac{dB_t(T)}{B_t(T)} = r_t dt + \vec{\Gamma}_t(T) d\vec{W}_t$$

and where the drift coefficient r_t, which is the short rate, is itself already (asymptotically) included in the curve input:

$$r_t = -\lim_{T \searrow t} \partial_T \ln \left[B_t(T) \right].$$

Also, it is worth recalling that the above interest rates result is, likewise, the structural consequence of no-arbitrage constraints, rather than modelling choices.

2.2.4 Synthesis and Overture

To summarise, we have so far expressed the ZDC (2.18) which, invoking regularity assumptions, we have then *specialised* to increasingly restrictive asymptotic domains: first in $\theta = 0$ with the IZDCs (2.28) and (2.29), then at the IATM point with the IATM identity (2.36).

A natural comment would be that other asymptotic domains deserve to be explored, in particular *extreme strikes* (as in $y = \pm\infty$), a topic which is covered for instance in [32, 33] or [34]. Although we have investigated this subject, this book will address it only from a numerical perspective, in Chap. 4.

Going back to the ZDC and its asymptotic corollaries, let us recall that our objective is to solve the direct and/or the inverse problem. In that respect we observe that the ZDC and Immediate ZDCs, while carrying more information (being valid in wider domains) than the IATM identity, seem to provide little information w.r.t. either problem. By contrast, based on the IATM Identity (2.36) and on Remark 2.4, it is tempting to assume that the SInsV model is actually *embedded* into the apparently richer SImpV framework. Indeed, the dynamics of the SImpV smile are specified in every individual point of the (y, θ) map, and can therefore describe complex deformation modes. If such was the case, then the inverse problem would become trivial, while the direct one would become definitively ill-posed.

In fact, this unilateral embedding *is* verified in certain conditions, dependent in particular on a low dimensionality. But we shall prove that, when it occurs, this inclusion is part of a wider *equivalence* (or bijection) which can be established between the two classes. Furthermore, and conversely, it turns out that the *direct* problem will in general be easier to solve, so that the embedding is in fact the other way round.

This apparent subordination of the SImpV framework to the SInsV class probably sounds counterintuitive. Its main justification is the very different kind of specification that we have used for each model. Indeed, the dynamics of the SInsV class are defined *in depth* via the chaos, while those of the SImpV framework are defined *in domain* by parametric processes. Let us now detail and contrast the two models in that respect.

Within the apparently simpler SInsV model, the value and the dynamics of the IATM volatility σ_t are *freely specifiable* using the Wiener expansion, to any required depth. So far, this is an SDE system describing adapted dynamics for only two financial instruments. To build the whole *market model* the arbitrage condition must *then* be invoked (Call prices as conditional expectations) which entirely determines the smile, both in its shape and dynamics. Unfortunately this information is, in general, not explicitly available.

By contrast, within the SImpV model the SDE specification is sparser and concerns the whole market, but it *must* be envisaged *along* its companion ZDC to carry any financial relevance, which brings a series of questions and remarks.

First, the dynamics of the SImpV coefficients ($\widetilde{b}, \widetilde{\nu}$ and $\overrightarrow{\widetilde{n}}$) have not been provided. So a generic-depth Wiener chaos expansion of these parametric processes seems *a fortiori* out of context. Hence, is a match to the SInsV class impossible, making both the direct and inverse problems nonsensical? The answer is no, as we will see that at the IATM point the *chaos* specification of $\widetilde{\Sigma}$ alone can translate into a *differential set* of all four SImpV functionals.

The second question is the degree of redundancy among these four SImpV functionals, in the perspective of *actual model parametrisation*: which are our degrees of freedom? Indeed, the ZDC is a stochastic PDE invoking multiple processes, so we must select a single dependent variable. The drift seems the best candidate, which explains why we presented the ZDC with $\widetilde{b}$ on the l.h.s. The endogenous coefficient $\widetilde{\nu}$ appears both squared and differentiated, while the exogenous coefficient $\overrightarrow{\widetilde{n}}$ only exposes its modulus, so that even in a bi-dimensional framework both are more difficult to infer. Finally, the shape functional $\widetilde{\Sigma}$ seems a non-starter, as having to solve dynamically a non-linear parabolic PDE is not an encouraging prospect.

The next subject is whether we have overlooked any more SImpV constraints. There are indeed some well-defined global restrictions, corresponding to the usual *validity conditions* of the smile (intra- and inter-expiry) satisfied statically but *a.s.* and at any time. We have intentionally not exploited these stochastic partial differential *inequalities*, because from an asymptotic perspective they do not bring additional *and* relevant information. They will, however, intervene in the whole-smile extrapolations covered in Chap. 4.

To link these two models and solve both the direct and the inverse problems, we must now manipulate further the main ZDC, and focus on the IATM point where the SImpV constraints are maximal. Indeed, the core principle of ACE is inductive: it involves cross-differentiating that stochastic PDE, and then taking its IATM limit under sufficient regularity assumptions.

2.3 Recovering the Instantaneous Volatility: The First Layer

After establishing some local constraints of the SImpV model, we show how to recover the associated SInsV model, then comment on and interpret this rich relationship.

2.3.1 Computing the Dynamics of σ_t

Our ultimate objective in this section is to offer a solution to the *inverse problem*, which is to *recover* the value and dynamics of the instantaneous stochastic volatility

σ_t, associated to a given SImpV model. In fact the achievements of this section will be relatively modest since, although the basics of the methods will be laid down, in effect we will only access the top level coefficients in the chaos expansion describing the dynamics of σ_t, which accounts for the mention of a "first layer".

First we derive a collection of IATM arbitrage constraints for the SImpV class.

Proposition 2.2 (IATM arbitrage constraints of the SImpV model: first layer) *Let us consider a given sliding SImpV model, as defined by* (2.11)–(2.12). *Then its dynamic coefficients are locally constrained to satisfy, at the IATM point* $(t, 0, 0)$*:*

$$\widetilde{\nu}(\star) = 2\sigma_t \widetilde{\Sigma}'_y(\star) \tag{2.37}$$

$$\widetilde{\nu}'_y(\star) = \widetilde{\Sigma}'^2_y(\star) + \frac{3}{2}\sigma_t \widetilde{\Sigma}''_{yy}(\star) - \frac{1}{2}\frac{\|\vec{\widetilde{n}}(\star)\|^2}{\sigma_t^2} \tag{2.38}$$

$$\widetilde{b}(\star) = 2\widetilde{\Sigma}'_\theta(\star) - \frac{1}{2}\sigma_t^2 \widetilde{\Sigma}''_{yy}(\star) + \sigma_t \widetilde{\nu}'_y(\star) - \frac{1}{2}\sigma_t \widetilde{\nu}(\star). \tag{2.39}$$

Proof Since we are not interested here in any θ-differential, we can take the limit of the ZDC in $\theta = 0$ first, before applying any y-differentiation. In other words, we can deal directly with the simpler IZDC instead.

Computation of $\widetilde{\nu}(t, 0, 0)$

Let us differentiate the Immediate Zero-Drift Condition (2.28) once w.r.t. y. Omitting the arguments by assuming that all functionals are taken in $(t, y, 0)$, we obtain

$$\begin{aligned}
0 = {} & 4\widetilde{\Sigma}^3 \widetilde{\Sigma}'_y - 2\sigma_t^2 \widetilde{\Sigma}\widetilde{\Sigma}'_y - 2\sigma_t \widetilde{\Sigma}\left[\widetilde{\nu} - \sigma_t \widetilde{\Sigma}'_y\right] \\
& - 2y\left[\sigma_t \widetilde{\Sigma}'_y\left(\widetilde{\nu} - \sigma_t \widetilde{\Sigma}'_y\right) + \sigma_t \widetilde{\Sigma}\left(\widetilde{\nu}'_y - \sigma_t \widetilde{\Sigma}''_{yy}\right) + \left(\widetilde{\nu} - \sigma_t \widetilde{\Sigma}'_y\right)^2 + \|\vec{\widetilde{n}}\|^2\right] \\
& - 2y^2\left[\left(\widetilde{\nu} - \sigma_t \widetilde{\Sigma}'_y\right)\left(\widetilde{\nu}'_y - \sigma_t \widetilde{\Sigma}''_{yy}\right) + \vec{\widetilde{n}}^{\perp}\vec{\widetilde{n}}'_y\right].
\end{aligned}$$

Evaluating this at $(\star) \triangleq (t, 0, 0)$ and using the IATM identity (2.36), we get

$$0 = 4\sigma_t^2 \widetilde{\Sigma}'_y(\star) - 2\sigma_t^2 \widetilde{\Sigma}'_y(\star) - 2\sigma_t\left[\widetilde{\nu}(\star) - \sigma_t \widetilde{\Sigma}'_y(\star)\right],$$

which after simplification proves (2.37).

Computation of $\widetilde{\nu}'_y(t, 0, 0)$

Differentiating the IZDC (2.28) twice w.r.t. y yields, with all functionals in $(t, y, 0)$:

$$\begin{aligned}
0 = {} & 12\widetilde{\Sigma}^2\, \widetilde{\Sigma}'^2_y + 4\widetilde{\Sigma}^3 \widetilde{\Sigma}''_{yy} - 4\sigma_t \widetilde{\Sigma}\widetilde{\nu}'_y + 2\sigma_t^2 \widetilde{\Sigma}\widetilde{\Sigma}''_{yy} - 2\left(\widetilde{\nu}^2 + \|\vec{\widetilde{n}}\|^2\right) \\
& - 2y\left[\sigma_t \widetilde{\Sigma}\widetilde{\nu}''_{yy} - 2\sigma_t \widetilde{\Sigma}'_y \widetilde{\nu}'_y - 3\sigma_t \widetilde{\Sigma}''_{yy}\widetilde{\nu} + \sigma_t^2 \widetilde{\Sigma}'_y \widetilde{\Sigma}''_{yy} - \sigma_t^2 \widetilde{\Sigma}\widetilde{\Sigma}'''_{yyy} \right.\\
& \left. \qquad + 4\left(\widetilde{\nu}\widetilde{\nu}'_y + \vec{\widetilde{n}}^{\perp}\vec{\widetilde{n}}'_y\right)\right]
\end{aligned}$$

$$-2y^2\left[\left(\widetilde{\mathrm{v}}'_y-\sigma_t\widetilde{\Sigma}''_{yy}\right)^2+\left(\widetilde{\mathrm{v}}-\sigma_t\widetilde{\Sigma}'_y\right)\left(\widetilde{\mathrm{v}}''_{yy}-\sigma_t\widetilde{\Sigma}'''_{yyy}\right)+\|\vec{\widetilde{n}}'_y\|^2+\vec{\widetilde{n}}^{\perp}\vec{\widetilde{n}}''_{yy}\right].$$

Evaluating this at $(\star)=(t,0,0)$, using (2.36) and (2.37), we get

$$0=4\sigma_t^2\,\widetilde{\Sigma}'^{\,2}_y(\star)+6\sigma_t^3\,\widetilde{\Sigma}''_{yy}(\star)-4\sigma_t^2\widetilde{\mathrm{v}}'_y(\star)-2\|\vec{\widetilde{n}}(\star)\|^2,$$

which after simplification gives (2.38).

Computation of $\widetilde{b}(t,0,0)$
Evaluating the Secondary IZDC (2.29) at the IATM point yields simply

$$\widetilde{\Sigma}^3\widetilde{b}(\star)=E(\star)+F'_\theta(\star). \tag{2.40}$$

Furthermore, the definition of $F(t,y,\theta)$ (2.21) leads to the differential

$$F'_\theta(t,y,\theta)=2\widetilde{\Sigma}^3\,\widetilde{\Sigma}'_\theta(\circ)-\sigma_t^2\,\widetilde{\Sigma}\,\widetilde{\Sigma}'_\theta(\circ)-y\,[\cdots]-\frac{1}{2}y^2\,[\cdots].$$

In particular, at the IATM point we have, using (2.36):

$$F'_\theta(\star)=2\widetilde{\Sigma}^3\,\widetilde{\Sigma}'_\theta(\star)-\widetilde{\Sigma}^3\,\widetilde{\Sigma}'_\theta(\star)=\widetilde{\Sigma}^3\,\widetilde{\Sigma}'_\theta(\star). \tag{2.41}$$

Substituting (2.41) into (2.40) we get

$$\widetilde{b}(\star)=2\widetilde{\Sigma}'_\theta(\star)-\frac{1}{2}\sigma_t^2\,\widetilde{\Sigma}''_{yy}(\star)+\sigma_t\widetilde{\mathrm{v}}'_y(\star)-\frac{1}{2}\sigma_t\widetilde{\mathrm{v}}(\star),$$

which proves (2.39) and concludes the proof. □

Proposition 2.2 does warrant further interpretation, as it illustrates the (over-) specification of the sliding stochastic implied volatility model. However, those comments will be postponed until Sect. 2.3.2 in order to give more insight into the recovery results themselves. Let us now move on to the inverse problem proper.

Theorem 2.1 (Recovery of the SInsV dynamics: the first layer) *Let us consider a given sliding SImpV model, as defined by* (2.11)–(2.12). *Then its local IATM specification is associated to a SInsV model, as defined by* (2.8)–(2.10), *whose value and dynamics are recovered as follows:*

$$d\sigma_t=a_{1,t}dt+a_{2,t}dW_t+\vec{a}^{\perp}_{3,t}d\vec{Z}_t, \tag{2.42}$$

with

$$a_{1,t} = 2\widetilde{\Sigma}'_{\theta}(\star) + \sigma_t^2 \widetilde{\Sigma}''_{yy}(\star) - \sigma_t^2 \widetilde{\Sigma}'_y(\star) + \sigma_t \, \widetilde{\Sigma}'^{\,2}_y (\star) - \frac{\| \vec{n}\,(\star)\|^2}{2\sigma_t} \tag{2.43}$$

$$a_{2,t} = 2\sigma_t \widetilde{\Sigma}'_y(\star) \tag{2.44}$$

$$\vec{a}_{\,3,t} = \vec{n}\,(\star) \tag{2.45}$$

$$a_{22,t} = 2\left[\widetilde{\Sigma}'_y(\star)\widetilde{\mathrm{v}}(\star) + \sigma_t \widetilde{\mathrm{v}}'_y(\star)\right]. \tag{2.46}$$

Note that we will shortly define the SInsV coefficients invoked here as the σ_t*-(2, 0) group (see Definition* 2.1 [p. 57]).

Proof The fundamental IATM identity (2.36) is static but valid *a.s.* and at any time t, while the parameters y and θ are constant. Therefore it also provides us with the following dynamics:

$$d\sigma_t = d\widetilde{\Sigma}(t, y = 0, \theta = 0) = \widetilde{b}(t, 0, 0)dt + \widetilde{\mathrm{v}}(t, 0, 0)dW_t + \vec{n}\,(t, 0, 0)^{\perp} d\vec{Z}_t. \tag{2.47}$$

By uniqueness of the decomposition and invoking Proposition 2.2 we get:

- The $a_{2,t}$ coefficient through (2.37), which provides (2.44).
- The $\vec{a}_{\,3,t}$ coefficient directly, which provides (2.45).

As for identifying the drift coefficient $a_{1,t}$, we wish to express it as a function of:

- The shape function $\widetilde{\Sigma}$ and its differentials, all taken at the origin point $(t, 0, 0)$;
- The exogenous coefficient $\vec{n}\,(t, 0, 0)$.

It suffices to replace both $\widetilde{\mathrm{v}}(\star)$ and $\widetilde{\mathrm{v}}'_y(\star)$ by their respective expressions (2.37) and (2.38) in the IATM drift expression (2.39) to obtain

$$\widetilde{b}(\star) = 2\widetilde{\Sigma}'_{\theta}(\star) - \frac{1}{2}\sigma_t^2 \widetilde{\Sigma}''_{yy}(\star) + \sigma_t \, \widetilde{\Sigma}'^{\,2}_y (\star) + \frac{3}{2}\sigma_t^2 \widetilde{\Sigma}''_{yy}(\star) - \frac{1}{2}\frac{\| \vec{n}\,(\star)\|^2}{\sigma_t} - \sigma_t^2 \widetilde{\Sigma}'_y(\star),$$

which after simplification provides (2.43). We have now covered all first-depth coefficients involved in the dynamics (2.42) of the instantaneous volatility σ_t. Moving on to the second depth, we then compute the dynamics of $a_{2,t}$ in order to extract the endogenous $a_{22,t}$ coefficient. From both sides of (2.37) and (2.44) it follows that

$$da_{2,t} = d\widetilde{\mathrm{v}}(\star) = [\cdot]\, dt + 2\widetilde{\Sigma}'_y(\star)\widetilde{\mathrm{v}}(\star)dW_t + 2\sigma_t \widetilde{\mathrm{v}}'_y(\star)dW_t + [\cdot]\, d\vec{Z}_t$$

and we get the final result

$$a_{22,t} = 2\left[\widetilde{\Sigma}'_y(\star)\widetilde{\mathrm{v}}(\star) + \sigma_t\widetilde{\mathrm{v}}'_y(\star)\right],$$

which concludes the proof. □

2.3.2 Interpretation and Comments

We shall first examine Proposition 2.2 in the perspective of SImpV model specification, before focusing on the Recovery Theorem 2.1, which answers the inverse problem. For practical reasons that shall become obvious in Chap. 3, we start by gathering the terms invoked by both results into two natural and consistent groups.

Definition 2.1 (*First Layer: the $\widetilde{\Sigma}$-(2,0) and σ_t-(2,0) groups*) The $\widetilde{\Sigma}$-(2,0) group comprises the following collection of IATM differentials:

$$\underbrace{\widetilde{\Sigma}(\star) \quad \widetilde{\Sigma}'_y(\star) \quad \widetilde{\Sigma}''_{yy}(\star) \quad \widetilde{\Sigma}'_\theta(\star)}_{\text{static coefficients}} \qquad \underbrace{\vec{n}(\star) \quad \widetilde{\mathrm{v}}(\star) \quad \widetilde{\mathrm{v}}'_y(\star) \quad \widetilde{b}(\star)}_{\text{dynamic coefficients}}$$

The σ_t-(2,0) group corresponds to the following coefficients of the SinsV model:

$$\sigma_t \qquad a_{1,t} \qquad a_{2,t} \qquad \vec{a}_{3,t} \qquad a_{22,t}$$

More generally, and as we shall confirm in the next section, the $\widetilde{\Sigma}$-(2,0) and σ_t-(2,0) collections constitute together what we will call the *first layer*. This terminology comes by reference to the successive *differentiation stages* of the ZDC, which are necessary to establish the asymptotic results (whether inverse or direct) as will be demonstrated in Chap. 3. In this case the suffix $(2, 0)$ indicates that the ZDC had to be differentiated twice w.r.t. y and not at all w.r.t. θ, which is clear from the proof of Proposition 2.2.

2.3.2.1 Specification of the Stochastic Implied Volatility Model

Let us first comment *globally* on the trio of arbitrage constraints (2.37)–(2.39), before covering them individually.

First we note that they represent only *necessary* conditions of no-arbitrage. It is indeed possible to express more conditions of the same type, invoking further y- and θ-differentials of the parametric processes $\widetilde{\Sigma}$, $\widetilde{b}$, $\widetilde{\mathrm{v}}$ and $\vec{n}$, also taken at the IATM point $(t, 0, 0)$: this will become clearer in the course of Chap. 3.

Note also that we have presented the three equations in such a way that the IATM drift and endogenous volatility are expressed as functions of only the local *static* differentials and of the exogenous coefficients $\vec{\vec{n}}$. This particular presentation is obviously artificial, and mainly serves the purpose of avoiding a chaos expansion at higher orders. In reality it consists in a simple, local manifestation of the ZDC (2.18), which itself stresses the growing over-specification of the SimpV model class, as we close on the IATM point.

Indeed, the model is defined by the four functionals $\widetilde{\Sigma}$, $\widetilde{b}$, $\widetilde{n}$ and $\vec{\vec{n}}$, and the NAO condition dictates that in the full domain (t, y, θ), one (and only one) of these functionals is redundant. Then in the Immediate sub-domain $(t, y, 0)$ the IZDC (2.28) and/or the Secondary IZDC (2.29) demonstrate that more restrictive conditions apply, involving a larger number of descriptive functionals: indeed, where θ is null the functionals $\widetilde{\Sigma}(t, y, 0)$ and $\widetilde{\Sigma}'_\theta(t, y, 0)$, for instance, are no longer redundant. Finally, at the IATM point $(t, 0, 0)$ the new Theorem 2.1 shows that the constraints become even stronger, involving a greater number of descriptors/functionals.

Therefore the presentation of Proposition 2.2 and of Theorem 2.1 is not unique, so that alternative expressions exist that will characterise a SImpV model, or at least its $\widetilde{\Sigma}$-(2,0) group of IATM differentials. Defining the model entirely through these differentials would indeed be natural, should the surface be designed using Taylor expansions for instance.[14] We observe that this group contains eight processes, which are themselves constrained by three no-arbitrage conditions, so that the specification only retains five degrees of freedom. Put simply, the following IATM differentials are interchangeable by pairs:

$$\widetilde{\mathrm{v}}(\star) \quad vs \quad \widetilde{\Sigma}'_y(\star) \qquad\qquad \widetilde{\mathrm{v}}'_y(\star) \quad vs \quad \widetilde{\Sigma}''_{yy}(\star) \qquad\qquad \widetilde{b}(\star) \quad vs \quad \widetilde{\Sigma}'_\theta(\star)$$

Note that in all three cases, we have equivalence between a static coefficient and a dynamic one, but the endogenous IATM coefficient $\vec{\vec{n}}(\star)$ is intrinsic and remains unmatched. In summary, the $\widetilde{\Sigma}$-(2,0) group can be uniquely and equivalently defined (and therefore the SImpV model characterised) by no less than eight equivalent, "minimal" configurations. Following these global considerations, let us comment individually on the three NAA constraints.

Remark 2.5 The IATM arbitrage constraint (2.37) shows that in any (sliding) stochastic implied volatility (SImpV) model, the IATM skew is HALF the Lognormal endogenous IATM coefficient.

This is certainly, once again, a strong design constraint for modellers. Equivalently, it means that if two models exhibit (possibly on purpose) the same IATM level *and* IATM skew, then the volatilities of that IATM level will necessary share the same endogenous component. This feature clearly has strong hedging implications, in particular in terms of Delta.

Let us now look at (2.38) which provides $\widetilde{\mathrm{v}}'_y(t, 0, 0)$. The latter IATM differential can be viewed in one of two ways: either as describing the variation of (endogenous)

[14] In fact, as we shall discuss in Sect. 2.4, this is generally a bad idea as one wishes that surface to be initially *valid*, and also to stay so dynamically.

volatility w.r.t. strike, or as the endogenous coefficient in the dynamics of the IATM skew, which we will adopt. Indeed, assuming enough regularity[15] we have

$$d\widetilde{\Sigma}'_y(\star) = \widetilde{b}'_y(\star)dt + \widetilde{\mathrm{v}}'_y(\star)dW_t + \overrightarrow{n}'_y(\star)^{\perp}d\overrightarrow{Z}_t. \tag{2.48}$$

In consequence, this term *will* determine a significant part of the risk for skew-based products such as ATM Binaries, Call Spreads, Risk Reversals or Collars. Furthermore, with respect to smile specification and behaviour, we observe that:

Remark 2.6 The IATM arbitrage constraint (2.38) shows that in any SImpV model, at the IATM point the endogenous Normal volatility of the skew $\widetilde{\mathrm{v}}'_y(\star)$ increases with the curvature $\widetilde{\Sigma}''_{yy}(\star)$ and with the square of the skew $\widetilde{\Sigma}'_y(\star)$ itself (all else equal). Note that the notion of curvature depends on the chosen variable: a smile which is convex in strike K is not necessarily so in log-moneyness y (see (B.3), p. 431).

Finally, and in a similar fashion, (2.39) shows how the drift of the IATM volatility depends on the exogenous volatility $\overrightarrow{n}(\star)$, but also on many static descriptors. In a way, it looks as if the IATM implied volatility is "riding the smile". Indeed, if the slope $\widetilde{\Sigma}'_\theta(\star)$ is positive, then it brings a positive component to the drift, just as if the IATM point was progressing along a static copy of the smile. In the same spirit, a convex smile (in y coordinates) will also bring a positive component: if the IATM point was riding a convex surface, then Itô's Lemma would generate a positive drift. However, interpreting the influence of the skew and of the exogenous volatility seems a bit more arduous for now.

2.3.2.2 Recovery of the Stochastic Instantaneous Volatility Model

The IATM Identity with (2.36) and the Recovery Theorem 2.1 with (2.42) provides, respectively, the value (2.42) and the first-depth dynamics of the instantaneous volatility σ_t. Therefore they do provide together the first elements of a solution for the inverse problem, as stated initially in Sect. 2.1.4. It is also comforting to note that these results are perfectly compatible with those of [12] (see Appendix E of [35]). Also on the subject of model correspondence, this recovery brings the following significant result:

Corollary 2.3 (Injectivity from $\widetilde{\Sigma}$-(2,0) to σ_t-(2,0)) *Starting from a given SImpV model and inferring the associated SInsV model, using the notations of Definition* 2.1 *we have that*

$$\textit{The function } \widetilde{\Sigma} - (2,0) \mapsto \sigma_t - (2,0) \textit{ is a.s. injective.}$$

[15] More on this point in the next section.

Note that it is mapping a group of five processes into another group of five processes, that $\widetilde{\Sigma}$-(2,0) presents equivalent configurations and that the non-nullity of $\widetilde{\Sigma}(\star)$ and σ_t has been assumed for convenience.

Proof Let us establish the property in a sequential fashion. Starting from any equivalent configuration of the $\widetilde{\Sigma}$-(2,0) group, $\widetilde{\Sigma}(\star)$ uniquely defines σ_t through (2.36). Then either $\widetilde{\Sigma}'_y(\star)$ or $\widetilde{\mathrm{v}}(\star)$ provides a_2 without ambiguity *via* respectively (2.44) or (2.37), while $\vec{a}_3$ stems directly from $\vec{\widetilde{n}}(\star)$ *via* (2.45). The coefficient a_{22} is now uniquely defined by $\widetilde{\mathrm{v}}'_y(\star)$, or alternatively from $\widetilde{\Sigma}''_{yy}(\star)$ by invoking (2.38), *via* (2.44). Finally, a_1 identifies with $\widetilde{b}(\star)$ which itself is now uniquely defined by $\widetilde{\Sigma}'_\theta(\star)$ as per (2.43) or (2.39).

Note that this inductive method is made possible by the *a.s.* positivity of σ_t, which is provided by the previous assumption (2.14), but this is an artificial constraint that can be lifted by considering the instantaneous variance instead of the volatility. □

At this point, the natural question to raise is whether we actually face a bijection between the two groups, which would establish the beginning of an equivalence between the SinsV and SimpV model classes themselves. We shall see in the coming Sect. 2.4.2 that the bijection does exist for Layer 1. However, and as will be discussed in Chap. 3, if the direct relationship is indeed injective (irrespective of the dimension) at any order, this is *a priori* not the case for the recovery problem.

Among all the minimal combinations for the $\widetilde{\Sigma}$-(2,0) group mentioned in Sect. 2.3.2.1, there is no escaping the inclusion of the exogenous coefficient $\vec{\widetilde{n}}(\star)$. However, for reasons that will be made clear in the sequel, it will prove especially useful to avoid the *full* specification of that exogenous coefficient, when possible. This is why we now present the following alternative expressions for the drifts and for the exogenous coefficients.

Corollary 2.4 (Alternative expressions within the Recovery Theorem 2.1) *The recovery of the SInsV drift can be expressed without the exogenous coefficient:*

$$a_{1,t} = 2\widetilde{\Sigma}'_\theta(\star) - \sigma_t^2\left[\widetilde{\Sigma}'_y(\star) + \frac{1}{2}\widetilde{\Sigma}''_{yy}(\star)\right] + \sigma_t\widetilde{\mathrm{v}}'_y(\star). \tag{2.49}$$

Furthermore, in the particular case of a scalar exogenous driver Z_t, where $\eta(\widetilde{n})$ denotes the sign of $\widetilde{n}(\star)$ and of $a_{3,t}$, we have the exogenous coefficients as

$$a_{3,t} = \eta(\widetilde{n})\sqrt{2}\sigma_t\left[\widetilde{\Sigma}'_y(\star)^2 + \frac{3}{2}\sigma_t\widetilde{\Sigma}''_{yy}(\star) - \widetilde{\mathrm{v}}'_y(\star)\right]^{\frac{1}{2}}. \tag{2.50}$$

Proof Expression (2.49) comes trivially from (2.39), while isolating the squared modulus $\|\vec{\widetilde{n}}(\star)\|^2$ on the left-hand side of (2.38) and then taking the square root on either side gives us (2.50). □

Turning now to the SInsV dynamics as described by (2.42), we find that they provide an interesting insight. Indeed, looking first at the drift of σ_t as per (2.43), we observe that a convex smile with a negative skew will create an increasing trend in volatility. This geometric situation happens to be extremely common, but the overall drift will depend also on the slope: a decreasing ATM volatility will counteract the previous effect and vice-versa.

As for the endogenous volatility of volatility, according to (2.44) its modulus will increase with the skew's, which also has strong trading implications as it relates a dynamic with a static option strategy.

In spite of, or rather because of the academic relevance of these results, it is important to stress the possible choices and difficulties one might encounter when confronting them in real-life markets. In particular, we have seen that it is possible to extract the SInsV coefficients from either static or dynamic descriptors of the smile: the question is naturally which is better.

In our view, the main criteria is the precision of such IATM market data. We have already mentioned (see Sect. 2.2.2) the possible difficulty of extrapolating the data in expiry, down to $\theta = 0$, which will affect the slope $\widetilde{\Sigma}'_\theta(\star)$. Furthermore, measuring a space differential such as the skew $\widetilde{\Sigma}'_y(\star)$ is subject to the liquidity of ITM and/or OTM strikes, whereas inferring dynamic coefficients, such as the endogenous vol of vol $\widetilde{\mathrm{v}}(\star)$, suffers from sampling error. In general, the latter error tends to dominate, hence it seems preferable to derive the dynamics of the SInsV model from the *shape* descriptors of the (sliding) smile. Unfortunately, and as previously mentioned, we cannot dispense with the exogenous coefficient $\overrightarrow{\widetilde{n}}(t, 0, 0)$, which is intrinsic to all $\widetilde{\Sigma}$-(2,0) configurations.

The second issue is obviously how to deal with conflicting information from both static and dynamic sources. One has to ponder whether the discrepancy falls within an acceptable noise range, whether the modelling of the dynamics is inappropriate (e.g. dimension of the drivers) or whether the market itself might not be arbitrage-free. However, most of these questions lie in the realm of model risk and statistical arbitrage, and are therefore left to further research.

It is naturally possible to interpret and use these results further, in a market-oriented approach. In particular, should we attempt to *specify* the shape and dynamics of the smile *ex ante*, in the fashion of stochastic *implied* volatility models, the Recovery Theorem 2.1 would give us strong guidance. We choose to postpone this discussion until Sect. 2.5 [p. 77], which is dedicated to the simultaneous interpretation of both recovery and first layer results.

2.4 Generating the Implied Volatility: The First Layer

We now turn to the *direct problem*, by assuming that the input SInsV model specifies the instantaneous volatility σ_t as per (2.8)–(2.9) while the shape and dynamics of the associated smile—described by (2.12)—are unknown. We aim at providing information on the associated SImpV model, pertaining either to its shape $\widetilde{\Sigma}(t, y, \theta)$ or to the SDE coefficients $\widetilde{b}(t, y, \theta)$, $\widetilde{\mathrm{v}}(t, y, \theta)$ and $\overrightarrow{\widetilde{n}}(t, y, \theta)$. As with the Recovery

Theorem 2.1, such information is focused on the IATM point $(t, 0, 0)$ but involves differentials in the y and θ directions.

Having brushed the subject of multi-dimensionality for the spot process, it seems reasonable to stick to the scalar case: therefore in the sequel of this chapter let us assume that $n_w = 1$.

2.4.1 Computing the Immediate ATM Differentials

Let us assume the dynamics of the *stochastic instantaneous volatility* model (2.8)–(2.10), although a_{21} and $\vec{a}_{23}$ will not be needed in this section.

Theorem 2.2 (Generating the first layer's IATM differentials) *Under the SInsV model defined by* (2.8)–(2.10) *the following local IATM differentials for the shape and dynamics of the sliding IV surface* $\widetilde{\Sigma}(t, y, \theta)$ *can be expressed. They constitute the* $\widetilde{\Sigma}$-(2, 0) *group, or first layer.*

- *Local differentials of the shape process:*

$$\widetilde{\Sigma}'_y(t, 0, 0) = \frac{1}{\sigma_t}\left[\frac{1}{2}a_2\right] \tag{2.51}$$

$$\widetilde{\Sigma}''_{yy}(t, 0, 0) = \frac{1}{\sigma_t^2}\left[\frac{1}{3}a_{22}\right] + \frac{1}{\sigma_t^3}\left[\frac{1}{3}\|\vec{a}_3\|^2 - \frac{1}{2}a_2^2\right] \tag{2.52}$$

$$\widetilde{\Sigma}'_\theta(t, 0, 0) = \sigma_t\left[\frac{1}{4}a_2\right] + \left[\frac{1}{2}a_1 - \frac{1}{6}a_{22}\right] \tag{2.53}$$

$$+\frac{1}{\sigma_t}\left[\frac{1}{8}a_2^2 + \frac{1}{12}\|\vec{a}_3\|^2\right]. \tag{2.54}$$

- *Local differentials of the dynamics processes:*

$$\widetilde{\mathrm{v}}(t, 0, 0) = a_2 \tag{2.55}$$

$$\widetilde{\mathrm{v}}'_y(t, 0, 0) = \frac{1}{\sigma_t}\left[\frac{1}{2}a_{22}\right] + \frac{1}{\sigma_t^2}\left[-\frac{1}{2}a_2^2\right] \tag{2.56}$$

$$\vec{\tilde{n}}(t, 0, 0) = \vec{a}_3. \tag{2.57}$$

As some readers might find it useful to express the static shape differentials w.r.t. strike K rather than log-moneyness y, we provide the two relevant expressions (skew and curvature) which come straight from applying (B.2) and (B.3):

Corollary 2.5 (IATM skew and curvature in absolute coordinates) *At the IATM point* $(t, S_t, K = S_t, T = t)$, *the skew and curvature can be expressed as*

$$\Sigma_K'(t, S_t, S_t, t) = \frac{a_2}{2S_t\sigma_t} \tag{2.58}$$

$$\Sigma_{KK}''(t, S_t, S_t, t) = \frac{1}{S_t^2}\left[\frac{1}{\sigma_t}\left[-\frac{a_2}{2}\right] + \frac{1}{\sigma_t^2}\left[\frac{a_{22}}{3}\right] + \frac{1}{\sigma_t^3}\left[\frac{\|\vec{a}_3\|^2}{3} - \frac{a_2^2}{2}\right]\right]. \tag{2.59}$$

It is naturally possible to accelerate the proof of Theorem 2.2 by making use of the results within the Recovery Theorem 2.1. We choose not to do so, for three reasons. The first is that we wish to bring into focus the similarities between the direct and inverse problem. Hence it makes sense to present them *in parallel* rather than sequentially. The second reason is that we will prove in the sequel (in particular, in Chap. 3) that these two problems are actually equivalent, within reasonable assumptions. Finally, the extensibility of the ACE methodology to higher orders will be established using the direct problem, and therefore the coming proof will serve as the basic sequence for further layers.

Proof (Theorem 2.2 and Corollary 2.5) As with the Recovery Theorem 2.1, **the first step** is to take the dynamics on both sides of the IATM Identity (2.36). The identification of the coefficients between (2.9) and (2.12) provides, *via* uniqueness of the martingale decomposition:

$$\widetilde{b}(\star) = a_{1,t} \qquad \widetilde{\mathrm{v}}(\star) = a_{2,t} \qquad \vec{\widetilde{n}}(\star) = \vec{a}_{3,t}$$

which proves (2.55) and (2.57) immediately.

The second step is to invoke the IATM arbitrage constraints of the SImpV model (the same that were necessary to prove the Recovery Theorem 2.1) which are gathered in Proposition 2.2. Then:

- combining first (2.37) with (2.55) provides the IATM skew as in (2.51);
- from (2.38) we have the IATM curvature $\widetilde{\Sigma}_{yy}''(\star)$ as an explicit function of $\widetilde{\mathrm{v}}_y'(\star)$;
- from (2.39) the IATM slope $\widetilde{\Sigma}_\theta'(\star)$ is an explicit function of $\widetilde{\Sigma}_{yy}''(\star)$ and/or $\widetilde{\mathrm{v}}_y'(\star)$,

which leaves us facing three variables for only two equations.

The third step is to derive new relationships by actually computing some dynamics from already established results: the strategy is to first express $\widetilde{\mathrm{v}}_y'(\star)$, which will then give us $\widetilde{\Sigma}_{yy}''(\star)$ and finally $\widetilde{\Sigma}_\theta'(\star)$.

Let us consider the formal dynamics of $\widetilde{\Sigma}(t, y, \theta)$ within the SImpV model, and assume that the regularity conditions required by Theorem 3.1.2 of [36] (p. 75) are satisfied. In other words, let us assume that we can apply the differentiation operator ∂_y on either side of (2.12) and express the corresponding flow, as was done with (2.48). Recall that we then get

$$d\widetilde{\Sigma}_y'(t, y, \theta) = \widetilde{b}_y'(t, y, \theta)dt + \widetilde{\mathrm{v}}_y'(t, y, \theta)dW_t + \vec{\widetilde{n}}_y'(t, y, \theta)^\perp d\vec{Z}_t. \tag{2.60}$$

We take this expression in $(t, 0, 0)$ and on the left-hand side replace $\widetilde{\Sigma}'_y(\star)$ by its expression (2.51) as a function of a_2 and σ_t. The dynamics of $a_{2,t}$ being defined by (2.10), those of the IATM skew come as

$$d\widetilde{\Sigma}'_y(\star) = d\left[\frac{a_2}{2\sigma_t}\right] = [\cdots]\,dt + \frac{1}{2}\left[\frac{a_{22}}{\sigma_t} - \frac{a_2^2}{\sigma_t^2}\right] dW_t + [\cdots]\,d\vec{Z}_t.$$

Then identifying the endogenous coefficients on either side of (2.60) we obtain

$$\widetilde{\mathrm{v}}'_y(t,0,0) = \frac{1}{\sigma_t}\left[\frac{1}{2}a_{22}\right] + \frac{1}{\sigma_t^2}\left[-\frac{1}{2}a_2^2\right]$$

which proves (2.56). We can then isolate the curvature $\widetilde{\Sigma}''_{yy}(\star)$ in (2.38) and substitute (2.51), (2.56) and (2.57) respectively in place of $\widetilde{\Sigma}'_y(\star)$, $\widetilde{\mathrm{v}}'_y(\star)$ and $\vec{\tilde{n}}(\star)$:

$$\begin{aligned}
\widetilde{\Sigma}''_{yy}(\star) &= -\frac{2}{3}\frac{1}{\sigma_t}\,\widetilde{\Sigma}'^{\,2}_y(\star) + \frac{1}{3}\frac{\|\vec{\tilde{n}}(\star)\|^2}{\sigma_t^3} + \frac{2}{3}\frac{1}{\sigma_t}\widetilde{\mathrm{v}}'_y(\star) \\
&= -\frac{2}{3}\frac{1}{\sigma_t}\left[\frac{a_2}{2\sigma_t}\right]^2 + \frac{1}{3}\frac{\|\vec{a}_3\|^2}{\sigma_t^3} + \frac{2}{3}\frac{1}{\sigma_t}\left[\frac{1}{\sigma_t}\left[\frac{1}{2}a_{22}\right] + \frac{1}{\sigma_t^2}\left[-\frac{1}{2}a_2^2\right]\right],
\end{aligned}$$

which after simplification proves (2.52). Finally, isolating the slope $\widetilde{\Sigma}'_\theta(\star)$ in the drift equation (2.39), before replacing $\widetilde{\Sigma}'_y(\star)$, $\widetilde{\Sigma}''_{yy}(\star)$, $\widetilde{\mathrm{v}}(\star)$ and $\widetilde{\mathrm{v}}'_y(\star)$ respectively by (2.51), (2.52), (2.55) and (2.56), we obtain

$$\begin{aligned}
\widetilde{\Sigma}'_\theta(\star) &= \frac{1}{2}\tilde{b}(\star) + \frac{1}{4}\sigma_t^2\,\widetilde{\Sigma}''_{yy}(\star) - \frac{1}{2}\sigma_t\widetilde{\mathrm{v}}'_y(\star) + \frac{1}{4}\sigma_t\widetilde{\mathrm{v}}(\star) \\
&= \frac{1}{2}a_1 + \frac{1}{4}\sigma_t^2\left[\frac{1}{\sigma_t^2}\left[\frac{1}{3}a_{22}\right] + \frac{1}{\sigma_t^3}\left[\frac{1}{3}\|\vec{a}_3\|^2 - \frac{1}{2}a_2^2\right]\right] \\
&\quad -\frac{1}{2}\sigma_t\left[\frac{1}{\sigma_t}\left[\frac{1}{2}a_{22}\right] + \frac{1}{\sigma_t^2}\left[-\frac{1}{2}a_2^2\right]\right] + \frac{1}{4}\sigma_t a_2,
\end{aligned}$$

which after simplification provides (2.53) and concludes the proof of Theorem 2.2.

In order to prove Corollary 2.5 it suffices to use the transition formulae: combining (2.51) and (B.2) proves (2.58), while applying (B.3) gives

$$\begin{aligned}
\Sigma''_{KK}(t, S_t, S_t, t) &= \frac{1}{S_t^2}\left[\widetilde{\Sigma}''_{yy}(\star) - \widetilde{\Sigma}'_y(\star)\right] \\
&= \frac{1}{S_t^2}\left[\frac{1}{\sigma_t^2}\left[\frac{1}{3}a_{22}\right] + \frac{1}{\sigma_t^3}\left[\frac{1}{3}\|\vec{a}_3\|^2 - \frac{1}{2}a_2^2\right] + \frac{1}{\sigma_t}\left[-\frac{1}{2}a_2\right]\right],
\end{aligned}$$

which matches (2.59) and concludes the overall proof. □

2.4.2 Interpretation and Comments

We now translate and extrapolate the Layer-1 results, i.e. Theorem 2.2 and Corollary 2.5, into theoretically meaningful and trading pertinent results. Naturally, the latter mainly concern the influence of each instantaneous coefficient $a_{i,t}$ on the shape and dynamics of the smile, as far as IATM differentials are concerned. Although these considerations are generally model-independent, whenever pertinent we compare our results with those available in the literature (whether exact or heuristic) for the specific class of pure local volatility models.

First we go through some general aspects of those Layer-1 results, before focusing on individual IATM differentials, in particular the (static) IATM skew and IATM curvature.

2.4.2.1 General Considerations on the First Layer

Let us first mention a peculiar feature of the $\widetilde{\Sigma}$-(2,0) group of IATM differentials, as presented by Theorem 2.2. As will be discussed in Sect. 2.5.1.1 [p. 77], it has significantly beneficial properties in terms of whole-smile extrapolation.

Remark 2.7 The first layer as per Theorem 2.2 provides the IATM differential of order 2 in strike, but only 1 in maturity (and no cross-terms, which will only appear in further layers). This distinct behaviour of the space and time differentials is clearly a consequence of Itô's formula. Not surprisingly, it is also a constant of our study: with higher-order layers, cross-terms might appear but increasing the order of differentiation w.r.t. θ by one will always drop the y-order by 2. We call this feature the "ladder effect", which will be formally established and graphically presented in Chap. 3 (see Fig. 3.1, p. 122).

In terms of model correspondence, a noticeable consequence of Theorem 2.2 is the establishment of the following result.

Corollary 2.6 (Injectivity from the σ_t-$(2,0)$ to the $\widetilde{\Sigma}$-$(2,0)$ group) *Starting from a given SInsV model and inferring the associated SImpV model, using the notations of Definition* 2.1 *we have that*

$$\textit{The function} \quad \sigma_t\text{-}(2,0) \ \mapsto \ \widetilde{\Sigma}\text{-}(2,0) \ \textit{is a.s. injective.}$$

As discussed previously in Sect. 2.3.2.1 [p. 57], *this statement has to be understood within the over-specification of the* $\widetilde{\Sigma}$-$(2,0)$ *group, which effectively possesses only five degrees of freedom.*

Proof To establish this result, again we proceed sequentially: starting with a given $\widetilde{\Sigma}$-(2,0) configuration, the IATM Identity (2.36) [p. 50] uniquely sets σ_t. Then (2.51) or equivalently (2.55) provides unambiguously $a_{2,t}$, while (2.57) determines $\overrightarrow{a}_{3,t}$. The coefficient $a_{22,t}$ is then uniquely defined by either (2.52) or (2.56), and finally (2.53) unambiguously sets $a_{1,t}$. □

Combining Corollaries 2.3 and 2.6 directly yields the following correspondence.

Corollary 2.7 (Bijectivity of the σ_t-(2,0) and $\widetilde{\Sigma}$-(2,0) groups)

The relationship $\sigma_t\text{-}(2,0) \longleftrightarrow \widetilde{\Sigma}\text{-}(2,0)$ *is a.s. bijective.*

We consider this result to be fundamental since, if upheld at all orders of differentiation (which it is, as will be proven in Sect. 3.1) it can establish complete correspondence between the SInsV and SImpV model classes. Indeed, assuming that the smile's static and dynamic functionals are analytic along both the y and θ coordinates (i.e. that they can be entirely determined by an infinite series of IATM differentials) then their specification is equivalent to writing "in depth" the chaos expansion of a generic stochastic instantaneous volatility model. But even without the still-to-be-proven extension to higher order, the σ_t-(2,0) $\longleftrightarrow$ $\widetilde{\Sigma}$-(2,0) correspondence can and will be used by itself, no later than in Sect. 2.5.3, which is dedicated to the "intuitive" re-parametrisation of SInsV models.

Turning now to more practical matters, as announced we shall now use the local volatility (LocVol) model class as an important benchmark, due to its simplicity but also to its academic relevance. Note that one can refer to [13] for an alternative approach to the class, in a dynamic smile context presenting similarities with our framework.

The class has previously been mentioned in Sect. 2.1.2.3 [p. 29] and we will maintain the same notation for its dynamics, as per (2.5). Recall also that although its practical usage is limited (due to flawed dynamics) its explanatory and demonstrative capacity is very strong. Also it has been thoroughly investigated in the literature, which provides a good opportunity for critically assessing our results using a simple case (more complex illustrations will follow).

All these features are mainly due to the minimal Markovian dimension, and to the alternative interpretation that they offer for the IV surface, as a conditional expectation and *via* Dupire. Evidently, since our results are generic and come in closed form, we must look for the same properties in the literature. However, to our knowledge there is no published result which is simultaneously generic, exact, explicit, valid in the full domain (or failing that in a neighborhood of the IATM point) and of a direct nature (i.e. providing $\widetilde{\Sigma}(\ldots)$ from $f(\cdot,\cdot)$). Hence we have selected three distinct approaches for their respective strengths.

The first natural candidate is certainly Dupire, which is expressed in price terms in [37] but can be re-parametrised using (sliding) implied volatility. The equivalent formulation can be found in [28][16] (p. 6) or adapted from [38] (p. 13) or even derived without difficulty from Dupire's formula. Using our usual notation $(\circ) = (t, y, \theta)$ it reads as:

$$f^2(T,K) = \frac{\widetilde{\Sigma}^4(\circ) + 2\theta\,\widetilde{\Sigma}^3\,\widetilde{\Sigma}'_\theta(\circ)}{\left[\widetilde{\Sigma}(\circ) - y\widetilde{\Sigma}'_y(\circ)\right]^2 - \frac{1}{4}\theta^2\widetilde{\Sigma}^4\,\widetilde{\Sigma}'^{\,2}_y(\circ) + \theta\,\widetilde{\Sigma}^3\,\widetilde{\Sigma}''_{yy}(\circ)}. \tag{2.61}$$

[16] Note that some non-official internet versions incorporate a minor typo in the denominator.

Note that this result is verified in the whole domain (t, y, θ), also that it falls into the inverse category (it provides the SinsV specification from the implied volatility) and finally that it is *exact*. Overall, in our context these properties make (2.61) better suited for verification purposes, rather than derivation of the direct results.

Furthermore, since our Layer-1 results are expressed at the IATM point, which is a subset of the immediate domain $\theta \equiv 0$, it becomes pertinent to bring forward another available result, provided by [30]. It states that in a local volatility model, the immediate implied volatility at any strike is the spatial *harmonic mean* of the local volatility between S_t and K. Importantly, this is a direct and exact result, which in our context reads as

$$\widetilde{\Sigma}(t, y, \theta = 0) = \left[\int_0^1 f^{-1}\left(t, S_t e^{sy}\right) ds \right]^{-1}, \tag{2.62}$$

which means that it provides the pure-y immediate differentials of the implied volatility surface, in particular at the IATM point.

Finally, we compare our results to an approximation, usually referred to as Gatheral's formula, that can be found in [39] and [38]. It relies on the very definition of local variance as a conditional expectation, and expresses the (stochastic) "forward implied variance" as a stochastic integral against dS_t, the integrand of which is written as an expansion around a particular "most probable" path (MPP). The Black implied variance itself is then taken as a time integral of that expression, but after selecting order zero for the expansion (see [38], p. 30). Within our framework, the formula reads as follows:

$$\Sigma^2(t, S_t; K, T) \approx \overline{\Sigma}^2(t, S_t; K, T) = \frac{1}{T-t} \int_t^T f^2\left(s, S_t \left(K/S_t\right)^{\frac{s-t}{T-t}}\right) ds. \tag{2.63}$$

The reasons we elected to test and compare Gatheral's formula against our asymptotic results are several. Firstly it constitutes an intuitive approach: the MPP is an easy concept, and the overall expression is linked to the notion of Brownian bridge, which is widely understood. Furthermore, the formula is popular among practitioners, not least because it provides an approximation of the implied volatility in the full domain, a feature which allows us to compute all IATM differentials of Theorem 2.2.

It is worth stressing that Gatheral's formula, unlike (2.61) and (2.62), and because it stems from a low-order expansion, is *not* an exact result. It is evident that it cannot capture with a single path the *whole* local volatility function until maturity, which *does* generate the marginal distribution. For all intents and purposes, it should therefore be considered in the current section as a heuristic.

In order to derive our asymptotic Layer-1 results, the first step with any model (class) is naturally to express the cast, i.e. the corresponding SInsV coefficients:

Lemma 2.2 (Instantaneous coefficients of the local volatility model) *In a pure local volatility model, defined by* (2.5)*, the SInsV coefficients are*

$$\sigma_t = f(t, S_t) \qquad a_{2,t} = S_t f(t, S_t) f_2'(t, S_t) \qquad a_{3,t} \equiv 0 \tag{2.64}$$

$$a_{1,t} = f_1'(t, S_t) + \frac{1}{2} S_t^2 f^2 f_{22}''(t, S_t) \tag{2.65}$$

$$a_{22,t} = S_t f(t, S_t) \left[(f \cdot f_2')(t, S_t) + S_t \, f_2'^{\,2}(t, S_t) + S_t (f \cdot f_{22}'')(t, S_t) \right]. \tag{2.66}$$

Proof For pure local volatility models, the cast immediately gives

$$\sigma_t = f(t, S_t).$$

Therefore the dynamics of σ_t become

$$d\sigma_t = \left[f_1'(t, S_t) + \frac{1}{2} f_{22}''(t, S_t) \langle dS_t \rangle \right] dt + f_2'(t, S_t) dS_t,$$

so that

$$a_{1,t} = f_1'(t, S_t) + \frac{1}{2} S_t^2 f^2 f_{22}''(t, S_t) \quad \text{and} \quad a_{2,t} = S_t f(t, S_t) f_2'(t, S_t),$$

obviously with $a_{3,t} = 0$. We then get the dynamics of $a_{2,t}$ as

$$da_2 = [\cdots] dt + \left[(f \cdot f_2')(t, S_t) + S_t \, f_2'^{\,2}(t, S_t) + S_t (f \cdot f_{22}'')(t, S_t) \right] dS_t,$$

so that

$$a_{22} = S_t f(t, S_t) \left[(f \cdot f_2')(t, S_t) + S_t \, f_2'^{\,2}(t, S_t) + S_t (f \cdot f_{22}'')(t, S_t) \right]. \qquad \square$$

Note that the exogenous coefficient $a_{3,t}$ is *a.s.* null; we use this opportunity to underline that, within our more general framework, we view local volatility models simply as a special instance of "purely endogenous" models. Those are characterised by a missing $\vec{Z}_t$ driver, which make of a_2 the "full" volatility of volatility. In that vein, another very simple type of endogenous model will be presented shortly (see Example 2.5, p. 108).

We can now derive the $\widetilde{\Sigma}$-(2,0) group of IATM differentials. We start with the level by combining (2.36) and (2.64):

Corollary 2.8 (IATM level in a local volatility model) *In a pure local volatility model defined by* (2.5) *the IATM level is simply given by*

$$\Sigma\,(t, S_t; K = S_t, T = t) = \widetilde{\Sigma}(t, 0, 0) = f(t, S_t). \tag{2.67}$$

Quite reassuringly, it is easily seen that all above-mentioned literature approaches (respectively (2.61), (2.62) and (2.63)) agree with the asymptotic result (2.36) on the IATM volatility level. Indeed, with (2.62) it suffices to take $y = 0$, and similarly with (2.61) the (assumed) non-negativity of both $f()$ and $\widetilde{\Sigma}()$ provides the desired result in $(t, 0, 0)$.

Finally, addressing Gatheral's heuristic, we start by re-expressing (2.63): defining $u \triangleq \frac{s-t}{T-t}$ and using sliding coordinates, we get the more practical formulation

$$\overline{\Sigma}^2(t, S_t; y, \theta) = \int_0^1 f^2\left(t + u\theta, S_t e^{uy}\right) du, \tag{2.68}$$

so that the same non-negativity argument brings (2.67).

2.4.2.2 Static Skew *vs* Endogenous Vol of Vol

Let us now examine the surprisingly simple result (2.51) which establishes a straightforward relationship between the IATM skew $\widetilde{\Sigma}'_y(\star)$ (a static, implied quantity) and the endogenous "vol of vol" $a_{2,t}$ (a dynamic, instantaneous coefficient). In simple terms, we observe that:

Remark 2.8 The direct result (2.51) shows that, in any SInsV model, the IATM skew is HALF the Lognormal endogenous volatility of volatility. The latter is defined by the convention

$$\frac{d\sigma_t}{\sigma_t} = \frac{a_1}{\sigma_t}dt + \frac{a_2}{\sigma_t}dW_t + \frac{\vec{a}_3}{\sigma_t}d\vec{Z}_t.$$

In qualitative terms, it is relatively easy to build some intuition as to why skew is an *increasing* function of the endogenous vol of vol, and furthermore shares its sign. Indeed, $a_{2,t}$ can be written as the quadratic co-variation between the underlying S_t and its volatility σ_t, when both are expressed using Lognormal conventions (formally, consider $\log(S_t)$ and $\log(\sigma_t)$):

$$\langle \frac{dS_t}{S_t}, \frac{d\sigma_t}{\sigma_t} \rangle = \sigma_t \frac{a_2}{\sigma_t} dt = a_2 dt.$$

That said, let us assume for instance that this instantaneous (effective) correlation a_2 is *negative*. Then, when the underlying S_t increases, "on average" its instantaneous Lognormal volatility will decrease, and vice-versa. For a given option maturity, this phenomenon will amplify down moves and curb upmoves, compared to an actual Lognormal process, tending (again, in law) to accumulate the marginal distribution[17]

[17] Obviously expressed under the martingale measure.

to the *left* of the money (S_t). This in turn increases the cumulative (Binary Put) value at-the-money: recalling how this cumulative is expressed as a function of the skew, we do obtain a *negative* Lognormal skew. Note that all volatilities (instantaneous and implied) must be considered under the same convention (here Lognormal) for this rough reasoning to be valid.

For a small number of specific models, including the pure local volatility class, this result has been known for some time, albeit usually expressed with differentials w.r.t. strike K. In some of these cases, the coefficient $a_{2,t}$ is referred to as the instantaneous *effective* correlation between the underlying and its volatility. This terminology owes to the fact that the co-variation we mentioned above can be generated by either correlated drivers and/or by a functional relationship, of the type $\sigma_t = f(S_t)$. We will see later in Sect. 2.5.2 how local volatility and correlation can be combined in a general class, and how to apply Theorem 2.2.

Let us now specialise the IATM skew formula (2.51) in the pure local volatility (LocVol) framework, and compare to results available in the literature. Our asymptotic results yield:

Corollary 2.9 (IATM skew in a local volatility model) *In a pure local volatility model, defined by* (2.5), *the IATM skew is given by*

$$\begin{aligned} \widetilde{\Sigma}'_y(t,0,0) &= \frac{1}{2} S_t f'_2(t,S_t) \\ \Sigma'_K(t,S_t;K=S_t,T=t) &= \frac{1}{2} f'_2(t,S_t). \end{aligned} \tag{2.69}$$

Proof It suffices to combine, respectively, (2.51) and (2.58) with (2.64) to obtain the desired result. □

In other words we have:

Remark 2.9 The direct result (2.51) shows that in local volatility models, the IATM skew of the Lognormal smile is HALF the IATM S-differential of the Lognormal local volatility function.

This result has been present in the literature for some time, and usually comes as a consequence of path integral approximations, of which the simplest form is the *mid-point method*. This rough proxy consists, for short-term options close to the money, in approximating the implied volatility at strike K and maturity T by the value of the *local* volatility, taken at the centre of the $[(t,S_t),(T,K)]$ segment. Note that the volatility convention is of course identical (Lognormal) in both the implied and local dynamics. Formally, we have that if

$$K = S_t + \Delta K \text{ with } \Delta K \ll 1 \qquad \text{and} \qquad T = t + \Delta t \text{ with } \Delta t \ll 1$$

then the approximation reads as

$$\Sigma(K,T) \approx \widehat{\Sigma}(K,T) \triangleq f\left(\frac{1}{2}(t+T), \frac{1}{2}(S_t+K)\right),$$

which happens to bring the exact IATM skew:

$$\Sigma_K'(S_t,t) \approx \widehat{\Sigma}_K'(S_t,t) = \frac{1}{2}f_2'(t,S_t).$$

The formula is naturally a very useful hedging tool, and over time has been presented in different forms and by several authors. In [40] it is brought as a heuristic, along with a couple of other "rules of thumb"; due to the pedagogic nature of the paper, the justification there is minimal. It has been made more rigorous in other, related papers (see references) but tends to rely on an intuitive idea: that the implied volatility for strike K is a path integral of the local volatility between the spot/forward S_t and K.

This is indeed the vein used by [38] to establish (2.63), albeit in a more elaborate fashion. Ir relies on the fact that in a Lognormal model, the MPP between (t, S_t) and (K,T) is a direct line in *log-space*, hence the integrand's argument in (2.63). Then for a short-expiry, close-to-the-money option, at first order the MPP becomes a direct line in the initial coordinates, while at zero order it can be approximated by the midpoint. Formally, it suffices to differentiate (2.68) once w.r.t. y and then to apply at the IATM point, obtaining sequentially

$$2\overline{\Sigma\Sigma_y'}(t,S_t;y,\theta) = \int_0^1 2ff_2'\left(t+u\theta, S_te^{uy}\right)S_tue^{uy}du \tag{2.70}$$

then

$$2\overline{\Sigma\Sigma_y'}(t,S_t;0,0) = 2S_tf(t,S_t)\,f_2'(t,S_t)\int_0^1 udu.$$

Finally, invoking (2.67) and (B.2) we get (2.69), as announced.

Apart from heuristics, (2.69) has also been established rigorously. For instance [41] uses a singular perturbation approach in the case where $f(s,x)$ is a function of separable variables. The resulting expansion is indeed taken around the midpoint, and differentiating once w.r.t. K provides the desired result. In the general context, we can obtain an exact formula from [30], as the skew is a pure-space differential. We simply differentiate both sides of (2.62) once w.r.t. to y to get

$$\widetilde{\Sigma}'_y(t, y, \theta = 0) = -\left[\int_0^1 f^{-1}\left(t, S_t e^{sy}\right) ds\right]^{-2} \left[\int_0^1 -f^{-2} f'_2\left(t, S_t e^{sy}\right) S_t e^{sy} s ds\right]. \tag{2.71}$$

Evaluating this expression at $(t, 0, 0)$, we have

$$\widetilde{\Sigma}'_y(t, 0, 0) = -f^2(t, S_t)\left[-S_t f^{-2}(t, S_t) f'_2(t, S_t) \int_0^1 s ds\right] = \frac{1}{2} S_t f'_2(t, S_t).$$

Note that, in order to check against an exact result, another alternative would be to invoke (2.61), which reassuringly provides the same answer.

2.4.2.3 Static Curvature *vs* Exogenous Vol of Vol

Let us now examine result (2.52), providing the curvature. We notice that the exogenous coefficient $\vec{a}_{3,t}$ has a systematic and positive effect on $\widetilde{\Sigma}''_{yy}(t, 0, 0)$. This feature is not surprising as it has been established for numerous model classes. In a Black model with an independent, stochastic volatility, the expression for Volga (C.2) combined with Jensen justifies that prices for ITM and OTM strikes will increase, relatively to the money: in terms of smile, this indeed translates into a curvature. More formally, following [42] it is proven in [43] that for a specific class of bi-dimensional diffusive models, with zero correlation (i.e. when a_2 and a_{22} are null), the implied volatility is symmetric and increasing with $|y = \ln(K/S_t)|$. Let us compare these results to our asymptotic approach, which gives:

Corollary 2.10 (IATM curvature in a local volatility model) *In a pure local volatility model, defined by* (2.5), *the IATM curvature is given by*

$$\widetilde{\Sigma}''_{yy}(t, 0, 0) = \frac{1}{3} S_t f'_2(t, S_t) - \frac{\mathbf{1}}{\mathbf{6}} S_t^2 f^{-1} f_2'^{\,2}(t, S_t) + \frac{1}{3} S_t^2 f''_{22}(t, S_t). \tag{2.72}$$

Proof Substituting (2.64) and (2.66) into (2.52) yields (dispensing with the argument):

$$\begin{aligned}\widetilde{\Sigma}''_{yy}(t, 0, 0) &= \frac{1}{3f^2} S_t f\left[f f'_2 + S_t f_2'^{\,2} + S_t f f''_{22}\right] - \frac{1}{2f^3} S_t^2 f^2 f_2'^{\,2} \\ &= \frac{1}{3} S_t f'_2(t, S_t) - \frac{1}{6} S_t^2 f^{-1} f_2'^{\,2}(t, S_t) + \frac{1}{3} S_t^2 f''_{22}(t, S_t).\end{aligned}$$ □

In order to check that result, we now compute the IATM curvature using [30].

Defining $(\bullet) \triangleq (t, S_t e^{sy})$ and differentiating (2.71) w.r.t. y gets us

$$\widetilde{\Sigma}''_{yy}(t, y, 0) = -2\left[\int_0^1 f^{-1}(\bullet)ds\right]^{-3}\left[\int_0^1 f^{-2}(\bullet)f_2'(\bullet)S_t e^{sy}s ds\right]^2 + \left[\int_0^1 f^{-1}(\bullet)ds\right]^{-2} S_t A(t, y),$$

where

$$A(t, y) \triangleq \int_0^1 \left[\left[-2f^{-3}\, f_2'^{\,2}(\bullet) + f^{-2}f_{22}''(\bullet)\right]S_t e^{2sy}s^2 + f^{-2}f_2'(\bullet)e^{sy}s^2\right]\cdot ds$$

Evaluating this expression at $(t, 0, 0)$, we obtain

$$\widetilde{\Sigma}''_{yy}(t, 0, 0) = 2f^3(t, S_t)S_t^2 f^{-4}\, f_2'^{\,2}(t, S_t)\left[\int_0^1 s ds\right]^2 + S_t f^2(t, S_t)\left[-2S_t f^{-3}\, f_2'^{\,2}(\star) + S_t f^{-2}f_{22}''(\star) + f^{-2}f_2'(\star)\right]\int_0^1 s^2 ds,$$

hence

$$\widetilde{\Sigma}''_{yy}(t, 0, 0) = \frac{1}{4}S_t^2 f^{-1}\, f_2'^{\,2}(t, S_t) + \frac{1}{3}\left[-2S_t^2 f^{-1}\, f_2'^{\,2}(\star) + S_t^2 f_{22}''(\star) + S_t f_2'(\star)\right],$$

which after simplification yields (2.72).

Alternatively, we can compare with the IATM curvature produced by Gatheral's formula (2.63). Using the following notations

$$(\circ) = (t, S_t; y, \theta) \qquad \text{and} \qquad (\diamond) = \left(t + u\theta, S_t e^{uy}\right)$$

we differentiate (2.70) w.r.t. y and obtain:

$$\overline{\Sigma}_y'^{\,2}(\circ) + \overline{\Sigma\Sigma}''_{yy}(\circ) = S_t\int_0^1 \left[\left(f_2'^{\,2} + ff_{22}''\right)(\diamond)S_t u^2 e^{2uy} + ff_2'(\diamond)u^2 e^{uy}\right]du.$$

Evaluating this expression at the IATM point $(\star) \triangleq (t, S_t; y = 0, \theta = 0)$, we obtain

$$\overline{\Sigma}_y'^{\,2}(\star) + \overline{\Sigma\Sigma}_{yy}''(\star) = \left[S_t^2\, f_2'^{\,2}(t, S_t) + S_t^2 f f_{22}''(t, S_t) + S_t f f_2'(t, S_t)\right] \int_0^1 u^2 du.$$

Substituting (2.67) and (2.69) and omitting argument (t, S_t) for f-differentials yields

$$\left[\frac{1}{2} S_t f_2'\right]^2 + f(t, S_t)\overline{\Sigma}_{yy}''(\star) = \frac{1}{3}\left[S_t^2\, f_2'^{\,2} + S_t^2 f f_{22}'' + S_t f f_2'\right],$$

so that finally the IATM curvature corresponding to Gatheral's formula comes as

$$\overline{\Sigma}_{yy}''(\star) = \frac{1}{3} S_t f_2'(t, S_t) + \mathbf{\frac{1}{12}} S_t^2 f^{-1}\, f_2'^{\,2}(t, S_t) + \frac{1}{3} S_t^2 f_{22}''(t, S_t). \tag{2.73}$$

Comparing (2.73) and (2.72) shows a discrepancy with the second coefficient (in bold). Indeed, the heuristics of (2.63) produce a *systematic positive bias* in the computation of the IATM curvature. This discrepancy might be due to the fact that Gatheral's formula is based on a zero-order expansion: it is entirely possible that further terms might bring the necessary correction, but this is left to further research.

2.4.2.4 Remaining IATM Differentials and General Remarks

Let us first examine the expression for the IATM slope (2.53). The fact that $\widetilde{\Sigma}_\theta'(\star)$ comes as an affine function of the instantaneous drift $a_{1,t}$ is a relief, as it supports the link between statics and dynamics. Beyond that property, the most obvious feature is the specific scaling of that coefficient. Indeed, it seems natural to normalise a_1, a_2 and a_3 by the initial value of the volatility, not as a Lognormal convention, but simply to distinguish magnitude from structural issues, i.e. to dissociate scale from shape. Then the IATM skew $\widetilde{\Sigma}_y'(\star)$ becomes dimensionless, while the next IATM y-differential $\widetilde{\Sigma}_{yy}''(\star)$ comes in σ_t^{-1}. Likewise $\widetilde{\mathrm{v}}_y'(\star)$, which is the only *dynamic* IATM strike-differential within the first layer, can be normalised as a single power of σ_t. We will see in Sect. 3.2.2 [p. 138] that this intuitive property (σ_t-homogeneity) is carried over to further *pure-strike* IATM differentials, both static and dynamic. By contrast, the IATM slope $\widetilde{\Sigma}_\theta'(\star)$ is σ_t-heterogeneous, therefore more complex and difficult to interpret. Therefore we revert back to the simpler framework of local volatility models.

Corollary 2.11 (IATM slope in a local volatility model) *In a local volatility model defined by* (2.5) [p. 30] *the IATM slope is*

$$\widetilde{\Sigma}_\theta'(t, 0, 0) = \frac{1}{2} f_1' + \frac{1}{12} S_t f^2 f_2' - \frac{1}{24} S_t^2 f\, f_2'^{\,2} + \frac{1}{12} S_t^2 f^2 f_{22}'' \tag{2.74}$$

where all differentials of $f(\cdot, \cdot)$ are taken at (t, S_t).

Proof Substituting all of Lemma 2.2 into (2.53), all f-differentials taken at (t, S_t), we get:

$$\begin{aligned}
\widetilde{\Sigma}'_\theta(t,0,0) &= \frac{1}{4}S_t f^2 f'_2 + \frac{1}{2}\left[f'_1 + \frac{1}{2}S_t^2 f^2 f''_{22}\right] \\
&\quad - \frac{1}{6}\left[S_t f^2 f'_2 + S_t^2 f\, {f'_2}^2 + S_t^2 f^2 f''_{22}\right] + \frac{1}{8}\frac{1}{f}S_t^2 f^2\, {f'_2}^2 \\
&= \frac{1}{2}f'_1 + \frac{1}{12}S_t f^2 f'_2 - \frac{1}{24}S_t^2 f\, {f'_2}^2 + \frac{1}{12}S_t^2 f^2 f''_{22}. \qquad (2.75)
\end{aligned}$$

□

Obviously we cannot verify the validity of this result against the exact result (2.62) as the latter is only defined in the immediate domain. This is where Dupire's formula (2.61) can be invoked, its explicit nature being overcome by its full domain of validity. Differentiating both sides once w.r.t. T, and omitting argument (t, y, θ) for all $\widetilde{\Sigma}$ differentials, we obtain

$$2ff'_1(T,K) = \frac{A'_\theta(\circ)}{B(\circ)} - \frac{A(\circ)B'_\theta(\circ)}{B^2(\circ)} \qquad (2.76)$$

with

$$A(\circ) = \widetilde{\Sigma}^4 + 2\theta\,\widetilde{\Sigma}^3\,\widetilde{\Sigma}'_\theta \qquad A'_\theta(\circ) = 6\widetilde{\Sigma}^3\,\widetilde{\Sigma}'_\theta + 6\theta\,\widetilde{\Sigma}^2\,{\widetilde{\Sigma}'_\theta}^2 + 2\theta\,\widetilde{\Sigma}^3\,\widetilde{\Sigma}''_{\theta\theta}$$

$$B(\circ) = \left[\widetilde{\Sigma} - y\widetilde{\Sigma}'_y\right]^2 - \frac{1}{4}\theta^2\,\widetilde{\Sigma}^4\,{\widetilde{\Sigma}'_y}^2 + \theta\,\widetilde{\Sigma}^3\,\widetilde{\Sigma}''_{yy}$$

$$\begin{aligned}
B'_\theta(\circ) &= 2\left[\widetilde{\Sigma} - y\widetilde{\Sigma}'_y\right]\left[\widetilde{\Sigma}'_\theta - y\widetilde{\Sigma}''_{y\theta}\right] - \frac{1}{2}\theta\,\widetilde{\Sigma}^4\,{\widetilde{\Sigma}'_y}^2 - \theta^2\,\widetilde{\Sigma}^3\,\widetilde{\Sigma}'_\theta\,{\widetilde{\Sigma}'_y}^2 \\
&\quad - \frac{1}{2}\theta^2\,\widetilde{\Sigma}^4\,\widetilde{\Sigma}'_y\,\widetilde{\Sigma}''_{y\theta} + \widetilde{\Sigma}^3\,\widetilde{\Sigma}''_{yy} + 3\theta\,\widetilde{\Sigma}^2\,\widetilde{\Sigma}'_\theta\,\widetilde{\Sigma}''_{yy} + \theta\,\widetilde{\Sigma}^3\,\widetilde{\Sigma}'''_{yy\theta}.
\end{aligned}$$

Therefore evaluating (2.76) at $(T = t, K = S_t)$ we get

$$2ff'_1(t,S_t) = 4\widetilde{\Sigma}\,\widetilde{\Sigma}'_\theta(\star) - \widetilde{\Sigma}^3\,\widetilde{\Sigma}''_{yy}(\star).$$

Isolating the IATM slope and then substituting both (2.67) and (2.72) we obtain

$$\begin{aligned}
\widetilde{\Sigma}'_\theta(\star) &= \frac{1}{2}f'_1(t,S_t) \\
&\quad + \frac{1}{4}f^2(t,S_t)\left[\frac{1}{3}S_t f'_2(t,S_t) - \frac{1}{6}S_t^2 f^{-1}\,{f'_2}^2(t,S_t) + \frac{1}{3}S_t^2 f''_{22}(t,S_t)\right],
\end{aligned}$$

which after simplification matches the desired result (2.74).

Note the similarity of the first term ($\frac{1}{2}f_1'$) with the IATM skew ($\frac{1}{2}f_2'$) and the mid-point method. Again it seems interesting to compare this exact result against Gatheral's heuristic formula. Differentiating (2.68) once w.r.t. θ and again defining $(\diamond) = (t + u\theta, S_t e^{uy})$ we obtain

$$\overline{\Sigma\Sigma}'_\theta(t, S_t; y, \theta) = \int_0^1 f(\diamond) f_1'(\diamond) u du.$$

Then evaluating this expression at ($y = 0, \theta = 0$) yields

$$f(t, S_t)\overline{\Sigma}'_\theta(t, S_t; 0, 0) = f f_1'(t, S_t) \int_0^1 u du \quad \text{so} \quad \overline{\Sigma}'_\theta(t, S_t; 0, 0) = \frac{1}{2} f_1'(t, S_t).$$

This expression clearly misses several terms, and ignores completely the role of the space coordinate in generating the IATM slope. However, we mitigate this structural shortcoming by the experience that, in practice, the magnitude of the ignored terms tends to be relatively small.

Let us now turn to the endogenous dynamics of the IATM skew, as described by (2.56). They show that $\widetilde{\mathrm{v}}'_y(\star)$ is totally independent from the exogenous specification (i.e. from $\overrightarrow{a}_{3,t}$ or $\overrightarrow{\overrightarrow{n}}(\star)$) which was not obvious when considering the recovery result (2.38) [p. 54]. Note also that $\widetilde{\mathrm{v}}'_y(\star)$ turns out to be a *dimensionless* quantity: it will remain unchanged if the volatility level σ_t is scaled by a constant λ. However, scaling the (endogenous) vol of vol $a_{2,t}$ will have a quadratic and negative effect on $\widetilde{\mathrm{v}}'_y(\star)$.

Note that in a LV model, if a proxy formula matches the IATM skew *as a process* (i.e. dynamically) then the latter's endogenous volatility will also agree, hence $\widetilde{\mathrm{v}}'_y(t, 0, 0)$ will match.

Now taking a step back and looking globally at (2.69)–(2.72) a very natural pattern is seen to emerge: indeed, these pure-strike differentials seem to be controlled sequentially by the successive space differentials of the local volatility function f. In order to put this impression into perspective, we bring forward the following remark.

Remark 2.10 The bijectivity established by Theorems 2.1 and 2.2 between the $\widetilde{\Sigma}$-(2,0) and σ_t-(2,0) groups can be modified for endogenous models. This is due to the exclusive relationship between $\overrightarrow{a}_{3,t}$ and $\overrightarrow{\overrightarrow{n}}(\star)$, characteristic of the constraints imposed on the SImpV model (see Sect. 2.3.2.1 [p. 57]). Specifically, we underline the following subset of the main bijection:

$$\left[\sigma_t, a_{1,t}, a_{2,t}, a_{22,t}\right] \quad \underset{\text{Endog. Models}}{\overset{\overrightarrow{a}_{3,t}\equiv 0}{\longleftrightarrow}} \quad \left[\widetilde{\Sigma}(\star), \widetilde{\Sigma}'_y(\star), \widetilde{\Sigma}''_{yy}(\star), \widetilde{\Sigma}'_\theta(\star)\right]$$

where we intentionally use only shape differentials on the right-hand side.

In particular, within LV models Corollaries 2.67, 2.69, 2.72 and 2.74 show that the IATM level, skew and curvature of the smile are sequentially and respectively controlled by the IATM space differentials of order 0, 1, 2 of the LV function.

By sequentially we mean that matching $f^{(i)}(t, S_t)$ is conditional on all $f^{(j)}(t, S_t)$ having been set, for $0 \leq j < i$. Likewise, the IATM slope is then controlled by the first time differential of the LV function. Although we haven't yet established the results to support this statement, in accordance with intuition this property is carried over to higher orders. In other words, with LV models the control over the IATM $\widetilde{\Sigma}$-differentials is established by the corresponding differentials of f.

2.5 Illustrations and Applications

In the course of solving the inverse and direct problem, our efforts so far have focused on establishing the raw asymptotic results, and providing a mathematical interpretation of the structural relationships between three groups of processes:

- The coefficients (in the chaos decomposition) of the stochastic instantaneous volatility model: σ_t, $a_{2,t}$, $a_{3,t}$, etc.
- The smile (IATM) shape descriptors: $\widetilde{\Sigma}'_y(\star)$, $\widetilde{\Sigma}''_{yy}(\star)$, etc.
- The smile (IATM) dynamics descriptors: $\widetilde{\nu}(\star)$, $\widetilde{n}(\star)$, $\widetilde{\nu}'_y(\star)$, etc.

We are coming now to a more applicative phase, where our previous results shall be examined with realistic modelling and trading concerns in mind. So far we have been considering SInsV and SImpV models "in parallel", not giving precedence to one over the other: simply because the mathematical framework exhibits such a symmetry between the two classes. But at the time of writing, stochastic *implied* volatility models are rarely used in practice. Therefore, the modelling concern will tend to be focused on the *instantaneous* model class.

The results we will use and interpret are those of Theorems 2.1 [p. 55] and 2.2 [p. 62]. We will group these Theorems, their corollaries and the exposed quantities under the overall denomination of "First Layer". Throughout this section, we make a deliberate effort to build some intuition, and also to continue introducing the topics to be developed in further chapters.

2.5.1 An Overview of Possible Applications

2.5.1.1 General Considerations

First of all, Remark 2.7 [p. 65] deserves additional comments, from the perspective of practical implementations. We observed that the IATM differentials provided by the first layer are of the second order in space (strike) but only of the first order in time (to maturity). This is naturally a consequence of Itô's Lemma, and this discrepancy will present itself at every order of differentiation, as shown later by the Ladder Effect (in Sect. 3.1). But this is also lucky for practitioners willing to extrapolate the smile shape. Indeed, most live market smiles do exhibit a lot more variation and irregularity in the strike dimension than in the expiry one: for instance curvature is a lot more pronounced,[18] so that we need $\widetilde{\Sigma}''_{yy}(\star)$ more than $\widetilde{\Sigma}''_{\theta\theta}(\star)$.

Also, one might wonder about the consequences of extrapolating the implied volatility far from the money, using Maclaurin series for instance. Again, in practice we are helped by the fact that Vega,[19] the sensitivity of price w.r.t. Lognormal implied volatility, vanishes when the strike goes to zero or to infinity. This might not have been be the case, or maybe not in such an obvious manner, had we used another implied parameter than Lognormal volatility. Beyond this mathematical argument let us recall that, for these far-from-the-money strikes, liquidity drops and therefore bid/ask spreads increase (at least in relative terms). This market feature does mitigate the fact that the attainable precision is comparatively lower in these regions.[20]

An attractive feature of this asymptotic approach is its generic nature: the fact that our results are produced for abstract stochastic volatility models, both instantaneous and implied. Also, as will be proven in Chap. 3, the methodology can be extended to any differential order.

Note also that with a single underlying and a scalar endogenous driver, we can directly apply our asymptotic results to the most common class of *mixtures* models, those combining *prices* with *fixed weights*. However, as a model class as well as a technical tool, mixtures constitute a wider family and do offer interesting (but not necessary satisfactory) properties: see [44, 45] and [46], among others. They are also strongly linked to the notion of a basket, which we will cover in Chap. 3, both in the constant and in the stochastic weights configurations (see Sect. 3.5 [p. 185]).

Despite these positive aspects, the asymptotic nature of the results brings structural limitations. Indeed, the model is "localised" at the IATM point. If, for instance, we are dealing with a SinsV model of the parametric diffusive class (the most common type) then all functional coefficients of the SDE will be seen through their partial differentials, taken at the initial point. True, if these functions are analytic, then pushing our method to an infinite order would theoretically solve the issue. Obviously in practice this is not an option, so that large variations of these maps w.r.t. time (non-stationarity) and/or space will degrade the output quality, at a given order of differentiation: ideally, we would like an integral approach as in [47].

There is, however, in the modelling community, a strong argument for stationary or time-homogeneous models (see [48] for instance). We support that view, within

---

[18] One could argue that this is a natural consequence of the higher liquidity of strike-based products (Butterflies, Strangles, etc.) compared to maturity spread products, which itself leads to the question of forward volatilities, but this is out of our current scope.

[19] $\mathscr{V} = S_t \mathscr{N}'(d_1)$.

[20] Note that the notions of precision and sensitivities must be clearly defined. In particular one might choose to focus either on absolute or on relative precision.

reason and while maintaining some flexibility,[21] because in practice stationary models tend to stabilise the calibration process and the hedge. Indeed, not relying on time-dependency to improve the calibration forces the modeller to develop deeper, more involved and hopefully more realistic dynamics. *A contrario*, heavily using time functions to fudge a good fit means that once the market has moved, the new calibrated time structure is likely to be very different. Although the later approach is easy to implement and will effectively provide an efficient calibration of the benchmark products, it will also create wild swings for the parameters and for most other products that are not included in the calibration set.

To some extent, the presence of *mean-reversion* will also require to obtain higher-order differentials, as it distorts the relationship between instantaneous and averaged dynamics. Another clear limitation of the method is that it does not envisage jumps: this is left for further research, as it requires a very different technical context.

Overall, when using these results we usually end up in one of two situations:

- either we exploit the raw, asymptotic equations linking IATM differentials and instantaneous coefficients; or
- we exploit these equations within *extrapolation* schemes, usually providing the whole smile, which tends to involve a large proportion of "engineering" skills.

Let us briefly detail these two approaches, before providing some examples.

2.5.1.2 Pure Asymptotic Applications: Qualitative Approaches

Let us first turn to asymptotic applications, starting with what we shall call *qualitative model design & analysis* and which clearly refers to the direct problem.

Most practitioners rely on stochastic instantaneous volatility models for pricing and hedging. It is therefore in their interest to better understand these models' behaviour, in particular the influence or cross-interferences of the various parameters (vol of vol, correlation, mean-reversion, etc.) or of specific functional forms (local volatility, time-dependent parameters, etc.) on the volatility surface, on the joint dynamics of the underlying with its instantaneous volatility, or with the smile, etc.

Such a precise understanding enables the agent to deliver a better hedge, and the modeller to customise an existing model or even design a new one *ex nihilo* in order to fulfil given trading needs. However, most currently used SInsV models[22] depend on numerical engines to price, whether it be Finite Differences or Monte-Carlo schemes, or even (Fast) Fourier Transform. Hence the difficulty with judging the impact of modelling choices on the smile, both in static and dynamic terms.

Alternatively, the approach that we advocate consists in focusing on the IATM region and using a low-level differential approach, by manipulating the usual and

---

[21] In particular in the absence of economic rationale, which could distinguish between time periods.

[22] With the noticeable exception of SABR.

meaningful smile descriptors which are level, skew, curvature, slope, etc. Clearly the $\widetilde{\Sigma}$-(2,0) group generated by Theorem 2.2 provides most of these. We will show in Chap. 3 that, providing simple and realistic assumptions, all IATM differentials can be expressed. In particular, Sect. 3.2 will give the most important ones, such as the twist and flattening (see the introduction of that chapter for a *typology* of the smile). Very shortly, Sect. 2.5.2 will be dedicated to an illustration of that approach. It will focus on the use of skew functions within stochastic volatility models, and in particular on the comparison of the Lognormal Displaced Diffusion and CEV instances.

Still in their pure asymptotic form, we can also use the formulas in an *inverse* manner. We can either exploit the recovery formulas provided by the Recovery Theorem 2.1, or try and invert the direct expressions of Theorem 2.2. The point is usually to re-parametrise, either totally or partially, a stochastic instantaneous volatility model by using meaningful, market-related quantities associated to the smile. As for model analysis, the rationale is the difficulty of appreciating the magnitude or impact of parameters in typical SinsV models, such as mean-reversion or vol of vol. Therefore the inverse method transforms the original model into an "intuitive" version, which is parametrised *via* its most pertinent *output*, the smile. Sect. 2.5.3 will illustrate this method on several simple examples.

2.5.1.3 Whole Smile Extrapolations and the Pertinence of Polynomials

When considering the direct problem, there is a great temptation to extrapolate the IATM differentials that have been computed (in particular those given by Theorem 2.2) to the whole smile, i.e. for all strikes and all maturities.

The natural method to exploit these results is to develop Taylor/Maclaurin series with both strike and maturity as variables. This is indeed the approach taken by [12] to describe the absolute surface $\Sigma(t, S_t, K, T)$, and we propose now to mimic that approach.

Within our framework, we denote by $\widetilde{\Sigma}^*(t, y, \theta)$ the polynomial approximation of the sliding implied volatility $\widetilde{\Sigma}(t, y, \theta)$, matched at the IATM point $(t, 0, 0)$. Let us invoke the IATM Identity (2.36) along with the static results of Theorem 2.2: (2.51), (2.52) and (2.53). We then write the Maclaurin series as

$$\begin{aligned}\widetilde{\Sigma}^*(t, y, \theta) = \sigma_t + y\frac{a_2}{2\sigma_t} + \frac{1}{2}y^2\frac{1}{\sigma_t^3}\left[\frac{1}{3}\sigma_t a_{22} - \frac{1}{2}a_2^2 + \frac{1}{3}\|\vec{a}_3\|^2\right] \\ + \theta\left[\frac{1}{2}a_1 + \frac{1}{4}\sigma_t a_2 + \frac{1}{8}\frac{a_2^2}{\sigma_t} + \frac{1}{12}\frac{\|\vec{a}_3\|^2}{\sigma_t} - \frac{1}{6}a_{22}\right].\end{aligned} \tag{2.77}$$

As for approximating the static *absolute* implied volatility surface, we can use Corollary 2.5, namely (2.36), (2.58), (2.59) and (2.53) to obtain

$$\Sigma^*(t,S_t,K,T) = \sigma_t + \left[\frac{K}{S_t} - 1\right]\frac{a_2}{2\sigma_t}$$
$$+ (T-t)\left[\frac{a_1}{2} + \frac{a_2}{4}\sigma_t + \frac{1}{8}\frac{a_2^2}{\sigma_t} + \frac{1}{12}\frac{\|\vec{a}_3\|^2}{\sigma_t} - \frac{a_{22}}{6}\right]$$
$$+ \frac{1}{2}\left[\frac{K}{S_t} - 1\right]^2 \left[\frac{1}{\sigma_t^3}\left[\frac{1}{3}\sigma_t a_{22} - \frac{1}{2}a_2^2 + \frac{1}{3}\|\vec{a}_3\|^2\right] - \frac{a_2}{2\sigma_t}\right]. \tag{2.78}$$

Quite reassuringly, these results are totally compatible with [12], as proven in Appendix E of [35]. It is important to note, however, that this expression cannot be used confidently for trading the vanillas, not necessarily because it constitutes only a low-level approximation,[23] but because it cannot *a priori* be guaranteed as valid, even statically. Besides, this is only part of the story, as the dynamic coefficients $\widetilde{b}$, $\widetilde{\nu}$ and $\vec{n}$ also need to ensure the validity of the surface in the future, as well as satisfy the ZDC everywhere. Finally, as *approximations of the dynamics*, they must tally the *dynamics of the (static) approximations*: in a nutshell, both groups have to be *consistent*.

At this stage of the study, we choose not to dwell on the various flavours and difficulties of these extrapolations. Some basic considerations will be discussed later in Sect. 4.1 and a practical application will be presented in Sect. 4.5. For now, it suffices to say that Taylor/Maclaurin series on the implied volatility and w.r.t. strike are certainly not the only option, and unfortunately are not as straightforward as we might hope. Also, we can already anticipate that the two hurdles that any approximation will have to tackle are, on one hand, the *validity* of the associated price surface (both statically and dynamically), and, on the other hand, the *precision* of that proxy w.r.t. the real model.

The fact that we are dealing with such approximations argues in itself against pure pricing applications: the risk of inconsistency and of arbitrage could indeed be significant. But on the other hand, these extrapolations can provide good initial guesses that can be used in static or dynamic calibration procedures: indeed, most *global* calibration procedures use an optimiser, which itself heavily invokes the actual pricer. More precisely, having selected a (large) collection of options $[K, T]$ for which market/target prices are available, then for a given set of model parameters the *market error* is defined as a metric[24] between the model option prices and the corresponding targets. The optimisation engine is tasked with minimising the market error, as a function of the model parameters, which are our (usually constrained) variables. Any such engine will therefore make numerous calls to the pricer, which itself for complex models usually consists in another numerical engine: FFT, PDE solver (finite differences or finite elements) or Monte-Carlo. That pricer will therefore be much slower than an extrapolation formula based on our asymptotic results, coming in closed form.

---

23 Higher-orders differentials should be required when far from the money.

24 Typically weighted squared differences, but some exact bootstrap is often involved.

However, the speed, the stability and possibly the convergence of the method will depend on a good starting point.[25] Therefore, by first running the optimisation engine on the approximate pricer we can attain a very good starting point, in a fraction of the time required by the nominal procedure. Also, before handing over to the real optimiser, the approximation formulas can provide good initial approximations of the Jacobian or of the Hessian. Indeed, this feature becomes very useful when the optimiser/root-finder relies on Gauss-Newton or Quasi-Newton algorithms (typically BFGS, Broyden, or Levenberg-Marquardt: see [49]). The proxy also enables us to address the possible instability issue, i.e. the existence of local minima: its speed enables it to be incorporated in dedicated methods, such as genetic algorithms.

In terms of fast calibration, note that the pure asymptotic results can also be used, as will be discussed in Sect. 2.5.3. Less precise, but faster than whole smile extrapolations, these asymptotic methods can therefore be employed to initialise the procedure described above, thus solving for the optimal parameters "in cascade".

2.5.1.4 Sensitivities and Hedge Ratios: Delta, Gamma, Vega & Co.

In both fields of application mentioned above, i.e. pure asymptotics and extrapolations, the IATM differential results in general, and the first layer in particular, grant access to more than just the smile's shape and dynamics. Indeed, they also provide hedging and risk information in the form of many *hedge ratios*, including Delta, Gamma, Vega, Volga and Vanna (see p. 44 for definitions).

As will shortly be discussed, in a market completed with options it is ultimately the *hedge strategy* that is the real determinant in terms of pricing and overall risk management. And in the context of unobservable and/or non-tradeable state variables, in particular with SV models (both instantaneous and implied), the very notion of hedge and its relationship to various price differentials becomes ambiguous.

Hedging with parametric diffusions

With these features in mind, we start by comparing the SImpV class to the dominant type of SInsV models, i.e. *parametric diffusions*. The ultimate purpose of this model class is to describe the dynamics of the (single) underlying S_t and *a priori* of no other financial quantity, especially not of options on S_t. To express those dynamics, the model is equipped with a finite number M of auxiliary *state variables* (typically some stochastic volatility) gathered under the notation $\overrightarrow{X_t}$. Restricting ourselves to Itô

[25] This is particularly the case for "scratch" calibration, where the parameters are completely unknown, but less so for re-calibration, which usually starts from the previous parameter set.

processes, this dynamic system is then driven by the following *Markovian* parametric diffusion:

$$dS_t = \vec{f}\left(t, S_t, \vec{X_t}; \vec{\beta}\right)^{\perp} d\vec{B_t} \tag{2.79}$$

$$d\vec{X_t} = \vec{g}\left(t, S_t, \vec{X_t}; \vec{\beta}\right) dt + \vec{\vec{h}}\left(t, S_t, \vec{X_t}; \vec{\beta}\right) d\vec{B_t}. \tag{2.80}$$

By convention, $\vec{B_t}$ is a standard Wiener process of finite dimension $N \leq M + 1$, with unit correlation matrix (i.e. with independent components) and which aggregates both the endogenous and exogenous drivers. The measure is chosen by convention as martingale for the underlying and for all European options considered. The individual components $X_{i,t}$, on the other hand, are not necessarily driftless and can typically be mean-reverting. Finally, by $\vec{\beta}$ we denote the finite set of parameters specifying the functional coefficients $f(\cdot)$, $g(\cdot)$ and $h(\cdot)$ which define the *parametric* diffusion.

The way in which the underlying S_t as well as its dynamics (2.79)–(2.80) are generated categorises the type of model considered, which usually come in one of two flavours:

- In *Factor Models*, a finite set of processes (often the components of $\vec{X_t}$ itself) represent abstract quantities, which with time t represent the actual state variables of the dynamic system. Consequently the underlying S_t is initially absent from the defining SDE system, which typically misses (2.80). The abstract factors are then mapped onto the underlying *via* a function. To fall back into our framework, this mapping function must be parametric[26] as with some Linear or Quadratic Gaussian models. We will not dwell on this class, which is sometimes also referred to as *Markov Functional*.
- In *Market Models*, S_t is a *market* instrument and its dynamics (2.79) are explicitly included. Formally, if $\vec{X_t}$ is empty then we have a local volatility model (*à la* Dupire) whereas in the case where $\vec{X_t} \equiv \sigma_t$ we may have a stochastic volatility model in the usual sense, such as with Heston or SABR.

Having defined the underlying's dynamics, we now turn to the price system of European options. The latter being uniquely determined by no-arbitrage assumptions under the chosen measure, European prices[27] will therefore come as deterministic (but not necessarily explicit) functions of:

- The Markovian state variables t, S_t and $\vec{X_t}$ (e.g. $[t, S_t, \sigma_t]$).
- The parameter set $\vec{\beta}$ specifying the diffusion (e.g. correlation, mean-reversion, vol of vol).
- The option parameters, usually strike K and expiry T.

[26] General Factor models strictly contain this limited setup, e.g. some use a non-parametric mapping.
[27] And therefore the associated marginal distributions.

The role of the auxiliary state variables $\overrightarrow{X_t}$ is partly to bring additional drivers to the smile. This is unlike pure local volatility (LV) models, for instance, which are strictly contained within *purely endogenous (PE) models*. In the latter, *all* components of $\overrightarrow{B_t}$ affect dS_t, which allow the shape and dynamics of the smile to be functions of the $X_{i,t}$ components. Since those processes are *a priori* non-observable and therefore non-tradeable, PE models imply *a priori* an incomplete market, as opposed to their LV subset.

The parametric diffusion family can of course be cast within our SInsV framework, as specified in Sect. 2.1.3.1 [p. 35]. But, although this class is noticeably simpler than the SImpV framework, it already exhibits ambivalent definitions for Greeks and hedges. Indeed, those can come either as partial differentials of the price w.r.t. state variables and parameters, or as individual weights in some replicating portfolio. Furthermore, although that portfolio value process is unique here, as an Itô integral it can be alternatively defined against the Brownian drivers, the state variables or some vanilla options. In the last two cases the weights are generally *not* unique, which is the downside of "hedging the drivers" through proxies.

Within the most generic framework of stochastic *implied* volatility models, the picture now gets even more complex as the notion of Greek becomes significantly ill-defined. Indeed, the Call price and/or the implied volatility processes are not necessarily deterministic functions of certain state processes (including t and S_t): recall that in the general case we simply dispense with state variables, apart from t and S_t (see Sect. 2.1.3.2). Accordingly, it is not *a priori* possible to identify a single such state variable that could pass as a "volatility" and be used to define Vega, Volga and Vanna. Also, differentiating price or volatility w.r.t. S_t becomes nonsensical in the general case: indeed we are now facing stochastic functionals of S_t, all driven by a multi-dimensional Wiener process which we know to be *a.s.* nowhere differentiable.

In summary, the differential definitions of Delta, Gamma and Vanna disappear, while the concept of Vega or Volga becomes obscure. As for the definition of those Greeks as replication weights against anything else than the drivers themselves, again there is no uniqueness. This is always the case when opting for a finite set of options as hedging instruments, but made even more pregnant as we are now dealing with a continuum of these. Whereas writing the process as a stochastic integral against some state variables is now only possible for S_t, which defines an extended Delta. Let us develop this last point, since Delta is by far the most important of the Greeks. Recall then that in the SImpV context, we have the dynamics of the Call from (2.22) [p. 44] as

$$\begin{aligned} dC(t, S_t, T, K) = & \left[\Delta S_t \sigma_t + \mathscr{V} \nu \sqrt{\theta}\right] dW_t + \mathscr{V} \sqrt{\theta}\, \overrightarrow{n}^{\perp} d\overrightarrow{Z}_t \\ & + \left[-\mathscr{V} \Sigma (2\sqrt{\theta})^{-1} + \mathscr{V} b \sqrt{\theta} + \frac{1}{2} \Gamma S_t^2 \sigma_t^2 \right. \\ & \left. + \frac{1}{2} \vartheta \left(\nu^2 + \| \overrightarrow{n} \|^2 \right) \theta + \Lambda S_t \sigma_t \nu \sqrt{\theta} \right] dt. \end{aligned}$$

The endogenous coefficient within brackets can be seen as a form of Delta because, in principle, it is only hedgeable with S_t:

$$\left[\Delta S_t \sigma_t + \mathscr{V}\nu\sqrt{\theta}\right] dW_t = \underbrace{S_t^{-1}\sigma_t^{-1}\left[\Delta S_t \sigma_t + \mathscr{V}\nu\sqrt{\theta}\right]}_{\text{Extended Delta}} dS_t.$$

However this extended Delta involves the Vega of Black's formula, which can be confusing. Also, it is *a priori* not measurable w.r.t. the filtration generated by the underlying, or even by the endogenous driver itself. Note, however, that this was already the case with parametric diffusion models.

We conclude that w.r.t. hedging and Greeks issues, our safest option is to limit ourselves to the parametric diffusion models (thankfully representing the overwhelming majority of practical implementations) and to define Greeks most simply as differentials w.r.t. the state variables t, S_t and $\overrightarrow{X_t}$ (the *hedges*) and w.r.t. the parameters $\overrightarrow{\beta}$, K and T (the *sensitivities*). We still have to link those Greeks to our IATM asymptotic results written on the sliding implied volatility. Although in principle we have access to the whole smile through extrapolations, we will focus on exact IATM results. In consequence, it is intuitive to anticipate that the limitations in scope of the first layer (or rather, the $\widetilde{\Sigma}$-(2,0) group: order 1 in maturity and 2 in strike) will necessarily have consequences.

Greeks *vs* the sliding smile

The most liquid options are usually those defined with fixed strike K and fixed expiry T (the *absolute* coordinates) and their hedging will (one way or another) invoke their price sensitivities. In our asymptotic context, some of the natural questions to address are therefore:

- Whether and how the Greeks (Delta, Gamma, Vega, etc.), defined here simply as price differentials, are affected by moving to the sliding coordinates y and θ;
- Which and how IATM *absolute* Greeks can be obtained, given the first layer results, i.e. with a specific set of IATM differentials for the *sliding* IV $\widetilde{\Sigma}$.

Our roadmap is the following: the first step will be to relate *absolute price* differentials to their *sliding implied volatility* counterparts. Then in a second phase we will identify *which differentials are affected* in the general domain. Finally, our third step will focus on the *IATM option* and derive the main absolute Greeks.

Remaining within the parametric diffusion framework (2.79)–(2.80) we have both the price and the IV as deterministic (but not necessarily explicit) functions $\underline{C}$ and $\underline{\Sigma}$ of the state variables, of the model parameters and of the option:

$$\underline{C}\left(t, S_t, \overrightarrow{X_t};\, \overrightarrow{\beta};\, K, T\right) = B\left(S_t, K, \sqrt{T-t}\,\underline{\Sigma}\left(t, S_t, \overrightarrow{X_t};\, \overrightarrow{\beta};\, K, T\right)\right).$$

A look at Appendix C confirms that all relevant differentials of the Call price w.r.t. its arguments can be obtained (without singularity issues) by combining the differentials of Black's formula and those of $\underline{\Sigma}$. We have hence answered the first question, and we also note that the same approach could be used with a *Normal* Implied Volatility and Bachelier's formula.

We now turn to the second question of how to extract the differentials of the absolute implied volatility $\underline{\Sigma}$ from those of its sliding version $\underline{\widetilde{\Sigma}}$, defined and denoted as follows:

$$\underline{\Sigma}(\Box) \triangleq \underline{\widetilde{\Sigma}}(\circ) \text{ with } (\Box) \triangleq \left(t, S_t, \overrightarrow{X_t};\, \overrightarrow{\beta};\, K, T\right) \text{ and } (\circ) \triangleq \left(t, S_t;\, \overrightarrow{X_t};\, \overrightarrow{\beta};\, y, \theta\right). \tag{2.81}$$

As discussed in Sect. 2.1.2.3 [p. 29], the move to sliding coordinates has *embedded* the two state variables t and S_t into $\underline{\widetilde{\Sigma}}$, now a *parametric* process. Clearly (2.81) shows that the differentials w.r.t. the auxiliary state variable $\overrightarrow{X_t}$ and w.r.t. the model parameter $\overrightarrow{\beta}$ are unaffected by that slide. In particular, within a classical SInsV model such as Heston, where the volatility process σ_t belongs to $\overrightarrow{X_t}$, the Vega and Volga as well as the sensitivity to correlation and vovol are unchanged. We are therefore left with four variables which, from Black's formula and from the definition of y and θ, can be grouped by symmetry into a *time* pair (time t and maturity T) and a space pair (underlying S_t and strike K). We obtain the relevant Greeks as:

- **Theta** $\quad \underline{\Sigma}'_t(\Box) = \underline{\widetilde{\Sigma}}'_t(\circ) - \underline{\widetilde{\Sigma}}'_\theta(\circ)$ (T-differentials are similarly obtained).
- **Delta** $\quad \underline{\Sigma}'_{S_t}(\Box) = \underline{\widetilde{\Sigma}}'_{S_t}(\circ) + \underline{\widetilde{\Sigma}}'_y(\circ)\dfrac{\partial y}{\partial S_t}$.
- **Gamma** $\quad \underline{\Sigma}''_{S_tS_t}(\Box) = \underline{\widetilde{\Sigma}}''_{S_tS_t}(\circ) + 2\underline{\widetilde{\Sigma}}''_{yS_t}(\circ)\dfrac{\partial y}{\partial S_t} + \underline{\widetilde{\Sigma}}''_{yy}(\circ)\left[\dfrac{\partial y}{\partial S_t}\right]^2 + \underline{\widetilde{\Sigma}}'_y(\circ)\dfrac{\partial^2 y}{\partial S_t^2}$.

The K-differentials would lead to similar equations, which are valid in the whole (K, T) domain. As announced, our third step is now to focus on the IATM point, where we shall use the abbreviations

$$(\blacklozenge) \triangleq \left(t, S_t, \overrightarrow{X_t};\, \overrightarrow{\beta};\, S_t, t\right) \qquad \text{and} \qquad (\star) \triangleq \left(t, S_t;\, \overrightarrow{X_t};\, \overrightarrow{\beta};\, 0, 0\right).$$

Evaluating the three differential expressions above at the IATM point, we get respectively

$$\begin{aligned}
&\text{IATM Theta} && \underline{\Sigma}'_t(\blacklozenge) = \underline{\widetilde{\Sigma}}'_t(\star) - \underline{\widetilde{\Sigma}}'_\theta(\star).\\
&\text{IATM Delta} && \underline{\Sigma}'_{S_t}(\blacklozenge) = \underline{\widetilde{\Sigma}}'_{S_t}(\star) - \frac{1}{S_t}\underline{\widetilde{\Sigma}}'_y(\star).\\
&\text{IATM Gamma} && \underline{\Sigma}''_{S_tS_t}(\blacklozenge) = \partial^2_{S_tS_t}\underline{\widetilde{\Sigma}}(\star) - \frac{2}{S_t}\partial_{S_t}\underline{\widetilde{\Sigma}}'_y(\star) + \frac{1}{S_t^2}\left[\underline{\widetilde{\Sigma}}'_y(\star) + \underline{\widetilde{\Sigma}}''_{yy}(\star)\right].
\end{aligned} \tag{2.82}$$

We can now conclude our answer to the second question with the following remark:

Remark 2.11 Let us denote by (t,T) the *time* pair and (S_t,K) the *space* pair. Given the three functionals $\underline{\widetilde{\Sigma}}'_\theta(\star)$, $\underline{\widetilde{\Sigma}}'_y(\star)$ and $\underline{\widetilde{\Sigma}}''_{yy}(\star)$ then we have any cross-differential of the absolute (traded) price, provided one of the following *exclusive* conditions is verified:

- The cumulative differential orders of the time *and* space pairs are both 0.
- The cumulative differential order is 1 for the time and 0 for the space pair.
- The cumulative differential order is 1 or 2 for the space and 0 for the time pair.

These Greeks include the Theta, Delta, Gamma and by consequence the Vanna cross-differential (since the volatility differentials are available).

Note that the time *vs* space discrepancy is naturally linked to Itô's Lemma, and is manifest at all higher differential orders with the *ladder effect* (see Fig. 3.1, p. 122). Finally, recall that should we wish to go beyond the differential definition of the Greeks and assess the hedge by transferring the dynamics themselves, we can use (2.16) [p. 42] and (2.22) [p. 44].

2.5.2 Illustration: Qualitative Analysis of a Classical SV Model Class

Later on, we will dedicate a large part of Chap. 4 to practical applications of the direct results to current real-life models, such as SABR, thus showing how Theorem 2.2, along with the higher-order extensions found in Sect. 3.2, can be used to produce model-specific IATM differentials and whole-smile approximations.

In this section, our ambitions are more modest, and they also take a different angle. Indeed, we show how the first layer (Theorem 2.2) alone can be invoked, not to approach a whole marginal distribution, but to predict the rough *qualitative* features of a smile, when it is generated by a given stochastic instantaneous volatility model. In other words, we emphasise the potential of the asymptotic results as a rapid *analysis* toolbox for such SInsV models.

To that intent, we select a rich and very popular SV model class, the Extended Skew Market Model, which incorporates a local volatility component. We then examine the basic statics and dynamics of its implied volatility surface, focusing on the close relationship between the skew and the local volatility function. We will see that we can extend the more academic results of pure local volatility models, with little additional work.

We then specialise these asymptotic results, and show that they allow fast and easy comparison of two very popular local volatility instances: the Lognormal Displaced Diffusion (LDD) and the Constant Elasticity of Variance (CEV). We show that our asymptotic results can provide fast answers to qualitative questions regarding the respective generated smiles, in particular the IATM level, skew as well as their dynamics. Overall, we show also that the computations involved are simple and repetitive, which suggests a possible automation.

2.5.2.1 Properties of Local Volatility "Skew" Functions

So far we have mainly been considering generic formulae, with only a few errands in the local volatility framework. We have seen, for instance, in Remark 2.10 [p. 76] that, when designing or specifying the parameters of a stochastic *instantaneous* volatility model, the triplet $\left[\sigma_t, a_{2,t}, a_{22,t}\right]$ would be sufficient to simultaneously control the IATM level, the skew and the curvature.

Although it has proven efficient to build some useful intuition, that local volatility class is nevertheless limited as a real-life hedging model. Indeed, in practice such pure local volatility models allow us to match any smooth and valid smile, but exhibit poor, if not downright dangerous, dynamic properties (see [11] for instance). Essentially for that reason, both academics and practitioners have engineered a class extending the local volatility models, and which currently dominates the modelling approach. These modern market models, such as SABR or FL-SV, combine both *local* and *stochastic* volatility by using a multiplicative perturbation process: we will call them "Extended Skew Market Models" for reasons that will be made clear shortly. But first of all, let us formalise the generic model class:

Definition 2.2 (*Extended Skew Market Model*)

$$\frac{dS_t}{S_t} = \alpha_t f(t, S_t) dW_t \qquad \text{or} \qquad dS_t = \alpha_t \varphi(t, S_t) dW_t, \tag{2.83}$$

$$d\alpha_t = h(t, \alpha_t)dt + \varepsilon g(t, \alpha_t) dB_t \qquad \text{with} \qquad \rho dt = \langle dW_t, dB_t \rangle. \tag{2.84}$$

This class clearly represents a subset of the generic *parametric diffusion* model (2.79)–(2.80). We chose not to treat the latter here simply because of its (potentially) higher Markovian dimension and its more implicit correlation structure. Those generate formulae which are quite complex and therefore tend to be counter-productive for our modest purpose of illustration.

The ESMM class includes, among others, Heston [2], Lewis [1], SABR [11], FL-SV [50] and FL-TSS [51]: they all share its principle of a local volatility model which is then multiplicatively perturbed by the α_t process. Note, however, that in the literature the dynamics specified *in lieu* of (2.84) tend to be those of the "variance" α_t^2 rather than α_t proper.

As to why we chose to present both Normal and Lognormal conventions, it is simply because both are equally often found in the literature: for instance Lognormal is used with Heston, whereas Normal is used with SABR and FL-SV. The "skew model" terminology can now be justified: it comes from the fact that some of these models only allow zero correlation between the underlying and its perturbation process: this is the case for FL-SV, by opposition to SABR. When thus restricted to independent perturbations, they need rely purely on their local volatility function to generate some IATM skew, hence the name.

Conversely, note that this Extended Skew Model naturally incorporates the case of a deterministic perturbation $\alpha(t)$, hence falling back onto the pure local volatility models seen previously.

The desire to present results simultaneously for the Normal and Lognormal conventions of the ESMM leads us to first establish a very low-level transition formula. This is also justified by the fact that, throughout this study, we will frequently be juggling between Lognormal and Normal conventions, wherever dynamics are concerned. For instance, our results so far have been expressed using the Black implied volatility (hence the Lognormal "baseline") while many models (including the ESMM class above) use Normal conventions instead. In Chap. 4, for example, high-order IATM differentials for SABR will require many such conversions. Another example of conversion between the Normal and Lognormal models can be found in Chap. 3, where baseline transfers are examined. For these reasons, establishing the following Lemma is clearly economical at the scale of the study, if a bit of an overkill for the current section.

Lemma 2.3 (Infinite-order differentiation of the Lognormal LV function) *Assume two C^n, $\mathbb{R} \to \mathbb{R}$ functions f and φ with $f(x) = x^{-1}\varphi(x)$. Then for $n \geq 0$,*

$$f^{(n)}(x) = \sum_{k=0}^{n} (-1)^k \frac{n!}{(n-k)!} x^{-(k+1)} \varphi^{(n-k)}(x).$$

Proof The result is clearly verified at the initial index $n = 0$.

Then by induction: we assume the property verified at index n, so that

$$\begin{aligned} f^{(n+1)} &= \sum_{k=0}^{n} \frac{(-1)^k n!}{(n-k)!} \left[(-1)(k+1) x^{-(k+2)} \varphi^{(n-k)}(x) + x^{-(k+1)} \varphi^{(n-k+1)}(x) \right] \\ &= \sum_{k=1}^{n+1} \frac{(-1)^k n! k}{(n-k+1)!} x^{-(k+1)} \varphi^{n-k+1)}(x) + \sum_{k=0}^{n} \frac{(-1)^k n!}{(n-k)!} x^{-(k+1)} \varphi^{(n+1-k)}(x), \end{aligned}$$

hence

$$\begin{aligned} f^{(n+1)} &= x^{-1} \varphi^{(n+1)}(x) + (-1)^{n+1} (n+1)! x^{-(n+2)} \varphi(x) \\ &\quad + \sum_{k=1}^{n} (-1)^k \left[\frac{n! k (n-k)! + n! (n+1-k)!}{(n+1-k)! (n-k)!} \right] x^{-(k+1)} \varphi^{(n+1-k)}(x). \end{aligned}$$

The bracket on the right-hand side can then be rewritten as

$$\left[\frac{n! k (n-k)! + n! (n+1-k)(n-k)!}{(n+1-k)!(n-k)!} \right] = \left[\frac{(n+1)!}{(n+1-k)!} \right]$$

and we obtain the desired expression at index $n+1$. □

In order to apply Theorem 2.2 the first step is generally to compute the native expressions for the coefficients of the SInsV model, as formatted in our framework by (2.8)–(2.10). Recall that those coefficients invoked in the first layer are only σ_t, $a_{1,t}$, $a_{2,t}$, $a_{3,t}$ and $a_{22,t}$.

Lemma 2.4 (Instantaneous coefficients of the ESMM) *Let us place ourselves within the Extended Skew Market Model defined by* (2.83)–(2.84). *In the following formulas, we assume the functions to be taken at the immediate point, i.e. at our current set of Markov variables* (t, S_t, α_t). *Accordingly, we use the following notations:*

$$f \triangleq f(t,S_t) \quad \varphi \triangleq \varphi(t,S_t) \quad g \triangleq g(t,\alpha_t) \quad h \triangleq h(t,\alpha_t) \quad \dot{\rho} \triangleq \sqrt{1-\rho^2}. \tag{2.85}$$

The instantaneous coefficients σ_t, a_2, a_3 *and* a_{22} *come as:*

- *Using the Lognormal convention for the ESMM:*

$$\sigma_t = \alpha_t f \qquad a_{2,t} = \sigma_t^2 S_t \frac{f_2'}{f} + \rho\varepsilon\sigma_t \frac{g}{\alpha_t} \qquad a_{3,t} = \dot{\rho}\varepsilon\sigma_t \frac{g}{\alpha_t} \tag{2.86}$$

$$a_{1,t} = \sigma_t \frac{f_1'}{f} + \frac{1}{2}\sigma_t^3 S_t^2 \frac{f_{22}''}{f} + \sigma_t \frac{h}{\alpha_t} + \rho\varepsilon\sigma_t g S_t f_2' \tag{2.87}$$

$$\begin{aligned} a_{22,t} = {} & \sigma_t^3 S_t \left[\frac{f_2'}{f} + S_t \left[\frac{f_2'}{f}\right]^2 + S_t \frac{f_{22}''}{f} \right] \\ & + 3\rho\varepsilon\sigma_t g S_t f_2' + \rho^2\varepsilon^2\sigma_t \frac{g}{\alpha_t} g_2'. \end{aligned} \tag{2.88}$$

- *Using the Normal convention for the ESMM:*

$$\sigma_t = \alpha_t S_t^{-1}\varphi \; a_{2,t} = \sigma_t^2 \left[S_t \frac{\varphi_2'}{\varphi} - 1 \right] + \rho\varepsilon\sigma_t \frac{g}{\alpha_t} \; a_{3,t} = \dot{\rho}\varepsilon\sigma_t \frac{g}{\alpha_t} \tag{2.89}$$

$$a_{1,t} = \sigma_t \frac{\varphi_1'}{\varphi} + \sigma_t^3 \left[1 - S_t \frac{\varphi_2'}{\varphi} + \frac{1}{2} S_t^2 \frac{\varphi_{22}''}{\varphi} \right] + \sigma_t \frac{h}{\alpha_t} + \rho\varepsilon\sigma_t g \left[\varphi_2' - S_t^{-1}\varphi \right]$$

$$\begin{aligned} a_{22,t} = {} & \sigma_t^3 \left[2 - 3S_t \frac{\varphi_2'}{\varphi} + S_t^2 \left[\frac{\varphi_2'}{\varphi}\right]^2 + S_t^2 \frac{\varphi_{22}''}{\varphi} \right] \\ & + 3\rho\varepsilon g\sigma_t \left[\varphi_2' - S_t^{-1}\varphi \right] + \rho^2\varepsilon^2 \frac{g}{\alpha_t} g_2' \sigma_t. \end{aligned} \tag{2.90}$$

Given the obvious filiation of the ESMM, it is natural, easy and reassuring to verify that these results match those obtained for pure LV models: see Lemma 2.2 [p. 68].

Proof (Lemma 2.4)
Using the Lognormal convention, the cast provides the instantaneous volatility:

$$\sigma_t = \alpha_t f(t, S_t). \tag{2.91}$$

Then, simply using Itô gives us its dynamics as

$$\begin{aligned} d\sigma_t = &\left[\alpha_t f_1' + \frac{1}{2}\alpha_t f_{22}''\alpha_t^2 S_t^2 f^2 + \rho\varepsilon\alpha_t g S_t f f_2'\right] dt \\ &+ \alpha_t f_2' \alpha_t S_t f dW_t + f h dt + f\varepsilon g\left[\rho dW_t + \sqrt{1-\rho^2} dZ_t\right]. \end{aligned}$$

Furthermore, applying Lemma 2.3 for $n = 1$ and $n = 2$, we easily get the conversions

$$f_2' = -S_t^{-2}\varphi + S_t^{-1}\varphi_2' \quad \text{and} \quad f_{22}'' = S_t^{-1}\varphi_{22}'' - 2S_t^{-2}\varphi_2' + 2S_t^{-3}\varphi. \tag{2.92}$$

Therefore, the first three instantaneous coefficients come respectively as

$$\begin{aligned} a_{1,t} &= \alpha_t f_1' + \frac{1}{2}\alpha_t^3 S_t^2 f^2 f_{22}'' + hf + \rho\varepsilon\alpha_t g S_t f f_2' \\ &= \alpha_t S_t^{-1}\varphi_1' + \frac{1}{2}\alpha_t^3\varphi^2\left[S_t^{-1}\varphi_{22}'' - 2S_t^{-2}\varphi_2' + 2S_t^{-3}\varphi\right] \\ &\quad + hS_t^{-1}\varphi + \rho\varepsilon\alpha_t g\varphi\left[-S_t^{-2}\varphi + S_t^{-1}\varphi_2'\right] \end{aligned}$$

and

$$\begin{aligned} a_{2,t} &= \alpha_t^2 S_t f(t, S_t) f_2'(t, S_t) + \rho\varepsilon g(t, \alpha_t) f(t, S_t) \\ &= \alpha_t^2\varphi(t, S_t)\left[-S_t^{-2}\varphi(t, S_t) + S_t^{-1}\varphi_2'(t, S_t)\right] + \rho\varepsilon g(t, \alpha_t) S_t^{-1}\varphi(t, S_t) \end{aligned}$$

and

$$a_{3,t} = \varepsilon\dot{\rho} g f = \varepsilon\dot{\rho} g S_t^{-1}\varphi.$$

We can then switch α_t and f with σ_t using (2.91), which provides the desired expressions. Note that we chose to conserve the arguments in the expression of the endogenous coefficient $a_{2,t}$, as we now need its dynamics:

$$\begin{aligned} da_{2,t} = &\left[\alpha_t^2\left[f f_2' + S_t\, f_2'^{\,2} + S_t f f_{22}''\right] + \rho\varepsilon g f_2'\right]\alpha_t S_t f dW_t \\ &+ \left[2S_t f f_2' \alpha_t + \rho\varepsilon g_2' f\right]\rho\varepsilon g dW_t + [\cdots] dt + [\cdots] dZ_t, \end{aligned}$$

where we again isolate the endogenous coefficient

$$a_{22,t} = \alpha_t^3 S_t f \left[f f_2' + S_t\, f_2'^{\,2} + S_t f f_{22}'' \right] + 3\rho\varepsilon\alpha_t g S_t f f_2' + \rho^2\varepsilon^2 g g_2' f.$$

Again switching σ_t with α_t and f in this expression gives the sought formula. We can also use the conversion formulas to obtain, under the Normal convention:

$$\begin{aligned} a_{22,t} &= 3\rho\varepsilon\alpha_t g\varphi \left[S_t^{-1}\varphi_2' - S_t^{-2}\varphi \right] + \rho^2\varepsilon^2 g g_2' S_t^{-1}\varphi \\ &\quad + \alpha_t^3\varphi \left[S_t^{-1}\varphi \left[\frac{\varphi_2'}{S_t} - \frac{\varphi}{S_t^2} \right] + S_t \left[\frac{\varphi_2'}{S_t} - \frac{\varphi}{S_t^2} \right]^2 + \varphi \left[\frac{\varphi_{22}''}{S_t} - 2\frac{\varphi_2'}{S_t^2} + 2\frac{\varphi}{S_t^3} \right] \right], \end{aligned}$$

which after simplification and switch gives (2.90). □

When performing this type of calculation, we recommend identifying as many recurring, symmetric or meaningful quantities as possible, as early as possible. Similarly, it is also a good idea to allocate *ad hoc* symbols to these quantities. It might not seem useful at the current level of the first layer, but when computing deeper coefficients (as needed by higher-order IATM differentials, see Sect. 3.2) they will come in handy. For examples of this "technique", see, for instance, the treatment of SABR in Chap. 4.

We can now move on to the IATM differentials, both static and dynamic.

Proposition 2.3 (IATM differentials of the ESMM) *Within the ESMM framework, defining* $(\star) \triangleq (t, 0, 0)$

- *The IATM skew is expressed as*

$$\widetilde{\Sigma}_y'(\star) = \widetilde{\Sigma}_{y,f}'(\star) + \widetilde{\Sigma}_{y,\rho}'(\star), \tag{2.93}$$

where

$$\widetilde{\Sigma}_{y,f}'(\star) \triangleq \frac{1}{2}\sigma_t S_t \frac{f'}{f} = \frac{1}{2}\sigma_t \left[S_t \frac{\varphi_2'}{\varphi} - 1 \right]$$

is the "local volatility" term, while the "correlation" term comes as

$$\widetilde{\Sigma}_{y,\rho}'(\star) \triangleq \frac{1}{2}\rho\varepsilon \frac{g}{\alpha_t}.$$

- *The IATM curvature is expressed as*

$$\widetilde{\Sigma}_{yy}''(\star) = \widetilde{\Sigma}_{yy,f}''(\star) + \widetilde{\Sigma}_{yy,\rho}''(\star),$$

with

$$\widetilde{\Sigma}''_{yy,f}(\star) \triangleq \frac{1}{3}\sigma_t \left[S_t \frac{f_2'}{f} - \frac{1}{2} S_t^2 \left[\frac{f_2'}{f} \right]^2 + S_t^2 \frac{f_{22}''}{f} \right]$$

$$\widetilde{\Sigma}''_{yy,\rho}(\star) \triangleq \frac{\varepsilon^2}{3} \frac{1}{\sigma_t} \left[\rho^2 \frac{g}{\alpha_t} g' + \left(1 - \frac{5}{2}\rho^2 \right) \frac{g^2}{\alpha_t^2} \right].$$

In Normal representation, the "local volatility" term becomes

$$\widetilde{\Sigma}''_{yy,f}(\star) = \frac{1}{6}\sigma_t \left[2S_t^2 \frac{\varphi_{22}''}{\varphi} - S_t^2 \left[\frac{\varphi_2'}{\varphi} \right]^2 + 1 \right].$$

- *The IATM slope is expressed as*

$$\widetilde{\Sigma}'_{\theta}(\star) = \widetilde{\Sigma}'_{\theta,f}(t,0,0) + \widetilde{\Sigma}'_{\theta,\rho}(t,0,0) + \frac{\sigma_t}{4}\rho\varepsilon g S_t f'$$

with

$$\widetilde{\Sigma}'_{\theta,f}(\star) \triangleq \frac{\sigma_t}{2} \frac{f_1'}{f} + \frac{\sigma_t^3}{12} \left[S_t \frac{f_2'}{f} - \frac{1}{2} S_t^2 \left[\frac{f_2'}{f} \right]^2 + S_t^2 \frac{f_{22}''}{f} \right]$$

$$\widetilde{\Sigma}'_{\theta,\rho}(\star) \triangleq \frac{\sigma_t^2}{4}\rho\varepsilon \frac{g}{\alpha_t} + \frac{\sigma_t}{12}\varepsilon^2 \left[1 + \frac{1}{2}\rho^2 \right] \left[\frac{g}{\alpha_t} \right]^2 - \frac{\sigma_t}{6}\rho^2\varepsilon^2 \frac{g}{\alpha_t} g' + \frac{\sigma_t}{2} \frac{h}{\alpha_t}.$$

The local volatility term rewrites, in Normal convention, as:

$$\widetilde{\Sigma}'_{\theta,f}(\star) = \frac{\sigma_t}{2} \frac{\varphi_1'}{\varphi} + \frac{\sigma_t^3}{24} \left[1 - S_t^2 \left[\frac{\varphi_2'}{\varphi} \right]^2 + 2S_t^2 \frac{\varphi_{22}''}{\varphi} \right].$$

- *As for the dynamic coefficients, the IATM volatility of volatility is*

$$\begin{cases} \widetilde{\nu}(\star) = \sigma_t^2 S_t \frac{f_2'}{f} + \rho\varepsilon\sigma_t \frac{g}{\alpha_t} = \underbrace{\sigma_t^2 \left[S_t \frac{\varphi_2'}{\varphi} - 1 \right]}_{\widetilde{\nu}_f(\star)} + \underbrace{\rho\varepsilon\sigma_t \frac{g}{\alpha_t}}_{\widetilde{\nu}_\rho(\star)} \\ \widetilde{n}(\star) = \dot{\rho}\varepsilon\sigma_t \frac{g}{\alpha_t} \end{cases}$$

while the endogenous volatility of the skew comes as

$$\widetilde{\mathrm{v}}_y'(\star) = \widetilde{\mathrm{v}}_{y,f}'(\star) + \widetilde{\mathrm{v}}_{y,\rho}'(\star) + \frac{1}{2}\rho\varepsilon g S_t f_2',$$

with

$$\widetilde{\mathrm{v}}_{y,f}'(\star) \triangleq \frac{1}{2}\sigma_t^2\left[S_t\frac{f_2'}{f} + S_t^2\frac{f_{22}''}{f}\right] = \frac{1}{2}\sigma_t^2\left[1 - S_t\frac{\varphi_2'}{\varphi} + S_t^2\frac{\varphi_{22}''}{\varphi}\right]$$

$$\widetilde{\mathrm{v}}_{y,\rho}'(\star) \triangleq \frac{1}{2}\rho^2\varepsilon^2\frac{g}{\alpha_t}\left[g_2' - \frac{g}{\alpha_t}\right].$$

Instead of heading thereupon into the proof (which is very straightforward) let us first comment on and interpret these results; we start with general considerations, before reviewing some of the IATM differentials on an individual basis.

First of all, it seems rather intuitive that the formulae describing the ESMM shall prove more complex when using the Normal convention. Indeed, recall that the generic dynamics for S_t, which define σ_t and therefore a_1, a_2, a_3 and a_{22}, are *Log*normal; likewise, the implied volatility $\widetilde{\Sigma}$ that we consider is also Lognormal. In other words, we have on one hand a "simple" (Lognormal) model used to define the implied volatility—the *baseline*—and on the other hand, a "complex" target model written along Normal lines: this dissimilarity is what complexifies the results. In Chap. 3, we will see that this also impacts the precision and usefulness of the asymptotic method. Nevertheless, we will also show that adapting the baseline to the target model is often possible. In particular the Normal baseline is frequently a natural option, and this would certainly be the case here.

Still on the presentation of those results, the main reason why we choose to position the differentials of f as numerators is to underline the inherent *scaling* of the problem. In other words, it stresses the fact that the IATM differentials correspond to *shape* factors of the local volatility function, while only the level depends on its scale.

The relative complexity of the formulas, in conjunction with the scaling aspect, point naturally to a change of variables. This is indeed no surprise, since the left-hand-sides of these equations represent (differentials of) functions of log-moneyness $\ln(K/S_t)$, while the right-hand sides are functions of S_t. We will use this feature in the sequel to rewrite some of the quantities in a more intuitive manner.

Evidently, it is easy to see that by neutralising the perturbation we fall back onto the expressions for pure local volatility models: refer to (2.67) for the level, (2.69) for the skew, (2.72) for the curvature, and (2.74) for the slope. Accordingly we find that, as was the case with pure local volatility models, once the perturbation process α_t is set then the differentials f, f_2', f_{22}'' and f_1' taken in (t, S_t) respectively control the IATM level, skew, curvature and slope.

But the fundamental feature to note, in all the IATM differentials expressions for the ESMM, is their clean decomposition between local volatility and correlation terms. Indeed, cross-terms only appear in the expressions for $\widetilde{\Sigma}'_\theta(\star)$ and $\widetilde{\mathrm{v}}'_y(\star)$. Moreover, these two terms are very similar and can easily be interpreted further, as will be discussed shortly. This decomposition implies that we can interpret many features of the model by referring to, respectively, pure local volatility models (Dupire) or pure correlation models (such as Heston). These two classes are extreme instances of the ESMM: the local volatility model is obtained with $\alpha_t \equiv 1$ or $\sigma_t \equiv f(t, S_t)$, while the pure correlation model corresponds to $f(t, S_t) \equiv 1$ or $\sigma_t \equiv \alpha_t$.

Let us now comment individually on the IATM differentials. First, with respect to the skew $\widetilde{\Sigma}'_y(\star)$, we remark that the correlation term $\widetilde{\Sigma}'_{y,\rho}$ is by definition unaffected by the presence, and therefore choice, of the local volatility function $f(\cdot)$. Consequently, when considering *skew vs local volatility* issues in this model framework, we will focus purely on $\widetilde{\Sigma}'_{y,f}$, which is equivalent to considering pure local volatility models.

That said, the correlation parameter does provide an additional degree of freedom which helps control the IATM skew. The Heston model for instance, relies solely on this correlation to create the descriptor, but can nevertheless be calibrated to fairly extreme market skews.

In fact, we observe it is actually the term $\rho\varepsilon g$, i.e. *the product of correlation with the Lognormal volatility of the perturbation*[28] which will provide that additional skew. In other words, the vol of vol "activates" and "compounds" the correlation, which is not to say that they cannot be dissociated, as proven by the terms $\widetilde{\Sigma}''_{yy,\rho}$ and $\widetilde{\Sigma}'_{\theta,\rho}(\star)$. Overall, this behaviour should come as no surprise to any modeller or practitioner with experience of Heston or SABR, for instance.

Conversely, when the correlated perturbation α_t is not available (as per FL-SV [50] or FL-TSS [51], which both use an *independent* perturbation), the skew function $f(\cdot)$ might have to be fairly extreme to match the market skews. It is possible to use a very steep $f(\cdot)$ around a certain point S_t^*, but not across a whole range of underlying's values. This means that the *conditional smile* will lose its skew, whereas there is no such constraint with $\widetilde{\Sigma}'_{y,\rho}(\star)$, which can consistently create skew for all values of S_t. Not surprisingly, there is a strong analogy with the smile curvature issue (local *vs* stochastic volatility) as discussed in [11].

In terms of calibration capacity, this is a main advantage of the SABR class, for instance. Indeed, this model can use correlation to control its skew, allowing us to specify the LV function to achieve other aims, such as controlling the backbone.

Turning now to the curvature $\widetilde{\Sigma}''_{yy}(\star)$, we observe the same decomposition between the effects of local volatility and correlation. This property is very convenient, both for model design and for static calibration. The relative complexity of the $\widetilde{\Sigma}''_{yy,f}(\star)$ term suggests aligning the variables on both sides: indeed, we have a differential w.r.t. log-moneyness y on the left-hand side, and w.r.t. S_t on the right-hand side. We elect to define a new local volatility function in sliding coordinates *via*

$$l(t, y) \triangleq f(t, S_t).$$

---

[28] $\varepsilon\alpha_t^{-1} g(\alpha_t)$, sometimes abusively called "vol of vol".

Then the transition formulae of Appendix B allow us to rewrite

$$\widetilde{\Sigma}''_{yy,f}(\star) = \frac{1}{3}\sigma_t\left[S_t\frac{1}{l}\left[\frac{-1}{S_t}\right]l'_2 - \frac{1}{2}S_t^2\left[\frac{1}{l}\left[\frac{-1}{S_t}\right]l'_2\right]^2 + S_t^2\frac{1}{l}\frac{1}{S_t^2}\left[l'_2 + l''_{22}\right]\right]$$
$$= \frac{1}{3}\sigma_t\left[\frac{l''_{22}}{l} - \frac{1}{2}\left[\frac{l'_2}{l}\right]^2\right],$$

which gains in coherence and compactness, but also lends itself to an easier interpretation. Note, in particular, that the second term corresponds to the squared skew, and also that the scaling property has been maintained. Indeed, the equation links the normalised curvature $\sigma_t^{-1}\widetilde{\Sigma}''_{yy}(\star)$ to the *shape* of l, and not its scale.

Looking at the expression for $\widetilde{\mathrm{v}}'_y$, notwithstanding a single cross term, we observe a similar decomposition between local volatility and correlation terms. Focusing first on the local volatility term, it appears that it corresponds to the endogenous volatility of the LV term of the skew. In other words, the LV-ρ split extends consistently to the dynamics, just as if we were dealing with a pure LV model:

$$d\widetilde{\Sigma}'_y(\star) \overset{\text{LV}}{\equiv} d\widetilde{\Sigma}'_{y,f}(\star) = [\cdot]\,dt + \frac{1}{2}\left[f'_2 + S_t f''_{22}\right]dS_t = [\cdot]\,dt + \widetilde{\mathrm{v}}'_{y,f}(\star)dW_t$$
$$\text{with}\quad \widetilde{\mathrm{v}}'_{y,f}(\star) \triangleq \frac{1}{2}\sigma_t\left[S_t f'_2 + S_t^2 f''_{22}\right].$$

To reduce the complexity of the $\widetilde{\mathrm{v}}'_{y,f}(\star)$ term, we can proceed in the same manner as before by changing to sliding variables. The new expression reads as

$$\widetilde{\mathrm{v}}'_{y,f}(\star) = \frac{1}{2}\sigma_t^2\left[S_t\left(-\frac{1}{S_t}\right)\frac{l'_2}{l} + S_t^2\frac{1}{S_t^2}\left[\frac{l'_2 + l''_{22}}{l}\right]\right] = \frac{1}{2}\sigma_t^2\frac{l''_{22}}{l}.$$

In other words, the convexity of the local volatility function creates positively correlated dynamics for the skew (and conversely if concave).

As for the correlation term $\widetilde{\mathrm{v}}'_{y,\rho}(\star)$, we again observe that the split between LV and ρ is extended to the dynamics, as if we were dealing with a pure correlation model such as Heston:

$$d\widetilde{\Sigma}'_y(\star) \overset{\text{Pure }\rho}{\equiv} d\widetilde{\Sigma}'_{y,\rho}(\star) = [\cdot]\,dt + \frac{1}{2}\rho\varepsilon\left[\frac{g'_2}{\alpha_t} - \frac{g}{\alpha_t^2}\right]d\alpha_t = [\cdot]\,dt + \widetilde{\mathrm{v}}'_{y,\rho}(\star)dW_t$$
$$\text{with}\quad \widetilde{\mathrm{v}}'_{y,\rho}(\star) \triangleq \frac{1}{2}\rho^2\varepsilon^2\frac{g}{\alpha_t}\left[g'_2 - \frac{g}{\alpha_t}\right].$$

In the $\widetilde{\mathrm{v}}'_{y,\rho}(\star)$ expression itself, we can interpret the bracket as representing the local "over-linearity" of the vol of vol function g. In practice, the function g is chosen so that

$g(0) = 0$ in order to maintain the non-negativity of the multiplicative perturbation α_t. It is also often taken as a power: typically $g(x) = x^\beta$ with $\beta \leq 1$ to ensure sub-linear growth and therefore existence of the solution for the SDE (2.84) (as per Itô's conditions). This is in particular the case for Heston ($\beta = -1$), SABR ($\beta = 1$) and most FL-SV or FL-TSS implementations (if $\psi(V) = V^\gamma$ for the variance then $\beta = 2\gamma - 1$). In that case we have the correlation term

$$\widetilde{\mathrm{v}}'_{y,\rho}(\star) = \frac{1}{2}\rho^2\varepsilon^2\alpha_t^{\beta-1}\left[\beta - 1\right]\alpha_t^{\beta-1} \leq 0. \tag{2.94}$$

This formula tells us that whenever g is a sub-linear power, the contribution of the effective correlation (i.e. *the product* $\rho\varepsilon g$) on the skew's endogenous volatility will be *negative*. In other words, it will tend to make the skew and the asset negatively correlated.

It would certainly be tempting to assume that an upward movement of the underlying S_t will (on average) be associated to a downward move of the IATM skew $\widetilde{\Sigma}'_y(\star)$. Unfortunately at this stage of the book we do not possess the required results to conclude so: in contrast to the underlying, the IATM skew *does* exhibit a drift coefficient $\widetilde{b}'_y(\star)$, and this finite variation term has *a priori* no reason to be null.

It would be an easy task to derive the expression for $\widetilde{b}'_y(\star)$ in the simple Heston framework for instance, but this falls outside the scope of this simple interpretative section. The generic formula, however, will be provided in Sect. 3.2: see (E.30) [p. 449].

Finally, let us link and interpret the two cross-terms appearing in the expressions of $\widetilde{\Sigma}'_\theta(\star)$ and $\widetilde{\mathrm{v}}'_y(\star)$. They are indeed almost identical, and can usefully be linked to the local volatility and correlation terms of the IATM *skew*, by noticing that

$$\rho\varepsilon g(t, \alpha_t) S_t f'_2(t, S_t) = 4\widetilde{\Sigma}'_{y,f}(\star)\widetilde{\Sigma}'_{y,\rho}(\star),$$

which explains why we did not deem it useful to provide either cross-term with Normal conventions. Having completed this interpretative review, let us now move on to the proof.

Proof (Proposition 2.3) Using the generic result (2.51) and the expressions for σ_t and a_2 as per (2.86) the skew comes as

$$\widetilde{\Sigma}'_y = \frac{1}{2}\sigma_t S_t \frac{f'_2}{f} + \frac{1}{2}\rho\varepsilon\frac{g}{\alpha_t} = \frac{1}{2}\sigma_t\left[S_t\frac{\varphi'_2}{\varphi} - 1\right] + \frac{1}{2}\rho\varepsilon\frac{g}{\alpha_t}.$$

For future use, we pre-compute

$$a_{2,t}^2 = \left[\sigma_t^2 S_t \frac{f'_2}{f} + \rho\varepsilon\sigma_t\frac{g}{\alpha_t}\right]^2 = \sigma_t^4 S_t^2\left[\frac{f'_2}{f}\right]^2 + \rho^2\varepsilon^2\sigma_t^2\left[\frac{g}{\alpha_t}\right]^2 + 2\rho\varepsilon\sigma_t^2 g S_t f'_2.$$

Then using the generic curvature formula (2.52) along with the specific expressions for a_3 and a_{22} given respectively by (2.86) and (2.88), the ESMM curvature comes as

$$\widetilde{\Sigma}''_{yy}(\star) = \frac{1}{3\sigma_t^2}\left[\sigma_t^3 S_t\left[\frac{f_2'}{f} + S_t\left[\frac{f_2'}{f}\right]^2 + S_t\frac{f_{22}''}{f}\right] + 3\rho\varepsilon\sigma_t g S_t f_2' + \rho^2\varepsilon^2\sigma_t\frac{g}{\alpha_t}g_2'\right]$$
$$+\frac{1-\rho^2}{3\sigma_t^3}\varepsilon^2\sigma_t^2\left[\frac{g}{\alpha_t}\right]^2 - \frac{1}{2\sigma_t^3}\left[\sigma_t^4 S_t^2\left[\frac{f_2'}{f}\right]^2 + \rho^2\varepsilon^2\sigma_t^2\left[\frac{g}{\alpha_t}\right]^2 + 2\rho\varepsilon\sigma_t^2 g S_t f_2'\right],$$

which, after simplifying and grouping the terms, yields

$$\widetilde{\Sigma}''_{yy}(\star) = \frac{\sigma_t}{3}\left[S_t\frac{f_2'}{f} - \frac{1}{2}S_t^2\left[\frac{f_2'}{f}\right]^2 + S_t^2\frac{f_{22}''}{f}\right] + \frac{\varepsilon^2}{3\sigma_t}\left[\rho^2\frac{g}{\alpha_t}g_2' + \left[1-\frac{5}{2}\rho^2\right]\frac{g^2}{\alpha_t^2}\right],$$

where only the first term depends on the local volatility function. With the Normal convention, this term becomes

$$\widetilde{\Sigma}''_{yy,f}(\star) = \frac{1}{3}\sigma_t\left[S_t\left[\frac{\varphi_2'}{\varphi} - \frac{1}{S_t}\right] - \frac{1}{2}S_t^2\left[\frac{\varphi_2'}{\varphi} - \frac{1}{S_t}\right]^2 + S_t^2\left[\frac{\varphi_{22}''}{\varphi} - \frac{2}{S_t}\frac{\varphi_2'}{\varphi} + \frac{2}{S_t^2}\right]\right]$$
$$= \frac{1}{6}\sigma_t\left[2S_t^2\frac{\varphi_{22}''}{\varphi} - S_t^2\left[\frac{\varphi_2'}{\varphi}\right]^2 + 1\right].$$

Let us now tackle the slope: substituting (2.86), (2.87) and (2.88) into (2.53), we get

$$\widetilde{\Sigma}'_{\theta}(\star) = \frac{\sigma_t}{4}\left[\sigma_t^2 S_t\frac{f_2'}{f} + \rho\varepsilon\sigma_t\frac{g}{\alpha_t}\right] + \frac{1}{12\sigma_t}(1-\rho^2)\varepsilon^2\sigma_t^2\left[\frac{g}{\alpha_t}\right]^2$$
$$+\frac{1}{2}\left[\sigma_t\frac{f_1'}{f} + \frac{1}{2}\sigma_t^3 S_t^2\frac{f_{22}''}{f} + \sigma_t\frac{h}{\alpha_t} + \rho\varepsilon\sigma_t g S_t f_2'\right]$$
$$-\frac{1}{6}\left[\sigma_t^3 S_t\left[\frac{f_2'}{f} + S_t\left[\frac{f_2'}{f}\right]^2 + S_t\frac{f_{22}''}{f}\right] + 3\rho\varepsilon\sigma_t g S_t f_2' + \rho^2\varepsilon^2\sigma_t\frac{g}{\alpha_t}g_2'\right]$$
$$+\frac{1}{8\sigma_t}\left[\sigma_t^4 S_t^2\left[\frac{f_2'}{f}\right]^2 + \rho^2\varepsilon^2\sigma_t^2\left[\frac{g}{\alpha_t}\right]^2 + 2\rho\varepsilon\sigma_t^2 g S_t f_2'\right].$$

After simplification and grouping the terms, we obtain

$$\widetilde{\Sigma}'_\theta(\star) = \underbrace{\frac{\sigma_t}{2}\frac{f_1'}{f} + \frac{\sigma_t^3}{12}\left[S_t\frac{f_2'}{f} - \frac{1}{2}S_t^2\left[\frac{f_2'}{f}\right]^2 + S_t^2\frac{f_{22}''}{f}\right]}_{\widetilde{\Sigma}'_{\theta,f}(\star)} + \frac{\sigma_t}{4}\rho\varepsilon g S_t f'$$

$$+\frac{\sigma_t^2}{4}\rho\varepsilon\frac{g}{\alpha_t} + \frac{\sigma_t}{12}\varepsilon^2\left[1+\frac{1}{2}\rho^2\right]\left[\frac{g}{\alpha_t}\right]^2 - \frac{\sigma_t}{6}\rho^2\varepsilon^2\frac{g}{\alpha_t}g' + \frac{\sigma_t}{2}\frac{h}{\alpha_t}.$$

Let us convert the first term using the Normal convention:

$$\widetilde{\Sigma}'_{\theta,f}(\star) = \frac{\sigma_t}{2}\frac{\varphi_1'}{\varphi} + \frac{\sigma_t^3}{12}\left[S_t\frac{\varphi_2' - S_t^{-1}\varphi}{\varphi} - \frac{1}{2}\left[S_t\frac{\varphi_2' - S_t^{-1}\varphi}{\varphi}\right]^2\right.$$
$$\left.\left. + S_t^2\left[\frac{\varphi_{22}'' - 2S_t^{-1}\varphi_2' + 2S_t^{-2}\varphi}{\varphi}\right]\right]\right],$$

which after simplification and grouping provides the desired result. Turning at last to the dynamic coefficients, we substitute (2.86) and (2.88) into (2.56) to obtain

$$\widetilde{\mathrm{v}}_y'(\star) = -\frac{1}{2\sigma_t^2}\left[\sigma_t^4 S_t^2\left[\frac{f_2'}{f}\right]^2 + \rho^2\varepsilon^2\sigma_t^2\left[\frac{g}{\alpha_t}\right]^2 + 2\rho\varepsilon\sigma_t^2 g S_t f_2'\right]$$
$$+\frac{1}{2\sigma_t}\left[\sigma_t^3 S_t\left[\frac{f_2'}{f} + S_t\left[\frac{f_2'}{f}\right]^2 + S_t\frac{f_{22}''}{f}\right] + 3\rho\varepsilon\sigma_t g S_t f_2' + \rho^2\varepsilon^2\sigma_t\frac{g}{\alpha_t}g_2'\right],$$

which, after simplification, yields

$$\widetilde{\mathrm{v}}_y'(\star) = \underbrace{\frac{1}{2}\sigma_t^2\left[S_t\frac{f_2'}{f} + S_t^2\frac{f_{22}''}{f}\right]}_{\widetilde{\mathrm{v}}_{y,f}'(\star)} + \frac{1}{2}\rho^2\varepsilon^2\frac{g}{\alpha_t}\left[g_2' - \frac{g}{\alpha_t}\right] + \frac{1}{2}\rho\varepsilon g S_t f_2'$$

with

$$\widetilde{\mathrm{v}}_{y,f}'(\star) = \frac{1}{2}\sigma_t^2\left[-S_t\frac{\varphi_2' - S_t^{-1}\varphi}{\varphi} + S_t^2\frac{\varphi_{22}'' - 2S_t^{-1}\varphi_2' + 2S_t^{-2}\varphi}{\varphi}\right]$$
$$= \frac{1}{2}\sigma_t^2\left[1 - S_t\frac{\varphi_2'}{\varphi} + S_t^2\frac{\varphi_{22}''}{\varphi}\right],$$

which concludes the proof. □

2.5.2.2 Application: Comparison of the LDD and CEV Skew Functions

We have established with Proposition 2.3 the Layer-1 results for the generic "Extended Skew" class: we apply them now to the two most popular forms of local volatility functions, the Lognormal Displaced Diffusion (LDD) and the Constant Elasticity of Variance (CEV). These two models are commonly and concurrently used in their pure local volatility version (see [52] for a comparison), i.e. with α_t as a deterministic function of time. We can easily extend them by adding a multiplicative perturbation as per (2.83)–(2.84) so that they fit into the ESMM framework. Following the Normal convention, the models are then specified with:

$$\begin{aligned}\textbf{LDD}: \quad \varphi_d(s,x) &= \lambda_d(s)\,(x+d(s)) \\ \textbf{CEV}: \quad \varphi_\beta(s,x) &= \lambda_\beta(s)x^{\beta(s)}.\end{aligned}$$

Both forms of local volatilities are commonly used, either on a standalone basis or within two-dimensional local/stochastic volatility models (such as SABR). Their role is essentially to simultaneously control the static IATM level and *skew*, exploiting the fact that each is specified by a pair of time functions: $\lambda_d(\cdot)$ and $d(\cdot)$ *vs* $\lambda_\beta(\cdot)$ and $\beta(\cdot)$. However, either local volatility function inevitably generates some unwanted and rather distinct side-effects. So that in the literature, as well as among practitioners, the argument for and against these two respective forms of skew functions has been a recurring issue of the modelling process. Rather surprisingly, few papers are available that provide a clean answer to these questions. For a prompt review of CEV properties (which is more technical than the LDD) a good reference is [53], while a comparison of the two models (based on a small-time expansion) is provided by [52].

In this context, we will show that the asymptotic results of Theorem 2.2 are well suited to settle (almost) effortlessly a large part of that old argument. Our objective is to justify rigorously how to achieve the joint level and skew fit, and in particular how to match one model onto the other. Naturally this will expose the limits of such an approach for the smile, both in static and dynamic terms. In order to ensure comparability, we assume *the same multiplicative perturbation process* α_t.

We show that once the static level/skew is achieved, the sliding IATM option will have the same volatility (both endogenous and exogenous coordinates) in both cases, and therefore that the fixed-strike IATM option will have the same Delta. However, we will also see that other important quantities will necessarily be different, and how to approximate these discrepancies. Of particular interest for trading are the static smile IATM curvature, the volatility of the skew, the drift of the sliding IATM option, and the Gamma of the fixed-strike IATM option.

As we have seen with Proposition 2.3, with regard to IATM level and skew the effects of correlation (and thus of the perturbation α_t) are dissociated from those of the local volatility. Therefore it makes sense to recall the structural differences between the two models in their pure LV form. It is well-known than when d and β are either constant or time functions, both pure local volatility models provide closed-form solutions for Calls and Puts, which is a large part of their appeal. However the solutions of their respective SDEs exhibit significantly distinct behaviours. Of particular concern are the existence and uniqueness of the CEV solution[29] and the support of the distribution for the LDD solution ($[-d, +\infty[$). Also, the (right) tail characteristics can be very different, affecting several quasi-vanilla products, such as CMS options.

The extension of the two models to a diffusive α_t inherits and enhances most of these differences. Furthermore, the ESMM decomposition between LV and perturbation effects ensures that few novel features are introduced. For instance, the backbone is controlled mainly by the LV function[30] and will therefore produce distinct *conditional* smiles, although the presence of stochastic volatility does ensure *some* Delta stickiness. Although the joint local/stochastic volatility context of the ESMM allows for further control of the smile properties, we will (artificially) consider that the primary role of $\varphi(\cdot, \cdot)$ is to control the IATM level and skew.

Our first move is obviously to unify the two local functions under the Extended Displaced CEV denomination, which is discussed for instance (in a time-homogeneous version) within [53].

Definition 2.3 (*The Extended Displaced CEV model*) Consider the EDCEV model to be specified as follows:

$$\begin{aligned} dS_t &= \alpha_t \varphi_{d,\beta}(t, S_t) dW_t & \text{with} \quad & \varphi_{d,\beta}(s, x) = \lambda(s)(x+d)^\beta \\ d\alpha_t &= h(t, \alpha_t)dt + \varepsilon g(t, \alpha_t) dB_t & \text{with} \quad & \langle dW_t, dB_t \rangle = \rho dt. \end{aligned} \tag{2.95}$$

We can then apply Proposition 2.3 directly in order to obtain the relevant IATM differentials. Again, due to the clear distinction between the correlation and local volatility effects, it is sufficient to express only the latter.

Corollary 2.12 (IATM level and skew of the EDCEV model) *Defining* $(\star) \triangleq (t, 0, 0)$ *the IATM level and skew are*

$$\widetilde{\Sigma}(\star) = \alpha_t \lambda(t) S_t^{-1} (S_t + d)^\beta \qquad \widetilde{\Sigma}'_{y,f}(\star) = \frac{1}{2} \widetilde{\Sigma}(\star) \left[\frac{\beta(t) S_t}{S_t + d(t)} - 1 \right], \tag{2.96}$$

which we can also express in strike units, using Corollary 2.5*:*

$$\Sigma'_{K,f}(LDD) = -\frac{1}{2} \frac{\sigma_t}{S_t} \frac{d}{S_t + d} \qquad\qquad \Sigma'_{K,f}(CEV) = -\frac{1}{2} \frac{\sigma_t}{S_t} (1 - \beta).$$

Consider CEV and LDD models with matching IATM levels and skew.

---

[29] For positive β: if $\beta < 1$ zero is an attainable boundary, for $\frac{1}{2} \le \beta \le 1$ it is an absorbing boundary, and for $0 < \beta < \frac{1}{2}$ there is no uniqueness of the solution unless a boundary condition is specified (absorption or reflection). Obviously for $\beta > 1$ the sub-linear growth Itô condition ensuring existence is not satisfied.

[30] $\widetilde{\Sigma}(\star) = \alpha_t S_t^{-1} \varphi(S_t)$.

Ignoring the IATM argument for $\widetilde{\Sigma}$ and $\widetilde{\Sigma}'_{y,f}$, we provide the transfer formulas:

$$\lambda_d = \lambda_\beta \beta S_t^{\beta-1} = \frac{1}{\alpha_t}\left[2\widetilde{\Sigma}'_{y,f} + \widetilde{\Sigma}\right] \quad \textit{and} \quad d = \frac{1-\beta}{\beta} = \frac{-2S_t\widetilde{\Sigma}'_{y,f}}{2\widetilde{\Sigma}'_{y,f} + \widetilde{\Sigma}},$$

$$\lambda_\beta = \lambda_d(S_t + d)S_t^{-\frac{S_t}{S_t+d}} = \frac{1}{\alpha_t}\widetilde{\Sigma} S_t^{2\widetilde{\Sigma}^{-1}\widetilde{\Sigma}'_{y,f}} \quad \textit{and} \quad \beta = \frac{S_t}{S_t + d} = 1 + 2\widetilde{\Sigma}^{-1}\widetilde{\Sigma}'_{y,f}.$$

Note that by normalising the displacement with $d(t) = c(t)S_t$ the *equivalence* between the two skew parameters becomes even more obvious. Indeed, the local volatility-induced IATM skew is then

$$\widetilde{\Sigma}'_{y,f}(t,0,0) = \frac{1}{2}\widetilde{\Sigma}(\star)\left[\frac{\beta(t)}{1+c(t)} - 1\right]. \tag{2.97}$$

Formulae (2.96) and (2.97) show that the E-LDD and E-CEV models can *a priori* be matched simultaneously in level and in (negative) skew. The latter is clearly bounded, even more so in the pure LDD and CEV cases, where for positive $\beta(\cdot)$ and $c(\cdot)$ the value of $-1/2\widetilde{\Sigma}(\star)$ cannot be exceeded (see [54] (p. 5) for instance). However, each model achieves these targets with distinct consequences on the marginal distributions. For example, recall that a CEV with $\beta = 1/3$ demands specification of the boundary conditions, while the equivalent LDD requires a displacement of $d = 2S_t$ to achieve the same skew: for most underlyings this support is considered to be unrealistic. By comparison, the EDCEV model allows us to reach the same negative skews in a healthier way, for instance by taking $\beta = 1/2$ and then using the displacement to complete. Overall, these results expose the limitations of pure local volatility models *vs* their (correlated) stochastic volatility extensions. Let us now move on to the proof.

Proof In this proof we will omit the arguments of $\widetilde{\Sigma}$ and $\widetilde{\Sigma}'_{y,f}$, which will be taken at $(t,0,0)$. Within (2.96) the IATM level is given by the cast, while the IATM skew comes directly from (2.93). Let us first consider an Extended LDD model with a given perturbation process α_t, and assume that it is matched to a given IATM level and skew. Then the skew match gives us

$$\widetilde{\Sigma}'_{y,f} = \frac{1}{2}\widetilde{\Sigma}\left[\frac{S_t}{S_t+d} - 1\right] \quad \text{hence} \quad \frac{S_t}{S_t+d} = \frac{2\widetilde{\Sigma}'_{y,f} + \widetilde{\Sigma}}{\widetilde{\Sigma}} \quad \text{thus} \quad d = \frac{-2S_t\widetilde{\Sigma}'_{y,f}}{2\widetilde{\Sigma}'_{y,f} + \widetilde{\Sigma}},$$

while the level match provides

$$\widetilde{\Sigma} = \alpha_t\lambda_d S_t^{-1}(S_t + d) \qquad \text{hence} \qquad \lambda_d = \frac{1}{\alpha_t}\widetilde{\Sigma}\frac{S_t}{S_t+d} = \frac{1}{\alpha_t}\left[2\widetilde{\Sigma}'_{y,f} + \widetilde{\Sigma}\right].$$

Let us now assume instead that this Extended LDD is matched to a given Extended CEV model. Then the skew match followed by the level match give us sequentially

$$\frac{1}{S_t+d} = \frac{\beta}{S} \Rightarrow d = \frac{1-\beta}{\beta} \qquad \text{then} \qquad \lambda_d\,[S_t+d] = \lambda_\beta S^\beta \Rightarrow \lambda_d = \beta\lambda_\beta S_t^{\beta-1}.$$

Conversely, let us then assume a CEV model, and suppose that it is matched to a given IATM level and skew. Then matching the skew gives us

$$\widetilde{\Sigma}'_{y,f} = \frac{1}{2}\widetilde{\Sigma}\,[\beta-1] \qquad \text{hence} \qquad \beta = 1 + 2\widetilde{\Sigma}^{-1}\widetilde{\Sigma}'_{y,f},$$

while the level match gives

$$\widetilde{\Sigma} = \alpha_t\lambda_\beta S_t^{\beta-1} \quad \Rightarrow \quad \lambda_\beta = \frac{1}{\alpha_t}\widetilde{\Sigma}\, S_t^{2\widetilde{\Sigma}^{-1}\widetilde{\Sigma}'_{y,f}}. \tag{2.98}$$

Suppose now that this Extended CEV is matched to a given Extended LDD model. Inverting the skew result in (2.96) before matching the level provides sequentially

$$\beta = \frac{S_t}{S_t+d} \qquad \text{then} \qquad \lambda_\beta = \lambda_d(S_t+d)S_t^{-\beta} = \lambda_d(S_t+d)S_t^{-\frac{S_t}{S_t+d}},$$

which concludes the proof. □

We have already noted in Sect. 2.3.2 that if two stochastic volatility models (whether SinsV or SImpV) provide matching IATM level and skew, then their endogenous volatilities a_2 (or $\widetilde{\mathrm{v}}(\star)$) are identical. In the general case, it is not true, however, that the total vol of vol $[a_2; a_3]$ or even its modulus $\sqrt{a_2^2+a_3^2}$ will be the same. As shown by the expression for a_3 within (2.89) however, this is the case for the Extended Skew Market Model, and in particular for the Extended LDD and CEV. This is important in terms of trading because these dynamics are those of the sliding IATM implied volatility and will therefore impact the *Delta* of the *fixed-strike* ATM Call. Indeed, as discussed in Sect. 2.5.1.4 and according in particular to (2.82), that Delta will be identical if both models are *dynamically* matched in level and in skew.

However, the lack of additional control parameters means that the above quantities represent the extent of any rigorous, IATM match between the two models. Indeed, differences will appear as soon as φ''_{22} or f''_{22} is invoked. In terms of instantaneous coefficients, this concerns a_1 and a_{22}, while in terms of IATM differentials it regards:

- the IATM curvature $\widetilde{\Sigma}''_{yy}(\star)$;
- the slope $\widetilde{\Sigma}'_\theta(\star)$;
- the endogenous volatility of the skew $\widetilde{\mathrm{v}}'_y(\star)$;
- the Gamma of the fixed-strike ATM option.

Overall, these are IATM results and a natural question to ask is whether the match of the EDCEV model is also possible for $\theta > 0$. The answer is that in general we can match ATM level and skew at a given maturity, because we have two targets (level and skew) and also two control parameters at our disposal. However, mathematically we cannot guarantee that this system of equations admits a (unique) solution.

2.5.3 *Second Illustration: Smile-Specification of SInsV Models*

In this section, we take the reverse approach from the previous illustration: starting form a given smile, we show how to build a stochastic instantaneous volatility model that would (approximatively) generate that surface. Taking that approach further, we make the case for the re-parametrisation of such models into what we call their "intuitive" versions. The latter consist in replacing one, several or all instantaneous parameters by more "trading-friendly", smile-related quantities. First we discuss some generalities relevant to that approach, before exposing several examples.

2.5.3.1 Intuitive Models: Principles

An intuitive modelling approach consists in an output-oriented re-parametrisation of an existing SinsV model. The idea is to expose that original model to the practitioner (typically a trader) by using parameters that relate to *observable* and *meaningful* quantities. The first qualifier limits us to liquid instruments, whose characteristics are either quoted (e.g. prices, rates, implied volatilities or correlations) or easily measurable (e.g. realised volatility, historical correlation). The second qualifier pertains to those quantities which have an actual influence on the agent's portfolio: they impact the prices, the hedges and/or the risk. Therefore, in order to link to the usual pricing, hedging and risk engines, we still need to internally express the native parameters driving the original SDE system.

Simple illustrations of this approach can already be found in practice-oriented literature. A classical instance consists in re-expressing volatility mean-reversion, in Heston's model for instance, as "frequencies" or time periods in the one-dimensional dynamics of the volatility. Indeed, it is difficult to gauge the impact of selecting, say, $\kappa = 0.583$, and especially so because of the interaction of that parameter with the vol of vol and the current level of volatility. It seems more practical however, and particularly for traders, to consider the associated (i.e. model-implied) half-life of an after-shock decay: because that quantity can readily be compared to historical data, simply by looking at a realised chart of ATM implied volatility.

Going back to the general approach, what our asymptotic results allow us is to extend and generalise this approach to a much larger scale, to very complex models, while keeping some useful flexibility. The market quantity that we select will of course be the smile, which is indeed an observable and measurable quantity, with primordial importance w.r.t. to prices, hedges, sensitivities and risk. In essence, we will *replace* some or all of the SinsV model native parameters, (such as vol of vol or correlation) by descriptors of the smile, (static and dynamic, such as curvature or skew) that they generate.

In a sense, this approach can be seen as the "poor man's stochastic implied volatility model". But, as we observed previously, the correspondence between the two classes (SinsV and SimpV) is very strong, so that adopting the above modelling strategy consists *de facto* in opting for a tradeoff: what we lose in freedom and precision[31] of the specification for the SimpV model, we gain in *validity*, as a SinsV model is intrinsically non-arbitrable.

An added benefit of the intuitive re-parametrisation is that it reduces the complexity of "extended" Vega computations. These usually consist in computing the sensitivity of the smile[32] w.r.t. to the model parameters, and then inverting (when possible) the Jacobian matrix, in order to evaluate the Vega/smile risk contained by exotic products. With the new representation, the process is simplified because this high-dimensional matrix is reduced or even diagonal. Indeed, the smile in question *is* now (a subset of) the parameter set.

Turning to practicalities, we start by characterising the static shape of the target smile in terms of several of its first IATM differentials: typically the ATM level, skew, curvature and slope for short expiry options. Alternatively, we can express or approximate these quantities with simple options, such as ATM Straddles, Strangles, Call Spreads, Butterflies, Maturity Spreads, etc. In order to compute the native parameters, we can either use the results of the Recovery Theorem 2.1 [p. 55] or invert the formulae of the First Layer Theorem 2.2 [p. 62].

In its principle, the intuitive approach belongs firmly to the inverse problem family; however we will tend to bring the results *via* the direct formulas of Theorem 2.2. The reason is that we demonstrate the method here by focusing on the *static* properties of the smile, which are the easiest to measure and comprehend. Obviously the method can also incorporate the dynamic coefficients: $\widetilde{b}$, $\widetilde{\nu}$, $\widetilde{\nu}'_y$ and $\widetilde{n}$ for the first layer, all taken in $(t, 0, 0)$. Also, there is a mathematical price to pay, a structural drawback to staying purely with static quantities. Indeed, we have seen with Corollary 2.7 that a bijectivity existed between the $\widetilde{\Sigma}$-(2,0) and σ_t-(2,0) groups. We have also mentioned in Sect. 2.4.2 that assuming bi-dimensionality and some regularity assumptions, a full bijection could be expected between, on one hand, some groups of instantaneous coefficients $(a_{i,t})$, and on the other hand, collections of IATM differentials. The thing is that those collections of differentials include both static and dynamic quantities. Granted, we have seen that some of these are interchangeable, since within the first layer for instance, we have the following equivalence pairs: $\widetilde{\nu}(\star)$ *vs* $\widetilde{\Sigma}'_y(\star)$, $\widetilde{\nu}'_y(\star)$ *vs* $\widetilde{\Sigma}''_{yy}(\star)$ and $\widetilde{b}(\star)$ *vs* $\widetilde{\Sigma}'_\theta(\star)$. But recall also that the dynamic coefficients relate either to the drift or to the endogenous driver. Hence $\vec{\widetilde{n}}\,(\star)$ cannot be swapped, so

[31] The method is only asymptotic, after all.

[32] The surface is usually represented either by a selection of points, or by a collection of parameters.

that removing it from the list of "intuitive" quantities will lose us some information, which we will have to compensate by making supplementary assumptions.

2.5.3.2 Intuitive Models: Examples

First, let us note that we have recently encountered such (potential) intuitive models. Indeed, Corollary 2.12 [p. 101] presents, respectively for the LDD and CEV models, the two native parameters as functions of the IATM level and skew. When limited to their pure local volatility versions, these two quantities define the models entirely. While in their extended (EDCEV) versions, only the perturbation (2.95) remains to be specified. In that case it is possible to use a mixed representation, with both native and implied (intuitive) parameters.

We now provide two distinct examples: note that these are purely "toy" models, or demonstrative instruments. As such we are not looking for high precision in a real-life model: this will be the object of Chap. 4.

The following example does not make use of a "real" SinsV model, but instead uses the generic formulation, the "cast" presented in Sect. 2.1.3.1. It demonstrates how easily we can generate desired surfaces, using sparse, low-dimensional specifications. In this case we have chosen to illustrate with an unusual, probably unheard of, *concave* smile shape. Indeed, result (2.53) implies that the smile generated by a SInsV model can theoretically exhibit negative convexity, at least in the vicinity of the origin $(t, 0, 0)$. In order to verify the validity of this prediction, we use the following toy model, which does not make use of any exogenous driver.

We have also seen that when using the asymptotic methodology, the SDE system describing the model is only accounted for *via* the values at the initial point. As such, this example underlines not only the efficiency but also the limitations of the asymptotic approach.

Example 2.4 (Endogenous SV model generating a concave smile) Assume the following stochastic instantaneous volatility model.

$$\frac{dS_t}{S_t} = \sigma_t dW_t \qquad \text{with} \qquad d\sigma_t = a_{2,t} dW_t \qquad \text{and} \qquad da_{2,t} = a_{22} dW_t,$$

where we choose the following initial values and parameter:

$$S_0 = 100 \qquad \sigma_0 = 0.2 \qquad a_{2,0} = 0 \qquad a_{22} = -1.$$

On a short-term basis, there is nothing drastically unrealistic with these dynamics. The level of volatility σ_t is consistent with usual market conditions, the vol of vol a_2 is null but very unstable and negatively correlated with the underlying: all in all, these parameters could economically correspond to a "tipping point" in the market. This configuration typically occurs just before an anticipated and important announcement

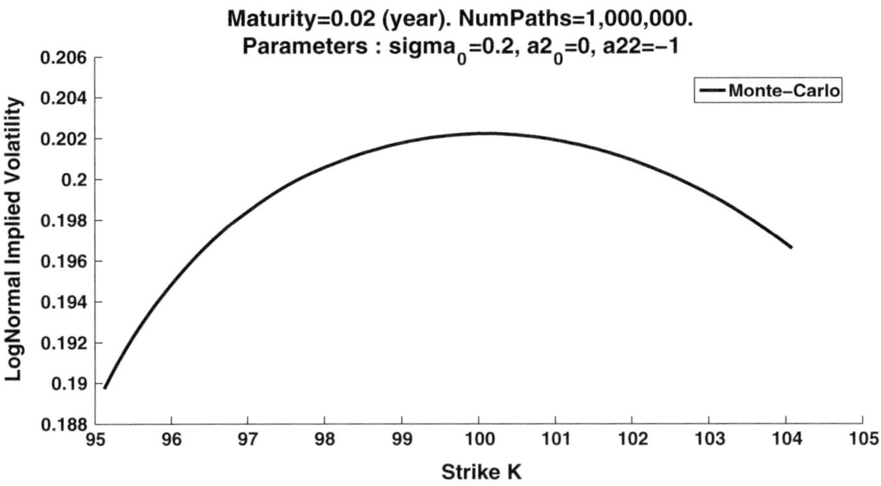

Fig. 2.3 LogNormal implied volatility: Monte-Carlo simulation with Euler scheme and $\Delta t = 10^{-3}$

which could swing either way: a central bank meeting or the release of a major economic indicator.

However, the resulting theoretical short-term smile is very unusual. Indeed, Theorem 2.2, in particular (2.58) and (2.59), predicts no skew and an IATM negative curvature of

$$\Sigma''_{KK}(0, S_t, S_t, 0) = \frac{1}{3} S_t^{-2} \sigma_t^{-2} a_{22,0} = \frac{1}{3} 100^{-2} 0.2^{-2} = 8.3310^{-4}$$

which, using a Maclaurin expansion for a strike of 95, represents a drop of $\approx 1\%$ Lognormal IV.

The actual smile is obtained by Monte-Carlo simulation, with a million paths, for a maturity of a week (2.10^{-2} year) and computed for strikes ranging 5 either side of the money: it is shown in Fig. 2.3.

Besides the possible economic interpretations, it is clear that the model served its purpose, as the Monte-Carlo simulation is in accordance with the theoretical prediction.[33]

Note that with such a high number of paths the standard error is negligible and therefore was not represented. For readers wishing to reproduce the test, note the issue of boundary conditions and the following identity:

$$a_{2,t} = a_{22} W_t \quad \text{hence} \quad d\sigma_t = a_{22} W_t dW_t \quad \text{thus} \quad \sigma_t = \tfrac{1}{2} a_{22} \left(W_t^2 - t \right).$$

[33] Indeed the level, skew and curvature are conform: check the 95 strike.

Of course, there exists an infinity of SinsV models that will provide the same cast and therefore the same IATM differentials, at least within Layer 1. This is due to the asymptotic nature of the method, which brings two restrictions:

- the processes/coefficients $a_{2,t}$, $a_{3,t}$, etc. are only considered at current time t;
- any IATM differential will only depend on a finite number of those coefficients.

The first restriction implies that, for instance, the method will be blind to a full time-dependency such as $\{\sigma(s)\}_{s\geq t}$: it will only consider the values $\partial^{(i)}_{s^i}\sigma(s)|_{s=t}$ until a finite differentiation order. The second restriction means, for instance, that if we were interested in $\widetilde{\Sigma}''_{yy}(t, 0, 0)$, then that curvature would be unaffected by an over-specification *in depth* of the dynamics of a_{22}. That specification, however, might lead to a significantly different model (in terms of dynamics, long-term marginal distribution, etc).

As a first illustration of these restrictions, it is of course possible to make the model of Example 2.4 even more realistic, especially for longer-term dynamics. We can, for instance, make the volatilities mean-reverting or Lognormal, thus avoiding some of the boundary issues. But all configurations providing the same instantaneous "cast" will generate very similar smiles, at least for such a short, one-week maturity. Accordingly, adding mean-reversion to volatility and vol of vol as per

$$d\sigma_t = \kappa\,(\theta - \sigma_t)\,dt + a_{2,t}dW_t \qquad\qquad da_{2,t} = \kappa_2\left(\theta_2 - a_{2,t}\right)dt + a_{22}dW_t$$

would not modify the IATM level, skew, and curvature. This is because, although this specification creates a (somewhat unusual) multi-scale SV model, it effectively provides an a_1 (and a_{21}) term, which in Layer 1 only affects the slope $\widetilde{\Sigma}'_\theta(t, 0, 0)$.

Similarly, and as a second illustration, since the toy model is purely endogenous we can obtain the same instantaneous cast with a pure local volatility model. As before, in order to match the current values of σ_t, $a_{2,t}$ and $a_{22,t}$ quoted above, it suffices[34] to specify a concave f as per

$$\frac{dS_t}{S_t} = f(S_t)dW_t \qquad \text{with} \qquad f(S_t) = 0.2 \quad f'(S_t) = 0 \quad f''(S_t) = -2.510^{-3}.$$

Clearly these conditions constrain very little the *global* behaviour of $f(\cdot)$ as a function, which implies through Dupire that all sorts of smiles could be generated when far from the immediate money. Assuming f is smooth and analytical, the only way through Dupire to extend our control over the smile would be to impose further IATM differentials $f^{(i)}(S_t)$. In our asymptotic framework, this would correspond to obtaining more IATM differentials *via* deeper coefficients $a_{i,t}$: this will be the object of Sects. 3.1 and 3.2.

Let us now illustrate the intuitive re-parametrisation with another toy model:

Example 2.5 (Lognormal model with Normal, correlated stochastic volatility) Let us consider the following bi-dimensional model:

[34] The simple proof is left to the reader, see Lemma 2.4, p. 90.

$$\frac{dS_t}{S_t} = \sigma_t dW_t \qquad \text{with} \qquad d\sigma_t = \varepsilon dB_t \qquad \text{and} \qquad \langle dW_t, dB_t \rangle = \rho dt \tag{2.99}$$

with $\varepsilon \geq 0$ some constant. The cast is easily deduced from the original model *via*

$$a_2 = \rho\varepsilon \qquad \text{and} \qquad a_3 = \sqrt{1-\rho^2}\varepsilon,$$

which justifies that both initial coefficients $a_{2,t}$ and $a_{3,t}$ should be positive.

Let us now assume that the agent is interested purely in the level, skew and curvature of the smile, and that he/she prefers expressing these differentials w.r.t. strike K, rather than log-moneyness y.[35] Also, the agent chooses to ignore the slope Σ_T' altogether, which justifies the lack of a drift (e.g. mean-reversion) in the volatility dynamics. The exposed model parameters are now Σ, Σ_K' and Σ_{KK}'' (all taken IATM).

Then applying Corollary 2.5, we get the IATM differentials:

$$\Sigma_K' = \frac{a_2}{2S_t\sigma_t} \qquad \text{and} \qquad \Sigma_{KK}'' = \frac{1}{S_t^2}\left[\frac{1}{\sigma_t}\left[-\frac{a_2}{2}\right] + \frac{1}{\sigma_t^3}\left[\frac{a_3^2}{3} - \frac{a_2^2}{2}\right]\right],$$

which we can invert, to express the instantaneous coefficients $a_{2,t}$ and $a_{3,t}$ as

$$a_2 = 2S_t\Sigma\Sigma_K' \qquad \text{and} \qquad a_3 = \sqrt{3}\left[S_t^2\Sigma^3\Sigma_{KK}'' + S_t\Sigma^3\Sigma_K' + 2S_t^2\Sigma^2\,{\Sigma_K'}^2\right]^{\frac{1}{2}}.$$

Note that we need to check the existence of a solution (i.e. that the quantity under the square root is positive) for the given set of IATM differentials. Note also that we could equivalently use the inverse results (2.44) and (2.50) (see pp. 56 and 60), providing they were first converted to K-differentials. Finally, we can use the cast to re-express the native model parameters as implied quantities with

$$\varepsilon = \sqrt{a_2^2 + a_3^2} \qquad \text{and} \qquad \rho = \frac{a_2}{\sqrt{a_2^2 + a_3^2}}.$$

Although our toy model (2.99) can be compared to Heston's [2], the dynamics of its volatility are Normal (as opposed to C.I.R.) and therefore no (semi-) closed form is *a priori* available. However, note that, were such a closed-form available for the price or even for the implied volatility, the intuitive approach would still be pertinent.

As mentioned above, the inversion technique is by no means always possible. In the current case, having excluded the exogenous dynamic coefficient $\vec{\vec{n}}$ ($\star$) from the "intuitive" quantities, we knew that (in the general case) we only had access to the *modulus* of $\vec{a}_{3,t}$. Therefore we had to rely on a bi-dimensional setup (Z_t is here scalar) along with the given, native constraint that $a_{3,t}$ was positive.

---

[35] Recall that with static quantities the absolute/sliding distinction is moot.

This illustrates the fact that, even with valid inputs (the IATM differentials), the existence and/or uniqueness of a solution might have to be enforced by applying supplementary (and possibly arbitrary) constraints. Formally, we first have to ensure that a bijection exists between the collection of IATM differentials that we take as input, and a group of instantaneous coefficients $(a_{i,t})$. This is where further assumptions or constraints might have to be applied, for instance if only shape differentials come as input, in order to establish the bijection. Then we have to ascertain that a bijective relationship also exists between that group of coefficients and the collection of native parameters that we wish to replace.

In practice this process rarely represents an issue, if only because the number of parameters and the order of differentiation are usually limited (typically within the first layer). This inversion technique will be illustrated in detail within Chap. 4, for more complex models such as SABR and FL-SV. We will see that it can also be a quick and powerful tool for the initial calibration, as it provides a close initial guess.

Note also that in this example it seems more appropriate to invert the "direct" results of Corollary 2.5 than to use the "inverse" formulas of the Recovery Theorem 2.1.

In our view, the fact that the asymptotic formulas lose pertinence and precision when far from the IATM point[36] does not hamper the relevance of the "intuitive modelling". The practitioner should not expect to parametrise the model by specifying *ex ante* and with precision any long-term smile. What he/she controls is the very short-term smile, which is certainly more natural than, for instance, gauging the magnitude of the vovol in Heston's model.

Note also that, as with the EDCEV model, a mixed representation using both implicit and native parameters is certainly conceivable. With model (2.99) we could alternatively specify

- either the IATM skew Σ_K' and the volatility of volatility ε; or
- the IATM curvature Σ_{KK}'' and the correlation ρ.

But as always, any such under-specification or excessive degree of freedom would be seen as an asset for modelling, but also as a liability for the calibration process.

2.5.3.3 Intuitive Models: Practical Usage

So far we have mentioned the principles of the approach, and broached the subject by means of a few simple examples. Beyond the theoretical aspects and all the choices that they offer, we have not discussed how to fit the methodology to market realities, and in particular which smile descriptors *could* or *should* in practice be taken as input. Nor have we offered a typical way of resolving the possible inversion issues that might arise from that choice: these are the goals of the current section. In order to remain as generic as possible, we do not specify an actual SInsV model, but instead we stay at the cast level. Note, however, that few of our following assumptions can

[36] See the discussion within Sect. 2.5.1.1, p. 77.

be seen as universal: those are just fairly common features of advanced markets, at the time of writing. We also make choices that are subjective: the point is to walk the reader through the thought process and provide general directions, nothing more definitive.

We wish to specify *ex ante* the most relevant features of the smile, both static and dynamic, through a stochastic instantaneous volatility model. In a trading context, "relevant" is often synonymous with "liquid" and in most single-underlying markets, the most traded assets will be the following, in order of frequency:

1. the underlying;
2. the short-dated ATM volatility level (Calls, Puts, Straddles);
3. the short-dated ATM skew (Binaries, Call spreads, Risk Reversals, Collars);
4. the short-dated ATM curvature (Butterflies, Strangles);
5. the short-dated ATM slope, providing enough liquidity in Calendar Spreads.

All in all, these products need to be matched at any given time, which gives us five *static* targets.

Although our presentation and methodology is both probabilistic and asymptotic, it is not incompatible with some statistical approaches. For practitioners, the modelling and calibration phases are usually torn between the implied and realised arguments: statistical arbitrage and hedging of exotics can seem at odds. Our results allow us to bridge part of that gap and combine both approaches, because they structurally link the statics and the dynamics.

Notwithstanding some dedicated, implied products such as volatility derivatives, statistics in our context means inference of processes, which in turn requires good-quality time series. These must be observable, financial quantities published at a sufficiently high frequency, which limits us therefore to the five static quantities quoted above. Incorporating the numeraire and risk premia defining the measure, it is possible to infer historically their dynamics, i.e. the drifts, volatilities and correlations. We already know, however, that in a SImpV context the drifts are redundant, and we will as usual consider vectorial volatilities, which incorporate the correlations.

When it comes to inferring those volatilities, the endogenous driver W_t should be extracted from the underlying's time series, which is the most liquid. As for exogenous components, the first hurdle is to decide on a dimension for the exogenous driver $\vec{Z}_t$. Given the list of static targets, the maximum useful dimension is 4; the actual dimension will usually be chosen after a Principal Component Analysis of those time series. In most markets though, only the short-dated *level*, and possibly the skew, present a liquidity comparable to the underlying's, and even then it comes at a lower frequency. Therefore all inference of exogenous volatilities will be less precise than endogenous ones. In our view, this statistical feature is structural and justifies not only the use of a single-dimensional Z_t (in this case), but also favouring endogenous volatility components over exogenous ones.

In accordance with these observations, and in view of our established asymptotic results, we will therefore limit ourselves to specifying:

6. the (scalar) volatility of the underlying;
7. the volatility (endog. *and* exog. components) of the short-dated ATM level;
8. the endogenous volatility component of the short-dated ATM skew.

In total, that brings us to nine targets in order to describe the SImpV-*via*-SInsV model. Obviously, the underlying is matched by construction, while σ_t and $\widetilde{\Sigma}(\star)$ are redundant. Furthermore, Recovery result (2.37) [p. 54] shows that when specifying the IV statics and dynamics, arbitrage constraints preclude the dissociation of the IATM level $\widetilde{\Sigma}(\star)$ and skew $\widetilde{\Sigma}_y'(\star)$ from the IATM endogenous volatility $\widetilde{\mathrm{v}}(\star)$. Similarly, result (2.38) establishes that $\widetilde{\mathrm{v}}_y'(\star)$ is entirely determined by $\widetilde{\Sigma}_y'(\star)$, $\widetilde{\Sigma}_{yy}''(\star)$ and $\vec{n}(\star)$. Together, these equations take us down to only four targets, but also with four instantaneous coefficients: $a_{1,t}$, $a_{2,t}$, $a_{3,t}$ and $a_{22,t}$.

As occurred previously with Example 2.5, the associated inverse problem can then be made well-posed, with the proviso that $a_{3,t}$ be taken scalar and constrained to a given sign[37]: it suffices to invoke (2.50) (see Corollary 2.4, p. 60).

So in principle we can infer a cast for the SInsV model; how this cast will determine the native parameters is an *ad hoc* issue, dependent on the chosen SInsV class, and might require the introduction of further constraints. More generally, the whole process depends on the collection of IATM differentials that was chosen, along with the parametrisation of the SInsV model. The problem can easily be ill-posed, for instance if we have more parameters than inputs. This kind of situation, on the other hand, is an appropriate context to introduce a more advanced usage of the inverse methodology, namely the *dynamics-influenced calibration*. Let us consider, for instance, the pure (re-)calibration issue of a given SInsV model. On one hand, assuming the underlying's path has been observed, the native parameters themselves can also be inferred (for instance, using likelihood estimators). On the other hand, dependent on which static and dynamic smile descriptors are available, the asymptotic methodology described above provides at least some of the $a_{i,t}$ coefficients, which determine or constrain the native parameters. We can then combine (hedging) or confront (arbitrage) both methods within a multi-objective optimisation process. How well the solution fits each objective is a measure of their compatibility, and therefore provides valuable information to both the arbitrageur and the hedger. The interest for statistical arbitrage is obvious. As for hedging, let us recall that a better dynamic stability of the parameters should also narrow down the tracking error and its variability.

In essence the principle is not new, but simply extended and accelerated by the asymptotic results. Extended because we can now throw in the *dynamics* of the smile, as opposed to only its shape. Accelerated because even the static calibration process is made easier and generic. All in all, we gather these variants of the calibration process under the denomination of *dynamics calibration*, which is strongly linked with but mitigates the process of *re-calibration*.

---

[37] Usually positive as in (2.99).

2.6 Conclusion and Overture

We consider that one of the main contributions of this study (summarised in Fig. 2.4 [p. 113]) is to make explicit the high degree of equivalence between the SInsV and SImpV classes. It is by now apparent that most practical applications will start with

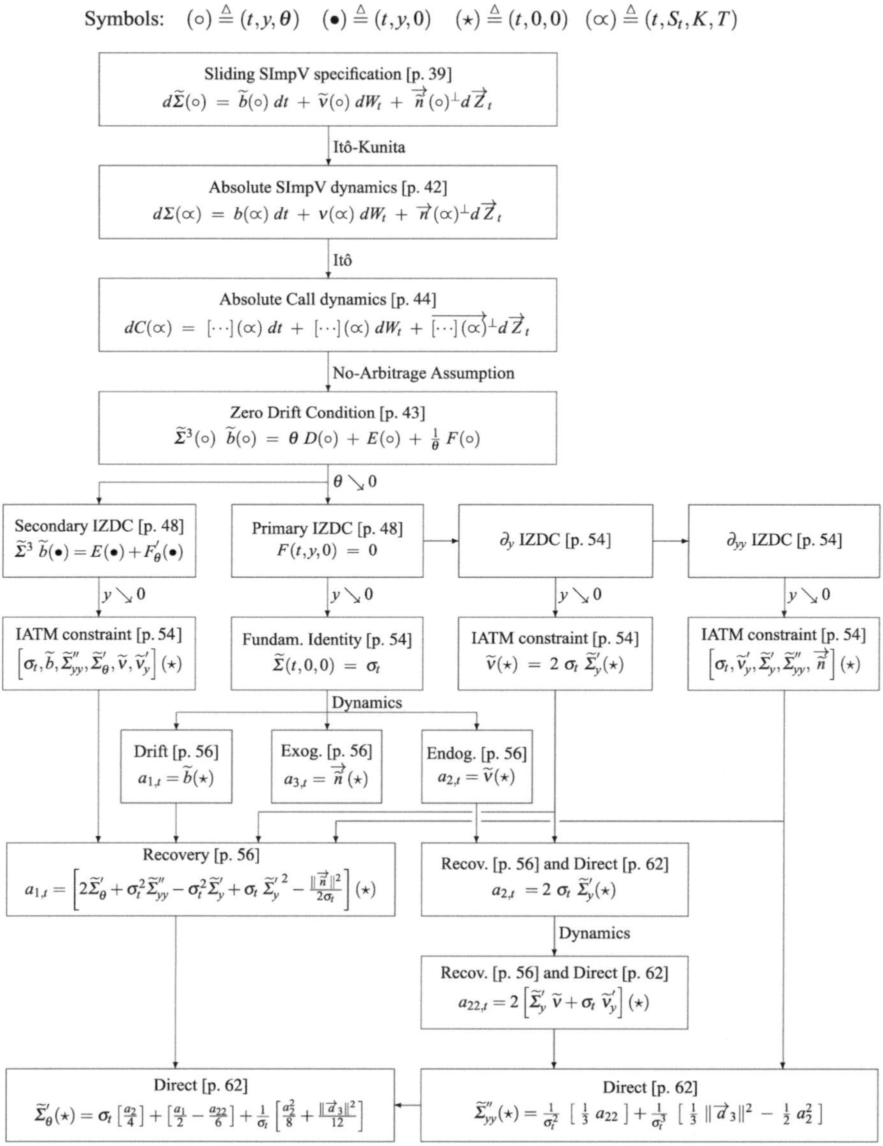

Fig. 2.4 Summary sketch of proofs within this chapter

SInsV models. In these cases the SImpV class becomes a kind of artefact, not destined to be used *per se* but simply to approximate the generated smile shape and dynamics. However, we would not discourage a stochastic implied volatility modelling approach. As mentioned earlier, this model class is rich and potent, although not without implementation difficulties. Overall we see it as one of the future solutions to deal with liquid option markets, but nonetheless we will tend to focus on the direct problem thereafter.

It is also clear that the subject is rich and deserves to be extended in several directions, if at all possible. A first interrogation relates to higher differentiation orders, and the possibility of a generic and programmable approach. In particular, can we compute some of the other meaningful differentials that describe the smile shape (and dynamics)? Another question is the impact of using a multi-dimensional framework, and the way it affects both the direct and inverse problems. Also, as discussed in Sect. 2.1.2.4 [p. 32], is it possible to extend the framework to term structures, and how does it affect the structure of the problem?

References

1. Lewis, A.L.: Option Valuation Under Stochastic Volatility. Finance Press, California (2000)
2. Heston, S.L.: A closed-form solution for options with stochastic volatility with applications to bond and currency options. Rev. Finance Stud. **6**(2), 327–343 (1993)
3. Black, F., Scholes, M.: The pricing of options and corporate liabilities. J. Polit. Econ. **3**(81), 637–654 (1973)
4. El Karoui, N., Jeanblanc-Picqu, M., Shreve, S.E.: Robustness of the black and scholes formula. Math. Finance **8**(2), 93–126 (1998)
5. Brace, A., Gatarek, D., Musiela, M.: The market model of interest rates dynamics. Math. Finance **7**(2), 127–155 (1997)
6. Schoenmakers, J.: Robust Libor Modelling and Pricing of Derivative Products. Financial Mathematics. Chapman & Hall, Boca Raton (2005)
7. Hafner, R.: Stochastic Implied Volatility. Number 545 in Lecture Notes in Economics and Mathematical Systems. Springer, Berlin (2004)
8. Gatheral, J.: Developments in Volatility Derivatives Modeling. Presentation at the Global Derivatives Trading and Risk Management. Merril Lynch (2007)
9. Gatheral, J.: Further Developments in Volatility Derivatives Modeling. Presentation at the Global Derivatives Trading and Risk Management. Merril Lynch (2008)
10. Derman, E.: Regimes of Volatility: Some Observations on the Variation of SP500 Implied Volatilities. Quantitative Strategies Research Notes, Goldman Sachs (1999)
11. Hagan, P.S., Kumar, D., Lesniewski, A.S., Woodward, D.E.: Managing smile risk. Wilmott Mag. **1**(1), 84–108 (2002)
12. Durrleman, V.: From implied to spot volatilities. Finance Stochast. **14**(2), 157–177 (2010)
13. Alexander, C., Nogueira, L.M.: Stochastic Local Volatility. In: Second IASTED International Conference, pp. 136–141 (2004)
14. Cont, R., da Fonseca, J.: Dynamics of implied volatility surfaces. Quant. Finance **2**, 45–60 (2002)
15. Fengler, M.R.: Semiparametric Modeling of Implied Volatility. Lecture Notes, Springer Finance (2005)
16. Ren, Y., Madan, D., Qian, M.: Calibrating and pricing with embedded local volatility models. Risk Mag. Lond. **20**(9), 138 (2007)

17. Heath, D., Jarrow, R.A., Morton, A.: Bond pricing and the term structure of interest rates: a new methodology for contingent claims valuation. Econometrica **60**, 77–105 (1992)
18. Schnbucher, P.J.: A market model for stochastic implied volatility. Philos. Trans. R. Soc. **357**(1758), 2071–2092 (1999)
19. Ledoit, O., Santa Clara, P.: Relative Pricing of Options with Stochastic Volatility. Technical Report, UCLA (1998)
20. Brace, A., Goldys, B., Klebaner, F., Womersley, R.: Market Model of Stochastic Implied Volatility with Application to the BGM Model. Technical Report, University of New South Wales (2001)
21. Bergomi, L.: Smile dynamics. Risk **17**(9), 117–123 (2004)
22. Bergomi, L.: Smile dynamics II. Risk **18**, 67–73 (2005)
23. Bergomi, L.: Smile dynamics III. Risk **21**, 90–96 (2008)
24. Bergomi, L.: Smile dynamics IV. Risk 94–100 (2009)
25. Rousseau, N.: How to Keep the Smile? Dynamic Vega Hedges and Volatility Derivatives. Ph.D. Thesis, Universite de Nice Sophia Antipolis (2007)
26. Hull, J., White, A.: Pricing interest-rate-derivatives securities. Rev. Finance Stud. **3**(4), 573–592 (1990)
27. Dupire, B.: Pricing with a smile. Risk **7**, 18–20 (1994)
28. Lee, R.W.: Implied volatility: statics, dynamics, and probabilistic interpretation. In: Recent Advances in Applied Probability. Springer, New York (2004)
29. Avellaneda, M., Lipkin, M.D.: A market-induced mechanism for stock pinning. Quant. Finance **3**, 417–425 (2003)
30. Beresticki, H., Busca, J., Florent, I.: Asymptotics and calibration of local volatility models. Quant. Finance **2**, 61–69 (2002)
31. Romano, M., Touzi, N.: Contingent claims and market completeness in a stochastic volatility model. Math. Finance **7**(4), 399–410 (1997)
32. Dragulescu, A.A., Yakovenko, V.M.: Probability distribution of returns in the Heston model with stochastic volatility. Quant. Finance **2**, 443–453 (2002)
33. Lee, R.W.: The moment formula for implied volatility at extreme strikes. Math. Finance **14**(3), 469–480 (2004)
34. Benaim, S., Friz, P.: Regular variation and smile asymptotics. Math. Finance **19**, 1–12 (2009)
35. Nicolay, D.: Volatility Dynamics. Ph.D. Thesis, Ecole Polytechnique (2011)
36. Kunita, H.: Stochastic Flows and Stochastic Differential Equations. Cambridge University Press, Cambridge (1990)
37. Dupire, B.: Pricing and Hedging with Smiles. Working paper, Paribas Capital Markets (1993)
38. Gatheral, J.: The Volatility Surface: A Practitioner's Guide. Wiley, New York (2006)
39. Gatheral, J.: Modeling the Implied Volatility Surface. Presentation at the Global Derivatives and Risk Management. Merril Lynch (2003)
40. Derman, E., Kani, I., Zou, J.Z.: The Local Volatility Surface: Unlocking the Information in Index Option Prices. Quantitative Strategies Research Notes, Goldman Sachs (1995)
41. Hagan, P.S., Woodward, D.E.: Equivalent black volatilities. Appl. Math. Finance **6**, 147–157 (1999)
42. Hull, J., White, A.: The pricing of options on assets with stochastic volatilities. J. Finance **42**, 281–300 (1987)
43. Renault, E., Touzi, N.: Option hedging and implied volatilities in a stochastic volatility model. Math. Finance **6**, 279–302 (1996)
44. Brigo, D., Mercurio, F., Rapisarda, F.: Lognormal-mixture dynamics and calibration to market volatility smiles. Int. J. Theor. Appl. Finance **5**, 427–446 (2002)
45. Alexander, C.: Normal mixture diffusion with uncertain volatility: modelling short- and long-term smile effects. J. Banking Finance **28**, 2957–2980 (2004)
46. Piterbarg, V.V.: Mixture of Models: A Simple Recipe for a ... Hangover? pp. 72–77, Wilmott Magazin (2005)
47. Fournie, E., Lebuchoux, J., Touzi, N.: Small noise expansion and importance sampling. Asymptotic Anal. **14**(4), 361–376 (1997)

48. Henrotte, P.: The case for time homogeneity. Wilmott Magazine, pp. 71–75 (2006)
49. Nocedal, J., Wright, S.J.: Numerical optimization. In: Springer Series in Operations Research, 2nd edn. Springer, New York (2006)
50. Andersen, L., Brotherton-Ratcliffe, R.: Extended Libor Market Models with Stochastic Volatility. Technical Report, Bank of America (2001)
51. Piterbarg, V.V.: Stochastic volatility model with time-dependent skew. Appl. Math. Finance **12**, 147–185 (2005)
52. Svoboda-Greenwood, S.: The Displaced Diffusion as an Approximation of the CEV. Technical Report, Oxford University, Oxford (2007)
53. Andersen, L., Andreasen, J.: Volatility skews and extension of the Libor market model. Appl. Math. Finance **7**, 1–32 (2000)
54. Lee, R.W., Wang, D.: Displaced lognormal volatility skews: analysis and applications to stochastic volatility simulations. Ann. Finance **8**, 159–181 (2012)

Chapter 3
Volatility Dynamics for a Single Underlying: Advanced Methods

Abstract This chapter is dedicated to generalising the ACE approach and its results. These extensions are performed in several directions, which offer some practical and/or some mathematical interest. First we describe the generic ACE methodology solving the direct problem at an arbitrary order. We then apply this algorithm to compute meaningful IATM differentials, all located within the second and third layers, which we can then exploit and interpret. Next we discuss alternative baselines to Black's model, and how to transfer IATM information between them. We then introduce a major structural extension, by considering the multi-dimensional framework. We follow the evolution of the inverse and direct problems, and illustrate these results with the case of generic baskets.

In Sect. 3.1 we extend the results of Chap. 2 concerning the *direct* problem to higher differentiation orders. In fact, we prove that any pure or cross-differential of the smile shape and dynamic maps, taken at the IATM point, can be computed from the dynamics of a generic stochastic instantaneous volatility model. Describing the associated Asymptotic Chaos Expansion methodology, we identify the *ladder effect* which specifies structurally the order of the computations and organise the differentials into well-defined groups called *layers*. We present the generic algorithm in a way that can clearly be automated, and restrict ourselves to the bi-dimensional case for simplicity.

In Sect. 3.2 we illustrate the generic methodology described above by computing the second and third layers. We start by justifying our choices, in particular the selection of IATM differentials which are financially and mathematically relevant. The computation itself is relegated to Appendix E, but we interpret those results, focusing on the significance of the dynamic coefficients and on the question of homogeneity. Furthermore, we show the impact of mean-reversion and vol-of-vol on the longer-term smile.

In Sect. 3.3 we continue extending the framework but in new directions. We start by decomposing the overall methodology into independent building blocks, and review how each of these can be generalised. Above all, we justify and examine the introduction of alternative *baselines* to replace the Lognormal model. We illustrate this point by running through the complete methodology for the *Normal* model, which shows striking similarities with the previous results. We characterise those baselines

D. Nicolay, *Asymptotic Chaos Expansions in Finance*,
Springer Finance Lecture Notes, DOI 10.1007/978-1-4471-6506-4_3

in terms of suitability and discuss several ways to make their implementation easier. In particular we present a generic but numerical method that allows us to *transfer* the results from one existing baseline to the next.

In Sect. 3.4 we extend the results of Chap. 2 on the first layer to multi-dimensional underlyings, drivers and dynamic coefficients. We discuss the benefits but also the difficulties involved, in particular with regard to solving the inverse problem, and propose several avenues for further research.

In Sect. 3.5 we illustrate the previous multi-dimensional results in the context of baskets, which is both mathematically rich and financially important. We treat the case of general stochastic weights as well as of fixed weights, with scalar but also with tensorial individuals. We show that the computational burden can be significantly reduced by adopting normalised notations, which in turn create and exploit an inductive structure.

Although we manage to extend the methodology in several directions and to a significant depth, it is clear that even in this single underlying context a number of questions remain unanswered. We obviously see these as potential for future research, and therefore as a good thing, which explains why some sections have been drafted in the shape of a discussion rather than of a definitive argument.

In terms of scope, and excluding the term structure framework developed in Part II, the current chapter represents the extent of the generalisations (conclusive or not) that we have officially considered in this study for the ACE methodology. Indeed, Sect. 3.2 and Chap. 4 will be focused on applications, with the former a direct instantiation of Sect. 3.1, and the latter a global illustration with specific complex models.

3.1 Higher-Order Expansions: Methodology and Automation

In this section, we focus on *the direct problem* in a bi-dimensional framework, i.e. when both the endogenous and exogenous drivers W_t and Z_t are (orthogonal) scalar Wiener processes. The extension of the methodology concerns the order of the attainable IATM differentiation, which is shown to be infinite in both the strike and maturity directions. Indeed, we prove that the input of the SInsV model, i.e. of the instantaneous volatility dynamics expressed in a chaos diffusion fashion, allows us to compute the differentials of the SImpV model (the implied volatility surface):

- taken at the IATM point $(t, 0, 0)$;
- both for the shape functional $\widetilde{\Sigma}$ and for its dynamic coefficients $\widetilde{b}$, $\widetilde{\nu}$ and $\widetilde{n}$;
- for any cross-differential order w.r.t. log-moneyness y and time-to-maturity θ.

It follows that, assuming benign convergence conditions, we can approximate the implied volatility surface at any required level of precision. Given our asymptotic approach, and without going into the assumption set introduced in Chap. 2, this assertion corresponds to making explicit the following link:

Definition 3.1 (*Direct problem in the generic bi-dimensional case*) Recall that in the generic bi-dimensional case, the SInsV and SImpV models are as usual both defined by the same underlying's dynamics

$$\frac{dS_t}{S_t} = \sigma_t dW_t$$

but differ by the specification of their own volatility, where $(\circ) \triangleq (t, y, \theta)$:

$$d\sigma_t = a_{1,t}dt + a_{2,t}dW_t + a_{3,t}dZ_t \qquad da_{\star,t} = a_{\star 1,t}dt + a_{\star 2,t}dW_t + a_{\star 3,t}dZ_t$$

$$d\widetilde{\Sigma}(t, y, \theta) = \widetilde{b}(\circ)dt + \widetilde{\nu}(\circ)dW_t + \widetilde{n}(\circ)dZ_t$$

so that in such a context the direct problem reads as

$$\left[\sigma_t, \left\{a_{\star,t}\right\}\right] \implies \begin{array}{c} \dfrac{\partial^n \partial^p}{\partial y^n \partial \theta^p} X(t, 0, 0) \quad \forall\, (n, p) \in \mathbb{N}^2 \\ \text{for} \quad X \in \left\{\widetilde{\Sigma}, \widetilde{b}, \widetilde{\nu}, \widetilde{n}\right\} \end{array}$$

where $\left\{a_{\star,t}\right\}$ must be understood as the current values of all chaos coefficients. Note that we are effectively dealing with processes on both sides.

It is worth remembering that the proof established for the first layer in Chap. 2 does *not* guarantee this generic property. In the literature, other expansion approaches are available but they are either limited at $n = 2$, or must be significantly and *ad hoc* altered to reach any further. This is typically the case for [1], for instance: differentiating the Call payoff twice leads to a Dirac function, and therefore any further y-order would require following a somewhat different route. By contrast, what we aim for is an "induction" approach, in the fashion of Fournie et al. [2], but for the completely generic setup presented so far. By contrast, [2] specialises to a single model class, a specific bi-dimensional Markovian model, but presents the advantage of an integral rather than differential approach, which is more appropriate for time-inhomogeneous or even time-discontinuous specifications.

Another interest in exposing the generic methodology is that it tends to reveal the deeper structure of the problem, especially its various symmetries. It does so in a fashion which is complementary to the simpler, low-order and low-depth computations of Chap. 2. It is indeed a common theme throughout this study that the generic and the specific investigations tend to feed each other.

But beyond its theoretical and structural importance, another reason to describe such an induction method is more applicative in nature, as we aim at enabling automation of the procedure. Indeed, the complexity of the computations, along with the mechanical and repetitive aspect of the process, create the perfect conditions for a computer program to take over. Of course, the latter does require a very simple formal calculus module, but this is not an issue.

The choice of a simple, bi-dimensional framework is justified by two concerns. As we will see later in Sect. 3.4, if the direct problem is not impaired by the dimensionality of the drivers, the inverse (or recovery) problem is deeply affected. In a nutshell, the reason for this feature is the poorer, scalar nature of the implied volatility *vs* the richer, vectorial structure of the instantaneous volatility. In other words, $\widetilde{\Sigma}$ tends to *aggregate* the information, which is then difficult to disentangle back. As will be shown in Sect. 3.4, there are clearly cases where the vectorial information can be recovered, but of course they require the consumption of more information in the form of higher-order differentials of the smile. In summary, the question of if and how the recovery is possible in a generic, multi-dimensional setup has been left for further research.

The second concern is that the multi-dimensionality does not bring much more to the direct problem, in terms of structure for the output. It does, however, render the formulae more complex and tends to clutter the picture.

In our view, these two reasons justify why we should stick to a bi-dimensional framework.

3.1.1 Tools and Roadmap

First, for the sake of completeness, let us formally define the functional space that covers the sliding processes that we will be considering:

Definition 3.2 We consider adapted processes on $\left(\Omega, \mathscr{F}, \{\mathscr{F}_t\}_{t\geq 0}, \mathbb{P}\right)$ where as before the filtration consists of the σ-algebra generated by the bi-dimensional driver, and parametrised by y and θ as per:

$$Proc \triangleq \left\{X_t(y,\theta) : \Omega \times [0, T^*] \times \mathbb{R} \times \mathbb{R}^+ \to \mathbb{R}\right\}.$$

Clearly this broad class encompasses the individual processes $\widetilde{\Sigma}(t, y, \theta)$, $\widetilde{b}(t, y, \theta)$, $\widetilde{\nu}(t, y, \theta)$ and $\widetilde{n}(t, y, \theta)$. But it also naturally contains *blocks* or polynomial combinations of these, which as sub-products of the ZDC differentiation are going to be our main objects.

Note that, in accordance with the bi-dimensionality condition, the target space is $\mathbb{R}$ and not some $\mathbb{R}^n$. As for the origin space, from the perspective of asymptotics we have included $\theta = 0$, which should be seen as a limit and therefore an abuse of proper notation.

Note also that we haven't imposed any regularity condition: this is because the sets of necessary and/or sufficient conditions are quite different between the shape process $\widetilde{\Sigma}$ and the dynamic coefficients $\widetilde{b}$, $\widetilde{\nu}$ and $\widetilde{n}$. Indeed, $\widetilde{\Sigma}$ and its differentials are destined to see their dynamics computed, which imposes specific constraints. This is not the case for the dynamics coefficients, which will only be differentiated.

Our rationale is therefore to keep a single process definition for *Proc* and introduce the various regularity packs *on demand* throughout the various proofs.

We can then introduce the classical differentiation operator

$$\nabla^{(m,p)} : Proc \longrightarrow Proc \quad \nabla^{(m,p)} X_t(y,\theta) \triangleq \frac{\partial^{m+p}}{\partial y^m \partial \theta^p} X(t,y,\theta).$$

In abridged form, we will thus denote the image of a given process by

$$X^{(m,p)} \nabla^{(m,p)} \triangleq X_t(y,\theta).$$

As for specific vocabulary, when $X^{(m,p)}(t,0,0)$ will be expressed as a function of a collection of chaos coefficients $\{a_{*,t}\}$, we will refer to that process as being *input specified*. Also, the useful *compounding properties* of this operator come trivially as

$$\nabla^{(m_1+m_2,p)} = \nabla^{(m_1,p)} \nabla^{(m_2,p)} \quad \text{and} \quad \nabla^{(m,p_1+p_2)} = \nabla^{(m,p_1)} \nabla^{(m,p_2)}.$$

As for links between the four target maps, we underline the following basic feature.

Remark 3.1 Once $\widetilde{\Sigma}^{(n,p)}(t,0,0)$ is input-specified then, assuming enough regularity, we can use Theorem 3.1.2 of Kunita [3] (p. 75) to formally swap the differentiation and diffusion operators, deriving the following dynamics:

$$\begin{aligned} d\left[\nabla^{(n,p)} \widetilde{\Sigma}(t,0,0)\right] &= a_{*1,t}dt + a_{*2,t}dW_t + a_{*3,t}dZ_t \\ &= \widetilde{b}^{(n,p)}(t,0,0)dt + \widetilde{\nu}^{(n,p)}(t,0,0)dW_t + \widetilde{n}^{(n,p)}(t,0,0)dZ_t. \end{aligned}$$

Starting with input expression, we obtain through Itô's Lemma the dynamics on the left-hand side. Therefore, by uniqueness of the decomposition we can identify the finite variation term and the endogenous and exogenous coefficients, which gives us $\widetilde{b}^{(n,p)}(t,0,0)$, $\widetilde{\nu}^{(n,p)}(t,0,0)$ and $\widetilde{n}^{(n,p)}(t,0,0)$ as function of the inputs.

We are now ready to start solving the problem set by Definition 3.1. Besides proving that the direct problem is *well-posed*, this section will also provide the *methodology* to be followed to access these quantities, as well as the first steps towards a computing automation.

Combining these multiple objectives in a compact fashion comes at the price of a rather unorthodox presentation. Indeed, we will prove various statements on the fly, as part of the methodology whose logic and progression sets the overall structure. It is interesting to note, however, that the latter revolves around the particular structure of the Zero-Drift Equation.

In light of Remark 3.1, we will focus the algorithm on the shape function $\widetilde{\Sigma}$ rather than on the dynamics coefficients $\widetilde{b}$, $\widetilde{\nu}$ and $\widetilde{n}$. In that spirit, we will consider the differentiation "matrix" of $\widetilde{\Sigma}$, i.e. $\widetilde{\Sigma}^{(m,p)}(t,0,0)$ with the m index for lines and p for columns: see Fig. 3.1 [p. 122] for a graphical representation. It must be made clear that the focus on the shape functional $\widetilde{\Sigma}$ is in no way an indication that we value static

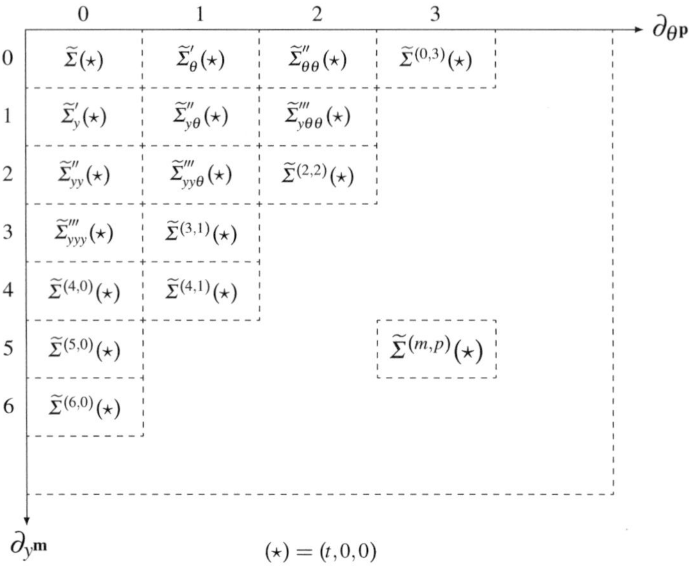

Fig. 3.1 Ladder effect in the differentiation matrix

considerations above dynamic ones. As shown by the IATM arbitrage constraints of the SImpV model, when taken at $(t, 0, 0)$ the links between the various differentials of $\widetilde{\Sigma}$, $\widetilde{b}$, $\widetilde{\nu}$ and $\widetilde{n}$ are very strong and it is nonsensical to define a definite hierarchy among them. But from the perspective of the direct problem, the redundancy is maximum, which explains our choice.

Accordingly, we divide the proof and therefore the methodology into two parts, in order to cover the whole matrix:

- In the first part (Sect. 3.1.2) we progress along the y axis on the *first* column. We use pure y-differentiations of the *Immediate* Zero-Drift Condition to compute any subset of that first column. In other words, we access

$$\widetilde{\Sigma}^{(k,0)} \quad \text{for} \quad 0 \le k \le m \quad (\text{with } m \in \mathbb{N}).$$

- In the second part (Sect. 3.1.3) we use cross-differentiations of the main Zero-Drift Condition to compute a subset of column p, based on previous computation of a specific set of associated sub-columns on the left (i.e. of lower θ-order):

$$\widetilde{\Sigma}^{(k,l)} \quad \text{with} \quad \begin{array}{c} 0 \le k \le m + 2(p-l) \\ 0 \le l \le p-1 \end{array} \quad \longrightarrow \quad \widetilde{\Sigma}^{(k,p)} \quad \text{with} \quad 0 \le k \le m.$$

The peculiar composition of the leftward sub-column set is known as the *ladder effect* and will be discussed later.

3.1.2 Computing the First Column of the Differentiation Matrix

In order to compute $\widetilde{\Sigma}^{(m,0)}(t,0,0)$ for all m, we must first gather the relevant smile differentials into a coherent set, which will be the very object on which our inductive process operates.

Definition 3.3 (*Process subset for the* $(m,0)$*-IZDC*) Assume $m \geq 0$ and let $\mathscr{S}^{IZDC}(m,0)$ be the following set of *Proc*-processes:

$$\mathscr{S}^{IZDC}(m,0) \triangleq \begin{bmatrix} \widetilde{\Sigma}^{(k,0)}(t,y,\theta) & 0 \leq k \leq m & \\ \widetilde{\nu}^{(k,0)}(t,y,\theta) & 0 \leq k \leq (m-1) & \textit{iff } m \geq 1 \\ \widetilde{n}^{(k,0)}(t,y,\theta) & 0 \leq k \leq (m-2) & \textit{iff } m \geq 2 \end{bmatrix}.$$

We also introduce the following **regularity pack** for any non-negative index m.

Assumption 3.1 (*Regularity assumption package for the* $(m,0)$*-layer*) The three parametric processes

$$\widetilde{\Sigma}^{(m,0)}(t,y,\theta) \quad \widetilde{\nu}^{(m-1,0)}(t,y,\theta) \quad \text{and} \quad \widetilde{n}^{(m-2,0)}(t,y,\theta)$$

admit a (finite) stochastic limit in the immediate domain $(t,y,0)$.

Then we have several constructive results w.r.t. the first column of the differentiation matrix (i.e. the pure strike differentials).

Lemma 3.1 (Derivation of the first column in a bi-dimensional model) *In a generic bi-dimensional framework specified by Definition* 3.1, *and assuming non-negative differentiation indices:*

(i) *Applying Itô to the input-expressions for*

$$\widetilde{\Sigma}^{(m-2,0)}(t,0,0) \quad \textit{and} \quad \widetilde{\Sigma}^{(m-1,0)}(t,0,0)$$

provides, respectively, the input-expressions of

$$\widetilde{\nu}^{(m-2,0)}(t,0,0) \quad \textit{and} \quad \widetilde{n}^{(m-1,0)}(t,0,0).$$

(ii) *Applying the* $\nabla^{(m,0)}$ *operator to the IZDC* (2.28) *while assuming the ad hoc regularity pack provides a non-linear (stochastic) ordinary differential equation*

- *linking* $\widetilde{\Sigma}^{(m,0)}$ *to the set* $\mathscr{S}^{IZDC}(\mathbf{m-1},0)$*;*
- *where all processes are taken in the immediate domain* $(t,y,\mathbf{0})$.

This equation constitutes another local arbitrage constraint of the SImpV model.

(iii) *Combining these two items recurrently provides the input expression of*

$$\widetilde{\Sigma}^{(m,0)}(t,0,0) \quad \forall m \geq 0.$$

Also, we prove the part of Remark 2.10 [p. 76] pertaining to local volatility models:

Lemma 3.2 (Differentiation of the first column in a local volatility model) *In an LV model defined by* (2.5) [p. 30] *the set of IATM pure-strike static differentials*

$$\left\{\widetilde{\Sigma}^{(i)}_{y^i}(\star)\right\}_{0\leq i\leq n}$$

is provided by the corresponding set of pure space f-differentials

$$\left\{f^{(i)}_{S^i_t}(t,S_t)\right\}_{0\leq i\leq n}.$$

Proof of Lemmas 3.1 and 3.2.
Recall that the IZDC (2.28) can be rewritten as

$$\text{(IZDC)} \qquad F(t,y,0) = 0 \tag{3.1}$$

where $F(t,y,\theta)$ has been defined by (2.21). Applying the $\nabla^{(m,0)}$ operator provides the $(m,0)$-IZDC in the form of

$$F^{(m,0)}(t,y,0) = 0 \quad \forall m \geq 1, \quad \forall y. \tag{3.2}$$

Our strategy is naturally to prove and develop result (iii) by induction, establishing the simple (i) and more qualitative (ii) as by-products. The local volatility result is run in parallel as it is closely related.

Initial check: for $m = 0$.
The fundamental IATM identity (2.36) [p. 50] and the first-layer result (2.55) [p. 62] along with (2.64) [p. 68] and (2.67) [p. 68] show that:

- applying $\nabla^{(0,0)}$ to the Immediate ZDC provides $\widetilde{\Sigma}^{(0,0)}(t,0,0)$ as a function of σ_t;
- $\widetilde{\Sigma}(t,0,0)$ is determined by $f(t,S_t)$ only.

Induction check: for $m \geq 1$.
The assumption is taken at level $m-1$ and states that

Ⓗ The whole $\mathscr{S}^{IZDC}(m-1,0)$ set taken in $(t,0,0)$ is input-specified.
Each component process of $\mathscr{S}^{IZDC}(m-1,0)$ is expressed
as polynomial in the input coefficients $\{a_{\star,t}\}$ and rational in σ_t.

Ⓗ$_L$ $\left\{\widetilde{\Sigma}^{(i)}_{y^i}(\star)\right\}_{0\leq i\leq m-1}$ is provided by $\left\{f^{(i)}_{x^i}(t,S_t)\right\}_{0\leq i\leq m-1}$.

We then introduce the regularity pack for the $(m, 0)$ layer, as per Assumption 3.1. Recalling Remark 3.1 [p. 121], we can now proceed with (i) and derive (through a straightforward application of Itô's Lemma) the dynamics of the input expressions for both $\widetilde{\Sigma}^{(m-2,0)}(t, 0, 0)$ and $\widetilde{\Sigma}^{(m-1,0)}(t, 0, 0)$, which by identification give us respectively the input expressions for

$$\widetilde{n}^{(m-2,0)}(t, 0, 0) \quad \text{and} \quad \widetilde{\nu}^{(m-1,0)}(t, 0, 0).$$

Note that the polynomial/rational property is inherited by the coefficients *via* Itô. Recall now from its definition (2.21) [p. 44] that

$$F(t, y, \theta) \triangleq \frac{1}{2}\widetilde{\Sigma}^4(\circ) - \frac{1}{2}\sigma_t^2\widetilde{\Sigma}^2(\circ) - y\sigma_t\widetilde{\Sigma}(\circ)\left[\widetilde{\nu}(\circ) - \sigma_t\widetilde{\Sigma}'_y(\circ)\right]$$
$$- \frac{1}{2}y^2\left[\left[\widetilde{\nu}(\circ) - \sigma_t\widetilde{\Sigma}'_y(\circ)\right]^2 + \|\vec{\widetilde{n}}\|^2(\circ)\right].$$

Introducing the following basic results for $n \geq 1$ and $f : \mathbb{R} \to \mathbb{R}$ n-differentiable:

$$\frac{\partial^n}{\partial x^n}\left[xf(x)\right] = nf^{(n-1)}(x) + xf^{(n)}(x) \tag{3.3}$$

and

$$\frac{\partial^n}{\partial x^n}\left[x^2f(x)\right] = 1_{\{n\geq 2\}}n(n-1)f^{(n-2)}(x) + 2nxf^{(n-1)}(x) + x^2f^{(n)}(x) \tag{3.4}$$

with the convention

$$f^{(k)}(x) = 0 \quad \text{if } k < 0$$

we can derive in the generic point the process $F^{(m,0)}(t, y, \theta)$ as

$$\nabla^{(m,0)}F(t, y, \theta) = y\,[\text{Ⓐ}] + \underbrace{\frac{1}{2}\sum_{k=0}^{m} C_m^k\,\widetilde{\Sigma}^{2\,(k,0)}\left[\widetilde{\Sigma}^{2\,(m-k,0)} - \sigma_t^2\right]}_{\text{Ⓑ}}$$
$$- m\sigma_t\sum_{k=0}^{m-1} C_{m-1}^k\,\widetilde{\Sigma}^{(m-k-1,0)}\left[\widetilde{\nu}^{(k,0)} - \sigma_t\widetilde{\Sigma}^{(k+1,0)}\right]$$
$$- 1_{\{m\geq 2\}}\frac{m(m-1)}{2}\sum_{k=0}^{m-2} C_{n-2}^k$$
$$\times\left[\left[\widetilde{\nu} - \sigma_t\widetilde{\Sigma}'_y\right]^{(k,0)}\left[\widetilde{\nu} - \sigma_t\widetilde{\Sigma}'_y\right]^{(m-k-2,0)} + \widetilde{n}^{(k,0)}\widetilde{n}^{(m-k-2,0)}\right] \tag{3.5}$$

where, on the right-hand side of (3.5), the first term in brackets Ⓑ invokes in a quadratic fashion all the processes in the set $\mathscr{S}^{IZDC}(m, 0)$, and these processes only.

As for the second term Ⓑ it can be further decomposed in (t, y, θ) as

$$
\begin{aligned}
&\sum_{k=0}^{m} C_m^k \; \widetilde{\Sigma}^{2\,(k,0)} \left[\widetilde{\Sigma}^{2\,(m-k,0)} - \sigma_t^2 \right] \\
&\quad = \sum_{k=0}^{m} C_m^k \left[\sum_{i=0}^{k} C_k^i \, \widetilde{\Sigma}^{(i,0)} \, \widetilde{\Sigma}^{(k-i,0)} \right] \left[\left[\sum_{j=0}^{m-k} C_{m-k}^j \, \widetilde{\Sigma}^{(j,0)} \, \widetilde{\Sigma}^{(m-k-j,0)} \right] - \sigma_t^2 \right] \\
&\quad = \sum_{\substack{i,j,k=0 \\ i \le k, j \le m-k}}^{m} C_m^k C_k^i C_{m-k}^j \left[\widetilde{\Sigma}^{(i,0)} \, \widetilde{\Sigma}^{(k-i,0)} \right] \left[\left[\widetilde{\Sigma}^{(j,0)} \, \widetilde{\Sigma}^{(m-k-j,0)} \right] - \sigma_t^2 \right].
\end{aligned}
$$

We can then combine the formal $(m, 0)$-IZDC (3.2) with the expression for $F^{m,0}$ (3.5) taken at $(t, y, 0)$, which produces the *explicit* $\nabla^{(m,0)}$ IZDC, valid in the Immediate region.

Then an inventory of the processes involved in (3.5) shows that we are faced with a functional of $\widetilde{\Sigma}^{(m,0)}(t, y, 0)$ and of the $\mathscr{S}^{IZDC}(m-1, 0)$ set taken at $(t, y, 0)$, and only those processes. The differential equation is obviously ordinary, since no θ-differential appears, and also quadratic, which completes the proof of result (ii).

Invoking the regularity Assumption 3.1 we can take the same explicit $\nabla^{(m,0)}$ IZDC now in $y = 0$. We end up with an ordinary equation which is affine in the only remaining unknown $\widetilde{\Sigma}^{(m,0)(t,0,0)}$. More precisely, according to (3.5)

- the linear coefficient is a polynomial of $\sigma_t = \widetilde{\Sigma}(t, 0, 0)$
- while the constant is polynomial in the $\{a_{\star,t}\}$ coefficients and rational in σ_t.

We conclude that $\widetilde{\Sigma}^{(m,0)(t,0,0)}$ is obtained as the required functional of the inputs so that the induction Ⓗ has been proven at the next level m, which proves (iii).

In the case of local volatility, still looking at (3.5) we recall that all exogenous terms are null while for $0 \le i \le m-1$ each $\widetilde{\nu}^{(i,0)}(\star)$ is the endogenous coefficient of $\widetilde{\Sigma}^{(i,0)}(\star)$. By applying Itô to the induction assumption $Ⓗ_L$ we find that the same population of IATM endogenous coefficients invokes the pure-space differentials $f_{x^i}^{(i)}$ for $0 \le i \le \mathbf{m}$ all taken in (t, S_t). Therefore solving for $\widetilde{\Sigma}^{(m,0)(t,0,0)}$ through the $\nabla^{(m,0)}$-IZDC proves $Ⓗ_L$ at the next index m and concludes the proof. □

3.1.3 *Computing Subsequent Columns of the Differentiation Matrix*

Here we are interested in $\widetilde{\Sigma}^{(m,p)}(t, 0, 0)$ for any m and for $p \ge 1$. But before coming to the main body of this section, we choose to establish a preliminary result, relative to the θ-expansion of cross-differentials for the functional $F(t, y, \theta)$:

Lemma 3.3 (Small-time expansion for the differentials of the $F(t, y, \theta)$ term) *We have, for $(m, p) \in \mathbb{N}^2$ and $y \in \mathbb{R}$,*

$$\sum_{i=0}^{p} C_p^i (-1)^i \frac{i!}{\theta^{i+1}} F^{(m,p-i)}(t, y, \theta) = \frac{1}{p+1} F^{(m,p+1)}(t, y, 0) + O(\theta).$$

Proof Let us develop the left-hand side using Taylor θ-expansions, invoking the θ-differentials of $F(t, y, \theta)$ up to order $p+1$:

$$\begin{aligned}
&\sum_{i=0}^{p} C_p^i \frac{(-1)^i i!}{\theta^{i+1}} F^{(m,p-i)}(\circ) \\
&= \sum_{i=0}^{p} C_p^i \frac{(-1)^i i!}{\theta^{i+1}} \left[\sum_{j=0}^{i+1} \frac{\theta^j}{j!} F^{(m,p-i+j)}(t, y, 0) + O(\theta^{i+2}) \right] \\
&= \sum_{i=0}^{p} \sum_{j=0}^{i+1} \frac{(-1)^i p!}{(p-i)! j!} \theta^{j-i-1} F^{(m,p-i+j)}(t, y, 0) + O(\theta).
\end{aligned}$$

We now introduce the new variables $k = p - i + j$ and $l = i - p + k$. Defining $(\bullet) = (t, y, 0)$, we isolate the terms in $F^{(m,p+1)}$ and use Fubini to re-write

$$\begin{aligned}
&\sum_{i=0}^{p} C_p^i (-1)^i \frac{F^{(m,p-i)}(\circ)}{\theta^{i+1}} \\
&= \sum_{i=0}^{p} \sum_{k=p-i}^{p} \frac{p!(-1)^i \theta^{k-p-1}}{(p-i)!(k+i-p)!} F^{(m,k)}(\bullet) \\
&\quad + \sum_{i=0}^{p} \frac{p!(-1)^i}{(p-i)!(i+1)!} F^{(m,p+1)}(\bullet) + O(\theta) \\
&= \sum_{k=0}^{p} \frac{(-1)^{p-k} p! \theta^{k-p-1}}{k!} F^{(m,k)}(\bullet) \sum_{l=0}^{k} \frac{k!(-1)^l}{(k-l)! l!} \\
&\quad - \frac{F^{(m,p+1)}(\bullet)}{p+1} \sum_{i=1}^{p+1} \frac{(-1)^i (p+1)!}{(p+1-i)! i!} + O(\theta).
\end{aligned}$$

Substituting the $(m, 0)$-IZDC we get $F^{(m,0)}(\bullet) \equiv 0$, which makes the first sum start from $k = 1$. Finally using Cramer formulae, we obtain the desired result:

$$\sum_{i=0}^{p} C_p^i (-1)^i i! \frac{F^{(m,p-i)}(\circ)}{\theta^{i+1}}$$

$$= \sum_{k=0}^{p} \frac{(-1)^{p-k} p! \theta^{k-p-1}}{k!} F^{(m,k)}(\bullet) \underbrace{\sum_{l=0}^{k} C_k^l (-1)^l}_{=0}$$

$$- \frac{F^{(m,p+1)}(\bullet)}{p+1} \sum_{i=\mathbf{1}}^{p+1} C_{p+1}^i (-1)^i + O(\theta)$$

$$= \frac{1}{p+1} F^{(m,p+1)}(\bullet) + O(\theta). \qquad \square$$

Let us now establish the main results and methodology involved in the computation of the static IATM differential $\widetilde{\Sigma}^{(m,p)}(t, 0, 0)$ for $p \geq 1$, in other words for any cell or column of the differentiation matrix. As before, we start with a qualitative result, which requires us to formally identify a specific group of processes:

Definition 3.4 (*Process subset for the (m, p)-ZDC*) For $m \in \mathbb{N}$ and $p \in \mathbb{N}^*$, let $\mathscr{S}^{ZDC}(m, p)$ be the following set of *Proc*-processes:

$$\mathscr{S}^{ZDC}(m,p) \triangleq \begin{bmatrix} \widetilde{\Sigma}^{(k,l)}(t,y,\theta) & 0 \leq k \leq m+2 & 0 \leq l \leq p & \\ \widetilde{\Sigma}^{(k,p+1)}(t,y,\theta) & 0 \leq k \leq m & & \\ \widetilde{b}^{(k,l)}(t,y,\theta) & 0 \leq k \leq m & 0 \leq l \leq p & \\ \widetilde{\nu}^{(k,l)}(t,y,\theta) & 0 \leq k \leq m+1 & 0 \leq l \leq p & \\ \widetilde{\nu}^{(k,p+1)}(t,y,\theta) & 0 \leq k \leq (m-1) & & \text{iff } m \geq 1 \\ \widetilde{n}^{(k,l)}(t,y,\theta) & 0 \leq k \leq m & 0 \leq l \leq p-1 & \text{iff } p \geq 1 \\ \widetilde{n}^{(k,l)}(t,y,\theta) & 0 \leq k \leq (m-2) & p \leq l \leq p+1 & \text{iff } m \geq 2 \end{bmatrix}$$

This leads us to the qualitative result:

Lemma 3.4 (IATM cross-differentials of the ZDC) *Applying the differentiation operator $\nabla^{(m,p)}$ to the Zero-Drift Condition (2.18) and then evaluating this stochastic PDE at the IATM point $(t, 0, 0)$ provides an equation that:*

- *involves only and all the processes in the $\mathscr{S}^{ZDC}(m, p)$ subset, taken in $(t, 0, 0)$;*
- *is polynomial in those processes;*
- *is affine in $\widetilde{\Sigma}^{(m,p+1)}(t, 0, 0)$.*

In order to keep the proof lighter, we now introduce a series of low-level definitions that it will require locally.

Definition 3.5 (*Process subsets for individual components of the* (m,p)*-ZDC*) Concerning the left-hand side of the (m,p)-ZDC we set

$$\mathscr{S}_L^{ZDC}(m,p) \triangleq \begin{bmatrix} \widetilde{\Sigma}^{(k,l)}(t,y,\theta) & 0 \le k \le m & 0 \le l \le p \\ \widetilde{b}^{(k,l)}(t,y,\theta) & 0 \le k \le m & 0 \le l \le p \end{bmatrix}.$$

With regard to the right-hand side, we start with the $D()$ component

$$\mathscr{S}_D^{ZDC}(m,p) \triangleq \begin{bmatrix} \widetilde{\Sigma}^{(k,l)}(t,y,\theta) & 0 \le k \le m+1 & 0 \le l \le p-1 & \text{iff } p \ge 1 \\ \widetilde{\nu}^{(k,l)}(t,y,\theta) & 0 \le k \le m & 0 \le l \le p-1 & \text{iff } p \ge 1 \\ \widetilde{n}^{(k,l)}(t,y,\theta) & 0 \le k \le m & 0 \le l \le p-1 & \text{iff } p \ge 1 \end{bmatrix}$$

before dealing with the $E()$ term

$$\mathscr{S}_E^{ZDC}(m,p) \triangleq \begin{bmatrix} \widetilde{\Sigma}^{(k,l)}(t,y,\theta) & 0 \le k \le m+2 & 0 \le l \le p \\ \widetilde{\Sigma}^{(k,p+1)}(t,y,\theta) & 0 \le k \le m & \\ \widetilde{\nu}^{(k,l)}(t,y,\theta) & 0 \le k \le m+1 & 0 \le l \le p \end{bmatrix}$$

and finally with the $F()$ component

$$\mathscr{S}_F^{ZDC}(m,p) \triangleq \begin{bmatrix} \widetilde{\Sigma}^{(k,l)}(t,y,\theta) & 0 \le k \le m & 0 \le l \le p+1 & \\ \widetilde{\nu}^{(k,l)}(t,y,\theta) & 0 \le k \le (m-1) & 0 \le l \le p+1 & \text{iff } m \ge 1 \\ \widetilde{n}^{(k,l)}(t,y,\theta) & 0 \le k \le (m-2) & 0 \le l \le p+1 & \text{iff } m \ge 2 \end{bmatrix}.$$

Proof For the sake of clarity we proceed in two steps: let us first differentiate the modified ZDC (2.18) m times w.r.t. y, keeping it in the generic point $(\circ) \triangleq (t,y,\theta)$. Formally, we end up with

$$\sum_{k=0}^{m} C_m^k \, \widetilde{\Sigma}^{3\,(m-k,0)} \, \widetilde{b}^{(k,0)}(\circ) = \theta D^{(m,0)}(\circ) + E^{(m,0)}(\circ) + \frac{1}{\theta} F^{(m,0)}(\circ).$$

Then we differentiate p times w.r.t. θ and obtain

$$\sum_{k=0}^{m} C_m^k \left[\sum_{l=0}^{p} C_p^l \; \widetilde{\Sigma}^{3\,(m-k,p-l)}(\circ)\widetilde{b}^{(k,l)}(\circ) \right]$$
$$= pD^{(m,p-1)}(\circ) + \theta D^{(m,p)}(\circ) + E^{(m,p)}(\circ) + \underbrace{\sum_{i=0}^{p} C_p^i (-1)^i \frac{i!}{\theta^{i+1}} F^{(m,p-i)}(\circ)}_{B}. \tag{3.6}$$

Assuming all invoked processes admit a finite stochastic limit in the immediate region $(t, y, 0)$, we can now take the limit of (3.6) when $\theta \searrow 0$. Invoking the preliminary Lemma 3.3 [p. 127] (with regard to the last term B on the r.h.s.) we obtain that **the (m, p)-ZDC taken in the region $(\bullet) \triangleq (t, y, 0)$ reads as**

$$\sum_{k=0}^{m} \sum_{l=0}^{p} C_m^k C_p^l \; \widetilde{\Sigma}^{3\,(m-k,p-l)} \widetilde{b}^{(k,l)}(\bullet)$$
$$= pD^{(m,p-1)}(\bullet) + E^{(m,p)}(\bullet) + \frac{1}{p+1} F^{(m,p+1)}(\bullet). \tag{3.7}$$

Let us now take (3.7) at the Immediate ATM point $(t, 0, 0)$ and list all the processes involved. Its left-hand side is a polynomial functional of all processes in the $\mathscr{S}_L^{ZDC}(m, p)$ subset (see Definition 3.5) since it is clear that for a generic process $X \in Proc$

$$X^{n\,(I,J)}(t, y, \theta) \text{ will only invoke } X^{(i,j)}(t, y, \theta) \text{ with } 0 \le i \le I \text{ and } 0 \le j \le J.$$

Furthermore, the developed expression is a polynomial functional of order n of these processes. On the right-hand side of (3.7) we now develop and analyse the three terms individually. First, recall from its Definition (2.19) [p. 43] that

$$D(t, y, \theta) \triangleq \frac{1}{8} \widetilde{\Sigma}^4 \left[\left(\widetilde{\nu} - \sigma_t \widetilde{\Sigma}'_y \right)^2 + \widetilde{n}^2 \right],$$

hence its cross-differential at the generic point (t, y, θ) is

$$D^{(m,p-1)}(t, y, \theta) = \nabla^{(0,p-1)} \frac{1}{8} \sum_{k=0}^{m} C_m^k \; \widetilde{\Sigma}^{4\,(k,0)} \left[\left(\widetilde{\nu} - \sigma_t \widetilde{\Sigma}'_y \right)^2 + \widetilde{n}^2 \right]^{(m-k,0)}$$
$$= \frac{1}{8} \sum_{k=0}^{m} \sum_{l=0}^{p-1} C_m^k C_{p-1}^l \; \widetilde{\Sigma}^{4\,(k,l)} \left[\left(\widetilde{\nu} - \sigma_t \widetilde{\Sigma}'_y \right)^2 + \widetilde{n}^2 \right]^{(m-k,p-1-l)},$$

so that $D^{(m,p-1)}(t,y,\theta)$ is a polynomial functional of all processes in the $\mathscr{S}_D^{ZDC}(m,p)$ subset. Concerning the E term, recall from its Definition (2.20) that

$$E(t,y,\theta) \triangleq \widetilde{\Sigma}^3 \left[\widetilde{\Sigma}'_\theta - \frac{1}{2}\sigma_t^2 \widetilde{\Sigma}''_{yy} + \sigma_t \widetilde{\nu}'_y - \frac{1}{2}\sigma_t \widetilde{\nu} \right]$$

leading to

$$E^{(m,p)}(t,y,\theta)$$
$$= \sum_{k=0}^{m} \sum_{l=0}^{p} C_m^k C_p^l \, \widetilde{\Sigma}^{3\,(m-k,p-l)} \left[\widetilde{\Sigma}^{(k,l+1)} - \frac{1}{2}\sigma_t^2 \widetilde{\Sigma}^{(k+2,l)} + \sigma_t \widetilde{\nu}^{(k+1,l)} - \frac{1}{2}\sigma_t \widetilde{\nu}^{(k,l)} \right],$$

so that $E^{(m,p)}(t,y,\theta)$ is a polynomial functional of all processes in subset $\mathscr{S}_E^{ZDC}(m,p)$, and of these processes only. Furthermore, note that the functional is affine in $\widetilde{\Sigma}^{(m,p+1)}(t,y,\theta)$. Turning to the F term we can θ-differentiate (3.5) [p. 125] to get, still at (t,y,θ):

$$\nabla^{(m,p+1)} F(t,y,\theta) = y[\cdot] + \frac{1}{2} \sum_{k=0}^{m} \sum_{l=0}^{p+1} C_m^k C_{p+1}^l \, \widetilde{\Sigma}^{2\,(k,l)} \left[\widetilde{\Sigma}^{2\,(m-k,p+1-l)} - \sigma_t^2 \right]$$
$$- 1_{\{m\geq 1\}} m\sigma_t \sum_{k=0}^{m-1} \sum_{l=0}^{p+1} C_{m-1}^k C_{p+1}^l \widetilde{\Sigma}^{(m-k-1,p+1-l)} \left[\widetilde{\nu}^{(k,l)} - \sigma_t \widetilde{\Sigma}^{(k+1,l)} \right]$$
$$- 1_{\{m\geq 2\}} \frac{m(m-1)}{2} \sum_{k=0}^{m-2} \sum_{l=0}^{p+1} C_{n-2}^k C_{p+1}^l A(t,y,\theta;k,l,m,p)$$

with

$$A(t,y,\theta;k,l,m,p) \triangleq \left[\widetilde{\nu} - \sigma_t \widetilde{\Sigma}'_y \right]^{(k,l)} \left[\widetilde{\nu} - \sigma_t \widetilde{\Sigma}'_y \right]^{(m-k-2,p+1-l)}$$
$$+ \widetilde{n}^{(k,l)} \widetilde{n}^{(m-k-2,p+1-l)}.$$

We conclude that $F^{(m,p+1)}(\mathbf{t},\mathbf{y},\theta)$ is a polynomial functional of all processes in the $\mathscr{S}_F^{ZDC}(m,p)$ subset, and that this expression is affine in $\widetilde{\Sigma}^{(m,p+1)}(t,y,\theta)$. Gathering the results above, we see that the (m,p)-ZDC (3.7), when taken at the IATM point $(t,0,0)$, invokes all processes in the set

$$\mathscr{S}^{ZDC} = \mathscr{S}_L^{ZDC} \bigcup \mathscr{S}_D^{ZDC} \bigcup \mathscr{S}_E^{ZDC} \bigcup \mathscr{S}_F^{ZDC}$$

and only these processes. Furthermore, since only $\mathscr{S}_F^{ZDC}$ contains $\widetilde{\Sigma}^{(m,p+1)}(t,0,0)$, that equation is affine in the latter variable, which concludes the proof. □

Definition 3.6 (*Layers and ladder constraint*) A layer is a specific set of processes, which are cross-differentials of $\widetilde{\Sigma}$, $\widetilde{b}$, $\widetilde{\nu}$ and $\widetilde{n}$ taken at $(t, 0, 0)$. This layer is defined by its order $m \geq 0$ so that we put

$$L_0 \triangleq \left\{\widetilde{\Sigma}(\star),\, \widetilde{\Sigma}'_y(\star),\, \widetilde{\Sigma}''_{yy}(\star),\, \widetilde{\Sigma}'_\theta(\star), \widetilde{b}(\star), \widetilde{\nu}(\star), \widetilde{\nu}'_y(\star), \widetilde{n}(\star)\right\}$$

for the first layer and

$$L_m \triangleq \begin{bmatrix} \widetilde{\Sigma}^{(m-2i+2,i)}(\star) & 0 \leq i \leq \frac{m}{2}+1 \\ \widetilde{b}^{(m-2i,i)}(\star) & 0 \leq i \leq \frac{m}{2} \\ \widetilde{\nu}^{(m-2i+1,i)}(\star) & 0 \leq i \leq \frac{m+1}{2} \\ \widetilde{n}^{(m-2i,i)}(\star) & 0 \leq i \leq \frac{m}{2} \end{bmatrix}$$

for the subsequent layers.

Note that $\mathscr{L}_0$ is a larger collection, since it serves as a base for the other sets, and that it corresponds to the $\widetilde{\Sigma}$-(2,0) group of IATM differentials of Definition 2.1 [p. 57]. All processes in all layers share the same distinctive geometry, which is easier to illustrate by focusing on the shape process $\widetilde{\Sigma}^{(k,p)}(\star)$. Indeed, as we progress by one order in the θ direction, we lose two orders in the y direction, until the y differentiation order reaches zero.

This is what we call the *ladder constraint*, the origin of which is of course Itô's Lemma, as will be shown shortly in the proof of Proposition 3.1. The only difference between the four processes is the height of that ladder: the y-order is highest for $\widetilde{\Sigma}$, then $\widetilde{\nu}$, then $\widetilde{n}$ and $\widetilde{b}$ together.

Note that the aggregate of all layers from 0 to m forms a connex set within the differentiation matrix, a fact that we will also rely on.

Proposition 3.1 (θ-progression and ladder constraint) *Sequential cross-differentiations of the Zero-Drift Condition* (2.18) *allow us to compute the following processes (assuming they are finite), still with* $(\star) \triangleq (t, 0, 0)$*:*

$$\widetilde{\Sigma}^{(k,p)}(\star) \qquad \widetilde{b}^{(k,p)}(\star) \qquad \widetilde{\nu}^{(k,p)}(\star) \qquad \widetilde{n}^{(k,p)}(\star) \qquad \text{for all } (k,p) \in \mathbb{N}^2$$

as polynomial/rational functions of the SInsV specification $\left[\sigma_t, \left\{a_{\star,t}\right\}\right]$*: any of these IATM processes comes as a sum of terms, each of the form*

$$\sigma_t^p \prod_{j=1}^{k} a_{i(j),t} \qquad \text{with } p \in \mathbb{N}, \tag{3.8}$$

where $i(j)$ *is a valid depth index. The methodology is layer-based, in the sense that the establishment of layer* L_m *requires all* L_k *layers with* $0 \leq k \leq m$*, and* $\bigcup_k L_k$ *covers the whole differentiation matrix. Furthermore, within a given layer the higher* θ*-orders require the lower ones to be established first.*

Proof We proceed naturally by induction on the order m of the layer.

Initial check: $m = 0$
Theorem 2.2 [p. 62] provides the input-expression of the first layer L_0 and satisfies the polynomial/rational property, which validates the result at the initial order.

Induction check: $m \geq 1$
We assume that the property is satisfied for all L_j with $0 \leq j \leq m-1$. Let us consider a single process in the layer L_m. From the process hierarchy within the layer and from Remark 3.1 [p. 121] we see that it is sufficient to reason on $\widetilde{\Sigma}^{(k,i)}(\star)$. Indeed, if that static process has the correct polynomial/rational structure, then (by Itô's Lemma) so will the dynamic processes $\widetilde{b}^{(k,i)}(\star)$, $\widetilde{\nu}^{(k,i)}(\star)$ and $\widetilde{n}^{(k,i)}(\star)$. Combining this feature with the induction assumption, we conclude that the property is already satisfied for the dynamic processes $\widetilde{b}^{(k,p)}(t,0,0)$, $\widetilde{\nu}^{(k,p)}(t,0,0)$ and $\widetilde{n}^{(k,p)}(t,0,0)$.

Also, we know from Lemma 3.1 [p. 123] that the first column is computable in the correct form. Let us therefore take that single process as

$$\widetilde{\Sigma}^{(m-2i+2,i)}(\star) \qquad \text{for some } i \text{ such that} \qquad \mathbf{1} \leq i \leq \frac{m}{2} + 1.$$

Our objective is to describe the methodology to express that process as the correct function of the inputs, and to identify the ladder constraint. Our first observation is that in terms of geometry and with regard to the existing layers, we can be in only one of two positions, depending on the indices m and i:

- either we aim at prolonging an existing column
- or we wish to start a new column.

Accordingly the remainder of the proof is organised along these two configurations.

Progressing from y-index m to $m+1$ (prolonging an existing sub-column)
Let us use (3.7) to produce the $(m-2i+2, i-1)$-ZDC in $(t, y, 0)$, which we then take at the IATM point $(t, 0, 0)$. Applying Lemma 3.4, we conclude that we have established a polynomial equation which is affine in the sought process $\widetilde{\Sigma}^{(m-2i+2,i)}$ and involves all processes within the subset

$$\mathscr{S}^{ZDC}(m-2i+2, i-1)$$

$$\triangleq \begin{bmatrix} \widetilde{\Sigma}^{(k,l)}(\star) & 0 \leq k \leq m-2i+4 & 0 \leq l \leq i-1 & \\ \widetilde{\Sigma}^{(k,i)}(\star) & 0 \leq k \leq m-2i+2 & & \\ \widetilde{b}^{(k,l)}(\star) & 0 \leq k \leq m-2i+2 & 0 \leq l \leq i-1 & \\ \widetilde{\nu}^{(k,l)}(\star) & 0 \leq k \leq m-2i+3 & 0 \leq l \leq i-1 & \\ \widetilde{\nu}^{(k,i)}(\star) & 0 \leq k \leq (m-2i+1) & & \text{iff } m \geq 2i-1 \\ \widetilde{n}^{(k,l)}(\star) & 0 \leq k \leq m-2i+2 & 0 \leq l \leq i-2 & \\ \widetilde{n}^{(k,l)}(\star) & 0 \leq k \leq (m-2i) & i-1 \leq l \leq i & \text{iff } m \geq 2i \end{bmatrix}$$

and only those processes (see Definition 3.4 [p. 128]). Since we can solve for $\widetilde{\Sigma}^{(m-2i+2,i)}(t,0,0)$ iff we have input-expressed all the other processes in that subset, and since from Remark 3.1 we must reason on $\widetilde{\Sigma}$ only, the necessary and sufficient prerequisite is therefore to have

$$\widetilde{\Sigma}^{(k,l)}(\star) \qquad \text{with} \qquad 0 \le k \le m-2i+4 \qquad \text{and} \qquad 0 \le l \le i-1$$

specified from the inputs. Since we have all the previous layers already, this comes down to requiring the single process $\widetilde{\Sigma}^{(m-2i+4,i-1)}$ which is one step on the left in the sought layer L_m. By induction we can work our way back to the first column, which proves that the property is satisfied.

Progressing from θ-index p to $p+1$ (starting a new column):
Since we are within the layer L_m, the configuration requires m to be even and the process that we want to input-specified is $\widetilde{\Sigma}^{(0,\frac{m}{2})}$. Let us put $p = \frac{m}{2}$ (clearly $p \ge 1$) and consider the $(0, p-1)$-ZDC, taken at $(t, 0, 0)$. Applying Lemma 3.4 [p. 128], we establish a polynomial equation which is affine in $\widetilde{\Sigma}^{(0,p)}$ (with a coefficient which is a polynomial of σ_t) and involves all the processes in subset

$$\mathscr{S}^{ZDC}(0,p-1) \triangleq \begin{bmatrix} \widetilde{\Sigma}^{(k,l)}(\star) & 0 \le k \le 2 & 0 \le l \le (p-1) \\ \widetilde{\Sigma}^{(0,p)}(\star) & & \\ \widetilde{b}^{(0,l)}(\star) & & 0 \le l \le (p-1) \\ \widetilde{\nu}^{(k,l)}(\star) & 0 \le k \le 1 & 0 \le l \le (p-1) \\ \widetilde{n}^{(0,l)}(\star) & & 0 \le l \le (p-2) \end{bmatrix}$$

and only these processes. Focusing as before on $\widetilde{\Sigma}$, we have that solving for $\widetilde{\Sigma}^{(0,p)}$ is now equivalent to expressing

$$\widetilde{\Sigma}^{(k,l)}(\star) \qquad \text{for} \qquad 0 \le k \le 2 \qquad \text{and} \qquad 0 \le l \le (p-1)$$

as functions of the inputs. Because the layers L_k for $0 \le k \le m-1$ have been provided by the induction, this means that we only need $\widetilde{\Sigma}^{(2,p-1)}(t,0,0)$. In turn this places us in the previous situation of extending a sub-column. This proves that the property is satisfied at the next level and therefore concludes the proof. □

The ladder effect is best described with a graph such as Fig. 3.1. It shows that the computation of $\widetilde{\Sigma}^{(m,p)}(t,0,0)$ requires the prior specification of $\widetilde{\Sigma}^{(m+2,p-1)}(t,0,0)$, which needs the expression of $\widetilde{\Sigma}^{(m+4,p-2)}(t,0,0)$, and so on. In other words, one sub-column of the differentiation matrix can only provide a sub-column for the next θ-order that is two cells shorter.

3.2 Higher-Order Expansions: Illustration and Interpretation

In this section we use ACE to express all IATM smile differentials deemed financially meaningful, for a wide model class covering most SInsV instances. First we select as input model the bi-dimensional, single-underlying framework defined in Sect. 2.1.3.1 [p. 35]. Then we identify which differentials—both static and dynamic—are relevant for pricing and hedging. Since all of these differentials are contained in the first, second and third layers (see Definition 3.6 [p. 132]) we apply the generic methodology presented in Sect. 3.1 [p. 118] and manually solve the direct problem, at the required order. We can then interpret some of the input-expressions obtained, and provide illustrations using concrete examples. For the sake of readability, the (lengthy but straightforward) computations have been externalised to Appendix E [p. 443].

3.2.1 Justification and Outline

Let us first justify the technical choices mentioned above.

3.2.1.1 Dimensionality

Let us note that the majority of SInsV models used in practice are indeed only bi-dimensional, in the sense that both their endogenous and exogenous drivers are scalar. Indeed, this specification is generally deemed sufficient to control the statics and dynamics of a single-maturity smile. Typical examples include Heston, SABR, FL-SV or the $3/2$ model.[1]

3.2.1.2 Direct Problem

It is clear that generating the smile from the instantaneous dynamics will *appear* more important to most practitioners than the *inverse* problem. This bias stems mainly from the calibration process, which usually involves the minimisation of a market error functional, itself making numerous calls to a (proxy) vanilla pricer. However, we stress that this observation takes nothing from the applicative potential of solving the inverse problem. Especially in a bi-dimensional setup, where recovery is easiest, as will be detailed in Sect. 3.4.3 [p. 176].

---

[1] See [4], for instance.

3.2.1.3 Manual Computation

We know from Sect. 3.1 how to formally produce IATM differentials up to any required order. Hence little theoretical ground is broken by the computations performed here. Nevertheless, detailing the inductive methodology underlines the exponential complexity of this analysis. It becomes clear that the sensible alternative to manual computation is the programming of a dedicated formal calculus engine, which is simple and achievable in a usual declarative language.

3.2.1.4 Selected IATM Differentials

As will shortly be justified, the most relevant differentials for financial *interpretation*, but also for *smile approximations*, are the *first, low-level ones*. Their selection will determine the required order for ACE, i.e. the number of layers that must be computed.

The first reason is that these differentials are easily associated with the prices, sensitivities or dynamics of the most simple and liquid options. The second reason is more mathematical: they are naturally the most important *descriptors* of the smile, when the latter is seen as a series expansion around the IATM point. As for practical smile approximations, the sensible alternative to a very high-order expansion (when computing manually) is to combine a low-level one with constraints on extreme strikes (the wings) and on extreme maturities. This last topic will be discussed in more detail within Chap. 4.

Now let us be more specific and select specific IATM differentials. For a practitioner, the knowledge of only *three* IATM static differentials is usually considered sufficient to *describe* the shape of the liquid smile (i.e. for short expiries and strikes close to the money). Also, when viewed as a finite difference, each of these differentials are *associated* to some liquid option(s):

- The IATM **level** $\widetilde{\Sigma}(t, 0, 0)$ is associated to Straddles.
- The IATM **skew** $\widetilde{\Sigma}'_y(t, 0, 0)$ corresponds to Risk Reversals or Call/Put Binaries.
- The IATM **curvature** $\widetilde{\Sigma}''_{yy}(t, 0, 0)$ is linked to Butterflies.

Since our approach focuses on dynamics, the coefficients driving these quantities are relevant as well. Given the bi-dimensional context, this will involve the following nine IATM dynamic differentials:

$$\widetilde{b}(\star) \quad \widetilde{\mathrm{v}}(\star) \quad \widetilde{n}(\star) \quad \widetilde{b}'_y(\star) \quad \widetilde{\mathrm{v}}'_y(\star) \quad \widetilde{n}'_y(\star) \quad \widetilde{b}''_{yy}(\star) \quad \widetilde{\mathrm{v}}''_{yy}(\star) \quad \widetilde{n}''_{yy}(\star). \tag{3.9}$$

Naturally, a static approximation based purely on the level, skew and curvature will lose too much precision at longer maturities: some markets (interest rates for instance) can be liquid for up to 30 years. Notwithstanding additional constraints on extreme maturities, it is therefore important to capture the *term structure* of the smile. Hence we will include the θ-differentials of the existing three static IATM differentials:

- The IATM **slope** $\widetilde{\Sigma}'_{\theta}(t,0,0)$.
- The IATM **twist** $\widetilde{\Sigma}''_{y\theta}(t,0,0)$.
- The IATM **flattening** $\widetilde{\Sigma}'''_{yy\theta}(t,0,0)$.

Additionally, certain areas of the smile far from the IATM point can be actively traded, and therefore warrant specific higher orders. The first of these areas is the "immediate" region, i.e. short expiries for all strikes, including deep ITM/OTM. For instance, a strike which is ATM at trade's inception is not necessarily close to the money at expiry. The second liquid area is the whole ATM line, up to long expiries. Therefore, the corresponding need for increased accuracy justifies the computation of:

- The IATM **hyperskew** $\widetilde{\Sigma}'''_{yyy}(t,0,0)$.
- The IATM **hypercurve** $\widetilde{\Sigma}^{(4')}_{y^4}(t,0,0)$.
- The IATM **arch** $\widetilde{\Sigma}''_{\theta\theta}(t,0,0)$.

Positioning all these IATM differentials within Fig. 3.1 [p. 122] shows that to address the above-mentioned concerns of financial interpretation and of approximation accuracy, we must implement ACE *up to and including the third layer.*

Moreover there are additional justifications—mathematical and numerical—for selecting these first, second and third layers. For instance, we will see in Sect. 3.3.3 [p. 164] how to *transfer* from the current *Lognormal* baseline, to a new implied volatility convention, such as *Normal* IV. We will describe a generic method to perform this *baseline transfer*, which uses finite differences but is also submitted to the ladder effect. As a consequence, in order to obtain the first layer at an appropriate numerical precision in the new baseline, we must already have the second and third layers in the initial baseline.

Another interest of the third layer is that it provides the IATM *arch* term $\widetilde{\Sigma}''_{\theta\theta}(\star)$, which is often not provided by other asymptotic results, developed *ad hoc* for some SInsV models. In [5] for instance, the implied volatility is provided as a *linear* function of time-to-maturity, which in practice not only hampers precision, but also leads to increased arbitrage opportunities (e.g. negative densities). As will be discussed in Sect. 4.2, Hagan et al.'s closed-form approximation *does* match the ACE results, for the eight other static differentials quoted above. However, obtaining the missing arch differential *via* that approach would require us to push the singular perturbation technique to higher orders, which is not straightforward.

3.2.1.5 Computation Outline

Out of the nine static targets, the first three—$\widetilde{\Sigma}(\star)$, $\widetilde{\Sigma}'_{y}(\star)$ and $\widetilde{\Sigma}''_{yy}(\star)$—have already been computed in Sect. 2.4. Similarly, we already have four of the nine dynamic differentials: $\widetilde{b}(\star), \widetilde{\nu}(\star), \widetilde{n}(\star)$ and $\widetilde{\nu}'_{y}(\star)$. Although the dynamic coefficients are important

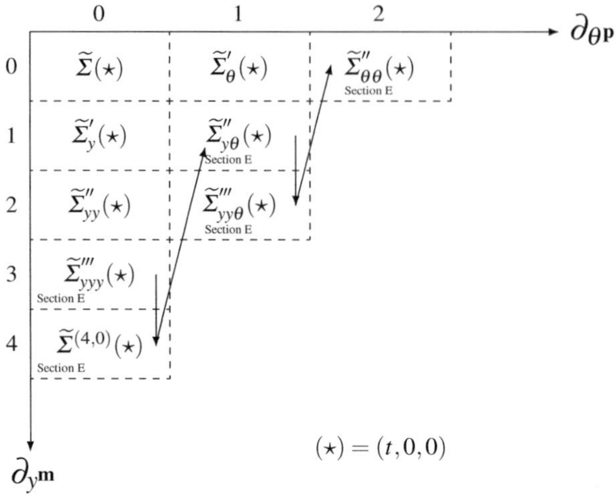

Fig. 3.2 Order of computation for layers 2 and 3

to us, for clarity we have organised the computations of Appendix E into five sections, each based on a *static* differential. However, instead of progressing layer by layer, we work column by column, in a fashion similar to the proofs and methodology of Sect. 3.1. The computation is therefore organised as per Fig. 3.2 and the ACE algorithm dictates that we must also express a specific collection of dynamic coefficients. This collection is constituted of list (3.9), and of the following group:

$$\widetilde{b}'''_{yyy}(\star) - \widetilde{\nu}''_{yyy}(\star) - \widetilde{n}'''_{yyy}(\star) \qquad \widetilde{b}'_{\theta}(\star) - \widetilde{\nu}'_{\theta}(\star) - \widetilde{n}'_{\theta}(\star) \qquad \widetilde{\nu}''_{y\theta}(\star).$$

Furthermore, for completeness we will also provide coefficients $\widetilde{b}''_{y\theta}(\star)$ and $\widetilde{n}''_{y\theta}(\star)$.

3.2.2 *Interpretation of the Results*

3.2.2.1 Static Maturity Differentials

The *twist* of the smile $\widetilde{\Sigma}''_{y\theta}(\star)$—provided by Proposition E.7 [p. 460]—is sometimes called the *term structure of skew*, in particular when the smile *does* correspond to a term structure of underlyings. This is the case of the Caplet smile for instance, which has motivated the FL-TSS model class [6] as an extension of FL-SV (see Sect. 4.4). It can also be applied to a single underlying framework and provides some control on the twist *via* a time-dependent skew function.

The *flattening* of the smile $\widetilde{\Sigma}'''_{yy\theta}(\star)$—given by Proposition E.9 [p. 468]—is a permanent concern of modellers and practitioners, mainly because it is symptomatic of both the volatility and mean-reversion of volatility. In Sect. 3.2.3 we will present a typical example of how that IATM differential provides information on the respective strengths of both parameters. In fact, the flattening *visually* observed can be misleading, as it depends on the moneyness employed, and in some markets it can actually be positive.

In that vein, there is a classical argument on the inappropriate *forward volatility* associated to a local volatility (LV) model that has been calibrated to a flattening smile. It should, however, be tempered by the fact that the concept of local volatility and the calibration process are both static in nature.

The flattening also gives an intuitive indication of the volatility diffusion, seen through the marginal distributions of the integrated variance process. As a simple example, let us consider a Lognormal model with an independent volatility:

$$\frac{dS_t}{S_t} = \sigma_t dW_t \qquad d\sigma_t = f(t, \sigma_t)dt + g(t, \sigma_t)dZ_t \qquad \text{with} \qquad \langle dW_t, dZ_t \rangle \equiv 0.$$

Then the smile curvature, at a given expiry T, is controlled by Jensen's inequality and by two functionals. The first functional is the Call price at each strike, given the (stochastic) integrated variance $\int_0^T \sigma_s^2 ds$, hence Black & Scholes formula. The second functional is the distribution of $\int_0^T \sigma_s^2 ds$. Therefore the flattening $\widetilde{\Sigma}'''_{yy\theta}(\star)$ tells us how this marginal distribution evolves with maturity T, which in turn informs us on the process σ_t.

3.2.2.2 Dynamic Coefficients

Combining the direct result (2.56) [p. 62] with Proposition E.1 [p. 449], we have now gathered input-expressions for the three IATM coefficients $\widetilde{b}'_y(\star)$, $\widetilde{\nu}'_y(\star)$ and $\widetilde{n}'_y(\star)$. Thus, given a generic bi-dimensional SInsV model, we can describe the *dynamics* of its IATM *skew* $\widetilde{\Sigma}'_y(t, 0, 0)$ exactly. As mentioned above, we can therefore approximate the dynamics of several *very* common option types. These are structures with prices essentially controlled by the IATM skew, such as long/short positions in Collars/Risk Reversals, Call/Put Strike-Spreads or Call/Put Binaries, provided they exhibit short maturities and are struck close to the money.

Similar interpretations can be given for other groups of dynamic coefficients. For instance, Proposition E.2 [p. 450] expresses the three IATM coefficients $\widetilde{b}''_{yy}(\star)$, $\widetilde{\nu}''_{yy}(\star)$ and $\widetilde{n}''_{yy}(\star)$, thus providing the full and exact dynamics of the IATM smile curvature $\widetilde{\Sigma}''_{yy}(t, 0, 0)$. In financial terms, this implies that we can approximate the dynamics of short-expiry ATM Butterfly options, whose payoff definition corresponds to a finite difference approximation for Σ''_{KK}.

Likewise, Proposition E.6 [p. 459] provides the dynamic differentials $\widetilde{b}'_\theta(\star)$, $\widetilde{\nu}'_\theta(\star)$ and $\widetilde{n}'_\theta(\star)$. These expressions give the dynamics of financial products which depend mainly on the IATM *slope* $\widetilde{\Sigma}'_\theta(\star)$ of the implied volatility surface. The most common of these products are calendar spreads, but some short-expiry barrier options could also be included.

3.2.2.3 The Question of Homogeneity

It is noteworthy that within the first three layers, expressions for the *pure-strike* IATM differentials—both static and dynamic—are *all* σ_t-homogeneous. In other words, once expressed using the normalised coefficients

$$c_{\star,t} \triangleq \sigma_t^{-1} a_{\star,t}$$

these input-expressions end up as a single power of the instantaneous volatility σ_t. For instance the expression of the hyperskew (E.38) (see Proposition E.3 [p. 452]) becomes

$$\widetilde{\Sigma}'''_{yyy}(\star) = \frac{1}{\sigma_t^2}\left[\frac{1}{4}c_{222} - \frac{3}{2}c_2c_{22} + \frac{3}{4}c_3c_{23} + \frac{1}{2}c_3c_{32} + \frac{3}{2}c_2^3 - \frac{5}{2}c_2c_3^2\right].$$

To be more specific, this property concerns the following computed expressions:

$$\widetilde{\Sigma}_y^{(1,0)}(\star)\ [\text{p. } 62] \qquad \widetilde{\Sigma}_{y^2}^{(2,0)}(\star)\ [\text{p. } 62] \qquad \widetilde{\Sigma}_{y^3}^{(3,0)}(\star)\ [\text{p. } 452] \qquad \widetilde{\Sigma}_{y^4}^{(4,0)}(\star)\ [\text{p. } 457]$$

$$\widetilde{b}'_y(\star) - \widetilde{\mathrm{v}}'_y(\star) - \widetilde{n}'_y(\star)\ [\text{pp. } 62, 449] \qquad \widetilde{b}''_{yy}(\star) - \widetilde{\mathrm{v}}''_{yy}(\star) - \widetilde{n}''_{yy}(\star)\ [\text{p. } 450]$$

$$\widetilde{b}'''_{yyy}(\star) - \widetilde{\mathrm{v}}''_{yyy}(\star) - \widetilde{n}'''_{yyy}(\star)\ [\text{p. } 453].$$

Conversely, *all* θ-differentials are heterogenous, spanning several σ_t-orders. Again for clarity, the concerned expressions are

$$\widetilde{\Sigma}'_\theta(\star)\ [\text{p. } 62] \qquad \widetilde{\Sigma}''_{y\theta}(\star)\ [\text{p. } 460] \qquad \widetilde{\Sigma}'''_{yy\theta}(\star)\ [\text{p. } 468] \qquad \widetilde{\Sigma}''_{\theta\theta}(\star)\ [\text{p. } 475]$$

$$\widetilde{b}'_\theta(\star) - \widetilde{\mathrm{v}}'_\theta(\star) - \widetilde{n}'_\theta(\star)\ [\text{p. } 459] \qquad \widetilde{b}''_{y\theta}(\star) - \widetilde{\mathrm{v}}''_{y\theta}(\star) - \widetilde{n}''_{y\theta}(\star)\ [\text{p. } 462].$$

This distinction on homogeneity between pure-strike and the rest of the IATM differentials was already apparent with the first layer, as discussed in Sect. 2.4.2.4 [p. 74]. We now present technical arguments explaining these properties—without providing a comprehensive proof—and why they should extend to all orders.

Let us recall that an important characteristic of ACE's generic methodology—as exposed in Sect. 3.1—is to alternate two phases: differentiation of the ZDC (or IZDC for the first column) and derivation of formal dynamics (via Itô) for some of the obtained static expressions. In order to establish the homogeneity (or lack thereof) of each IATM differential, we must therefore examine how each of the two phases propagates this property.

Let us start by investigating the SDE phase. We know from Proposition 3.1 that each static IATM differential $\widetilde{\Sigma}^{(n,p)}(\star)$ comes as a sum of terms, each of formal type

$$(3.8)\ [\text{p. } 132] \qquad \sigma_t^p \prod_{j=1}^{k} a_{i(j),t} \qquad \text{with } p \in \mathbb{N}.$$

The σ_t-order of this product is $p+k$, and it is easily seen that the coefficients of its dynamics will exhibit the same formal σ_t-order. We will call this type of process σ_t-*stable*. Therefore, in the general case the σ_t-homogeneity (resp. heterogeneity) of the IATM static differentials will be carried over to its dynamic coefficients $\widetilde{b}^{(n,p)}(\star)$, $\widetilde{\nu}^{(n,p)}(\star)$ and $\widetilde{n}^{(n,p)}(\star)$.

Remark 3.2 Note, incidentally, that Lemma E.1 [p. 443] also confirms that for any σ_t-stable formal process x_t, the ratio x_t/σ_t^n for $n \geq 1$ shares the same property. Indeed, assuming the generic SDE

$$dx_t = z_1 dt + z_2 dW_t + z_3 dZ_t$$

and $n \in \mathbb{N}^*$, then the Lemma gives the *normalised* dynamics of the ratio as

$$d\left[\frac{x_t}{\sigma_t^n}\right] = \frac{1}{\sigma_t^n}\left[z_1 - n\left(c_2 z_2 + c_3 z_3 + x_t c_1\right) + \frac{1}{2}n(n+1)x_t\left(c_2^2 + c_3^2\right)\right] dt$$
$$+ \frac{1}{\sigma_t^n}\left[z_2 - n x_t c_2\right] dW_t + \frac{1}{\sigma_t^n}\left[z_3 - n x_t c_3\right] dZ_t.$$

Let us now examine the PDE phase, first in the simpler case of the first column, i.e. for pure-strike differentials. We know from Lemma 3.1 [p. 123] that to express $\widetilde{\Sigma}^{(m,0)}(\star)$ we need only some $\widetilde{\Sigma}^{(i,0)}(\star)$, $\widetilde{\nu}^{(j,0)}(\star)$ and $\widetilde{n}^{(k,0)}(\star)$ with i, j, k strictly below m. By induction it follows that all IATM pure-strike differentials will be homogeneous.

For cross-differentials, Proposition 3.1 [p. 132] shows that IATM differentials w.r.t. θ will *a priori* be heterogenous. Indeed, all these quantities depend (directly or not) in a polynomial/rational way on the slope $\widetilde{\Sigma}'_\theta(\star)$, which is not σ_t-homogeneous. Again invoking the σ-stability argument, by induction heterogeneity will be contagious. One reason why the above explanation does not constitute a comprehensive proof is that unforeseen simplifications, which can potentially bring back the homogeneity, cannot be ruled out.

This behaviour w.r.t. σ_t-homogeneity can be attributed to Itô's Lemma, but also to our choice of parametrisations for the dynamics of both the instantaneous and implied stochastic volatilities. Note, however, that the alternative parametrisation used by [7] (implied variance) does not remove or simplify the problem.

3.2.2.4 Consistency

In the computation of the hyperskew—see Proposition E.3 [p. 452]—the concept of (3, 0)-IZDC should be seen as a formal notation. Indeed, since the (Primary) IZDC (2.1) [p. 25] is expressed in the immediate region $(t, y, 0)$, it cannot accept any differential w.r.t. θ. The only interest of this notation is therefore to ensure some consistency with the usual ZDC cross-differentials.

It can be checked (see Appendix E of [8]) that all *further layers* expressions obtained in [7] are compatible with those presented here. This statement concerns the hyperskew $\widetilde{\Sigma}'''_{yyy}(\star)$ and the twist $\widetilde{\Sigma}''_{y\theta}(\star)$.

3.2.3 Illustration of the Maturity Effect

With the previous IATM differentials we have gained quite some insight into the higher-order maturity differentials. However, these results are complex and therefore difficult to interpret, which suggests that our parametrisation of the problem might not be optimal. Nevertheless, we propose to illustrate these new results by applying them to some given, realistic (instantaneous) local-stochastic volatility model.

To simplify the resulting formulae and thus make easier their interpretation, we forgo any skew problematic. This implies that the local volatility part of the model must be Lognormal, and also that the correlation must be null (see Sect. 2.5.2 for more details) resulting in a purely exogenous volatility. We will therefore be interested mainly in the smile level and in its curvature, with a particular concern for their dependency with respect to the maturity θ. In other words, we will look at

$$\widetilde{\Sigma}(\star) \qquad \widetilde{\Sigma}''_{yy}(\star) \qquad \widetilde{\Sigma}'_{\theta}(\star) \qquad \widetilde{\Sigma}'''_{yy\theta}(\star) \qquad \widetilde{\Sigma}''_{\theta\theta}(\star).$$

In order to provide more control over the effect of maturity, it is common practice to impose some mean-reversion either on the instantaneous volatility or on the instantaneous variance (e.g. Heston). Choosing the former option for simplicity's sake, and selecting a Normal vol of vol for the same reason, we get the dynamics

$$\begin{cases} \dfrac{dS_t}{S_t} = \sigma_t dW_t & (3.10) \\ & \text{with} \quad W_t \perp Z_t \\ d\sigma_t = \kappa\left[\overline{\sigma} - \sigma_t\right]dt + \varepsilon dZ_t. & (3.11) \end{cases}$$

Proposition 3.2 (IATM static differentials of a mean-reverting SV model) *The SInsV model* (3.10)–(3.11)*-yields the following IATM static differentials:*

$$\widetilde{\Sigma}(\star) = \sigma_t \tag{3.12}$$

$$\widetilde{\Sigma}'_{\theta}(\star) = \frac{1}{2}\kappa(\overline{\sigma} - \sigma_t) + \frac{1}{12}\frac{1}{\sigma_t}\varepsilon^2 \tag{3.13}$$

$$\begin{aligned} \widetilde{\Sigma}''_{\theta\theta}(\star) = & -\frac{1}{12}\sigma_t\varepsilon^2 + \frac{1}{12}\frac{1}{\sigma_t}\left[\kappa^2(\overline{\sigma} - \sigma_t)^2 - \kappa\varepsilon^2\right] \\ & -\frac{1}{3}\kappa^2(\overline{\sigma} - \sigma_t) - \frac{1}{6}\kappa\frac{1}{\sigma_t^2}(\overline{\sigma} - \sigma_t)\varepsilon^2 + \frac{7}{80}\frac{1}{\sigma_t^3}\varepsilon^4 \end{aligned} \tag{3.14}$$

in regard of the level, and

$$\widetilde{\Sigma}''_{yy}(\star) = \frac{1}{3}\frac{\varepsilon^2}{\sigma_t^3} \tag{3.15}$$

$$\widetilde{\Sigma}'''_{yy\theta}(\star) = -\frac{1}{4}\kappa\frac{1}{\sigma_t^3}\varepsilon^2 - \frac{5}{12}\kappa\frac{1}{\sigma_t^4}(\overline{\sigma} - \sigma_t)\varepsilon^2 + \frac{4}{45}\frac{1}{\sigma_t^5}\varepsilon^4 \tag{3.16}$$

with respect to the curvature.

We now proceed with the quite straightforward proof, before commenting qualitatively on these results. We will be interested, in particular, in the impact of each of the model parameters on the level, curvature and their maturity evolution.

Proof Casting this model into our generic (bi-dimensional) framework it follows that any coefficient with a 2 in its tag will be null:

$$0 = a_2 = a_{12} = a_{21} = a_{22} = a_{23} = a_{32}$$

$$0 = a_{122} = a_{212} = a_{221} = a_{222} = a_{223} = a_{232} = a_{322} = a_{2222}$$

while at 1st and 2nd depth the relevant drift and/or exogenous coefficients are

$$a_1 = \kappa(\overline{\sigma} - \sigma_t) \qquad a_3 = \varepsilon$$

and

$$a_{11} = -\kappa^2(\overline{\sigma} - \sigma_t) \qquad a_{13} = -\kappa\varepsilon \qquad a_{31} = 0 \qquad a_{33} = 0.$$

Then applying (2.36) [p. 50] and (2.52) [p. 62] directly we get, respectively,

$$\widetilde{\Sigma}(\star) = \sigma_t \qquad \text{and} \qquad \widetilde{\Sigma}''_{yy}(\star) = \frac{\varepsilon^2}{3\sigma_t^3},$$

which proves (3.12) and (3.15). Turning to the slope, it is equally fast to apply (2.53) [p. 62], which gives

$$\widetilde{\Sigma}'_{\theta}(\star) = \frac{1}{2}\kappa(\overline{\sigma} - \sigma_t) + \frac{1}{12}\frac{1}{\sigma_t}\varepsilon^2 \tag{3.17}$$

and proves (3.13). As for the flattening, starting from (E.58) [p. 468] we get formally, by removing all null coefficients:

$$\widetilde{\Sigma}'''_{yy\theta}(\star) = \frac{1}{4}\frac{1}{\sigma_t^3}a_3a_{13} - \frac{5}{12}\frac{1}{\sigma_t^4}a_1a_3^2 + \frac{4}{45}\frac{1}{\sigma_t^5}a_3^4.$$

By replacing with the model input we get immediately

$$\widetilde{\Sigma}'''_{yy\theta}(\star) = -\frac{1}{4}\kappa\frac{1}{\sigma_t^3}\varepsilon^2 - \frac{5}{12}\kappa\frac{1}{\sigma_t^4}(\overline{\sigma} - \sigma_t)\varepsilon^2 + \frac{4}{45}\frac{1}{\sigma_t^5}\varepsilon^4,$$

which proves (3.16). Finally addressing the arch, applying (E.62) [p. 475] gives, again by removing all null coefficients:

$$\widetilde{\Sigma}''_{\theta\theta}(\star) = -\frac{\sigma_t}{12}a_3^2 + \frac{1}{3}a_{11} + \frac{1}{\sigma_t}\left[\frac{1}{12}a_1^2 + \frac{1}{12}a_3 a_{13}\right] - \frac{1}{6}\frac{1}{\sigma_t^2}a_1 a_3^2 + \frac{7}{80}\frac{1}{\sigma_t^3}a_3^4.$$

Replacing by the model inputs, we get

$$\begin{aligned}\widetilde{\Sigma}''_{\theta\theta}(\star) &= -\frac{1}{12}\sigma_t\varepsilon^2 - \frac{1}{3}\kappa^2(\overline{\sigma} - \sigma_t)\\ &\quad + \frac{1}{12}\frac{1}{\sigma_t}\left[\kappa^2(\overline{\sigma} - \sigma_t)^2 - \kappa\varepsilon^2\right] - \frac{1}{6}\kappa\frac{1}{\sigma_t^2}(\overline{\sigma} - \sigma_t)\varepsilon^2 + \frac{7}{80}\frac{1}{\sigma_t^3}\varepsilon^4,\end{aligned}$$

which proves (3.14) and concludes the proof. □

The curvature being simpler than the level ones, they lend themselves to an easier interpretation. Therefore we will comment upon the former in the general case, before applying a classical simplification that will allow us to gauge both differential sets with some ease.

Using (3.15) and (3.16) we re-scale the maturity-sensitivity of the curvature to get:

$$\widetilde{\Sigma}'''_{yy\theta}(\star) = \widetilde{\Sigma}''_{yy}(\star)\left[\frac{4}{15}\frac{1}{\sigma_t^2}\varepsilon^2 - \frac{1}{4}\kappa\left(3 + \frac{5}{\sigma_t}(\overline{\sigma} - \sigma_t)\right)\right]. \tag{3.18}$$

In order to simplify matters, it is customary to assume that the initial volatility σ_t is identical to its long-term mean $\overline{\sigma}$:

Assumption 3.2 Within model (3.10)–(3.11) we assume that

$$\sigma_t = \overline{\sigma}.$$

Note that this assumption is quite unrealistic, and that used dynamically it runs contrary to the desire for a stationary model. Even in a purely static and implicit approach, imposing this restriction deprives the modeller of an additional degree of freedom, able to control the short-term evolution of the smile. Nevertheless, in our case it is precisely to mask the latter effect that we opt for this simplifying assumption.

The combination of (3.15) and (3.18) therefore rewrites as

$$\widetilde{\Sigma}''_{yy}(\star) = \frac{1}{3}\frac{\varepsilon^2}{\sigma_t^3} \qquad \text{and} \qquad \widetilde{\Sigma}'''_{yy\theta}(\star) = \widetilde{\Sigma}''_{yy}(\star)\left[\frac{4}{15}\frac{1}{\sigma_t^2}\varepsilon^2 - \frac{3}{4}\kappa\right].$$

In this form, the model is very demonstrative of the fairly intuitive influence of the two main parameters, i.e. mean-reversion and vol of vol. Indeed, κ has no effect on the IATM curvature but will decrease it as maturity progresses. In contrast, ε alone will create a positive IATM curvature which it will also tend to increase with maturity. Observe that adopting Assumption 3.2 is *not* equivalent to downgrading the model to non-mean-reverting dynamics for the volatility, i.e. taking $\kappa = 0$. Indeed, this is yet another proof that although the method is asymptotic and can apprehend any coefficient only at current time t, because it is using a dynamic approach it will capture some deeper effects.

The overall impact, i.e. whether we will observe either a flattening or a deepening of the smile, depends on the respective numerical values of κ and ε:

$$\widetilde{\Sigma}'''_{yy\theta}(\star) \geq 0 \qquad \Longleftrightarrow \qquad \kappa \leq \frac{16}{45}\left[\frac{\varepsilon}{\sigma_t}\right]^2.$$

Note that the role played by the IATM level σ_t can be regarded as a scaling of the various quantities. This is not to say that the influence is trivial to analyse, so we will not dwell much longer on this particular issue. It suffices to mention that the comparison of κ with a term proportional to $[\varepsilon/\sigma_t]^2$ is coherent with the homogeneity of the problem. Indeed, the mean-reversion is expressed in a Lognormal fashion, while the volatility comes with a Normal convention. Therefore, in terms of diffusion coefficients, κ should be compared to ε/σ_t. But then, in terms of dynamics and quadratic variation (which is all that matters eventually), in such a diffusion the drift coefficient must come along the *square* of the volatility.

Still with Assumption 3.2 the level formulae (3.12)–(3.14) simplify to

$$\widetilde{\Sigma}(\star) = \sigma_t \qquad \widetilde{\Sigma}'_{\theta}(\star) = \frac{\varepsilon^2}{12\sigma_t} \qquad \widetilde{\Sigma}''_{\theta\theta}(\star) = \frac{7\varepsilon^4}{80\sigma_t^3} - \frac{\sigma_t\varepsilon^2}{12} - \frac{\kappa\varepsilon^2}{12\sigma_t}.$$

We observe again a clear opposition between the vol of vol and the mean-reversion, with some restrictions. Indeed, ε will clearly increase the IATM slope $\widetilde{\Sigma}'_{\theta}(\star)$, while the mean-reversion has no bearing on it. At the next θ-order, clearly κ will tend to tame that slope as maturity increases. The influence of ε however, is more complex in that respect, as we have

$$\widetilde{\Sigma}''_{\theta\theta}(\star) \geq 0 \qquad \Longleftrightarrow \qquad \kappa + \sigma_t^2 \leq \frac{21}{20}\left[\frac{\varepsilon}{\sigma_t}\right]^2$$

which is difficult to interpret. At last removing the mean-reversion tells us a simpler story, in that the arch is non-negative *if and only if* $\varepsilon \geq \sqrt{20/21}\sigma_t^2$.

Overall, it is interesting to note that a high enough vol of vol will create positive and θ-increasing curvature and slope, while the mean-reversion will counteract this evolution in maturity.

3.3 Framework Extensions and Generalisation

Having established a generic method for higher orders, in this section we take a step back in order and explore the other extension possibilities for the ACE methodology. We start in Sect. 3.3.1 by decomposing the latter into several independent *building blocks* or *ingredients*. We thus establish a general configuration, of which we have only explored one instance so far.

This breakdown naturally suggests several avenues for extensions, more or less along each block's function, and we discuss each of them briefly in Sects. 3.3.1.1, 3.3.1.2 and 3.3.1.3. However, among these prospects, in our opinion one stands out as quite promising, and we choose to investigate it further. This extension (or rather replacement) is the selection of a better-suited "simple" model, which is called the *baseline*, and which for now has been geometric Brownian motion. It allows us to re-parametrise the price surface via an implied parameter, and the suitability of that baseline must be understood w.r.t. a given target model.

As a significant example, in Sect. 3.3.2 we run through the complete process again but within the *Normal* baseline this time, underlining the structural differences and more importantly the striking similarities with the Lognormal case.

Then in Sect. 3.3.3 we describe a series of simple techniques that can be used in order to transform the IATM differentials (both static and dynamic) from one baseline to another, without having to go through the derivation of some Zero-Drift Condition and the subsequent process. As a general concept, we have named this approach the *baseline transfer*, but it comes in several flavours, all with their specific pros and cons.

3.3.1 Building Blocks and Available Extensions

Let us first list the *ingredients* required by the methodology in order to provide its results for both the inverse and direct problems. So far we have required:

1. Market-wise, an underlying S_t, a (possibly implicit) numeraire N_t and a continuum of options for the (K, T) field, whose payoff is paid at T for a value of

$$\Phi^N(X_T; K, T) \triangleq N_T \Phi(X; K, T)$$

where X is measurable w.r.t G_T with $G_u \triangleq \sigma(S_u)$. Previously these were Call options:

$$\Phi(X_T; K, T) \triangleq [S_T - K]^+ .$$

2. A baseline, i.e. a (simple) instantaneous model whose dynamics involve a parameter ψ. Typically this model will belong to a local volatility family:

$$\text{(baseline model)} \quad dS_t = f(t, S_t, \psi) dB_t. \tag{3.19}$$

Furthermore the baseline must satisfy the following two properties:

- A closed-form formula is available for the deflated price of these options (previously the Black formula)

$$B(t, S_t; T, K; \psi) = \mathbb{E}^N \left[\Phi(X_T; K, T) | \mathscr{F}_t \right];$$

- the single parameter ψ (previously the Lognormal volatility) ensures a bijection with the deflated price, enabling an implicit re-parametrisation of the price mapping:

$$C(t, S_t; K, T) \Longleftrightarrow \Psi(t, S_t; K, T).$$

It follows that the baseline model *can* use several parameters (e.g. a Lognormal displaced diffusion), but that one of them must be selected for the role of ψ.

3. A (possibly complex) stochastic instantaneous volatility model $\mathscr{M}$, which can be cast in a chaos into the baseline by making the ψ parameter stochastic. When the baseline is a local volatility model, for instance, then we would write

$$\text{(target model)} \qquad \begin{aligned} dS_t &= f(t, S_t, \psi_t) dW_t \\ d\psi_t &= a_{1,t} dt + a_{2,t} dWt + \vec{a}_{3,t}^{\perp} d\vec{Z}_t \end{aligned} \tag{3.20}$$

with the subsequent Wiener expansion on the $a_{i,t}$ coefficients.

4. A moneyness, i.e. a bijective function transforming the native option parameters K and T into a pair of variables z and ξ, possibly parametrised by the observable state variables t and S_t:

$$(z, \xi) = m_{[t, S_t]}(K, T).$$

So far we have been using the log-moneyness $y = \ln(K/S_t)$ and time-to-maturity $\theta = T - t$, since those are supposed or expected to represent the main invariance modes of the smile. We aim at choosing the coordinates ensuring maximum stability, and in practice we will often take $\xi = \theta$.

So what are the possible avenues for extending the methodology? Well, as announced we can try and modify each of the main *ingredients* above. However, providing new complex models (typically as input for the direct problem) is a given, since it is the whole point of our generic approach, which leaves us with three directions.

Modifying the moneyness is certainly a possibility, the principles of which have been discussed in Sect. 2.1.2.3. We provide more insight, however, in Sect. 3.3.1.1 and recall that the choice of moneyness must often be adapted to the baseline in order to simplify the main Zero Drift Condition.

Alternatively we an use payoffs that are different from European Calls, but again this choice will be closely related to the choice of the baseline. Indeed, there must be an available formula (and it must be simple enough to derive a ZDC) and ψ must define a bijection, so that there aren't so many choices. Note that, for extrapolation purposes, for instance, we could settle for an approximate formula. These points will be discussed in Sect. 3.3.1.2.

We observe that the baseline issue seems to be at the centre of these choices. Furthermore, we think that there is a significant potential for performance improvement in adapting that ingredient to the complex model $\mathscr{M}$ at hand. This justifies the closer attention that we brought to Sects. 3.3.1.3, 3.3.2 and 3.3.3, where we discuss, respectively, the topics of baseline choice, Normal baseline and baseline transfer.

3.3.1.1 Extending the Moneyness

Recall that, as discussed in Sect. 2.1.2.3 [p. 29], we have made the *arbitrary* choice of re-parameterising the price and implied volatility surfaces, from the natural K and T variables to the log-moneyness $y = \ln(K/S_t)$ and the time-to-maturity $\theta = T - t$. Granted, such a change of coordinates is mainly cosmetic, as it does not affect the *nature* of our asymptotic results. However it does have two important consequences:

- It allows the modeller to enforce some *stationarity* in the surface dynamics, if he/she thinks such a feature representative of historic/anticipated market moves. For instance, by choosing the log-moneyness he/she implicitly assumes that the smile is roughly sticky-Delta. Also, it provides some symmetry in a problem which is structurally one-sided since S_t must stay positive.
- It modifies the sliding Zero Drift Condition, which is also dependent on the choice of baseline. Here y and θ happen to fit that specific Lognormal model, since they are the natural variables of the (normalised) Black-Scholes formula. In turn, this selection contributes to making the ZDC a reasonably simple and therefore manageable (stochastic) PDE.

A logical question is therefore whether a different choice of moneyness could improve the stationarity and/or the ZDC. As previously mentioned (in Sects. 2.1.2.3 and 3.3.1) the moneyness must ensure a bijection with the original coordinates (K, T) and can only use the observable state variables t and S_t as parameters in that transformation. Furthermore, on the latter point, note that only S_t can be absorbed, since t is needed to define a process.

Classically we tend to focus on the strike dimension (with variations such as the proportional moneyness K/S_t or the simpler moneyness $K - S_t$) while the time-to-maturity $(T-t)$ seems to remain the usual way of transposing T. There are exceptions to this trend though, in particular when deterministic volatilities are involved. This subject will be brushed in Sect. 3.3.1.3, dedicated to baseline changes, and will stress the level of ZDC simplification that the choice of good sliding variables can bring (see (3.21) *vs* (3.22) [p. 153]).

We underline, though, that there is *a priori* no imperative either to absorb S_t or to split the K and T transforms. However, in order to ensure a bijection the only practical solution does seem to be the mapping to another *pair* of variables, as opposed to a singleton or a triplet. In full generality we can have either

$$(t, S_t, K, T) \overset{m}{\longleftrightarrow} (t, S_t, z, \xi) \qquad \text{or} \qquad (t, S_t, K, T) \overset{m}{\longleftrightarrow} (t, z, \xi)$$

as long as these functions do establish bijections. The term *moneyness* then refers to the bijection $m()$ as a whole, not just the strike dimension. For instance, the mappings

$$(t, S_t, K, T) \overset{m}{\longleftrightarrow} \left(t, S_t, K^3, \sqrt{T-t}\right)$$
$$\text{or}$$
$$(t, S_t, K, T) \overset{m}{\longleftrightarrow} \left(t, \frac{K}{S_t}\sqrt{T-t}, \ln(T-t)\right)$$

would be valid choices, albeit probably not very effective ones. It is interesting to note that by defining the moneyness we establish in particular the fundamentals of how the ATM implied volatility evolves with S_t, the extreme case being a constant or deterministic sliding IV. In that respect, the introduction of a structural *backbone* in the transform can be done without difficulty, since any well-behaved function of S_t is appropriate and does not have to be bijective.

In conclusion, a good choice of moneyness is obviously dependent on the baseline model, but also on the specific real-life dynamics of the vanilla market considered. A healthy starting point would be to select a linear transformation of the (K, T) coordinates, running a PCA[2]-type inference over the historical dynamics of the smile. Indeed, that procedure minimises the residual variance which is a reasonable measure of stationarity. In that vein, [9] provides a good idea of what statistical analysis of a liquid smile can entail.

3.3.1.2 Extending the Payoff

With respect to the option field, we have so far limited ourselves to European Calls, and therefore we might want to vary the payoff and/or the exercise type.

Focusing on the payoff, it is clear that we could equivalently use Puts or Straddles, whose Black prices are monotonic: they increase with volatility at a given fixed strike. The implied volatility map would be identical to the one associated with the Calls, which can be seen either through parity or by examining the ZDC.

For the Puts it is demonstrative to follow the latter alternative: we observe that the SDE for Put dynamics is still (2.22) [p. 44], but where the Greeks are those of a Black Put. However, it leads to the Zero-Drift Condition (2.23) [p. 44] involving only the Vega, Gamma, Volga and Vanna, which all happen to be identical to their Call versions.

The same monotonicity argument obviously stands for a Strangle, because its payoff involves a position *simultaneously* long (or short) a Call and a Put (albeit

[2] Principal Component Analysis.

of different strikes).[3] Conversely, the bijectivity criteria makes Risk Reversals, Butterflies, Call/Put Spreads or Binaries difficult if not impossible to use, at least with a Lognormal baseline, because they combine both long and short positions on Calls and Puts. It follows that a sufficient condition to ensure the bijection (of prices w.r.t. the baseline parameter) is clearly to limit ourselves to *cones*, i.e. linear combinations of Calls and Puts where all the weights share the same sign.

Another avenue is to modify the type of exercise for the option, since our only requirements were for the random variable X to be measurable w.r.t. the paths of S_u between current time t and payment date T. In consequence, it is theoretically possible to envisage American, barrier or Asian options,[4] depending on the baseline model. Clearly with most baselines, and certainly for the Lognormal one, in the best of cases we would have to rely on an *approximation* for the price (as opposed to a true closed form). In the sequel we will discuss this proxy alternative in more practical terms, so for now it suffices to say that it degrades the performance of the methodology but not its principles.

The main attraction of this line of work is of course to expand the domain of (quasi-) vanilla options whose dynamics can be approximated under the *complex* model $\mathcal{M}$, because in so doing, we increase our ability to hedge a specific exotic payoff. However, irrespective of the mathematical implications, this prospect is currently limited simply by the narrow spectrum of options which are traded with sufficient liquidity.

3.3.1.3 Extending the Baseline

In this section we justify the interest of opening our methodology to new baseline models, and cover the case of such candidates presenting a closed-form formula, either exact or approximate. Among the models in the former situation, we stress the important role played by the *Normal* baseline, which will be treated in Sect. 3.3.2. Conversely, we also introduce the notion of *baseline transfer*, which is most appropriate when no simple enough closed-form is available, and that will be covered in Sect. 3.3.3.

In slight contrast with the payoff and moneyness ingredients, the choice of the baseline model suggests significant improvements for the methodology. Indeed, for most practitioners the main use of our asymptotics lie in the direct problem. More specifically, the focus will be on whole-smile static extrapolation as a module of the calibration process, and will therefore rely on the *speed* and *precision* of the method.

The speed is unlikely to be an issue, at least compared to full-blown numerical schemes such as FD or Monte-Carlo, although obviously the faster the better. As for the precision, defined either with the absolute or with the relative convention, our

---

[3] This would be *a priori* useful in FX markets, for instance, where the smile level is quoted w.r.t. Straddles, and the curvatures in terms of Strangles.

[4] Note that mid-curve options are already treatable within our framework, but nowhere near liquid enough: see p. 276.

opinion is that it should be based on the price rather than on the implied volatility. In order to justify this claim, we point to the strong link between the hedging and calibration processes involved in the trading of structured products. When the agent uses vanilla options in the replicating or hedging portfolio, either statically or dynamically, he/she must ensure that their initial *value* is accurately represented in the model. Indeed the most pertinent dynamics to consider are ultimately those of *assets*, rather than re-parametrisations such as *rates* or *volatilities*, if only for reasons of martingale representation. It is therefore the *price* of the vanillas that will impact the structured product's price process, the tracking error and ultimately the P&L. There is also a more numerical aspect to this issue, which is the impact on the *market error* functional. Indeed, because of the Vega *vs* strike profiles provided by most realistic baseline models, using implied volatility will artificially focus the calibration process on certain parts of the smile. And in doing so, it will distort any discretionary weighting scheme put in place either for hedging or for risk management purposes.

It follows that the quality of the price approximation depends trivially on two related aspects:

- The sensitivity of the price P w.r.t. the implied parameter Ψ.
- The quality of that implied parameter's approximation.

Let us write the (absolute) pricing error simply as a function of the parameter error:

$$\Delta P(t, S_t, T, K) = \partial_\theta B(t, S_t, T, K, \Theta(t, S_t, K, T)) \cdot \Delta\Psi(t, S_t, K, T).$$

Obviously we would like some criteria to help us choose between several possible baselines, whenever such an opportunity presents itself. Our view is that it is *a priori* not obvious to differentiate between two baselines on the basis of the above price sensitivity, and our reasoning is again backed by the Vega profiles.

Comparing the Lognormal and Normal cases, for instance, it is clear from the respective sensitivities (C.1) [p. 433] and (C.4) [p. 435] that not only does each model show significant variation in strike, but that these profiles are different between the two models. In consequence, the optimal choice would have to be made on a strike-by-strike basis, which tends to negate the notion of whole-smile extrapolation and global dynamics, raises the issue of intra-smile compatibility and probably affects the speed significantly.

The second aspect—the implied parameter error—is in our view closely related to the similarity between the baseline and the complex models. To illustrate this point, let us envisage as a complex model the following instantaneous volatility specification:

$$dS_t = \sigma_t dW_t \qquad \text{where} \qquad d\sigma_t = \sigma_t \nu dZ_t \qquad \text{with} \qquad \langle dW_t, dZ_t\rangle = 0.$$

It seems intuitive that, as vol of vol ν increases from 0, this model and therefore its associated option prices will further and further deviate from their pure Normal counterparts. We could indeed derive an expansion in the perturbation parameter ν, based upon the heat equation, similarly to the approaches of Lewis [4] (Chap. 3) or [10].

In this toy case, it is equally clear that using a Normal baseline will provide better results, at a given order, than employing a Lognormal baseline. Taking matters to the extreme, if $\nu = 0$ then the Normal baseline will provide exact results at expansion order 0, whereas the Lognormal baseline will require an infinite order to achieve the same result. Even more structurally, it seems futile to use a Lognormal reference in order to approximate a marginal distribution whose support extends in the negative axis. It appears that there is benefit to be gained from expanding the collection of available baseline models.

It is possible to significantly improve our baseline specification, while staying within the broad Lognormal framework and recycling some of our existing work. Indeed, and as discussed before, one of the target's model features that can limit the method's efficiency is a low level of time-homogeneity or *stationarity*.

Although we argue that stationarity is a healthy feature in general, practical models *do* employ time-dependent parameters and diffusion coefficients, if only to match the ATM level across all expiries. The ACE methodology, as exposed so far, will asymptotically *predict* the future behaviour of any deterministic parameter, such as a volatility $\sigma(s)$. It will do so through its time-differentials at current time t ($\sigma'(t)$, $\sigma''(t)$, etc.) which will be invoked by successive IATM θ-differentials. In consequence we will have to compute higher layers than with a constant volatility, in order to reach the same precision on a smile extrapolation. For the same reason, a non-continuous deterministic volatility in general and a piecewise constant specification in particular, would significantly impair that pure Lognormal baseline.

The natural solution to this issue is the *deterministic volatility Lognormal baseline*:

$$\frac{dS_t}{S_t} = \sigma \upsilon(t)\, dW_t.$$

We allocate the implied parameter ψ to the *scale factor* σ, while $\upsilon(t)$ becomes a *shape function*. Call options are priced under that baseline model with the same normalised Black functional as usual, replacing the constant by the accumulated variance. Anticipating the next steps, we propose two (trivially equivalent) ways to rewrite (2.3) [p. 25]:

$$\begin{aligned}
&\text{either} \quad C(t, S_t, T, K) = N_t C^{BS}\left(S_t, K, \Sigma(t, S_t, K, T).\sqrt{\xi(t, T)}\right) \\
&\text{or} \quad C(t, S_t, T, K) = N_t C^{BS}\left(S_t, K, \Sigma(t, S_t, K, T).\Upsilon(t, T).\sqrt{T-t}\right) \\
&\text{with} \quad \Upsilon^2(t, T) \triangleq \tfrac{1}{T-t}\xi(t, T) \quad \text{and} \quad \xi(t, T) \triangleq \textstyle\int_t^T \upsilon^2(s)\, ds.
\end{aligned}$$

Writing the Call dynamics through Itô will generate new terms, compared to (2.22) [p. 44], due to the combined presence of the accumulated variance in the BS formula and of $\upsilon(t)$ in the underlying dynamics. However, the *apparent* complexity of these dynamics will depend on which maturity re-parametrisation we choose for the sliding IV representation.

For instance if we stick to the time-to-maturity θ then the Call dynamics will read as

$$\begin{aligned} dC(t, S_t, T, K) = \mathscr{V}\Sigma(\Upsilon_t^{'}\sqrt{\theta} - \frac{\Upsilon}{2\sqrt{\theta}})dt + \Delta S_t\sigma_t\upsilon(t)dW_t \\ + \frac{1}{2}\Gamma S_t^2\sigma_t^2\upsilon^2(t)dt + \mathscr{V}\Upsilon\sqrt{\theta}\left[bdt + \nu dW_t + \vec{n}^{\perp}d\vec{Z}_t\right] \\ + \frac{1}{2}\vartheta\Upsilon^2\theta\left[\nu^2 + \|\vec{n}\|^2\right]dt + \Lambda\Upsilon\sqrt{\theta}S_t\sigma_t\upsilon(t)\nu dt. \end{aligned} \tag{3.21}$$

Alternatively, choosing ξ as a sliding maturity variable leads to these Call dynamics:

$$\begin{aligned} dC(t, S_t, T, K) = & -\mathscr{V}\Sigma(2\sqrt{\xi})^{-1}\upsilon^2(\mathbf{t})dt + \Delta S_t\sigma_t\upsilon(\mathbf{t})dW_t \\ & + \mathscr{V}\sqrt{\xi}\left[bdt + \nu dW_t + \vec{n}^{\perp}d\vec{Z}_t\right] + \frac{1}{2}\Gamma S_t^2\sigma_t^2\upsilon^2(\mathbf{t})dt \\ & + \frac{1}{2}\vartheta\xi\left[\nu^2 + \|\vec{n}\|^2\right]dt + \Lambda\sqrt{\xi}S_t\sigma_t\upsilon(t)\nu dt. \end{aligned} \tag{3.22}$$

In both cases, those of the new/replacement terms appearing in the drift will in turn modify the ZDC, which is the starting point of the methodology. But in comparing, respectively, (3.21) and (3.22) with 2.22, we observe that choosing ξ instead of θ will conserve much more of the original ZDC structure. Indeed, apart from the first term, the new expression is achieved simply by replacing θ by ξ and scaling the instantaneous volatility by $\upsilon(t)$. This is indeed the technique employed in [1] and can be interpreted as a time change. Also, it is a good example of how setting another kind of moneyness function can simplify the ZDC.

Continuing any further with the derivation would fall outside of the scope of this study, but there is *a priori* no new structural difficulty involved. Furthermore, it is possible to check the associated computations, by downgrading the framework to the constant volatility case. Naturally the Immediate (and in particular the IATM) smile differentials will still invoke the derivatives of $\upsilon(\cdot)$ taken in t, so that the shape function must be differentiable at that point. This is, however, a small price to ask and an intrinsic constraint of asymptotic methods.

Overall, this new baseline allows us to manage many more real-life, complex models. In particular, the Extended Skew Market Model class (see Definition 2.2 [p. 88]) usually employs a skew function $f(\cdot)$ with separate variables, so that the overall dynamics rewrite as

$$\begin{aligned} \frac{dS_t}{S_t} &= \alpha_t f_1(t) f_2(S_t) dW_t \qquad && \text{where} \quad f_1(\cdot) \text{ can be discontinuous,} \\ d\alpha_t &= h(t, \alpha_t)dt + \varepsilon g(t, \alpha_t)dB_t \quad && \text{with} \quad \rho dt = \langle dW_t, dB_t\rangle. \end{aligned}$$

By allocating the baseline's shape function $\upsilon(s)$ to the deterministic volatility $f_1(s)$ of the complex model, we end up with the cast

$$\sigma_t = \alpha_t f_2(S_t)$$

and can therefore tackle the deterministic/discontinuous volatility specification. Obviously we have made the baseline clearly specific to the complex model, but the whole process can be made abstract with regard to the shape function $\upsilon(\cdot)$, hence the approach *does* correspond to our objective of a generic methodology. Finally, getting ahead of our narrative, we note a striking similarity between, on one hand, this *constant vs deterministic* comparison of the single underlying setup, and, on the other hand, the *term-structure vs term-by-term* argument in the maturity-dependent framework of Chap. 5 (see Sect. 5.4.2.2 [p. 319]).

Deviating further from the Lognormal case, it is clear that we could work with the displaced version of that model, i.e. the LDD local volatility model, which we have already mentioned in Sect. 2.5.2.2. On the other side of the LDD we also find the Normal model, which clearly deserves particular attention.

Indeed, many models used in practice, and in particular those within the stochastic instantaneous volatility class, are parametrised according to the Normal convention rather than the Lognormal one. This is due in part to the very steep ATM market skews experienced over the last decade, but also to a desired/observed behaviour for the ATM Delta. In that spirit, the CEV-SABR model, for instance, (see Sect. 4.2) can see either its correlation ρ or its power β calibrated to historical time series data, while the other parameter is selected to match the ATM market skew. In current markets, this approach can result in a β as low as 0.2, and sometimes lower. It is clear that such a model is much closer, in its statics and in its dynamics, to a Normal baseline than to a Lognormal one.

This observation justifies in itself the investigation of the application of the methodology to the Normal baseline, which is the object of Sect. 3.3.2. However, we recognise further, more mathematical reasons to investigate the Normal case. In particular, we consider that using this convention tends to align the dynamics of the underlying S_t with those of its instantaneous volatility σ_t and of its subsequent dynamic coefficients $a_{\star,t}$. And establishing such a recurring scheme seems *a priori* to be a good start if we are to detect inductive structures in the generic expressions for the IATM differentials.

Let us go back to the general need for more baseline models, and the subject of feasibility in particular. If we were to run through our full methodology for *each* of these simple models, then we would be structurally limited to those associated to closed-form pricing. This is due to the fact that we need to derive, differentiate and then take to the limits the main Zero Drift Condition. This ZDC will be jointly determined by each baseline dynamics, by its closed form pricing formula and by the choice of moneyness. Leaving the latter point aside, it follows that the feasibility or difficulty to proceed will depend on the simplicity of both the baseline SDE and closed form.

In practice, imposing the existence of the closed form would seriously limit the number of usable baseline models, even more so since the formula might not be simple enough for the ZDC. To illustrate this point, we consider the most complex (*sic*) *market model*[5] available with closed-form pricing. This is the Displaced Constant Elasticity of Variance (CEV) local volatility model, which we have already encountered in Chap. 2:

$$dS_t = \sigma\,(d + S_t)^\beta\, dW_t.$$

In terms of closed-form pricing, the displacement d brings no additional difficulty, hence without loss of generality we will take $d = 0$ and downgrade to a pure CEV. The Call formula then becomes

$$0 < \beta < 1 \qquad C(t, S_t, K, T) = S_t\left[1 - \chi^2(a, b+2, c)\right] - K\chi^2(c, b, a) \quad (3.23)$$

with

$$a = \frac{K^{2(1-\beta)}}{(1-\beta)^2\sigma^2(T-t)} \qquad b = \frac{1}{1-\beta} \qquad c = \frac{S_t^{2(1-\beta)}}{(1-\beta)^2\sigma^2(T-t)},$$

where $\chi^2(x, y, z)$ denotes the non-central χ^2 cumulative distribution at value x, for non-centrality parameter y and z degrees of freedom. See Beckers [11] or Schroder [12] for a rigorous study, while this parametrisation corresponds to [13] p. 457. Notwithstanding the various mathematical regularity issues presented by this model[6] we observe that although the model is quite poor, the pricing formula involves the χ^2 function, which itself involves the Gamma function and its lower incomplete version (see Press et al. [14] Sect. 6.2 for details on the relevant implementation aspects). Therefore, although *a priori* feasible, establishing the ZDC might prove very involved and thus not practical.

We are likely to face the situation where a given model might appear particularly well suited as a baseline (mainly because of its proximity to the target model) without any closed form solution at hand for the chosen option type. However, *if* an *approximation* formula were available instead for that baseline, *and if* that formula were simple enough to establish the ZDC, then we would be in a position to deploy the methodology. Of course the overall precision would suffer from the proxy aspect, but this approach offers the prospect to combine Asymptotic Chaos Expansions with other approximation techniques, such as singular perturbations [5], Markovian projection [15], etc.

This line of thought offers new perspectives, in particular the possibility of using a stochastic volatility model for the baseline. Assuming we have a reasonably simple

---

[5] We note for future reference that there are several other available classes of baselines, in particular in the factor model class. There is, for instance, substantial literature on various quadratic Gaussian parametrisations, but transferring our methodology to this environment requires a brand new derivation.

[6] For instance, the boundary conditions associated to the uniqueness of a solution when $0 < \beta \leq \frac{1}{2}$, or the existence/integrability constraints when $\beta > 1$.

proxy at our disposal then this price function would be a deterministic function of (t, S_t, K, T, ψ) but also of the additional state variables (including the volatility). It is simple to accommodate these features within the existing framework, by assimilating the new initial values to superfluous parameters, which are then *hidden* in the pricing functional $B(t, S_t; T, K; \psi)$. However, this approach also leads to new mathematical questions, such as the stability of the ZDC with respect to the closed-form approximation.

It remains that either in exact or approximate form, extending the method to new baselines is dependent on the availability of *some* closed-form for that specific model. We introduce, therefore, a default approach that does not rely on establishing each idiosyncratic ZDC. Instead its principle is to carry over the results for one central baseline to the new ones, and that notion of *baseline transfer* will be covered in Sect. 3.3.3.

3.3.2 An Important Example: The Normal Baseline via Its ZDC

The object of this section is to illustrate the possibility of changing the baseline model and of rebuilding accordingly our methodology from scratch, which serves two purposes. The first goal is to gauge the level of complexity of such an approach, in the sense that we should be able to recycle a large part of the results of Chap. 2 established for the Lognormal baseline. The second aim is to compare the actual results obtained (the ZDC, the IZDC, the IATM differentials, etc.) with those obtained for that Lognormal case. In particular, we would like to distinguish the systemic features of the method from the idiosyncracies of each specific baseline.

In several respects the Gaussian framework is naturally simpler than the Lognormal one, and one could argue that it would have been a better choice for the foundation Chap. 2. However, most of the literature related to local and stochastic volatility, along with the associated pricing and approximation methods, has been written according to the Lognormal convention. In consequence we would have lost a certain degree of comparability.

Without surprise we have the dynamics of the Normal baseline as

$$dS_t = \sigma \, dW_t,$$

and the Call pricing solution comes from (C.3) as

$$C(t, S_t, K, T) = B^n\left(K - S_t, \sigma\sqrt{T-t}\right)$$

where B^n is the *normalised Bachelier formula*:

$$B^n(z, v) = v\mathscr{G}\left(\frac{z}{v}\right) - z\left[1 - \mathscr{N}\left(\frac{z}{v}\right)\right]. \tag{3.24}$$

Let us now run through the full methodology, retracing our steps quickly through Sects. 2.1, 2.2, 2.3 and 2.4.

Framework (parallel of Sect. 2.1 [p. 24])
As before, we assume a continuum of Call option prices, represented in absolute and sliding coordinates respectively by $C(t, S_t, K, T)$ and $C(t, z, \theta)$ where the new moneyness reads:

$$z = K - S_t \qquad \text{while} \qquad \theta = T - t.$$

Using Bachelier's formula, we translate these two maps into *Normal* implied volatility surfaces $\Sigma(t, S_t, K, T)$ and $\widetilde{\Sigma}(t, y, \theta)$. For good measure, we check easily from (C.4) that $\frac{\partial B^n}{\partial v} \geq 0$ and therefore the Normal implied volatility indeed provides a bijection to the price.

On one hand, the SInsV model is defined by the SDE system

$$\begin{cases} dS_t = \sigma_t dW_t, \\ d\sigma_t = a_{1,t}dt + a_{2,t}dW_t + \vec{a}^{\perp}_{3,t} d\vec{Z}_t, \\ da_{2,t} = a_{21,t}dt + a_{22,t}dW_t + \vec{a}^{\perp}_{23,t} d\vec{Z}_t. \end{cases} \tag{3.25}$$

On the other hand, the sliding SImpV model is described by

$$\begin{cases} dS_t = \sigma_t dW_t, \\ d\widetilde{\Sigma}(t, z, \theta) = \widetilde{b}(t, z, \theta)dt + \widetilde{\nu}(t, z, \theta)dW_t + \vec{n}(t, z, \theta)^{\perp} d\vec{Z}_t, \end{cases} \tag{3.26}$$

where, as usual, W_t and $\vec{Z}_t$ are orthogonal drivers.

Derivation of the Zero-Drift Conditions (parallel of Sect. 2.2 [p. 42])
Our first task is to derive the dynamics of the absolute volatility, which come as:

Lemma 3.5 (Dynamics of the absolute IV surface with a Normal baseline) *In the framework defined by* (3.26) *the dynamics of the absolute IV surface is*

$$d\Sigma(t, S_t, K, T) = b(t, S_t, K, T)dt + \nu(t, S_t, K, T)dW_t + \vec{n}(t, S_t, K, T)^{\perp} d\vec{Z}_t,$$

with

$$b(t, S_t, K, T) = \widetilde{b}(t, z, \theta) - \widetilde{\Sigma}'_{\theta}(t, z, \theta) + \frac{1}{2}\sigma_t^2 \widetilde{\Sigma}''_{zz}(t, z, \theta) - \sigma_t \widetilde{\nu}'_z(t, z, \theta), \tag{3.27}$$

$$\nu(t, S_t, K, T) = \widetilde{\nu}(t, z, \theta) - \sigma_t \widetilde{\Sigma}'_z(t, z, \theta), \tag{3.28}$$

$$\vec{n}(t, S_t, K, T) = \vec{n}(t, z, \theta). \tag{3.29}$$

Proof Invoking Itô-Kunita's Theorem A.1 with $\vec{\alpha}_t = \binom{K-S_t}{\theta}$ we get

$$\begin{aligned} d\Sigma(t, S_t, K, T) &= \widetilde{b}(t, z, \theta)dt + \widetilde{\nu}(t, z, \theta)dW_t + \vec{n}(t, z, \theta)^{\perp} d\vec{Z}_t \\ &\quad - \widetilde{\Sigma}'_z(t, z, \theta)dS_t + \widetilde{\Sigma}'_{\theta}(t, z, \theta)d\theta + \frac{1}{2}\widetilde{\Sigma}''_{zz}(t, z, \theta)d\langle S_t \rangle - \widetilde{\nu}'_z(t, z, \theta)\sigma_t dt, \end{aligned}$$

hence

$$d\Sigma(t, S_t, K, T) = \widetilde{b}(t, z, \theta)dt + \widetilde{\nu}(t, z, \theta)dW_t + \overrightarrow{\widetilde{n}}(t, z, \theta)^{\perp} d\overrightarrow{Z}_t$$
$$- \widetilde{\Sigma}'_z(t, z, \theta)\sigma_t dW_t - \widetilde{\Sigma}'_\theta(t, z, \theta)dt + \frac{1}{2}\widetilde{\Sigma}''_{zz}(t, z, \theta)\sigma_t^2 dt - \widetilde{\nu}'_z(t, z, \theta)\sigma_t dt,$$

and the result follows by grouping the finite and non-finite variation terms. □

We are now capable of expressing the dynamics of the actual assets, hence the next step is the main Zero Drift Condition itself.

Proposition 3.3 (Zero Drift Condition with a Normal baseline) *With a Normal baseline the Zero Drift Condition imposed on the SImpV model* (3.26) *in the general domain* $(\circ) \stackrel{\Delta}{=} (t, z, \theta)$ *is*

$$\widetilde{\Sigma}^3(\circ)\widetilde{b}(\circ) = E(\circ) + \frac{1}{\theta}F(\circ) \tag{3.30}$$

with

$$E(\circ) = \widetilde{\Sigma}^3\left[\widetilde{\Sigma}'_\theta - \frac{1}{2}\sigma_t^2\widetilde{\Sigma}''_{zz} + \sigma_t\widetilde{\nu}'_z\right],$$

$$F(\circ) = \frac{1}{2}\widetilde{\Sigma}^2\left[\widetilde{\Sigma}^2 - \sigma_t^2\right] - z\widetilde{\Sigma}\sigma_t\left[\widetilde{\nu} - \sigma_t\widetilde{\Sigma}'_z\right] - \frac{1}{2}z^2\left[\left(\widetilde{\nu} - \sigma_t\widetilde{\Sigma}'_z\right)^2 + \|\overrightarrow{\widetilde{n}}\|^2\right].$$

Comparing the Normal ZDC (3.30) to its Lognormal equivalent (2.18) [p. 43] we can make a first series of observations on this structural result.

- There is no $D(\circ)$ term, i.e. no term in θ. This comes from the new expression for the Volga and will affect some of the higher-order θ-differentials, such as the arch $\widetilde{\Sigma}''_{\theta\theta}(\star)$ (see (E.63) [p. 475]). Indeed, successive maturity-differentiations of the ZDC involve term D with no θ in factor, so that it survives at the IATM point.
- The $E(\circ)$ term is almost identical to (2.20) [p. 44], except for its last sub-term which is missing, and taking into account the change of sliding variable from y to z. This will quickly impact the IATM differentials as term E is invoked *as soon as* we differentiate w.r.t. θ, starting with $\widetilde{\Sigma}'_\theta$.
- The $F(\circ)$ term is identical (again, replacing y with z) which should lead to a replica of the Immediate ZDC. In turn, this suggests that all the pure-strike differentials should be identical to the Lognormal case.

It is remarkable that the two terms which are almost unchanged are those which are guaranteed to give a finite limit in $\theta = 0$, provided the regularity assumptions are valid. This leads to the conclusion that the two implied volatility maps, each corresponding to their own Normal/Lognormal SInsV model, should have the same Immediate smile but will then start to differ as maturity increases. However, we will need the actual expressions of the IATM differentials in order to gauge with more precision that divergence.

Proof of Proposition 3.3.
We now must compute the dynamics of the absolute (rebased) Calls. Denoting by

$$(\propto) \triangleq (t, S_t, K, T) \qquad \text{and} \qquad (\ltimes) = \left(K - S_t,\, \Sigma\,(t, S_t, K, T)\,\sqrt{T-t}\right)$$

respectively the absolute and sliding arguments, we can apply Itô's Lemma to the Bachelier formula and write (still with $\theta \triangleq T - t$)

$$\begin{aligned} dC(\propto) &= \frac{\partial B^n}{\partial v}(\ltimes)\Sigma(\propto)\frac{-1}{2\sqrt{\theta}}dt + \frac{\partial B^n}{\partial z}(\ltimes)\,(-1)\,dS_t + \frac{\partial B^n}{\partial v}(\ltimes)\sqrt{\theta}d\Sigma(\propto) \\ &\quad + \frac{\partial^2 B^n}{\partial z \partial v}(\ltimes)\left(-\sqrt{\theta}\right)\langle dS_t, d\Sigma_t\rangle + \frac{1}{2}\frac{\partial^2 B^n}{\partial z^2}(\ltimes)\langle dS_t\rangle + \frac{1}{2}\frac{\partial^2 B^n}{\partial v^2}\theta\langle d\Sigma_t\rangle \\ &= \frac{\partial B^n}{\partial v}(\ltimes)\Sigma(\propto)\frac{-1}{2\sqrt{\theta}}dt + \frac{\partial B^n}{\partial v}(\ltimes)\sqrt{\theta}b(\propto)dt - \frac{\partial^2 B^n}{\partial z \partial v}(\ltimes)\sqrt{\theta}\sigma_t \mathrm{v}(\propto)dt \\ &\quad + \frac{1}{2}\frac{\partial^2 B^n}{\partial z^2}(\ltimes)\sigma_t^2 dt + \frac{1}{2}\frac{\partial^2 B^n}{\partial v^2}\theta\left[\mathrm{v}^2 + \|\vec{n}\|^2\right](\propto)dt + [\cdot]\,dW_t + [\cdot]\,d\vec{Z}_t. \end{aligned}$$

The relevant differentials of the Bachelier formula can be found in Sect. C.2. The zero-drift condition therefore reads, omitting the argument $(\propto)$:

$$\begin{aligned} 0 &= \mathscr{G}\left(\frac{K-S_t}{\Sigma\sqrt{\theta}}\right)\left[-\frac{\Sigma}{2\sqrt{\theta}} + b\sqrt{\theta}\right] - \frac{K-S_t}{\Sigma^2\theta}\mathscr{G}\left(\frac{K-S_t}{\Sigma\sqrt{\theta}}\right)\left(-\sqrt{\theta}\right)\sigma_t\mathrm{v} \\ &\quad + \frac{1}{2}\frac{1}{\Sigma\sqrt{\theta}}\mathscr{G}\left(\frac{K-S_t}{\Sigma\sqrt{\theta}}\right)\sigma_t^2 + \frac{1}{2}\frac{(K-S_t)^2}{\Sigma^3\theta^{\frac{3}{2}}}\mathscr{G}\left(\frac{K-S_t}{\Sigma\sqrt{\theta}}\right)\theta\left[\mathrm{v}^2 + \|\vec{n}\|^2\right], \end{aligned}$$

which after simplification gives

$$0 = -\frac{1}{2}\Sigma + b(T-t) + \frac{K-S_t}{\Sigma^2}\sigma_t\mathrm{v} + \frac{1}{2}\frac{\sigma_t^2}{\Sigma} + \frac{1}{2}\frac{(K-S_t)^2}{\Sigma^3}\left[\mathrm{v}^2 + \|\vec{n}\|^2\right].$$

We replace the absolute coefficients b, v and $\vec{n}$ with their sliding versions $\widetilde{b}$, $\widetilde{\mathrm{v}}$ and $\vec{\widetilde{n}}$, taken at (t, z, θ) as per (3.27)–(3.29). Then isolating the drift gives

$$\begin{aligned} \widetilde{b} &= \left[\widetilde{\Sigma}'_\theta - \frac{1}{2}\sigma_t^2\widetilde{\Sigma}''_{zz} + \sigma_t\widetilde{\mathrm{v}}'_z\right] \\ &\quad + \frac{1}{\theta}\left[\frac{1}{2}\widetilde{\Sigma} - \frac{1}{2}\frac{\sigma_t^2}{\widetilde{\Sigma}} - \frac{z}{\widetilde{\Sigma}^2}\sigma_t\left[\widetilde{\mathrm{v}} - \sigma_t\widetilde{\Sigma}'_z\right] - \frac{1}{2}\frac{z^2}{\widetilde{\Sigma}^3}\left[\left(\widetilde{\mathrm{v}} - \sigma_t\widetilde{\Sigma}'_z\right)^2 + \|\vec{\widetilde{n}}\|^2\right]\right]. \end{aligned}$$

Finally, scaling by $\widetilde{\Sigma}^3$ and putting $(\circ) = (t, z, \theta)$ gives (3.30), concluding the proof. □

As a corollary we can express the Immediate Zero Drift Condition. Having previously transposed the regularity assumptions of the Lognormal case, we end up with

$$F(t, z, 0) = 0$$

so that the Primary IZDC becomes, with $(\bullet) = (t, z, 0)$

$$0 = \frac{1}{2}\widetilde{\Sigma}^2(\bullet)\left[\widetilde{\Sigma}^2(\bullet) - \sigma_t^2\right] - z\widetilde{\Sigma}(\bullet)\sigma_t\left[\widetilde{\mathrm{v}}(\bullet) - \sigma_t\widetilde{\Sigma}'_z(\bullet)\right]$$

$$\text{(IZDC)} \qquad -\frac{1}{2}z^2\left[\left(\widetilde{\mathrm{v}}(\bullet) - \sigma_t\widetilde{\Sigma}'_z(\bullet)\right)^2 + \|\vec{\widetilde{n}}(\bullet)\|^2\right]. \tag{3.31}$$

We note that (3.31) is naturally "identical" to its Lognormal counterpart (2.28) [p. 48] and also that we get the usual lower bound for the convergence speed.

Finally, evaluating at $z = 0$ we get the same IATM identity as

$$\widetilde{\Sigma}(t, 0, 0) = \sigma_t. \tag{3.32}$$

Recovering the instantaneous volatility (parallel of Sect. 2.3 [p. 53]) Our first step is to establish the IATM arbitrage constraints of the SImpV model.

Proposition 3.4 (SImpV model's IATM constraints with a Normal baseline) *With a Normal baseline, the SImpV IATM arbitrage constraints* (3.26) *become*

$$\begin{cases} \widetilde{\mathrm{v}}(\star) = 2\sigma_t\widetilde{\Sigma}'_z(\star), & (3.33)\\[2ex] \widetilde{\mathrm{v}}'_z(\star) = \widetilde{\Sigma}'^{\,2}_z(\star) + \dfrac{3}{2}\sigma_t\widetilde{\Sigma}''_{zz}(\star) - \dfrac{1}{2}\dfrac{\|\vec{\widetilde{n}}(\star)\|^2}{\sigma_t^2}, & (3.34)\\[2ex] \widetilde{b}(\star) = 2\widetilde{\Sigma}'_\theta(\star) + \sigma_t\,\widetilde{\Sigma}'^{\,2}_z(\star) + \sigma_t^2\widetilde{\Sigma}''_{zz}(\star) - \dfrac{1}{2}\dfrac{\|\vec{\widetilde{n}}(\star)\|^2}{\sigma_t}. & (3.35) \end{cases}$$

Note that the Normal IATM drift $\widetilde{b}(t, 0, 0)$ is different by a single term from its Lognormal counterpart (2.39) [p. 54]. As will be seen in the proof, this is simply because of a different expression for the functional $E(t, y, \theta)$.

Proof To obtain $\widetilde{\mathrm{v}}(t, 0, 0)$ and $\widetilde{\mathrm{v}}'_z(t, 0, 0)$ we can differentiate the IZDC once and then twice w.r.t. z, i.e. compute $\nabla^{1,0}F(t, z, \theta)$ and $\nabla^{2,0}F(t, z, \theta)$. Alternatively we can exploit the similarity of the Normal IZDC with its Lognormal equivalent, transposing (2.37) and (2.38) [p. 54]. Either way we end up with (3.33). Invoking the

IZDC again, we get the IATM drift by first using L'Hopital's rule and re-writing the ZDC within a small-θ expansion:

$$\widetilde{b}(t, z, \theta) = E(t, z, \theta) + F'_\theta(t, z, 0) + O(\theta). \tag{3.36}$$

We can then compute $\nabla^{0,1}F(t, z, \theta)$ or simply use again the similarity to obtain that

$$F'_\theta(\star) = \widetilde{\Sigma}^3 \widetilde{\Sigma}'_\theta(\star).$$

Evaluating (3.36) at $(\star)$, replacing $F'_\theta(\star)$ by its expression, and simplifying by $\widetilde{\Sigma}^3(\star)$ we get

$$\widetilde{b}(\star) = \widetilde{\Sigma}^{-3}(\star)\left[E(\star) + F'_\theta(\star)\right] = 2\widetilde{\Sigma}'_\theta(\star) - \frac{1}{2}\sigma_t^2\widetilde{\Sigma}''_{zz}(\star) + \sigma_t\widetilde{\mathrm{v}}'_z(\star).$$

Finally, replacing $\widetilde{\mathrm{v}}'_z(\star)$ with (3.33) we have

$$\widetilde{b}(\star) = 2\widetilde{\Sigma}'_\theta(\star) - \frac{1}{2}\sigma_t^2\widetilde{\Sigma}''_{zz}(\star) + \sigma_t\left[\widetilde{\Sigma}'^{\,2}_z(\star) + \frac{3}{2}\sigma_t\widetilde{\Sigma}''_{zz}(\star) - \frac{1}{2}\frac{\|\vec{n}(\star)\|^2}{\sigma_t^2}\right]$$

and after simplification comes (3.33), which concludes the proof. □

We can deal with the recovery itself, again invoking the *ad hoc* regularity packages.

Theorem 3.1 (Recovery with the Normal baseline) *A sliding stochastic Normal implied volatility model defined by* (3.26) *is associated to a stochastic instantaneous Normal volatility model* (3.25) *as per*

$$d\sigma_t = a_{1,t}dt + a_{2,t}dW_t + \vec{a}^{\perp}_{3,t}d\vec{Z}_t$$

with

$$a_{1,t} = 2\widetilde{\Sigma}'_\theta(\star) + \sigma_t\widetilde{\Sigma}'^{\,2}_z(\star) + \sigma_t^2\widetilde{\Sigma}''_{zz}(\star) - \frac{1}{2}\frac{\|\vec{n}(\star)\|^2}{\sigma_t}, \tag{3.37}$$

$$a_{2,t} = 2\sigma_t\widetilde{\Sigma}'_z(\star), \tag{3.38}$$

$$\vec{a}_{3,t} = \vec{n}(\star), \tag{3.39}$$

$$a_{22,t} = 2\left[\widetilde{\Sigma}'_z\widetilde{\mathrm{v}}(\star) + \sigma_t\widetilde{\mathrm{v}}'_z(\star)\right]. \tag{3.40}$$

Comparing, respectively, (2.43) with (3.37), (2.44) with (3.38), (2.45) with (3.39), and (2.46) with (3.40), we observe that the equivalence with the Recovery Theorem (2.42) [p. 55] is complete. In other words, the fact that we chose in each case the moneyness corresponding precisely to the relevant closed form (i.e. y or z along θ) aligns the ZDC sufficiently for the first layer to be identical. This should actually not come as a surprise, considering that the respective pricing PDEs for the Normal and Lognormal models are equivalent, providing the exact same change of variable.

Proof Taking the dynamics on both sides of (3.32) we obtain

$$d\sigma_t = d\widetilde{\Sigma}(t, z=0, \theta=0) = \widetilde{b}(t,0,0)dt + \widetilde{\nu}(t,0,0)dW_t + \overrightarrow{\widetilde{n}}(t,0,0)^{\perp} d\overrightarrow{Z}_t$$

then by identification:

- The exogenous coefficient (3.39) is obtained straight away.
- The endogenous coefficient (3.38) is obtained *via* (3.33).
- The drift coefficient (3.37) is obtained through (3.33).

Invoking (3.38) we get its dynamics by Itô as

$$da_{2,t} = d\left[2\sigma_t \widetilde{\Sigma}'_z(\star)\right] = [\cdots]\,dt + 2\left[\widetilde{\Sigma}'_z(\star)a_2 + \sigma_t \widetilde{\nu}'_z(\star)\right]dW_t + \overrightarrow{[\cdots]}^{\perp} dZ_t$$

so that substituting (3.33) again proves (3.40) and concludes the proof. □

Generating the implied volatility (parallel of Sect. 2.4) [p. 61] We can now address the direct problem, which tends to make the comparison easier.

Theorem 3.2 (First layer IATM differentials in the Normal baseline) *A Normal SInsV model* (3.25) *is associated as follows to a sliding Normal SImpV model* (3.26), *at the IATM point* $(\star) = (t, z=0, \theta=0)$ *and for the first layer:*

Static differentials:

$$\widetilde{\Sigma}(\star) = \sigma_t, \tag{3.41}$$

$$\widetilde{\Sigma}'_z(\star) = \frac{a_2}{2\sigma_t}, \tag{3.42}$$

$$\widetilde{\Sigma}''_{zz}(\star) = \frac{1}{\sigma_t^2}\left[\frac{1}{3}a_{22}\right] + \frac{1}{\sigma_t^3}\left[\frac{1}{3}\|\vec{a}_3\|^2 - \frac{1}{2}a_2^2\right], \tag{3.43}$$

$$\widetilde{\Sigma}'_\theta(\star) = \left[\frac{1}{2}a_1 - \frac{1}{6}a_{22}\right] + \frac{1}{\sigma_t}\left[\frac{1}{8}a_2^2 + \frac{1}{12}\|\vec{a}_3\|^2\right]. \tag{3.44}$$

Dynamic differentials:

$$\widetilde{\nu}(\star) = a_2, \tag{3.45}$$

$$\widetilde{\nu}'_z(\star) = \frac{1}{\sigma_t}\left[\frac{1}{2}a_{22}\right] - \frac{1}{\sigma_t^2}\left[\frac{1}{2}a_2^2\right], \tag{3.46}$$

$$\overrightarrow{\widetilde{n}}(\star) = \vec{a}_3. \tag{3.47}$$

We note that all expressions except the static IATM slope are the exact z-equivalents of those in Theorem 2.2 [p. 62], established for the Lognormal baseline. We had anticipated this feature for the static pure-strike differentials $\widetilde{\Sigma}'_z(\star)$ and $\widetilde{\Sigma}''_{zz}(\star)$ but the fact that $\widetilde{\mathrm{v}}'_z$ behaves identically should not come as a surprise. Indeed, the primary IZDC is identical and it is that equation which by further z-differentiation will provide all the pure-strike differentials, for static *and for dynamic* coefficients. We should therefore expect $\widetilde{\mathrm{v}}''_{zz}$, $\vec{\widetilde{n}}'_z$, etc. to come with the same expressions as in the Lognormal case.

In particular, we observe that σ_t is still present at the denominator of these expressions, so that this feature has clearly nothing to do with the convention used for the asset dynamics. It seems more related to the convention we have taken to write the dynamics of the instantaneous *volatility*, rather than the instantaneous *variance*, and is yet another indicator that improvements are certainly possible in our parametrisation.

Looking at the IATM slope expression (3.44) and comparing to the Lognormal version (2.53) [p. 62] we note that the first term in σ_t has disappeared. Again this comes as no surprise, since we remarked earlier on a missing component in the E term. It is indeed that difference which has been carried on through the IATM drift $\widetilde{b}$. It is interesting to remark as well that the sign and magnitude of the difference between these two expressions is quite well understood. Indeed, we have

$$\widetilde{\Sigma}'^{\,LN}_{\theta}(t, y = 0, 0) - \widetilde{\Sigma}'^{\,N}_{\theta}(t, z = 0, 0) = \frac{1}{4}\sigma_t^{LN} a_{2,t}^{LN} = \frac{1}{2}\,\widetilde{\Sigma}^{LN\,2}\,\widetilde{\Sigma}'^{\,LN}_{y}(t, y = 0, 0).$$

Hence (3.44) tells us that the difference in slope between the two environments will depend on the sign of the IATM skew. We must keep in mind, however, that each coefficient (i.e. σ_t and $a_{\star,t}$) has a different significance in the two frameworks. In particular, we are *not* looking at the implied Normal and Lognormal smiles of the *same* SInsV model.

Proof of Theorem 3.2.
The IATM Identity (3.32) proves (3.41), while taking its dynamics, then identifying the finite and non-finite variation terms provides

$$\widetilde{b}(\star) = a_1 \qquad \widetilde{\mathrm{v}}(\star) = a_2 \qquad \vec{\widetilde{n}}(\star) = \vec{a}_3 \tag{3.48}$$

which proves (3.45) and (3.47). Then combining the former with the IATM SImpV constraint (3.33) leads to (3.42). Taking the dynamics of that IATM skew expression we obtain formally

$$d\left[\frac{a_2}{2\sigma_t}\right] = d\widetilde{\Sigma}'_z(t, z, \theta) = \widetilde{b}'_z(t, z, \theta)dt + \widetilde{\mathrm{v}}'_z(t, z, \theta)dW_t + \vec{\widetilde{n}}'_z(t, z, \theta)^{\perp} d\vec{Z}_t.$$

Applying Itô on the l.h.s. and susbstituting the IATM constraint (3.33) on the r.h.s. we get by identification of the endogenous term

$$\frac{1}{2\sigma_t}\left[a_{22} - \frac{a_2^2}{\sigma_t}\right] = \widetilde{\mathrm{v}}'_z(\star) = \widetilde{\Sigma}'^{\,2}_z(\star) + \frac{3}{2}\sigma_t \widetilde{\Sigma}''_{zz}(\star) - \frac{1}{2}\frac{\|\vec{\widetilde{n}}(\star)\|^2}{\sigma_t^2},$$

which proves (3.46). Substituting (3.42) and (3.47) into that expression we get the IATM curvature (3.43). We can now move on to $\widetilde{\Sigma}'_\theta(\star)$ by replacing the IATM drift $\widetilde{b}(\star)$ in (3.48) using the SImpV IATM constraint (3.33):

$$a_1 = 2\widetilde{\Sigma}'_\theta(\star) + \sigma_t\, \widetilde{\Sigma}'^{\,2}_z(\star) + \sigma_t^2 \widetilde{\Sigma}''_{zz}(\star) - \frac{1}{2}\frac{\|\overrightarrow{n}(\star)\|^2}{\sigma_t}.$$

Isolating the slope and then substituting (3.42)–(3.43) and (3.47) we get

$$\begin{aligned}\widetilde{\Sigma}'_\theta(\star) &= \frac{1}{2}a_1 - \frac{1}{2}\sigma_t\, \widetilde{\Sigma}'^{\,2}_z(\star) - \frac{1}{2}\sigma_t^2 \widetilde{\Sigma}''_{zz}(\star) + \frac{1}{4}\frac{\|\overrightarrow{n}(\star)\|^2}{\sigma_t}\\ &= \frac{1}{2}a_1 - \frac{1}{2}\sigma_t\frac{a_2^2}{4\sigma_t^2} - \frac{1}{2}\sigma_t^2\left[\frac{1}{\sigma_t^2}\left[\frac{a_{22}}{3}\right] + \frac{1}{\sigma_t^3}\left[\frac{\|\overrightarrow{a}_3\|^2}{3} - \frac{a_2^2}{2}\right]\right] + \frac{1}{4}\frac{\|\overrightarrow{a}_3\|^2}{\sigma_t},\end{aligned}$$

which after simplification yields (3.44) and concludes the proof. □

3.3.3 The Generic Baseline Transfer

As discussed previously in Sect. 3.3.1.3, it is desirable to increase the number of available baselines, since a high degree of similarity between that simple model and the complex target $\mathscr{M}$ should improve extrapolation performance. Having now run the full ACE methodology through both the Normal and Lognormal baselines, we can gauge the level of technicality as well as the limitations associated to this approach.

In particular, it is clear that few baseline models can be used in practice, and for those that can, the most involved part of the method is to establish the specific Zero-Drift Condition. Then, subject to adequate regularity assumptions as well as overall structure of this PDE, we can proceed with the generic solution to the direct problem, as described in Sect. 3.1. In algorithmic terms, combining an *ad hoc* differentiation module for the ZDC with a simple Itô calculus formal engine allows us to automate the computation of all-order IATM (n, p)-differentials of the smile, both static and dynamic.

The obvious question is therefore whether we can avoid going through the specific ZDC of each model, either because it is very complex or because it cannot be derived: typically there might not be any closed-form solution, either exact or approximate. In the sequel we discuss an alternative possibility, or rather a suite of possibilities, which be available depending on the baseline choice. These options share the notion of a *central* baseline from which we can transfer the final IATM information to other baselines, from one implied parameter map to the next. In other words, this approach bypasses the need for the ZDC.

3.3.3.1 General Considerations and Methodology

Having selected an option type (e.g. Calls) parametrised by (K, T), the first step is to select that single, central baseline (arguably either Normal or Lognormal) providing either a closed form, a semi-closed form or an approximation formula. We then focus on developing the required *generic* differentials, once and for all, under this master model. By generic we understand expressions such as those of Theorem 2.2 or of Sect. 3.2, invoking purely σ_t and the $a_{\star,t}$ coefficients, and whose computation can be automated as per Sect. 3.1. When confronted with a new complex model, we cast its dynamics in the SInsV chaos framework (2.8)–(2.9) and populate the $a_{\star,t}$ coefficient collection: again, a formal calculus tool might be helpful. Let us now introduce a different baseline, and the following abstract model notations:

$$M_1 \quad \text{is the central baseline} \qquad M_2 \quad \text{is the new baseline}$$

$$\mathcal{M} \quad \text{is the target complex model}$$

For illustration purposes let us take the baselines in the local volatility model class:

$$(M_i) \qquad\qquad dS_t = f_i(t, S_t, \psi_i)dW_t \qquad i \in \{1, 2\},$$

where ψ_1 and ψ_2 are the respective parameters ensuring a bijection with the option price. Accordingly, let us denote the corresponding baseline pricing formulae by B_1 and B_2:

$$\text{Under} M_i \qquad\qquad P_i(K, T; t, S_t) = B_i(K, T; t, S_t; \psi_i) \qquad i \in \{1, 2\}.$$

Note that we do not impose either formula to be *exact* as it does not affect the *principle* of the approach. In the sequel, for simplicity's sake, we will omit the initial conditions arguments, starting with t and S_t, for all models M_1, M_2 and $\mathcal{M}$. Now we can formally re-parametrise the option price under the complex model, using each baseline formula and implied parameter:

$$\text{Under } \mathcal{M} \qquad\qquad P_{\mathcal{M}}(K, T) = B_i\left(K, T, \Psi_{i,\mathcal{M}}(K, T)\right) \qquad i \in \{1, 2\}.$$

Hence the relationship that we are aiming to make explicit is

$$\frac{\partial^{m+p}}{\partial K^m \partial T^p}\Psi_{1,\mathcal{M}}(K = S_t, T = t) \qquad \Longrightarrow \qquad \frac{\partial^{m+p}}{\partial K^m \partial T^p}\Psi_{2,\mathcal{M}}(K = S_t, T = t)$$

for a given set of (m, p)-differentials. Before tackling this subject let us recall that:

- we cannot transfer the IATM information analytically through the price;
- the implied parameter transition can be either functional or purely numerical.

The first restriction comes from the irregularity of the price function, in the whole Immediate domain ($T = t$) where the distribution is a Dirac, and at the IATM point, in particular where the function is not even C^1 w.r.t K. This is valid for any baseline and means that the chain-rule on the price, which (for instance) at the first order in strike reads as

$$\begin{aligned}\partial_K P_{\mathcal{M}}(K,T) &= \partial_1 B_i\left(K,T,\Psi_{i,\mathcal{M}}(K,T)\right)\\ &\quad + \partial_3 B_i\left(K,T,\Psi_{i,\mathcal{M}}(K,T)\right)\partial_K \Psi_{i,\mathcal{M}}(K,T)\end{aligned}$$

cannot be used by induction to match the respective differentials of Ψ_1 and Ψ_2. This again boils down to the implied volatility having a regularisation effect on the price, which makes it a good re-parametrisation tool.

The second comment refers to the two (non-exclusive) configurations of the baseline transfer. The first situation occurs when a functional relationship exists between $\Psi_1()$ and $\Psi_2()$, while the second corresponds to the general case and uses a numerical approach to provide differential *values* as opposed to generic expressions.

First case: the parameter-to-parameter functional
Certain pairs of baselines (central and new) provide us with the best-case scenario where we can *represent the smile associated to the central baseline model within the new baseline* using a function $\Psi_{2,1}(K,T,\psi_1)$. In other words, if a given value ψ_1 specifies the central model then

$$\text{Under } M_1 \quad P_1(K,T) = B_2\left(K,\Psi_{2,1}(K,T,\psi_1)\right).$$

We can then write the price in the complex model $\mathcal{M}$ as

$$P_{\mathcal{M}}(K,T) = B_1\left(K,T,\Psi_{1,\mathcal{M}}(K,T)\right) = B_2\left(K,\Psi_{2,1}(K,T,\Psi_{1,\mathcal{M}}(K,T))\right)$$

and the bijective property now ensures that

$$\Psi_{2,\mathcal{M}}(K,T) = \Psi_{2,1}(K,\Psi_{1,\mathcal{M}}(K,T)). \tag{3.49}$$

It is now possible to differentiate (3.49), to take that expression at the IATM point, and then to inject the generic formulae for the IATM differentials of $\mathcal{M}$'s *central* smile. Note that in this ideal case the computational price is mainly concentrated in the initial expansion. Note also that $\Psi_{2,1}$ can be either exact or approximate, the latter case simply degrading the output expressions.

A textbook example of an *exact* link is the one that exists between two displaced versions of the same model family. Starting, for instance, from a given local volatility class we can extend it as per

$$dS_t = f(t,S_t,\psi)dW_t \qquad \Longrightarrow \qquad dS_t = f(t,S_t+d,\psi)dW_t.$$

Of course this simplistic technique can be applied to more complex classes, such as local-stochastic volatility models. The fact is that by re-introducing the underlying into the argument we can write

$$B_{d_2}(S_t; K, T; \psi) = B_{d_1}(S_t + d_2 - d_1; K + d_2 - d_1, T; \psi),$$

hence
$$\Psi_{d_2}(S_t; K, T) = \Psi_{d_1}(S_t + d_2 - d_1; K + d_2 - d_1, T).$$

Although it is possible to extend further by applying an affine transform, and/or to apply the same modifications to the time variable, the interest of this observation is somewhat limited. Firstly, because any space displacement modifies the support of the marginals, and secondly since the closed-form is likely to be given for the whole displaced family at once. We will convene that overall, an *exact* correspondence is quite rarely found for the whole smile, leading us not to discuss this case any further. Of more general interest is the case of the Immediate smile of local volatility models. Indeed, we have result (2.62) [p. 67] from [16] that gives us the *Lognormal* implied volatility for any member of that class. Therefore if the master baseline is Normal or CEV, for instance, we can then move *all pure-strike differentials* to the Lognormal baseline.

Alternatively, in several classical cases where no exact link is known, a proxy analytical conversion between the two baseline parametrisations is nevertheless available, usually obtained *via* expansion techniques. Let us illustrate this interesting situation with the Normal/Lognormal as well as the CEV/Lognormal configurations:

- In the (very likely) case of a Lognormal (B) model expressed in a Normal (N) baseline we have from [5], through some expansions, that

$$\sigma_N(K) \approx \sigma_B \sqrt{S_t K} \frac{1 + \frac{1}{24}\log^2\left(\frac{S_t}{K}\right) + \frac{1}{1920}\log^4\left(\frac{S_t}{K}\right)}{1 + \frac{1}{24}\left[1 - \frac{1}{120}\log^2\left(\frac{S_t}{K}\right)\right]\sigma_B^2(T-t) + \frac{1}{5760}\sigma_B^4(T-t)^2}.$$

- In the situation of a CEV (C) model (with power coefficient β) expressed in a Lognormal baseline then [17] gives us similarly that

$$\sigma_B(K) \approx \sigma_C S_{av}^{\beta-1}\left[1 + \frac{(1-\beta)(2+\beta)}{24}\left[\frac{S_t - K}{S_{av}}\right]^2 + \frac{(1-\beta)^2}{24}\sigma_C^2(T-t)S_{av}^{2\beta-2}\right],$$

$$\text{with} \quad S_{av} \triangleq \frac{1}{2}(S_t + K). \tag{3.50}$$

To summarise this first case, our opinion is that although the functional link is not always available, when the opportunity presents itself it is worth investigating that route, even in the approximate case. The rationale being that it does provide *generic* IATM differential expressions, the usefulness of which does not stop at a given set of parameters or even at a specific complex (input) model. By contrast, the method that we present now is by nature a proxy, but presents the advantage of being systematically applicable.

Second case: the finite differences approach

This generic baseline transfer method establishes a relationship between the Ψ_1- and Ψ_2-IATM differentials which is both *approximate* and *numerical*. However, such a summary description deserves a number of comments:

- The fact that it delivers only the differentials is not an issue: after all, this is the necessary and sufficient information for the transfer to be achieved.
- The link being numerical, i.e. value to value as opposed to functional, we cannot obtain *generic* formulae for the IATM differentials in the new baseline. Instead, we will have to run through this module every time the complex model $\mathscr{M}$ changes, but also every time its parameters are altered. On the other hand, the module is generic: it only needs to know about M_1 and M_2 (and not $\mathscr{M}$) which can be input in an abstract manner.
- The accuracy of the computation is controlled by finite differences (FD) parameters, thus the highest attainable precision will mainly be limited by platform,[7] language[8] or implementation[9] specifics.

The method is based on a couple of features. The first one is the fact that, although the prices cannot be used as a media in the *Immediate* region, a very small time-to-maturity is usually sufficient to eliminate that issue (notwithstanding the numerical limitations mentioned above). The second feature is the inherent structure of the layers, as described in Sect. 3.1, which means that to increase the θ-order of any IATM cross-differential by one, we need either one or two more layers (see Fig. 3.1 [p. 122]).

The simplest way to introduce this numerical baseline transfer is to work through an example. Let us therefore take M_1 and M_2, respectively, as the Lognormal and the Normal baselines, and denote the associated implied volatilities by $\Sigma(K,T) = \Psi_1(K,T)$ and $\Gamma(K,T) = \Psi_1(K,T)$. For a given set of parameters of the complex model $\mathscr{M}$, we assume that the first, second and third layers have been computed for the central, Lognormal implied volatility Σ. Recall that the involved (exact) static IATM differentials are

$$\Sigma(\star) \qquad \Sigma'_K(\star) \qquad \Sigma''_{K^2}(\star) \qquad \Sigma'''_{K^3}(\star) \qquad \Sigma^{(4')}_{K^4}(\star)$$

$$\Sigma'_T(\star) \qquad \Sigma''_{KT}(\star) \qquad \Sigma'''_{K^2T}(\star) \qquad \Sigma''_{TT}(\star)$$

whose generic expressions are provided in Sect. 3.2. For the benefit of the FD scheme, we then start by sampling the (K,T) space locally, in the vicinity of the IATM point. For the sake of argument we select the K and T variables, but this choice is asymptotically irrelevant. Similarly we opt for a grid which is regular in each direction and controlled by a single scale parameter ε, i.e. $\delta_K = \varepsilon.\Delta_K$ and $\delta_T = \varepsilon.\Delta_T$. The scheme itself is chosen classically as central in strike, and of course forward in maturity. Using a Taylor expansion, we then compute the $\Sigma(K,T)$ values associated

[7] Typically the *machine epsilon* and more generally underflow issues.

[8] For instance, float *vs* double.

[9] Selecting a good cumulative Normal approximation is usually important.

Table 3.1 Pointwise expansion precision for $\Sigma_t(K,T)$

	$t+\delta_T$	$t+2\delta_T$
$S_t+\delta_K$	$o(\varepsilon^3)$	$o(\varepsilon^3)$
S_t	$o(\varepsilon^2)$	$o(\varepsilon^2)$
$S_t-\delta_K$	$o(\varepsilon^3)$	$o(\varepsilon^3)$

to the non-immediate part of that grid, and plot their respective precision order. With a slight abuse of notation we get Table 3.1.

We can then convert these volatilities into prices using the Black function, and then back into *Normal* implied volatilities $\Gamma_t(K,T)$ using a Bachelier inverter. Since each pricing formula uses the square root of its own volatility, the precision orders on the Normal $\Sigma_t(K,T)$ and Lognormal $\Gamma_t(K,T)$ grids are identical.

The last phase consists in computing the IATM differentials of the Γ map through the FD scheme. For demonstrative purposes we artificially break down this phase into two steps and keep using visual representations rather than writing the associated system of Maclaurin series.

We first reconstruct the Immediate values for the Normal implied volatility. Using finite differences on the rest of the grid, we get Table 3.2.

But the actual precision attained can be significantly improved. Indeed, we have the IATM level $\Gamma(\star)$ in generic form, so in particular we have its *exact* value here.

This is true here for the Normal baseline, but also for any regular local volatility model: comparing (3.19) and (3.20) we find that, after cast, the arbitrage argument will lead invariably to the IATM level expressed as $\Psi_{i,\mathscr{M}}(K=S_t, T=t)=\psi_t$.

Although we do not provide a formal proof here, it is reasonable to expect the same behaviour for even more elaborate baselines, outside of the local volatility class. Assuming then that we do have the IATM level of the implied parameter in the new baseline, it is natural to use that information in order to apply a further *correction* to the raw FD method. Using the FD scheme again we can now approximate the static IATM differentials of the first layer for the Normal implied volatility, with the respective precisions presented in Table 3.3.

Irrespective of this level adjustment, and due to the dependency of the method on the ladder structure, it is clear that the availability of further layers for the central baseline will increase the overall precision. This is indeed a good thing since, being able to dedicate all our efforts solely to the central baseline, we should be able to compute once and for all a high number of these, in generic form.

Table 3.2 Pointwise expansion precision for $\Gamma_t(K,T)$

	t	$t+\delta_T$	$t+2\delta_T$
$S_t+\delta_K$	$o(\varepsilon^3)$	$o(\varepsilon^3)$	$o(\varepsilon^3)$
S_t	$o(\varepsilon^2)$	$o(\varepsilon^2)$	$o(\varepsilon^2)$
$S_t-\delta_K$	$o(\varepsilon^3)$	$o(\varepsilon^3)$	$o(\varepsilon^3)$

Table 3.3 Overall expansion precision for $\Gamma_t(K,T)$

Γ	Γ'_K	Γ''_{KK}	Γ'_T
Exact	$o(\varepsilon^2)$	$o(\varepsilon)$	$o(\varepsilon)$

Applications of the baseline transfer

The systematic capacity to execute a baseline transfer offers a wealth of opportunities. The FD method in particular can be applied to any *pseudo-baseline* model whose pricing function is either *semi-closed* (usually in integral form) or even fully numerical (Monte-carlo, but more realistically FD or tree).Clearly we can also employ it with *real* baselines providing a closed-form formula, either exact or approximate, in the case where that function is deemed too complex for establishing a ZDC. Among all these prospects, in our view a few seem to be worthy of future research, and therefore we describe them briefly now.

Within the baseline models providing closed-form exact pricing, the CEV is an interesting case. Firstly, because it is a candidate for several of the baseline extension methods mentioned above:

- Through a proper ZDC (see (3.23)).
- By a simpler approximate ZDC (through [17] or [5]).
- Through an approximate baseline transfer (see (3.50)).
- By the FD approach.

Secondly, because it would *a priori* be a good baseline for the very popular CEV-SABR model (see Chap. 4). Even more so since the CEV versions using displacement and/or time-dependent parameters do fall into the same category.

We continue our investigation with baselines providing semi-closed-form exact pricing, usually *via* an integral or transform expression, among which the Heston model (2.6)–(2.7) [p. 34] stands out as a prime candidate. For instance, it could be considered as an appropriate baseline for the following complex model class

$$\begin{cases} dS_t = \sqrt{v_t} f(t, S_t) dW_t \\ dv_t = \kappa\left(\overline{v} - v_t\right) dt + \varepsilon \sqrt{v_t} dB_t \quad \text{with} \quad \langle dW_t, dB_t \rangle = \rho dt \end{cases}$$

as an alternative to a pure local volatility baseline. In the same family as Heston, the 3/2 model is defined in [4] with

$$\begin{cases} \dfrac{dS_t}{S_t} = \sqrt{v_t} dW_t \\ dv_t = \kappa\left(\overline{v} v_t - v_t^2\right) dt + \varepsilon v_t^{3/2} dB_t \quad \text{with} \quad \langle dW_t, dB_t \rangle = \rho(v_t) dt \end{cases}$$

where a semi-closed form solution is provided in the shape of a generalised transform, which certainly warrants further investigation.

But since we now have the FD alternative for the transfer, it may also be worth examining baselines that provide pricing through different, fast numerical schemes. For instance, with a CEV-SABR model whose perturbation α_t is independent ($\rho = 0$), the Call price is given by

$$C(t, S_t, K, T) = \mathbb{E}\left[CEV(t, S_t, K, T; \sigma, \beta) | \sigma = \sqrt{\frac{1}{T-t} \int_t^T \alpha_s^2 ds} \right]$$

which is also computed in semi-closed form, by numerical integration[10] of the CEV formula (3.23) with deterministic volatility, against the density of the functional $\int_t^T \alpha_s^2 ds$. Since the multiplicative perturbation α_t is Lognormal, very good approximations can be found for that marginal. In summary, in the perspective of approximating the smile generated by a full CEV-SABR ($\rho \neq 0$) this baseline should *a priori* give us better results than the pure CEV model, but at a higher computational cost.

Going even further, it follows by transitivity that we can relate *two complex models* through the smiles provided by their respective implied parameters, (provided they each offer such a bijection). Granted, the overall precision would suffer from the numerical schemes on both sides, but this is nevertheless an interesting perspective, as it allows us to compare both marginal *surfaces* in static and dynamic terms.

Overall, an important consideration when choosing first a baseline, and then an expansion variable, is the specific support of the marginal distribution. For instance, if the complex model has positive support, then any expansion of the implied *Normal* volatility will require that functional to converge to 0 when the strike does. This point must be kept in mind when dealing with local volatility coefficients, typically skew functions of stochastic volatility models, since they will usually determine that support.

On another tack, note that we have been focusing on transferring the static IATM differentials, but that the dynamic coefficients $\nabla^{(m,p)}\widetilde{\mathrm{v}}(\star)$ and $\nabla^{(m,p)}\overrightarrow{\widetilde{n}}(\star)$ can also be investigated. In particular, if we have the *generic* expression for $\widetilde{\Sigma}^{(m,p)}$ in the new baseline then we can apply Itô and obtain the desired coefficients.

Note finally that this baseline transfer discussion justifies in part the choice of Black's model as our initial baseline (see Remark 2.1, p. 41) as it shows that such a choice is not binding, and that the results obtained can be transferred.

3.4 Multi-dimensional Extensions, or the Limitations of Recovery

The aim of this section is to extend the previous results to the multi-dimensional case where *both* endogenous drivers $\overrightarrow{W}_t$ and $\overrightarrow{Z}_t$, along with the instantaneous volatilities σ_t, $\overrightarrow{\widetilde{\mathrm{v}}}(t, y, \theta)$ and $\overrightarrow{\widetilde{n}}(t, y, \theta)$, are all vectorial processes. In particular we show that in that configuration the recovery of the SInsV dynamics, associated to a given SImpV model, quickly gains in complexity.

First, let us comment on the modelling pertinence of this extension. As we have discussed before (see Sect. 2.1.3.3 [p. 40]), with only a scalar underlying S_t in our setup the sole modulus $\|\overrightarrow{\sigma_t}\|$ conveys all the relevant information, so that the multi-dimensional configuration seems futile. However, the situation is different when the underlying itself becomes multi-dimensional, which can correspond to a number of financial applications, from cross-currency to basket options *via* spread options.

---

[10] Obviously single-dimensional, typically using a Gaussian quadrature.

First, we will rapidly extend the expressions obtained in Chap. 2, namely the Zero drift Conditions, the IATM constraints of the SImpV model, the Recovery theorem and the First Layer theorem. We shall see that the structural differences are obvious, which is due to the fact that the implied volatility is a scalar which therefore cannot convey as easily all the information contained in vectorial processes. These differences will mainly affect our ability to solve the inverse problem, but we shall see that such issues are not necessarily unsurmountable.

Later we will apply these new results to a number of practical issues. We shall focus on the rich case of baskets (see Sect. 3.5) which we will also exploit in the interest rates environment of Chaps. 6 and 7.

The methodology employed and the results obtained being understandably similar to the scalar case, it seems reasonable to take a shortcut and simply retrace our steps through Sects. 2.1 (the framework), 2.2 (the ZDCs), 2.3 (the recovery) and 2.4 (the direct problem).

It is also justified to skip the proofs which are easily derived from the scalar case. We will, however, pay particular attention to stressing the changes in both the problem and in the results, and propose some interpretations when possible.

Another reason for such a rapid coverage is that this section can be regarded simply as a "primer" for Part II, where the multidimensional setup will again be extended to term structures, and treated in a comprehensive way. In other words, the results hereafter can be attained quickly by simply *degrading* those presented in Chap. 5.

3.4.1 Framework

Let us start by defining our new extended setup. The *chaos* dynamics of the generic SInsV model—down to depth two—are

$$\begin{cases} \dfrac{dS_t}{S_t} = \vec{\sigma}_t^{\perp} d\vec{W}_t, & (3.51) \\[2ex] d\vec{\sigma}_t = \vec{a}_{1,t}dt + \vec{\vec{a}}_{2,t}d\vec{W}_t + \vec{\vec{a}}_{3,t}d\vec{Z}_t, & (3.52) \\[2ex] d\vec{\vec{a}}_{2,t} = \vec{\vec{a}}_{21,t}dt + \vec{\vec{\vec{a}}}_{22,t}d\vec{W}_t + \vec{\vec{\vec{a}}}_{23,t}d\vec{Z}_t, & (3.53) \end{cases}$$

while the SImpV dynamics are now set up as

$$\begin{cases} \dfrac{dS_t}{S_t} = \vec{\sigma}_t^{\perp} d\vec{W}_t \quad \text{and with} \quad (\circ) \triangleq (t, y, \theta) \\[2ex] d\widetilde{\Sigma}(\circ) = \widetilde{b}(\circ)dt + \vec{\widetilde{\nu}}(\circ)^{\perp} d\vec{W}_t + \vec{\widetilde{n}}(\circ)^{\perp} d\vec{Z}_t, \end{cases} \tag{3.54}$$

where both drivers are still orthogonal and each with a unit correlation matrix. As in the scalar case we understate that the underlying is deflated by the numeraire and we have chosen the corresponding measure to express those dynamics.

3.4.2 Derivation of the Zero-Drift Conditions

Let us start with the main Zero-Drift Condition, analogous to (2.18).

Proposition 3.5 (Zero Drift Condition in the multidimensional case) *Let us consider the vectorial framework* (3.51)–(3.54).

Then we have the following Zero Drift Condition, valid in the whole domain:

$$\widetilde{\Sigma}^3\widetilde{b}(t,y,\theta) = \theta D(t,y,\theta) + E(t,y,\theta) + \frac{1}{\theta}F(t,y,\theta), \tag{3.55}$$

with, defining $(\circ) \triangleq (t,y,\theta)$,

$$\begin{cases} D(t,y,\theta) = \frac{1}{8}\widetilde{\Sigma}^4(\circ)\left[\|\vec{\mathrm{v}} - \widetilde{\Sigma}'_y\vec{\sigma}_t\|^2 + \|\vec{n}\|^2\right](\circ), \\ E(t,y,\theta) = \widetilde{\Sigma}^3(\circ)\left[\widetilde{\Sigma}'_\theta(\circ) - \frac{1}{2}\|\vec{\sigma}_t\|^2\widetilde{\Sigma}''_{yy}(\circ) + \vec{\sigma}_t^\perp\vec{\mathrm{v}}'_y(\circ) - \frac{1}{2}\vec{\sigma}_t^\perp\vec{\mathrm{v}}(\circ)\right], \\ F(t,y,\theta) = \frac{1}{2}\widetilde{\Sigma}^4(\circ) - \frac{1}{2}\|\vec{\sigma}_t\|^2\widetilde{\Sigma}^2(\circ) - y\widetilde{\Sigma}(\circ)\vec{\sigma}_t^\perp\left[\vec{\mathrm{v}}(\circ) - \widetilde{\Sigma}'_y(\circ)\vec{\sigma}_t\right] \\ \qquad -\frac{1}{2}y^2\left[\|\vec{\mathrm{v}} - \widetilde{\Sigma}'_y\vec{\sigma}_t\|^2 + \|\vec{n}\|^2\right](\circ). \end{cases}$$

Proof Following closely the scalar case, our first step is to express the dynamics of the absolute implied volatility Σ through Itô-Kunita. Indeed, we have

$$dy = \frac{1}{2}\|\vec{\sigma}_t\|^2dt - \vec{\sigma}_t^\perp d\vec{W}_t \qquad \text{and} \qquad \langle dy\rangle = \|\vec{\sigma}_t\|^2dt.$$

Therefore, evaluating all sliding quantities at the generic point $(\circ) \triangleq (t,y,\theta)$, the dynamics of the absolute $\Sigma(t,S_t,K,T)$ surface are

$$d\Sigma = \widetilde{b}dt + \vec{\mathrm{v}}^\perp d\vec{W}_t + \vec{n}^\perp d\vec{Z}_t + \widetilde{\Sigma}'_y dy + \widetilde{\Sigma}'_\theta d\theta + \frac{1}{2}\widetilde{\Sigma}''_{yy}\langle dy\rangle - \vec{\sigma}_t^\perp\vec{\mathrm{v}}'_y dt,$$

which simplifies into

$$d\Sigma(t,S_t,K,T) = \underbrace{\left[\vec{\mathrm{v}}(\circ) - \widetilde{\Sigma}'_y(\circ)\vec{\sigma}_t\right]}_{\vec{\mathrm{v}}(t,S_t,K,T)}{}^\perp d\vec{W}_t + \underbrace{\vec{n}(t,y,\theta)}_{\vec{n}(t,S_t,K,T)}{}^\perp d\vec{Z}_t$$

$$+ \underbrace{\left[\widetilde{b}(\circ) - \widetilde{\Sigma}'_\theta(\circ) + \frac{1}{2}\|\vec{\sigma}_t\|^2\left[\widetilde{\Sigma}'_y + \widetilde{\Sigma}''_{yy}\right](\circ) - \vec{\sigma}_t^\perp\vec{\mathrm{v}}'_y(\circ)\right]}_{b(t,S_t,K,T)} dt. \tag{3.56}$$

We can now compute the dynamics of C_t/N_t where C_t is the *absolute* Call and N_t the numeraire. We apply Itô to the normalised Black formula $C^{BS}\left(S_t, K, \sqrt{\theta}\,\Sigma\right)$, focusing on the finite variation terms:

$$\begin{aligned} dC^{BS} &= \overrightarrow{[\cdot]}^{\perp} d\vec{W}_t + \overrightarrow{[\cdot]}^{\perp} d\vec{Z}_t \\ &\quad + \left[\mathcal{V} \left[\theta^{\frac{1}{2}} b - \frac{1}{2} \theta^{\frac{-1}{2}} \Sigma \right] + \frac{1}{2} \Gamma S_t^2 \|\vec{\sigma}_t\|^2 + \frac{1}{2} \theta \Lambda \left[\|\vec{\mathrm{v}}\|^2 + \|\vec{n}\|^2 \right] \right. \\ &\quad \left. + \vartheta \theta^{\frac{1}{2}} S_t \vec{\sigma}_t^{\perp} \vec{\mathrm{v}} \right] dt. \end{aligned}$$

Invoking the martingale argument and substituting the Black Greeks (see Appendix C) we get

$$\begin{aligned} 0 = \mathcal{V} \left[-\frac{1}{2} \theta^{-\frac{1}{2}} \Sigma + \theta^{\frac{1}{2}} b \right] + \frac{1}{2} \theta \mathcal{V} \left[y^2 \Sigma^{-3} \theta^{-\frac{3}{2}} - \tfrac{1}{4}\, \Sigma \theta^{\frac{1}{2}} \right] \left[\|\vec{\mathrm{v}}\|^2 + \|\vec{n}\|^2 \right] \\ + \frac{1}{2} \mathcal{V} S_t^{-2} \Sigma^{-1} \theta^{-\frac{1}{2}} S_t^2 \|\vec{\sigma}_t\|^2 + \mathcal{V} S_t^{-1} \left[\frac{1}{2} + y \Sigma^{-2} \theta^{-1} \right] \theta^{\frac{1}{2}} S_t \vec{\sigma}_t^{\perp} \vec{\mathrm{v}}. \end{aligned}$$

Considering the strict positivity of $\mathcal{V}$ and of θ, we simplify that expression into

$$\begin{aligned} b = \frac{1}{2} \theta^{-1} \Sigma + \left[\frac{1}{8} \Sigma \theta - \frac{1}{2} y^2 \Sigma^{-3} \theta^{-1} \right] \left[\|\vec{\mathrm{v}}\|^2 + \|\vec{n}\|^2 \right] \\ - \frac{1}{2} \Sigma^{-1} \theta^{-1} \|\vec{\sigma}_t\|^2 - \left[\frac{1}{2} + y \Sigma^{-2} \theta^{-1} \right] \vec{\sigma}_t^{\perp} \vec{\mathrm{v}}. \end{aligned}$$

Using (3.56), we replace the absolute functionals by their sliding counterparts:

$$\begin{aligned} \widetilde{b} &= \widetilde{\Sigma}'_{\theta} - \frac{1}{2} \|\vec{\sigma}_t\|^2 \left[\widetilde{\Sigma}'_y + \widetilde{\Sigma}''_{yy} \right] + \vec{\sigma}_t^{\perp} \vec{\widetilde{\mathrm{v}}}'_y + \frac{1}{2} \theta^{-1} \widetilde{\Sigma} - \frac{1}{2} \widetilde{\Sigma}^{-1} \theta^{-1} \|\vec{\sigma}_t\|^2 \\ &\quad + \left[\frac{1}{8} \widetilde{\Sigma} \theta - \frac{1}{2} y^2 \widetilde{\Sigma}^{-3} \theta^{-1} \right] \left[\|\vec{\widetilde{\mathrm{v}}} - \widetilde{\Sigma}'_y \vec{\sigma}_t\|^2 + \|\vec{\widetilde{n}}\|^2 \right] \\ &\quad - \left[\frac{1}{2} + y \widetilde{\Sigma}^{-2} \theta^{-1} \right] \vec{\sigma}_t^{\perp} \left[\vec{\widetilde{\mathrm{v}}} - \widetilde{\Sigma}'_y \vec{\sigma}_t \right]. \end{aligned}$$

Simplifying and re-scaling by $\widetilde{\Sigma}^3(\circ)$ proves (3.55), concluding the proof. □

Let us now leave the general field (t, y, θ) to focus on the immediate domain $(t, y, 0)$. In order to establish the following results, we need to ensure the existence of limits when $\theta \searrow 0$ for several processes. This is where a regularity package equivalent to Assumption 2.2 [p. 48] needs to be invoked. Since it consists in a simple translation to a vectorial framework for $\widetilde{\mathrm{v}}'_{\theta}$ and $\widetilde{\mathrm{v}}''_{y\theta}$, we shall dispense with making it explicit. We can then move on to the Immediate Zero-Drift Condition, which sees the scalar result (2.28) becoming:

Corollary 3.1 (Immediate Zero Drift Conditions in the vectorial case) *As a consequence of the ZDC* (3.55), *the sliding SImpV model* (3.51)–(3.54) *is constrained by these two equivalent IZDCs:*

In the immediate domain $(\bullet) = (t, y, 0)$ *we have a.s.:*

- *The Primary IZDC:*

$$0 = \widetilde{\Sigma}^4(\bullet) - \|\vec{\sigma}_t\|^2 \widetilde{\Sigma}^2(\bullet) - y 2\widetilde{\Sigma}(\bullet)\vec{\sigma}_t^{\perp}\left[\vec{\mathrm{v}}(\bullet) - \widetilde{\Sigma}'_y(\bullet)\vec{\sigma}_t\right]$$
$$- y^2\left[\|\vec{n}(\bullet)\|^2 + \|\vec{\mathrm{v}}(\bullet) - \widetilde{\Sigma}'_y(\bullet)\vec{\sigma}_t\|^2\right]. \tag{3.57}$$

- *The Secondary IZDC:*

$$\widetilde{\Sigma}^3\widetilde{b}(t, y, 0) = E(t, y, 0) + F'_\theta(t, y, 0). \tag{3.58}$$

Proof Again we follow very closely the proof for the scalar case, since the dimensionality issue does not affect its principles. The (multi-dimensional) Immediate Regularity Assumption 2.2 applied to the ZDC (3.55) implies in particular that

$$\lim_{\theta \searrow 0} F(t, y, \theta) = 0,$$

which proves the Primary IZDC (3.57) and provides a lower bound for the convergence speed. Using a small-θ expansion on that same term F gives that

$$lim_{\theta \searrow 0}\frac{1}{\theta}F(t, y, \theta) = F'_\theta(t, y, 0).$$

Then injecting that result into the ZDC (3.55) and invoking Assumption 2.2 proves the Secondary IZDC (3.58) and concludes the proof. □

We observe that *all* the coefficients corresponding to the non-finite variation are now vectorial, so that any product involving a pair of these becomes a scalar product, while quadratic terms now come up as squared moduli. But apart from these points, the structure remains very similar to the single-dimensional case. Finally, we can bring forward the IATM Identity, where (2.36) becomes:

Corollary 3.2 (IATM Identity in the multi-dimensional case) *Evaluating* (3.57) *at* $y = 0$ *we have a.s. at the Immediate ATM point* $(t, 0, 0)$

$$\|\vec{\sigma}_t\| = \widetilde{\Sigma}(t, 0, 0). \tag{3.59}$$

This result supports Remark 2.2 [p. 41] regarding the modulus of $\vec{\sigma}_t$ and its positivity. It also highlights the structural loss of information, with regard to the individual components of the vectorial instantaneous stochastic volatility $\vec{\sigma}_t$, that comes from using an aggregated, scalar implied parameter such as a volatility, or even a price for that matter.

Let us illustrate this point in the simple context of a bi-dimensional volatility $\vec{\sigma_t}$. To that end we assume a pair of underlyings, each with its own scalar volatility, and which can be associated to generate a third scalar underlying:

$$S_t = f\left(S_{1,t}, S_{2,t}\right) \qquad \text{with} \qquad \begin{aligned} dS_{1,t}/S_{1,t} &= \overrightarrow{\sigma_{1,t}}^{\perp} d\vec{W}_t, \\ dS_{2,t}/S_{2,t} &= \overrightarrow{\sigma_{2,t}}^{\perp} d\vec{W}_t. \end{aligned}$$

This kind of functional setup covers, for instance, the cross-currency framework of [18]. It can also be generalised to higher dimensions in order to include baskets, which will be covered shortly. Considering the new underlying S_t, we assume an associated smile $\widetilde{\Sigma}$ and we are interested in its volatility $\vec{\sigma_t}$. The dimensionality imposes that, should we wish to recover the full information on $\vec{\sigma_t}$, we will require two independent sources. Using, in particular, a polar representation instead of a cartesian one, we can interpret the IATM Identity (3.59) as providing only the *modulus* of the instantaneous volatility. In other words, we are still missing its *angle*, which represents its allocation along each driver: intuitively, that information is conditioned by the instantaneous *correlation* between S_1 and S_2. At this stage, it also seems natural that the missing data should come from a further product, and the respective smiles of S_1 and S_2 are prime candidates.

3.4.3 Recovering the Instantaneous Volatility: The First Layer

First, let us establish the equivalent of Proposition 2.2.

Proposition 3.6 (IATM constraints on the SImpV model in the vectorial case) *A sliding SImpV model defined by* (3.51)–(3.54) *has its dynamic coefficients constrained to satisfy the following, at the IATM point* $(t, 0, 0)$*:*

$$\vec{\sigma}_t^{\perp} \vec{\mathrm{v}}(\star) = 2\widetilde{\Sigma}^2 \widetilde{\Sigma}'_y(\star), \tag{3.60}$$

$$\vec{\sigma}_t^{\perp} \vec{\mathrm{v}}'_y(\star) = \frac{3}{2}\widetilde{\Sigma}^2 \widetilde{\Sigma}''_{yy}(\star) + 3\widetilde{\Sigma}\, \widetilde{\Sigma}'^{\,2}_y(\star) - \frac{\|\vec{\mathrm{v}}(\star)\|^2 + \|\vec{n}(\star)\|^2}{2\widetilde{\Sigma}(\star)}, \tag{3.61}$$

$$\widetilde{b}(\star) = 2\widetilde{\Sigma}'_\theta(\star) - \frac{1}{2}\widetilde{\Sigma}^2 \widetilde{\Sigma}''_{yy}(\star) - \frac{1}{2}\vec{\sigma}_t^{\perp} \vec{\mathrm{v}}(\star) + \vec{\sigma}_t^{\perp} \vec{\mathrm{v}}'_y(\star). \tag{3.62}$$

Proof Differentiating the new IZDC (3.57) once w.r.t. y, then evaluating it at $(t, 0, 0)$ and invoking the IATM identity (3.59) leads to (3.60). Similarly, differentiating the new IZDC twice w.r.t. y, then evaluating it at $(t, 0, 0)$ and using the IATM identity leads to (3.61). Applying a small-θ-expansion to the ZDC (3.55) leads to the same symbolic equation as (2.33), where now

$$F'_\theta(\circ) = 2\widetilde{\Sigma}^3\,\widetilde{\Sigma}'_\theta(\circ) - \|\vec{\sigma}_t\|^2\,\widetilde{\Sigma}\,\widetilde{\Sigma}'_\theta(\circ) - y\,[\ldots] - \frac{1}{2}y^2\,[\ldots].$$

Evaluating that expression at $(t, 0, 0)$ and invoking the IATM identity (3.59) then yields (3.62). □

Comparing Proposition 3.6 [p. 176] with its scalar counterpart, which is Proposition 2.2 [p. 54], it appears that although the drift expressions (2.39) and (3.62) are very similar, by contrast in (3.60) and in (3.61) no further simplification is possible. In other words, we can no longer isolate, respectively, $\vec{\mathrm{v}}\,(t, 0, 0)$ and $\vec{\mathrm{v}}'_y(\star)$, as was the case with (2.37) and (2.38). Indeed, we can rewrite (3.60), for instance, as

$$\vec{\sigma}_t^{\perp}\left[\vec{\mathrm{v}}\,(\star)\right] = \vec{\sigma}_t^{\perp}\left[2\widetilde{\Sigma}'_y(\star)\,\vec{\sigma}_t\right],$$

but clearly that equation does not ensure that both the l.h.s. and r.h.s. brackets are *a.s.* equal. In other words, due to the aggregation of information brought by the scalar implied volatility, the only constraints applied to $\vec{\mathrm{v}}\,(\star)$ and $\vec{\mathrm{v}}'_y(\star)$ concern their moduli and their *angle* w.r.t. the instantaneous volatility $\vec{\sigma}_t$. In the prospect of recovery, however, it is possible to turn this argument around and extract more information on $\vec{\sigma}_t$, as illustrated in the following remark.

Remark 3.3 Let us assume that $\vec{W}_t$ is of dimension two, which means that overall our model is at least tri-dimensional. We omit arguments t and $(t, 0, 0)$ and use cartesian coordinates for the following notations:

$$\vec{\sigma}_t \triangleq \begin{bmatrix}\sigma_1\\ \sigma_2\end{bmatrix} \qquad \vec{\mathrm{v}}\,(\star) \triangleq \begin{bmatrix}\mathrm{v}_1\\ \mathrm{v}_2\end{bmatrix} \qquad \vec{\mathrm{v}}'_y(\star) \triangleq \begin{bmatrix}\mathrm{v}'_1\\ \mathrm{v}'_2\end{bmatrix} \qquad \vec{\vec{M}} \triangleq \begin{bmatrix}\mathrm{v}_1 & \mathrm{v}_2\\ \mathrm{v}'_1 & \mathrm{v}'_2\end{bmatrix}.$$

Then combining (3.60) with (3.61) yields the system:

$$\vec{\vec{M}}\begin{bmatrix}\sigma_1\\ \sigma_2\end{bmatrix} = \begin{bmatrix}I_1\\ I_2\end{bmatrix},$$

where I_1 and I_2 are expressions provided purely by the IATM SImpV model:

$$\begin{cases} I_1 = 2\widetilde{\Sigma}^2\,\widetilde{\Sigma}'_y(\star),\\[4pt] I_2 = \dfrac{3}{2}\,\widetilde{\Sigma}^2\,\widetilde{\Sigma}''_{yy}(\star) + 3\widetilde{\Sigma}\,\widetilde{\Sigma}'^{\,2}_y(\star) - \dfrac{1}{2}\widetilde{\Sigma}^{-1}\left[\|\vec{\mathrm{v}}\|^2 + \|\vec{\vec{n}}\|^2\right](\star). \end{cases}$$

Therefore, assuming $\overset{\Rightarrow}{M}$ is not singular, in this simple case it is possible to entirely recover $\vec{\sigma}_t$. Note that the IATM identity (3.59) must also be verified, amounting to

$$\sigma_1^2 + \sigma_2^2 = \widetilde{\Sigma}^2(\star).$$

Note that this equation, combined exclusively with *either* (3.60) *or* (3.61), is *a priori* not sufficient to reconstruct the vectorial volatility by solving for σ_1 and σ_2.

As for higher endogenous dimensions, it seems rather natural to anticipate that each successive differentiation of the Primary IZDC (3.57) should bring another equation of the type of (3.60) and (3.61), to be added to the existing system. Indeed, $\nabla^{(n)}$ IZDC taken at the IATM point should involve $\vec{\sigma}_t^{\perp} \vec{\mathrm{v}}\, y^{n(n)}(\star)$, which suggests that an n-dimensional instantaneous volatility $\vec{\sigma}_t$ could be recovered. We leave for further research, however, the corresponding formal study, which should establish the generic shape of the differentiated IZDC and of the equation system, and if possible some partial or complete characterisation of the latter's solvability.

Note, finally, that this remark is extended to a more powerful version when working in the interest rates framework: see Remark 5.5 [p. 306].

We will see that, notwithstanding the technique presented above in Remark 3.3, the dimensionality feature *does* alter the recovery, into which we now move.

Theorem 3.3 (Recovering the modulus' dynamics in the vectorial case) *A given sliding SImpV model, as defined by* (3.51)–(3.54), *is associated to a SInsV model, defined by* (3.51)–(3.53) *and for which we have the following modulus dynamics:*

$$\begin{aligned} d\|\vec{\sigma}_t\| = \; & \vec{\mathrm{v}}(\star)^{\perp} d\vec{W}_t + \vec{n}(\star)^{\perp} d\vec{Z}_t \\ & + \Bigg[2\widetilde{\Sigma}'_\theta(\star) + \|\vec{\sigma}_t\|^2 \left[\widetilde{\Sigma}''_{yy} - \widetilde{\Sigma}'_y \right](\star) + 3\|\vec{\sigma}_t\|\; \widetilde{\Sigma}'^{\,2}_y(\star) \\ & \quad - \frac{\|\vec{\mathrm{v}}(\star)\|^2 + \|\vec{n}(\star)\|^2}{2\|\vec{\sigma}_t\|} \Bigg] dt. \end{aligned} \tag{3.63}$$

Proof Assuming sufficient regularity to do so, we first take the dynamics on both sides of (3.59). Then replacing the last two terms of (3.62), respectively, by (3.60) and (3.61) yields after simplification:

$$\widetilde{b}(\star) = 2\widetilde{\Sigma}'_\theta(\star) + \widetilde{\Sigma}^2 \left[\widetilde{\Sigma}''_{yy} - \widetilde{\Sigma}'_y \right](\star) + 3\widetilde{\Sigma}\; \widetilde{\Sigma}'^{\,2}_y(\star) - \frac{\|\vec{\mathrm{v}}\|^2 + \|\vec{n}\|^2}{2\widetilde{\Sigma}}(\star). \tag{3.64}$$

Injecting IATM identity (3.59) gives the drift as in (3.63) and concludes the proof. □

However, the actual dynamics that we seek are not those of the modulus $\|\vec{\sigma}_t\|$ but those describing the vectorial instantaneous volatility $\vec{\sigma}_t$, which are inherently richer. This explains why the recovery is only partial, which we formalise with the following corollary.

Corollary 3.3 (Partial recovery of the first layer in the vectorial case) *As a consequence of Theorem* 3.3, *the dynamics of the SInsV model* (3.51) *can be partially recovered from the sliding SImpV model with*

$$\vec{\vec{a}}_{2,t}\vec{\sigma}_t = \widetilde{\Sigma}(\star)\,\vec{\mathrm{v}}(\star), \tag{3.65}$$

$$\vec{\vec{a}}_{3,t}\vec{\sigma}_t = \widetilde{\Sigma}(\star)\,\vec{n}(\star), \tag{3.66}$$

$$\begin{aligned}\vec{\sigma}_t^{\perp}\vec{a}_{1,t} = &-\frac{1}{2}\left[\|\vec{\vec{a}}_{2,t}\|^2 + \|\vec{\vec{a}}_{3,t}\|^2\right]\\ &+ 2\widetilde{\Sigma}\,\widetilde{\Sigma}'_{\theta}(\star) + \widetilde{\Sigma}^3\left[\widetilde{\Sigma}''_{yy}(\star) - \widetilde{\Sigma}'_y(\star)\right] + 3\widetilde{\Sigma}^2\,\widetilde{\Sigma}'^{\,2}_y(\star),\end{aligned} \tag{3.67}$$

$$\begin{aligned}\overrightarrow{\left[a_{22\,\mathbf{ij}k}\,\sigma_{\mathbf{i}}\,\sigma_{\mathbf{j}}\right]_t} = &-\vec{\vec{a}}^{\,\perp}_{2,t}\left[\vec{\vec{a}}_{2,t} + \vec{\vec{a}}^{\,\perp}_{2,t}\right]\vec{\sigma}_t\\ &+ 2\left[3\widetilde{\Sigma}^2\widetilde{\Sigma}'_y\,\vec{\mathrm{v}}(\star) + \widetilde{\Sigma}^3\,\vec{\mathrm{v}}'_y(\star)\right].\end{aligned} \tag{3.68}$$

Proof For ease of computation, let us first square the IATM identity (3.59), before computing its dynamics. Developing the right-hand side we obtain

$$d\widetilde{\Sigma}^2(\star) = \left[2\widetilde{\Sigma}\widetilde{b}(\star) + \|\vec{\mathrm{v}}\|^2 + \|\vec{n}\|^2\right](\star)dt + 2\widetilde{\Sigma}\,\vec{\mathrm{v}}(\star)^{\perp}d\vec{W}_t + 2\widetilde{\Sigma}\,\vec{n}(\star)^{\perp}d\vec{Z}_t.$$

In the drift bracket above, replacing the term $\widetilde{b}(t,0,0)$ by its expression (3.64) gives

$$\begin{aligned}d\widetilde{\Sigma}^2(\star) = &\left[4\widetilde{\Sigma}\,\widetilde{\Sigma}'_{\theta}(\star) + 2\widetilde{\Sigma}^3(\star)\left[\widetilde{\Sigma}''_{yy}(\star) - \widetilde{\Sigma}'_y(\star)\right] + 6\widetilde{\Sigma}^2\,\widetilde{\Sigma}'^{\,2}_y(\star)\right]dt\\ &+ 2\widetilde{\Sigma}\,\vec{\mathrm{v}}(\star)^{\perp}d\vec{W}_t + 2\widetilde{\Sigma}\,\vec{n}(\star)^{\perp}d\vec{Z}_t.\end{aligned} \tag{3.69}$$

In parallel, we can also compute the dynamics of $\|\vec{\sigma}_t\|^2$ directly from the SInsV specification. Using (D.14) [p. 442] and (D.13) [p. 441] we get

$$\begin{aligned}d\|\vec{\sigma}_t\|^2 = &\left[2\vec{\sigma}_t^{\perp}\vec{a}_1 + \|\vec{\vec{a}}_{2,t}\|^2 + \|\vec{\vec{a}}_{3,t}\|^2\right]dt\\ &+ 2\vec{\sigma}_t^{\perp}\vec{\vec{a}}_{2,t}d\vec{W}_t + 2\vec{\sigma}_t^{\perp}\vec{\vec{a}}_{3,t}d\vec{Z}_t.\end{aligned} \tag{3.70}$$

From the uniqueness of Itô's decomposition we can identify terms between (3.69) and (3.70). We obtain, respectively, (3.65) and (3.66) for the non-finite variation terms, while from the drift term comes

$$
\begin{aligned}
&2\vec{\sigma}_t^{\perp}\vec{a}_{1,t} + \|\overset{\Rightarrow}{a}_{2,t}\|^2 + \|\vec{a}_{3,t}\|^2 \\
&\quad = 4\widetilde{\Sigma}\,\widetilde{\Sigma}'_{\theta}(\star) + 2\widetilde{\Sigma}^3\left[\widetilde{\Sigma}''_{yy} - \widetilde{\Sigma}'_{y}\right](\star) + 6\widetilde{\Sigma}^2\,\widetilde{\Sigma}'^{\,2}_{y}(\star)
\end{aligned}
\tag{3.71}
$$

which after simplification yields (3.67). We now compute the right product of (3.65) by $\vec{\sigma}_t$ and substitute (3.60) to obtain

$$
\vec{\sigma}_t^{\perp}\overset{\Rightarrow}{a}_{2,t}\vec{\sigma}_t = \widetilde{\Sigma}(\star)\,\vec{\mathrm{v}}\,(\star)^{\perp}\vec{\sigma}_t = 2\widetilde{\Sigma}^3\,\widetilde{\Sigma}'_{y}(\star). \tag{3.72}
$$

We derive the dynamics either side of (3.72), focusing on endogenous components.

- On the r.h.s. we get

$$
d\left[2\widetilde{\Sigma}^3\,\widetilde{\Sigma}'_{y}(\star)\right] = 2\left[3\widetilde{\Sigma}^2\,\widetilde{\Sigma}'_{y}\,\vec{\mathrm{v}} + \widetilde{\Sigma}^3\,\vec{\mathrm{v}}'_{y}\right](\star)^{\perp}d\vec{W}_t + [\cdots]\,dt + \overrightarrow{[\cdots]}^{\perp}d\vec{Z}_t.
$$

- On the l.h.s., using (D.17) we obtain

$$
\begin{aligned}
d\left[\vec{\sigma}_t^{\perp}\overset{\Rightarrow}{a}_{2,t}\vec{\sigma}_t\right] = {} & \vec{\sigma}_t^{\perp}\overset{\Rightarrow}{a}{}_{2,t}^{\perp}\overset{\Rightarrow}{a}_{2,t}d\vec{W}_t + \vec{\sigma}_t^{\perp}\left[\overset{\Rrightarrow}{a}_{22,t}d\vec{W}_t\vec{\sigma}_t + \overset{\Rightarrow}{a}_{2,t}\overset{\Rightarrow}{a}_{2,t}d\vec{W}_t\right] \\
& + \overrightarrow{[\cdots]}d\vec{Z}_t + [\cdots]\,dt,
\end{aligned}
$$

  so that, identifying both sides and using modified Einstein notation we get

$$
\vec{\sigma}_t^{\perp}\left[\overset{\Rightarrow}{a}{}_{2,t}^{\perp} + \overset{\Rightarrow}{a}_{2,t}\right]\overset{\Rightarrow}{a}_{2,t} + \overrightarrow{\left[a_{22\,\mathbf{ij}k}\,\sigma_{\mathbf{i}}\,\sigma_{\mathbf{j}}\right]}_t^{\perp} = 2\left[3\widetilde{\Sigma}^2\,\widetilde{\Sigma}'_{y}\,\vec{\mathrm{v}} + \widetilde{\Sigma}^3\,\vec{\mathrm{v}}'_{y}\right](\star)^{\perp},
$$

  which after simplification and transposition gives (3.68) and concludes the proof. □

In view of the previous Remark 3.3, one might wonder whether the same technique could be applied in order to recover $\overset{\Rightarrow}{a}_{2,t}$, for instance. Without a convincing case to present, though, we prefer to leave this argument for further research. It is interesting to note, however, that for these recovery issues, the endogenous direction should be expected to appear first, as it is structurally privileged.

3.4.4 Generating the Implied Volatility: The First Layer

Moving on to the direct problem, it becomes interesting to exploit the *angle vs modulus* arguments developed above, and to introduce the following scaled coefficients.

Definition 3.7 The normalised instantaneous volatility is represented by

$$\vec{u}_t \triangleq \frac{\vec{\sigma}_t}{\sigma}, \qquad \text{where} \qquad \sigma \triangleq \|\vec{\sigma}_t\|.$$

Similarly, we can scale the relevant tensorial coefficients invoked in the dynamics of $\vec{\sigma}_t$, which is consistent since the convention used to write those has been chosen as Normal:

$$\vec{c}_{1,t} \triangleq \frac{\vec{a}_{1,t}}{\|\vec{\sigma}_t\|} \qquad \vec{\vec{c}}_{2,t} \triangleq \frac{\vec{\vec{a}}_{2,t}}{\|\vec{\sigma}_t\|} \qquad \vec{\vec{c}}_{3,t} \triangleq \frac{\vec{\vec{a}}_{3,t}}{\|\vec{\sigma}_t\|} \qquad \vec{\vec{\vec{c}}}_{22,t} \triangleq \frac{\vec{\vec{\vec{a}}}_{22,t}}{\|\vec{\sigma}_t\|}.$$

Theorem 3.4 (Generation of the first layer in the multi-dimensional case) *Under the SInsV model defined by* (3.51)–(3.53), *the IATM differentials of the sliding SImpV model within the* $\widetilde{\Sigma}$-(2, 0) *group (the first layer) can be expressed as follows:*

Local differentials of the shape process:

$$\widetilde{\Sigma}'_y(\star) = \frac{1}{2}\vec{u}_t^{\perp}\vec{\vec{c}}_{2,t}\vec{u}_t, \tag{3.73}$$

$$\begin{aligned}\widetilde{\Sigma}''_{yy}(\star) = \sigma^{-1}\Bigg[&-\frac{3}{2}\left[\vec{u}^{\perp}\vec{\vec{c}}_2\vec{u}\right]^2 + \frac{1}{3}\vec{u}^{\perp}\left[\vec{\vec{c}}_2^{\perp}\vec{\vec{c}}_2 + \vec{\vec{c}}_2\vec{\vec{c}}_2 + \vec{\vec{c}}_2\vec{\vec{c}}_2^{\perp}\right]\vec{u} \\ &+ \frac{1}{3}\vec{u}^{\perp}\vec{\vec{c}}_3\vec{\vec{c}}_3^{\perp}\vec{u} + \frac{1}{3}\left[u_{\mathbf{i}}u_{\mathbf{j}}u_{\mathbf{k}}\, c_{22\,\mathbf{ijk}}\right]\Bigg],\end{aligned} \tag{3.74}$$

$$\widetilde{\Sigma}'_\theta(\star) = \sigma^2\left[\frac{1}{4}\vec{u}^{\perp}\vec{\vec{c}}_2\vec{u}\right] + \sigma A_t, \tag{3.75}$$

with

$$\begin{aligned}A_t \triangleq\ &\frac{1}{2}\vec{u}^{\perp}\vec{c}_1 + \frac{1}{4}\|\vec{\vec{c}}_2\|^2 + \frac{1}{4}\|\vec{\vec{c}}_3\|^2 + \frac{3}{8}\left[\vec{u}^{\perp}\vec{\vec{c}}_2\vec{u}\right]^2 - \frac{1}{6}\vec{u}^{\perp}\vec{\vec{c}}_3\vec{\vec{c}}_3^{\perp}\vec{u} \\ &- \frac{1}{6}\vec{u}^{\perp}\left[\vec{\vec{c}}_2^{\perp}\vec{\vec{c}}_2 + \vec{\vec{c}}_2\vec{\vec{c}}_2 + \vec{\vec{c}}_2\vec{\vec{c}}_2^{\perp}\right]\vec{u} - \frac{1}{6}\left[u_{\mathbf{i}}u_{\mathbf{j}}u_{\mathbf{k}}\, c_{22\,\mathbf{ijk}}\right].\end{aligned}$$

Local differentials of the dynamics processes:

$$\vec{\mathrm{v}}(\star) = \vec{\vec{a}}_2^{\perp}\vec{u}, \tag{3.76}$$

$$\vec{n}(\star) = \vec{\vec{a}}_3^{\perp}\vec{u}, \tag{3.77}$$

$$\overrightarrow{\mathrm{v}}'_y(\star) = -\frac{3}{2}\left[\vec{u}^{\perp}\vec{\vec{c}}_2\vec{u}\right]\vec{\vec{c}}_2^{\perp}\vec{u} + \frac{1}{2}\vec{\vec{c}}_2^{\perp}\left[\vec{\vec{c}}_2 + \vec{\vec{c}}_2^{\perp}\right]\vec{u} + \frac{1}{2}\overrightarrow{\left[u_{\mathbf{i}}\, c_{22\,\mathbf{ij}k}\, u_{\mathbf{j}}\right]}. \quad (3.78)$$

Before going into the proof, let us comment on these results. First, for the benefit of readers interested in the tensorial aspects of the problem, in particular the numerical and (model) versatility concerns, we provide the following re-expressions of the terms invoked by Theorem 3.4:

$$\left[\vec{u}^{\perp}\vec{\vec{c}}_2\vec{u}\right]\vec{\vec{c}}_2^{\perp}\vec{u} = \overrightarrow{\left[u_{\mathbf{i}}u_{\mathbf{j}}u_{\mathbf{k}}\, c_{2\,\mathbf{ij}}\, c_{2\,\mathbf{k}l}\right]} \qquad \vec{u}^{\perp}\vec{\vec{c}}_3\,\vec{\vec{c}}_3^{\perp}\vec{u} = \left[u_{\mathbf{i}}u_{\mathbf{j}}\, c_{3\,\mathbf{ik}}\, c_{3\,\mathbf{jk}}\right]$$

$$\vec{\vec{c}}_2^{\perp}\left[\vec{\vec{c}}_2 + \vec{\vec{c}}_2^{\perp}\right]\vec{u} = \overrightarrow{\left[u_{\mathbf{i}}\left(c_{2\,\mathbf{ij}} + c_{2\,\mathbf{ji}}\right) c_{2\,\mathbf{j}k}\right]} \qquad \vec{u}^{\perp}\vec{\vec{c}}_2\vec{u} = \left[u_{\mathbf{i}}u_{\mathbf{j}}\, c_{2\,\mathbf{ij}}\right].$$

Furthermore, we have trivially that $\|\vec{\vec{c}}_*\|^2 = \left[c_{*\,\mathbf{ij}}^{\,2}\right]$ and also we note the most complex term, that we shall revisit shortly:

$$\vec{u}^{\perp}\left[\vec{\vec{c}}_2^{\perp}\vec{\vec{c}}_2 + \vec{\vec{c}}_2\vec{\vec{c}}_2 + \vec{\vec{c}}_2\vec{\vec{c}}_2^{\perp}\right]\vec{u} = \left[u_{\mathbf{i}}u_{\mathbf{j}}\left(c_{2\,\mathbf{ki}}\, c_{2\,\mathbf{kj}} + c_{2\,\mathbf{ik}}\, c_{2\,\mathbf{kj}} + c_{2\,\mathbf{ik}}\, c_{2\,\mathbf{jk}}\right)\right].$$

The natural next move is to check that those expressions fall back onto the scalar ones when the dimension collapses, which it does. We can then ponder on homogeneity aspects, where the use of normalised quantities becomes useful. It shows that the three dynamic coefficients are *dimensionless*, while the IATM skew and curvature show consistency and suggest the existence of a pattern for pure-y differentials.[11] On the other hand, the IATM slope $\widetilde{\Sigma}'_\theta(\star)$ exhibits two distinct units: beyond the inherent cohabitation of drift terms and squared non-finite-variation terms, which is due to Itô, we see this feature as another hint that a different parametrisation might provide some structural information.

We now turn to symmetry issues, in which respect any distinct treatment of the endogenous matrix $\vec{\vec{a}}_{2,t}$ and of its transpose $\vec{\vec{a}}_{2,t}^{\perp}$ seems surprising. This occurs for only one term, the most involved expression mentioned above, but which can be symmetrised by using (D.16):

$$\vec{u}^{\perp}\left[\vec{\vec{c}}_2^{\perp}\vec{\vec{c}}_2 + \vec{\vec{c}}_2\vec{\vec{c}}_2 + \vec{\vec{c}}_2\vec{\vec{c}}_2^{\perp}\right]\vec{u}$$
$$= \vec{u}^{\perp}\left[\frac{1}{2}\vec{\vec{c}}_2\vec{\vec{c}}_2 + \frac{1}{2}\vec{\vec{c}}_2^{\perp}\vec{\vec{c}}_2^{\perp} + \vec{\vec{c}}_2\vec{\vec{c}}_2^{\perp} + \vec{\vec{c}}_2^{\perp}\vec{\vec{c}}_2\right]\vec{u}.$$

[11] More precisely, that the unit of $\widetilde{\Sigma}^{(n)}_{y^n}(\star)$ should be $\|\vec{\sigma}_t\|^{1-n}$. Certainly in the bi-dimensional setting, as was shown in Sect. 3.2, this property is verified for $\widetilde{\Sigma}'_y(\star)$, $\widetilde{\Sigma}''_{yy}(\star)$, $\widetilde{\Sigma}'''_{yyy}(\star)$ and $\widetilde{\Sigma}^{(4')}_{y^4}(\star)$.

In our view, these remarks confirm that within the ACE methodology, whether scalar or multi-dimensional, symmetries and induction structures are very common. For the sake of interpretation as well as to simplify the computations, these features should be identified and exploited as soon as possible in the process, which would probably require the introduction of dedicated notations.

Proof of Theorem 3.4.
The first step is to square the IATM Identity (3.59) and then to compute the dynamics of both sides, as was done for the recovery:

- On one hand, the identification of the non-finite variation terms between (3.69) and (3.70) combined with the normalised notations of Definition 3.7 immediately provides the coefficients through (3.76) and (3.77).
- On the other hand, identifying the drifts provides (3.71) which will be used later to obtain the IATM slope.

The second step is to invoke the IATM arbitrage constraints of the SImpV model. First we combine (3.60) with (3.76) and then normalise all coefficients, which provides the IATM skew through (3.73). Then we observe that (3.61) gives the IATM curvature $\widetilde{\Sigma}'_y(\star)$ as an explicit function of $\overrightarrow{\mathrm{v}}'_y(\star)$ while (3.62) gives the slope as an explicit function of $\widetilde{\Sigma}'_y(\star)$. In consequence, we embark upon expressing $\overrightarrow{\mathrm{v}}'_y(\star)$.

To that end, the third step consists in computing the dynamics of the skew expression (3.73). For simplicity we start with the original (scaled) coefficients:

$$d\left[2\widetilde{\Sigma}^3\,\widetilde{\Sigma}'_y(\star)\right] = d\left[\vec{\sigma}_t^{\perp}\,\overset{\Rightarrow}{a}_{2,t}\,\vec{\sigma}_t\right].$$

Using (D.17) [p. 442], omitting non-endogenous terms and arguments gives

$$2\left[3\widetilde{\Sigma}^2\widetilde{\Sigma}'_y\,\overrightarrow{\mathrm{v}} + \widetilde{\Sigma}^3\,\overrightarrow{\mathrm{v}}'_y\right]^{\perp} d\overrightarrow{W}_t = \vec{\sigma}^{\perp}\overset{\Rightarrow}{a}{}_2^{\perp}\left[d\vec{\sigma}\right] + \vec{\sigma}^{\perp}\left[\left[d\overset{\Rightarrow}{a}_2\right]\vec{\sigma} + \overset{\Rightarrow}{a}_2\left[d\vec{\sigma}\right]\right] + \cdots$$

Developing that expression, we note that according to (D.9) [p. 440] we have

$$\left[d\overset{\Rightarrow}{a}_2\right]\vec{\sigma} = \left[\overset{\Rrightarrow}{a}_{22}\,d\overrightarrow{W} + \cdots\right]\vec{\sigma} = \overrightarrow{\overrightarrow{\left[a_{22\,\mathbf{i}\mathbf{j}k}\sigma_{\mathbf{j}}\right]}}\,d\overrightarrow{W} + \cdots$$

The uniqueness of Itô's decomposition allows us to identify the endogenous terms:

$$2\left[3\widetilde{\Sigma}^2\widetilde{\Sigma}'_y\,\overrightarrow{\mathrm{v}} + \widetilde{\Sigma}^3\,\overrightarrow{\mathrm{v}}'_y\right]^{\perp} = \vec{\sigma}^{\perp}\left[\overset{\Rightarrow}{a}{}_2^{\perp} + \overset{\Rightarrow}{a}_2\right]\overset{\Rightarrow}{a}_2 + \overrightarrow{\left[\sigma_{\mathbf{i}}\,a_{22\,\mathbf{ij}k}\,\sigma_{\mathbf{j}}\right]}^{\perp},$$

which, after invoking (3.76) and (3.73), yields

$$\widetilde{\Sigma}^3 \vec{\vec{\mathrm{v}}}'_y = -\frac{3}{2}\widetilde{\Sigma}^2 \left[\vec{u}^{\perp}\vec{\vec{c}}_2\vec{u}\right]\vec{\vec{a}}_2^{\perp}\vec{u} + \frac{1}{2}\vec{\vec{a}}_2^{\perp}\left[\vec{\vec{a}}_2 + \vec{\vec{a}}_2^{\perp}\right]\vec{\sigma}$$
$$+\frac{1}{2}\overrightarrow{\left[\sigma_{\mathbf{i}}\, a_{22\,\mathbf{ij}k}\, \sigma_{\mathbf{j}}\right]}.$$

Eventually the normalisation leads us to

$$\vec{\vec{\mathrm{v}}}'_y = -\frac{3}{2}\left[\vec{u}^{\perp}\vec{\vec{c}}_2\vec{u}\right]\vec{\vec{c}}_2^{\perp}\vec{u} + \frac{1}{2}\vec{\vec{c}}_2^{\perp}\left[\vec{\vec{c}}_2 + \vec{\vec{c}}_2^{\perp}\right]\vec{u} + \frac{1}{2}\overrightarrow{\left[u_{\mathbf{i}}\, c_{22\,\mathbf{ij}k}\, u_{\mathbf{j}}\right]}$$
$$= -\frac{3}{2}\overrightarrow{\left[u_{\mathbf{i}}\, c_{2\,\mathbf{ij}}\, u_{\mathbf{j}}\, c_{2\,\mathbf{i}j}\, u_{\mathbf{i}}\right]} + \frac{1}{2}\overrightarrow{\left[c_{2\,\mathbf{j}i}\left(c_{2\,\mathbf{jk}} + c_{2\,\mathbf{kj}}\right) u_{\mathbf{k}}\right]} + \frac{1}{2}\overrightarrow{\left[u_{\mathbf{i}}\, c_{22\,\mathbf{ij}k}\, u_{\mathbf{j}}\right]},$$

which after re-ordering proves (3.78). We can now rewrite (3.61) as

$$\widetilde{\Sigma}''_{yy}(\star) = \frac{2}{3}\widetilde{\Sigma}^{-2}(\star)\vec{\sigma}_t^{\perp}\vec{\vec{\mathrm{v}}}'_y(\star) - 2\widetilde{\Sigma}^{-1}\,\widetilde{\Sigma}'^{\,2}_y(\star)$$
$$+\frac{1}{3}\widetilde{\Sigma}^{-3}(\star)\left[\|\vec{\vec{\mathrm{v}}}\|^2 + \|\vec{\vec{n}}\|^2\right](\star).$$

Substituting (3.73), (3.76)–(3.78) and omitting the arguments yields

$$\widetilde{\Sigma}''_{yy}(\star) = -2\widetilde{\Sigma}^{-1}\frac{1}{4}\left[\vec{u}^{\perp}\vec{\vec{c}}_2\vec{u}\right]^2 + \frac{1}{3}\widetilde{\Sigma}^{-1}\left[\vec{u}^{\perp}\vec{\vec{c}}_2\vec{\vec{c}}_2^{\perp}\vec{u} + \vec{u}^{\perp}\vec{\vec{c}}_3\vec{\vec{c}}_3^{\perp}\vec{u}\right]$$
$$+\frac{2}{3}\widetilde{\Sigma}^{-2}\vec{\sigma}_t(t)^{\perp}\left[\frac{-3}{2}\left[\vec{u}^{\perp}\vec{\vec{c}}_2\vec{u}\right]\vec{\vec{c}}_2^{\perp}\vec{u} + \frac{1}{2}\vec{\vec{c}}_2^{\perp}\left[\vec{\vec{c}}_2 + \vec{\vec{c}}_2^{\perp}\right]\vec{u}\right.$$
$$\left.+\frac{1}{2}\overrightarrow{\left[u_{\mathbf{i}}\, c_{22\,\mathbf{ij}k}\, u_{\mathbf{j}}\right]}\right].$$

Using (D.16) we develop this expression into

$$\widetilde{\Sigma}\,\widetilde{\Sigma}''_{yy}(\star) = -\left[\vec{u}^{\perp}\vec{\vec{c}}_2\vec{u}\right]^2 + \frac{1}{3}\vec{u}^{\perp}\left[\vec{\vec{c}}_2^{\perp}\vec{\vec{c}}_2 + \vec{\vec{c}}_2\vec{\vec{c}}_2\right]\vec{u} + \frac{1}{3}\left[u_{\mathbf{i}}u_{\mathbf{j}}u_{\mathbf{k}}\, c_{22\,\mathbf{ijk}}\right]$$
$$-\frac{1}{2}\left[\vec{u}^{\perp}\vec{\vec{c}}_2\vec{u}\right]^2 + \frac{1}{3}\vec{u}^{\perp}\vec{\vec{c}}_2\vec{\vec{c}}_2^{\perp}\vec{u} + \frac{1}{3}\vec{u}^{\perp}\vec{\vec{c}}_3\vec{\vec{c}}_3^{\perp}\vec{u},$$

which after simplification proves (3.74). Finally, in order to get the slope we recall the drift identity (3.71) which rewrites as

$$4\widetilde{\Sigma}\,\widetilde{\Sigma}'_{\theta}(\star)$$
$$= 2\vec{\sigma}_t^{\perp}\vec{a}_1 + \|\vec{\vec{a}}_2\|^2 + \|\vec{\vec{a}}_3\|^2 - 2\widetilde{\Sigma}^3\widetilde{\Sigma}''_{yy}(\star) + 2\widetilde{\Sigma}^3\widetilde{\Sigma}'_y(\star) - 6\widetilde{\Sigma}^2\,\widetilde{\Sigma}'^{\,2}_y(\star).$$

Substituting (3.73) and (3.74) we get

$$
\begin{aligned}
4\widetilde{\Sigma}\,\widetilde{\Sigma}'_{\theta}(\star) = {} & 2\vec{\sigma}_t^{\perp}\vec{a}_1 + \|\overset{\Rightarrow}{a}_2\|^2 + \|\overset{\Rightarrow}{a}_3\|^2 + 3\widetilde{\Sigma}^2\left[\vec{u}^{\perp}\overset{\Rightarrow}{c}_2\vec{u}\right]^2 \\
& - \frac{2}{3}\widetilde{\Sigma}^2\vec{u}^{\perp}\left[\overset{\Rightarrow}{c}_2^{\perp}\overset{\Rightarrow}{c}_2 + \overset{\Rightarrow}{c}_2\overset{\Rightarrow}{c}_2 + \overset{\Rightarrow}{c}_2\overset{\Rightarrow}{c}_2^{\perp}\right]\vec{u} - \frac{2}{3}\widetilde{\Sigma}^2\vec{u}^{\perp}\overset{\Rightarrow}{c}_3\overset{\Rightarrow}{c}_3^{\perp}\vec{u} \\
& - \frac{2}{3}\widetilde{\Sigma}^2\left[u_{\mathbf{i}}u_{\mathbf{j}}u_{\mathbf{k}}\ c_{22\,\mathbf{ijk}}\right] + \widetilde{\Sigma}^3\vec{u}^{\perp}\overset{\Rightarrow}{c}_2\vec{u} - \frac{3}{2}\widetilde{\Sigma}^2\left[\vec{u}^{\perp}\overset{\Rightarrow}{c}_2\vec{u}\right]^2,
\end{aligned}
$$

which after simplification proves (3.75) and concludes the proof. □

3.5 Illustration of the Vectorial Framework: The *Basket Case*

Before introducing the notion of a maturity-dependent underlying in Part II, we now extend the previous results by focusing on the basket case,[12] i.e. considering baskets of underlyings, which naturally bring forward the multi-dimensional context.

After exposing several motivational arguments, we set a generic framework for all baskets. We then list the various configurations that can be met in practice, and select the two types that are most relevant for this study: the *coefficient* and *asset* baskets. Afterwards we can state our objectives, which consist in computing the first layer's $\widetilde{\Sigma}$-(2,0) IATM differentials associated to the smile of the asset basket. Given the results of the previous sections, we know that obtaining those differentials only requires deriving the σ-(2,0) dynamic coefficients of the same asset basket. But since its volatility structure is multi-dimensional, this implies in turn computing the dynamics of the coefficient basket. We can then specialise the results to fixed-weights asset baskets and propose interesting interpretations in both cases.

3.5.1 *Motivation*

We consider that the motivation for this intermediate step (i.e. before moving on to term structures) can be broken down into four broad categories:

1. Basket products represent genuine and common features of the financial world, which come in many forms: indexes, custom underlyings for exotic options, or even spreads. Furthermore, European options written on those are also frequent and can be very liquid.
2. In the modelling world, the mixture approaches have recently regained in popularity. They can be schematically split into two categories: either mixtures of prices

[12] Pun intended.

or mixtures of underlyings. We will see that the (most common) first case can be treated directly using our scalar framework. The second case, however, a.k.a. "basket of models" is both more interesting and more involved. If one considers a (single) underlying whose dynamics are defined by a mixture of SDEs, then applying our asymptotic approach naively can prove lengthy (the computing cost can be linear in the number of component models). Hence it is useful to propose generic formulas and some qualitative interpretations for that case.

3. The basket provides a good illustration of the added complexity brought by the multi-dimensionality, as described previously. Furthermore, this is achieved without any maturity dependency, and we shall see later that in practice it is the former factor that drives the complexity of the computations (but not of the conceptual framework).
4. Lastly, the generic results obtained in this section will prove very useful to accelerate several computations in the coming part of the study dedicated to maturity-dependent frameworks. Indeed, many term structure instruments, such as swap rates for instance, can be either written or approximated as baskets (i.e. with either stochastic or fixed weights).

Let us quickly expand on the first point, i.e. the motivation for baskets as existing financial instruments: the notion is fairly widespread in modern finance, although not always in an explicit fashion. For instance, we can easily relate to baskets of stocks, whether they define an equity index (SP500, CAC40, FTSE100, etc.) or the underlying for an equity derivative product (e.g. "Mountain Range" options; these are usually OTC custom baskets mainly used for correlation trading). Note that in the former case, the weights tend to be fixed, but looking at cross-currency and quanto products for instance shows that the weights themselves can be stochastic. But baskets can be generalised and found in various asset classes. They appear with inflation products, since the CPI[13] is by definition a typical "household" consumption basket. They also show up in credit derivatives, which involve baskets of names, and, as mentioned, they intervene with interest rates derivatives such as CMS spreads or swap rates.[14] This list is far from exhaustive, and simply illustrates the familiarity of the concept.

It is therefore no surprise that in the literature, the topic of financial baskets has also attracted some attention. Generally the objective of the authors is to relate the individual dynamics, often implied from the individual smiles, to those of the basket. The trend is to focus on fixed weights and naturally to focus on the correlation structure. The methods employed can broadly be categorised, as usual, between closed-form and numerical, the latter involving the full available spectrum: Monte-Carlo, trees, PDE schemes, Fourier transforms, etc. Naturally, in the context of our study we will focus mainly on analytical methods. Note that baskets are often seen as spatial averages, while arithmetic Asian options represent time averages; for that reason, many articles treat both subjects, with a definite popularity for blending them

[13] Consumer Price Index.

[14] Indeed, as we will recall in Chap. 7, a par swap rate can be written as a basket of Libor rates, but with stochastic (albeit relatively stable) weights.

into *Asian baskets*. Let us mention only a few of the contributions in this domain. A popular topic is the establishment of upper and lower limits for prices of European basket options. In [19] several bounds are computed, in closed and semi-closed forms *via* linear programming, but only for specific cases. By contrast, in [20] some lower and upper bounds are computed in a very general case, where little information is available w.r.t. the individual components, by using duality methods and replication arguments.

Another technique which has attracted a lot of attention is the use of moment matching. In [21] the authors use such methods in order to approximate the basket's marginal distribution in the sense of two specific distances, including Kullback-Leibler, and focus on simple dynamics of the shifted Lognormal family. Other moment-based methods, often based on specific classes of parametric distribution functions, can be found in [22], for instance.

An interesting approach can be found in [23], an equity basket with fixed weights is considered, where each individual stock admits local volatility dynamics and with constant correlations. The aim is to provide pricing approximations for options on the basket, and the key ingredient is to rely on the *most likely configuration* of the individual stock prices corresponding to a given basket level. In turn, the authors link the local volatility of the basket to the individual local volatilities, before moving on to implied volatilities. One can find many more approaches in the literature, including, for instance, expansion techniques, or even geodesics.

Finally, let us mention [24] (Chap. 3), which in the context of an LMM framework investigates the *freezing* approximation, used to proxy the volatility of a basket of Lognormal underlyings. Exploiting an asymptotic expansion approach, it proposes corrective terms to the *frozen* price formula: we will discuss these subjects again in Sect. 7.6 [p. 400].

Let us now expand on the second point by discussing further the mixture modelling approach. In mathematical finance, the term applies to several distinct pricing methodologies. The most common of these fall into the broader category of *mixture of prices*, or mixture of marginals. In a nutshell, it consists in considering a (usually finite) number of simple processes, weighting and adding up their marginal distributions, and then considering those to belong to the underlying. The second category is sometimes called *mixture of assets* and is more rarely found. It assumes the underlying itself to be a mixture of simpler processes, whose dynamics can be correlated. In terms of establishing the marginal distributions, this second method will evidently be more complex, as it naturally invokes convolution properties. Let us present in more detail both approaches and how our methodology can be applied to them.

Concerning the mixture of prices methodology, an easy introduction and insightful comparison can be found in Piterbarg [25]. In particular, it shows that a naive undertaking can easily bring inconsistency, in the form of arbitrage opportunities. We now discuss the most popular approaches found in the literature.

The first *price mixing* method that we present is fully consistent but uses the mixture indirectly, as a tool to get fast European option prices. Developed in Brigo et al. [26] and in several other papers from the same authors, its final target is actually a pure local volatility model. Its principle is to fit a whole implied volatility surface[15] as best as possible, with a finite collection of displaced Lognormal processes.[16] It uses a fixed-weights, convex combination of the latter, so that all European prices are obtained in closed form. Given this projection of the marginals onto the individual displaced Black-Scholes models, a Dupire-like argument then provides the local volatility model associated to the mixture.

In summary, this corresponds to a *parametric* local volatility approach, but exhibiting the additional property of easily providing vanilla prices. This Dupire model is now calibrated to the smile, and can be used dynamically to price and hedge exotic, path-dependent options. Note that the methodology can easily be extended, and the model's dynamic behaviour improved (although at a higher numerical cost) by adding a single and independent stochastic volatility perturbation, shared by all individual components and therefore by the underlying.

This leads us to the price mixing second approach, whose principle can be tracked back to such independent, stochastic volatility models (e.g. FL-SV). Let us consider, for instance, a Lognormal model with an orthogonal, multiplicative volatility perturbation. When pricing a European Call in that model, we can simply condition the expectancy w.r.t. individual volatility paths, and then integrate Black's formula against the latter's distribution. In that integration, the pertinent variable, or functional of the path, is simply the effective variance $\zeta(t,T) = \int_t^T \sigma_s^2 ds$, whose distribution can usually be well approximated by a sparse, discrete measure $\{(\zeta_i, p_i)\}_{1\leq i\leq N}$. This means that the Call is priced by convex combination as

$$C_t(K,T) = \sum_{i=1}^{N} p_i(t,T)\; Black(K, \sigma^2 = (T-t)^{-1}\zeta_i(t,T)).$$

Using this approximation for European options of a given expiry is valid. What is not consistent, however, as pointed out in Piterbarg [25], is to generalise the above pricing approach indiscriminately. Indeed, several authors propose to use a pricing methodology such as

$$V(\Phi) = \sum_{i=1}^{N} p_i V_i(\Phi), \tag{3.79}$$

where Φ represents the payoff to evaluate and V_i is its price in each simple, individual model. In Gatarek [27], for instance, the author proposes to mix displaced Lognormal instances of the Libor Market Model, in order to capture skew and curvature. Each individual model is therein considered parametrised by its own term-structure of time-dependent volatilities and of constant displacements. Since each model produces

[15] Actually, a series of smiles for benchmark expiries and therefore the associated marginal distributions.

[16] All sharing the same initial value as the underlying.

its own Caplet prices in closed form, it is then proposed to proceed by a convex combination of these models.

Unfortunately, this approach presents the drawback of not being consistent with a dynamic model. Taking our previous perturbed Lognormal example, it is clear that the discrete $\{(\zeta_i, p_i)\}$ distribution will *a priori* change with the Call's expiry T, so that (3.79) would not be valid to price another European option, let alone an exotic option. Even if we could find a dynamic process for the volatility that would make that distribution stationary, we would have to start with an *uncertain* volatility. Furthermore, provided a solution exists, we have no reason to believe that it is unique (actually, quite the contrary). This is problematic, since leaving the dynamic process undetermined does not allow us to price path-dependent options. Finally, even if we were to forgo any idea of a dynamic process for our underlying, [25] exhibits a simple example of path-dependent payoff (a Compound Put, or Bermudan option) for which the pricing methodology of (3.79) is not self-consistent.

All in all, if the weights are fixed and if we are only interested in the direct problem, i.e. extrapolating the smile shape and dynamics from the specification of the model's dynamics, then our asymptotic methodology applies seamlessly to all these *mixture of prices* approaches. Indeed, it suffices to extrapolate the implied volatility for each individual sub-model, to deduce the corresponding Call prices for all strikes and therefore the marginal densities, and then to proceed with the weighted sum. Clearly, this class of models does not warrant a specific treatment in comparison with the single-underlying, scalar framework of Chap. 2.

On the other hand, the *mixture of underlyings* approach does require the multi-dimensional results, and deserves more attention. This more complex framework builds on the notion of mixtures of stochastic *processes*, which is a well-documented field both in the probabilistic and statistical domains. Simply put, we equip ourselves with a finite collection of N individual processes, each explicitly defined by its SDE, and then take their weighted sum to obtain our scalar underlying. For instance, we can set:

$$\frac{dS_{i,t}}{S_{i,t}} = \overrightarrow{\sigma_{i,t}}^{\perp} d\overrightarrow{W}_t \quad \text{for } 1 \leq i \leq N \qquad \text{and then} \qquad S_t = \sum_{i=1}^{N} \lambda_{i,t} S_{i,t},$$

where the dynamics of the individual volatilities $\overrightarrow{\sigma_{i,t}}$ and of the weights $\lambda_{i,t}$ will have to be specified, and can involve an exogenous driver. In most practical cases the weights are taken as constants, and the individual dynamics are such that closed forms are available for the pricing of vanilla options. Typically, the $S_{i,t}$ are taken as normals or lognormals, but one can throw in some local and/or stochastic local volatility, or even jumps.

The bottom line is that S_t is a legitimate underlying with rich dynamics, so that pricing and hedging of vanilla and exotic (including path-dependent) options can (theoretically) be achieved consistently. Note, however, that the Markovian dimension of the problem will usually stay high (at N), so that numerical methods will become expensive. We now show how to apply our asymptotic results in such a context of *mixture of underlyings* modelling.

3.5.2 Framework and Objectives

It serves our purpose well to define a basket in full generality, as

$$\mathbf{M}_t \triangleq \sum_{i=1}^{N} \lambda_{i,t} \mathbf{X}_{i,t},$$

where, in full generality, $\mathbf{M}_t$ along with all individual $\lambda_{i,t}$ and $\mathbf{X}_{i,t}$ are *tensorial* continuous Itô processes of compatible orders and dimensions. In other words, all $\lambda_{i,t}\mathbf{X}_{i,t}$ products make sense, they share the same order and dimensions. In practice, the weights come overwhelmingly as scalar, and this is the convention we will take in the sequel. As a consequence, the basket M_t and its individual constituents $\mathbf{X}_{i,t}$ will share the same tensorial order and dimensions.

One might wonder why we could possibly be interested in other-than-scalar baskets $\mathbf{M}$, since the focus of this study is on liquid options surfaces $C(K, T)$ for single underlyings. The reason is that when considering a basket of underlyings, its volatility can often also be written as a basket of the individual volatilities. Because we will require the dynamics of this volatility in order to apply our methodology, it makes sense to address these two related issues simultaneously.

Still on the structural level, and besides the dimensionality question for these processes, the basket definition will require us to select a further feature: whether the dynamics of the weights $\lambda_{i,t}$ incorporate an exogenous driver $d\vec{Z}_t$. In other words, whether the composition of the mixture can be influenced by a source independent of the individuals' dynamics. In fact, such a framework is closely linked to stochastic volatility issues, as it suffices to consider a Lognormal model with independent stochastic volatility, which indeed can be seen as a mixture (this is usually referred to as the *mixing argument* of Hull and White: see [28]).

Having now defined the structure of the basket itself, then in order to apply our methodology and to present our results we also need to specify the conventions, i.e. Normal or Lognormal (LogN), used to describe the dynamics of the above processes and of their volatilities (diffusion coefficients). Clearly when the quantities are tensorial, the only meaningful representation is Normal. In the scalar case, and in our generic framework, where no assumptions are made on the processes, evidently all representations are equivalent in principle. But in practice, we mainly use as inputs for $\mathbf{X}_{i,t}$ some bi-dimensional instantaneous stochastic volatility (market) models (such as Heston, SABR or FL-SV) which are written under specific but limited conventions: classically, they use either the Normal or LogN, for both the component $\mathbf{X}_{i,t}$ and the perturbation. To a lesser extent, this also tends to be true for the weights $\lambda_{i,t}$. Furthermore, the expansions output concerns the basket M_t, and for the dynamics of the latter to be readily interpretable, they should fall into one of these two mainstream formats. All in all, selecting one of these conventions does not structurally affect the results, but is motivated by convenience from the reader's perspective.

Faced with so many cases and subcases, we need to make some choices and thus restrict our framework. Therefore we select a pair of configurations, that we will name the "asset basket"and the "coefficient basket". Before describing and justifying these

Table 3.4 Main basket configurations

			$X_\star$ and M are scalar			$X_\star$ and M are tensorial
			$X_\star \sim$ LogN		$X_\star \sim$ Normal	
			M $\sim$ LogN	M $\sim$ Normal	M $\sim$ LogN	
Fixed Weights						
Stochastic weights	Endogenous dynamics	LogN	*"Asset"*			
		Normal				
	Generic dynamics	LogN				*"Coefficient"*
		Normal				

configurations, we present Table 3.4, which demonstrates the variety of the basket population, and which places our two chosen configurations in context. Note that this table is in no way exhaustive, as for instance it does not specify conventions for the volatility processes.

Let us now make precise the notions of "asset"and "coefficient" baskets, as well as justifying their choice. Note that together, these two configurations will cover most of our needs in Part II of this study, which focuses on interest rates and term structures.

1. **The *asset basket***
   This configuration is geared towards the situation where the individuals $\mathbf{X}_i$ and the basket $\mathbf{M}$ are traded assets, represented as scalars, but of course it can be applied to other purposes. We treat this all-scalar case simply because its practical importance is highest. We cover it with stochastic weights (mainly in preparation for dealing with swap rates, as mentioned above), and then specialise the result to the fixed-weights situation (due to its preponderance as a market instrument). Also, the dynamics of these weights are deemed purely endogenous, in order to preserve the same property for the basket $\mathbf{M}$. As for the conventions, we elect to write the setup with:

   - The Lognormal convention for the basket $\mathbf{M}$ and the individual constituents $\mathbf{X}_i$. This is mainly to simplify comparisons and to be consistent with the single-underlying framework treated previously in Chap. 2.
   - The Lognormal convention for the weights λ_i. This is first of all to improve the applicability of our results, since such weights are often defined as ratios of assets (see Part II). Then there are psychological/conventional reasons, since weights tend to remain positive in most financial baskets. Note that the *convention* itself does not prevent the weights from changing sign if necessary, as can be the case when they are defined as spreads.
   - The Normal convention for all volatility processes. Again this is for consistency with Chap. 2, but also to simplify computations. Finally, recall that the Lognormal convention becomes useless with vectors and higher-order tensors.

2. **The *coefficient basket***
   This configuration is designed primarily for linear combinations of volatility vectors, but treated with more generality it also applies to other Itô coefficients,

of higher tensorial order, in view of our necessary Wiener chaos expansion. Therefore we deal here with the case where the basket $\mathbf{M}$ and the constituents $\mathbf{X}_i$ are tensors, whereas the weights λ_i remain scalar.

- The dynamics of $\mathbf{M}$ and $\mathbf{X}_i$ are naturally represented under the Normal convention, but they now incorporate the exogenous driver.
- The dynamics of λ_i are written in Lognormal fashion, but they too involve $\vec{Z}_t$.

Our *objective* is to express the first layer of IATM differentials for the smile of the *asset basket*. In view of Theorem 3.4 [p. 181], in order to obtain these differentials (both static and dynamic) it suffices to express the dynamics of the basket M_t within a generic SInsV model, using the usual chaos expansion. More precisely, all we need is to compute four of the (tensorial) coefficients invoked by those dynamics: $\vec{\sigma}_t$, $\vec{a}_{1,t}$, $\vec{\vec{a}}_{2,t}$, $\vec{\vec{a}}_{3,t}$ and $\vec{\vec{\vec{a}}}_{22,t}$, in other words the σ-(2,0) group.

We choose, however, not to present *all* the $\widetilde{\Sigma}$-(2,0) differential expressions within the general case. Indeed, those formulas are very involved and therefore not really subject to interpretation. Nevertheless, we will provide some qualitative analysis for specific cases, namely for the static (shape) descriptors in the case of fixed weights asset baskets.

3.5.3 The Coefficient Basket

This section is intended to deliver a purely technical result, and also to be used again in later chapters, in particular those in Part II dedicated to interest rates modelling: we choose therefore to present it as a Lemma.

Lemma 3.6 (Dynamics of the stochastic-weights coefficient basket) *Let $\mathbf{X}_t$ be a tensorial Itô process of order n, with the following dynamics:*

$$d\mathbf{X}_t = \mathbf{A}_{1,t}dt + \mathbf{A}_{2,t}d\vec{W}_t + \mathbf{A}_{3,t}d\vec{Z}_t.$$

Note that $\mathbf{A}_{1,t}$ is of order n, $\mathbf{A}_{2,t}$ and $\mathbf{A}_{3,t}$ of order $n+1$, and dimensions must agree. Let w_t be a scalar Itô process with the following dynamics (using a Lognormal convention):

$$\frac{dw_t}{w_t} = c_{1,t}dt + \overrightarrow{c_{2,t}}^{\perp}d\vec{W}_t + \overrightarrow{c_{3,t}}^{\perp}d\vec{Z}_t.$$

Then the dynamics of the product $[w_t\mathbf{X}_t]$ are

$$\begin{aligned} d\left[w_t\mathbf{X}_t\right] = w_t\left[\mathbf{A}_{1,t} + c_{1,t}\mathbf{X}_t + \mathbf{A}_{2,t}\overrightarrow{c_{2,t}} + \mathbf{A}_{3,t}\overrightarrow{c_{3,t}}\right]dt \\ + w_t\left[\mathbf{A}_{2,t} + \mathbf{X}_t \otimes \overrightarrow{c_{2,t}}\right]d\vec{W}_t + w_t\left[\mathbf{A}_{3,t} + \mathbf{X}_t \otimes \overrightarrow{c_{3,t}}\right]d\vec{Z}_t \end{aligned} \quad (3.80)$$

where ⊗ denotes the outer product.[17] *Now let us consider a finite collection of such pairs, defining the tensorial basket:*

$$\exists \left\{ \left(w_{i,t}, \mathbf{X}_{i,t} \right) \right\}_{1 \le i \le N} \qquad \textit{and} \qquad \mathbf{M}_t \triangleq \sum_{i=1}^{N} w_{i,t} \mathbf{X}_{i,t}.$$

We then have the basket dynamics as

$$d\mathbf{M}_t = \mathbf{C}_{1,t} dt + \mathbf{C}_{2,t} d\overrightarrow{W}_t + \mathbf{C}_{3,t} d\overrightarrow{Z}_t, \tag{3.81}$$

with

$$\mathbf{C}_{1,t} \triangleq \sum_{i=1}^{N} w_{i,t} \left[\mathbf{A}_{i,1,t} + c_{i,1,t} \mathbf{X}_{i,t} + \mathbf{A}_{i,2,t} \overrightarrow{c_{i,2,t}} + \mathbf{A}_{i,3,t} \overrightarrow{c_{i,3,t}} \right],$$

$$\mathbf{C}_{2,t} \triangleq \sum_{i=1}^{N} w_{i,t} \left[\mathbf{A}_{i,2,t} + \mathbf{X}_{i,t} \otimes \overrightarrow{c_{i,2,t}} \right],$$

$$\mathbf{C}_{3,t} \triangleq \sum_{i=1}^{N} w_{i,t} \left[\mathbf{A}_{i,3,t} + \mathbf{X}_{i,t} \otimes \overrightarrow{c_{i,3,t}} \right].$$

Note the natural induction feature, in the sense that the coefficients $\mathbf{C}_{1,t}$, $\mathbf{C}_{2,t}$ and $\mathbf{C}_{3,t}$ are themselves written as tensorial baskets on the same collection of weights. Observe also that $\mathbf{M}_t$ and $\mathbf{C}_{1,t}$ are naturally of order n, while $\mathbf{C}_{2,t}$ and $\mathbf{C}_{3,t}$ are of order $n+1$.

Proof By a simple application of Itô's Lemma, coordinate by coordinate, we get that

$$\begin{aligned} d\left[w_t \mathbf{X}_t \right] &= \mathbf{X}_t w_t \left[c_{1,t} dt + \overrightarrow{c_{2,t}}^{\perp} d\overrightarrow{W}_t + \overrightarrow{c_{3,t}}^{\perp} d\overrightarrow{Z}_t \right] \\ &\quad + w_t \left[\mathbf{A}_{1,t} dt + \mathbf{A}_{2,t} d\overrightarrow{W}_t + \mathbf{A}_{3,t} d\overrightarrow{Z}_t \right] \\ &\quad + w_t \left[\mathbf{A}_{2,t} \overrightarrow{c_{2,t}} + \mathbf{A}_{3,t} \overrightarrow{c_{3,t}} \right] dt, \end{aligned}$$

which, after grouping finite and non-finite variation terms and introducing the outer product, proves (3.80). We can now move on to the tensorial basket:

$$\begin{aligned} d\mathbf{M}_t &= \sum_{i=1}^{N} w_{i,t} \left[\mathbf{A}_{i,1,t} + c_{i,1,t} \mathbf{X}_{i,t} + \mathbf{A}_{i,2,t} \overrightarrow{c_{i,2,t}} + \mathbf{A}_{i,3,t} \overrightarrow{c_{i,3,t}} \right] dt \\ &\quad + \sum_{i=1}^{N} w_{i,t} \left[\mathbf{A}_{i,2,t} + \mathbf{X}_{i,t} \otimes \overrightarrow{c_{i,2,t}} \right] d\overrightarrow{W}_t + \sum_{i=1}^{N} w_{i,t} \left[\mathbf{A}_{i,3,t} + \mathbf{X}_{i,t} \otimes \overrightarrow{c_{i,3,t}} \right] d\overrightarrow{Z}_t, \end{aligned}$$

which proves (3.81). □

[17] Whereas the tensorial inner product, such as $\mathbf{A}_{2,t} \overrightarrow{c_{2,t}}$, is left implicit.

3.5.4 The Asset Basket in the General Case

As mentioned above, this type covers a large proportion of the baskets that one meets in practice. Let us start by formalising the setup and presenting our notations.

Definition 3.8 We consider the stochastic-weights asset basket with *a.s.* non-null price:

$$M_t = \sum_{i=1}^{N} \lambda_{i,t} S_{i,t} \qquad \text{and} \qquad M_u \neq 0 \; \forall u \geq t, \; a.s. \tag{3.82}$$

We assume the existence of a measure $\mathbb{Q}$ as M_t-martingale, and consider a $\mathbb{Q}$-Brownian endogenous driver $\vec{W}_t$, along with an independent exogenous driver $\vec{Z}_t$. We then express the dynamics of both weights and individuals using the Lognormal convention:

$$\frac{dS_{i,t}}{S_{i,t}} = \mu_{i,t} dt + \vec{\sigma_{i,t}}^{\perp} d\vec{W}_t \qquad \text{and} \qquad \frac{d\lambda_{i,t}}{\lambda_{i,t}} = \eta_{i,t} dt + \vec{\gamma_{i,t}}^{\perp} d\vec{W}_t.$$

We specify the chaos dynamics for the two groups of instantaneous volatilities as:

$$d\vec{\sigma_{i,t}} = \vec{a}_{i,1,t} dt + \vec{\vec{a}}_{i,2,t} d\vec{W}_t + \vec{\vec{a}}_{i,3,t} d\vec{Z}_t,$$

$$d\vec{\vec{a}}_{i,2,t} = [\cdot]\, dt + \vec{\vec{\vec{a}}}_{i,22,t} d\vec{W}_t + [\cdot]\, d\vec{Z}_t,$$

$$d\vec{\gamma_{i,t}} = \vec{\alpha}_{i,1,t} dt + \vec{\vec{\alpha}}_{i,2,t} d\vec{W}_t + \vec{\vec{\alpha}}_{i,3,t} d\vec{Z}_t,$$

$$d\vec{\vec{\alpha}}_{i,2,t} = [\cdot]\, dt + \vec{\vec{\vec{\alpha}}}_{i,22,t} d\vec{W}_t + [\cdot]\, d\vec{Z}_t.$$

Note that the individual components $S_{i,t}$ and the weights $\lambda_{i,t}$ are not assumed martingale under the chosen measure, and that we have not specified any dynamics for their respective drifts $\mu_{i,t}$ and $\eta_{i,t}$. In fact, these coefficients are immaterial for our purposes, as will be shown later. Having defined our setup, let us establish the expressions for the instantaneous coefficients constituting the basket's σ-(2,0) group.

Proposition 3.7 (Chaos diffusion of the asset basket up to the first layer) *Let us place ourselves within the framework of Definition* 3.8 *for the asset basket. We then introduce the normalised weight processes, which sum to unity:*

$$\omega_{i,t} = \frac{\lambda_{i,t} S_{i,t}}{M_t} = \frac{\lambda_{i,t} S_{i,t}}{\sum_{k=1}^{N} \lambda_{k,t} S_{k,t}} \qquad \textit{and} \qquad \sum_{i=1}^{N} \omega_{i,t} = 1. \tag{3.83}$$

We can now re-express the basket dynamics with its own volatility:

$$\frac{dM_t}{M_t} = \vec{\sigma_t}^{\perp} d\vec{W}_t \qquad \textit{with} \qquad \vec{\sigma_t} = \sum_{i=1}^{N} \omega_{i,t} \left[\overrightarrow{\sigma_{i,t}} + \overrightarrow{\gamma_{i,t}}\right], \tag{3.84}$$

while the martingale condition reads as

$$0 = \sum_{i=1}^{N} \omega_{i,t} \left[\mu_{i,t} + \eta_{i,t} + \overrightarrow{\sigma_{i,t}}^{\perp} \overrightarrow{\gamma_{i,t}}\right]. \tag{3.85}$$

Continuing the Wiener chaos, the basket's instantaneous volatility dynamics are

$$\begin{aligned} d\vec{\sigma_t} &= \vec{a}_{1,t} dt + \vec{\vec{a}}_{2,t} d\vec{W}_t + \vec{\vec{a}}_{3,t} d\vec{Z}_t, \\ d\vec{\vec{a}}_{2,t} &= [\cdot]\, dt + \vec{\vec{\vec{a}}}_{22,t} d\vec{W}_t + [\cdot]\, d\vec{Z}_t, \end{aligned}$$

where

$$\begin{aligned} \vec{a}_{1,t} \triangleq\; & -\vec{\vec{a}}_{2,t} \vec{\sigma_t} \\ & + \sum_{i=1}^{N} \omega_{i,t} \left[\left[\vec{a}_{i,1,t} + \vec{\alpha}_{i,1,t}\right] + \left[\vec{\vec{a}}_{i,2,t} + \vec{\vec{\alpha}}_{i,2,t}\right] \left[\overrightarrow{\sigma_{i,t}} + \overrightarrow{\gamma_{i,t}}\right]\right], \end{aligned} \tag{3.86}$$

$$\vec{\vec{a}}_{2,t} \triangleq -\vec{\sigma_t}^{2\otimes} + \sum_{i=1}^{N} \omega_{i,t} \left[\left[\vec{\vec{a}}_{i,2,t} + \vec{\vec{\alpha}}_{i,2,t}\right] + \left[\overrightarrow{\sigma_{i,t}} + \overrightarrow{\gamma_{i,t}}\right]^{2\otimes}\right], \tag{3.87}$$

$$\vec{\vec{a}}_{3,t} \triangleq \sum_{i=1}^{N} \omega_{i,t} \left[\vec{\vec{a}}_{i,3,t} + \vec{\vec{\alpha}}_{i,3,t}\right] \tag{3.88}$$

and ignoring time-dependency (for compactness)

$$\begin{aligned} \vec{\vec{\vec{a}}}_{22} =\; & -\vec{\vec{a}}_2 \overset{\leftrightarrow}{\otimes} \vec{\sigma} + \sum_{i=1}^{N} w_i \left(\vec{\vec{a}}_{i,2} + \vec{\vec{\alpha}}_{i,2}\right) \overset{\leftrightarrow}{\otimes} \left(\vec{\sigma_i} + \vec{\gamma_i}\right) \\ & + \sum_{i=1}^{N} w_i \left[\left(\vec{\vec{\vec{a}}}_{i,22} + \vec{\vec{\vec{\alpha}}}_{i,22}\right) + \left[(\sigma_i + \gamma_i - \sigma)_k \left(a_{i,2} + \alpha_{i,2}\right)_{jl}\right]\right] \\ & + \sum_{i=1}^{N} w_i \left[\left(\vec{\sigma_i} + \vec{\gamma_i}\right)^{2\otimes} - \left(\vec{\sigma_i} + \vec{\gamma_i}\right) \otimes \vec{\sigma}\right] \otimes \left(\vec{\sigma_i} + \vec{\gamma_i}\right), \end{aligned} \tag{3.89}$$

where we use modified Einstein notation (see Appendix D), an obvious extension of the external product ($2\otimes$), *as well as the commutative outer product introduced in* ($D.6$) [p. 439].

Let us comment on these assumptions and results, before going into the proof.

The fact that all individual components $S_{i,t}$ might not be martingale under $\mathbb{Q}$, or might even each possess their own private martingale measure, is not a mere matter of overstretched generalisation: it does happen in practice. This is indeed the case when considering a swap rate as a basket of Libor rates.

As for the original weights $\lambda_{i,t}$, there is no reason why they should be assets themselves, but they are usually built from either assets or underlyings, typically as rational functions. We note that their dynamics can be quite complex, and intervene in a similar fashion to those of the true individual underlyings $S_{i,t}$. Indeed, M_t is simply a sum of pair products, so that we can anticipate a *duality structure* between the underlyings and their weights. What concerns us directly, though, in terms of smile approximation, is the fact that the $\lambda_{i,t}$ family can contribute significantly to the overall basket's volatility.

For that reason, we have made the choice of *not* incorporating the exogenous driver $\overrightarrow{Z}_t$ in the weights' dynamics. Note that this restriction is not binding, since the following proofs could easily be adapted if necessary. For our purposes, however, it would bring unwarranted complexity, and we try to abide by the notion that the endogenous driver $\overrightarrow{W}_t$ concerns assets and underlyings, while the exogenous driver $\overrightarrow{Z}_t$ is allocated to volatilities and smiles. In fact, all we ask is for those volatility-related processes to access driver components that are orthogonal to the underlyings' and weights' dynamics. That point being granted, it is always possible to redistribute the components between $\overrightarrow{W}_t$ and $\overrightarrow{Z}_t$, and to write the framework as we did.

With respect to the normalised weights now, observe that in the fairly common case where all individual weights $\lambda_{i,t}$ and components $S_{i,t}$ remain positive, then so does the basket, and the normalised weights $\omega_{i,t}$ live in the $]0, 1[$ range. Note that the normalised weights are well-defined at time t if and only if $M_t \neq 0$. We have asked for this property to be maintained in the future ($M_u \neq 0 \ \forall u \geq t$) and *a.s.*, but this is purely for consistency. Indeed, since we are targeting small-time asymptotics, and since these normalised weights are only a tool to express the IATM differentials, it would suffice to write that at current time $M_t \neq 0$.

The practical selection of $\mathbb{Q}$ as the M_t-martingale measure is not necessarily trivial, mainly because the weights $\lambda_{i,t}$ do possess dynamics of their own. The technicality of this point will be discussed further in the course of the proof, but it is important to stress the more fundamental implications now. First of all, it is safe to say that, in most cases, it proves relatively easy to assert suitable martingale measures for each of the individual components $S_{i,t}$. Similarly, let us assume that the original weights' dynamics are provided driftless. Now, a common measure $\mathbb{Q}$ can always be chosen, at the cost of defining unequivocally the drifts $\mu_{i,t}$ and $\eta_{i,t}$. Naively, it seems reasonable that by "spanning" the measure $\mathbb{Q}$ we should eventually find the one that satisfies the martingale condition ... but this intuition is no formal guarantee of existence. The martingale condition is therefore entirely relying on the initial assumption that *there exists* such a martingale measure. In practice this point is no issue, as the basket is often delivered from the start along with its own, dedicated measure: certainly this is the case with the swap rate. Note also that the selection of $\mathbb{Q}$ would normally be associated with a change of numeraire, but we choose not to dwell on this aspect in the current section.

Turning to the expressions for the coefficients themselves, we postpone any real interpretation to Sect. 3.5.6 as we are still missing an important notion. It is already possible, however, by comparing the formulas for $\overset{\Rightarrow}{a}_{2,t}$ and $\overset{\Rrightarrow}{a}_{22,t}$, to note that the complexification associated to the depth of the coefficient is magnified by the dimensional issue. This confirms the intuition that in the case of baskets, obtaining high-order IATM differentials will be significantly more expensive. Comparing the same two coefficients, we note, however, that an induction structure starts to appear, which mitigates somewhat the previous argument.

We have used several tools from linear algebra in order to simplify our expressions, and also to stress the symmetries and inductive structures. Some readers might ask why we have not pushed this logic further in order to gain in compactness. Indeed, it is possible to introduce tensorial quantities such as

$$\overrightarrow{S}_t \triangleq \begin{bmatrix} S_{1,t} \\ S_{2,t} \\ \vdots \\ S_{N,t} \end{bmatrix} \qquad \overrightarrow{\Lambda}_t \triangleq \begin{bmatrix} \lambda_{1,t} \\ \lambda_{2,t} \\ \vdots \\ \lambda_{N,t} \end{bmatrix} \qquad \overset{\Rightarrow}{\xi}_t \triangleq \begin{bmatrix} \overrightarrow{\sigma_{1,t}}^{\perp} \\ \overrightarrow{\sigma_{2,t}}^{\perp} \\ \vdots \\ \overrightarrow{\sigma_{N,t}}^{\perp} \end{bmatrix}$$

which allow us to define the basket effortlessly *via* $M_t = \overrightarrow{\Lambda}_t^{\perp} \overrightarrow{S}_t$ instead of the classical summation (3.82). The issue is simply dynamics: indeed, if a single collection (e.g. $\{S_{i,t}\}_{1\le i\le N}$ or $\{\omega_{i,t}\}_{1\le i\le N}$) has its evolution defined through a Lognormal convention (which is the case for both the coefficient and asset baskets), then the methodology becomes awkward. For instance, writing the dynamics of the individuals involves the introduction of the unit (diagonal) matrix $\overset{\Rightarrow}{\mathbb{I}}$ in

$$d\overrightarrow{S}_t = \overset{\Rightarrow}{\mathbb{I}}\,\overrightarrow{S}_t \overset{\Rightarrow}{\xi}_t d\overrightarrow{W}_t.$$

In addition, the latter requires the introduction of supplementary notations and concepts linked to Itô calculus applied to tensorial quantities: although we would gain in *compactness*, it is unclear whether *clarity* would benefit. Furthermore, using the summation explicitly makes it easier to grasp the discrete probability interpretation soon to be introduced (see Sect. 3.5.6).

Proof of Proposition 3.7
We start by considering the measure $\mathbb{Q}$ as unspecified, and destined to be constrained later on. The dynamics of a weighted individual are simply given by Itô as

$$d\left[\lambda_{i,t}S_{i,t}\right] = \lambda_{i,t}S_{i,t}\left[\mu_{i,t}dt + \overrightarrow{\sigma_{i,t}}^{\perp} d\overrightarrow{W}_t\right] + \lambda_{i,t}S_{i,t}\left[\eta_{i,t}dt + \overrightarrow{\gamma_{i,t}}^{\perp} d\overrightarrow{W}_t\right]$$
$$+ \lambda_{i,t}S_{i,t}\overrightarrow{\sigma_{i,t}}^{\perp}\overrightarrow{\gamma_{i,t}}dt,$$

hence

$$\frac{d\left[\lambda_{i,t}S_{i,t}\right]}{\lambda_{i,t}S_{i,t}} = \left[\mu_{i,t} + \eta_{i,t} + \overrightarrow{\sigma_{i,t}}^{\perp}\overrightarrow{\gamma_{i,t}}\right]dt + \left[\overrightarrow{\sigma_{i,t}} + \overrightarrow{\gamma_{i,t}}\right]^{\perp} d\overrightarrow{W}_t.$$

Which provides the Lognormal dynamics for the whole basket as

$$\frac{dM_t}{M_t} = \sum_{i=1}^{N} \frac{\lambda_{i,t}S_{i,t}}{M_t}\left[\left[\mu_{i,t} + \eta_{i,t} + \overrightarrow{\sigma_{i,t}}^{\perp}\overrightarrow{\gamma_{i,t}}\right]dt + \left[\overrightarrow{\sigma_{i,t}} + \overrightarrow{\gamma_{i,t}}\right]^{\perp} d\overrightarrow{W}_t\right].$$

Using normalised weights, we write the basket dynamics with its own coefficients:

$$\frac{dM_t}{M_t} = \mu_t dt + \overrightarrow{\sigma_t}^{\perp} d\overrightarrow{W}_t$$

with

$$\mu_t = \sum_{i=1}^{N} \omega_{i,t}\left[\mu_{i,t} + \eta_{i,t} + \overrightarrow{\sigma_{i,t}}^{\perp}\overrightarrow{\gamma_{i,t}}\right] \quad \text{and} \quad \overrightarrow{\sigma_t} = \sum_{i=1}^{N} \omega_{i,t}\left[\overrightarrow{\sigma_{i,t}} + \overrightarrow{\gamma_{i,t}}\right]. \tag{3.90}$$

The basket is *a priori* not martingale under the generic measure $\mathbb{Q}$ considered so far. In such a (classical) situation we basically have two options: either we modify $\mathbb{Q}$ by computing through Girsanov the drift that would exactly compensate for μ_t, or we recall that $\mathbb{Q}$ was left unspecified in the first place and select it *ad hoc*. In the latter case, we assume that we have initially chosen the M_t-martingale measure, and then report all subsequent changes to the specifications of $\mu_{i,t}$ and $\eta_{i,t}$. These two alternatives are of course mathematically equivalent, and not surprisingly we solve this moot argument by choosing the (easier) second option. Whichever way we choose, we assume now that $\mu_t \equiv 0$ i.e.

$$\frac{dM_t}{M_t} = \overrightarrow{\sigma_t}^{\perp} d\overrightarrow{W}_t,$$

which proves (3.84) and (3.85).

From their definition, the Lognormal dynamics of the normalised weights are

$$\begin{aligned}
\frac{d\omega_{i,t}}{\omega_{i,t}} &= \left[\mu_{i,t} + \eta_{i,t} + \overrightarrow{\sigma_{i,t}}^{\perp}\overrightarrow{\gamma_{i,t}} - \mu_t\right]dt + \left[\overrightarrow{\sigma_{i,t}} + \overrightarrow{\gamma_{i,t}} - \overrightarrow{\sigma_t}\right]^{\perp}\left[d\overrightarrow{W}_t - \overrightarrow{\sigma_t}\right] \\
&= \left[\left[\left(\mu_{i,t} + \eta_{i,t} + \overrightarrow{\sigma_{i,t}}^{\perp}\overrightarrow{\gamma_{i,t}}\right) - \mu_t\right] - \left[\left(\overrightarrow{\sigma_{i,t}} + \overrightarrow{\gamma_{i,t}}\right) - \overrightarrow{\sigma_t}\right]^{\perp}\overrightarrow{\sigma_t}\right]dt \\
&\quad + \left[\left(\overrightarrow{\sigma_{i,t}} + \overrightarrow{\gamma_{i,t}}\right) - \overrightarrow{\sigma_t}\right]^{\perp} d\overrightarrow{W}_t.
\end{aligned}$$

We note that the basket's volatility $\vec{\sigma_t}$ is itself a basket of volatilities, which falls into the framework of the "coefficient basket" covered previously. We can therefore make use of Lemma 3.6, noting that here the weights' dynamics present no exogenous component ($\overrightarrow{c_{i,3,t}} \equiv 0, \forall i, \forall t$).

Applying the Lemma first to $\vec{\sigma_t}$, we have the dynamics:

$$d\vec{\sigma_t} = \vec{a}_{1,t}dt + \vec{\vec{a}}_{2,t}d\vec{W}_t + \vec{\vec{a}}_{3,t}d\vec{Z}_t.$$

Making explicit the non-finite variation coefficients, the Lemma provides

$$\begin{aligned}\vec{\vec{a}}_{2,t} &= \sum_{i=1}^{N} \omega_{i,t} \left[\left[\vec{\vec{a}}_{i,2,t} + \vec{\vec{\alpha}}_{i,2,t}\right] + \left[\overrightarrow{\sigma_{i,t}} + \overrightarrow{\gamma_{i,t}}\right] \otimes \left[\left(\overrightarrow{\sigma_{i,t}} + \overrightarrow{\gamma_{i,t}}\right) - \vec{\sigma_t}\right]\right] \\ &= \sum_{i=1}^{N} \omega_{i,t} \left[\left[\vec{\vec{a}}_{i,2,t} + \vec{\vec{\alpha}}_{i,2,t}\right] + \left[\overrightarrow{\sigma_{i,t}} + \overrightarrow{\gamma_{i,t}}\right]^{2\otimes}\right] - \vec{\sigma_t}^{2\otimes}\end{aligned}$$

and

$$\vec{\vec{a}}_{3,t} = \sum_{i=1}^{N} \omega_{i,t} \left[\vec{\vec{a}}_{i,3,t} + \vec{\vec{\alpha}}_{i,3,t}\right],$$

which proves (3.87) and (3.88). As for the drift coefficient, it comes as

$$\vec{a}_{1,t} \triangleq \sum_{i=1}^{N} \omega_{i,t} \left[\left[\vec{a}_{i,1,t} + \vec{\alpha}_{i,1,t}\right] + \left[\vec{\vec{a}}_{i,2,t} + \vec{\vec{\alpha}}_{i,2,t}\right]\left[\left(\overrightarrow{\sigma_{i,t}} + \overrightarrow{\gamma_{i,t}}\right) - \vec{\sigma_t}\right] + \vec{H}_{i,t}\right],$$

with

$$\vec{H}_{i,t} \triangleq \left[\underbrace{\mu_{i,t} + \eta_{i,t} + \overrightarrow{\sigma_{i,t}}^{\perp}\overrightarrow{\gamma_{i,t}} - \mu_t} - \left[\left(\overrightarrow{\sigma_{i,t}} + \overrightarrow{\gamma_{i,t}}\right) - \vec{\sigma_t}\right]^{\perp} \vec{\sigma_t}\right] \left[\overrightarrow{\sigma_{i,t}} + \overrightarrow{\gamma_{i,t}}\right].$$

Using the weights property (3.83) and the basket martingale condition (3.85), we note that the under-braced part of this expression vanishes due to the summation:

$$\begin{aligned}&\sum_{i=1}^{N} \omega_{i,t} \left[\left[\mu_{i,t} + \eta_{i,t} + \overrightarrow{\sigma_{i,t}}^{\perp}\overrightarrow{\gamma_{i,t}}\right] - \mu_t\right] \\ &\quad = \sum_{i=1}^{N} \omega_{i,t} \left[\mu_{i,t} + \eta_{i,t} + \overrightarrow{\sigma_{i,t}}^{\perp}\overrightarrow{\gamma_{i,t}}\right] - \mu_t \sum_{i=1}^{N} \omega_{i,t} \\ &\quad = 0.\end{aligned}$$

Hence the drift coefficient is

$$\vec{a}_{1,t} = \|\vec{\sigma_t}\|^2\vec{\sigma_t} + \sum_{i=1}^{N}\omega_{i,t}\left[\left[\vec{a}_{i,1,t} + \vec{\alpha}_{i,1,t}\right] - \left[\left(\overrightarrow{\sigma_{i,t}} + \overrightarrow{\gamma_{i,t}}\right)^{\perp}\vec{\sigma_t}\right]\left[\overrightarrow{\sigma_{i,t}} + \overrightarrow{\gamma_{i,t}}\right]\right.$$
$$\left. + \left[\vec{\vec{a}}_{i,2,t} + \vec{\vec{\alpha}}_{i,2,t}\right]\left[\left(\overrightarrow{\sigma_{i,t}} + \overrightarrow{\gamma_{i,t}}\right) - \vec{\sigma_t}\right]\right]. \tag{3.91}$$

Invoking some elementary properties of the outer product (see (D.4) and (D.5) [p. 438]) we can extract and re-formulate some of the right-hand side:

$$\|\vec{\sigma_t}\|^2\vec{\sigma_t} - \sum_{i=1}^{N}\omega_{i,t}\left[\left[\left(\overrightarrow{\sigma_{i,t}} + \overrightarrow{\gamma_{i,t}}\right)^{\perp}\vec{\sigma_t}\right]\left[\overrightarrow{\sigma_{i,t}} + \overrightarrow{\gamma_{i,t}}\right] + \left[\vec{\vec{a}}_{i,2,t} + \vec{\vec{\alpha}}_{i,2,t}\right]\vec{\sigma_t}\right]$$

$$= \vec{\sigma_t}^{2\otimes}\vec{\sigma_t} - \sum_{i=1}^{N}\omega_{i,t}\left[\left(\overrightarrow{\sigma_{i,t}} + \overrightarrow{\gamma_{i,t}}\right)^{2\otimes}\vec{\sigma_t} + \left[\vec{\vec{a}}_{i,2,t} + \vec{\vec{\alpha}}_{i,2,t}\right]\vec{\sigma_t}\right] = -\vec{\vec{a}}_{2,t}\vec{\sigma_t}.$$

Substituting into (3.91) we get

$$\vec{a}_{1,t} = -\vec{\vec{a}}_{2,t}\vec{\sigma_t} + \sum_{i=1}^{N}\omega_{i,t}\left[\left(\vec{a}_{i,1,t} + \vec{\alpha}_{i,1,t}\right) + \left(\vec{\vec{a}}_{i,2,t} + \vec{\vec{\alpha}}_{i,2,t}\right)\left(\overrightarrow{\sigma_{i,t}} + \overrightarrow{\gamma_{i,t}}\right)\right],$$

which proves (3.86). Now in order to obtain the expression for $\vec{\vec{\vec{a}}}_{22,t}$, we need apply Lemma 3.6 again, but this time to the basket

$$\vec{\vec{a}}_{2,t} = \sum_{i=1}^{N}\omega_{i,t}\mathbf{X}_{i,t}$$

with

$$\mathbf{X}_{i,t} = \left[\vec{\vec{a}}_{i,2,t} + \vec{\vec{\alpha}}_{i,2,t}\right] + \left[\overrightarrow{\sigma_{i,t}} + \overrightarrow{\gamma_{i,t}}\right] \otimes \left[\left(\overrightarrow{\sigma_{i,t}} + \overrightarrow{\gamma_{i,t}}\right) - \vec{\sigma_t}\right]. \tag{3.92}$$

The Lemma gives us then that

$$\vec{\vec{\vec{a}}}_{22,t} = \sum_{i=1}^{N}w_{i,t}\left[\mathbf{A}_{i,2,t} + \mathbf{X}_{i,t} \otimes \left[\left(\overrightarrow{\sigma_{i,t}} + \overrightarrow{\gamma_{i,t}}\right) - \vec{\sigma_t}\right]\right], \tag{3.93}$$

where $\mathbf{A}_{i,2,t}$ is the endogenous coefficient of $\mathbf{X}_{i,t}$, which we compute now. To that end, we get very easily the dynamics of its first term:

$$d\left[\vec{\vec{a}}_{i,2,t} + \vec{\vec{\alpha}}_{i,2,t}\right] = [\cdot]\,dt + \left[\vec{\vec{\vec{a}}}_{i,22,t} + \vec{\vec{\vec{\alpha}}}_{i,22,t}\right]d\vec{W}_t + [\cdot]\,d\vec{Z}_t.$$

For the second term, Appendix result (D.10) gives us immediately

$$d\left(\overrightarrow{\sigma_{i,t}}+\overrightarrow{\gamma_{i,t}}\right)^{2\otimes}=[\cdots]\,dt+[\cdots]\,d\overrightarrow{Z}_t$$
$$+\left[\left[\left(\sigma_{i,t}+\gamma_{i,t}\right)_k\left(a_{i,2,t}+\alpha_{i,2,t}\right)_{jl}\right]+\left(\overrightarrow{\sigma_{i,t}}+\overrightarrow{\gamma_{i,t}}\right)\otimes\left(\overrightarrow{\overrightarrow{a}}_{i,2,t}+\overrightarrow{\overrightarrow{\alpha}}_{i,2,t}\right)\right]d\overrightarrow{W}_t$$

and

$$d\left[\left[\overrightarrow{\sigma_{i,t}}+\overrightarrow{\gamma_{i,t}}\right]\otimes\overrightarrow{\sigma_t}\right]=[\cdots]\,dt$$
$$+[\cdots]\,d\overrightarrow{Z}_t+\left[\left[\sigma_{t_k}\left(a_{i,2,t}+\alpha_{i,2,t}\right)_{jl}\right]+\left(\overrightarrow{\sigma_{i,t}}+\overrightarrow{\gamma_{i,t}}\right)\otimes\overrightarrow{\overrightarrow{a}}_{2,t}\right]d\overrightarrow{W}_t.$$

So that, overall, the endogenous coefficient comes as

$$\mathbf{A}_{i,2,t}=\left[\overrightarrow{\overrightarrow{a}}_{i,22,t}+\overrightarrow{\overrightarrow{\alpha}}_{i,22,t}\right]+\left[\left(\sigma_{i,t}+\gamma_{i,t}-\sigma_t\right)_k\left(a_{i,2,t}+\alpha_{i,2,t}\right)_{jl}\right]$$
$$+\left[\overrightarrow{\sigma_{i,t}}+\overrightarrow{\gamma_{i,t}}\right]\otimes\left[\overrightarrow{\overrightarrow{a}}_{i,2,t}+\overrightarrow{\overrightarrow{\alpha}}_{i,2,t}-\overrightarrow{\overrightarrow{a}}_{2,t}\right].$$

Putting this expression into (3.93) and omitting the time argument, we obtain

$$\overrightarrow{\overrightarrow{a}}_{22}=\sum_{i=1}^{N}w_i\left[\left[\overrightarrow{\overrightarrow{a}}_{i,22}+\overrightarrow{\overrightarrow{\alpha}}_{i,22}\right]+\overrightarrow{\overrightarrow{K}}_{i,t}\right]$$

with

$$\overrightarrow{\overrightarrow{K}}_{i,t}=\left[\left(\sigma_i+\gamma_i-\sigma_t\right)_k\left(a_{i,2}+\alpha_{i,2}\right)_{jl}\right]+\left(\overrightarrow{\sigma_i}+\overrightarrow{\gamma_i}\right)\otimes\left[\overrightarrow{\overrightarrow{a}}_{i,2}+\overrightarrow{\overrightarrow{\alpha}}_{i,2}-\overrightarrow{\overrightarrow{a}}_{2}\right]$$
$$+\left[\overrightarrow{\overrightarrow{a}}_{i,2}+\overrightarrow{\overrightarrow{\alpha}}_{i,2}\right]\otimes\left[\left(\overrightarrow{\sigma_i}+\overrightarrow{\gamma_i}\right)-\overrightarrow{\sigma_t}\right]$$
$$+\left[\overrightarrow{\sigma_i}+\overrightarrow{\gamma_i}\right]\otimes\left[\left(\overrightarrow{\sigma_i}+\overrightarrow{\gamma_i}\right)-\overrightarrow{\sigma_t}\right]^{2\otimes}.$$

In the first line of the expression for $\overrightarrow{\overrightarrow{K}}_{i,t}$, we extract from the last term that

$$\sum_{i=1}^{N}w_i\left(\overrightarrow{\sigma_i}+\overrightarrow{\gamma_i}\right)\otimes\overrightarrow{\overrightarrow{a}}_2=\overrightarrow{\sigma}_t\otimes\overrightarrow{\overrightarrow{a}}_2.$$

Furthermore, in the second line of that expression we have

$$\left[\overrightarrow{\overrightarrow{a}}_{i,2,t}+\overrightarrow{\overrightarrow{\alpha}}_{i,2,t}\right]\otimes\left[\left(\overrightarrow{\sigma_{i,t}}+\overrightarrow{\gamma_{i,t}}\right)-\overrightarrow{\sigma_t}\right]+\left[\overrightarrow{\sigma_{i,t}}+\overrightarrow{\gamma_{i,t}}\right]\otimes\left[\left(\overrightarrow{\sigma_{i,t}}+\overrightarrow{\gamma_{i,t}}\right)-\overrightarrow{\sigma_t}\right]^{2\otimes}$$

$$= \left[\left[\vec{\vec{a}}_{i,2,t} + \vec{\vec{\alpha}}_{i,2,t}\right] + \left[\vec{\sigma}_{i,t} + \vec{\gamma}_{i,t}\right]^{2\otimes} - \left(\vec{\sigma}_{i,t} + \vec{\gamma}_{i,t}\right) \otimes \vec{\sigma}_t\right] \otimes \left(\vec{\sigma}_{i,t} + \vec{\gamma}_{i,t}\right)$$

$$- \underbrace{\left[\left[\vec{\vec{a}}_{i,2,t} + \vec{\vec{\alpha}}_{i,2,t}\right] + \left[\vec{\sigma}_{i,t} + \vec{\gamma}_{i,t}\right]^{2\otimes} - \left(\vec{\sigma}_{i,t} + \vec{\gamma}_{i,t}\right) \otimes \vec{\sigma}_t\right]}\otimes \vec{\sigma}_t .$$

We identify the under-braced term above with (3.92), hence after summation

$$\vec{\vec{a}}_{22} = \sum_{i=1}^{N} w_i \left[\left(\vec{\vec{a}}_{i,22} + \vec{\vec{\alpha}}_{i,22}\right) + \left[(\sigma_i + \gamma_i - \sigma)_k \left(a_{i,2} + \alpha_{i,2}\right)_{jl}\right]\right]$$

$$+ \sum_{i=1}^{N} w_i \left[(\vec{\sigma}_i + \vec{\gamma}_i) \otimes \left(\vec{\vec{a}}_{i,2} + \vec{\vec{\alpha}}_{i,2}\right) + \left(\vec{\vec{a}}_{i,2} + \vec{\vec{\alpha}}_{i,2}\right) \otimes (\vec{\sigma}_i + \vec{\gamma}_i)\right]$$

$$+ \sum_{i=1}^{N} w_i \left[(\vec{\sigma}_i + \vec{\gamma}_i)^{2\otimes} - (\vec{\sigma}_i + \vec{\gamma}_i) \otimes \vec{\sigma}\right] \otimes (\vec{\sigma}_i + \vec{\gamma}_i)$$

$$- \left[\vec{\vec{a}}_2 \otimes \vec{\sigma} + \vec{\sigma} \otimes \vec{\vec{a}}_2\right],$$

which proves (3.89) and concludes the proof. □

3.5.5 The Asset Basket Specialised to Fixed Weights

We now restrict our *asset basket* framework, as presented in Definition 3.8, to fixed (constant) weights λ_i. The basket could correspond to an index price process, for instance, or to a simple mixture model. Whichever way, we assume the existence of an option price continuum written on the basket, itself re-parametrised *via* Lognormal implied volatility. Then we have the following results for the small-time asymptotics of that smile:

Corollary 3.4 (Chaos dynamics of the fixed-weights asset basket) *Given the framework of Definition* 3.8 *where the weights $\lambda_{i,t}$ are considered constant $\forall(i,t)$, the dynamic coefficients of the chaos expansion for the basket's dynamics (a.k.a the σ-(2,0) group) are*

$$\vec{\sigma}_t = \sum_{i=1}^{N} \omega_{i,t} \vec{\sigma}_{i,t}, \tag{3.94}$$

$$\vec{a}_{1,t} \triangleq -\vec{\vec{a}}_{2,t} \vec{\sigma}_t + \sum_{i=1}^{N} \omega_{i,t} \left[\vec{a}_{i,1,t} + \vec{\vec{a}}_{i,2,t} \vec{\sigma}_{i,t}\right], \tag{3.95}$$

$$\vec{\vec{a}}_{2,t} \triangleq -\vec{\sigma_t}^{2\otimes} + \sum_{i=1}^{N} \omega_{i,t} \left[\vec{\vec{a}}_{i,2,t} + \overrightarrow{\sigma_{i,t}}^{2\otimes} \right], \tag{3.96}$$

$$\vec{\vec{a}}_{3,t} \triangleq \sum_{i=1}^{N} \omega_{i,t} \vec{\vec{a}}_{i,3,t}, \tag{3.97}$$

$$\vec{\vec{\vec{a}}}_{22} = -\vec{\vec{a}}_2 \overset{\leftrightarrow}{\otimes} \vec{\sigma} + \sum_{i=1}^{N} w_i \vec{\vec{a}}_{i,2} \overset{\leftrightarrow}{\otimes} \vec{\sigma_i}$$

$$+ \sum_{i=1}^{N} w_i \left[\vec{\vec{\vec{a}}}_{i,22} + \left[(\sigma_i - \sigma)_k \; a_{i,2\,jl} \right] + \vec{\sigma_i}^{3\otimes} - \vec{\sigma_i} \otimes \vec{\sigma} \otimes \vec{\sigma_i} \right]. \tag{3.98}$$

Therefore the IATM level and skew are simply given by

$$\widetilde{\Sigma}(\star) = \left\| \sum_{i=1}^{N} \omega_{i,t} \overrightarrow{\sigma_{i,t}} \right\|, \tag{3.99}$$

$$\widetilde{\Sigma}_y'(\star) = \frac{1}{2} \| \vec{\sigma_t} \|^{-3} \vec{\sigma_t}^{\perp} \left[\sum_{i=1}^{N} \omega_{i,t} \vec{\vec{a}}_{i,2,t} + \sum_{i=1}^{N} \omega_{i,t} \left(\overrightarrow{\sigma_{i,t}}^{2\otimes} - \vec{\sigma_t}^{2\otimes} \right) \right] \vec{\sigma_t}$$

$$= -\frac{1}{2} \| \vec{\sigma_t} \| + \frac{1}{2} \| \vec{\sigma_t} \|^{-3} \vec{\sigma_t}^{\perp} \left[\sum_{i=1}^{N} \omega_{i,t} \left[\vec{\vec{a}}_{i,2,t} + \overrightarrow{\sigma_{i,t}}^{2\otimes} \right] \right] \vec{\sigma_t}. \tag{3.100}$$

Note that since we are considering Lognormal dynamics for the components and for the basket, although the actual weights (the σ-(2,0) group) are constant, the normalised weights $\omega_{i,t}$ themselves are still stochastic. Also, since the basket is essentially a finite sum of pair products, our computations would clearly be easier if all dynamics were taken with the *Normal* convention.

This is a tempting approach, especially in light of the baseline transfer methods presented in Sect. 3.3.3, which ensure seamless navigation between both the Normal and Lognormal conventions. However, although feasible this method eventually turns out to be more complex, at least at the order with which we are concerned here, i.e. the first layer.

Interpreting the IATM level (3.94) is very straightforward: indeed we now have the basket's instantaneous volatility simply as the (normalised)-weighted sum of the individual volatilities. In other words, that volatility is itself a tensorial basket, but with stochastic weights. Note also that the Markovian dimension is not *a priori* reduced: even with constant individual volatilities $\overrightarrow{\sigma_{i,t}}$, the state variables would still include each single component $S_{i,t}$ through the normalised weights $\omega_{i,t}$. Overall, the result seems fairly intuitive and makes expressing the IATM level for the basket's smile very easy, as per (3.99). Note finally that, had we *a contrario* considered *Normal* conventions for the individuals' and basket's dynamics,[18] the weights would have been identical between the static definition and the dynamic expressions, hence fixed.

The expression for the IATM skew (3.101), however, is a bit more involved. It can naturally be broken down into two weighted sums, again with the same normalised weights. The first of these we identify as the "natural" component, since it involves the weighted average of the individual endogenous coefficients $\overset{\Rightarrow}{a}_{i,2,t}$. The second sum can be interpreted as a dispersion term, which incorporates both the decorrelation and the differences in modulus between the instantaneous volatilities $\sigma_{i,t}$ of the individual underlyings $S_{i,t}$. Note that in the simple but frequent (at least in the literature) case of constant, deterministic or even independent individual volatilities, the individual skews and the first term are null, while the basket's smile itself still exhibits a skew. In order to improve this first interpretation and to extend beyond this limited configuration, we will shortly introduce new concepts, in Sect. 3.5.6.

Proof The assumption of constant weights $\lambda_{i,t}$ translates into

$$\forall i, t \qquad \eta_{i,t} \equiv 0 \qquad \overrightarrow{\gamma_{i,t}} \equiv 0 \qquad \overrightarrow{\alpha}_{i,1,t} \equiv 0 \qquad \overset{\Rightarrow}{\alpha}_{i,2,t} \equiv 0 \qquad \overset{\Rightarrow}{\alpha}_{i,3,t} \equiv 0 \qquad \overset{\Rrightarrow}{\alpha}_{i,22,t} \equiv 0.$$

Injecting this information into (3.84), (3.86)–(3.89) respectively proves the corresponding expressions (3.94)–(3.98). Having obtained the instantaneous coefficients, we can now move on to the IATM coefficients. We have the IATM level from (3.59), hence (3.99), and according to (3.73) [p. 181] the IATM skew comes as

$$\begin{aligned}
\widetilde{\Sigma}'_y(t,0,0) &= \frac{1}{2}\|\overrightarrow{\sigma_t}\|^{-3}\overrightarrow{\sigma_t}^{\perp}\overset{\Rightarrow}{a}_{2,t}\overrightarrow{\sigma_t}\\
&= \frac{1}{2}\|\overrightarrow{\sigma_t}\|^{-3}\overrightarrow{\sigma_t}^{\perp}\left[\sum_{i=1}^{N}\omega_{i,t}\left[\overset{\Rightarrow}{a}_{i,2,t}+\overrightarrow{\sigma_{i,t}}^{2\otimes}\right]-\overrightarrow{\sigma_t}^{2\otimes}\right]\overrightarrow{\sigma_t},
\end{aligned}$$

which proves (3.101) and concludes the proof. □

3.5.6 Interpretation and Applications

We now come back to the general asset basket with stochastic weights, as developed in Sect. 3.5.4. Our intention is to provide more intuitive interpretations for the instantaneous coefficients of the basket, and therefore for the associated IATM differentials.

---

[18] By denoting $dS_t^i = \overrightarrow{\gamma_{i,t}}^{\perp} d\overrightarrow{W}_t$ and $dM_t = \overrightarrow{\gamma_t}^{\perp} d\overrightarrow{W}_t$.

3.5.6.1 A Discrete Measure Associated to the Basket Structure

We first make a reasonably restrictive assumption about the structure of the basket, before introducing a new, discrete probability measure.

Assumption 3.3 This hypothesis corresponds to the non-negativity of the normalised weights:

$$\omega_{i,t} \geq 0 \quad \forall i, \forall t.$$

Note that this situation can be enforced by ensuring the non-negativity of the weighted individuals, since by assumption we have $M_t > 0$:

$$\lambda_{i,t} S_{i,t} \geq 0 \Rightarrow \omega_{i,t} \geq 0 \quad \forall i, \forall t,$$

which is a natural occurrence in most financial baskets (typically indexes, long portfolios, etc.). We can now introduce the main interpretative tool.

Definition 3.9 (*Discrete measure associated to the asset basket*) Under Assumption 3.3 we have

$$\omega_{i,t} \in [0, 1]\, \forall i \qquad \text{and} \qquad \sum_{i=1}^{N} \omega_{i,t} = 1,$$

therefore the collection of normalised weights $\{\omega_{i,t}\}_{1 \leq i \leq N}$ defines a *discrete probability measure* over any indexed collection of N variables, including any family of tensorial processes $\mathbf{X}_{\cdot,t}$.

3.5.6.2 Interpretation in the General Case

We can re-express the basket's drift μ_t and volatility $\vec{\sigma_t}$ from (3.90) into expectancies, under the discrete measure system:

$$\mu_t = \mathbb{E}^{\omega}\left[\mu_{\cdot,t} + \eta_{\cdot,t} + \overrightarrow{\sigma_{\cdot,t}}^{\perp}\overrightarrow{\gamma_{\cdot,t}}\right] \qquad \text{and} \qquad \vec{\sigma_t} = \mathbb{E}^{\omega}\left[\overrightarrow{\sigma_{\cdot,t}} + \overrightarrow{\gamma_{\cdot,t}}\right].$$

This goes to show, through Jensen's inequality, that any convex functional of instantaneous volatility will be lower when applied to the basket, as opposed to being averaged over the individual components. This is satisfied, in particular, by the norm, which confirms the usual and intuitive property that the basket's IATM implied volatility is lower than the weighted average of individual IATM smile levels. This is indeed a simple instance of diversification.

The same interpretation stands true for the first depth coefficients, so that we can adapt the expressions (3.87) and (3.88) for the non-finite variation coefficients respectively into

$$\vec{\vec{a}}_{2,t} = -\vec{\sigma_t}^{2\otimes} + \mathbb{E}^{\omega}\left[\left(\vec{\vec{a}}_{\cdot,2,t} + \vec{\vec{\alpha}}_{\cdot,2,t}\right) + \left[\overrightarrow{\sigma_{\cdot,t}} + \overrightarrow{\gamma_{\cdot,t}}\right]^{2\otimes}\right]$$

and

$$\overrightarrow{\overrightarrow{a}}_{3,t} = \mathbb{E}^{\omega}\left[\overrightarrow{\overrightarrow{a}}_{\cdot,3,t} + \overrightarrow{\overrightarrow{\alpha}}_{\cdot,3,t}\right].$$

It is then interesting to rewrite the endogenous coefficient $\overrightarrow{\overrightarrow{a}}_{2,t}$ as

$$\overrightarrow{\overrightarrow{a}}_{2,t} = \mathbb{E}^{\omega}\left[\left(\overrightarrow{\overrightarrow{a}}_{\cdot,2,t} + \overrightarrow{\overrightarrow{\alpha}}_{\cdot,2,t}\right)\right] + \underbrace{\mathbb{E}^{\omega}\left[\left[\overrightarrow{\sigma_{\cdot,t}} + \overrightarrow{\gamma_{\cdot,t}}\right]^{2\otimes}\right] - \vec{\sigma}_t^{\,2\otimes}}_{\mathbb{E}^{\omega}\left[\left[\overrightarrow{\sigma_{\cdot,t}}+\overrightarrow{\gamma_{\cdot,t}}\right]^{2\otimes}\right]-\left[\mathbb{E}^{\omega}\left[\overrightarrow{\sigma_{\cdot,t}}+\overrightarrow{\gamma_{\cdot,t}}\right]\right]^{2\otimes}}.$$

The under-braced term appears naturally as the usual covariance matrix of a random N-dimensional variable, which we denote by $\mathbb{V}^{\omega}_{\otimes}[\cdots]$. Thus we can write $\overrightarrow{\overrightarrow{a}}_{2,t}$ as

$$\overrightarrow{\overrightarrow{a}}_{2,t} = \mathbb{E}^{\omega}\left[\overrightarrow{\overrightarrow{a}}_{\cdot,2,t} + \overrightarrow{\overrightarrow{\alpha}}_{\cdot,2,t}\right] + \mathbb{V}^{\omega}_{\otimes}\left[\overrightarrow{\sigma_{\cdot,t}} + \overrightarrow{\gamma_{\cdot,t}}\right].$$

As for the drift coefficient, omitting time-dependency for clarity, (3.86) becomes

$$\begin{aligned}\vec{a}_1 \triangleq \|\vec{\sigma}\|^2\vec{\sigma} + \mathbb{E}^{\omega}\Big[\left[\vec{a}_{\cdot,1} + \vec{\alpha}_{\cdot,1}\right] - \left[(\vec{\sigma_\cdot} + \vec{\gamma_\cdot})^{\perp}\vec{\sigma}\right]\left[\vec{\sigma_\cdot} + \vec{\gamma_\cdot}\right] \\ + \left[\overrightarrow{\overrightarrow{a}}_{\cdot,2} + \overrightarrow{\overrightarrow{\alpha}}_{\cdot,2}\right]\left[(\vec{\sigma_\cdot} + \vec{\gamma_\cdot}) - \vec{\sigma}\right]\Big].\end{aligned}$$

Turning to the expression for $\overrightarrow{\overrightarrow{\overrightarrow{a}}}_{22,t}$, we observe that it is more involved and difficult to decipher. It exhibits, however, some reassuring symmetry properties, in particular the appearance of double products of the type

$$\overrightarrow{\overrightarrow{a}}_2 \otimes \vec{\sigma} + \vec{\sigma} \otimes \overrightarrow{\overrightarrow{a}}_2.$$

This is the point at which the dimensionality becomes an issue in terms of interpretation, and we shall leave it at that for the time being.

3.5.6.3 Interpretation in the Case of Fixed Weights

In order to extend this interpretation to the IATM differentials, we choose first to restrict our framework to constant weights. Note, however, that, because of the basket's zero drift condition (3.85), the *drift* of the weights disappears from the first layer coefficients. Therefore the following results could be obtained if the weights were deterministic functions of time $\lambda_i(t)$. Making use of the discrete measure introduced above, we can re-express the first-depth coefficients given by Corollary 3.4:

$$\vec{\sigma}_t = \mathbb{E}^{\omega}\left[\overrightarrow{\sigma_{\cdot,t}}\right], \qquad \overrightarrow{\overrightarrow{a}}_{2,t} = \mathbb{E}^{\omega}\left[\overrightarrow{\overrightarrow{a}}_{\cdot,2,t}\right] + \mathbb{V}^{\omega}_{\otimes}\left[\overrightarrow{\sigma_{\cdot,t}}\right], \qquad \overrightarrow{\overrightarrow{a}}_{3,t} = \mathbb{E}^{\omega}\left[\overrightarrow{\overrightarrow{a}}_{\cdot,3,t}\right],$$

$$\vec{a}_{1,t} = -\left[\mathbb{E}^\omega\left[\vec{\vec{a}}_{\cdot,2,t}\right] + \mathbb{V}^\omega_\otimes\left[\vec{\sigma}_{\cdot,t}\right]\right]\mathbb{E}^\omega\left[\vec{\sigma}_{\cdot,t}\right] + \mathbb{E}^\omega\left[\vec{a}_{\cdot,1,t}\right] + \mathbb{E}^\omega\left[\vec{\vec{a}}_{\cdot,2,t}\vec{\sigma}_{\cdot,t}\right].$$

So presented, they lend themselves to more interpretations, depending on our focus. For instance, in the expression for $\vec{a}_{1,t}$ we can extract the term

$$\mathbb{E}^\omega\left[\vec{\vec{a}}_{\cdot,2,t}\vec{\sigma}_{\cdot,t}\right] - \mathbb{E}^\omega\left[\vec{\vec{a}}_{\cdot,2,t}\right]\mathbb{E}^\omega\left[\vec{\sigma}_{\cdot,t}\right],$$

which can be seen as another type of covariance, albeit non-commutative. Moving on to the IATM differentials, the IATM skew is

$$\widetilde{\Sigma}'_y(t,0,0) = \frac{1}{2}\|\vec{\sigma}_t\|^{-3}\vec{\sigma}_t^{\perp}\left[\mathbb{E}^\omega\left[\vec{\vec{a}}_{\cdot,2,t}\right] + \mathbb{V}^\omega_\otimes\left[\vec{\sigma}_{\cdot,t}\right]\right]\vec{\sigma}_t. \tag{3.101}$$

It seems difficult to draw any conclusion when the differential is presented in this form. Let us assume, for instance, that all individual underlyings exhibit an identical IATM level for their implied volatility surface. Then the same IATM level for the basket must be lower, due to the dispersion in individual volatilities, in other words the diversification effect. If we now tried to compare the basket's skew with the individual skews, that scaling issue would significantly complicate the picture. For that reason, our next natural move is to normalise (3.101) by using the notations of Definition 3.7 [p. 181]. We end up with

$$\widetilde{\Sigma}'_y(t,0,0) = \frac{1}{2}\|\vec{\sigma}_t\|^{-1}\left[\vec{u}_t^{\perp}\mathbb{E}^\omega\left[\vec{\vec{a}}_{\cdot,2,t}\right]\vec{u}_t + \vec{u}_t^{\perp}\mathbb{V}^\omega_\otimes\left[\vec{\sigma}_{\cdot,t}\right]\vec{u}_t\right].$$

The first term is the most natural, in the sense that it is similar to an individual IATM skew as per (3.73) [p. 181]. Note that *a priori* we cannot normalise *inside* the expectancy:

$$\|\vec{\sigma}_t\|^{-1}\mathbb{E}^\omega\left[\vec{\vec{a}}_{\cdot,2,t}\right] \neq \mathbb{E}^\omega\left[\vec{\vec{c}}_{\cdot,2,t}\right] = \mathbb{E}^\omega\left[\|\vec{\sigma}_{\cdot,t}\|^{-1}\vec{\vec{a}}_{\cdot,2,t}\right].$$

Also, this first term is different from the weighted average of individual skews:

$$\frac{1}{2}\|\vec{\sigma}_t\|^{-1}\vec{u}_t^{\perp}\mathbb{E}^\omega\left[\vec{\vec{a}}_{\cdot,2,t}\right]\vec{u}_t \neq \mathbb{E}^\omega\left[\frac{1}{2}\vec{u}_{\cdot,t}^{\perp}\vec{\vec{c}}_{\cdot,2,t}\vec{u}_{\cdot,t}\right] = \mathbb{E}^\omega\left[\widetilde{\Sigma}'_{y,\cdot,t}\right].$$

As for the second, corrective term, it is associated to the covariance of the collection of instantaneous volatilities $\sigma_{\cdot,t}$. As such, it reflects the dispersion of that family, both in direction and magnitude. Note that, since $\mathbb{V}^\omega_\otimes\left[\vec{\sigma}_{\cdot,t}\right]$ is a genuine covariance matrix, it stays semi-definite positive, hence the whole corrective term remains non-negative. The influence of dispersion, however, is more difficult to assess, as it will have opposite effects on the inverse modulus $\|\vec{\sigma}_t\|^{-1}$ and on the variance $\mathbb{V}^\omega_\otimes\left[\vec{\sigma}_{\cdot,t}\right]$.

An easy way to bypass this issue is simply to change our variable to the squared implied volatility (closer to variance) and therefore write that

$$\left[\widetilde{\Sigma}^2\right]'_y(t,0,0) = \vec{u}_t^{\perp}\mathbb{E}^\omega\left[\vec{\vec{a}}_{\cdot,2,t}\right]\vec{u}_t + \vec{u}_t^{\perp}\mathbb{V}^\omega_\otimes\left[\vec{\sigma}_{\cdot,t}\right]\vec{u}_t.$$

A specific instance of the fixed-weights basket occurs when all individual components are deemed independent. This subcase is of particular interest, as it corresponds essentially to a mixture: as mentioned in Sect. 3.5.1, this configuration can be attained asymptotically (when $N \to +\infty$) *via* a stochastic volatility model using an independent perturbation, such as FL-SV.

We can even specialise this subcase further, and apply these formulae to an extreme, albeit very common instance of basket. Let us consider the situation where all individual components S_i are indeed independent, but also iso-weighted and identically distributed[19] (i.i.d.). In our case, we will have to assume an identical initial value of S_t as well as identical dynamics (at least asymptotically in t) using independent drivers. Assuming a very large N, as well as enough regularity and integrability for the processes, we know then from the Central Limit Theorem (CLT) that the basket's marginal distribution will tend to a Gaussian. The question is whether and how our asymptotic formulae reflect this behaviour.

To simplify, let us assume that each individual underlying is Lognormal, with identical instantaneous volatility σ. Then we have $\lambda_i = \omega_{i,t} = N^{-1} \forall(i,t)$, so that (using the canonical base for $\mathbb{R}^N$) the instantaneous volatilities of the individuals and of the basket come respectively as

$$\overrightarrow{\sigma_{i,t}} \triangleq \sigma \left[00 \cdots \underset{i}{1} \cdots 0 \right]^{\perp} \qquad \text{and} \qquad \overrightarrow{\sigma_t} = \frac{\sigma}{N} [11 \cdots 1]^{\perp}.$$

Thus, according to our asymptotic results, the IATM *level* of the basket's smile is

$$\widetilde{\Sigma}(t,0,0) = \|\overrightarrow{\sigma_t}\| = \frac{\sigma}{\sqrt{N}}.$$

On the other hand, each individual $S_{i,t}$ process generates a Lognormal marginal distribution at time $T > t$, whose standard deviation is $S_t \sqrt{e^{\sigma^2 T} - 1}$. Therefore, and according to the CLT, the marginal distribution of the basket, for the same date T, converges in law (as $N \nearrow +\infty$) towards a Gaussian random variable with mean S_t and standard deviation $S_t \sqrt{e^{\sigma^2 T} - 1} N^{-\frac{1}{2}}$.

As a consequence we have the asymptotic price of the T-expiry ATM Call written on the basket, which we can also expand for short expiries, at the first order:

$$P_{atm}(T) \underset{N \nearrow +\infty}{\longrightarrow} S_t \sqrt{e^{\sigma^2 T} - 1} \, (2\pi N)^{-\frac{1}{2}} = S_t \sigma \sqrt{T} \, (2\pi N)^{-\frac{1}{2}} + o(\sqrt{T}).$$

[19] We specify this property in the weak sense, but dynamically: in other words, all components follow the same SDE but respond to independent drivers, generating the same, but orthogonal, marginal laws.

However, we can also express this price *via* the ATM Black implied volatility $\Sigma_{atm}(T)$, and proceed with the same expansion:

$$P_{atm}(T) = S_t \left[\mathcal{N}\left(\frac{1}{2}\Sigma_{atm}(T)\sqrt{T}\right) - \mathcal{N}\left(-\frac{1}{2}\Sigma_{atm}(T)\sqrt{T}\right)\right]$$
$$= S_t \Sigma_{atm}(T)\sqrt{T}\,(2\pi)^{-\frac{1}{2}} + o(\sqrt{T}).$$

By comparing both expressions of P_{atm} for short expiries, and taking the limit in $T = t$, we conclude that

$$\widetilde{\Sigma}(t, 0, 0) \underset{N\to+\infty}{\longrightarrow} \frac{\sigma}{\sqrt{N}},$$

which *does* match our asymptotic results.

There is obviously room for some more interpretation in the basket framework. Even on the pure computational front, tools such as the discrete probability measure help in identifying symmetries, inductions schemes and simplifications.

In the context of this study, we will see the basket results used again in Part II, which deals with term structures. Indeed, the archetypal example of a derivative product that can be seen as a basket is the par swap rate. But the overall application field of basket concepts in the interest rates world is in fact much, much wider.

References

1. Osajima, Y.: The asymptotic expansion formula of implied volatility for dynamic SABR model and FX hybrid model. University of Tokyo, Graduate School of Mathematical Sciences, Report (2006)
2. Fournie, E., Lebuchoux, J., Touzi, N.: Small noise expansion and importance sampling. Asymptotic Anal. **14**(4), 361–376 (1997)
3. Kunita, H.: Stochastic Flows and Stochastic Differential Equations. Cambridge University Press, Cambridge (1990)
4. Lewis, A.L.: Option Valuation Under Stochastic Volatility. Finance Press, Newport Beach (2000)
5. Hagan, P.S., Kumar, D., Lesniewski, A.S., Woodward, D.E.: Managing smile risk. Wilmott Mag. **1**(September), 84–108 (2002)
6. Piterbarg, V.V.: Stochastic volatility model with time-dependent skew. Appl. Math. Finance **12**, 147–185 (2005)
7. Durrleman, V.: From implied to spot volatilities. Finance Stochast. **14**(2), 157–177 (2010)
8. Nicolay, D.: Volatility dynamics. Ph.D. thesis, Ecole Polytechnique (2011)
9. Cont, R., da Fonseca, J.: Dynamics of implied volatility surfaces. Quant. Finance **2**, 45–60 (2002)
10. Fouque, J.-P., Papanicolau, G., Sircar, K.R.: Derivatives in Financial Markets with Stochastic Volatility. Cambridge University Press, Cambridge (2000)

11. Beckers, S.: The constant elasticity of variance model and its implications for option pricing. J. Finance **35**(3), 661–673 (1980)
12. Schroder, M.: Computing the constant elasticity of variance option pricing formula. J. Finance **44**(1), 211–219 (1989)
13. Hull, J.C.: Options, Futures, and Other Derivatives, 5th edn. Finance Series. Prentice Hall, Upper Saddle River (2003)
14. Press, W.H., Teukolsky, S.A., Vetterling, W.T., Flannery, B.P.: Numerical Recipes in C, 2nd edn. Cambridge University Press, Cambridge (1992)
15. Piterbarg, V.V.: Markovian projection for volatility calibration. Risk Mag. **20**, 84–89 (2007)
16. Berestycki, H., Busca, J., Florent, I.: Asymptotics and calibration of local volatility models. Quant. Finance **2**, 61–69 (2002)
17. Hagan, P.S., Woodward, D.E.: Equivalent Black volatilities. Appl. Math. Finance **6**, 147–157 (1999)
18. Durrleman, V., El Karoui, N.: Coupling smiles. Quant. Finance **8**, 573–590 (2008)
19. d'Aspremont, A., El Ghaoui, L.: Static arbitrage bounds on basket option prices. Math. Program. **106**, 467–489 (2006)
20. Laurence, P., Wang, T.H.: What's a basket worth? Risk Mag. **17**, 73–77 (2004)
21. Brigo, D., Mercurio, F., Rapisarda, F., Scotti, R.: Approximated moment-matching dynamics for basket-options simulation. Working paper, Banca IMI (2002)
22. Posner, S.E., Milevsky, M.A.: Valuing exotic options by approximating the SPD with higher moments. J. Financial Eng. **7**, 109–125 (1998)
23. Avellaneda, M., Boyer-Olson, D., Busca, J., Fritz, P.: Reconstructing the smile. Risk Mag. **15**(10), 84–108 (2002)
24. d'Aspremont, A.: Interest Rate model calibration and risk-management using semidefinite programming. Ph.D. thesis, Ecole Polytechnique (2003)
25. Piterbarg, V.V.: Mixture of models: a simple recipe for a ... hangover? Wilmott Mag. **13**, 72–77 (2005)
26. Brigo, D., Mercurio, F., Rapisarda, F.: Lognormal-mixture dynamics and calibration to market volatility smiles. Int. J. Theor. Appl. Finance **5**, 427–446 (2002)
27. Gatarek, D.: LIBOR market model with stochastic volatility. Technical Report, Deloitte and Touche (2003)
28. Hull, J., White, A.: The pricing of options on assets with stochastic volatilities. J. Finance **42**, 281–300 (1987)

Chapter 4
Practical Applications and Testing

Abstract In this chapter we turn to even more practical considerations, by applying ACE results to some popular stochastic (instantaneous) volatility models, namely the SABR and FL-SV classes. We start by discussing the financial, practical and numerical issues involved. We then derive the chaos dynamics of each model, up to the third layer, stressing the technical benefits of staying model-generic and of exploiting induction. We can then express the desired static IATM differentials, which we subsequently use in either direct or inverse mode. In inverse fashion, we use those quantities to illustrate an "intuitive" model re-parametrisation of the generic SABR class. In direct mode, we test the flexibility and quality of static smile approximations provided by ACE for the CEV-SABR model, compared to Hagan et al's benchmark.

First in Sect. 4.1 we review general considerations, focusing mainly on the direct problem (due to its importance for practitioners). In particular, we discuss the practicality and expected performance of a Taylor expansion as a smile proxy, destined to be used for calibration and/or within a numerical scheme,[1] or even for hedging.

Then in Sect. 4.2 we apply these principles to the generic SABR model, which is presented in Sect. 4.2.1 both in a trading and mathematical perspective. In Sect. 4.2.2 we derive the coefficients of the chaos diffusion which are required by all static IATM differentials exposed in Sect. 3.2. In Sect. 4.2.3 we use some of these coefficients to establish bilateral relationships between the model parameters and the IATM level, skew and curvature.

We then turn in Sect. 4.3 to the more common CEV instance of the SABR class, beginning naturally in Sect. 4.3.2 by producing the relevant coefficients of the chaos diffusion. In Sect. 4.3.3 we link explicitly the model parameters to the IATM smile, and in Sect. 4.3.4 we show that Hagan's formula (see [1]) does verify these IATM differentials.

In Sect. 4.4 we address another rich stochastic instantaneous volatility model, the FL-SV class. We present the model in Sect. 4.4.1, along with its research background and financial rationale. We introduce a more generic version of the class, called Extended FL-SV, that will be the basis of our work. Then in Sect. 4.4.2 we present

[1] In order to improve Monte-Carlo speed or the exercise boundary definition.

D. Nicolay, *Asymptotic Chaos Expansions in Finance*,
Springer Finance Lecture Notes, DOI 10.1007/978-1-4471-6506-4_4

as an exercise the computation of all static IATM differentials within the first layer. This involves three steps (setting up the induction, deriving the chaos coefficients and applying our direct problem results), for which solutions are provided in Appendix E.

Finally in Sect. 4.5 we illustrate these results with a few simulations. In Sect. 4.5.1 we discuss the various possibilities for the testing phase and justify our choices. We select a model (CEV-SABR) as well as an expansion type, and adopt a testing protocol focused on assessing both precision and validity, in absolute terms and compared to Hagan. Then in Sect. 4.5.2 we graphically present and analyse the results of this testing protocol. In Sect. 4.5.3 we briefly expose our conclusions as well as avenues for further research and experimentation.

Note that in the sequel we could have used the results of the Extended Skew Market Model (ESMM) (see Sect. 2.5.2) in order to generate some or all of the chaos coefficients for SABR and FL-SV. We chose not to do so, essentially because the chaos coefficients computed for the ESMM in Lemma 2.4 [p. 90] were only those of the first layer. Therefore we were faced with two choices: either continue with the ESMM chaos computation, or transfer only these first coefficients to each model, and then compute their dynamics on an *ad hoc* basis.

The first solution is very costly, and the formulae generated by the ESMM for high-depth coefficients very involved. To be convinced of this fact, it suffices to have a look at the simpler Extended FL-SV results given later on (see Lemma E.5 [p. 482]). That option is therefore not adapted to the current applicative chapter, although we do not exclude investigating the matter in further research.

The second solution *is* feasible, and indeed we did apply it initially. However, it turned out to be actually slower than the direct approach. Besides, it made the induction technique (which is obvious in our notation) more difficult to implement as it lacked the initialisation phase.

Finally, it is worth mentioning that the model *classes* being examined in this chapter are fairly generic, and can be customised by either a single (SABR) or five (FL-SV) functionals.

4.1 General Considerations on Practical Applications

In many respects, the current chapter can be seen as an extension of Sect. 2.5, which also focused on the practical illustrations and applications of the theoretical results obtained for the first layer. Indeed, the latter section exploited the Extended Skew Market Model (see Definition 2.2 [p. 88]) which as a very wide two-dimensional model class embraced both SABR and FL-SV.

However, our scope is now both narrower and deeper than that of Sect. 2.5, for several reasons:

- We now have at our disposition a generic method to compute the direct problem, i.e. both static and dynamic IATM differentials, at any required order (Sect. 3.1). In particular, we have expressed in Sect. 3.2 the second and third layers, which

provide the most useful descriptors of the smile. Our intention here is to provide all the information necessary to exploit these higher-order IATM differentials, as opposed to Sect. 2.5.2 [p. 87], which limited itself to the first layer.
- In order to remain manageable, the natural tradeoff to computing such involved formulae is to decrease the complexity of the model itself. Ideally we would have liked to present all IATM differentials, for the second and third layer, in the ESMM framework. This task presents no technical difficulty but the resulting formulae are quite imposing, mitigating or even negating the illustration purpose of the current chapter. In consequence (and regrettably) we will specialise our computations to the specific models at hand (SABR and FL-SV) from the beginning.
- Another difference with Sect. 2.5 is that the latter presented succinctly a quite large array of possible applications (see Sect. 2.5.1 [p. 77]), from pure asymptotics (using both the direct and inverse approaches) to helping with hedging issues, *via* full-smile extrapolations. In contrast, the current chapter will focus mainly on pure asymptotics, with an emphasis on the direct problem, although we will spend some time on reversing the first layer IATM differentials back into the model parameters. Note that the latter approach is not *stricto sensu* equivalent to solving the inverse problem.

Although our focus in the current chapter will be on pure asymptotics for a trio of models, we also wish to discuss in more generality the potential and practical use of these formulae, when applied to other classes of dynamics. These comments should be seen as complementing or extending those of Sect. 2.5.

One might wonder how our asymptotic and generic formulae actually compare in practice, compared to *ad hoc* techniques developed for a given model class. There are numerous and rigorous candidate methods in the latter category, usually approximate in nature and often following asymptotic approaches[2] (see [2, 3] or [4], for instance) but not always (for example [5] or [6]). Our take is that these different approaches should not be seen as exclusive but as complementary, if at all possible. In that spirit, what makes the asymptotic chaos expansion method[3] remarkable is its level of compatibility with most other approaches. Indeed, it provides pure differentials which are *raw* results, as opposed to a functional approximation such as [1]. In that respect its results can easily be used to compare, adjust or replace all or part of a full-smile approximation, for instance. A typical example would be to correct Gatheral's heuristic formula at the second order (which is itself rooted in a low-order asymptotic expansion) with the correction for local volatility models found in Sect. 2.4.2.3 [p. 72].

An important feature to take into account is the actual *stationarity* of the (input) SInsV model. Indeed, the asymptotic chaos expansion, because it is differential and not integral in nature, relies on the assumption that the current dynamics are

---

[2] Often these methods will use singular perturbation and/or expansions on the vol of vol parameter, which creates the deviation from the Normal or Lognormal model, which itself provides either a manageable infinitesimal generator or the closed-form formula. Pure chaos as in [2] is less frequently found.

[3] As described initially in [7] and then in this study.

representative of what they will be throughout the life of the option. The pertinence of the IATM differentials, or rather the precision of any functional full-smile extrapolation will depend on the degree of *stationarity* and time homogeneity of the process. Although an instantaneous volatility model with time-dependent (and even discontinuous) volatility can be managed within our asymptotic methodology (see Sect. 3.3.1.3) the same cannot necessarily be said for correlation or vovol. Obviously the method will cope well with the smooth, deterministic evolution of some parameter $\mu(s)$. It will do so in the same way (and probably even better) than it would if that parameter was actually stochastic. But the asymptotic algorithm envisages the dynamics of $\mu(s)$ only from what it "knows" at initial time t. Therefore if $s \to \mu(s)$ is smooth, then successive orders of the expansion will approximate this map better and better, by calling in higher and higher derivatives (taken at the origin) i.e. $\frac{\partial^n \mu}{\partial s^n}|_{s=t}$. Basically this corresponds to approximating the $\mu(s)$ function with polynomials of increasing order in the variable $(s-t)$. This is fine for a smooth function, but obviously if $\beta(s)$ were to be piecewise constant, for instance, then *a priori* the associated derivatives and therefore the long-term smile approximations would necessary be of a lower quality.

This kind of limitation is not to be confused with some other classical configurations, where the initial dynamics might be misleading and lead the modeller to undue pessimism about the performance potential of the method. A typical example would be a Lognormal process with mean-reverting multiplicative instantaneous volatility (such as Heston) and where the initial value of the latter is exactly the long-term mean. Clearly a_1, a_{11}, a_{111} and so on are all null, but we know from Sect. 3.2.3 [p. 142] that deeper layers will take into account the mean-reversion.[4]

The choice of a good *baseline model* is important but usually not difficult, essentially because there aren't so many candidate models available. Obviously the baseline must be associated to a closed-form solution,[5] but within the few possible choices we should use one that closely resembles the input SInsV dynamics, or that generates similar smiles. Indeed, it is fairly intuitive that if we start closer to the solution (in terms of statics and perhaps dynamics too) then a lower number of differentials is needed in order to achieve the same precision. For a CEV-SABR with low β (or very negative correlation), for instance, which generates large amounts of skew, it would make sense to use the Gaussian model as a baseline, rather than the Lognormal one. Conversely, in a pseudo-Lognormal setup (with high β and small effective correlation $\rho\nu$) the Black model would seem more appropriate. This is all very well and intuitive, but what should we do then with a Heston model, which, depending on correlation (and therefore vovol), can exhibit either a low or high skew? A first answer would be to choose whichever baseline produces the lowest endogenous vovol a_2 for the input SInsV model, once cast. But this matter is still very much an open subject and will clearly depend on the objective function of the modeller: which part of the model is he/she concentrating on, which static or dynamic properties is he/she

---

[4] If κ was null then we would have $a_{12} = 0$, which is not the case.

[5] See Sect. 3.3 [p. 146] for a more in-depth discussion of the various extensions and alternatives to the mainstream methodology.

looking for? For instance, if one considers a SABR model and insists on the support of the distribution being faithful, then a specific displaced diffusion baseline might be chosen, etc.

Selecting an appropriate *moneyness*, or more conventionally a pair of variables in bijection with strike K and maturity T, can also be both important and difficult. As discussed in Chaps. 2 and 3, the main purpose of this change of variable is to enforce (or benefit from) a given type of stationarity, in the smile this time. Other methodologies do employ this type of technique: in [2], for instance, the time-to-maturity variable is taken as the accumulated variance $\int_t^T \sigma_s^2 ds$, which corresponds both to a sliding parametrisation and to a time change.

But these variables (say y and θ, for instance) are those of the framework, not necessarily the best ones to conduct a series expansion on, if one chooses to do so. As discussed in Sect. 2.5.1.3, a straight Maclaurin (or Taylor) series expansion is not necessarily the best option. There are other functional families on which to project the asymptotic information, one of which will be discussed shortly. For those readers interested in extrapolation methods and their respective efficiency in a more general context, we refer to [8]. Nevertheless, for the time being we will stay with Taylor series as they are fairly demonstrative of the wings issue. Indeed, it is the behaviour of the extrapolation for very low and very high strikes that will be its main quality criteria. We see two main and related issues for the extrapolated smile: the first is the *validity* of the marginal density associated to it, and the second is its *precision* with regard to the true IV generated by the input SInsV model.

The validity criteria is very difficult to gauge visually for a given smile, as a positive C''_{KK} translates into a complex, non-linear PDE for the implied volatility. However, negative densities tend to appear on the wings first, and on the left in particular (this is certainly the case with the Lognormal formula of [1]). There is indeed a symmetric bound on the implied volatility, imposed by arbitrage constraints, imposing that $\widetilde{\Sigma}(t, y, \theta)$ should not grow faster than $\sqrt{|y|}$ at a given expiry (see [9], for instance). It follows that if we can control the wings then we should improve matters across the whole strike range.

The precision of the method is perfect at the IATM point, hence the question is the speed at which it degrades as both $|y|$ and θ increase. Again, for many practical models we have good knowledge of the wing behaviour: this is the case for local volatility models (see [5], for instance) but for wider classes as well, including stochastic volatility (see [9–11], for instance). The problem is of course similar for high expiries, as exact or approximate results can be found in the literature. For large θ asymptotics in mean-reverting models, for instance, see [12, 13]. When theoretical results are not available, traders will often express a *view* for the extreme strike and extreme maturity behaviour.

Since we wish to control the asymptotics in those two directions, the idea of a polynomial in y and θ is not very palatable. Let us present alternative methods: Since the problem is similar but more demonstrative in K than in T, we will focus on the strike dimension.

If we stay with Taylor series then higher orders bring more precision and the only degree of freedom is the choice of the expansion variable. Note that we can apply a transform to either y or $\widetilde{\Sigma}$, whichever is more convenient (this is mathematically equivalent). Keeping in mind that we are now touching heuristics, we propose a few solutions that have been successfully tested. The first method is to adapt the expanded *functional* according to the chosen order. For instance, with a fourth-order expansion we can expand $\widetilde{\Sigma}^8$ on y and then re-scale down, ensuring that the $\sqrt{|y|}$ barrier is not breached. Another method consists in choosing an expansion variable $z(y)$ that will provide sub-square root growth when taken at any power: clearly $z = \ln(1 + |y|)$ is a good starting point. Again another choice for $z(y)$ is a variable that itself exhibits a limit when $|y| \to +\infty$, forcing the Taylor series to converge to a fixed value in the wings. Ideally that limit on $z(\pm\infty)$ should be controllable so that the IV limit itself can be set. In that respect, variations on the theme of decreasing exponentials or of the cumulative Normal can be tried, but the Delta should be avoided as it leads to some delicate fixed-point issues.

An important point to note is that we do not have to use the same expansion for positive and negative y. Indeed, as long as the junction is smoothly established at-the-money there should be no issue. In particular, if the distribution support provides a lower bound (typically for an *a.s.* positive underlying), then on the left-hand side we can expand simply on strike K.

An altogether more rigorous and generic approach is to employ Padé approximants, which use rational functions whereas Taylor employs polynomials:

$$f(x) \simeq R(x) = \frac{p_0 + p_1 x + p_2 x^2 + \cdots + p_m x^m}{1 + q_1 x + q_2 x^2 + \cdots + q_n x^n}.$$

Note that by construction this family contains the polynomials, and in the same manner as Taylor the p_i and q_i coefficients are selected to ensure matching of the first differentials at the origin

$$f^{(i)}(0) = R^{(i)}(0) \qquad \forall 0 \leq i \leq m + n.$$

Although the matching algorithm is more involved, additional control can be provided to ensure given asymptotes when $x \to \pm\infty$, so that it seems an overall more attractive alternative. For a detailed description of some numerical implementation aspects, refer to [14] (Sect. 5.12).

However, whether with Padé or Taylor, the classical issues of series expansions do remain, in particular w.r.t. their actual *radius of convergence*. Consider, for instance, a Taylor expansion of $f(x) = x^n$ around $x = 0$ when n is high. Only one differential is non-null, at order n, but this is sufficient to reconstruct the whole function. The question is therefore whether the differentials will have been computed to the required order. It is of course possible to project the IATM information onto other functional families than rational and polynomial functions. But the feasibility of the IATM match, as well as the performance in the high-$|y|$ and/or high-θ regions, is likely to be *ad hoc* rather than generic (which is a main focus of this study). Whereas with

Padé or Taylor, we know that (within the convergence radius) a high enough order will give us any required degree of precision, and we know that this order is attainable in a generic fashion (as illustrated in Sect. 3.1).

With regard to smile validity, it is tempting to re-parametrise the surface back into prices, since the marginal is much easier to control from the Call price surface. At a single expiry for instance and with $\underline{S}$ the floor for S_t, we must simply enforce the following four criteria for $C(K)$:

$$C(\underline{S}) = S_t \qquad C'_K(\underline{S}) = -1 \qquad C(K) \underset{K\nearrow+\infty}{\searrow} 0 \qquad C''_{KK} \geq 0 \quad \forall K.$$

Imposing a valid surface is not much more difficult, so that would seem to be the best way to proceed. Unfortunately, our method is asymptotic, and the Black function does not lend itself well to strike differentiation at the IATM point. Indeed, recall that this was one of the main advantages of using implied volatility: as a regularisation function.

There is another avenue, though, which is to work either on the implied cumulative or density, but these techniques fall outside the scope of this study. It is nevertheless worth noting that an expansion on $\widetilde{\Sigma}$ can indeed be considered as a mixture of implied volatilities, but that this not equivalent to a mixture of marginals, processes or underlyings.

Finally, let us mention the *dynamic coherence* aspect, as we have only reasoned in static terms so far. We first note that the static validity conditions must be enforced at current time, but also dynamically. However, in order to apply dynamics ($\widetilde{b}$, $\vec{\nu}$ and $\vec{n}$) one can approximate them statically as well. To simplify matters, we assume a SInsV model with state vector $\vec{x_t} = (t, S_t, \alpha_t)$ driven by two scalar drivers W_t and Z_t. We reason with an absolute rather than sliding representation, and use a Taylor expansion on strike. We note then as follows the relevant static approximations (b is redundant by arbitrage) at the first order:

$$\Sigma \approx \Sigma^* \triangleq \lambda_0^\Sigma(\vec{X_t}) + \lambda_1^\Sigma(\vec{X_t})K \qquad \nu \approx \nu^* \triangleq \lambda_0^\nu(\vec{X_t}) + \lambda_1^\nu(\vec{X_t})K$$

$$n \approx n^* \triangleq \lambda_0^n(\vec{X_t}) + \lambda_1^n(\vec{X_t})K.$$

The question is then to ensure consistency between, on one hand, ν^* and n^* , and, on the other hand, the dynamic coefficients of

$$d\Sigma^* = d\left[\lambda_0^\Sigma(\vec{X_t})\right] + Kd\left[\lambda_1^\Sigma(\vec{X_t})\right].$$

We know that this coherence is established at the IATM point, but we should enforce it throughout the smile. However, this level of concern falls outside the scope of the current study.

4.2 Application to the Generic SABR Class

The SABR model class was introduced in [1], initially for the IR vanilla market, and quickly became popular due to a combination of several factors. Its first attraction was the capacity to match the traded smile, showing significant skew and curvature, so that its CEV instance has now become the market standard to mark European Swaptions (and Caplets to a lesser extent). The second novelty was its capacity to provide adequate and clear dynamics for the smile, combining both local and correlated stochastic volatility to ensure a good degree of Delta stickiness, along with a controllable backbone. The third feature (noticeably well exposed in [1]) was the simple one-on-one role played by each of the parameters and initial values, on both the statics and dynamics of the smile. Indeed it is not rare today to hear practitioners mention *correlation* when they actually mean *skew*, and so on. Finally, the model was delivered with a closed-form approximation for vanilla options. The most commonly used is the Lognormal implied volatility approximation for CEV-SABR (see (4.68) [p. 240]), but [1] actually contains three formulae (price, Lognormal and Normal IV) which are provided for any local volatility function. Developed using a singular perturbation technique, they can (and will), unfortunately, generate negative implied densities on the left-hand side.

4.2.1 Presentation of the Model

The SABR class is often deemed to belong to the "local-stochastic volatility" family and describes the dynamics of the single underlying under its associated martingale measure as

$$(\mathbf{SABR}) \qquad \begin{cases} dS_t = \alpha_t f(S_t)\, dW_t & (4.1) \\ & \langle dW_t, dB_t\rangle = \rho\, dt \\ d\alpha_t = \nu\, \alpha_t\, dB_t & (4.2) \end{cases}$$

with $f(\cdot)$ a local volatility function satisfying the usual regularity conditions. Note that α_t provides a multiplicative perturbation that scales up and down the backbone[6] for any expiry $T \geq t$. The latter is almost entirely determined by the product $\alpha_t S_t^{-1} f(S_t)$ whereas the smile exhibits a strong dependency on correlation ρ and vovol ν. Note that the dynamics of the perturbation α_t are Lognormal, and this lack of mean-reversion for the instantaneous volatility/variance increase significantly the terminal variance of S_T. Apart from being unrealistic and capable of causing explosions (see [15]) or integrability issues, these dynamics also create difficulties for

[6] The backbone is defined as the function $S_t \to \Sigma(K = S_t, T)$. In the absence of parameter T this function will refer to a very short time-to-maturity.

numerical schemes. To mitigate this issue some time-dependent parameters can be used, and [1] caters for those under the label of *dynamic* SABR. However, other authors have actually extended the model with a volatility mean-reversion into the λ-SABR class (see [6]).

The SABR model has been the object of active academic research. Several alternative approximations have therefore been published, ranging from customisation of the existing Hagan et al's formulae to radically different approaches: let us name but a few. In [16] Hagan's *price* formula is modified using a local time technique. Interestingly, although the resulting approximation presents issues of the same magnitude as Hagan's Lognormal IV (LNIV) formula, these are endemic to different areas of the smile. In consequence, a combination of both approaches can be envisaged. In the short paper [17], the fundamental results of [18] are exploited to identify some inconsistent behaviour in Hagan's LNIV formula for the CEV-SABR instance, but the proposed corrections only mitigate the issue of negative density. In [2] some Wiener chaos is used but in an integral fashion, and although the method is not presented as generic in order, it is very clearly laid out and could probably be used on other model classes. The same author has also explored hybrid versions of SABR and the related expansions. Last but not least, let us mention the hyperbolic geometry approach taken in [6].

4.2.2 Coefficients of the Chaos Dynamics

Let us pool all the available input expressions for the static IATM differentials, in other words $\widetilde{\Sigma}^{(m,p)}$ in the first, second and third layers:

$$\widetilde{\Sigma}(\star)(2.36) \qquad \widetilde{\Sigma}'_y(\star)(2.51) \qquad \widetilde{\Sigma}''_{yy}(\star)(2.52) \qquad \widetilde{\Sigma}'_\theta(\star)(2.53)$$
$$\widetilde{\Sigma}''_{y\theta}(\star)(\text{E}.38) \qquad \widetilde{\Sigma}'''_{yy\theta}(\star)(\text{E}.46) \qquad \widetilde{\Sigma}''_{\theta\theta}(\star)(\text{E}.62).$$

A survey of the involved chaos dynamic coefficients gives us the following set:

Definition 4.1 (*Chaos coefficients for the first three static layers*)

$$\begin{array}{ccccccccccc}
 & & & & & \sigma_t & & & & & \\
 & a_1 & & & & a_2 & & & & a_3 & \\
a_{11} & a_{12} & a_{13} & & a_{21} & a_{22} & a_{23} & & a_{31} & a_{32} & a_{33} \\
 & a_{122} & & a_{212} & a_{221} & a_{222} & a_{223} & a_{232} & & a_{322} & \\
 & & & & & a_{2222} & & & & &
\end{array}$$

We aim at expressing all of these 21 coefficients as functions of the model parameters, however these involve, on top of the constants (β, ρ, ν) and the initial values (S_t, α_t), the local volatility function $f(\cdot)$. It is clear that with our Lognormal baseline the function of interest will soon become $g(x) \triangleq x^{-1}f(x)$, or rather the successive derivatives $g^{(n)}(x)$ of this new expression. We could express these fully, but the formulae are already quite complex, and this is a low-level computation. Besides, Lemma 2.3 [p. 89] provides all the necessary tools, hence we have opted for semi-compact notations that work by induction, and that we now introduce.

Definition 4.2 For the LV function $f(\cdot)$ regular enough, we set the following functional notations:

$$
\begin{array}{lll}
g(x) \triangleq f(x)x^{-1} & g_1(x) \triangleq f(x)g'(x) & g_2(x) \triangleq f^2(x)g''(x) \\
g_{11}(x) \triangleq f(x)g_1'(x) & g_{12}(x) \triangleq f^2(x)g_1''(x) & g_{21}(x) \triangleq f(x)g_2'(x) \\
g_{22}(x) \triangleq f^2(x)g_2''(x) & g_{111}(x) \triangleq f(x)g_{11}'(x) & g_{112}(x) \triangleq f^2(x)g_{11}''(x) \\
g_{121}(x) \triangleq f(x)g_{12}'(x) & g_{211}(x) \triangleq f(x)g_{21}'(x) & g_{1111}(x) \triangleq f(x)g_{111}'(x).
\end{array}
$$

In turn, these allow us to define the $G^*_{\star,t}$ processes of the SABR model:

$$
\begin{array}{lll}
G^2_{1,t} \triangleq \alpha_t^2 g_1(S_t) & G^3_{2,t} \triangleq \alpha_t^3 g_2(S_t) & G^3_{11,t} \triangleq \alpha_t^3 g_{11}(S_t) \\
G^4_{12,t} \triangleq \alpha_t^4 g_{12}(S_t) & G^4_{21,t} \triangleq \alpha_t^4 g_{21}(S_t) & G^4_{111,t} \triangleq \alpha_t^4 g_{111}(S_t) \\
G^5_{22,t} \triangleq \alpha_t^5 g_{22}(S_t) & G^5_{112,t} \triangleq \alpha_t^5 g_{112}(S_t) & G^5_{121,t} \triangleq \alpha_t^5 g_{121}(S_t) \\
G^5_{211,t} \triangleq \alpha_t^5 g_{211}(S_t) & G^5_{1111,t} \triangleq \alpha_t^5 g_{1111}(S_t) &
\end{array}
$$

where the simple logic of the tagging relies on the following rules:

$$
g_{\star 1}(x) \triangleq f(x)g_\star'(x) \qquad g_{\star 2}(x) \triangleq f^2(x)g_\star''(x) \qquad G^n_{\star,t} \triangleq \alpha_t^n g_\star(S_t).
$$

We can now provide semi-compact expressions for the relevant chaos coefficients.

Lemma 4.1 (Chaos dynamics up to 3rd layer for the generic SABR model) *In the Generic SABR model* (4.1)–(4.2) *the chaos coefficients are as follows.*

First-level coefficients:

$$\sigma_t = \alpha_t g(S_t) \tag{4.3}$$

$$a_{1,t} = \rho\nu G_1^2 + \frac{1}{2}G_2^3 \tag{4.4}$$

$$a_{2,t} = \rho\nu\sigma_t + G_1^2 \tag{4.5}$$

$$a_{3,t} = \dot{\rho}\nu\sigma_t \tag{4.6}$$

Second-level coefficients:

$$a_{11,t} = \rho\nu^3 G_1^2 + 2\rho^2\nu^2 G_{11}^3 + \frac{3}{2}\nu^2 G_2^3 + \frac{1}{2}\rho\nu G_{12}^4 + \frac{3}{2}\rho\nu G_{21}^4 + \frac{1}{4}G_{22}^5 \tag{4.7}$$

$$a_{12,t} = 2\rho^2\nu^2 G_1^2 + \rho\nu G_{11}^3 + \frac{3}{2}\rho\nu G_2^3 + \frac{1}{2}G_{21}^4 \tag{4.8}$$

$$a_{13,t} = 2\rho\dot{\rho}\nu^2 G_1^2 + \frac{3}{2}\dot{\rho}\nu G_2^3 \tag{4.9}$$

$$a_{21,t} = \rho\nu a_1 + \nu^2 G_1^2 + 2\rho\nu G_{11}^3 + \frac{1}{2}G_{12}^4 \tag{4.10}$$

$$a_{22,t} = \rho\nu a_2 + 2\rho\nu G_1^2 + G_{11}^3 \tag{4.11}$$

$$a_{23,t} = \rho\nu a_3 + 2\dot{\rho}\nu G_1^2 \tag{4.12}$$

$$a_{31,t} = \dot{\rho}\nu a_1 \tag{4.13}$$

$$a_{32,t} = \dot{\rho}\nu a_2 \tag{4.14}$$

$$a_{33,t} = \dot{\rho}\nu a_3 \tag{4.15}$$

Third-level coefficients:

$$\begin{aligned} a_{122,t} = {} & 4\rho^3\nu^3 G_1^2 + 5\rho^2\nu^2 G_{11}^3 + \rho\nu G_{111}^4 \\ & + \frac{9}{2}\rho^2\nu^2 G_2^3 + \frac{7}{2}\rho\nu G_{21}^4 + \frac{1}{2}G_{211}^5 \end{aligned} \tag{4.16}$$

$$\begin{aligned} a_{212,t} = {} & \rho\nu a_{12} + \nu^2\left[2\rho\nu G_1^2 + G_{11}^3\right] \\ & + 2\rho\nu\left[3\rho\nu G_{11}^3 + G_{111}^4\right] + \frac{1}{2}\left[4\rho\nu G_{12}^4 + G_{121}^5\right] \end{aligned} \tag{4.17}$$

$$\begin{aligned} a_{221,t} = {} & \rho\nu a_{21} + 2\rho\nu^3 G_1^2 + \nu^2\left(3 + 4\rho^2\right) G_{11}^3 \\ & + 3\rho\nu G_{111}^4 + \rho\nu G_{12}^4 + \frac{1}{2}G_{112}^5 \end{aligned} \tag{4.18}$$

$$a_{222,t} = \rho\nu a_{22} + 4\rho^2\nu^2 G_1^2 + 5\rho\nu G_{11}^3 + G_{111}^4 \tag{4.19}$$

$$a_{223,t} = \rho\nu a_{23} + 4\rho\dot{\rho}\nu^2 G_1^2 + 3\dot{\rho}\nu G_{11}^3 \tag{4.20}$$

$$a_{232,t} = \rho\nu a_{32} + 4\rho\dot{\rho}\nu^2 G_1^2 + 2\dot{\rho}\nu G_{11}^3 \tag{4.21}$$

$$a_{322,t} = \dot{\rho}\nu a_{22} \tag{4.22}$$

And finally the single fourth-level coefficient:

$$a_{2222,t} = \rho\nu a_{222} + 8\rho^3\nu^3 G_1^2 + 19\rho^2\nu^2 G_{11}^3 + 9\rho\nu G_{111}^4 + G_{1111}^5. \tag{4.23}$$

An immediately obvious feature is that the correlation parameter ρ, as well as its counterpart $\dot{\rho}$, will never be found without the vol of vol ν in factor. This feature is well-known by practitioners, has been noted before in Chap. 2 and is due to the artificial decomposition imposed by the model dynamics (4.1)–(4.2). Indeed, once rewritten with a pair of orthogonal drivers, both $\rho\nu$ and $\dot{\rho}\nu$ become the diffusion coefficients of α_t, hence by induction they will stay in such form throughout the chaos.

Instead of heading into the proof, let us first establish a simple preliminary result:

Lemma 4.2 (Dynamics of an elementary H_t^n block for the generic SABR model) *Let us consider a C^2 function $h(\cdot)$ and a process S_t driven by the SABR dynamics* (4.1)–(4.2). *We use the following notations:*

$$\dot{\rho} \triangleq \sqrt{1-\rho^2} \qquad h_1(x) \triangleq f(x)h'(x) \qquad h_2(x) \triangleq f^2(x)h''(x)$$

$$H_t^n \triangleq \alpha_t^n h(S_t) \qquad H_{1,t}^n \triangleq \alpha_t^n h_1(S_t) \qquad H_{2,t}^n \triangleq \alpha_t^n h_2(S_t)$$

Note that the above-defined use of superscripts forbids the use of powers.

Then we can express the dynamics of H_t^n, with the obvious simplified notation:

$$dH_t^n = \left[\frac{1}{2}n(n-1)\nu^2 H_t^n + n\rho\nu H_{1,t}^{n+1} + \frac{1}{2}H_{2,t}^{n+2}\right]dt$$
$$+ \left[n\rho\nu H_t^n + H_{1,t}^{n+1}\right]dW_t + \left[n\dot{\rho}\nu H_t^n\right]dZ_t.$$

Note that out of the six terms above, three consist of a constant times the original expression H_t^n. The remaining ones respect the generic form $\alpha_t^n h^*(S_t)$ where $h^*(\cdot)$ is a simple transform of h using its first two differentials along with the local volatility function f. This stable structure will enable us to speed up and simplify the computations by heavily using induction.

Proof By Itô, we get simply:

$$d\left[\alpha_t^n h(S_t)\right] = h(S_t)\left[n\alpha_t^{n-1}d\alpha_t + \frac{1}{2}n(n-1)\alpha_t^{n-2}\langle d\alpha_t\rangle\right]$$
$$+\alpha_t^n\left[h'(S_t)dS_t + \frac{1}{2}h''(S_t)\langle dS_t\rangle\right] + \langle d\alpha_t^n, dh(S_t)\rangle,$$

hence

$$d\left[\alpha_t^n h(S_t)\right] = \left[n\rho\nu\alpha_t^n h(S_t) + \alpha_t^{n+1}h'(S_t)f(S_t)\right]dW_t + \left[n\dot{\rho}\nu\alpha_t^n h(S_t)\right]dZ_t$$
$$+\left[\frac{1}{2}n(n-1)\nu^2\alpha_t^n h(S_t) + n\rho\nu\alpha_t^{n+1}h'(S_t)f(S_t) + \frac{1}{2}\alpha_t^{n+2}h''(S_t)f^2(S_t)\right]dt.$$

Replacing with the given notations, we obtain the desired result. □

We are now equipped to prove the main Lemma.

Proof of Lemma 4.1.
Using our new notations, the casting into Lognormal dynamics gives us immediately

$$\sigma_t = \alpha_t g(S_t) = G^1$$

which proves (4.3). Then using Lemma 4.2 we get the dynamics of σ_t as

$$d\sigma_t = \left[\rho\nu G_1^2 + \frac{1}{2}G_2^3\right]dt + \left[\rho\nu G^1 + G_1^2\right]dW_t + \left[\dot{\rho}\nu G^1\right]dZ_t$$

so that we have the first-level coefficients as

$$a_{1,t} = \rho\nu G_1^2 + \frac{1}{2}G_2^3 \qquad a_{2,t} = \rho\nu G^1 + G_1^2 = \rho\nu\sigma_t + G_1^2 \qquad a_{3,t} = \dot{\rho}\nu G^1 = \dot{\rho}\nu\sigma_t$$

which proves (4.4)–(4.6). As for second-level coefficients, let us now compute $da_{1,t}$ using the same method. We get

$$dG_1^2 = \left[\nu^2 G_1^2 + 2\rho\nu G_{11}^3 + \frac{1}{2}G_{12}^4\right]dt + \left[2\rho\nu G_1^2 + G_{11}^3\right]dW_t + \left[2\dot{\rho}\nu G_1^2\right]dZ_t$$

and

$$dG_2^3 = \left[3\nu^2 G_2^3 + 3\rho\nu G_{21}^4 + \frac{1}{2}G_{22}^5\right]dt + \left[3\rho\nu G_2^3 + G_{21}^4\right]dW_t + \left[3\dot{\rho}\nu G_2^3\right]dZ_t$$

so that

$$\begin{aligned}
a_{11,t} &= \rho\nu\left[\nu^2 G_1^2 + 2\rho\nu G_{11}^3 + \frac{1}{2}G_{12}^4\right] + \frac{1}{2}\left[3\nu^2 G_2^3 + 3\rho\nu G_{21}^4 + \frac{1}{2}G_{22}^5\right] \\
&= \rho\nu^3 G_1^2 + 2\rho^2\nu^2 G_{11}^3 + \frac{3}{2}\nu^2 G_2^3 + \frac{1}{2}\rho\nu G_{12}^4 + \frac{3}{2}\rho\nu G_{21}^4 + \frac{1}{4}G_{22}^5, \\
a_{12,t} &= \rho\nu\left[2\rho\nu G_1^2 + G_{11}^3\right] + \frac{1}{2}\left[3\rho\nu G_2^3 + G_{21}^4\right] \\
&= 2\rho^2\nu^2 G_1^2 + \rho\nu G_{11}^3 + \frac{3}{2}\rho\nu G_2^3 + \frac{1}{2}G_{21}^4, \\
a_{13,t} &= \rho\nu\left[2\dot{\rho}\nu G_1^2\right] + \frac{1}{2}\left[3\dot{\rho}\nu G_2^3\right] = 2\rho\dot{\rho}\nu^2 G_1^2 + \frac{3}{2}\dot{\rho}\nu G_2^3,
\end{aligned}$$

which proves (4.7)–(4.9). Let us establish the dynamics of $a_{2,t}$ in the same way. All terms are already computed, hence

$$a_{21,t} = \rho\nu a_1 + \nu^2 G_1^2 + 2\rho\nu G_{11}^3 + \frac{1}{2} G_{12}^4$$

$$a_{22,t} = \rho\nu a_2 s + 2\rho\nu G_1^2 + G_{11}^3$$

$$a_{23,t} = \rho\nu a_3 + 2\dot{\rho}\nu G_1^2$$

which proves (4.10)–(4.12). Finally the dynamics of $a_{3,t}$ are

$$a_{31,t} = \dot{\rho}\nu a_1 \qquad a_{32,t} = \dot{\rho}\nu a_2 \qquad a_{33,t} = \dot{\rho}\nu a_3$$

which proves (4.13)–(4.15). Let us now tackle the third-level coefficients, starting with a_{122}. Still using Lemma 4.2 we get both elementary dynamics

$$\begin{aligned} dG_{11}^3 &= \left[3\nu^2 G_{11}^3 + 3\rho\nu G_{111}^4 + \frac{1}{2} G_{112,t}^5\right] dt \\ &\quad + \left[3\rho\nu G_{11}^3 + G_{111}^4\right] dW_t + \left[3\dot{\rho}\nu G_{11}^3\right] dZ_t, \\ dG_{21}^4 &= \left[6\nu^2 G_{21}^4 + 4\rho\nu G_{211}^5 + \frac{1}{2} H_{212}^6\right] dt \\ &\quad + \left[4\rho\nu G_{21}^4 + G_{211}^5\right] dW_t + \left[4\dot{\rho}\nu G_{21}^4\right] dZ_t, \end{aligned}$$

therefore

$$\begin{aligned} a_{122} &= 2\rho^2\nu^2 \left[2\rho\nu G_1^2 + G_{11}^3\right] + \rho\nu \left[3\rho\nu G_{11}^3 + G_{111}^4\right] \\ &\quad + \frac{3}{2}\rho\nu \left[3\rho\nu G_2^3 + G_{21}^4\right] + \frac{1}{2}\left[4\rho\nu G_{21}^4 + G_{211}^5\right] \\ &= 4\rho^3\nu^3 G_1^2 + 5\rho^2\nu^2 G_{11}^3 + \rho\nu G_{111}^4 + \frac{9}{2}\rho^2\nu^2 G_2^3 + \frac{7}{2}\rho\nu G_{21}^4 + \frac{1}{2} G_{211}^5, \end{aligned}$$

which proves (4.16). We move on to a_{212} and start with similar dynamics

$$\begin{aligned} dG_{12}^4 &= \left[6\nu^2 G_{12}^4 + 4\rho\nu G_{121,t}^5 + \frac{1}{2} G_{122,t}^6\right] dt \\ &\quad + \left[4\rho\nu G_{12}^4 + G_{121,t}^5\right] dW_t + \left[4\dot{\rho}\nu G_{12}^4\right] dZ_t \end{aligned}$$

leading to

$$a_{212} = \rho\nu a_{12} + \nu^2 \left[2\rho\nu G_1^2 + G_{11}^3\right] + 2\rho\nu \left[3\rho\nu G_{11}^3 + G_{111}^4\right] + \frac{1}{2}\left[4\rho\nu G_{12}^4 + G_{121,t}^5\right],$$

which proves (4.17). Now let us derive the dynamics of $a_{22,t}$ to obtain $a_{221,t}$, $a_{222,t}$ and $a_{223,t}$. Having already differentiated each term, we can express immediately:

$$\begin{aligned}
a_{221} &= \rho\nu a_{21} + 2\rho\nu\left[\nu^2 G_1^2 + 2\rho\nu G_{11}^3 + \frac{1}{2}G_{12}^4\right] + \left[3\nu^2 G_{11}^3 + 3\rho\nu G_{111}^4 + \frac{1}{2}G_{112,t}^5\right] \\
&= \rho\nu a_{21} + 2\rho\nu^3 G_1^2 + \nu^2\left(3 + 4\rho^2\right)G_{11}^3 + 3\rho\nu G_{111}^4 + \rho\nu G_{12}^4 + \frac{1}{2}G_{112}^5, \\
a_{222} &= \rho\nu a_{22} + 2\rho\nu\left[2\rho\nu G_1^2 + G_{11}^3\right] + \left[3\rho\nu G_{11}^3 + G_{111}^4\right] \\
&= \rho\nu a_{22} + 4\rho^2\nu^2 G_1^2 + 5\rho\nu G_{11}^3 + G_{111}^4, \\
a_{223} &= \rho\nu a_{23} + 2\rho\nu\left[2\dot{\rho}\nu G_1^2\right] + \left[3\dot{\rho}\nu G_{11}^3\right] = \rho\nu a_{23} + 4\rho\dot{\rho}\nu^2 G_1^2 + 3\dot{\rho}\nu G_{11}^3,
\end{aligned}$$

which proves (4.18)–(4.20). Then we get a_{232} in a similar fashion:

$$a_{232} = \rho\nu a_{32} + 2\dot{\rho}\nu\left[2\rho\nu G_1^2 + G_{11}^3\right] = \rho\nu a_{32} + 4\rho\dot{\rho}\nu^2 G_1^2 + 2\dot{\rho}\nu G_{11}^3,$$

which proves (4.21), and the same method applies to a_{322}:

$$a_{322} = \dot{\rho}\nu a_{22},$$

which proves (4.22). Finally, let us compute the only fourth-level coefficient $a_{2222,t}$ by differentiating G_{111}^4:

$$dG_{111}^4 = [\cdots]\,dt + \left[4\rho\nu G_{111}^4 + G_{1111}^5\right]dW_t + [\cdots]\,dZ_t,$$

hence we get

$$\begin{aligned}
a_{2222} &= \rho\nu a_{222} + 4\rho^2\nu^2\left[2\rho\nu G_1^2 + G_{11}^3\right] + 5\rho\nu\left[3\rho\nu G_{11}^3 + G_{111}^4\right] + \left[4\rho\nu G_{111}^4 + G_{1111}^5\right] \\
&= \rho\nu a_{222} + 8\rho^3\nu^3 G_1^2 + 19\rho^2\nu^2 G_{11}^3 + 9\rho\nu G_{111}^4 + G_{1111}^5,
\end{aligned}$$

which proves (4.23) and concludes the proof. □

4.2.3 Mapping the Model and the Smile

As one would expect, it appears that the high-order terms are fairly involved, notwithstanding a large number of similarities and symmetries. But our first layer or σ-(2,0) coefficients, namely σ_t, $a_{1,t}$, $a_{2,t}$, $a_{3,t}$ and $a_{22,t}$, seem reasonably simple. Since these quantities are necessary and sufficient to describe the IATM level, skew and curvature (along with the slope), which themselves represent the most important *smile descriptors*, we should focus on expressing the latter as a function of all the parameters. We then note that when the local volatility function $f(\cdot)$ is fixed (e.g. fixed β in the CEV-SABR case) there are three main smile quantities that we can infer or observe, while three model parameters remain (α_t, ρ and ν).

As one would hope and as will be proven shortly, with reasonable assumptions this relationship happens to be bijective. In other words, given the shape of the smile in the vicinity of the money, hence a good proxy for the IATM level, skew and curvature, we can find the corresponding three SABR parameters. This is of significant practical importance, since it provides good initial guesses on which to base a calibration engine and therefore reduces the computing time while improving stability. In practice, one would provide three prices (or implied volatilities) typically and respectively struck at

$$K = S_t - dK \qquad K = S_t \qquad \text{and} \qquad K = S_t + dK \qquad \text{with} \qquad dK \ll S_t.$$

It suffices then to proxy the differentials by the finite difference to obtain, through these initial guesses, a very good fit to the SABR smile. Let us now make this approach more explicit.

4.2.3.1 Direct Problem: From Model to Smile Shape

Here we are looking at generating the IATM level, skew and curvature from the model inputs.

Proposition 4.1 (Static first layer strike-differentials of the generic SABR model) *In the generic SABR model (4.1)–(4.2) the IATM level, skew and curvature can be expressed as functions of:*

- *the initial condition α_t and two model parameters ν and ρ;*
- *a collection of differential transforms of the local volatility f, all taken in S_t.*

With $(\star) \triangleq (t, 0, 0)$ *they are given by*

$$\widetilde{\Sigma}(\star) = \alpha_t g(S_t), \tag{4.24}$$

$$\widetilde{\Sigma}'_y(\star) = \frac{1}{2}\rho\nu + \frac{1}{2}\alpha_t \frac{g_1}{g}(S_t), \tag{4.25}$$

$$\widetilde{\Sigma}''_{yy}(\star) = \frac{1}{\alpha_t g(S_t)}\left[\frac{1}{3}\nu^2 - \frac{1}{2}\rho^2\nu^2 + \alpha_t^2\left[\frac{1}{3}\frac{g_{11}}{g} - \frac{1}{2}\frac{g_1^2}{g^2}\right](S_t)\right]. \tag{4.26}$$

Note that (4.26) implies that the sign of the IATM curvature, in other words whether a smile is convex for options struck close to the money and with a short expiry, is indeed controlled by both the correlation ρ and vol of vol ν. The local volatility being fixed, their combined impact on the curvature will be positive iff

$$|\rho| \le \sqrt{\frac{2}{3}} \simeq 81.65\,\%.$$

Such a high correlation might be high in the interest rates vanilla world, where SABR has its roots and a large number of its followers, but in full generality, and in particular in other markets, it *is* possible to obtain a *concave* contribution.

Proof Combining (2.36) with (4.3) and (2.51) with (4.3)–(4.5) we obtain the IATM level and skew respectively as

$$\widetilde{\Sigma}(\star) = \alpha_t g(S_t)$$

and

$$\widetilde{\Sigma}'_y(\star) = \frac{a_2}{2\sigma_t} = \frac{1}{2}\rho\nu + \frac{\alpha_t^2 g_1(S_t)}{2\alpha_t g(S_t)} = \frac{1}{2}\rho\nu + \frac{1}{2}\alpha_t\frac{g_1}{g}(S_t),$$

which proves (4.24) and (4.25). Similarly, combining (2.52) with (4.3)–(4.6) we get the IATM curvature as

$$\begin{aligned}
\widetilde{\Sigma}''_{yy}(\star) &= \frac{1}{\sigma_t^2}\left[\frac{1}{3}a_{22}\right] + \frac{1}{\sigma_t^3}\left[\frac{1}{3}a_3^2 - \frac{1}{2}a_2^2\right] \\
&= \frac{1}{3\alpha_t^2 g^2(S_t)}\left[\rho\nu(\rho\nu\alpha_t g(S_t) + \alpha_t^2 g_1(S_t)) + 2\rho\nu\alpha_t^2 g_1(S_t) + \alpha_t^3 g_{11}(S_t)\right] \\
&\quad + \frac{1}{\alpha_t^3 g^3(S_t)}\left[\frac{1}{3}\rho^2\nu^2\alpha_t^2 g^2(S_t) - \frac{1}{2}\left(\rho\nu\alpha_t g(S_t) + \alpha_t^2 g_1(S_t)\right)^2\right].
\end{aligned}$$

After simplification this expression becomes

$$\widetilde{\Sigma}''_{yy}(t,0,0) = \frac{1}{\alpha_t g(S_t)}\left[\frac{1}{3}\nu^2 - \frac{1}{2}\rho^2\nu^2 + \alpha_t^2\left(\frac{1}{3}\frac{g_{11}(S_t)}{g(S_t)} - \frac{1}{2}\frac{g_1^2(S_t)}{g^2(S_t)}\right)\right],$$

which proves (4.26) and concludes the proof. □

We are aware that these results might not appeal to some readers in their current form. Since the object of the current chapter is to maximise the applicability potential, we provide the same formulas using the strike K variable (as opposed to log-moneyness y) and the initial f function (instead of its g-transforms).

Corollary 4.1 (Static first layer strike-differentials of the CEV-SABR model) *For the generic SABR model (4.1)–(4.2) the level, skew and curvature of the smile taken at the point* $(\propto) \triangleq (t, S_t, K = S_t, T = t)$ *are*

$$\Sigma(\propto) = \alpha_t S_t^{-1} f(S_t), \tag{4.27}$$

$$\Sigma_K'(\propto) = \frac{1}{2}\rho\nu S_t^{-1} + \frac{1}{2}\alpha_t \left[S_t^{-1} f'(S_t) - S_t^{-2} f(S_t)\right], \tag{4.28}$$

$$\begin{aligned}\Sigma_{K^2}''(\propto) = &\frac{1}{\alpha_t}\nu^2 S_t^{-1} f^{-1}(S_t)\left[\frac{1}{3} - \frac{1}{2}\rho^2\right] - \frac{1}{2}\rho\nu S_t^{-2} \\ &+\alpha_t\left[\frac{2}{3}S_t^{-3} f(S_t) - \frac{1}{2}S_t^{-2} f'(S_t) - \frac{1}{6}S_t^{-1} f^{-1} f'^2(S_t)\right. \\ &\left.+\frac{1}{3}S_t^{-1} f''(S_t)\right].\end{aligned} \tag{4.29}$$

Proof We have the functional relationships

$$\begin{aligned} g &= x^{-1} f, \\ g_1 &= fg' = f\left[-x^{-2} f + x^{-1} f'\right] = x^{-1} f\left[f' - x^{-1} f\right], \\ g_{11} &= fg_1' = ff'\left[-x^{-2} f + x^{-1} f'\right] + f^2\left[2x^{-3} f - x^{-2} f' - x^{-2} f' + x^{-1} f''\right] \\ &= fx^{-1}\left[2x^{-2} f^2 - 3x^{-1} ff' + f'^2 + ff''\right]. \end{aligned}$$

We then have

$$\begin{aligned}\frac{1}{3}\frac{g_{11}}{g}(x) - \frac{1}{2}\frac{g_1^2}{g^2}(x) &= \frac{1}{3}\left[2x^{-2} f^2 - 3x^{-1} ff' + f'^2 + ff''\right] - \frac{1}{2}\left[f' - x^{-1} f\right]^2 \\ &= \frac{1}{6}x^{-2} f^2 - \frac{1}{6}f'^2 + \frac{1}{3}ff''.\end{aligned}$$

Therefore the IATM level and skew are re-expressed as

$$\widetilde{\Sigma}(\star) = \alpha_t S_t^{-1} f(S_t) \quad \text{and} \quad \widetilde{\Sigma}_y'(\star) = \frac{1}{2}\rho\nu + \frac{1}{2}\alpha_t\left[f'(S_t) - S_t^{-1} f(S_t)\right]$$

while the IATM curvature is

$$\widetilde{\Sigma}_{yy}''(\star) = \frac{S_t}{\alpha_t f(S_t)}\left[\frac{1}{3}\nu^2 - \frac{1}{2}\rho^2\nu^2 + \frac{\alpha_t^2}{6}\left[S_t^{-2} f^2(S_t) - f'^2(S_t) + 2ff''(S_t)\right]\right],$$

which proves the level result (4.27). Note that all three results match the IATM differentials of the Extended Skew Market Model established in Proposition (2.3).

Furthermore (B.2) and (B.3) [p. 431] provide the transition formulae as

$$\frac{\partial \Sigma}{\partial K}(t, S_t, K, T) = \frac{1}{K} \widetilde{\Sigma}'_y(t, y, \theta) \text{ and}$$
$$\frac{\partial^2 \Sigma}{\partial K^2}(t, S_t, K, T) = \frac{1}{K^2} \left[\widetilde{\Sigma}''_{yy} - \widetilde{\Sigma}'_y \right](t, y, \theta)$$

which gives (4.28) straight away, while the curvature is given by

$$S_t^2 \Sigma''_{K^2}(\propto) = \frac{S_t}{\alpha_t f(S_t)} \left[\frac{1}{3} \nu^2 - \frac{1}{2} \rho^2 \nu^2 \right] - \frac{1}{2} \alpha_t \left[f'(S_t) - S_t^{-1} f(S_t) \right]$$
$$- \frac{1}{2} \rho \nu + \frac{1}{6} \alpha_t \left[S_t^{-1} f(S_t) - S_t f^{-1} f'^2(S_t) + 2 S_t f''(S_t) \right],$$

which after simplification yields (4.29) and concludes the proof. □

As was noted with the Extended Skew Market Model, at this low level of differentiation the dissociation between the effects of the stochastic volatility (correlation ρ and vol of vol ν) and of the local volatility (derivatives of f) is blatant. Recall that in order to remove the $f(S_t)$ terms it suffices to invoke the IATM level as per (4.27).

4.2.3.2 Inverse Problem and Initial Guesses

Trying to revert the smile generation process is a natural endeavour for practitioners: after all, this is the essence of calibration. However, this is not *stricto sensu* solving the inverse problem, for a couple of reasons.

The first way in which we deviate from the Recovery Theorem 2.1 [p. 55] is that, on one hand, we start purely from static differentials[7] and, on the other hand, we end up not with SInsV formal coefficients (such as a_2, a_3, etc.) but with model parameters (ρ, ν) and initial values (α_t).

The second noticeable difference is that the possibility of reconstructing the model parameters and initial values is quite specific to SABR, and might not be feasible with another stochastic volatility model. Unlike the Recovery theorem, which guarantees a generic result. In fact, the class of practical models for which some or all of the static first layer IATM differentials can be mapped is quite large, but clearly the parametrisation must be sparse.

Proposition 4.2 (Inverse parametrisation of the generic SABR model) *In the generic SABR model* (4.1)–(4.2) *we can infer the initial condition* α_t*, along with the two parameters* ρ *and* ν*, from the IATM level, skew and curvature. The inversion formulae are*

---

[7] Although the IATM SImpV constraints ensure a large amount of redundancy between static and dynamics: see Proposition 2.2 [p. 54].

$$\alpha_t = \frac{\Sigma}{g} \tag{4.30}$$

$$\nu = \left[3\Sigma \left[S_t^2 \Sigma_{KK}^{''} + S_t \Sigma_K^{'} \right] + \frac{3}{2} \left[2S_t \Sigma_K^{'} - \frac{g_1}{g^2} \Sigma \right]^2 + \frac{\Sigma^2}{g^2} \left[\frac{3}{2} \frac{g_1^2}{g^2} - \frac{g_{11}}{g} \right] \right]^{\frac{1}{2}} \tag{4.31}$$

$$\rho = \frac{1}{\nu} \left[2S_t \Sigma_K^{'} - \frac{g_1}{g^2} \Sigma \right] if \nu \neq 0 \quad whereas\ \rho = 0\ otherwise \tag{4.32}$$

where all IATM differentials have been taken in $(t, S_t.K = S_t, T = t)$ *and the g-functions are expressed in* S_t.

Note that the strike differentials have been expressed w.r.t. the strike K, as opposed to log-moneyness y.

Proof By inverting the level equation (4.24) we trivially get α_t as in (4.30).

Then from the skew equation (4.24) we extract the *effective correlation* product

$$\rho\nu = 2S_t \Sigma_K^{'} - \frac{g_1}{g^2} \Sigma \tag{4.33}$$

which proves (4.32). Furthermore, we can rewrite the curvature equation (4.26) as

$$\Sigma \left[S_t^2 \Sigma_{KK}^{''} + S_t \Sigma_K^{'} \right] = \frac{1}{3}\nu^2 - \frac{1}{2}\rho^2\nu^2 + \frac{\Sigma^2}{g^2} \left(\frac{1}{3} \frac{g_{11}}{g} - \frac{1}{2} \frac{g_1^2}{g^2} \right).$$

Substituting (4.33) into this equation and isolating ν^2 we get

$$\begin{aligned} \nu^2 = 3\Sigma \left[S_t^2 \Sigma_{KK}^{''} + S_t \Sigma_K^{'} \right] + \frac{3}{2} \left[2S_t \Sigma_K^{'} - \frac{g_1}{g^2} \Sigma \right]^2 \\ - 3\frac{\Sigma^2}{g^2} \left[\frac{1}{3} \frac{g_{11}}{g} - \frac{1}{2} \frac{g_1^2}{g^2} \right]. \end{aligned} \tag{4.34}$$

Taking the square root finally gives us (4.31) and concludes the proof. □

Note that (4.34) guarantees the non-negative sign of the right-hand side, which is only natural as that r.h.s. identifies the excess curvature which is not created by the local volatility.

4.3 Application to the CEV-SABR Model

This model and the associated Hagan et al's formula, along with their various extensions and customisations, is the workhorse of the interest rates vanilla world. It is obviously used with other asset classes, such as exchange rates for instance, where it can be competing against (displaced) Heston or more recent stochastic volatility classes. However, this is by no means a term-structure model, as it requires a single martingale underlying, so that the issues raised tend to be similar.

4.3.1 Presentation of the Model

The local volatility function $f(\cdot)$ provides an important degree of freedom for the generic SABR model, but a large proportion of practitioners prefer to use the simple CEV instance, which was the illustration case of [1]. The model is given by

$$\textbf{(CEV-SABR)} \quad \begin{cases} dS_t = \alpha_t S_t^\beta dW_t & (4.35) \\ \qquad\qquad\qquad \text{with} \quad \langle dW_t, dB_t \rangle = \rho\, dt \\ d\alpha_t = \nu \alpha_t dB_t. & (4.36) \end{cases}$$

Although it might appear simplistic, there are several advantages to this version of SABR. The first one is that it is based upon an existing, well-understood local volatility model (the Constant Elasticity of Variance). Since the latter can provide significant skew while maintaining a positive support for the distribution, it provides a good base for calibration. In that respect, and in light of Sect. 2.5.2.2 [p. 100], we recall that the CEV can obviously be displaced, which means that it contains and supersedes the displaced Lognormal diffusion.

Another attractive feature is its simplicity, since the local volatility function is controlled by the single parameter β. Furthermore, *in the liquid strike region* of interest, the influence of β is simple (monotonous) on the most important smile descriptors, in particular the level, the skew and the ATM Delta. In a calibration perspective this might appear constraining, but for hedging such sparsity enhances intuition and therefore efficiency. More globally, and from a trading perspective, the β power parameter influences the following features of the smile:

- It determines the support of the distribution, in particular its lower bound (which can be non-zero). In turn this floor influences the smile: in the rate environment for instance, it can be slightly negative for long maturities. This means that the *Lognormal* smile must structurally go to infinity as $K \searrow 0$, while the *Normal* smile must converge to zero at the distribution's floor.
- It sets the fundamental static smile shape, the *local volatility* and sticky-strike part of it. This basic profile will then be perturbed (skewed and curved) by the stochastic volatility.
- It is the almost unique determinant of the backbone, i.e. the ATM level as a function of the underlying S_t. In turn, the backbone provides the most important part of the smile dynamics, and is therefore a vital information for hedging.

In consequence, the β power parameter can be chosen to fulfil one of several roles. Some practitioners set it close to zero to provide a near-Normal ATM skew and Delta, while others select a high β to generate a high right-hand wing. Indeed, in the interest rates world for instance, a thick right tail is usually necessary in order to provide the high market prices of Constant Maturity Swaps.

In our view it is because of its practical importance, and of the simplicity of its local volatility specification, that the CEV-SABR deserves an almost distinct treatment from the generic version.

4.3.2 Coefficients of the Chaos Dynamics

The specific CEV-SABR case can be treated either as an instance of the more generic SABR case exposed above, or from scratch by deriving the chaos dynamics *ad hoc*. It happens that the relative simplicity of the CEV local volatility function—which is an important part of the model's appeal—makes the second option easier than the first. Let us therefore derive the relevant coefficients for that model.

Lemma 4.3 (Chaos dynamics up to 3rd layer for the CEV-SABR model) *In the CEV-SABR model* (4.35) *the chaos dynamics' coefficients are:*

- *First-level coefficients:*

$$\sigma_t = \alpha_t S_t^{\beta-1} \tag{4.37}$$

$$a_{1,t} = (\beta-1)\sigma_t^2\left[\rho\nu + \frac{1}{2}(\beta-2)\sigma_t\right] \tag{4.38}$$

$$a_{2,t} = \sigma_t\left[\rho\nu + (\beta-1)\sigma_t\right] \tag{4.39}$$

$$a_{3,t} = \sigma_t\sqrt{1-\rho^2}\nu \tag{4.40}$$

- *Second-level coefficients:*

$$\begin{aligned} a_{11,t} = {} & [(\beta-1)\rho\nu]\left[2\sigma_t a_1 + a_2^2 + a_3^2\right] \\ & + \left[\frac{3}{2}(\beta-2)(\beta-1)\right]\left[\sigma_t^2 a_1 + \sigma_t\left(a_2^2+a_3^2\right)\right] \end{aligned} \tag{4.41}$$

$$a_{12,t} = [2(\beta-1)\rho\nu]\,\sigma_t a_2 + \left[\frac{3}{2}(\beta-1)(\beta-2)\right]\sigma_t^2 a_2 \tag{4.42}$$

$$a_{13,t} = [2(\beta-1)\rho\nu]\,\sigma_t a_3 + \left[\frac{3}{2}(\beta-1)(\beta-2)\right]\sigma_t^2 a_3 \tag{4.43}$$

$$a_{21,t} = [\rho\nu + 2(\beta-1)\sigma_t]\,a_1 + (\beta-1)\left(a_2^2+a_3^2\right) \tag{4.44}$$

$$a_{22,t} = [\rho\nu + 2(\beta-1)\sigma_t]\,a_2 \tag{4.45}$$

$$a_{23,t} = [\rho\nu + 2(\beta-1)\sigma_t]\,a_3 \tag{4.46}$$

$$a_{31,t} = \sqrt{1-\rho^2}\nu a_1 \tag{4.47}$$
$$a_{32,t} = \sqrt{1-\rho^2}\nu a_2 \tag{4.48}$$
$$a_{33,t} = \sqrt{1-\rho^2}\nu a_3 \tag{4.49}$$

- *Third-level coefficients:*

$$\begin{aligned} a_{122} &= \left[\frac{3}{2}(\beta-2)(\beta-1)\right]\sigma_t^2 a_{22} + 3(\beta-2)(\beta-1)a_2^2\sigma_t \\ &\quad + [2(\beta-1)\rho\nu]\, a_{22}\sigma_t + [2(\beta-1)\rho\nu]\, a_2^2 \end{aligned} \tag{4.50}$$
$$\begin{aligned} a_{212,t} &= 2(\beta-1)a_1a_2 + [\rho\nu + 2(\beta-1)\sigma_t]\, a_{12} \\ &\quad + 2(\beta-1)a_2a_{22} + 2(\beta-1)a_3a_{32} \end{aligned} \tag{4.51}$$

$$a_{221,t} = 2(\beta-1)(a_1a_2 + a_2a_{22} + a_3a_{23}) + [\rho\nu + 2(\beta-1)\sigma_t]\, a_{21} \tag{4.52}$$
$$a_{222,t} = 2(\beta-1)a_2^2 + [\rho\nu + 2(\beta-1)\sigma_t]\, a_{22} \tag{4.53}$$
$$a_{223,t} = 2(\beta-1)a_2a_3 + [\rho\nu + 2(\beta-1)\sigma_t]\, a_{23} \tag{4.54}$$
$$a_{232,t} = 2(\beta-1)a_2a_3 + (\rho\nu + 2(\beta-1)\sigma_t)\, a_{32} \tag{4.55}$$
$$a_{322,t} = \sqrt{1-\rho^2}\nu a_{22} \tag{4.56}$$

And finally:

$$a_{2222,t} = 6(\beta-1)a_2a_{22} + [\rho\nu + 2(\beta-1)\sigma_t]\, a_{222}. \tag{4.57}$$

Proof First, let us rewrite the perturbation specification (4.35) as

$$d\alpha_t = \rho\nu\alpha_t dW_t + \sqrt{1-\rho^2}\nu\alpha_t dZ_t \qquad \text{with} \qquad W_t \perp Z_t$$

leading to

$$\frac{dS_t}{S_t} = \sigma_t dW_t \qquad \text{with} \qquad \sigma_t = \alpha_t S_t^{\beta-1},$$

which proves (4.37). Then let us turn to the dynamics of σ_t which come as

$$\begin{aligned} d\sigma_t &= S_t^{\beta-1}\left[\rho\nu\alpha_t dW_t + \sqrt{1-\rho^2}\nu\alpha_t dZ_t\right] + (\beta-1)\alpha_t S_t^{\beta-2}\alpha_t S_t^{\beta} dW_t \\ &\quad + \frac{1}{2}(\beta-1)(\beta-2)\alpha_t S_t^{\beta-3}\alpha_t^2 S_t^{2\beta} dt + (\beta-1)S_t^{\beta-2}\rho\nu\alpha_t\alpha_t S_t^{\beta} dt \end{aligned}$$

therefore

$$a_{1,t} = (\beta-1)\sigma_t^2\left[\rho\nu + \frac{1}{2}(\beta-2)\sigma_t\right],$$

$$a_{2,t} = \sigma_t \left[\rho\nu + (\beta - 1)\sigma_t\right], \qquad a_{3,t} = \sigma_t\sqrt{1-\rho^2}\nu,$$

which proves (4.38) and (4.39). For later use, let us recall that

$$d\sigma_t^2 = \left[2\sigma_t a_1 + a_2^2 + a_3^2\right] dt + \left[2\sigma_t a_2\right] dW_t + \left[2\sigma_t a_3\right] dZ_t.$$

We focus now on the dynamics of $a_{1,t}$. Using (4.38) we get

$$\begin{aligned} da_1 &= \frac{1}{2}(\beta-2)(\beta-1)\left[2\sigma_t a_2^2 + 2\sigma_t a_3^2\right] dt \\ &\quad + \frac{1}{2}(\beta-2)(\beta-1)\sigma_t^2\left[a_1 dt + a_2 dW_t + a_3 dZ_t\right] \\ &\quad + (\beta-1)\left[\rho\nu + \frac{1}{2}(\beta-2)\sigma_t\right] d\sigma_t^2, \end{aligned}$$

so that

$$\begin{aligned} a_{11} &= \left[(\beta-1)\rho\nu\right]\left[2\sigma_t a_1 + a_2^2 + a_3^2\right] \\ &\quad + \left[\frac{3}{2}(\beta-2)(\beta-1)\right]\left[\sigma_t^2 a_1 + \sigma_t\left(a_2^2 + a_3^2\right)\right] \\ a_{12} &= \left[2(\beta-1)\rho\nu\right]\sigma_t a_2 + \left[\frac{3}{2}(\beta-1)(\beta-2)\right]\sigma_t^2 a_2 \\ a_{13} &= \left[2(\beta-1)\rho\nu\right]\sigma_t a_3 + \left[\frac{3}{2}(\beta-1)(\beta-2)\right]\sigma_t^2 a_3 \end{aligned}$$

which proves (4.41)–(4.43). Similarly (4.39) gives the dynamics of $a_{2,t}$:

$$da_2 = \left[\rho\nu + 2(\beta-1)\sigma_t\right]\left[a_1 dt + a_2 dW_t + a_3 dZ_t\right] + (\beta-1)\left[a_2^2 + a_3^2\right] dt$$

giving us

$$\begin{aligned} a_{21} &= \left[\rho\nu + 2(\beta-1)\sigma_t\right] a_1 + (\beta-1)\left(a_2^2 + a_3^2\right) \\ a_{22} &= \left[\rho\nu + 2(\beta-1)\sigma_t\right] a_2 \\ a_{23} &= \left[\rho\nu + 2(\beta-1)\sigma_t\right] a_3. \end{aligned}$$

From (4.40) we obtain the simpler dynamics of $a_{3,t}$ as

$$a_{31,t} = \sqrt{1-\rho^2}\nu a_1 \qquad a_{32,t} = \sqrt{1-\rho^2}\nu a_2 \qquad a_{33,t} = \sqrt{1-\rho^2}\nu a_3$$

which proves (4.47)– (4.49). Let us now turn to third-level coefficients, starting with (4.42), for which we write the dynamics of $a_{12,t}$ as

$$\begin{aligned} da_{12} &= \frac{3}{2}(\beta-2)(\beta-1)\left[a_2 d\sigma_t^2 + \sigma_t^2 da_2\right] \\ &\quad + 2(\beta-1)\rho\nu\left[a_2 d\sigma_t + \sigma_t da_2\right] + [\ldots]\,dt, \end{aligned}$$

which gives

$$da_{12} + [\ldots]\,dt + [\ldots]\,dZ_t = a_{122} dW_t$$

with

$$\begin{aligned} a_{122} &= 2(\beta-1)\rho\nu\left[a_2^2 + \sigma_t a_{22}\right] + \frac{3}{2}(\beta-2)(\beta-1)\left[a_2\left(2\sigma_t a_2\right) + \sigma_t^2 a_{22}\right] \\ &\quad + 3(\beta-2)(\beta-1)a_2^2\sigma_t + \frac{3}{2}(\beta-2)(\beta-1)a_{22}\sigma_t^2 + 2(\beta-1)\rho\nu a_2^2 \\ &\quad + 2(\beta-1)\rho\nu a_{22}\sigma_t \end{aligned}$$

and therefore

$$\begin{aligned} a_{122} &= \left[\frac{3}{2}(\beta-2)(\beta-1)\right]\sigma_t^2 a_{22} + 3(\beta-2)(\beta-1)a_2^2\sigma_t \\ &\quad + \left[2(\beta-1)\rho\nu\right]a_{22}\sigma_t + \left[2(\beta-1)\rho\nu\right]a_2^2, \end{aligned}$$

proving (4.50). Looking now at a_{21} as expressed in (4.44), its dynamics are

$$\begin{aligned} da_{21} &= 2(\beta-1)a_1 d\sigma_t + \left[\rho\nu + 2(\beta-1)\sigma_t\right]da_1 \\ &\quad + 2(\beta-1)a_2 da_2 + 2(\beta-1)a_3 da_3 + [\ldots]\,dt, \end{aligned}$$

hence

$$\begin{aligned} da_{21} &= 2(\beta-1)a_1 a_2 dW_t + \left[\rho\nu + 2(\beta-1)\sigma_t\right]a_{12} dW_t + 2(\beta-1)a_2 a_{22} dW_t \\ &\quad + 2(\beta-1)a_3 a_{32} dW_t + [\cdot]\,dt + [\cdot]\,dZ_t, \end{aligned}$$

therefore

$$a_{212,t} = 2(\beta-1)a_1 a_2 + \left[\rho\nu + 2(\beta-1)\sigma_t\right]a_{12} + 2(\beta-1)a_2 a_{22} + 2(\beta-1)a_3 a_{32},$$

which validates (4.51). Similarly, using (4.45) we get the dynamics of $a_{22,t}$ as

$$da_{22,t} = 2(\beta-1)a_2 d\sigma_t + \left[\rho\nu + 2(\beta-1)\sigma_t\right]da_2 + 2(\beta-1)(a_2 a_{22} + a_3 a_{23})dt$$

therefore

$$\begin{aligned}a_{221,t} &= 2(\beta-1)(a_1a_2+a_2a_{22}+a_3a_{23})+[\rho\nu+2(\beta-1)\sigma_t]\,a_{21}\\a_{222,t} &= 2(\beta-1)a_2^2+[\rho\nu+2(\beta-1)\sigma_t]\,a_{22}\\a_{223,t} &= 2(\beta-1)a_2a_3+[\rho\nu+2(\beta-1)\sigma_t]\,a_{23},\end{aligned}$$

which proves (4.52), (4.53) and (4.54). Turning to $a_{23,t}$, using (4.46) we get

$$da_{23,t} = 2(\beta-1)a_3d\sigma_t+[\rho\nu+2(\beta-1)\sigma_t]\,da_3+2(\beta-1)(a_2a_{32}+a_3a_{33})dt,$$

therefore

$$a_{232,t} = 2(\beta-1)a_2a_3+[\rho\nu+2(\beta-1)\sigma_t]\,a_{32},$$

which proves (4.55). In a simpler way, we express the dynamics of $a_{32,t}$ to obtain

$$a_{322,t} = \sqrt{1-\rho^2}\nu a_{22},$$

giving us (4.56). Eventually, we turn to the dynamics of $a_{222,t}$, i.e.

$$\begin{aligned}da_{222,t} &= 2(\beta-1)\,[2a_2da_2]+2(\beta-1)a_{22}d\sigma_t+[\rho\nu+2(\beta-1)\sigma_t]\,da_{22}+(.)dt\\&= [4(\beta-1)a_2a_{22}+2(\beta-1)a_2a_{22}+[\rho\nu+2(\beta-1)\sigma_t]\,a_{222}]\,dW_t\\&\quad+[\cdots]\,dt+[\cdots]\,dZ_t,\end{aligned}$$

which allows us to express

$$a_{2222,t} = 6(\beta-1)a_2a_{22}+[\rho\nu+2(\beta-1)\sigma_t]\,a_{222},$$

which validates (4.57) and concludes the proof. □

4.3.3 *Mapping the Model and the Smile Shape*

We are now equipped to express any IATM differential of the first, second or third layer as a function of the model parameters (ρ and ν) and initial values (α_t and S_t).

4.3.3.1 Direct Problem: From Model to Smile Shape

We first present a simple result, usually sufficient to calibrate moderate maturities.

Proposition 4.3 (First layer static IATM differentials of the CEV-SABR model) *In the CEV-SABR model* (4.35) *the smile's static IATM differentials are expressed in* *terms of the model parameters as follows. We provide these results both in sliding* $(\star) = (t, y = 0, \theta = 0)$ *and absolute* $(\propto) = (t, S_t, K = S_t, T = t)$ *coordinates.*

The IATM level:

$$\Sigma(\propto) = \widetilde{\Sigma}(\star) = \alpha_t S_t^{\beta-1}. \tag{4.58}$$

The IATM skew:

$$\begin{aligned}\widetilde{\Sigma}'_y(\star) &= \frac{1}{2}\left[\rho\nu + (\beta-1)\alpha_t S_t^{\beta-1}\right],\\ \Sigma'_K(\propto) &= \frac{1}{2S_t}\left[\rho\nu + (\beta-1)\alpha_t S_t^{\beta-1}\right].\end{aligned} \tag{4.59}$$

The IATM curvature:

$$\widetilde{\Sigma}''_{yy}(\star) = \alpha_t S_t^{\beta-1}\left[\frac{1}{6}(\beta-1)^2\right] + \alpha_t^{-1} S_t^{1-\beta}\left[\left(\frac{1}{3}-\frac{1}{2}\rho^2\right)\nu^2\right], \tag{4.60}$$

$$\Sigma''_{KK}(\propto) = \frac{1}{S_t^2}\left[\left[\frac{1}{3}-\frac{1}{2}\rho^2\right]\frac{\nu^2}{\alpha_t}S_t^{1-\beta} - \frac{1}{2}\rho\nu + \frac{1}{6}\left[\beta^2-5\beta+4\right]\alpha_t S_t^{\beta-1}\right]. \tag{4.61}$$

The IATM slope:

$$\begin{aligned}\Sigma'_T(\propto) &= \widetilde{\Sigma}'_\theta(\star)\\ &= \alpha_t^3 S_t^{3\beta-3}\left[\frac{1}{24}(\beta-1)^2\right] + \alpha_t^2 S_t^{2\beta-2}\left[\frac{1}{4}\rho\nu\beta\right]\\ &\quad + \alpha_t S_t^{\beta-1}\left[\left(\frac{1}{12}-\frac{1}{8}\rho^2\right)\nu^2\right].\end{aligned} \tag{4.62}$$

These formulae provide us with a clear understanding of each parameter's influence on the short-expiry smile, which is one of the positive features of SABR. If we fix β, then in terms of control we can allocate α_t to the level of the smile, ρ to the skew and ν to the curvature.

Usually β is chosen to generate a given backbone (IATM level as a function of the underlying S_t) and/or to control the smile shape at a characteristic expiry. What Proposition 4.3 shows us, though, is how a certain degree of control could be applied to the IATM slope $\widetilde{\Sigma}'_\theta(\star)$ as well.

Proof The IATM level (4.58) comes directly from (4.37) and (2.36). The IATM skew is expressed by combining (2.51) with (4.37) and (4.39):

$$\widetilde{\Sigma}'_y(t,0,0) = \frac{a_2}{2\sigma_t} = \frac{1}{2\sigma_t}\sigma_t\left[\rho\nu + (\beta-1)\sigma_t\right] = \frac{1}{2}\left[\rho\nu + (\beta-1)\alpha_t S_t^{\beta-1}\right].$$

Invoking (B.2) we obtain σ'_K which completes the proof of (4.59). Turning to the curvature, we invoke (4.39) and (4.40) to write

$$a_2^2 = \sigma_t^2 \left[\rho\nu + (\beta-1)\sigma_t\right]^2 = \sigma_t^4(\beta-1)^2 + \sigma_t^3 2\rho\nu(\beta-1) + \sigma_t^2\rho^2\nu^2,$$
$$a_3^2 = \sigma_t^2(1-\rho^2)\nu^2,$$

therefore

$$\frac{1}{3}a_3^2 - \frac{1}{2}a_2^2 = \sigma_t^4\left[-\frac{1}{2}(\beta-1)^2\right] + \sigma_t^3\left[-\rho\nu(\beta-1)\right] + \sigma_t^2\left[\frac{1}{3}\left(1-\frac{5}{2}\rho^2\right)\nu^2\right].$$

Furthermore, the second-depth coefficient is

$$\begin{aligned} a_{22} &= \left[\rho\nu + 2(\beta-1)\sigma_t\right] a_2 = \sigma_t\left[\rho\nu + 2(\beta-1)\sigma_t\right]\left[\rho\nu + (\beta-1)\sigma_t\right] \\ &= \sigma_t^3\left[2(\beta-1)^2\right] + \sigma_t^2\left[3(\beta-1)\rho\nu\right] + \sigma_t\left[\rho^2\nu^2\right]. \end{aligned}$$

Therefore the IATM curvature formula (2.52) becomes

$$\begin{aligned} \widetilde{\Sigma}''_{yy}(\star) &= \frac{1}{3\sigma_t^2}a_{22} + \frac{1}{\sigma_t^3}\left[\frac{1}{3}a_3^2 - \frac{1}{2}a_2^2\right] \\ &= \sigma_t\left[\frac{1}{6}(\beta-1)^2\right] + \frac{1}{\sigma_t}\left[\frac{1}{3}\left(1-\frac{3}{2}\rho^2\right)\nu^2\right]. \end{aligned}$$

Hence

$$\widetilde{\Sigma}''_{yy}(t,0,0) = \alpha_t S_t^{\beta-1}\left[\frac{1}{6}(\beta-1)^2\right] + \frac{1}{\alpha_t}S_t^{1-\beta}\left[\frac{1}{3}\left(1-\frac{3}{2}\rho^2\right)\nu^2\right],$$

which proves (4.60). Then from (B.3) we get

$$\begin{aligned} \Sigma''_{KK} = \frac{1}{K^2}\left[\widetilde{\Sigma}''_{yy} - \widetilde{\Sigma}'_{y}\right] = \frac{1}{K^2}\Bigg[&\alpha_t S_t^{\beta-1}\left[\frac{1}{6}(\beta-1)^2\right] + \alpha_t^{-1}S_t^{1-\beta}\left[\frac{1}{3} - \frac{1}{2}\rho^2\right]\nu^2 \\ &-\frac{1}{2}\left[\rho\nu + (\beta-1)\alpha_t S_t^{\beta-1}\right]\Bigg], \end{aligned}$$

hence

$$\Sigma''_{KK} = \frac{1}{K^2}\left[\left(\frac{1}{3} - \frac{1}{2}\rho^2\right)\nu^2\alpha_t^{-1}S_t^{1-\beta} - \frac{1}{2}\rho\nu + \frac{1}{6}(\beta-1)(\beta-4)\alpha_t S_t^{\beta-1}\right],$$

which proves (4.61). Finally, turning to the IATM slope we first provide

$$\begin{aligned} \frac{1}{8}a_2^2 + \frac{1}{12}a_3^2 &= \frac{1}{8}\sigma_t^2\left[\rho^2\nu^2 + (\beta-1)^2\sigma_t^2 + 2\rho\nu(\beta-1)\sigma_t\right] + \frac{1}{12}\sigma_t^2\left(1-\rho^2\right)\nu^2 \\ &= \sigma_t^4\left[\frac{1}{8}(\beta-1)^2\right] + \sigma_t^3\left[\frac{1}{4}\rho\nu(\beta-1)\right] + \sigma_t^2\left[\frac{1}{12}\left(1+\frac{1}{2}\rho^2\right)\nu^2\right] \end{aligned}$$

before expressing the IATM differential from (2.53) as

$$
\begin{aligned}
\widetilde{\Sigma}_\theta'(\star) &= \sigma_t\left[\frac{1}{4}a_2\right]+\left[\frac{1}{2}a_1-\frac{1}{6}a_{22}\right]+\frac{1}{\sigma_t}\left[\frac{1}{8}a_2^2+\frac{1}{12}a_3^2\right]\\
&= \sigma_t\left[\frac{1}{4}\sigma_t\left[\rho\nu+(\beta-1)\sigma_t\right]\right]+\frac{1}{2}(\beta-1)\sigma_t^2\left[\rho\nu+\frac{1}{2}(\beta-2)\sigma_t\right]\\
&\quad-\frac{1}{6}\left[\sigma_t^3\left[2(\beta-1)^2\right]+\sigma_t^2\left[3(\beta-1)\rho\nu\right]+\sigma_t\left[\rho^2\nu^2\right]\right]\\
&\quad+\sigma_t^3\left[\frac{1}{8}(\beta-1)^2\right]+\sigma_t^2\left[\frac{1}{4}\rho\nu(\beta-1)\right]+\sigma_t\left[\frac{1}{12}\left(1+\frac{1}{2}\rho^2\right)\nu^2\right].
\end{aligned}
$$

Gathering terms according to powers of σ_t and simplifying we obtain

$$
\widetilde{\Sigma}_\theta'(\star) = \sigma_t^3\left[\frac{1}{24}(\beta-1)^2\right]+\sigma_t^2\left[\frac{1}{4}\rho\nu\beta\right]+\sigma_t\left[\frac{1}{12}\left(1-\frac{3}{2}\rho^2\right)\nu^2\right],
$$

which we can re-express as

$$
\begin{aligned}
\widetilde{\Sigma}_\theta'(\star) &= \alpha_t^3 S_t^{3\beta-3}\left[\frac{1}{24}(\beta-1)^2\right]\\
&\quad+\alpha_t^2 S_t^{2\beta-2}\left[\frac{1}{4}\rho\nu\beta\right]+\alpha_t S_t^{\beta-1}\left[\frac{1}{12}\left(1-\frac{3}{2}\rho^2\right)\nu^2\right],
\end{aligned}
$$

which proves (4.62) and concludes the proof. □

4.3.3.2 Inverse Problem and Initial Guesses

Since CEV-SABR is by far the most popular instance of the SABR class, the capacity to invert the IATM smile back into the model parameters and initial value is all the more useful. Again, these formulae benefit from the simplicity of the local volatility function.

Lemma 4.4 (Inverse parametrisation of the CEV-SABR model) *In the CEV-SABR model* (4.35) *the model parameters ρ and ν, and the initial value α_t, are deducted from IATM level, skew and curvature as follows. The sliding Lognormal implied volatility $\widetilde{\Sigma}$ (resp. the absolute LNIV Σ) and its differentials are taken at the IATM point $(t, y=0, \theta=0)$ (resp. $(t, S_t, K=S_t, T=t)$).*

$$
\begin{aligned}
\alpha_t &= \Sigma S_t^{1-\beta} = \widetilde{\Sigma} S_t^{1-\beta}, &(4.63)\\
\nu^2 &= 6\,\widetilde{\Sigma}_y'^{\,2}+(\beta-1)^2\widetilde{\Sigma}^2+\widetilde{\Sigma}\left[3\widetilde{\Sigma}_{yy}''+6(1-\beta)\widetilde{\Sigma}_y'\right] &(4.64)\\
&= (\beta-1)^2\Sigma^2+6S_t^2\,\Sigma_K'^{\,2}+(9-6\beta)S_t\Sigma\Sigma_K'+3S_t^2\Sigma\Sigma_{KK}'', &(4.65)\\
\rho\nu &= \left[2\widetilde{\Sigma}_y'+(1-\beta)\widetilde{\Sigma}\right]=\left[2S_t\Sigma_K'+(1-\beta)\Sigma\right]. &(4.66)
\end{aligned}
$$

Proof We have from 4.58 that

$$\alpha_t = \widetilde{\Sigma}(\star)S_t^{(1-\beta)},$$

which proves (4.63). The IATM skew formula 4.59 gives us that

$$\rho\nu = 2\widetilde{\Sigma}'_y + (1-\beta)\alpha_t S_t^{\beta-1}$$

and (B.2) completes the proof of (4.66). Furthermore, from the curvature expression (4.60) we have that

$$\nu^2 = \frac{3}{2}(\rho\nu)^2 + 3\alpha_t S_t^{\beta-1}\left[\widetilde{\Sigma}''_{yy} - \frac{1}{6}(\beta-1)^2\alpha_t S_t^{\beta-1}\right]. \tag{4.67}$$

We solve this system by substituting (4.66) into (4.67), which gives us

$$\begin{aligned}\nu^2 &= \frac{3}{2}\left[2\widetilde{\Sigma}'_y + (1-\beta)\alpha_t S_t^{\beta-1}\right]^2 + 3\alpha_t S_t^{\beta-1}\widetilde{\Sigma}''_{yy} - \frac{1}{2}(\beta-1)^2\left[\alpha_t S_t^{\beta-1}\right]^2\\ &= 6\,\widetilde{\Sigma}'^{\,2}_y + (\beta-1)^2\widetilde{\Sigma}^2 + \widetilde{\Sigma}\left[3\widetilde{\Sigma}''_{yy} + 6(1-\beta)\widetilde{\Sigma}'_y\right]\end{aligned}$$

and proves (4.64). Now replacing the sliding skew and curvature by their absolute counterparts through (B.2) and (B.3) we obtain

$$\begin{aligned}\nu^2 &= 6S_t^2\,\Sigma'^{\,2}_K + (\beta-1)^2\Sigma^2 + \Sigma\left[3\left(S_t^2\Sigma''_{KK} + S_t\Sigma'_K\right) + 6(1-\beta)K\Sigma'_K\right]\\ &= (\beta-1)^2\Sigma^2 + 6S_t^2\,\Sigma'^{\,2}_K + (9-6\beta)S_t\Sigma\Sigma'_K + 3S_t^2\Sigma\Sigma''_{KK},\end{aligned}$$

which proves (4.65) and concludes the proof. □

4.3.4 Compatibility with Hagan et al.

Since CEV-SABR and the corresponding Hagan et al. formula are still respectively the most used by practitioners and the best-known approximation available, it is worth checking whether the latter provides the correct IATM differentials, at least for the first layer.

Corollary 4.2 (Compatibility of ACE with Hagan's formula for CEV-SABR) *The singular perturbation formula provided in* [1] *for CEV-SABR, e.g.*

$$\widetilde{\Sigma}^H(t,y,\theta) = \widetilde{\Sigma}_1^H\,\widetilde{\Sigma}_2^H\,\widetilde{\Sigma}_3^H(t,y,\theta) \tag{4.68}$$

with

$$\widetilde{\Sigma}_1^H(\beta; \alpha_t, S_t; K) \triangleq \frac{\alpha_t}{(KS_t)^{\frac{1-\beta}{2}}\left[1+\frac{(1-\beta)^2}{24}y^2+\frac{(1-\beta)^4}{1920}y^4\right]}$$

$$\widetilde{\Sigma}_2^H(\beta, \nu, \rho; \alpha_t, S_t; K) \triangleq \left[\frac{z}{\xi(z)}\right]$$

$$\widetilde{\Sigma}_3^H(\beta, \nu, \rho; \alpha_t, S_t; K, T-t) \triangleq \left[1+\left[\frac{(1-\beta)^2\,\alpha_t^2}{24\,(KS_t)^{1-\beta}}+\frac{\rho\beta\nu\alpha_t}{4\,(KS_t)^{\frac{1-\beta}{2}}}+\frac{2-3\rho^2}{24}\nu^2\right]\theta\right]$$

where $z = -\frac{\nu}{\alpha}(KS_t)^{\frac{1-\beta}{2}}\,y$ *and* $\xi(z) = \log\left(\frac{\sqrt{1-2\rho z+z^2}+z-\rho}{1-\rho}\right)$

provides the same results as our asymptotic chaos expansion methods for the static differentials of the first layer, i.e. *the IATM level $\widetilde{\Sigma}(\star)$, skew $\widetilde{\Sigma}_y'(\star)$, curvature $\widetilde{\Sigma}_{yy}''(\star)$ and slope $\widetilde{\Sigma}_\theta'(\star)$.*

Proof The first and second order pure-strike differentials are actually provided by the authors through the following expansion:

$$\widetilde{\Sigma}^H(t, y, \theta) = \frac{\alpha_t}{S_t^{1-\beta}}\left[1-\frac{1}{2}(1-\beta-\rho\lambda)y\right.$$
$$\left.+\frac{1}{2}\left[(1-\beta)^2+(2-3\rho^2)\lambda^2\right]y^2\right]+o(y^2)$$

with $\lambda = \frac{\nu}{\alpha_t}S_t^{1-\beta}$ so that

$$\widetilde{\Sigma}^H(t, 0, 0) = \alpha_t S_t^{\beta-1}. \tag{4.69}$$

As for the IATM skew:

$$\widetilde{\Sigma}_y'^H(t, 0, 0) = -\frac{1}{2}\alpha_t S_t^{\beta-1}1-\beta-\rho\lambda = \frac{1}{2}\rho\nu+(\beta-1)\alpha_t S_t^{\beta-1}. \tag{4.70}$$

And the IATM curvature:

$$\begin{aligned}\widetilde{\Sigma}_{yy}''^{\,H}(t, 0, 0) &= \frac{1}{6}\alpha_t S_t^{\beta-1}\left[(1-\beta)^2+(2-3\rho^2)\lambda^2\right]\\ &= \alpha_t S_t^{\beta-1}\left[\frac{1}{6}(\beta-1)^2\right]+\frac{1}{\alpha_t}S_t^{1-\beta}\left[\frac{1}{6}(2-3\rho^2)\nu^2\right].\end{aligned} \tag{4.71}$$

Comparing (4.69), (4.70) and (4.71) respectively to (4.58), (4.59) and (4.60), we observe a match.

In order to compute Hagan's ATM slope $\widetilde{\Sigma}'^{H}_{\theta}(t,0,0)$ we have to work on the original formula. Note that the limit of z when $y \searrow 0$ is null, which makes the second bracket in Hagan's formula (4.68) undetermined. Therefore we must start with the small-z expansion

$$\log\left[\frac{\sqrt{1-2\rho z+z^2}+z-\rho}{1-\rho}\right] = z + O(z^2),$$

which enables us to write the limit of the second bracket simply as

$$\lim_{y\to 0} \quad \frac{z}{\xi(z)} = 1$$

and therefore the IATM slope in $(\star) \triangleq (t,0,0)$ is

$$\begin{aligned}\widetilde{\Sigma}'^{H}_{\theta}(\star) &= \frac{\alpha_t}{S_t^{1-\beta}}\left[\frac{(1-\beta)^2\,\alpha_t^2}{24S_t^{2-2\beta}} + \frac{\rho\beta\nu\alpha_t}{4S_t^{1-\beta}} + \frac{2-3\rho^2}{24}\nu^2\right]\\ &= \alpha_t^3 S_t^{3\beta-3}\left[\frac{(\beta-1)^2}{24}\right] + \alpha_t^2 S_t^{2\beta-2}\left[\frac{1}{4}\rho\beta\nu\right] + \alpha_t S_t^{\beta-1}\left[\frac{1}{12}\left(1-\frac{3}{2}\rho^2\right)\nu^2\right],\end{aligned}$$

which matches (4.62). □

4.4 Application to the FL-SV Class (Exercise)

The FL-SV acronym stands for *Forward Libor-Stochastic Volatility* which indicates that (like SABR) this model was initially developed for the interest rates environment. More precisely, it was targeted at the term structure of caplets and from then on, *via* several necessary approximations, to the swaption grid. It was initially introduced in [19] and later complemented by [20], presented as an improvement to the local volatility extensions of the Libor Market Model. Those have been formalised in [21], and allow control of both the term-structure level and the skew of the smile, by clearly allocating the former to a deterministic volatility and the latter to a (local volatility) skew function.

The FL-SV extension consists in adding an exogenous, multiplicative perturbation to the volatility, responsible for producing a convexity in the smile that will stay *Delta sticky*. It is then clear that one of the main strengths of this model lies in its clear one-on-one parameter *vs* smile allocation. But any model must also be calibrated, hence [19] provides two closed-form approximations of the static smile. These are based on a fairly complex small-time expansion of a Lognormal volatility proxy, using log-moneyness as the variable. The authors observe that the growth rate of the wings is very high and therefore introduce a heuristic *dampening* of the wings. In fact, without that adjustment the usual arbitrage conditions (the growth rate of $\sqrt{y}$) are typically breached.

4.4.1 Presentation of the Model

The bi-dimensional SDE system reads as

$$\textbf{FL-SV}\begin{cases} dS_t = \varphi(S_t)\sqrt{V_t}\gamma(t)\,dW_t & \\ & \text{with} \\ dV_t = \kappa\left(\overline{V} - V_t\right)\,dt + \varepsilon\psi(V_t)\,dZ_t & \end{cases} \quad \begin{matrix} V_0 = 1 \\ \\ W_t \perp Z_t. \end{matrix} \tag{4.72}$$

Note that in order to normalise the problem, we can also write $\varphi(x) = \sigma\varphi^*(x)$ with $\varphi^*(S_t) = 1$. The same convention can evidently be applied to $\gamma(\cdot)$ and $\psi(\cdot)$. This technique clearly dissociates scale from shape and simplifies the interpretation. However, it does so in an artificial way, as the parametrisation is not stationary, hence we will maintain the original formulation.

The model has been quite successful with practitioners, inasmuch as that when SV-LMM frameworks are implemented, it tends to be along these lines. But this class has also been well received by academics, since it is quite intuitive, and has been both extended and modified. In [22], a time-dependency is introduced in $\varphi(t, S_t)$ which allows the generation of a *term-structure of skew*, hence the FL-TSS denomination. The calibration method focuses on projecting the model onto a time-homogeneous Markovian process, a method which is generalised in [23].

In [24], a low-dimensional, separable volatility HJM model (a stochastic volatility extension of the Cheyette class) is parametrised to approximate the dynamics of the FL-SV, and again several *ad hoc* calibration approaches are discussed.

In order for us to analyse this model *via* the ACE methodology, we first proceed with a simplification step. In fact the latter consists in *extending* the model by considering

$$\textbf{Extended FL-SV}\quad\begin{cases} dS_t = \gamma(t)b\,(V_t)\,\varphi(S_t)dW_t & V_0 = 1 \\ dV_t = \kappa c\,(V_t)\,dt + \varepsilon\psi\,(V_t)\,dZ_t & W_t \perp Z_t. \end{cases} \tag{4.73}$$

We will assume any required degree of regularity for the five functions involved in the model, so that all differentials invoked are meaningful. Note that it would not take much to transform this framework into an extended FL-TSS model, but the added complexity is undesirable here. We now proceed to the derivation of the chaos dynamics for this Extended FL-SV class.

4.4.2 Derivation Exercise

We start by establishing some elementary results which will help exploit the induction, then we derive the relevant chaos dynamics coefficients, and finally we express the static IATM differentials of the first layer.

Step 1: induction setup

Let us first introduce a few notations that will help in creating a recurring scheme.

Definition 4.3 In the context of the Extended FL-SV framework (4.73) we replace $\varphi(\cdot)$ by

$$g(x) \triangleq \varphi(x)x^{-1} \tag{4.74}$$

and define several functions derived from $\gamma(\cdot)$, $b(\cdot)$ and $g(\cdot)$ by induction:

$$\begin{array}{lll} n \in \{0,1,2,3,4,5\} & m \in \{0,1,2,3\} & p \in \{0,1,2\} \\ b_0(x) \triangleq b(x) & \gamma_0(x) \triangleq \gamma(x) & g_0(x) \triangleq g(x) \\ b_{n1}(x) \triangleq b_n^{'}\psi(x) & \gamma_{m1}(x) \triangleq \gamma_m^{'}(x) & g_{p1}(x) \triangleq g_p^{'}\varphi(x) \\ b_{n2}(x) \triangleq b_n^{''}\psi^2(x) & \gamma_{m2}(x) \triangleq \gamma_m\gamma(x) & g_{p2}(x) \triangleq g_p^{''}\varphi^2(x) \\ b_{n3}(x) \triangleq b_n b(x) & \gamma_{m3}(x) \triangleq \gamma_m\gamma^2(x) & \\ b_{n4}(x) \triangleq b_n b^2(x) & & \\ b_{n5}(x) \triangleq b_n^{'}c(x) & & \end{array}$$

In the sequel, we will omit index 0 for simplicity.

Note that these notations are simple but arbitrary. Hence they can certainly be modified and improved by the reader. We can now establish the basic induction result.

Exercise 4.1 Following the notations of Definition 4.3 and exploiting the induction, express the dynamics of the following **elementary block**

$$[\gamma_m \, b_n \, g_p]$$

where m, n and p represent abstract multi-digit indices in $\mathbb{N}$.

The solution to Exercise 4.1 can be found on p. 481.

Step 2: computation of the chaos coefficients

We can now make explicit the chaos dynamics in a reasonably compact fashion.

Exercise 4.2 Using the output of Exercise 4.1, express the chaos coefficients listed in Definition 4.1 [p. 219] for the FL-SV model, as a function of the native parameters.

The solution to Exercise 4.2 is given on p. 482.

Step 3: expression of some static IATM differentials

We now have all the information to produce the static IATM differentials of the first layer.

Exercise 4.3 Using the results of Exercise 4.2, express the IATM differentials of the first layer, namely

$$\widetilde{\Sigma}(\star) \qquad \widetilde{\Sigma}'_y(\star) \qquad \widetilde{\Sigma}''_{yy}(\star) \qquad \widetilde{\Sigma}'_\theta(\star)$$

as functions of the native FL-SV parameters.

The solution to Exercise 4.3 is provided on p. 487.

4.5 Numerical Implementation and Testing

In this section we illustrate part of the considerations and techniques, discussed in Sect. 4.1, pertaining to whole-smile extrapolations based on ACE results. To do so, we also exploit some of the results derived in the later Sects. 4.2–4.4 dedicated to specific models.

4.5.1 Testing Environment and Rationale

4.5.1.1 Selecting a Model

In order to test our extrapolation techniques, a single model first has to be selected. Indeed, given the focus and constraints of this study, it seems more appropriate to explore various configurations and parameter sets for a well-chosen class, rather than superficially covering a number of different models. Among the three classes covered in this chapter, **CEV-SABR** appears to be the best choice, and for several reasons:

- The SABR class presents marginal distributions with much more variance that FL-SV, due to the lack of mean-reversion. Since our extrapolations should be tested w.r.t. the *tails* that they generate, and therefore the associated high-order moments, SABR represents a more demonstrative environment.
- SABR offers the benefit of a closed-form approximation formula, which has been obtained by a singular perturbation technique. Although the latter can be categorised in the same broad asymptotic family as ACE, the associated extrapolation presents fewer similarities with our method than the approach of [19] for FL-SV. Thus it brings even more contrast, which is a desirable feature in our testing context.
- Within the SABR class, and as mentioned before, the CEV instance is by far the most popular. As such a standard market model, it presents more interest both for

practitioners and academics. This means that the market-calibrated parameter sets are already available for testing and comparison, and also that several other published approximation methods can be used instead of Hagan's formula, if desired.

4.5.1.2 Defining the Objectives

For reasons of financial interest as well as practicality, we are interested in the direct problem, more specifically the static shape of the smile (rather than its dynamic properties). Since we know the IATM differentials to be *exact*, we aim at testing the quality of various extrapolation techniques, following the general principles mentioned in Sect. 4.1.

Our applicative concern is naturally the calibration process, hence our focus will be on significant expiries but in the *liquid* range, which is usually one or two standard deviations either side of the ATM point. In that respect, far-from-the-money regions and especially the left-hand side should be examined essentially along *validity* criteria. In other words, in these areas we should evaluate the implied distribution rather than the volatility. But when we *do* gauge the IV extrapolation, our measure of quality should be the *real* CEV-SABR model proxied *via* a *very precise numerical scheme*. Therefore our protocol requires a numerical method providing a clear error control, while there is no significant constraint for speed. In such a specific *testing and academic* environment, as opposed to a pricing and hedging framework, Monte-Carlo seems more appropriate than a PDE scheme (typically finite differences).

4.5.1.3 Specifying the Scheme and the Constants

Given the style and space constraints of this book, our testing strategy is to select a realistic, given set of model parameters as reference, which will be the *central configuration*. We examine how the model and the approximations behave under these conditions, then observe and analyse the influence on these features of each relevant parameter. The central configuration is chosen to satisfy several criteria, but common sense dictates that it should be roughly midway through each parameter range, as well as representative of the model specificities and of the benchmark's artifacts. Furthermore, we elected a configuration which is compatible with recent and sustainable IR market conditions, in particular the level, skew and curvature of the smile. Since the CEV model is scalable, we have fixed the underlying's initial value at $\mathbf{S_t} = \mathbf{100}$. Obviously this choice does *not* fit the interest rates environment, but it is the usual academic convention as it allows easy comparison.

The central power has been set at $\beta = \mathbf{0.5}$ as a compromise between a Normal and a Lognormal local volatility/backbone. The initial value $\alpha_\mathbf{t}$ has been chosen as a function of S_t and β so as to produce an ATM Lognormal Implied Volatility (**LNIV**) of approximately **20** %. Correlation has been chosen negative, in accordance with most historical time series analyses. It has been set at $\rho = -\mathbf{50}\,\%$ again as an arbitrary midway point, but also in order to generate realistic ATM skews given the chosen β power. The last model parameter, the volatility of volatility, has been set at $\nu = \mathbf{30}\,\%$ simply to create a smile curvature that is compatible with current markets, at the chosen expiry.

The latter Time-To-Maturity has been fixed at $T - t = 10Y$, which indeed falls well outside of the common range for short-expiry *asymptotics* and therefore should exacerbate the limitations of our extrapolation. Note, however, that, for lack of an alternative methodology, it is common for practitioners to rely on this approximation class for their vanilla products. To provide some perspective on the subject, note that in the interest rates environment, for instance, those liquid products (essentially swaptions, but implicitly caplets also) can be found at up to 30 or 40 years' expiry.

As for the benchmark Monte-Carlo scheme, we will be using the following:

- a Marsaglia's Ziggurat generator for the Brownian increments, with a period of 2^{62} which is more than sufficient for our needs;
- a common time step Δt of 10^{-2} (slightly less than four days in calendar terms);
- a log-Euler scheme for both S_t and α_t, justified by a β of 0.5;
- a common threshold of 10^{-3} for S_t and α_t, with a one-step only absorption rule.

Unless specified otherwise, we will simulate 10^6 paths and plot the confidence interval at 3 standard deviations. As the only variance reduction technique, we use S_T for the control variate. Due to the large variance and specific l.h.s. behaviour of the marginal, this significantly improves the implied volatility precision for low strikes, as demonstrated by Fig. 4.1.

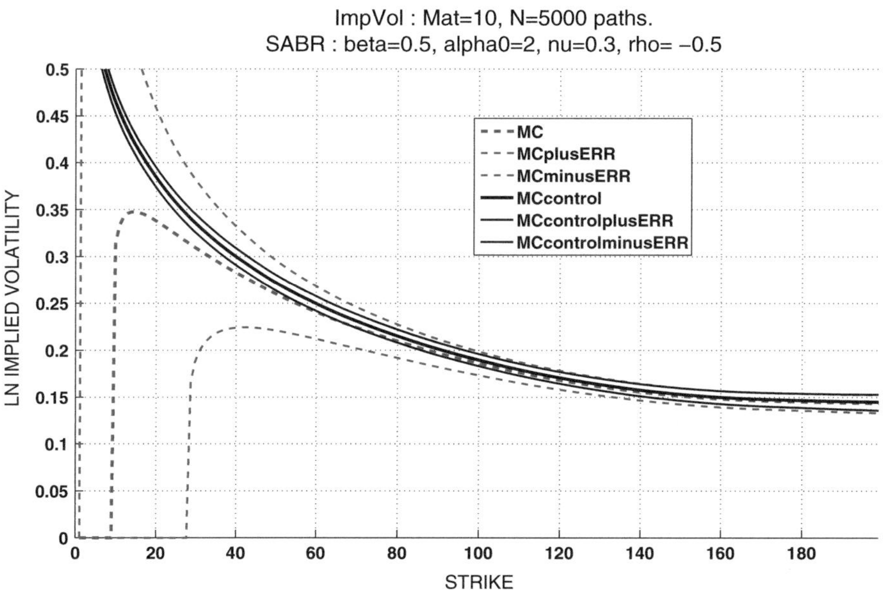

Fig. 4.1 Section 4.5.2.1 [p. 251]. LNIV by MC: pure (*dotted blue*) *vs* controlled (*plain black*)

Table 4.1 CEV-SABR testing: central configuration

β[a]	α_t^{b}	ρ[a]	ν[a]	$(T-t)$[c]	NbPath[d]
0.5	2	−50 %	30 %	10Y	1,000,000

[a] Model param
[b] Model init
[c] Option param
[c] Numerical param

Combining the model and the scheme parameters, the **central configuration** is summarised in Table 4.1 and we will use the same presentation for each graph.

Let us now describe in more detail the type of ACE approximation used, the sequence of configurations to be tested, as well as the types of output to be plotted and analysed.

4.5.1.4 Selecting an Expansion Type

Our first task is to select an order (keeping in mind the ladder constraint) and a variable for the expansion, among the several possibilities discussed in Sect. 4.1. Our primary concern is to employ a variable that at least does not *guarantee* a breach of validity/arbitrage in either of the tails. Recalling that it is possible to treat separately (to *split*) the left and right-hand side of the expansion, the above concern therefore leaves us with the following choices:

- On the left and/or on the right, an expansion of order $N \leq 4$ on the log-moneyness $y = \ln(K/S_t)$ but for the functional $\widetilde{\Sigma}^{2N}$, which we then re-scale down in order not to exceed the $\sqrt{y}$ growth rule.
- On the left and/or on the right, an expansion of order $N \leq 4$ on a variable z that grows as the logarithm of y:

$$z = \begin{cases} -\ln(1-y) & \text{for } y \leq 0, \\ \ln(1+y) & \text{for } y \geq 0. \end{cases}$$

  Indeed, and as discussed in Sect. 4.1, this variable satisfies the growth condition at any order. Note that as a function of the strike or of the underlying, which are the native space variables, $z(K)$ or $z(S_t)$ is very close to an *iterated logarithm.*
- On the left only, an expansion on strike K (combined with any of the above choices on the right).

The first re-scaling alternative provides good results on the left-hand side, albeit quite similar to the z option. It also creates some unfortunate artefacts on the right, in the form of really obvious inflexion points. As for a left-K/right-z solution, Figs. 4.3 and 4.4 allow a comparison with the pure-z approximation. They show that the Monte-Carlo (as well as Hagan's approximation) exhibits a specific shape for the LNIV, namely a surge due to a strictly positive asymptote for the density, which only the z expansion can match. As for the implied density, although the z and K expansions feature the same qualitative behaviour (a surge as $K \searrow 0$) it is significantly less pronounced for the former.

In consequence we opt for the z variable, the order of which remains to be set. Indeed, the question of whether we should be using the *maximum* available number of differentials, those included in layers 1, 2 and 3, is not trivial. It is a well-known feature of polynomial interpolations and extrapolations, that when limited by a *finite* order, the performance can be counterintuitive. The optimal expansion order will effectively depend on the function considered, on the chosen range for the variable, and obviously on the specific measure of precision employed.

In the current context, our quality criteria are complex (they concern volatility *and* density) but the strike range *is* well-defined (see below). Hence we compare the results obtained with three distinct sets of differentials: those involved in layer 1 alone, then those in the union of layers 1 and 2, and finally the whole group (layers 1 to 3). This comparison is plotted in Fig. 4.2 in terms of LNIV error w.r.t. the benchmark Monte-Carlo, and fortunately shows that in terms of precision the highest layer is indeed the most appropriate. Note that in a typical fashion, the first and second alternatives could not clearly be ranked in that respect. Furthermore, and as a sanity check, Fig. 4.7 shows that the density associated to this maximum order expansion cannot *a priori* be rejected.

4.5.1.5 Testing Plan and Rationale

Starting from this realistic central configuration, in full generality we would like to assess the influence of each available input, i.e. the initial values of the state variable (S_t and α_t), the model parameters (β, ρ and ν) and the option specifiers (K and T). This is how we choose to proceed:

- We will plot the relevant information (see below) for the selected **strike range of** $[0, 2 * S_t]$, which makes our focus proportional to the initial value S_t. Furthermore, the CEV local volatility function ensures that a scaling of S_t by λ is translated into a scaling of α_t by $\lambda^{\beta-1}$. This redundancy implies that there is no point in modifying the initial value S_t.
- The impact of modifying the β power will be examined by switching to a Lognormal setup ($\beta = 1$). The change in backbone (ATM LNIV as a function of S_t) is simple and well understood (see [1]) therefore we will not represent it. However, the modification of the smile is significant, especially the increase of the ATM level and the moderation of the skew. Therefore in order to maintain approximately the same at-the-money implied volatility, which enhances comparability, we will decrease α_t to 0.2.
- The influence of the initial perturbation α_t will be measured by increasing its value from 2.0 to $\alpha_t = \mathbf{3.0}$, bringing the ATM Lognormal Implied Volatility roughly from 20 to 30 %. This is still a realistic configuration, and although both Hagan and

ACE provide formulae which are proportional to α_t, it shows how asymptotics on the *implied volatility* cope with a fattening of the tails for the *marginal* distribution.
- The effect of correlation ρ will be observed by setting $\rho = \mathbf{0}$. In this uncorrelated case the joint dynamics are simpler, the skew and overall smile asymmetry less pronounced, so that the relative importance of the local volatility and of the vol of vol ν is enhanced.
- The impact of the vol of vol ν will be examined similarly, by imposing $\nu = \mathbf{0}$, which provides the best insight on the sole influence of the β parameter.
- Finally, the influence of Time-To-Maturity $T - t$ will be measured by adopting $T = 5Y$ for a first benign case, and then $T = 20Y$ for a more demanding example.

We still need to select the relevant outputs. Since our concern is first the precision and then the validity of the expansion, we will observe and plot all or some of the following graphs:

- The Lognormal implied volatility, by comparison with the two benchmarks which are the numerical scheme and Hagan's formula. To that end we will plot the smiles in parallel and, when pertinent, the difference between the analytic expressions (ACE and Hagan) and the numerical scheme.
- The density, when deemed relevant, since a visual check usually provides a good assessment of the approximation's validity. Again, we will simultaneously graph ACE, Hagan and the Monte-Carlo scheme.

To avoid confusion, we summarise in a table the parameter configuration corresponding to each figure, with the parameters deviating from the central configuration in bold. Also, whenever possible, we maintain the same scales and graph conventions in order to facilitate comparison.

Note that this protocol aims at providing only a rough idea of the performance of the approximation methods. In the context of the current study we cannot afford to spend much more time on the subject. *A contrario* the same comparison in a production perspective would be more thorough, and in particular it would involve plotting the *sensitivities* of the smile to each of the parameters, in various configurations. A typical requirement would be to graph the correlation and vol of vol:

$$K \longrightarrow \frac{\partial \Sigma}{\partial \rho} \qquad \text{and} \qquad K \longrightarrow \frac{\partial \Sigma}{\partial \nu}.$$

The overall shape of these functionals for the model are simple and well-documented (see [1], for instance) so we will not plot them.

4.5.2 Tests Data and Results

Most tests below cover the effect of modifying a single parameter or feature, compared to the central configuration. For each test we will first present a table containing the whole configuration, therefore allowing the reader to reproduce the results if necessary. We will then interpret the results based on the output graphs (starting on p. 247) which illustrate the relevant impact.

4.5.2.1 Justifying the Choice of Expansion Variable, Order and Benchmark

Figures 4.1 [p. 247] and 4.2 [p. 251] demonstrate, respectively, the benefit of using the forward as MC control variate, and of increasing the expansion order. They use the central configuration but with a much lower number of paths to be demonstrative, respectively 5,000 and 10,000 (see Table 4.2).

Figures 4.3 and 4.4 [p. 252] illustrate and justify the choice of expansion variable on the left-hand side of the smile. They compare a pure-z approximation with a left-K/right-z alternative (whose splitting point is At-The-Money) using the central configuration (see Table 4.3).

These topics will be discussed within Sect. 4.5.2.2 below.

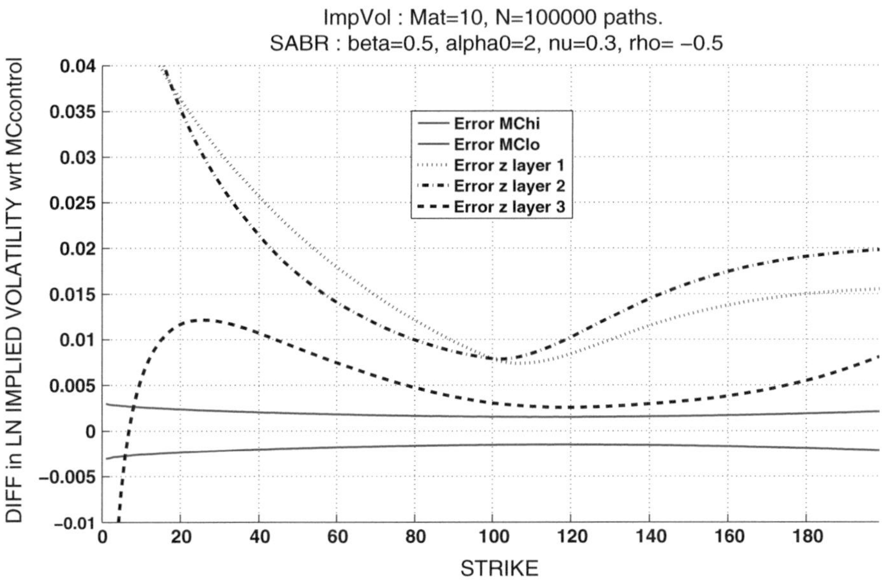

Fig. 4.2 Section 4.5.2.1 [p. 251]. Influence of the expansion order on the LNIV error

Table 4.2 CEV-SABR: configuration for testing the expansion order and control variate

	β	α_t	ρ (%)	ν (%)	$(T-t)$	NbPath
Figure 4.1	0.5	2	−50	30	10Y	**5,000**
Figure 4.2	0.5	2	−50	30	10Y	**10,000**

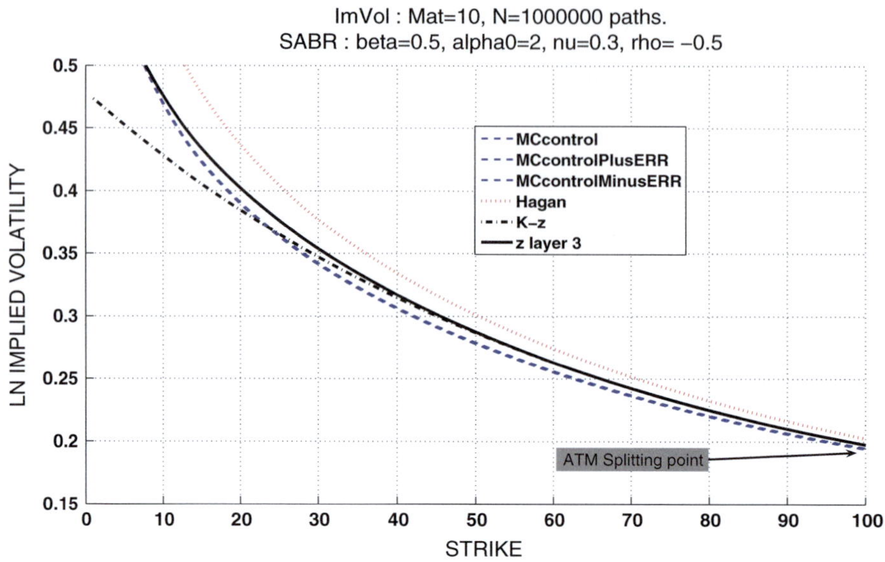

Fig. 4.3 Section 4.5.2.1 [p. 251]. Impact of expansion variable on LNIV: pure z *vs* mixed $K\&z$

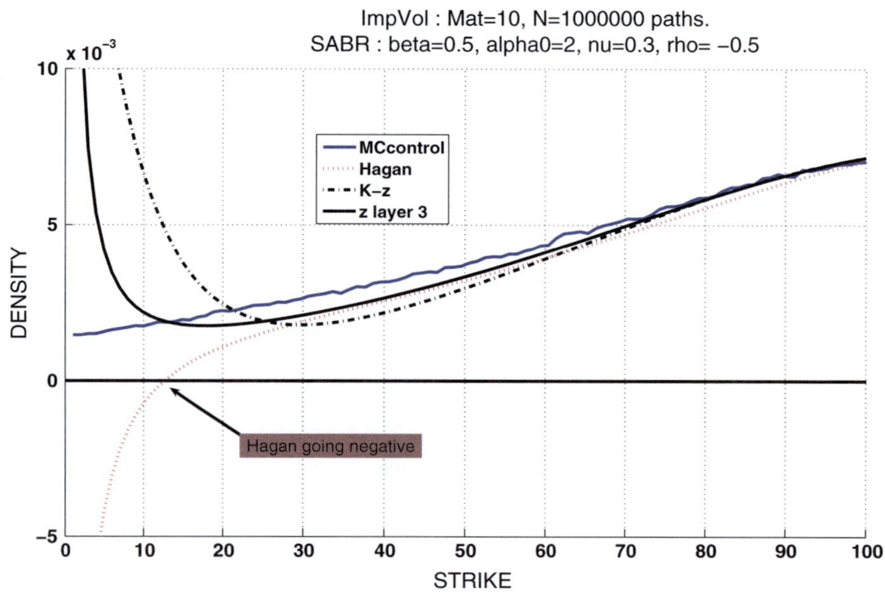

Fig. 4.4 Section 4.5.2.1 [p. 251]. Impact of expansion variable on density: pure z *vs* mixed $K\&z$

Table 4.3 CEV-SABR: configuration for testing the expansion variable

	β	α_t	ρ	ν	$(T-t)$	NbPath
Figures 4.3 and 4.4	0.5	2	$-50\,\%$	$30\,\%$	10Y	1,000,000

4.5.2.2 Reference Configuration

Figures 4.5 [p. 253], 4.6 [p. 254], 4.7 [p. 254] and 4.8 [p. 255] demonstrate the extrapolation's performance for the central parameter configuration (see Table 4.4), establishing the main qualities and shortfalls of the approximation. This performance can be assessed on its own, and in comparison to the benchmark Monte-Carlo and/or to Hagan's formula.

In particular, Figs. 4.5 and 4.6 show that within the liquid range and for a significant expiry of 10 years, Hagan presents a smile shape which is stable and consistent with the model. However, it overestimates the Lognormal IV across the whole strike range, and increasingly so as K nears zero. In our view, this is mainly due to a combination of two factors in formula (4.68):

- The geometric average $(KS_t)^{\frac{1-\beta}{2}}$ in the first multiplicative term's denominator.
- The specific expression for the third and last multiplicative term $[1+[\cdots]\theta]$.

A quick examination of the proof in [1] shows that alternatives can be used for the average (typically the arithmetic one) while maintaining the singular perturbation at

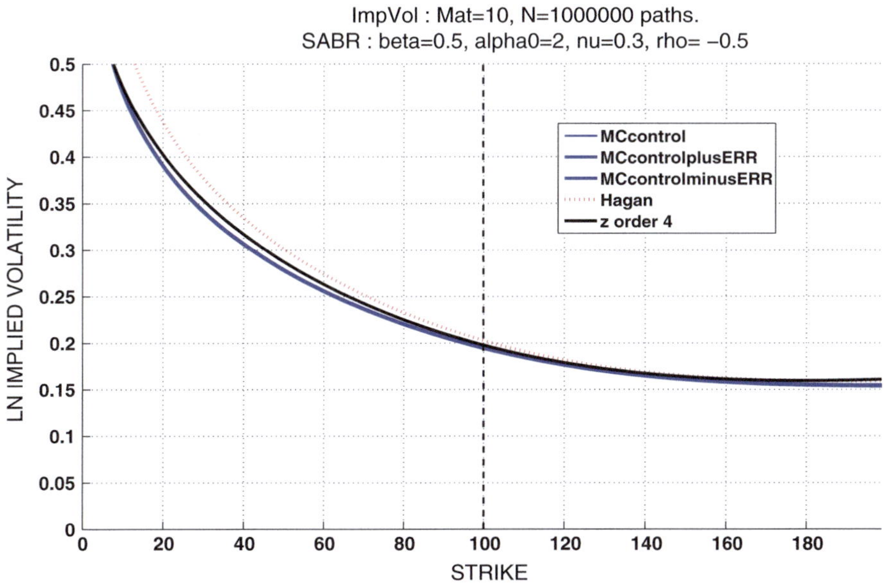

Fig. 4.5 Section 4.5.2.2 [p. 253]. LNIV of central config. Layer-3 expansion on z *vs* Hagan

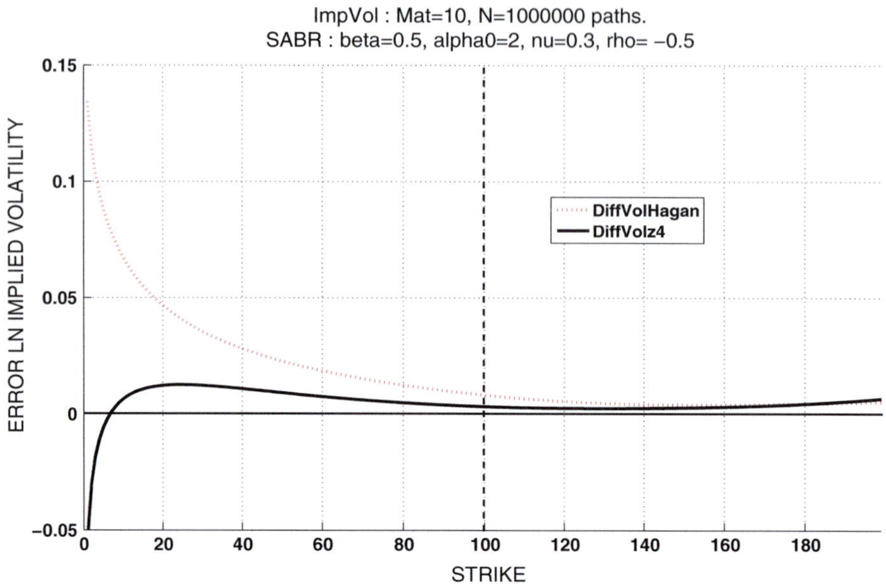

Fig. 4.6 Section 4.5.2.2 [p. 253]. LNIV error of central config. Layer-3 expansion on z *vs* Hagan

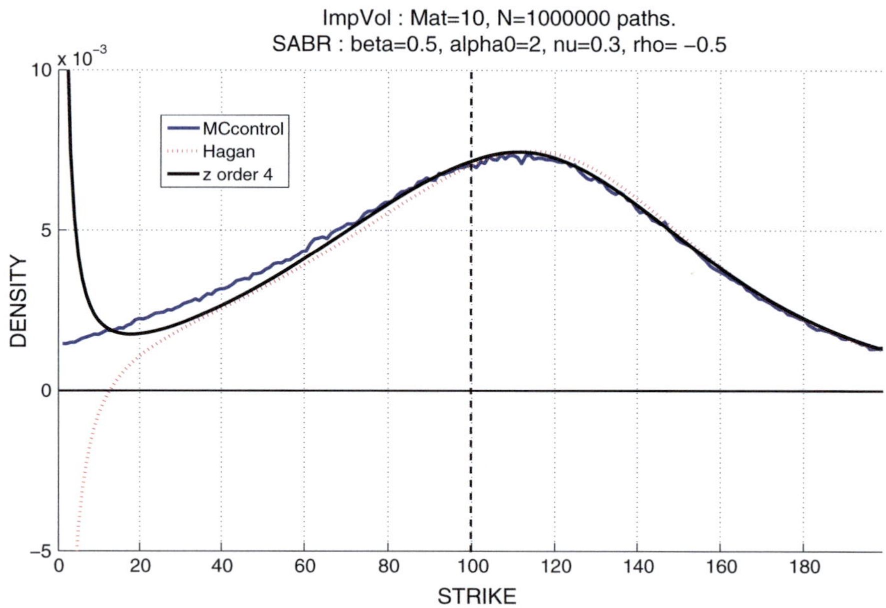

Fig. 4.7 Section 4.5.2.2 [p. 253]. Density of central config. Layer-3 expansion on z *vs* Hagan

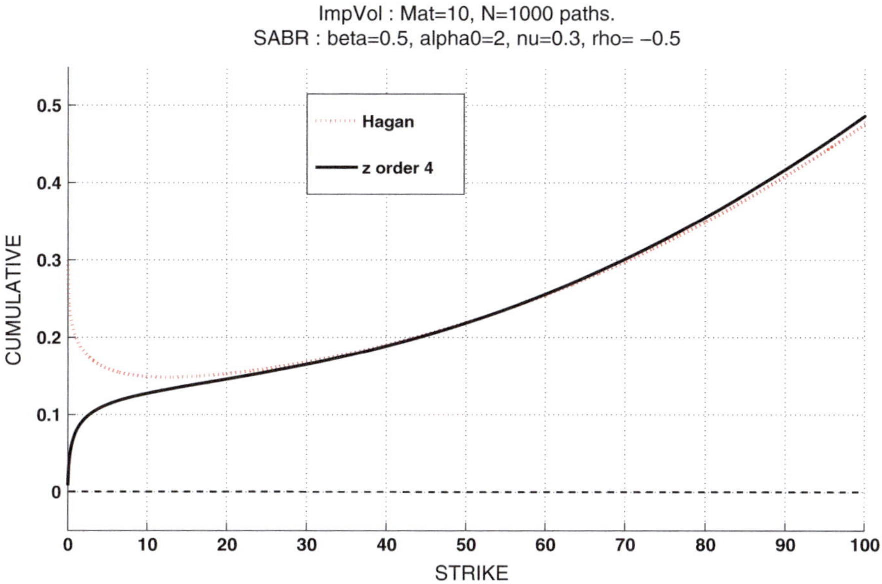

Fig. 4.8 Section 4.5.2.2 [p. 253]. Cumulative of central config. Layer-3 expansion on z *vs* Hagan

Table 4.4 CEV-SABR: testing the performance of the central configuration

	β	α_t	ρ	ν	$(T-t)$	NbPath
Figures 4.5, 4.6, 4.7, 4.8	0.5	2	−50 %	30 %	10Y	1,000,000

the same order of precision. They *can* mitigate the left-hand side behaviour, but they also tend to develop their own specific issues, so that we are not aware of any choice ensuring a systematic improvement on Hagan.[8]

As for the third, maturity-correction term, its linearity in θ tends to overestimate the ATM implied volatility. Indeed most SABR configurations exhibit an *ATM vs maturity* profile which is concave, a feature that requires the $\widetilde{\Sigma}''_{\theta\theta}(\star)$ IATM differential to be captured. The fact that Hagan's formula is linear in term-to-maturity is inherent to the expansion order at which it has been computed, so that to achieve a better long-maturity fit we would need to extend the proof in [1] significantly.

As an alternative, we could choose a new expression for the θ variable that would ensure concavity, such as $C(1-e^{-\lambda(T-t)})$, for instance. Note that this offers the added benefit of capping the volatility in the long term, a feature that we have so far overlooked. This choice does not *have* to be a heuristic either, since results on the long-term implied volatility of some stochastic volatility models are available (see [12] and [13], for instance).

[8] Which is not to say that it has not been published or that this avenue should be abandoned.

In comparison, the layer-3 z-expansion exhibits the same bias, but in a much milder fashion, while exhibiting an unwelcome inversion for very low strikes. The implied density graph (Fig. 4.7) shows interestingly that, as $K \searrow 0$:

- The *real* Monte-carlo density admits a non-null asymptote.
- The z expansion exhibits an infinite asymptote.
- Hagan approximation corresponds to a negative density.

It is important to recall that, although the infinite asymptote of the z expansion seems quite unusual, it does not *a priori* breach arbitrage conditions and the validity of the marginal distribution. To be convinced of this point, it suffices to examine Fig. 4.11 [p. 258], which proves that the empirical Monte-Carlo density *can* exhibit a surge as well.

Alternatively, one can examine Fig. 4.8 which plots the cumulative for both Hagan and the z expansion, again for the central parameter configuration. It follows that the z-density is indeed integrable, since the cumulative starts from zero.

By comparison, Hagan's approximation is clearly invalid and arbitrable below $K \approx 12$, as shown by the density and by the cumulative (which exhibits a strictly positive initial value) and is certainly problematic as early as $K = 40$.

Placing ourselves, for instance, in an interest rates context with $S_t = 5\,\%$, this would mean that any binary spread below 50 bp would be negative, and this is actually a (non-liquid) traded range.[9]

In fact, and as will be observed later, the z approximation *can* produce negative densities on the left, just as Hagan does. Indeed, we have chosen the z variable to prevent a single and specific arbitrage condition, i.e. the convergence of the Call price in $K = 0$ and $K = +\infty$. Therefore it satisfies no more than a *necessary* validity criteria, and does not guarantee *a priori* the positivity of the density.

4.5.2.3 Influence of the Power β: The Lognormal Case

Figures 4.9 [p. 257], 4.10 [p. 257] and 4.11 [p. 258] show the expansion's performance for a Lognormal version of SABR. Note that the α_t initial condition has been modified, in order to maintain roughly the same ATM Lognormal Implied Volatility as in the central configuration case (see Table 4.5).

The role of β has been described in some detail in Sect. 4.3.1 [p. 231]. With regard to the interpretation of the current graphs, it is worth recalling that this parameter affects the support of the distribution and determines the basic smile shape. Also, it sets the backbone and therefore has a very strong influence on the smile dynamics. It is also worth mentioning that a β close to 1 (typically 0.9) is not uncommon with practitioners wishing to set a very thick right-hand tail.

By comparing Figs. 4.6 and 4.10 we observe that the precision of Hagan's formula improves slightly in the Lognormal case. As for the z expansion, it is even more

[9] On EUR markets for instance, quotes for zero-strike floors on nominal rates can be obtained from the brokers.

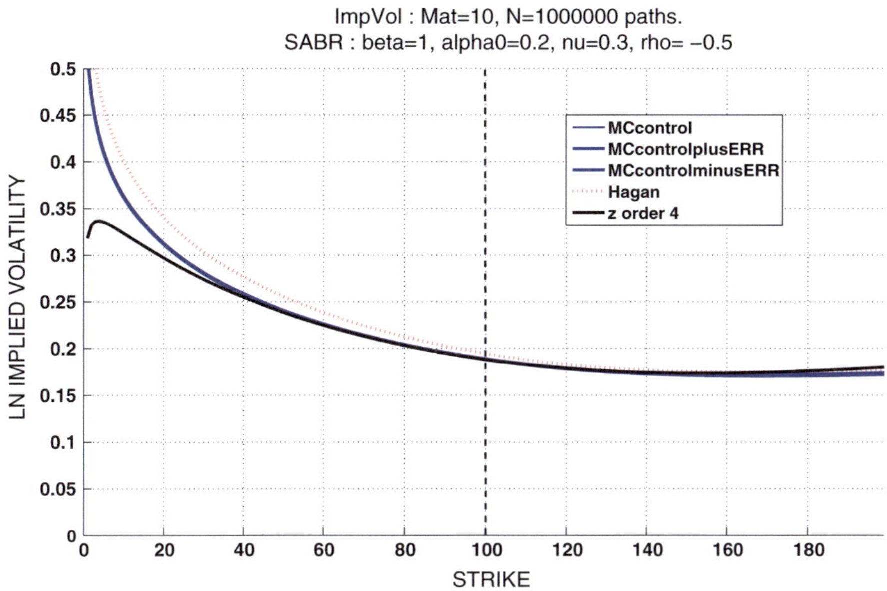

Fig. 4.9 Section 4.5.2.3 [p. 256]. LNIV with a lognormal beta: layer-3 expansion on z *vs* Hagan

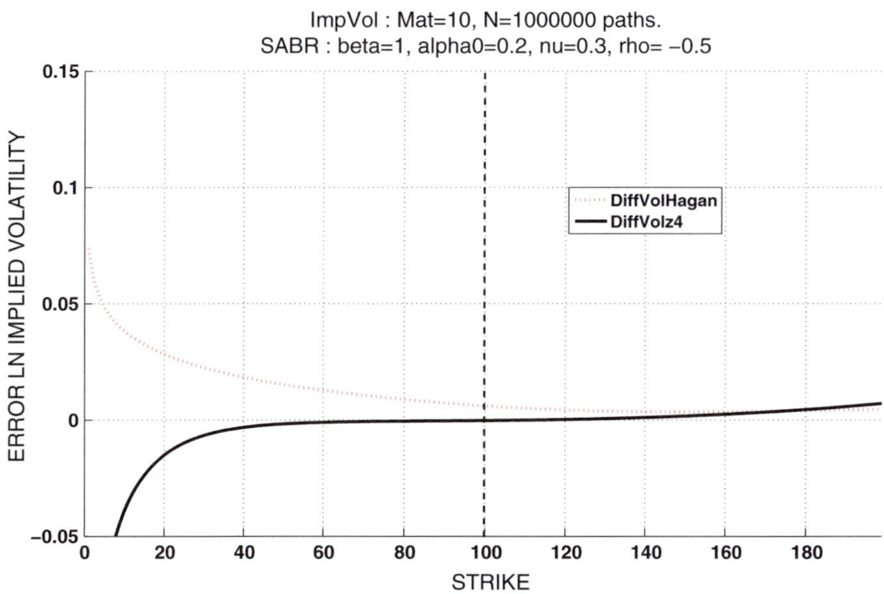

Fig. 4.10 Section 4.5.2.3 [p. 256]. LNIV error with a lognormal beta: expansion on z *vs* Hagan

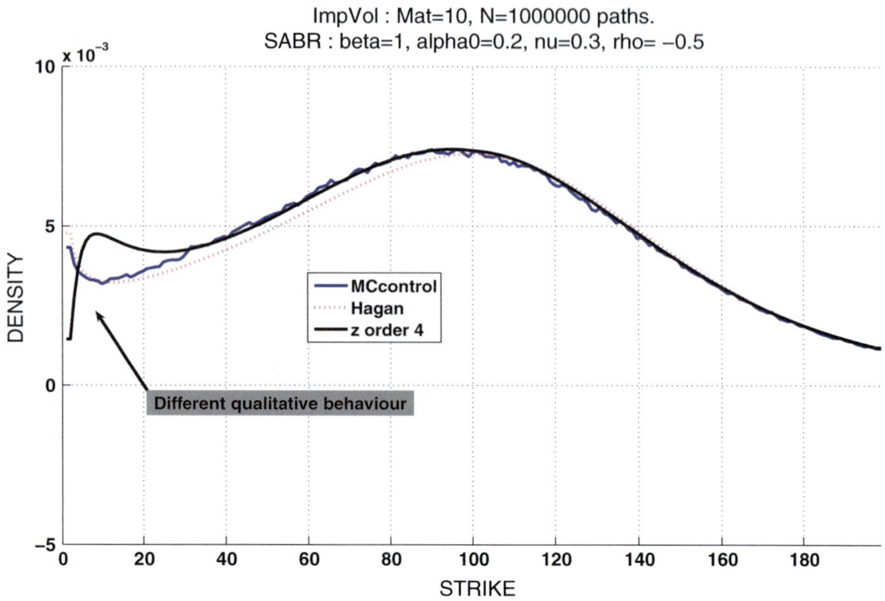

Fig. 4.11 Section 4.5.2.3 [p. 256]. Density with a lognormal beta: expansion on z *vs* Hagan

Table 4.5 CEV-SABR: testing the lognormal case (LNIV, density)

	β	α_t	ρ	ν	$(T-t)$	NbPath
Figures 4.9, 4.10, 4.11	**1.0**	**0.2**	−50 %	30 %	10Y	1,000,000

accurate now, at least in the liquid range: the difference with the benchmark is immaterial between 60 and 140 % of the money. This can be attributed to the fact that the target model (SABR) is now very close to our chosen Lognormal baseline. However, the undershooting happens sooner on the left, which also corresponds to a qualitative modification of the density's behaviour.

Indeed, Fig. 4.11 shows that the benchmark's density is now surging just before $S_t = 0$, albeit with an apparent finite limit. This behaviour is well reproduced by Hagan, but not so by the z-expansion, which after providing a secondary mode decreases down to zero, as would a Lognormal distribution.

In the light of these results, it seems fair to say that the Lognormal environment is quite favourable to both approximations.

4.5.2.4 Influence of a Higher Volatility α_t

Figures 4.12 [p. 259], 4.13 [p. 259] and 4.14 [p. 260] demonstrate the influence on the expansion of increasing α_t (and thus with it, the ATM Lognormal Implied Volatility) as per Table 4.6.

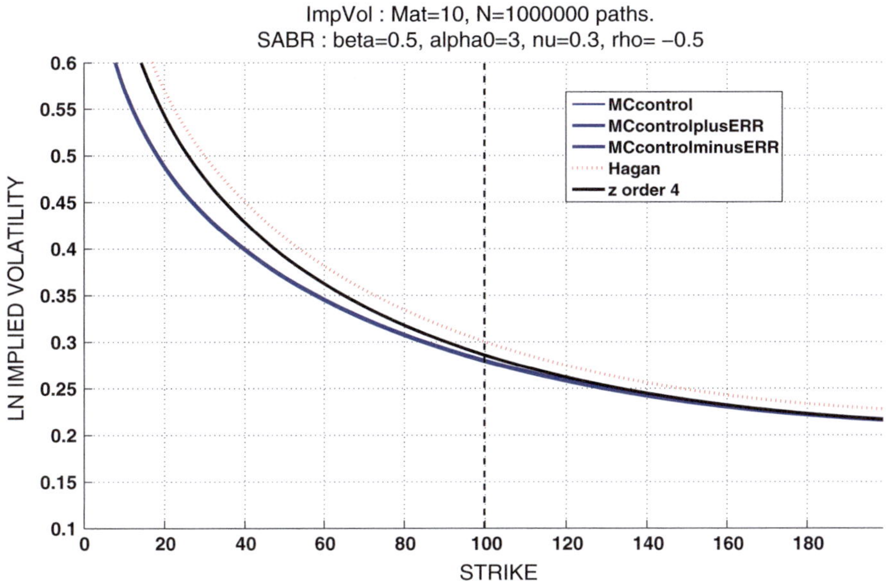

Fig. 4.12 Section 4.5.2.4 [p. 258]. LNIV with high α_t: layer-3 expansion on z *vs* Hagan

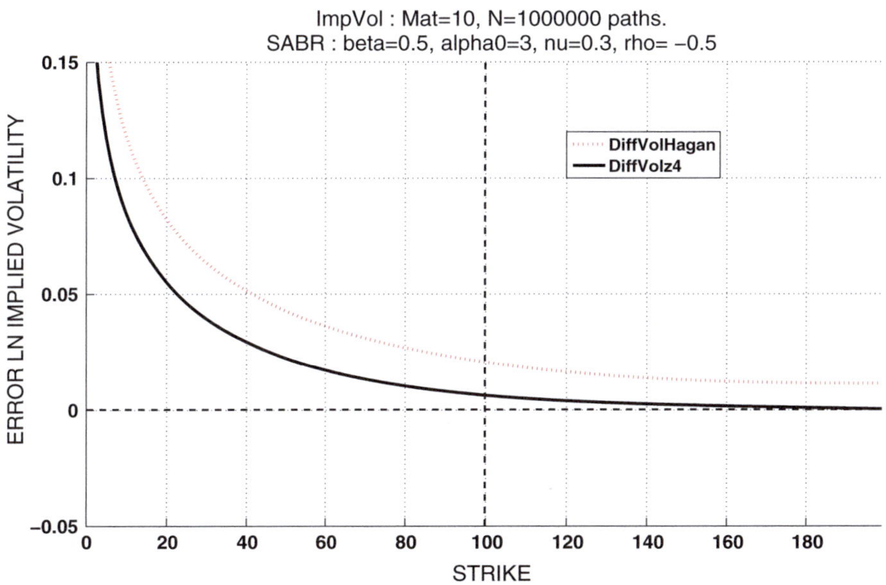

Fig. 4.13 Section 4.5.2.4 [p. 258]. LNIV error with high α_t: expansion on z *vs* Hagan

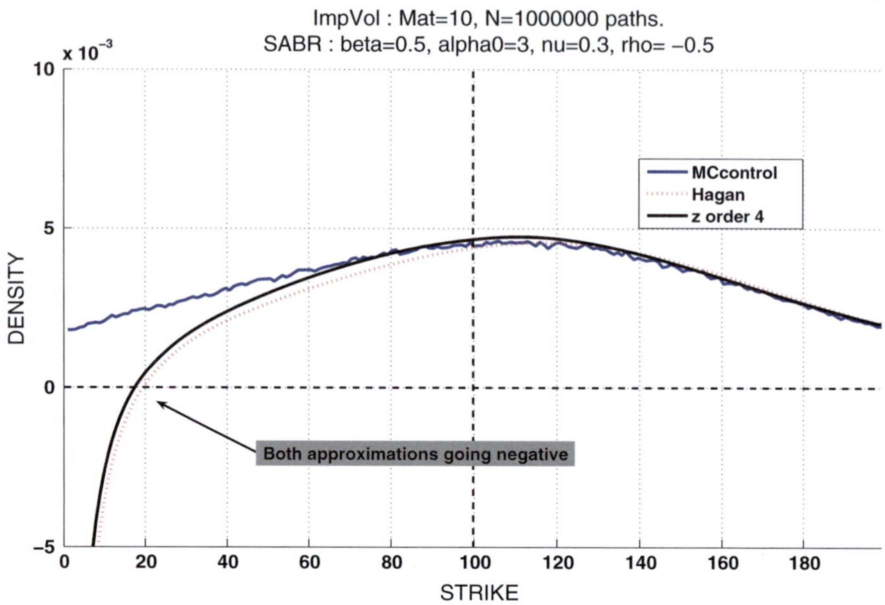

Fig. 4.14 Section 4.5.2.4 [p. 258]. Density with high α_t: expansion on z *vs* Hagan

Table 4.6 CEV-SABR: testing the influence of α_t (LNIV, density)

	β	α_t	ρ	ν	$(T-t)$	NbPath
Figures 4.12, 4.13, 4.14	0.5	**0.3**	−50 %	30 %	10Y	1,000,000

Our first observation is that, at such high expiry, the LNIV (as well as the NIV) is visibly not proportional to α_t. This (potentially counter-intuitive) property is due in part to the instantaneous SV dynamics, but is also observed in a pure CEV setting. Indeed, Dupire's LNIV formula (see (2.61) [p. 66]) shows that for any local volatility (LV) model, the relationship between the LV function and the LNIV for non-null expiries is *not* linear.

This is typically where ACE can provide qualitative insight. Considering the first layer's IATM static differentials (4.58)–(4.62) we observe that α_t is invoked with several powers. Killing the SV with $\rho = \nu = 0$ does make the (second-order) Immediate smile proportional to α_t but the slope $\widetilde{\Sigma}'_\theta(\star)$ still exhibits a cubic dependency. The consequence is an increased LNIV error for both Hagan and the z-expansion, with the latter still ahead (see Fig. 4.13).

The situation is more complex for the density, which is differentiated from the price, itself far from linear w.r.t. the IV. In Fig. 4.14 we observe a fatter density (note the lower mode and the similar left asymptote) but also that *both* approximations now plunge into negative territory, with only a marginal advantage to the z-expansion.

Overall, such high volatility does not suit either approximation well. However, since in a pure Lognormal setting their precision would be perfect, we can safely conclude that the volatility only *exacerbates* their existing shortfalls. This is clear in terms of LNIV precision, but we can also observe that the left-hand side density behaviour for Hagan, and even more so for the z-expansion, is quite extreme and unpredictable.

4.5.2.5 Influence of ρ: The Non-correlated Case

Figures 4.15 [p. 261], 4.16 [p. 262] and 4.17 [p. 262] illustrate the impact of a zero correlation between the underlying and its "stochastic volatility" α_t (see Table 4.7).

We observe that the lack of correlation has—as expected—clearly affected the skew (Fig. 4.15) but does not seem to change the performance of Hagan's approximation much, either in terms of LNIV precision (Fig. 4.16) or with regard to its density (see Fig. 4.17).

By contrast the z-expansion seems to suffer from the uncorrelated environment, as its precision is now decreased, although its bias is now systematically positive (the inversion has disappeared). Also, and more importantly, *both* approximations now have a plunging density, although the z expansion seems slightly more benign.

This qualitative change in behaviour for the z approximation cannot intuitively be attributed to the asymptotic results themselves: the model is simpler, closer to the baseline and therefore the same layers *should* brings a higher degree of IATM

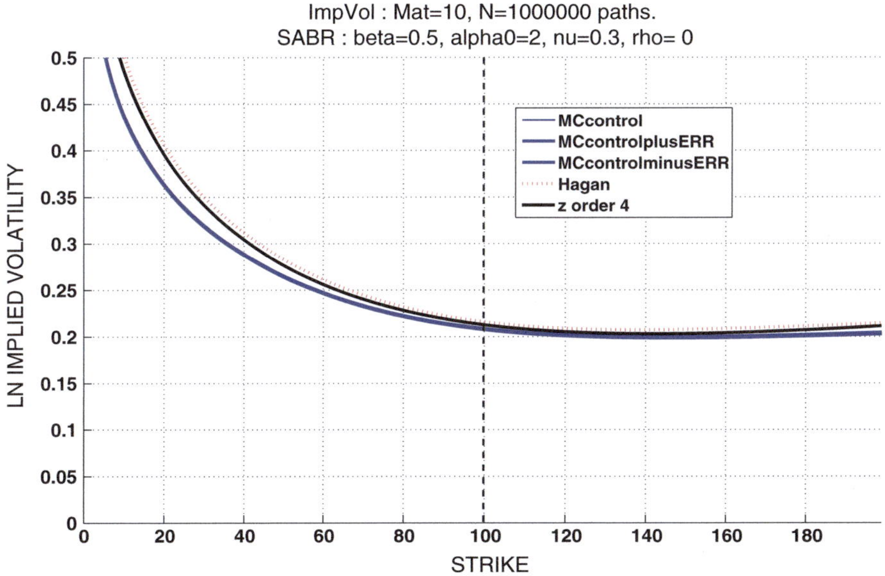

Fig. 4.15 Section 4.5.2.5 [p. 261]. LNIV with no correlation: layer-3 z *vs* Hagan

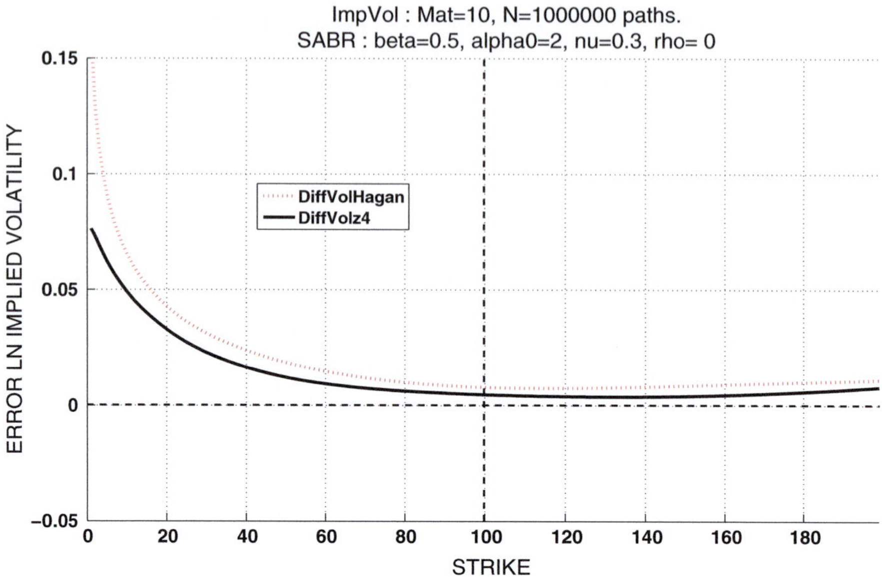

Fig. 4.16 Section 4.5.2.5 [p. 261]. LNIV error with no correlation: layer-3 z *vs* Hagan

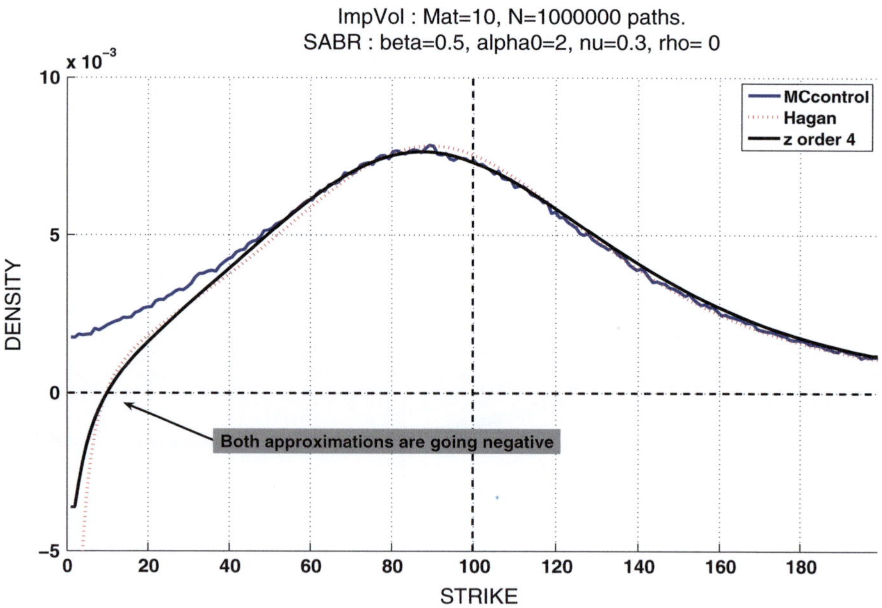

Fig. 4.17 Section 4.5.2.5 [p. 261]. Density with no correlation: layer-3 z *vs* Hagan

Table 4.7 CEV-SABR: testing the influence of ρ (LNIV, density)

	β	α_t	ρ	ν	$(T-t)$	NbPath
Figures 4.15, 4.16, 4.17	0.5	2	**0 %**	30 %	10Y	1,000,000

precision. The most probable explanation is that we are facing an idiosyncracy of the z variable and/or of the Lognormal baseline, which seem to favour a certain asymmetry. Irrespective of the real causes, it is now apparent that the very-far left-hand side of that z-expansion is difficult to predict.

Having neutralised the correlation and therefore made the volatility purely exogenous, we now consider an extreme (and in a way inverse) case where the stochastic perturbation is neutralised.

4.5.2.6 Influence of the Vol of Vol ν: The Pure Local Volatility Case

Figures 4.18 [p. 263], 4.19 [p. 264] and 4.20 [p. 264] demonstrate the impact of taking volatility of volatility as zero (see Table 4.8). First, let us recall that a zero vol of vol neutralises the correlation, which explains why this case does correspond to a pure LV setting.

We note that both approximations are now very accurate, with a slight advantage to Hagan below $K = 20$. This superior accuracy can be observed in the LNIV error graph (Fig. 4.19) and even more so on the density plot (Fig. 4.20). This is

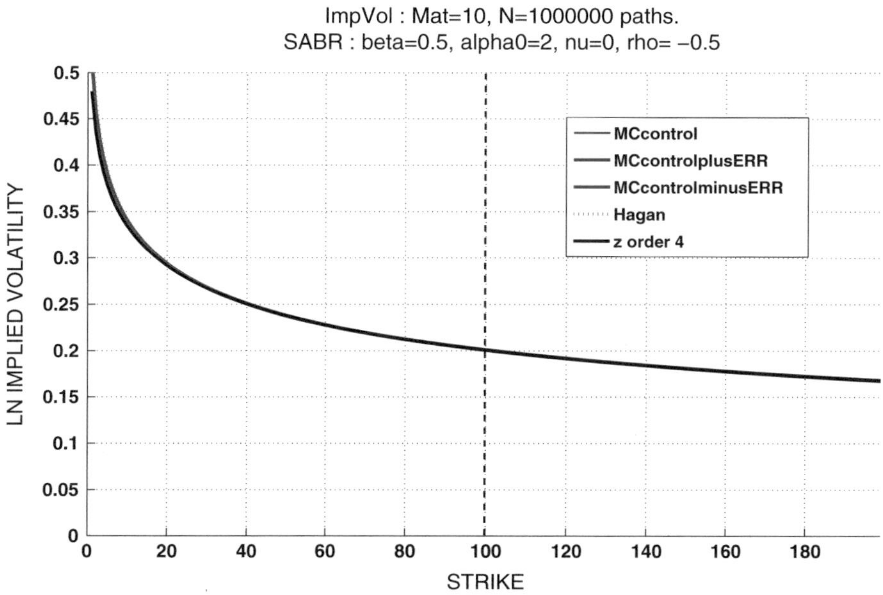

Fig. 4.18 Section 4.5.2.6 [p. 263]. LNIV with no SV: layer-3 expansion on z *vs* Hagan

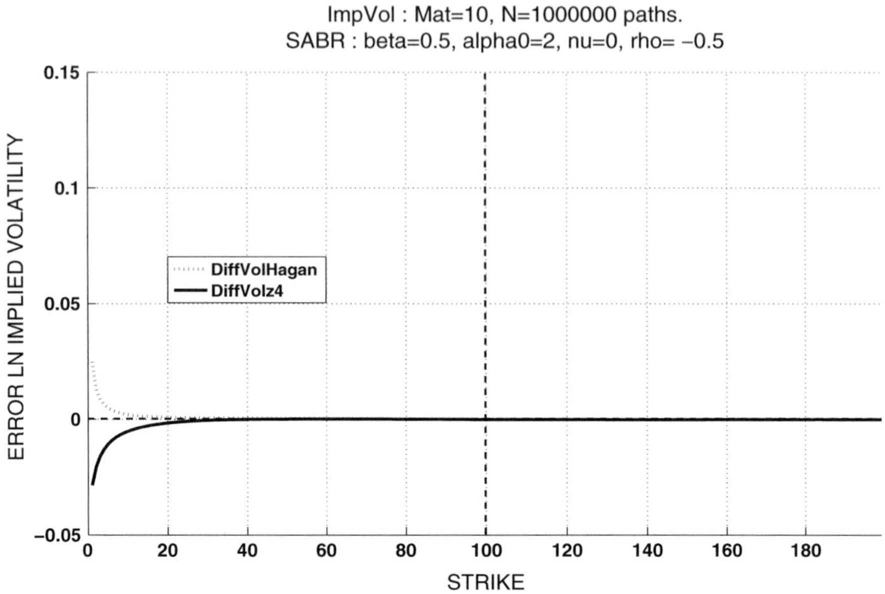

Fig. 4.19 Section 4.5.2.6 [p. 263]. LNIV error with no SV: layer-3 expansion on *z vs* Hagan

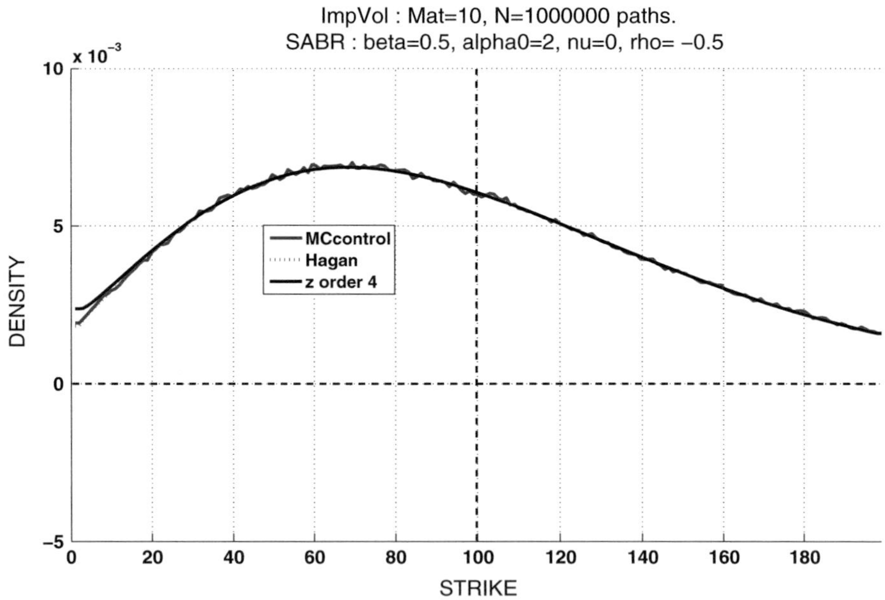

Fig. 4.20 Section 4.5.2.6 [p. 263]. Density with no SV: layer-3 expansion on *z vs* Hagan

Table 4.8 CEV-SABR: testing the influence of ν (LNIV, density)

	β	α_t	ρ	ν	$(T-t)$	NbPath
Figures 4.18, 4.19, 4.20	0.5	2	$-50\,\%$	**0 %**	10Y	1,000,000

not surprising, considering that Hagan's formula extends the pure-CEV results of [25]. Note that we could have used the CEV closed-form solution of (3.23) [p. 155]. However, the low variance of this CEV model, combined with using a control variate and that many paths, makes this MC scheme sufficiently precise.

The point to take is that in such a pure LV setting, the *generic* ACE methodology produces results which are comparable to those of an *ad hoc* and quite involved method. In fact, the precision attained is sufficiently high that it would be interesting to compare these approximations to a finite differences PDE scheme.It is well-known that for pricing in a bi-dimensional model, the PDE approach outperforms Monte-Carlo in terms of speed. However, the precision of a PDE scheme can be difficult to measure (seesaw effect in particular), whereas in a Monte-Carlo framework the error is stochastic but well controlled. Also, the PDE scheme relies on an appropriate implementation of the boundary conditions which, for instance, in a CEV with a low β, is not necessarily trivial. Finally, results on the marginal density associated to a PDE scheme are rarely provided, and we are not aware of a comprehensive study for the case of local volatility models.

Coming back to our testing concern, we can summarise by stating that the z-expansion clearly fares better with a lower vol of vol.

4.5.2.7 Influence of the Expiry *T*

First we modify the central configuration with a benign, lower expiry $T-t=5Y$, which provide Figs. 4.21 [p. 266], 4.22 [p. 266] and 4.23 [p. 267]. Then we analyse a more extreme case by adopting a higher expiry $T-t=20Y$, as shown in Figs. 4.24 [p. 267], 4.25 [p. 268] and 4.26 [p. 268]. These configurations are summarised in Table 4.9.

The lower expiry exhibits the same qualitative effect as the pure LV: a significant increase in performance, for both approximations, with regard to LNIV (Fig. 4.22) as well as to density (Fig. 4.23). Note that in the latter case we can observe, for the same parameter configuration, the three most common and *a priori* valid near-zero behaviours. Indeed, the Monte-Carlo exhibits a non-zero asymptote, while Hagan converges to zero in Lognormal fashion, and finally the z-expansion surges.

Conversely, increasing maturity seems to degrade the approximations, as shown by the loss of precision (Fig. 4.25) and by *some* negative density displayed by both (see Fig. 4.26). In the case of the z expansion this is just an excursion, whereas Hagan goes and stays negative as early as $35\,\%$ of S_t. In fact, Hagan's density is impaired across the whole strike range, since the binary spread prices that it gives are significantly inaccurate *immediately to the left of the forward.*

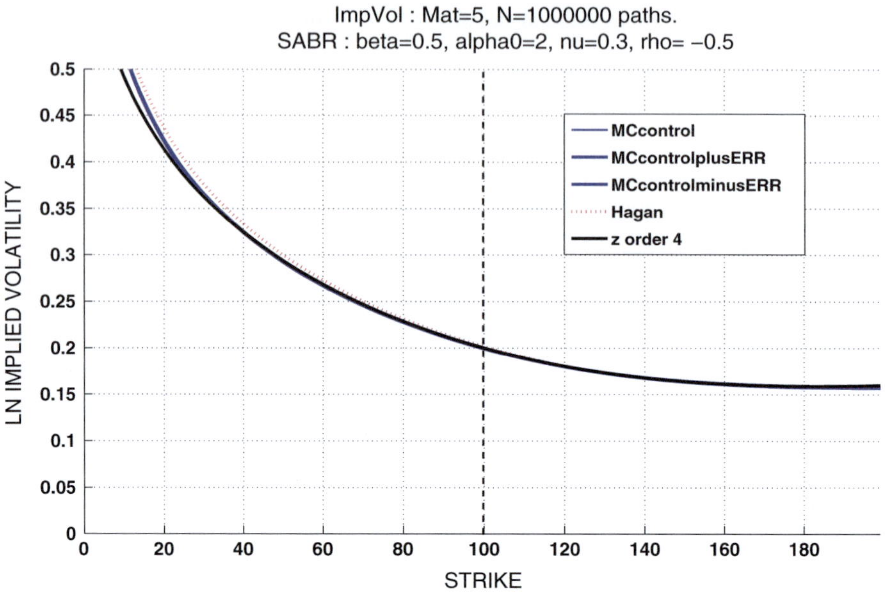

Fig. 4.21 Section 4.5.2.7 [p. 265]. LNIV with low expiry $T = 5Y$: layer-3 expansion on z *vs* Hagan

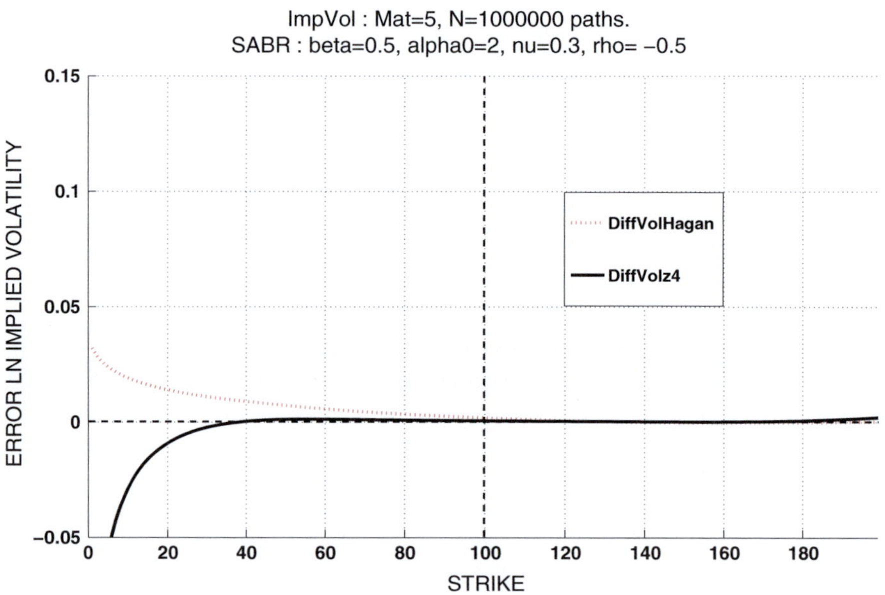

Fig. 4.22 Section 4.5.2.7 [p. 265]. LNIV error with low expiry $T = 5Y$: layer-3 z *vs* Hagan

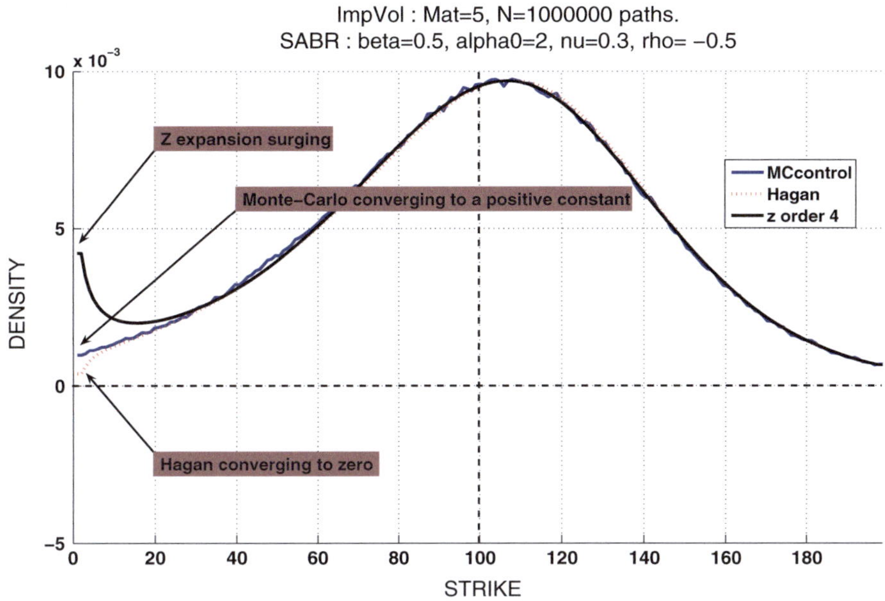

Fig. 4.23 Section 4.5.2.7 [p. 265]. Density with low expiry $T = 5Y$: expansion on z *vs* Hagan

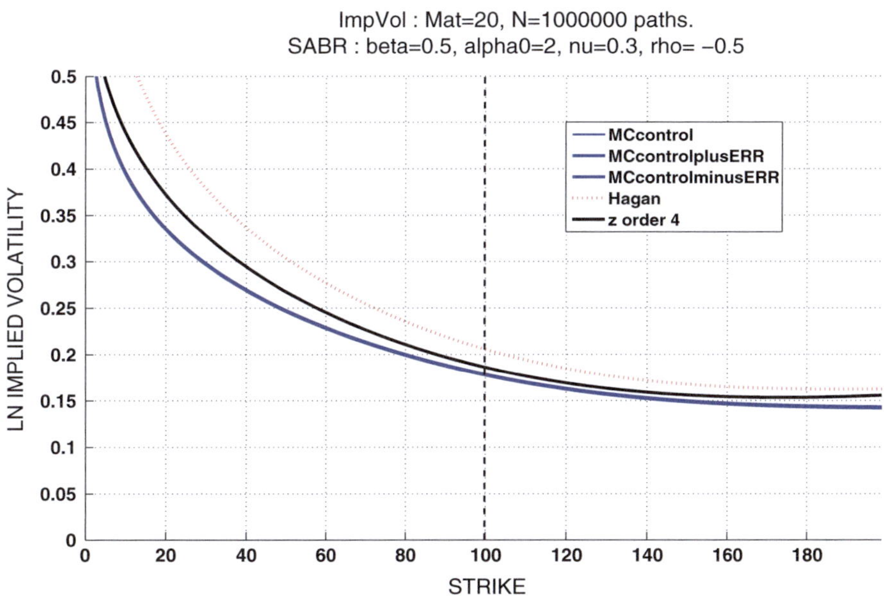

Fig. 4.24 LNIV with high expiry $T = 20Y$: layer-3 z *vs* Hagan

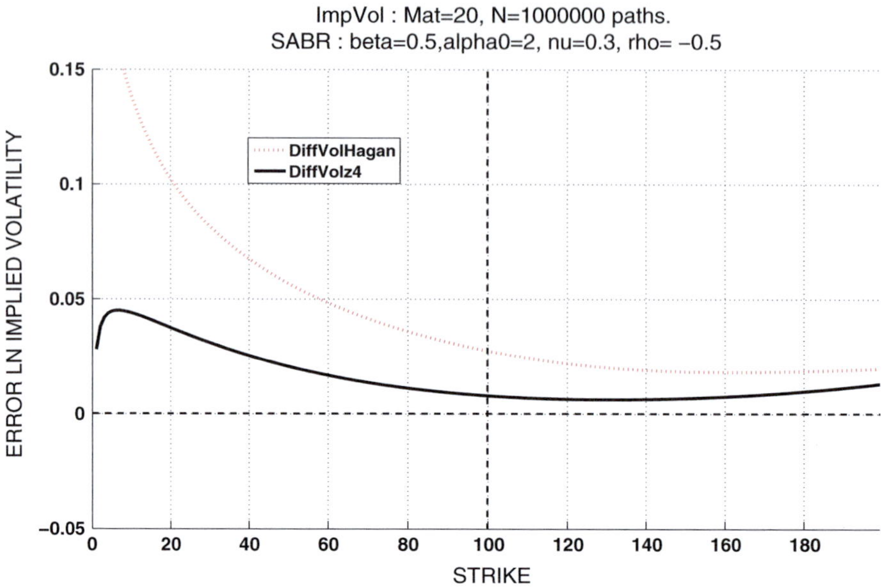

Fig. 4.25 Section 4.5.2.7 [p. 265]. LNIV error with high expiry $T = 20Y$: layer-3 z *vs* Hagan

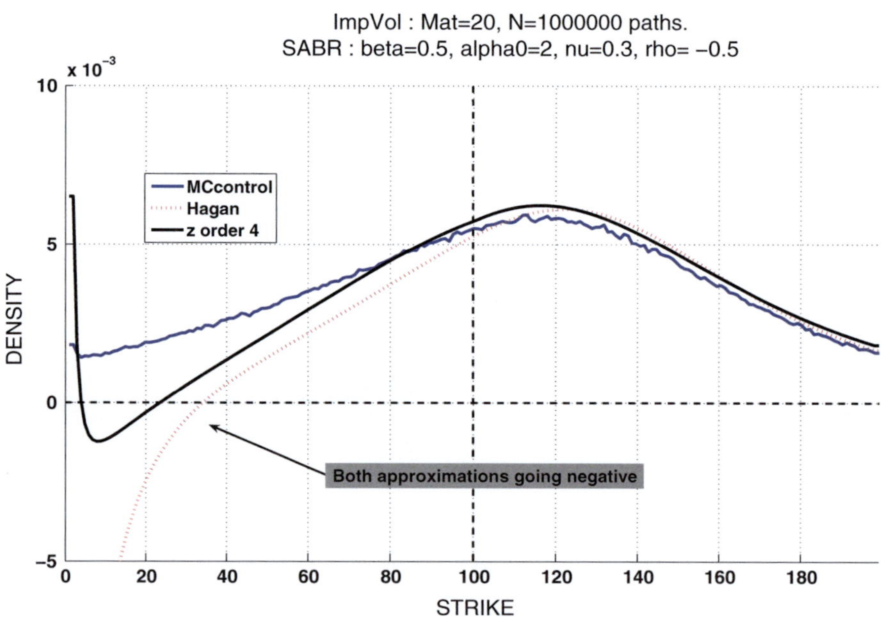

Fig. 4.26 Section 4.5.2.7 [p. 265]. Density with high expiry $T = 20Y$: layer-3 z *vs* Hagan

Table 4.9 CEV-SABR: testing the influence of a lower/higher maturity T (LNIV, density)

	β	α_t	ρ (%)	ν (%)	$(T-t)$	NbPath
Figures 4.21, 4.22, 4.23	0.5	2	−50	30	**5Y**	1,000,000
Figures 4.24, 4.25, 4.26	0.5	2	−50	30	**20Y**	1,000,000

On both counts (density and smile), when expiry varies the advantage remains to the z expansion. This is especially obvious in terms of precision, hence for the calibration process. However, in order to be used for full-smile pricing the excursions in negative territory would need to be prevented.

4.5.3 Conclusions

Around the chosen central configuration (see Table 4.1 p. 248) the z-expansion seems to be systematically more precise in the liquid range, but its bias can be quite unpredictable for very low strikes. In terms of validity, the behaviour of the density close to zero also seems difficult to anticipate, as it can switch easily between the various asymptotic modes (apparently $-\infty$, 0 and $+\infty$). Although the negative density areas seem to be more limited than with Hagan, they can still occur within reasonable parameter sets.

By comparison, although the performance of Hagan is overall lower, it seems to be reasonably robust and to behave in a more consistent manner across the parameter space. This feature could be attributed to the singular perturbation approach, where the asymptotic focus is on a PDE (hence global) rather than on a single point on the smile. This is not to say that both approaches cannot be made equivalent in certain cases, but that would be on the pure asymptotic results. In fact, Hagan's formula is in itself an *extrapolation*, since it comes by taking $\varepsilon = 1$.

We observe both approximations behaving better in a low volatility, low vol of vol environment. Indeed, fat-tails distributions (and especially a non-zero left asymptote) seem difficult to handle when the IV and the baseline are Lognormal.

Arguably, the better behaviour of ACE in the liquid domain can be attributed mainly to using the *arch* differential $\widetilde{\Sigma}''_{\theta\theta}(\star)$ (see Fig. 4.2). This view suggests a new, *mixed* approach to using ACE results for whole-smile extrapolations. Indeed, we can *combine* ACE's pure asymptotic results with global results, by *correcting* (as opposed to replacing) existing methods, such as Hagan.

There are many options in that line of work. For instance, we could correct Hagan's level by adding the *arch*, probably with a new, bounded variable θ. Also, since ACE behaves better with stationary models, piecewise-constant parametrisations will *a priori* degrade its performance. We have shown in Chap. 3 that time-dependent volatilities could be handled, but a time-dependent correlation or LV would be more challenging. Therefore we could pre-process the model with the Markovian projection method of [23], hence averaging the parameters into "efficient" constants, before applying ACE.

References

1. Hagan, P.S., Kumar, D., Lesniewski, A.S., Woodward, D.E.: Managing smile risk. Wilmott Mag. **1**, 84–108 (2002)
2. Osajima, Y.: The asymptotic expansion formula of implied volatility for dynamic SABR model and FX hybrid model. University of Tokyo, Graduate school of mathematical sciences, Report (2006)
3. Fouque, J.-P., Papanicolau, G., Sircar, K.R.: Derivatives in Financial Markets with Stochastic Volatility. Cambridge University Press, Cambridge (2000)
4. Fournie, E., Lebuchoux, J., Touzi, N.: Small noise expansion and importance sampling. Asymptotic Anal. **14**(4), 361–376 (1997)
5. Beresticki, H., Busca, J., Florent, I.: Asymptotics and calibration of local volatility models. Quant. Financ. **2**, 61–69 (2002)
6. Henry-Labordere, P.: A general asymptotic implied volatility for stochastic volatility models. Technical report, Barclays Capital (Apr 2005)
7. Durrleman, V.: From implied to spot volatilities. Finance Stochast. **14**(2), 157–177 (2006) (Springer, Cambridge)
8. Liu, E.H.L.: Fundamental Methods of Numerical Extrapolation with Applications. Mitopencourseware. Massachusetts Institute of Technology, Cambridge (2006)
9. Lee, R.W.: The moment formula for implied volatility at extreme strikes. Math. Financ. **14**(3), 469–480 (2004)
10. Benaim, S., Friz, P.: Regular variation and smile asymptotics. Math. Financ. **19**, 1–12 (2009)
11. Dragulescu, A.A., Yakovenko, V.M.: Probability distribution of returns in the Heston model with stochastic volatility. Quant. Financ. **2**, 443–453 (2002)
12. Fouque, J.-P., Papanicolau, G., Sircar, K.R.: Financial modeling in a fast mean-reverting stochastic volatility environment. Asia-Pacific Financ. Markets **6**, 37–48 (1999)
13. Fouque, J.-P., Papanicolau, G., Sircar, K.R.: Mean-reverting stochastic volatility. SIAM J. Control Optim. **31**, 470–493 (2000)
14. Press, W.H., Teukolsky, S.A., Vetterling, W.T., Flannery, B.P.: Numerical Recipes in C, 2nd edn. Cambridge University Press, New York (1992)
15. Andersen, L.B.G., Piterbarg, V.V.: Moment explosions in stochastic volatility models. Financ. Stochast. **11**, 29–50 (2006)
16. Benhamou, E., Croissant, O.: Local time for the SABR model. Connection with the complex Black-Scholes and application to CMS and spread options. Working paper, IXIS CIB, Sep (2007)
17. Obloj, J.: Fine-tune tour smile. Correction to Hagan & al. Technical report,Imperial College, London (Mar 2008)
18. Beresticki, H., Busca, J., Florent, I.: Computing the implied volatility in stochastic volatility models. Commun. Pure Appl. Math. **LVII**, 1352–1373 (2004)
19. Andersen, L., Brotherton-Ratcliffe, R.: Extended libor market models with stochastic volatility. Technical report, Bank of America (2001)
20. Andersen, L., Andreasen, J.: Volatile volatilities. Risk Mag. **15**, 163–168 (2002)
21. Andersen, L., Andreasen, J.: Volatility skews and extension of the Libor market model. Appl. Math. Financ. **7**, 1–32 (2000)
22. Piterbarg, V.V.: Stochastic volatility model with time-dependent skew. Appl. Math. Financ. **12**, 147–185 (2005)
23. Piterbarg, V.V.: Markovian projection for volatility calibration. Risk Mag. **20**, 84–89 (2007)
24. Andreasen, J.: Back to the future. Risk Mag. **18**, 104–109 (2005)
25. Hagan, P.S., Woodward, D.E.: Equivalent black volatilities. Appl. Math. Financ. **6**, 147–157 (1999)

Part II
Term Structures

As we have seen in the fundamental Chap. 2 (see Sect. 2.1.2.4) the modelling of a term structure framework in general and the Interest Rate (IR) derivatives environment in particular, cannot fall into the simple context developed in Part I for a single asset. Therefore the tools we have been developing so far would equally prove insufficient. Evidently, one could consider each underlying individually (e.g. each forward swap or forward Libor rate) and would then be facing a term-by-term or "point-by-point" calibration approach, thereby forfeiting most of the inter-maturities dependency. Alternatively, what we envisage now is a way to model the shape and *joint* dynamics[1] of the *whole* caplet or swaption surface.[2]

To illustrate this point, let us consider two different swap underlyings—say $1Y5Y$ and $2Y7Y$—each with an associated swaption smile (i.e. the implied volatility for all strikes, at the unique expiry). Then the results of Part I would allow us to approximate *separately*, and under two different measures, the shape and dynamics of these two smiles. Hence these results would provide no information on how the smiles move *together*. However, in order to price and hedge a structured product dependent on *both* forward swap rates (and/or their volatilities) the determination of these joint dynamics is essential.

This is typically the type of problem addressed by Part II. We introduce a new level of complexity to ACE, by considering a term structure as underlying. In an abstract framework we establish the new ZDC and compute the first layer, analysing how the multi-dimensional equations are altered.

Then we apply these results to an interest rate framework, more specifically some generic Stochastic Volatility HJM and LMM (Libor Market Model) setups. In particular, we show how the input chaos dynamics of these models allow us to approximate the implied volatility surface of bond options, caplets and physical swaptions.

---

1 The choice of measure will be discussed later.

2 By swaption surface we usually understate the swaption cube for all strikes and expiries, but for a given fixed tenor. Whether it be in the caplet or swaption case, the accrual—say 3M, or 6M Libor—also has to be fixed.

Chapter 5
Volatility Dynamics in a Term Structure

Abstract In this chapter we start by considering a generalised framework, encompassing in particular the Caplets and Swaptions markets, but potentially applicable to other products. This is made possible because these payoffs, as well as the martingale method used to price them, are very similar. Hence the main requirement is to find the correct numeraire and pricing measure. The difference with the single underlying setting of Part I is that we are now dealing with a *collection of underlyings*, for instance the forward Libor or forward par swap rates. Each of these underlying has its own numeraire, is martingale under the associated measure, and defines a specific strike-continuum of vanilla options. Hence we end up with associated *collections of numeraires, measures and options*. All these families are parametrised by their own list of maturities, which we will naturally extend to a common maturity continuum. We end up naturally with a term structure (TS) framework, and in solving the direct and indirect problems we will point to the structural difference simpler single-underlying environment of Part I.

The architecture of this chapter has been designed to follow as much as possible the structure of Chap. 2. This choice enables us to immediately detect the differences between the two frameworks, in particular the new terms. We have also paid particular attention to describing and interpreting these new terms, providing an intuitive insight whenever practical.

In Sect. 5.1, we build the framework by defining the three collections, the implied and instantaneous stochastic volatility models, and set the objectives.

In Sect. 5.2, we express the Main Zero-Drift Condition (ZDC), define our regularity assumptions and develop the Immediate ZDC.

In Sect. 5.3, we establish the Recovery Theorem, which deals with solving the inverse problem.

In Sect. 5.4, we provide one of the most useful result of this chapter, from a practical point of view. Indeed, we solve for the direct problem, which is intensively invoked for calibration purposes. Technically, we express the first layer of Immediate At-The-Money differentials for the smile.

With regard to the association between different maturities, we start by considering a fairly general case, before restricting ourselves to the practical situation presented by Caplet and Swaptions.

D. Nicolay, *Asymptotic Chaos Expansions in Finance*,
Springer Finance Lecture Notes, DOI 10.1007/978-1-4471-6506-4_5

Finally, in Sect. 5.5 we propose further research and conclude on the applicability of the method.

5.1 Framework and Objectives

In this section we set the context and the objectives for the term structure study. Although our aims are similar, the introduction of the maturity dimension requires some more attention as its implications reach deep into the proofs.

5.1.1 Numeraires, Underlyings and Options

We start by considering the three distinct families of processes defining our term structure framework: underlyings, numeraires and European Call options. We assume that each collection is indexed (or parametrised) by a *shared* maturity $T \in]0, T_{max}]$ with respect to which it provides a bijective *continuum*. Schematically, we denote these families as in Fig. 5.1.

Without surprise, the underlyings will define the reference for the options, and the numeraire assets will define the discounting and appropriate measure through the usual valuation argument. Besides the maturity indexation, we now associate functionally all three collections as follows:

- **Underlying dynamics**
  For each underlying there exists a numeraire, and therefore a corresponding measure, under which it becomes a martingale. Formally, and using a Lognormal convention:

$$\begin{cases} dX_t(T) = X_t(T)\overrightarrow{\sigma}_t(T)^\perp \, d\overrightarrow{W}_t^{N(T)} & (5.1) \\ d\overrightarrow{W}_t^{N(T)} \triangleq d\overrightarrow{W}_t - \overrightarrow{\lambda}_t^N(T)\, dt & (5.2) \\ dN_t(T) = N_t(T)\, r_t\, dt + N_t(T)\, \overrightarrow{\lambda}_t^N(T)^\perp \, d\overrightarrow{W}_t & (5.3) \end{cases}$$

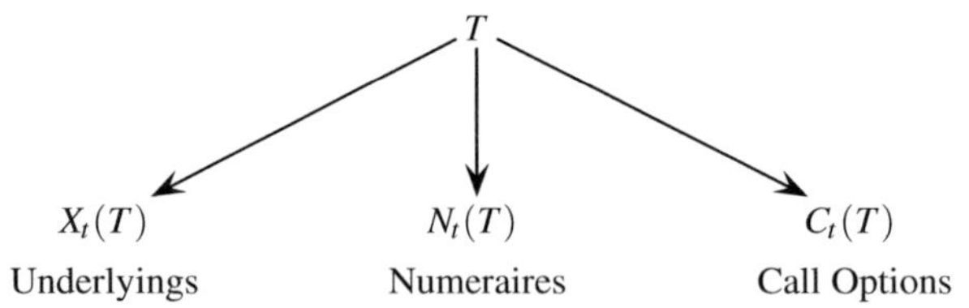

Fig. 5.1 Shared maturity indexation

where $\overrightarrow{W}_t$ is a Wiener process under $\mathbb{Q}$, the risk-neutral measure. Note that specifying the numeraire and therefore its volatility structure $\overrightarrow{\lambda}_t^N(T)$ is entirely equivalent to defining the term structure of martingale measures through the (same) associated risk premia.

- **Option specification**
  For each underlying there exists a strike-continuum[1] of European Call options, whose cash payoff sequence is either defined by or equivalent by arbitrage to the following

$$h\ \left(X_T(T), N_{\eta(T)}(T),\ K\right) \triangleq N_{\eta(T)}(T)\left[X_T(T) - K\right]^+ \quad \text{paid at time } \eta(T). \tag{5.4}$$

Note that the *fixing date* for the payoff, i.e. the date at which the underlying's value is observed, has been chosen as the index T. The payment date $\eta(T)$ however, *must* correspond to the argument of the numeraire process as invoked in the payoff definition.

With regard to the payoff, a good example of an *equivalent* cashflow sequence comes with the case of a plain vanilla physical swaption. Indeed, we can take the dates of observation and of payment (resp. T and $\eta(T)$) identical to the same swaption expiry and swap start date (T_0). We would then have to discount the N individual ($\mathscr{F}_{T_0}$-measurable) net coupons down to T_0 using the annuity in factor, which falls back exactly onto the payoff definition (5.4).

This leads us to comment further on the date and maturity implications of these definitions. The maturity T is used primarily to parametrise and therefore link our three collections: it can be seen as a way to index the whole framework at once. As such, the *bijective property* allows us to use that indexation very formally if required. We can, for instance, shift T by a constant in order to map one of our three families onto a real-life object, whose indexation style might be more conventional. Typically, in the case of a Caplet smile and for a given accrual δ, we use $T \rightarrow T+\delta$. In principle we could envisage more complex indexation schemes, such as affine or functional mappings. Although feasible, they would, however, complexify the equations without *a priori* much practical interest.

The fact that the payoff observation date comes specifically as T is simply a matter of convention, a way to anchor the configuration by allocating the reference maturity T to *some* physical reality. This explains the possibly confusing situation where the same symbol T appears both as the time argument and as a parameter of a process. We could very well define some function $\rho(T)$ as the fixing date instead.

This leads us to the payment date $\eta(T)$, effectively made redundant by the presence of the numeraire in the payoff definition. Indeed, the coming proofs will simply use

---

[1] The continuous nature of any European option family is understood as the permanent availability of a common bid/ask tradeable price, for every maturity and for every strike.

the usual valuation argument, discounting by the *same* numeraire value. Furthermore, the underlying volatility life to be considered in order to value a European option is schematically from valuation to fixing, not to payment. In consequence, the $\eta(\cdot)$ function does not actually provide an additional degree of freedom, but conversely we could consider *mid-curve* options within the same framework. Accordingly we will try to get rid of the $\eta(\cdot)$ mapping whenever possible, which is made easier by the given possibility of selecting an alternative, equivalent payoff sequence.

Note also that since we have imposed the maturity T to be deterministic, our setup excludes *a priori* the introduction of a stopping time for instance, as would be necessary for a Bermudan or an American option.

Finally, note that we do not impose the underlyings $X_t(T)$ to be *tradeable* assets, as opposed to the numeraires $N_t(T)$. The reason is that—as we will see shortly—the various proofs require only that the individual vanilla option written on each underlying be tradeable (and therefore martingale), not the underlying itself. This distinction is important, in particular, for Interest Rates (IR) underlyings, such as the Forward Libor *rate* and the Forward Swap *rate*. It is noticeable that in the IR world, one cannot (Delta-) hedge itself with the underlying *per se*, but instead must do so with the contract that it represents (e.g. the FRA or the swap). Their numeraire however, respectively the forward zero-coupon and the forward annuity, are proper tradeable assets. If there is some tradability issue here, it is actually more a modelling constraint than anything else: the *continuum* assumption ignores that liquidity generally gets thinner as both maturity and tenor increase.

5.1.2 Absolute and Sliding Implied Volatilities

Having selected a volatility risk premium and therefore a martingale measure, these options admit a non-arbitrage price. The "dollar" price of these options is denoted $V_t(X_t(T), K, T)$, and this price surface is then mapped or re-parametrised into an implied volatility surface, using:

$$V_t(X_t(T), T, K) = N_t(T) \cdot C^{BS}\left(X_t(T), K, \Sigma_t(X_t(T), T, K) \cdot \sqrt{T-t}\right) \quad (5.5)$$

with $C^{BS}(f, k, v, b)$ the deflated price, written with the time-normalised Black-Scholes functional (which, for our purposes, is identical to the normalised Black taken with a unit zero-coupon) defined by

$$C^{BS}(f, k, v) = f\,\mathcal{N}(d_1) - k\,\mathcal{N}(d_2) \quad (5.6)$$

$$\text{with} \quad d_{1/2}(f, k, v) = \frac{-y}{v} \pm \frac{1}{2}v \quad \text{and} \quad y(f, k) = \ln(k/f).$$

Which is to say that using the classical no-arbitrage and change-of-numeraire arguments (see [1], for instance) we have

$$C^{BS}\left(X_t(T), K, \Sigma_t(X_t(T), T, K)\cdot\sqrt{T-t}\right) = \mathbb{E}^{\mathbb{Q}^{N(T)}}\left[\frac{(X_T(T)-K)^+}{N_T(T)}|\mathcal{F}_t\right].$$

We define the "sliding" counterparts associated to these "absolute" quantities by

$$\theta \triangleq T-t \tag{5.7}$$

$$\begin{aligned}\widetilde{X}_t(\theta) &\triangleq X_t(T)\\ y &\triangleq \ln\left(K/\widetilde{X}_t(\theta)\right)\end{aligned} \tag{5.8}$$

$$\begin{aligned}\widetilde{V}_t\,(y,\theta) &\triangleq V_t(X_t(T), K, T)\\ \widetilde{\Sigma}_t\,(y,\theta) &\triangleq \Sigma_t(X_t(T), K, T)\end{aligned}$$

Remark 5.1 Note that a significant difference with the single underlying case (see Sect. 2.1) is that now the underlying is also made sliding in time. Note in particular that the log-moneyness y, still defined with respect to the underlying $\widetilde{X}$, is also sliding with time. However, although the numeraire is necessarily sliding as well, as soon as we use the Black formula this feature is only perceived implicitly through the deflation. Indeed, as will be recalled [p. 289], it is the quantity $V_t(X_t(T), T, K)/N_t(T)$ which is martingale under $\mathbb{Q}^{N(T)}$ [1]. In other words, the numeraire aspect is effectively "hidden" by only considering the Black functional re-parametrisation.

5.1.3 The Two Stochastic Volatility Models

In a similar fashion to the single underlying setup, we now independently present the stochastic instantaneous volatility and the stochastic implied volatility, before establishing their connection. This is again the first step before we can prove that the former specification is indeed included in the latter. Both models share the same underlying dynamics already defined by the SDE system (5.1)–(5.3), and differ by definition in the type of volatility dynamics that they describe.

5.1.3.1 The (Term Structure) Stochastic Instantaneous Volatility Model

Since we are now dealing with a term structure of instantaneous volatility $\vec{\sigma}_t(T)$, we might want to specify the dynamics of that field on its own, irrespective of the collection of underlyings it is associated to, and therefore using a different measure. This is the case, in particular, if the modeller wishes to incorporate information from other sources than just the observed static vanilla prices. Examples include historical dynamics of these prices, or, when available, prices of volatility derivatives products. Such rich market data can clearly be extended to a continuum and formalised as a term structure, with specific dynamics and maturity-dependent risk premia.

With that important motivation in mind, we elect to use a *specific endogenous driver* for the instantaneous volatility, a driver whose drift must obviously be maturity-dependent and *a priori* distinct from the one used for the underlying dynamics (5.1). In the sequel, we will refer to this feature as *endogenous driver (ED) disalignment*.

Therefore, using a chaos expansion specification, similar to the one used in Part I, we adopt the following notations for our framework:

$$d\vec{\sigma}_t(T) = \vec{a}_{1,t}(T)\,dt + \vec{\vec{a}}_{2,t}(T)\,d\vec{W}_t^{\sigma(T)} + \vec{\vec{a}}_{3,t}(T)\,d\vec{Z}_t \tag{5.9}$$

$$\text{with} \quad d\vec{W}_t^{\sigma(T)} \triangleq d\vec{W}_t - \vec{\lambda}_t^{\sigma}(T)\,dt.$$

Similarly, we write the dynamics of the "endogenous" coefficient matrix $\vec{\vec{a}}_{2,t}(T)$ as

$$d\vec{\vec{a}}_{2,t}(T) = \vec{\vec{a}}_{21,t}(T)\,dt + \vec{\vec{\vec{a}}}_{22,t}(T)\,d\vec{W}_t^{\sigma(T)} + \vec{\vec{\vec{a}}}_{23,t}(T)\,d\vec{Z}_t. \tag{5.10}$$

We assume (in the most general case) that all diffusion *tensorial* coefficients[2] are themselves fully stochastic and maturity-dependent. Note, in particular, that $\vec{\vec{a}}_2$ and $\vec{\vec{a}}_3$, the "volatility" coefficients of $\vec{\sigma}_t(T)$, take their values respectively in $\mathbb{R}^{n_w \times n_w}$ and $\mathbb{R}^{n_w \times n_z}$. This implies that, at each step of the stochastic "chaos" expansion that awaits us (and bears strong similarities with the single underlying case) we will increase the coefficient tensorial order by one. Fortunately, exploring the "first layer" will not require going deeper than depth 2 in the chaos expansion.[3] It turns out that, in practice and compared to the single-asset case, this dimensionality issue can prove more difficult to manage that a reasonably specified term-dependency.

Remark 5.2 Note that we could also define the *sliding instantaneous volatility* with

$$\vec{\vec{\sigma}}_t(\theta) = \vec{\sigma}_t(t+\theta) \tag{5.11}$$

[2] Meaning $\vec{a}_{1,t}(T)$, $\vec{\vec{a}}_{2,t}(T)$, $\vec{\vec{a}}_{3,t}(T)$, $\vec{\vec{\vec{a}}}_{22,t}(T)$ and all subsequent offspring.

[3] With the tensorial family $\vec{\vec{\vec{a}}}_{22,t}(T)$.

and that its dynamics are then

$$d\vec{\vec{\sigma}}_t(\theta) = \left[\vec{\vec{\sigma}}'_T(t, t+\theta) + \vec{\vec{a}}_{1,t}(t+\theta)\right] dt \qquad (5.12)$$
$$+ \vec{\vec{\vec{a}}}_{2,t}(t+\theta)^{\perp} \, d\vec{W}_t^{\sigma(T)} + \vec{\vec{\vec{a}}}_{3,t}(t+\theta)^{\perp} \, d\vec{Z}_t.$$

This volatility would naturally be associated to the sliding underlying, however this new notion would not bring anything to our current focus, and besides we opted for a fully *absolute* description of the *underlying* dynamics, as will be justified later in Sect. 5.1.3.3.

5.1.3.2 The (Term Structure) Stochastic Implied Volatility Model

All the arguments cited above, establishing the necessity to freely specify the dynamics of the term structure of instantaneous volatility, under whichever measure we see fit, also apply to the implied volatility surface. Ultimately, this approach boils down to considering that underlyings, volatilities and marginal distributions consist in associated but distinct dynamic entities. Eventually, the choice of the measure used to describe the dynamics of any of these should rest with the modeller.

As in the single underlying case, the implied volatility dynamics are specified as

$$d\tilde{\Sigma}_t(y,\theta) = \tilde{b}_t(y,\theta)\, dt + \vec{\vec{\nu}}_t(y,\theta)^{\perp} \, d\vec{W}_t^{\tilde{\Sigma}(\theta)} + \vec{\vec{n}}_t(T)^{\perp} \, d\vec{Z}_t \qquad (5.13)$$

with $d\vec{W}_t^{\tilde{\Sigma}(\theta)} \triangleq d\vec{W}_t - \vec{\lambda}_t^{\tilde{\Sigma}}(\theta)\, dt,$

where θ and y are respectively defined by (5.7) and (5.8).

We note that there is little difference with the single-underlying case. The term-dependency was already present then, although its was brought solely by the option definition, and not the underlying. As for the dimensionality of the drivers, the same remarks as in Sect. 5.1.3.1 apply when it comes to increasing the expansion order and the associated difficulties.

For further use we also introduce the trivial absolute notations:

$$\vec{\lambda}_t^{\Sigma}(T) \triangleq \vec{\lambda}_t^{\tilde{\Sigma}}(T-t) \qquad \text{and} \qquad d\vec{W}_t^{\Sigma(T)} \triangleq d\vec{W}_t - \vec{\lambda}_t^{\Sigma}(T)\, dt.$$

5.1.3.3 Comments and Comparison of the Two Models

It is noteworthy that the volatility dynamics under consideration are, respectively, *absolute* for the instantaneous volatility, and *sliding* for the implied volatility. This feature is in fact non-binding, and the reason for this choice of presentation is to simplify the task of end-users in general, and modellers/practitioners in particular.

Indeed, practical models (as seen through the dynamics of the underlying assets) are usually specified in absolute terms, as is the case for the Libor and Swap Market Models. Whereas implied volatility surfaces are usually observed and modelled as "sticky", meaning that their shape "moves along" with the underlying and with time. Consequently, these surfaces are usually better represented in sliding terms.

Conversely, one could argue that, on one hand, the quality of re-calibration is better assessed by observing the stationarity of the parametrisation (i.e. in sliding terms for the underlying), in particular the term-structure of volatility. And that, on the other hand, hedging instruments are usually available only in absolute terms: the liquid strikes are fixed!

All these different focuses are actually honored throughout this study, since the transitions between sliding and absolute dynamics are explicitly expressed in the course of the proofs.

Arguably, the level of abstraction and generalisation adopted in this setup is high and subjective, especially the allowance for endogenous driver disalignment. In consequence, it appears desirable to justify these choices by expressing the rationale that underpins them.

As for selecting a distinct endogenous driver for the volatilities in (5.9) and (5.13), it is obviously equivalent to keeping a $\mathbb{Q}^{N(T)}$-Wiener process and altering the drifts. For instance, we could trivially re-express (5.9) as

$$\begin{aligned} d\vec{\sigma}_t(T) = &\left[\vec{a}_{1,t}(T) + \vec{\vec{a}}_{2,t}(T)\left(\vec{\lambda}_t^N(T) - \vec{\lambda}_t^\sigma(T)\right)\right] dt \\ &+ \vec{\vec{a}}_{2,t}(T)\, d\vec{W}_t^{N(T)} + \vec{\vec{a}}_{3,t}(T)\, d\vec{Z}_t. \end{aligned}$$

But modifying the drift coefficient in such a fashion might prove counter-productive in the long term. Indeed, the nature of the chaos expansion we use implies that, as the expansion order increases, eventually we will have to compute the dynamics of the finite variation term above. Unfortunately, we know from Part I that the complexity of the coefficients tend to increase exponentially with each level.

This remark applies, in particular, to LMM models, where it might be tempting to express all dynamics under a single, non-maturity-dependent measure, such as the terminal measure. The repercussions on the drifts of the remaining FL rates are well-known (see [2], for instance), so that the resulting diffusions are quite involved and difficult to simulate precisely.

Although there is some sense in describing the FL rates dynamics under the measure associated to one of them, the rationale of using that same measure on the volatility dynamics is far less convincing. Besides, computing the dynamics of the consequent drifts would prove very tedious indeed, and increase the computation risk.

The bottom line is that complexifying either the drift or the measure is—as usual—a matter of choice. Again, our approach is to allow maximum flexibility to the end users of our study, and to enable them to match their model to our notations with minimum work on their part (and therefore a reduced risk of error).

Beyond such concern for end-user usefulness and practicality, let us remind ourselves that one objective of the method is to provide (an approximation of) the joint dynamics of all points on the implied volatility surface. Considering, for instance, the Caplet smile, the choice of the $(T+\delta)$-forward neutral measure is a must in order to obtain term-by-term martingale dynamics, but it is dictated by the very definition of the underlying, the Forward Libor rate. When it comes to representing the dynamics of the volatilities (either instantaneous or implied), there is no such unique and obvious choice for a pertinent measure.

This flexibility we allow, however, cannot go without consequences. Any endogenous driver disalignment will a priori generate additional terms in most results thereafter, as will be observed and discussed in Sect. 5.3.

Playing the devil's advocate, the next logical question is therefore: "could we have gone further, and chosen another new measure for the endogenous driver of further coefficients, such as $\vec{\vec{a}}_{2,t}(T)$ for instance?" In principle, the answer is yes, and the added complexity would have been manageable at this stage (Layer 1). But the interest of doing so would have been moot at best.

Firstly, because in practice there is no market (yet!) for the term structure of "volatility of volatility", the dynamics of the skew or the dynamics of the curvature, which are the main quantities affected by $\vec{\vec{a}}_{2,t}(T)$.

Also because one of the aims of this chapter is to demonstrate the *mechanisms* of the method in the maturity-dependent framework, and to underline the differences with the single underlying framework of Part I. From that perspective, the effects of endogenous driver disalignment will seem very clear at this level of generality.
Last but not least, it happens that current published interest rates models are relatively limited and unambitious when it comes to volatility specification.[4]

In order to match as closely as possible the configuration of the most popular IR models, in particular the various local/stochastic volatility versions of the LMM, as well as for the reader's convenience, we will provide whenever useful a simplified, "aligned" version of our results by taking

$$\vec{\lambda}_t^{\sigma}(T) \equiv \vec{\lambda}_t^{N}(T) \equiv \vec{\vec{\lambda}}_t^{\widetilde{\Sigma}}(t+\theta).$$

5.1.3.4 Several Technical Assumptions

Before proceeding with the proof, we introduce technical, positivity assumptions that will facilitate our computations. Respectively,

Assumption 5.1 Almost surely, $\quad X_t(T) > 0 \quad \forall (t,T) \in \mathbb{R}^{+2}.$

Assumption 5.2 Almost surely, $\quad \|\vec{\sigma}_t(T)\| > 0, \quad \forall (t,T) \in \mathbb{R}^{+2}.$

Assumption 5.3 Almost surely, $\quad \widetilde{\Sigma}(t,y,\theta) > 0 \quad \forall (t,y,\theta) \in \mathbb{R}^{+} \times \mathbb{R} \times \mathbb{R}^{+*}.$

---

[4] Which is coherent with the low liquidity of the IR volatility market. At the time of writing, "vol bonds", "vol swaps" or "vol caps" are certainly not vanilla products and even less advertised as such.

Note that the same comments as in Remark 2.2, p. 41, still hold, underlining the non-binding character of these assumptions. They mainly consist in realistic limitations aimed at simplifying the coming calculus.

5.1.4 The Objectives

Drawing a parallel with the single underlying case, we aim at relating the value and dynamics of two distinct volatility processes: on one hand, the (vectorial) instantaneous volatility $\vec{\sigma}_t(T)$ of the underlying family, and, on the other hand, the sliding implied volatility surface $\widetilde{\Sigma}_t(y,\theta)$ of the vanilla options. The parallel with Part I stops here, though, for two reasons. First, because $\vec{\sigma}_t(T)$ is no longer a scalar, but a mapping of $\mathbb{R}^+$ into $\mathbb{R}^n$, and second, because the signification of $\widetilde{\Sigma}_t(y,\theta)$—albeit not its dimensionality—has changed: to every maturity is now associated a different underlying.

Modestly, we do not seek exact relationships valid on the whole field (y,θ). Instead, we will be content with approximations, provided the error can be controlled and asymptotically vanish. We will see that we attain our goal by linking, on one hand, the chaos decomposition of $\vec{\sigma}_t$ with its T-differentials (all taken in $T = t$) and, on the other hand, $\widetilde{\Sigma}_t$ with its cross y- and θ-differentials (all taken in $(y = 0, \theta = 0)$).

As for vocabulary, the process of extracting information about the (term structure of) instantaneous volatility from the input of the implied volatility will still be called the "inverse" problem (the theoretical calibration issue). Reciprocally, inferring some information about the implied volatility given the instantaneous volatility, will again be called the "direct" problem.

But in contrast to the single underlying case, we will see that the inverse problem is under-determined in the general case. Therefore the usual calibration method, which consists of a series of direct (model to smile) procedures,[5] seems now to be not another, but the (only) sensible alternative.

5.2 Derivation of the Zero-Drift Conditions

We now embark on establishing the fundamental element of the method, namely the PDE linking the shape and dynamics of the implied volatility mapping with the term structure of instantaneous volatility, across the whole space (y,θ). From this Zero-drift Condition we then also derive an Immediate version valid in small time (i.e. when $\theta \searrow 0$).

The methodology we employ is straightforward and follows in the steps of Sect. 2.2, with the exception of the benign Lemma 5.1 that we cover shortly afterwards and which expresses the dynamics of the sliding underlying.

[5] Determined via a trial-and-error method or a more complex optimisation routine. These optimisation procedures are typically non-trivial because the "market error" value function is *a priori* non-convex w.r.t. the model parameters.

5.2.1 The Main Zero-Drift Condition

For ease of presentation, we will thereafter allocate specific notations to the two relevant argument sets, used respectively for the implied volatility surface and the Black function:

$$(\dagger) \triangleq (X_t(T), T, K) \qquad \text{and} \qquad (\ddagger) \triangleq \left(X_t(T), K, \Sigma_t(X_t(T), T, K) \cdot \sqrt{T-t}\right).$$

Let us first establish some intermediate results, starting with the above-mentioned dynamics of the sliding underlying and of the absolute implied volatility. For convenience, we elect to express these SDEs using the risk-neutral driver $\overrightarrow{W}_t$, but this technical detail is irrelevant for the sequel. Our further aim is indeed to provide an expression (the deflated Call dynamics) which happens to structurally combine several distinct drivers. It makes practical sense, then, to cast all dynamics beforehand under a single measure, if only to clarify the computations. We arbitrarily chose this common measure as $\mathbb{Q}$, but again this is completely insignificant in the wider picture.

Lemma 5.1 (Dynamics of the absolute IV surface in the TS framework) *Using the risk-neutral driver, the dynamics of the sliding underlying are*

$$d\widetilde{X}_t(\theta) = \left[\frac{\partial \widetilde{X}_t}{\partial \theta}(\theta) - \widetilde{X}_t(\theta)\, \overrightarrow{\sigma}_t(t+\theta)^{T\perp} \overrightarrow{\lambda}_t^N(t+\theta)\right] dt + \widetilde{X}_t(\theta)\, \overrightarrow{\sigma}_t(t+\theta)^{\perp}\, d\overrightarrow{W}_t \tag{5.14}$$

while the dynamics of the absolute implied volatility are expressed as:

$$d\Sigma_t(\dagger) = b(\dagger)\, dt + \overrightarrow{\mathrm{v}}(\dagger)^{\perp}\, d\overrightarrow{W}_t + \overrightarrow{n}(\dagger)^{\perp}\, d\overrightarrow{Z}_t \tag{5.15}$$

with

$$b(\dagger) = b_1(\dagger) + b_2(\dagger) \tag{5.16}$$

$$b_1(\dagger) = \widetilde{b}(\circ) - \widetilde{\Sigma}'_\theta(\circ) - \widetilde{\Sigma}'_y(\circ)\, \frac{\widetilde{X}'_\theta}{\widetilde{X}}(t,\theta) - \overrightarrow{\widetilde{\mathrm{v}}}'_y(\circ)^{\perp} \overrightarrow{\sigma}_t(t+\theta) + \frac{1}{2}\left[\widetilde{\Sigma}'_y(\circ) + \widetilde{\Sigma}''_{yy}(\circ)\right] \|\overrightarrow{\sigma}_t(t+\theta)\|^2 \tag{5.17}$$

$$b_2(\dagger) = -\overrightarrow{\widetilde{\mathrm{v}}}(\circ)^{\perp} \overrightarrow{\lambda}_t^{\widetilde{\Sigma}}(\theta) + \widetilde{\Sigma}'_y(\circ)\, \overrightarrow{\sigma}_t(t+\theta)^{\perp} \overrightarrow{\lambda}_t^N(t+\theta) \tag{5.18}$$

$$\overrightarrow{\mathrm{v}}(\dagger) = \overrightarrow{\widetilde{\mathrm{v}}}(\circ) - \widetilde{\Sigma}'_y(\circ)\, \overrightarrow{\sigma}_t(t+\theta) \tag{5.19}$$

$$\overrightarrow{n}(\dagger) = \overrightarrow{\widetilde{n}}(\circ). \tag{5.20}$$

We remark straight away that our simple change of coordinates, from absolute to sliding, has already generated several new and interesting terms. Comparing (5.15) above with its counterpart (2.16) [p. 42] in the single-underlying case, we note that concerning the drift:

- In (5.17) all terms are the natural equivalents of the scalar case, except

$$\widetilde{\Sigma}'_y(\circ)\,\frac{\widetilde{X}'_\theta}{\widetilde{X}}(t,\theta)$$

  which accounts for the change in implied volatility due to a switch in the actual underlying, itself caused by a move along the θ axis. This quantity is called the *slope compensation* term and will be discusses in more detail on p. 288.
- In (5.18) all the terms are new, since the numeraire and measure aspect is now not only explicit but maturity-dependent. Also, should we align the endogenous measures for the underlying and the implied volatility, we would then have

$$b2(\dagger) \underset{\vec{\lambda}_t^{\widetilde{\Sigma}} \equiv \vec{\lambda}_t^N}{\Rightarrow} -\left[\vec{\mathrm{v}}(\circ) - \widetilde{\Sigma}'_y(\circ)\,\vec{\sigma}_t(t+\theta)\right]^{\perp} \vec{\lambda}_t^N(t+\theta) = -\vec{\mathrm{v}}(\dagger)^{\perp}\vec{\lambda}_t^N(t+\theta)$$

  and thus the dynamics of the absolute implied volatility would simply come as

$$d\Sigma_t(\dagger) = b_1(\dagger)\,dt + \vec{\mathrm{v}}(\dagger)^{\perp}\,d\vec{W}_t^{N(T)} + \vec{n}(\dagger)^{\perp}\,d\vec{Z}_t.$$

Proof Let us start with dynamics of the *sliding* underlying $\widetilde{X}_t(\theta)$. Having defined the dynamics of the *absolute* underlying as

$$dX_t(T) = X_t(T)\,\vec{\sigma}_t(T)^{\perp}\left[d\vec{W}_t - \vec{\lambda}_t^N(T)\,dt\right]$$

those of the sliding version come as

$$\begin{aligned} d\widetilde{X}_t(\theta) = dX_t(t+\theta) = &\left[\frac{\partial X_t}{\partial T}(t+\theta) - X_t(t+\theta)\,\vec{\sigma}_t(t+\theta)^{\perp}\vec{\lambda}_t^N(t+\theta)\right] dt \\ &+ X_t(t+\theta)\,\vec{\sigma}_t(t+\theta)^{\perp}\,d\vec{W}_t. \end{aligned}$$

Therefore re-expressing with the *sliding* underlying yields:

$$\begin{aligned} d\widetilde{X}_t(\theta) = &\left[\frac{\partial \widetilde{X}_t}{\partial\theta}(\theta) - \widetilde{X}_t(\theta)\,\vec{\sigma}_t(t+\theta)^{\perp}\vec{\lambda}_t^N(t+\theta)\right] dt \\ &+ \widetilde{X}_t(\theta)\,\vec{\sigma}_t(t+\theta)^{\perp}\,d\vec{W}_t, \end{aligned}$$

which proves (5.14). We can now move on to the dynamics of the *absolute* implied volatility. Starting with the dynamics of the sliding implied volatility function $\widetilde{\Sigma}$, we can use the Itô-Kunita formula (Theorem A.1) by selecting

$$\vec{\alpha}_t \longleftarrow \begin{bmatrix} \widetilde{X}_t(\theta) \\ \theta \end{bmatrix}.$$

Again defining $(\circ) \triangleq (t, y, \theta)$ and recalling that

$$\Sigma_t\left(X_t(T), T, K\right) = d\widetilde{\Sigma}\left(t, \ln\left(K/\widetilde{X}_t(\theta)\right), \theta\right),$$

we get

$$\begin{aligned} d\Sigma_t\left(X_t(T), T, K\right) = {} & \widetilde{b}(\circ)\, dt + \vec{\widetilde{\mathrm{v}}}(\circ)^{\perp}\left[d\vec{W}_t - \vec{\widetilde{\lambda}}_t^{\widetilde{\Sigma}}(\theta)\, dt\right] \\ & + \vec{\widetilde{n}}(\circ)^{\perp}\, d\vec{Z}_t - \frac{\widetilde{\Sigma}'_y(\circ)}{\widetilde{X}_t(\theta)}\, d\widetilde{X}_t(\theta) + \widetilde{\Sigma}'_\theta(\circ)\, d\theta \\ & + \frac{1}{2\,\widetilde{X}_t^{\,2}(\theta)}\left[\widetilde{\Sigma}'_y(\circ) + \widetilde{\Sigma}''_{yy}(\circ)\right] d\langle\widetilde{X}_t(\theta)\rangle \\ & - \frac{1}{\widetilde{X}_t(\theta)}\,\vec{\widetilde{\mathrm{v}}}'_y(\circ)^{\perp}\left[\widetilde{X}_t(\theta)\vec{\sigma}_t(t+\theta)\right] dt. \end{aligned}$$

Substituting the dynamics for the sliding underlying we get

$$\begin{aligned} & d\Sigma_t\left(X_t(T), T, K\right) \\ & \quad = \widetilde{b}(\circ)\, dt - \vec{\widetilde{\mathrm{v}}}(\circ)^{\perp}\vec{\widetilde{\lambda}}_t^{\widetilde{\Sigma}}(\theta)\, dt + \vec{\widetilde{\mathrm{v}}}(\circ)^{\perp}\, d\vec{W}_t + \vec{\widetilde{n}}(\circ)^{\perp}\, d\vec{Z}_t \\ & \quad - \frac{\widetilde{\Sigma}'_y(\circ)}{\widetilde{X}_t(\theta)}\left[\frac{\partial\widetilde{X}_t}{\partial\theta}(\theta) - \widetilde{X}_t(\theta)\,\vec{\sigma}_t(t+\theta)^{\perp}\vec{\lambda}_t^{N}(T)\right] dt \\ & \quad - \widetilde{\Sigma}'_y(\circ)\,\vec{\sigma}_t(t+\theta)^{\perp}\, d\vec{W}_t - \widetilde{\Sigma}'_\theta(\circ)\, dt + \frac{1}{2}\left[\widetilde{\Sigma}'_y(\circ) + \widetilde{\Sigma}''_{yy}(\circ)\right] \\ & \quad \times \|\vec{\sigma}_t(t+\theta)\|^2\, dt - \vec{\widetilde{\mathrm{v}}}'_y(\circ)^{\perp}\vec{\sigma}_t(t+\theta)\, dt. \end{aligned}$$

Finally regrouping the finite and non-finite variation terms provides (5.15), (5.16), (5.19), (5.20) and concludes the proof. □

Having expressed the dynamics of the absolute implied volatility $\Sigma_t\left(X_t(T), T, K\right)$ now allows us to develop our main result, the **Main Zero-Drift Condition**.

Proposition 5.1 (Zero Drift Condition in the generic TS framework) *The shape and dynamics functionals specifying the sliding implied volatility model (5.1)–(5.3) and (5.13) are constrained by arbitrage. Specifically, in the generic point* $(\circ) = (t, y, \theta) \in \mathbb{R}^+ \times \mathbb{R} \times \mathbb{R}^{+*}$ *the drift must satisfy the following Main Zero-Drift Condition:*

$$\widetilde{\Sigma}^3(\circ)\,\widetilde{b}(\circ) = \theta\, D(\circ) + E(\circ) + \frac{1}{\theta} F(\circ) \tag{5.21}$$

with

$$D(\circ) = \frac{1}{8}\,\widetilde{\Sigma}^4(\circ)\left[\|\vec{\widetilde{\mathrm{v}}}(\circ) - \widetilde{\Sigma}'_y(\circ)\vec{\sigma}_t(t+\theta)\|^2 + \|\vec{\widetilde{n}}(\circ)\|^2\right]$$

$$\begin{aligned}
\widetilde{\Sigma}^{-3}E(\circ) = {} & \widetilde{\Sigma}'_\theta(\circ) - \frac{1}{2}\widetilde{\Sigma}''_{yy}(\circ)\|\vec{\sigma}_t(t+\theta)\|^2 \\
& - \frac{1}{2}\vec{\widetilde{\mathrm{v}}}(\circ)^\perp\vec{\sigma}_t(t+\theta) + \vec{\widetilde{\mathrm{v}}}'_y(\circ)^\perp\,\vec{\sigma}_t(t+\theta) \\
& \underbrace{+\widetilde{\Sigma}'_y(\circ)\frac{\widetilde{X}'_\theta}{\widetilde{X}}(t,\theta)}_{I(\circ)}\ \underbrace{+\vec{\widetilde{\mathrm{v}}}(\circ)^\perp\left[\vec{\lambda}_t^{\widetilde{\Sigma}}(\theta) - \vec{\lambda}_t^N(t+\theta)\right]}_{K(\circ)}
\end{aligned} \tag{5.22}$$

$$\begin{aligned}
F(\circ) = {} & \frac{1}{2}\widetilde{\Sigma}^4(\circ) - \frac{1}{2}\widetilde{\Sigma}^2(\circ)\,\|\vec{\sigma}_t(t+\theta)\|^2 \\
& -y\,\widetilde{\Sigma}(\circ)\,\vec{\sigma}_t(t+\theta)^\perp\left[\vec{\widetilde{\mathrm{v}}}(\circ) - \widetilde{\Sigma}'_y(\circ)\vec{\sigma}_t(t+\theta)\right] \\
& -\frac{1}{2}y^2\left[\|\vec{\widetilde{\mathrm{v}}}(\circ) - \widetilde{\Sigma}'_y(\circ)\vec{\sigma}_t(t+\theta)\|^2 + \|\vec{\widetilde{n}}(\circ)\|^2\right].
\end{aligned} \tag{5.23}$$

Observe that this term-structure, multi-dimensional expression matches the single-underlying formulae, whether scalar (2.18) [p. 43] *or multi-dimensional* (3.55) [p. 173], *except for terms* $I(\circ)$ *and* $K(\circ)$ *appearing in expression (5.22) for* $E(\circ)$.

Note also that the self-consistency of term $I(\circ)$ *is ensured by using the sliding ratio* $\widetilde{X}'_\theta\widetilde{X}^{-1}(t,\theta)$, *which is simply equal to the absolute quantity* $X'_T X^{-1}(t, T)$.

The first natural step is clearly to compare this new ZDC (5.21), established in the term structure framework, to its equivalents (2.18) and (3.55) in the single underlying environment.

Similarities with the Single-Underlying Case

First of all, is this new result consistent with the previous, simpler formulas? Indeed, the single underlying case can easily be seen as a subset of the current framework: it suffices to ignore all dimensionality (only for the scalar case) and maturity-dependence (in both cases) in the underlying and in its instantaneous volatility, and then to align all endogenous drivers. Applying these reductions produces, thankfully, precisely the single-underlying ZDCs.

Drawing a parallel with Part I, we can identify and interpret easily some quantities of interest.

We draw first the now familiar term $\vec{\vec{\mathrm{v}}}(\circ) - \widetilde{\Sigma}'_y(\circ)\vec{\vec{\sigma}}_t(t+\theta)$ which is present in D and F, and nothing else than $\vec{\mathrm{v}}$, the endogenous coefficient of the *absolute* smile Σ. Again, it is easy to see that its second part is compensating the change in log-moneyness induced by the underlying's movement.

It follows immediately that the term $\| \vec{\vec{\mathrm{v}}}(\circ) - \widetilde{\Sigma}'_y(\circ)\vec{\vec{\sigma}}_t(t+\theta)\|^2 + \| \vec{\vec{n}}(\circ)\|^2$, which constitutes most of the $D(t, y, \theta)$ expression, and is also present in $F(t, y, \theta)$, represents exactly the quadratic variation of the absolute implied volatility Σ: again this property is transferred seamlessly from the single underlying case.

Analysis of the New Terms

It seems, therefore, that the real novelty regarding this equation resides in the underlined terms $I(\circ)$ and $K(\circ)$ in (5.22), which account, respectively, for the term structure of the underlying collection, and for the specification of measures used in the dynamics of the different volatility maps.

Starting with term $K(\circ)$, we realise that it could already have been present in the single-asset case, albeit without the maturity dependence. This is where we observe the impact of the simplification choices we made in Part I, in order then to match closely the specification of popular single-asset stochastic volatility models. Specifically, the fact that we elected to express the dynamics of the volatility with the same endogenous driver as the underlying were effectively "hiding" the possible misalignment. Again, mathematically the expressions are just as valid and general, since the volatility drift $a_{1,t}$ can be made to contain that information. But in terms of modelling, in particular when it comes to specifying the volatility risk premiums, the spirit is noticeably different.

Let us now turn to the more structural term $I(t, y, \theta)$, which was introduced into the implied volatility dynamics by the change from absolute to sliding coordinates, as described by (5.15) and in particular (5.17) [p. 283]. This brand new quantity within the parametric process $E(\circ)$ can in fact be re-expressed in several ways, each leading to a distinct interpretation:

- As a first attempt, we can change all quantities to their absolute counterparts by using (B.1) and hence obtain

$$
\begin{aligned}
I(t,y,\theta) &= \widetilde{\Sigma}'_y(t,y,\theta)\,\frac{\widetilde{X}'_\theta}{\widetilde{X}}(t,\theta) = -X_t(T)\;\partial_X \Sigma_t\left(X_t(T),T,K\right)\;\frac{\partial_T X_t(T)}{X_t(T)} \\
&= -\partial_X \Sigma_t\left(X_t(T),T,K\right)\;\partial_T X_t(T).
\end{aligned}
$$

Then, noticing that the slope $\widetilde{\Sigma}'_\theta(\circ)$ also appears in the same term $E(\circ)$, we can legitimately group them together and argue that their sum $\widetilde{\Sigma}'_\theta(\circ) + I(\circ)$ represents the total variation in implied volatility due to an infinitesimal step in time-to-maturity θ (or maturity T, which are both *parameters*), but *compensated* for the associated "slide" of the underlying. In other words, it represents the slope as if the underlying were static as opposed to maturity-dependent.
The fact that we are changing the underlying will at the first order generate a change in the level of the implied volatility, in proportion to the local "Delta" Σ'_X of the implied volatility surface. This situation is somewhat similar to what we have already observed above for the coefficient $\overrightarrow{v}$ and in the single underlying environment: the movements in the *absolute* implied volatility Σ, for a given strike/expiry, were a combination of the dynamics of $\widetilde{\Sigma}$, the *sliding* implied volatility, and of the movements of both time and the underlying S_t. Simply because a movement of the underlying automatically repositions a given strike somewhere else on the y axis.
In other words, this corrective or compensation term can be interpreted as the deterministic change in the sliding implied volatility, generated by the underlying "riding down" the maturity curve. Besides, we happen to know that the local "Delta" of the implied volatility Σ'_X is strongly linked to its local skew Σ'_K. This is easily seen, in particular for at-the-money options within pure local volatility models, as discussed, for instance, in [3]. Therefore, in the case of a Caplet smile, for instance, the combination of a skewed implied volatility surface with a non-flat forward rate curve will mechanically generate by itself some deterministic dynamics in the smile.

- Independently, for any positive constant λ we can re-express term $I(t,y,\theta)$ as

$$
\begin{aligned}
I(t,y,\theta) &= \widetilde{\Sigma}'_y(t,y,\theta)\,\frac{\widetilde{X}'_\theta}{\widetilde{X}}(t,\theta) = \widetilde{\Sigma}'_y(t,y,\theta)\,\partial_\theta \ln\left[\lambda\,\widetilde{X}(t,\theta)\right] \\
&= \widetilde{\Sigma}'_y(t,y,\theta)\,\partial_T \ln\left[\lambda\,X(t,T)\right].
\end{aligned}
\tag{5.24}
$$

Therefore, if by any chance $X(t,T)$ were to represent $B_t(T,T+\delta)$, denoting as usual the t-value of the forward zero-coupon for the period $[T,T+\delta]$, then with $\lambda = 1$ we would have, for $(\circ) \triangleq (t,y,\theta)$,

$$
\begin{aligned}
I(\circ) &= -\widetilde{\Sigma}'_y(\circ)\;\partial_T\left[(T-t)L_t(T,T+\delta)\right] = -\widetilde{\Sigma}'_y(\circ)\;\partial_T\left[\int_T^{T+\delta} f_t(s)\,ds\right] \\
&= -\widetilde{\Sigma}'_y(\circ)\left[f_t(T+\delta) - f_t(T)\right],
\end{aligned}
$$

where $L_t(T, T+\delta)$ denotes the linear (Libor) forward rate, and $f_t(T)$ the instantaneous forward rate. Recall that in this configuration $\widetilde{\Sigma}$ represents the (hypothetical) δ-period ZC option (sliding) smile, which we can effectively link to Bond options with the same frequency. In summary, we can consider that we have roughly expressed term I_θ as a product of the Bond options skew and the slope of the yield curve.

In light of these two interpretations, we will call I_θ the "slope compensation" term.

Proof of Proposition 5.1.
We know that

$$\frac{V_t(X_t(T), T, K)}{N_t(T)} = C^{BS}\left(X_t(T), K, \Sigma_t(X_t(T), T, K) \cdot \sqrt{T-t}\right) \quad (5.25)$$

is martingale under $\mathbb{Q}^{N(T)}$, since both numerator and denominator are traded assets.

We can therefore apply Itô's Lemma to the functional on the r.h.s. side to obtain:

$$\begin{aligned} d\left[\frac{V_t(X_t(T), T, K)}{N_t(T)}\right] &= C_f^{BS'}(\ddagger)\, dX_t(T) \\ &+ \frac{1}{2} C_{vv}^{BS''}(\ddagger)\theta \langle d\Sigma_t(\dagger)\rangle + \frac{1}{2} C_{ff}^{BS''}(\ddagger)\langle dX_t(T)\rangle \\ &+ C_v^{BS'}(\ddagger)\left[\sqrt{\theta} d\Sigma_t(\dagger) - \frac{\Sigma_t(\dagger)}{2\sqrt{\theta}}\, dt\right] \\ &+ C_{fv}^{BS''}(\ddagger)\sqrt{\theta}\langle dX_t(T), d\Sigma_t(\dagger)\rangle. \end{aligned} \quad (5.26)$$

Using the Black function differentials of Appendix C, this result becomes:

$$\begin{aligned} & d\left[\frac{V_t(X_t(T), T, K)}{N_t(T)}\right] \\ &= \mathcal{N}(d_1)\left[X_t(T)\, \vec{\sigma}_t(T)^\perp\, d\vec{W}_t - X_t(T)\, \vec{\sigma}_t(T)^\perp \vec{\lambda}_t^N(T)\, dt\right] \\ &+ \frac{1}{2}\sqrt{\theta} X_t(T)\Sigma(\dagger)\mathcal{N}'(d_1)\left[\frac{ln^2\left(K/X_t(T)\right)}{\Sigma(\dagger)^4\theta^2} - \frac{1}{4}\right]\theta\left[\|\vec{v}(\dagger)\|^2 + \|\vec{n}(\dagger)\|^2\right] dt \\ &+ \frac{1}{2}\frac{\mathcal{N}'(d_1)}{X_t(T)\Sigma(\dagger)\sqrt{\theta}} X_t(T)^2 \|\vec{\sigma}_t(T)\|^2\, dt \\ &+ X_t(T)\mathcal{N}'(d1)\left[\sqrt{\theta}\left(b(\dagger)\, dt + \vec{v}(\dagger)^\perp\, d\vec{W}_t + \vec{n}(\dagger)^\perp\, d\vec{Z}_t\right) - \frac{\Sigma(\dagger)}{2\sqrt{\theta}}\, dt\right] \\ &+ \mathcal{N}'(d_1)\left[\frac{1}{\Sigma(\dagger)^2\theta} ln\left(\frac{K}{X_t(T)}\right) + \frac{1}{2}\right]\sqrt{\theta}\, X_t(T)\vec{\sigma}_t(T)^\perp \vec{v}(\dagger)\, dt \end{aligned}$$

with

$$d_1 = \frac{1}{\sqrt{\theta}\Sigma(\dagger)} ln\left(\frac{X_t(T)}{K}\right) + \frac{1}{2}\sqrt{\theta}\Sigma(\dagger) = \frac{-y}{\sqrt{\theta}\,\Sigma(\dagger)} + \frac{1}{2}\sqrt{\theta}\,\Sigma(\dagger).$$

Note that this expression is well-defined since we are considering $\theta > 0$ and also making Assumption 5.3. We can re-write after Itô decomposition:

$$d\left[\frac{V_t(\dagger)}{N_t(T)}\right] = A(\dagger)\,dt + \vec{B}(\dagger)^{\perp}\,d\vec{W}_t + \vec{C}(\dagger)^{\perp}\,d\vec{Z}_t \tag{5.27}$$

with

$$\begin{aligned} A(\dagger) = & -\mathcal{N}(d_1)X_t(T)\,\vec{\sigma}_t(T)^{\perp}\,\vec{\lambda}_t^N(T) + \mathcal{N}'(d_1)\frac{X_t(T)\|\vec{\sigma}_t(T)\|^2}{2\Sigma(\dagger)\sqrt{\theta}} \\ & + \mathcal{N}'(d_1)\sqrt{\theta}X_t(T)b(\dagger) - \mathcal{N}'(d_1)\frac{X_t(T)\Sigma(\dagger)}{2\sqrt{\theta}} \\ & + \mathcal{N}'(d_1)\sqrt{\theta}X_t(T)\left[\frac{y}{\Sigma(\dagger)^2\theta} + \frac{1}{2}\right]\vec{\sigma}_t(T)^{\perp}\vec{v}(\dagger) \\ & + \frac{1}{2}\mathcal{N}'(d_1)\theta^{\frac{3}{2}}X_t(T)\Sigma(\dagger)\left[\frac{y^2}{\Sigma(\dagger)^4\theta^2} - \frac{1}{4}\right]\left[\|\vec{v}(\dagger)\|^2 + \|\vec{n}(\dagger)\|^2\right] \end{aligned}$$

and

$$\vec{B}(\dagger) = \mathcal{N}(d_1)X_t(T)\vec{\sigma}_t(T) + \mathcal{N}'(d_1)\sqrt{\theta}X_t(T)\vec{v}(\dagger)$$

and

$$\vec{C}(\dagger) = \mathcal{N}'(d_1)\sqrt{\theta}X_t(T)\vec{n}(\dagger).$$

After changing the reference measure to $\mathbb{Q}^{N(T)}$, (5.27) becomes

$$\begin{aligned} d\left[\frac{V_t(\dagger)}{N_t(T)}\right] = & A(\dagger)\,dt + \vec{B}(\dagger)^{\perp}\left[d\vec{W}_t^{N(T)} + \vec{\lambda}_t^N(T)\,dt\right] \\ & + \vec{C}(\dagger)^{\perp}\,d\vec{Z}_t. \end{aligned} \tag{5.28}$$

The martingale property (5.25) is therefore expressed as:

$$0 = A(\dagger) + \vec{B}(\dagger)^{\perp}\vec{\lambda}_t^N(T)$$

which we expand into:

$$
\begin{aligned}
0 = &-\mathcal{N}(d_1)X_t(T)\,\vec{\sigma}_t(T)^{\perp}\,\vec{\lambda}_t^N(T) + \mathcal{N}'(d_1)\frac{X_t(T)\|\vec{\sigma}_t(T)\|^2}{2\Sigma(\dagger)\sqrt{\theta}} \\
&+ \mathcal{N}'(d_1)\sqrt{\theta}X_t(T)b(\dagger) - \mathcal{N}'(d_1)\frac{X_t(T)\Sigma(\dagger)}{2\sqrt{\theta}} \\
&+ \mathcal{N}'(d_1)\sqrt{\theta}X_t(T)\left[\frac{y}{\Sigma^2(\dagger)\theta} + \frac{1}{2}\right]\vec{\sigma}_t(T)^{\perp}\vec{\mathrm{v}}(\dagger) \\
&+ \mathcal{N}(d_1)X_t(T)\vec{\sigma}_t(T)^{\perp}\vec{\lambda}_t^N(T) + \mathcal{N}'(d_1)\sqrt{\theta}X_t(T)\vec{\mathrm{v}}(\dagger)^{\perp}\vec{\lambda}_t^N(T) \\
&+ \frac{1}{2}\mathcal{N}'(d_1)\theta^{\frac{3}{2}}X_t(T)\Sigma(\dagger)\left[\frac{y^2}{\Sigma^4(\dagger)\theta^2} - \frac{1}{4}\right]\left[\|\vec{\mathrm{v}}\|^2(\dagger) + \|\vec{n}\|^2(\dagger)\right].
\end{aligned}
$$

Observing that (trivially) $\mathcal{N}'(d_1) > 0$, using Assumption 5.1 and remembering that we are only considering strictly positive θ, we can divide both sides by $\sqrt{\theta}X_t(T)$ $\mathcal{N}'(d_1) > 0$ and thus simplify into:

$$
\begin{aligned}
0 = &\frac{1}{2}\theta\Sigma(\dagger)\left[\frac{y^2}{\Sigma^4(\dagger)\theta^2} - \frac{1}{4}\right]\left[\|\vec{\mathrm{v}}(\dagger)\|^2 + \|\vec{n}(\dagger)\|^2\right] + \frac{\|\vec{\sigma}_t(T)\|^2}{2\Sigma(\dagger)\theta} + b(\dagger) \\
&- \frac{\Sigma(\dagger)}{2\theta} + \left(\frac{y}{\Sigma^2(\dagger)\theta} + \frac{1}{2}\right)\vec{\sigma}_t(T)^{\perp}\vec{\mathrm{v}}(\dagger) + \vec{\mathrm{v}}(\dagger)^{\perp}\vec{\lambda}_t^N(T).
\end{aligned}
$$

We can now replace absolute quantities with their sliding counterparts according to (5.16), (5.19) and (5.20). Omitting the argument (t, y, θ) for clarity, we get

$$
\begin{aligned}
0 = &\frac{1}{2}\theta\,\widetilde{\Sigma}\left[\frac{y^2}{\widetilde{\Sigma}^4\theta^2} - \frac{1}{4}\right]\left[\|\vec{\widetilde{\mathrm{v}}} - \widetilde{\Sigma}'_y\vec{\sigma}_t(t+\theta)\|^2 + \|\vec{\widetilde{n}}\|^2\right] + \frac{\|\vec{\sigma}_t(t+\theta)\|^2}{2\widetilde{\Sigma}\theta} \\
&+ \widetilde{b} - \widetilde{\Sigma}'_\theta - \vec{\widetilde{\mathrm{v}}}^{\perp}\vec{\lambda}_t^{\widetilde{\Sigma}}(\theta) + \widetilde{\Sigma}'_y\left[\vec{\sigma}_t(t+\theta)^{\perp}\vec{\lambda}_t^N(t+\theta) - \frac{\widetilde{X}'_\theta}{\widetilde{X}}(t,\theta)\right] \\
&+ \frac{1}{2}\left[\widetilde{\Sigma}''_{yy} + \widetilde{\Sigma}'_y\right]\|\vec{\sigma}_t(t+\theta)\|^2 + \left[\frac{y}{\widetilde{\Sigma}^2\theta} + \frac{1}{2}\right]\vec{\sigma}_t(t+\theta)^{\perp}\left[\vec{\widetilde{\mathrm{v}}} - \widetilde{\Sigma}'_y\vec{\sigma}_t(t+\theta)\right] \\
&- \vec{\widetilde{\mathrm{v}}}'^{\perp}_y\vec{\sigma}_t(t+\theta) - \frac{\widetilde{\Sigma}}{2\theta} + \left[\vec{\widetilde{\mathrm{v}}} - \widetilde{\Sigma}'_y\vec{\sigma}_t(t+\theta)\right]^{\perp}\vec{\lambda}_t^N(t+\theta).
\end{aligned}
$$

Simplifying, arranging the terms according to the power of θ, and re-scaling by $\widetilde{\Sigma}^3(t, y, \theta)$ proves (5.21), which concludes the proof. □

5.2.2 The Immediate Zero Drift Condition

Let us now focus on the immediate smile. An important point to observe is that in doing so, we are effectively considering a single asset, the *immediate underlying* $\widetilde{X}(0)$. If θ represented the maturity of a forward quantity, for instance if $\widetilde{X}(\theta)$ were a forward rate of a given period/tenor, then $\widetilde{X}(0)$ would simply be the *spot* rate for that same period/tenor.

As in Part I, we now introduce some sufficient regularity assumptions:

Assumption 5.4 (*Immediate regularity*) The following processes

(i) $\widetilde{\Sigma}\ \widetilde{\Sigma}'_\theta\ \widetilde{\Sigma}'_y\ \widetilde{\Sigma}''_{yy}\ \widetilde{b}\ \vec{\widetilde{\nu}}\ \vec{\widetilde{\nu}}'_y\ \vec{\widetilde{n}}$

(ii) $\widetilde{X}\ \widetilde{X}'_\theta$

(iii) $\vec{\sigma}\ \vec{\lambda}^N\ \vec{\lambda}^{\widetilde{\Sigma}}$

admit a finite (stochastic) limit when $\theta \searrow 0$, respectively in (t, y, θ) for group (i), (t, θ) for pair (ii) and $(t, t + \theta)$ for group (iii). Moreover, we denote these limits by using the extended arguments, respectively $(t, y, 0)$, $(t, 0)$ and (t, t).

For simplicity, these assumptions have been chosen to be strong (sufficient) but they are not *necessary*. For instance, the proof does not require the existence of individual limits for $\vec{\lambda}^N$ and $\vec{\lambda}^{\widetilde{\Sigma}}$: only that their difference $\vec{\lambda}^N - \vec{\lambda}^{\widetilde{\Sigma}}$ admit a limit in $\theta = 0$.

We can now move on to the result, the **Immediate Zero-Drift Condition**:

Corollary 5.1 (Immediate Zero Drift Conditions in the TS framework) *In the immediate domain* $(\bullet) \triangleq (t, y, 0)$, *the local (IATM) shape and dynamics of the implied volatility can be linked to the immediate instantaneous volatility with the two equivalent Immediate Zero Drift Conditions (IZDCs):*

- *The Primary IZDC:*

$$\begin{aligned} F(\bullet) = 0 = \widetilde{\Sigma}^2(\bullet) &\left[\|\vec{\sigma}_t(t)\|^2 - \widetilde{\Sigma}^2(\bullet) \right] \\ &+ 2y\, \widetilde{\Sigma}(\bullet) \vec{\sigma}_t(t)^\perp \left[\vec{\widetilde{\nu}}(\bullet) - \widetilde{\Sigma}'_y(\bullet) \vec{\sigma}_t(t) \right] \\ &+ y^2 \left[\| \vec{\widetilde{\nu}}(\bullet) - \widetilde{\Sigma}'_y(\bullet) \vec{\sigma}_t(t) \|^2 + \| \vec{\widetilde{n}}(\bullet) \|^2 \right]. \end{aligned} \tag{5.29}$$

- *The Secondary IZDC:*

$$\widetilde{\Sigma}^3(t, y, 0)\, \widetilde{b}(t, y, 0) = E(t, y, 0) + F'_\theta(t, y, 0). \tag{5.30}$$

Remark 5.3 Note that, similarly to what was observed in the single-asset case in Remark 2.3, p. 49, it is the Secondary IZDC which corresponds to the ZDC taken into the immediate domain $(t, y, 0)$, and which represents its true asymptotics. From that angle, the Primary IZDC consists rather in a useful and compact consequence of the structural result.

Proof As in Part I, let us denote the immediate argument $(t, y, 0)$ by $(\bullet)$. The Assumption set 5.4 implies that the processes D, E, and F all admit, pointwise in y and $a.s.$ in w, individual finite limits in $(\bullet)$. In light of the main ZDC (5.1), this means that term $\frac{1}{\theta}F(t, y, \theta)$ converges to a finite limit. In turn, this implies that necessarily

$$\lim_{\theta\searrow 0} F(t, y, \theta) = 0$$

and imposes a lower bound on the convergence speed ("faster than θ"). Then invoking (5.23), which defines term $F(t, y, \theta)$, we get

$$\begin{aligned} 0 = &\frac{1}{2}\,\|\vec{\sigma}_t(t)\|^2\,\widetilde{\Sigma}^2(\bullet) - \frac{1}{2}\,\widetilde{\Sigma}^4(\bullet) + y\,\widetilde{\Sigma}(\bullet)\;\vec{\sigma}_t(t)^{\perp}\left[\vec{\mathrm{v}}\,(\bullet) - \widetilde{\Sigma}'_y(\bullet)\,\vec{\sigma}_t(t)\right] \\ &+\frac{1}{2}\,y^2\left[\,\|\vec{\mathrm{v}}\,(\bullet) - \widetilde{\Sigma}'_y(\bullet)\vec{\sigma}_t(t)\|^2 + \|\,\vec{n}\,(\bullet)\|^2\right], \end{aligned} \tag{5.31}$$

which immediately provides the Primary IZDC (5.29). In turn, using a simple Taylor expansion we obtain

$$F(t, y, \theta) = F(t, y, 0) + \theta\,F'_\theta(t, y, 0) + O(\theta^2) = \theta\,F'_\theta(t, y, 0) + O(\theta^2). \tag{5.32}$$

Alternatively, using L'Hopital's rule:

$$\lim_{\theta\searrow 0}\,\frac{1}{\theta}F(t, y, \theta) = F'_\theta(t, y, 0). \tag{5.33}$$

The main ZDC (5.1) can therefore be rewritten in the general domain as

$$\widetilde{\Sigma}^3(\circ)\,b(\circ) = \theta\,D(t, y, \theta) + E(t, y, \theta) + F'_\theta(t, y, 0) + O(\theta).$$

Its then suffice to invoke the regularity conditions of Assumption 5.4 to take the limit of that expression in $\theta = 0$: we obtain the Secondary IZDC (5.30) which concludes the proof. □

Again, our first natural move is to check the consistency of this equation with its counterpart in the single underlying case.

Remark 5.4 This time round the comparison with the non-TS framework is straightforward. By removing only the maturity dependency in the volatility structure we get the multifactor version (3.57), p. 175, and by further ignoring the dimensionality we fall back onto our original IZDC (2.28) [p. 48]. Note however that the instantaneous volatility being invoked is also "immediate", as its maturity argument is t. Similarly, the log-moneyness y incorporates the immediate sliding underlying $\widetilde{X}_t(t)$.

But this time around, we do not have to contend with the slope compensation term I_θ or any driver disalignment issue such as term K_θ. This is due to the fact that the IZDC only calls on term F, and not on process E. In particular, the immediate numeraire $N_t(t)$, associated to the dynamics of the immediate sliding underlying $\widetilde{X}_t(0)$, does not appear in the new IZDC.

In conclusion, looking at small-time quantities and dynamics tends to significantly neutralise the term-structure aspect, at least without further differentiation. As we will see later, this property will disappear at higher orders of precision (i.e. when further θ-differentiating the ZDC in order to generate immediate ATM differentials of the smile).

Still on the matter of interpreting and building our intuition, we present here a subjective re-writing of the Primary Immediate Zero Drift Condition. The purpose of this syntax is to underline the strong links between two groups:

- the skew, the endogenous coefficient $\vec{\nu}$ and the exogenous coefficient $\vec{n}$;
- the polynomial decomposition of the *Immediate* smile shape.

Let us reformulate the immediate (or small-time) zero-drift condition (5.29). First, let us recall that the endogenous coefficient of the *absolute* implied volatility appears in two instances, according to (5.19) [p. 283]. We then modify the convention used to describe the dynamics of the latter, and adopt a Lognormal convention. We can then introduce the two Lognormal absolute coefficients:

$$\overrightarrow{\nu^{\text{LN}}} = \frac{\vec{\nu}}{\Sigma} \quad \text{and} \quad \overrightarrow{n^{\text{LN}}} = \frac{\vec{n}}{\Sigma}.$$

The IZDC (5.29) can then be re-written as

$$\begin{aligned} 0 = \widetilde{\Sigma}^2(\bullet) \left[\|\vec{\sigma}_t(t)\|^2 - \widetilde{\Sigma}^2(\bullet) \right] + 2\, y\, \widetilde{\Sigma}^2(\bullet)\, \vec{\sigma}_t(t)^{\perp} \overrightarrow{\nu^{\text{LN}}}(t, X_t(t), K) \\ + y^2\, \widetilde{\Sigma}^2(\bullet) \left[\|\overrightarrow{\nu_t^{\text{LN}}}(X_t(t), t, K)\|^2 + \|\overrightarrow{n_t^{\text{LN}}}(X_t(t), t, K)\|^2 \right]. \end{aligned}$$

Simplifying and reorganising the expression we get eventually

$$\widetilde{\Sigma}^2(\bullet) = \|\, \vec{\sigma}_t(t) + y \cdot \overrightarrow{\nu^{\text{LN}}}(t, X_t(t), K)\, \|^2 + \|\, y \cdot \overrightarrow{n_t^{\text{LN}}}(X_t(t), t, K)\, \|^2.$$

Reduced in such a way, the IZDC lends itself easily to interpretation. The *static* shape of the immediate smile, hence a function of log-moneyness y (or equivalently of strike K), is the combination of two *dynamic* and *orthogonal* components, respectively purely endogenous and purely exogenous. Furthermore, we observe that the *Lognormal* dynamic coefficients $\overrightarrow{\nu^{ln}}$ and $\overrightarrow{n_t^{ln}}$ are accounted for *via* their *Lognormal* torque,[6] i.e. with a leverage of $y = \ln(K/X_t(t))$. Besides, it appears that on the right-hand side the y coordinate is only taken into account in conjunction with the dynamics. In other words, both the endogenous and exogenous coefficient maps act as *relative* volatilities compared to the IATM dynamics.

Besides the decomposition, it is interesting to note that the implied volatility appears through its squared modulus. This is, of course, another way to present the aggregation of the directional information into a scalar quantity. But this triviality underlines that, at a generic point (t, y, θ), and when it comes to marginal distributions, the term of interest is usually more the total variance $\widetilde{\Sigma}^2(t, y, \theta)\,(T - t)$ than the "baseline model" implied parameter $\widetilde{\Sigma}$. However, having placed ourselves in the immediate domain, we can ignore the time-to-maturity altogether.

Finally, although this re-expression is focused on the immediate smile, there is every reason to believe that its principle extends to the rest of the (y, θ) field.

5.2.3 The IATM Identity

Having been through the same progression in both the scalar and multi-dimensional single-underlying cases, we can state the result without further work.

Corollary 5.2 (The IATM Identity in a Term Structure Framework) *Let us consider the sliding implied volatility model defined by* (5.1)–(5.3) *and (5.13). Then the static modulus of the (absolute) zero-expiry instantaneous stochastic volatility, defined by* (5.1)–(5.3) *and* (5.9), *can be recovered from the implied volatility surface:*

At the Immediate ATM (IATM) point $(\star) = (t, 0, 0)$ *we have a.s.*

$$\| \vec{\sigma}_t(t) \| = \widetilde{\Sigma}\,(t, 0, 0). \tag{5.34}$$

Proof Selecting $y = 0$ in the IZDC (5.31), we indeed recover from the Immediate ATM implied volatility smile the modulus of the instantaneous volatility of the immediate underlying $X_t(t)$, which proves the fundamental identity (5.34). □

It is clear that the "static" result (5.34) concerning the recovery of the instantaneous volatility is entirely consistent with its single underlying counterpart (2.36) [p. 50] and even more so with its multi-dimensional version (3.59) [p. 175]. We underline two slight differences, however, which arise from the very structure of the instantaneous volatility $\vec{\sigma}_t(T)$:

- The multi-dimensionality implies that we cannot recover the full information (meaning each individual coordinate) on the volatility of the immediate underlying $\vec{\sigma}_t(t)$ from the Immediate ATM implied volatility $\widetilde{\Sigma}(t, y, 0)$. Indeed, the

[6] a.k.a. "moment" or "couple".

implied volatility is obviously scalar, which in a sense "aggregates" the directional information contained in the multi-dimensional $\vec{\sigma}_t(t)$. Hence the implied volatility appear to be blind to any sort of correlation structure within our current underlying family or between the latter (or more precisely, the immediate underlying) and any other process (even partially) driven by $\vec{W}_t$ and $\vec{Z}_t$. However, this statement has to be mitigated by the knowledge of scalar products (5.35) and (5.36), as will be discussed shortly.

- The dependence of the instantaneous volatility $\vec{\sigma}_t(t+\theta)$ on the time-to-maturity θ means that the information recovered only concerns the "initial value" or "level" of the map $\theta \to \vec{\sigma}_t(t+\theta)$. In other terms, we are looking purely at the volatilities of the immediate underlying. How much this information tells us about the rest of the instantaneous volatility structure is a matter of modelling. It remains that the objective of recovering the whole map $T \to \vec{\sigma}_t(T)$ is not attainable using only Theorem 5.1. As will be discussed later on, it is possible to access all T-differentials of $\vec{\sigma}_t(T)$ taken in $T = t$, using inductive θ-differentiation of the main ZDC (5.21). But in order for that information to provide the whole of $\vec{\sigma}_t(T)$, we obviously need the latter to be an *analytic* function of the maturity T, which excludes piecewise affine specifications. In conclusion, the limitations we had w.r.t. the stationarity of smooth time-dependency of the model extend to the maturity dependency.

5.3 Recovering the Instantaneous Volatility

Our aim in this section is to tackle the *inverse problem*, which is to infer or recover information about the *stochastic instantaneous volatilities* (seen as model parametric inputs) from the shape $\widetilde{\Sigma}(t, y, \theta)$ and from the dynamics $\vec{\mathrm{v}}(t, y, \theta)$ and $\vec{n}(t, y, \theta)$ of the implied volatility surface. By "volatilities" we obviously refer to the underlyings' $\vec{\sigma}_t(T)$, but also possibly to the term structure of volatility for the numeraire collection $\vec{\lambda}_t^N(T)$, as well as other risk premia used to define the relevant measures.

In the end, it is only a matter of defining which elements of the model are given as inputs, and which ones need to be solved for. We will see that, depending on the configuration, the problem might be (partially) solvable or downright ill-defined.

Let us start by stating the main result, before commenting on it and then presenting the proof.

5.3.1 Establishing the Main Result

Following the outline of Sects. 2.3 and 3.4.3, let us first focus on the information inherent to the Stochastic Implied Volatility model, without any connection yet to the Instantaneous framework.

Proposition 5.2 (IATM arbitrage constraints of the SImpV model: first layer) *Let us consider a given stochastic sliding implied volatility model, as defined by* (5.1)–(5.3) *and (5.13). Then its dynamic coefficients are locally constrained to satisfy, at the IATM point* $(t, 0, 0)$*:*

$$\vec{\mathrm{v}}(\star)^{\perp}\vec{\sigma}_t(t) = 2\,\widetilde{\Sigma}^2\,\widetilde{\Sigma}'_y(\star) \tag{5.35}$$

$$\vec{\mathrm{v}}'_y(\star)^{\perp}\vec{\sigma}_t(t) = \frac{3}{2}\,\widetilde{\Sigma}^2\,\widetilde{\Sigma}''_{yy}(\star) + 3\,\widetilde{\Sigma}\,\widetilde{\Sigma}'^{\,2}_y(\star) - \frac{\left[\|\vec{\mathrm{v}}\|^2 + \|\vec{n}\|^2\right]}{2\,\widetilde{\Sigma}}(\star) \tag{5.36}$$

$$\widetilde{b}(\star) = 2\,\widetilde{\Sigma}'_\theta(\star) - \frac{1}{2}\,\widetilde{\Sigma}^2\,\widetilde{\Sigma}''_{yy}(\star) - \frac{1}{2}\,\vec{\mathrm{v}}(\star)^{\perp}\vec{\sigma}_t(t) + \vec{\mathrm{v}}'_y(\star)^{\perp}\,\vec{\sigma}_t(t) \\ \underbrace{+I_0 + J_0 + K_0}_{\textit{new TS terms}} \tag{5.37}$$

where the new terms, specific to the Term-Structure framework, are

$$I_0 = \widetilde{\Sigma}'_y(\star)\frac{\widetilde{X}'_\theta}{\widetilde{X}}(t,0), \quad J_0 = -\frac{\vec{\sigma}_t(t)^{\perp}\,\vec{\sigma}'_T(t,t)}{\|\vec{\sigma}_t(t)\|}, \quad K_0 = \vec{\mathrm{v}}(\star)^{\perp}\left[\vec{\lambda}_t^{\widetilde{\Sigma}}(0) - \vec{\lambda}_t^{N}(t)\right].$$

Note that, similarly to the remark made in Proposition 5.1, *the sliding ratio* $\widetilde{X}'_\theta\widetilde{X}^{-1}(t,0)$ *is equal to the absolute quantity* $X'_T X^{-1}(t,t)$*. Note also that two of the new terms have been introduced earlier through (5.22)* [p. 286]*:* $I_0 = I(\star)$ *and* $K_0 = K(\star)$.

Proof

Step 1/3: computation of the first product

$\vec{\mathrm{v}}(\star)^{\perp}\vec{\sigma}_t(t)$ Let us differentiate the Primary IZDC (5.29) once w.r.t. y; we obtain

$$\begin{aligned}
\overbrace{0 = 2\,\|\vec{\sigma}_t(t)\|^2\,\widetilde{\Sigma}\,\widetilde{\Sigma}'_y(\bullet) - 4\,\widetilde{\Sigma}^3\,\widetilde{\Sigma}'_y(\bullet) + 2\widetilde{\Sigma}(\bullet)\,\vec{\sigma}_t(t)^{\perp}\left[\vec{\mathrm{v}}(\bullet) - \widetilde{\Sigma}'_y(\bullet)\vec{\sigma}_t(t)\right]}\\
+ 2\,y\,\widetilde{\Sigma}'_y(\bullet)\,\vec{\sigma}_t(t)^{\perp}\left[\vec{\mathrm{v}}(\bullet) - \widetilde{\Sigma}'_y(\bullet)\vec{\sigma}_t(t)\right]\\
+ 2\,y\,\widetilde{\Sigma}(\bullet)\,\vec{\sigma}_t(t)^{\perp}\left[\vec{\mathrm{v}}'_y(\bullet) - \widetilde{\Sigma}''_{yy}(\bullet)\vec{\sigma}_t(t)\right]\\
+ 2y\left[\|\vec{\mathrm{v}}(\bullet) - \widetilde{\Sigma}'_y(\bullet)\vec{\sigma}_t(t)\|^2 + \|\vec{n}(\bullet)\|^2\right]\\
+ 2\,y^2\left[\left[\vec{\mathrm{v}}(\bullet) - \widetilde{\Sigma}'_y(\bullet)\vec{\sigma}_t(t)\right]^{\perp}\left[\vec{\mathrm{v}}'_y(\bullet) - \widetilde{\Sigma}''_{yy}(\bullet)\vec{\sigma}_t(t)\right] + \vec{n}(\bullet)^{\perp}\vec{n}'_y(\bullet)\right]
\end{aligned} \tag{5.38}$$

where the first (over-braced) term simplifies as

$$2\|\vec{\sigma}_t(t)\|^2 \widetilde{\Sigma}\,\widetilde{\Sigma}'_y(\bullet) - 4\widetilde{\Sigma}^3\,\widetilde{\Sigma}'_y(\bullet) + 2\widetilde{\Sigma}(\bullet)\vec{\sigma}_t(t)^\perp\left[\vec{\mathrm{v}}\,(\bullet) - \widetilde{\Sigma}'_y(\bullet)\vec{\sigma}_t(t)\right]$$
$$= 2\widetilde{\Sigma}(\bullet)\,\vec{\sigma}_t(t)^\perp\,\vec{\mathrm{v}}\,(\bullet) - 4\widetilde{\Sigma}^3\,\widetilde{\Sigma}'_y(\bullet).$$

Evaluating (5.38) at $y = 0$ yields, in view of the IATM identity (5.34):

$$0 = 2\,\widetilde{\Sigma}(\star)\,\vec{\sigma}_t(t)^\perp\,\vec{\mathrm{v}}\,(\star) - 4\,\widetilde{\Sigma}^3\,\widetilde{\Sigma}'_y(\star)$$

which proves (5.35).

Step 2/3: computation of the second product $\vec{\mathrm{v}}\,'_y(\star)^\perp\,\vec{\sigma}_t(t)$

We differentiate the IZDC (5.31) twice w.r.t. y and obtain:

$$\begin{aligned}0 = {} & 2\,\widetilde{\Sigma}'_y(\bullet)\,\vec{\sigma}_t(t)^\perp\,\vec{\mathrm{v}}\,(\bullet) + 2\,\widetilde{\Sigma}(\bullet)\,\vec{\sigma}_t(t)^\perp\,\vec{\mathrm{v}}\,'_y(\bullet)\\ & - 4\left[3\widetilde{\Sigma}^2\,\widetilde{\Sigma}'^{\,2}_y + \widetilde{\Sigma}^3\,\widetilde{\Sigma}''_{yy}\right](\bullet) + 2\,\widetilde{\Sigma}'_y(\bullet)\vec{\sigma}_t(t)^\perp\left[\vec{\mathrm{v}}\,(\bullet) - \widetilde{\Sigma}'_y(\bullet)\vec{\sigma}_t(t)\right]\\ & + 2\,\widetilde{\Sigma}(\bullet)\vec{\sigma}_t(t)^\perp\left[\vec{\mathrm{v}}\,'_y(\bullet) - \widetilde{\Sigma}''_{yy}(\bullet)\vec{\sigma}_t(t)\right]\\ & + 2\left[\|\vec{\mathrm{v}}\,(\bullet) - \widetilde{\Sigma}'_y(\bullet)\vec{\sigma}_t(t)\|^2 + \|\vec{n}\,(\bullet)\|^2\right] + y\,[\cdots].\end{aligned}$$

Evaluating this equation at $y = 0$, using (5.34) and (5.35) yields:

$$\begin{aligned}0 = {} & -12\,\widetilde{\Sigma}^2\,\widetilde{\Sigma}'^{\,2}_y(\star) + 4\,\widetilde{\Sigma}(\star)\,\vec{\sigma}_t(t)^\perp\,\vec{\mathrm{v}}\,'_y(\star)\\ & - 6\,\widetilde{\Sigma}^3\,\widetilde{\Sigma}''_{yy}(\star) + 2\left[\|\vec{\mathrm{v}}\,(\star)\|^2 + \|\vec{n}\,(\star)\|^2\right]\end{aligned}$$

which proves (5.36).

Step 3/3: computation of the IATM drift $\widetilde{b}(t, 0, 0)$

Let us recall the Secondary IZDC (5.30), p. 292:

$$\widetilde{\Sigma}^3(t, y, 0)\,\widetilde{b}(t, y, 0) = E(t, y, 0) + F'_\theta(t, y, 0).$$

In the above expression, the very definition of F (5.23) leads to

$$F'_\theta(t, y, \theta) = \frac{1}{2}\,\widetilde{\Sigma}'_\theta(\circ) - \frac{\partial}{\partial\theta}\left[\frac{\|\vec{\sigma}_t(t+\theta)\|^2}{2\widetilde{\Sigma}(\circ)}\right] + y\,[\cdot]$$

which is *a priori* non-null and where we express

$$\frac{\partial}{\partial\theta}\left[\frac{\|\vec{\sigma}_t(t+\theta)\|^2}{\widetilde{\Sigma}(\circ)}\right] = \frac{\partial}{\partial\theta}\left[\frac{\vec{\sigma}_t(t+\theta)^{\perp}\vec{\sigma}_t(t+\theta)}{\widetilde{\Sigma}(\circ)}\right]$$

$$= \frac{1}{\widetilde{\Sigma}^2(\circ)}\Big[2\vec{\sigma}_t(t+\theta)^{\perp}\,\vec{\sigma}'_T(t,t+\theta)\,\widetilde{\Sigma}(\circ)$$

$$- \|\vec{\sigma}_t(t+\theta)\|^2\,\widetilde{\Sigma}'_\theta(\circ)\Big]$$

so that

$$F'_\theta(t,y,\theta) = \frac{1}{2}\widetilde{\Sigma}'_\theta(\circ) - \frac{\vec{\sigma}_t(t+\theta)^{\perp}\,\vec{\sigma}'_T(t,t+\theta)}{\widetilde{\Sigma}(\circ)} + \frac{\|\vec{\sigma}_t(t+\theta)\|^2}{2\,\widetilde{\Sigma}^2(\circ)}\widetilde{\Sigma}'_\theta(\circ) + y\,[\cdot]$$

$$= \frac{1}{2}\left[1 + \frac{\|\vec{\sigma}_t(t+\theta)\|^2}{\widetilde{\Sigma}^2(\circ)}\right]\widetilde{\Sigma}'_\theta(\circ) - \frac{\vec{\sigma}_t(t+\theta)^{\perp}\,\vec{\sigma}'_T(t,t+\theta)}{\widetilde{\Sigma}(\circ)} + y\,[\cdot]\,.$$

Evaluating this expression at $(t,0,0)$ and invoking (5.34) gives

$$F'_\theta(\star) = \frac{1}{2}\left[1 + \frac{\|\vec{\sigma}_t(t)\|^2}{\|\vec{\sigma}_t(t)\|^2}\right]\widetilde{\Sigma}'_\theta(\star) - \frac{1}{\|\vec{\sigma}_t(t)\|}\vec{\sigma}_t(t)^{\perp}\,\vec{\sigma}'_T(t,t)$$

$$= \widetilde{\Sigma}'_\theta(\star) - \frac{\vec{\sigma}_t(t)^{\perp}\,\vec{\sigma}'_T(t,t)}{\|\vec{\sigma}_t(t)\|}.$$

Therefore, evaluating the Secondary IZDC (5.30) at $(t,0,0)$ we obtain the drift as

$$\widetilde{b}(t,0,0) = \widetilde{\Sigma}^{-3}(\star)\,\big[\,E(\star) + F'_\theta(\star)\big]$$

hence

$$\widetilde{b}(t,0,0) = \widetilde{\Sigma}'_\theta(\star) - \frac{1}{2}\widetilde{\Sigma}''_{yy}(\star)\|\vec{\sigma}_t(t)\|^2 - \frac{1}{2}\vec{\widetilde{\nu}}(\star)^{\perp}\vec{\sigma}_t(t) + {\vec{\widetilde{\nu}}'_y(\star)}^{\perp}\,\vec{\sigma}_t(t)$$

$$+ \widetilde{\Sigma}'_y(\star)\frac{\widetilde{X}'_\theta}{\widetilde{X}}(t,0) + \vec{\widetilde{\nu}}(\star)^{\perp}\Big[\vec{\widetilde{\lambda}}{}^{\widetilde{\Sigma}}_t(0) - \vec{\widetilde{\lambda}}{}^{N}_t(t)\Big]$$

$$+ \widetilde{\Sigma}'_\theta(\star) - \frac{\vec{\sigma}_t(t)^{\perp}\,\vec{\sigma}'_T(t,t)}{\|\vec{\sigma}_t(t)\|}.$$

Finally, simplifying and invoking (5.34) provides (5.37) and concludes the proof. □

Theorem 5.1 (Recovery of the modulus dynamics in a TS framework) *We denote by* $(\star) = (t, 0, 0)$ *the immediate ATM sliding argument and consider the sliding implied volatility model defined by* (5.1)–(5.3) *and (5.13). Then the dynamics of the (absolute) zero-expiry instantaneous stochastic volatility, defined by* (5.1)–(5.3) *and (5.9), can be partially recovered. They can be expressed as a function of the IATM shape, of the IATM driving coefficients* ($\vec{\nu}$ *and* $\vec{n}$) *and of the Immediate term structure for the underlying and for its instantaneous volatility.*

$$d \, \| \vec{\sigma}_t(t) \| \triangleq \left[2\widetilde{\Sigma}'_\theta + \widetilde{\Sigma}^2 \left[\widetilde{\Sigma}''_{yy} - \widetilde{\Sigma}'_y \right] + 3\widetilde{\Sigma} \, \widetilde{\Sigma}'^{\,2}_y - \frac{\left[\| \vec{\nu} \|^2 + \| \vec{n} \|^2 \right]}{2\widetilde{\Sigma}} \right] (\star) \, dt$$
$$+ \left[I_0 + J_0 + K'_0 \right] dt + \vec{\nu}(\star)^\perp \, d\vec{W}_t^{\sigma(t)} + \vec{n}(\star)^\perp \, d\vec{Z}_t \qquad (5.39)$$

where

$$K'_0 = \vec{\nu}(\star)^\perp \left[\vec{\lambda}_t^\sigma(t) - \vec{\lambda}_t^N(t) \right].$$

Proof The IATM identity (5.34) remaining true *a.s.* for any time t, we can take the dynamics of both its sides to obtain

$$d \, \| \vec{\sigma}_t(t) \| = d \; \widetilde{\Sigma} \, (t, 0, 0) \qquad (5.40)$$

which implies, according to our stochastic implied volatility framework (5.13):

$$\begin{aligned} d \, \| \vec{\sigma}_t(t) \| &= \widetilde{b}(\star) \, dt + \vec{\nu}(\star)^\perp \, d\vec{W}_t^{\widetilde{\Sigma}(0)} + \vec{n}(\star)^\perp \, d\vec{Z}_t \\ &= \widetilde{b}(\star) \, dt + \vec{\nu}(\star)^\perp \left[d\vec{W}_t^{\sigma(t)} + \vec{\lambda}_t^\sigma(t) \, dt - \vec{\lambda}_t^{\widetilde{\Sigma}}(0) \, dt \right] \\ &\quad + \vec{n}(\star)^\perp \, d\vec{Z}_t \\ &= \left[\widetilde{b}(\star) + \vec{\nu}(\star)^\perp \left[\vec{\lambda}_t^\sigma(t) - \vec{\lambda}_t^{\widetilde{\Sigma}}(0) \right] \right] dt \\ &\quad + \vec{\nu}(\star)^\perp \, d\vec{W}_t^{\sigma(t)} + \vec{n}(\star)^\perp \, d\vec{Z}_t . \end{aligned} \qquad (5.41)$$

In the drift term, we invoke the expression for $\widetilde{b}(\star)$ given by the SImpV constraint (5.37), where we replace $\|\vec{\sigma}_t(t)\|$ along with the products $\vec{\nu}(\star)^\perp \vec{\sigma}_t(t)$ and $\vec{\nu}'_y(\star)^\perp \vec{\sigma}_t(t)$ by their respective expressions (5.34), (5.35) and (5.36) to obtain:

$$\begin{aligned}\widetilde{b}(\star) &= 2\widetilde{\Sigma}'_\theta(\star) - \frac{1}{2}\widetilde{\Sigma}^2\widetilde{\Sigma}''_{yy}(\star) - \widetilde{\Sigma}^2\,\widetilde{\Sigma}'_y(\star) + \frac{3}{2}\,\widetilde{\Sigma}^2\,\widetilde{\Sigma}''_{yy}(\star)\\ &\quad - \frac{\left[\|\overrightarrow{\mathrm{v}}\|^2 + \|\overrightarrow{n}\|^2\right]}{2\widetilde{\Sigma}}(\star) + 3\widetilde{\Sigma}\,\widetilde{\Sigma}'^{\,2}_y(\star) + \widetilde{\Sigma}'_y\frac{\widetilde{X}'_\theta}{\widetilde{X}}(t,0)\\ &\quad - \frac{\overrightarrow{\sigma}_t(t)^\perp\,\overrightarrow{\sigma}'_T(t,t)}{\widetilde{\Sigma}(\star)} + \overrightarrow{\mathrm{v}}(\star)^\perp\left[\overrightarrow{\lambda}^{\widetilde{\Sigma}}_t(0) - \overrightarrow{\lambda}^N_t(t)\right].\end{aligned}$$

After simplification the Immediate ATM drift is

$$\begin{aligned}\widetilde{b}(\star) &= 2\widetilde{\Sigma}'_\theta(\star) + \widetilde{\Sigma}^2\left[\widetilde{\Sigma}''_{yy} - \widetilde{\Sigma}'_y\right](\star) + 3\widetilde{\Sigma}\,\widetilde{\Sigma}'^{\,2}_y(\star)\\ &\quad - \frac{\left[\|\overrightarrow{\mathrm{v}}\|^2 + \|\overrightarrow{n}\|^2\right]}{2\widetilde{\Sigma}}(\star) + I_0 + J_0 + \overrightarrow{\mathrm{v}}(\star)^\perp\left[\overrightarrow{\lambda}^{\widetilde{\Sigma}}_t(0) - \overrightarrow{\lambda}^N_t(t)\right].\end{aligned} \tag{5.42}$$

Note that this equation is simply a direct consequence of the arbitrage constraints of the SImpV model, as per Proposition 5.2. Substituting that expression into (5.41) and invoking the uniqueness of Itô's decomposition (in order to identify finite and non-finite variation terms) we prove (5.39) and conclude the proof. □

Obviously, the ideal recovery would have meant obtaining the full map of the instantaneous volatility, i.e. $T \to \overrightarrow{\sigma}_t(T)$. Clearly the asymptotic nature of the method seems to suggest that our information should be restricted to the Immediate point $T = t$. However, we can reasonably expect (by analogy with the single-asset case, especially the ladder effect) that successive differentiations of the ZDC should involve the T-derivatives of the map, still taken at the Immediate point. If that approach gave us $\overrightarrow{\sigma}'_T(t, T = t)$, then $\overrightarrow{\sigma}''_{TT}(t, T = t)$, etc., then provided that the map itself is analytic we could theoretically extrapolate: in other words, we would recover an approximation of the full map, up to any desired degree of precision.

The issue, however, is that the dynamics that we recover through Theorem 5.1 pertain to the *modulus* $\|\overrightarrow{\sigma}_t\|$, when those that we ideally seek describe the vectorial instantaneous volatility $\overrightarrow{\sigma}_t$, and are inherently richer. Consequently, the recovery is only partial, which we shall formalise with the following Corollary.

Corollary 5.3 (Partial recovery of the first layer in the Term Structure case) *As a consequence of* Theorem 5.1, *the dynamics of the SInsV model* (5.1), *(5.9) and (5.10) can be partially recovered from the sliding SImpV model as per:*

$$\overrightarrow{a}_{2,t}(t)\overrightarrow{\sigma}_t(t) = \widetilde{\Sigma}(\star)\,\overrightarrow{\mathrm{v}}(\star) \tag{5.43}$$

$$\overrightarrow{a}_{3,t}(t)\overrightarrow{\sigma}_t(t) = \widetilde{\Sigma}(\star)\,\overrightarrow{n}(\star) \tag{5.44}$$

$$\vec{\sigma}_t(t)^\perp \vec{a}_{1,t}(t) = -\frac{1}{2}\left[\|\vec{\vec{a}}_{2,t}\|^2 + \|\vec{\vec{a}}_{3,t}\|^2\right](t) + 2\,\widetilde{\Sigma}\,\widetilde{\Sigma}'_\theta(\star)$$
$$+\,3\widetilde{\Sigma}^2\,\widetilde{\Sigma}'^{\,2}_y(\star) + \widetilde{\Sigma}^3\left[\widetilde{\Sigma}''_{yy} - \widetilde{\Sigma}'_y\right](\star)$$
$$+\,2\widetilde{\Sigma}(\star)\left[I_0 + J_0 + K'_0\right] \tag{5.45}$$

$$\overrightarrow{\left[a_{22\,\mathbf{ij}k}\ \sigma_{\mathbf{i}}\ \sigma_{\mathbf{j}}\right]}(t) = -\vec{\vec{a}}_{2,t}(t)^\perp\left[\vec{\vec{a}}_{2,t} + \vec{\vec{a}}^{\,\perp}_{2,t}\right](t)\,\vec{\sigma}_t(t)$$
$$+2\left[3\,\widetilde{\Sigma}^2\widetilde{\Sigma}'_y\,\vec{\mathrm{v}} + \widetilde{\Sigma}^3\,\vec{\mathrm{v}}'_y\right](\star). \tag{5.46}$$

Proof We square the IATM Identity (5.2) and take its dynamics on both sides. The r.h.s. expands into

$$d\,\widetilde{\Sigma}^2(\star) = \left[2\widetilde{\Sigma}(\star)\widetilde{b}(\star) + \|\vec{\mathrm{v}}(\star)\|^2 + \|\vec{n}(\star)\|^2\right]dt$$
$$+\,2\widetilde{\Sigma}(\star)\,\vec{\mathrm{v}}(\star)^\perp\,d\vec{W}^{\widetilde{\Sigma}(0)}_t + 2\widetilde{\Sigma}(\star)\,\vec{n}(\star)^\perp\,d\vec{Z}_t.$$

We replace $\widetilde{b}(t,0,0)$, in the drift bracket above, by its expression (5.42), which is induced purely by the arbitrage constraints on the SImpV model:

$$d\,\widetilde{\Sigma}^2(\star) = \mu_t\,dt + 2\widetilde{\Sigma}(\star)\,\vec{\mathrm{v}}(\star)^\perp\,d\vec{W}^{\widetilde{\Sigma}(0)}_t + 2\widetilde{\Sigma}(\star)\,\vec{n}(\star)^\perp\,d\vec{Z}_t \tag{5.47}$$

with

$$\mu_t = 4\widetilde{\Sigma}\,\widetilde{\Sigma}'_\theta(\star) + 2\widetilde{\Sigma}^3\left[\widetilde{\Sigma}''_{yy} - \widetilde{\Sigma}'_y\right](\star) + 6\widetilde{\Sigma}^2\,\widetilde{\Sigma}'^{\,2}_y(\star) + 2\widetilde{\Sigma}(\star)\left[I_0 + J_0 + K_0\right].$$

Independently, on the l.h.s. we can compute the dynamics of $\|\vec{\sigma}_t(t)\|^2$ directly from the SInsV specification. According to (D.13) [p. 441] and (D.14) [p. 442], we have

$$d\,\|\vec{\sigma}_t(t)\|^2 = \left[2\vec{\sigma}_t(t)^\perp\vec{a}_{1,t}(t) + \|\vec{\vec{a}}_{2,t}(t)\|^2 + \|\vec{\vec{a}}_{3,t}(t)\|^2\right]dt$$
$$+\,2\,\vec{\sigma}_t(t)^\perp\,\vec{\vec{a}}_{2,t}(t)\,d\vec{W}^{\sigma(t)}_t + 2\,\vec{\sigma}_t(t)^\perp\,\vec{\vec{a}}_{3,t}(t)\,d\vec{Z}_t. \tag{5.48}$$

From the uniqueness of Itô's decomposition we can identify terms between (5.47) and (5.48). First, we recall that

$$d\vec{W}^{\sigma(t)}_t = d\vec{W}_t - \vec{\lambda}^{\sigma}_t(t)\,dt = d\vec{W}^{\widetilde{\Sigma}(0)}_t + \vec{\lambda}^{\widetilde{\Sigma}}_t(0)\,dt - \vec{\lambda}^{\sigma}_t(t)\,dt.$$

Then for the non-finite variation terms, we obtain (5.43) and (5.44) w.r.t. respectively the endogenous and the exogenous driver. As for the drift term, we have

$$2\vec{\sigma}_t^{\perp}\vec{a}_1 + \|\vec{\vec{a}}_2\|^2 + \|\vec{\vec{a}}_3\|^2 + 2\vec{\sigma}_t(t)^{\perp}\vec{\vec{a}}_{2,t}(t)\left[\vec{\vec{\lambda}}_t^{\widetilde{\Sigma}}(0) - \vec{\lambda}_t^{\sigma}(t)\right]$$
$$= 4\widetilde{\Sigma}\,\widetilde{\Sigma}'_\theta(\star) + 2\widetilde{\Sigma}^3\left[\widetilde{\Sigma}''_{yy} - \widetilde{\Sigma}'_y\right](\star) + 6\widetilde{\Sigma}^2\,\widetilde{\Sigma}'^{\,2}_y(\star) + 2\widetilde{\Sigma}(\star)\left[I_0 + J_0 + K_0\right], \tag{5.49}$$

which after simplification provides (5.45). We now compute the scalar product of (5.43) by $\vec{\sigma}_t(t)$ and again substitute (3.60)—which comes from the SImpV intrinsic constraints—to obtain

$$\vec{\sigma}_t(t)^{\perp}\vec{\vec{a}}_{2,t}(t)\vec{\sigma}_t(t) = \widetilde{\Sigma}(\star)\,\vec{\vec{\mathrm{v}}}(\star)^{\perp}\vec{\sigma}_t(t) = 2\,\widetilde{\Sigma}^3\,\widetilde{\Sigma}'_y(\star). \tag{5.50}$$

We derive the dynamics on both sides of (3.72), focusing on the endogenous term.

- On the r.h.s. we get

$$d\left[2\,\widetilde{\Sigma}^3\,\widetilde{\Sigma}'_y(\star)\right] = 2\left[3\,\widetilde{\Sigma}^2\widetilde{\Sigma}'_y\,\vec{\vec{\mathrm{v}}}(\star) + \widetilde{\Sigma}^3\,\vec{\vec{\mathrm{v}}}'_y(\star)\right]^{\perp} d\vec{W}_t^{\widetilde{\Sigma}(0)}$$
$$+\,[\cdots]\,dt + \overrightarrow{[\cdots]}^{\perp}\,d\vec{Z}_t.$$

- On the l.h.s., using (D.17) we obtain

$$d\left[\vec{\sigma}_t(t)^{\perp}\vec{\vec{a}}_{2,t}(t)\vec{\sigma}_t(t)\right]$$
$$= +\overrightarrow{[\cdots]}d\vec{Z}_t + \vec{\sigma}_t(t)^{\perp}\vec{\vec{a}}_{2,t}(t)^{\perp}\vec{\vec{a}}_{2,t}(t)\,d\vec{W}_t^{\sigma(t)}$$
$$+\,\vec{\sigma}_t(t)^{\perp}\left[\vec{\vec{\vec{a}}}_{22,t}(t)\,d\vec{W}_t^{\sigma(t)}\,\vec{\sigma}_t(t) + \vec{\vec{a}}_{2,t}(t)\,\vec{\vec{a}}_{2,t}(t)\,d\vec{W}_t^{\sigma(t)}\right] + [\cdots]\,dt,$$

so that, identifying both sides and using modified Einstein notation, we get

$$\vec{\sigma}_t(t)^{\perp}\left[\vec{\vec{a}}_{2,t}(t)^{\perp} + \vec{\vec{a}}_{2,t}(t)\right]\vec{\vec{a}}_{2,t}(t) + \overrightarrow{\left[a_{22\,\mathbf{ij}k}\,\sigma_{\mathbf{i}}\,\sigma_{\mathbf{j}}\right]_t}^{\perp}(t)$$
$$= 2\left[3\,\widetilde{\Sigma}^2\,\widetilde{\Sigma}'_y\,\vec{\vec{\mathrm{v}}}(\star) + \widetilde{\Sigma}^3\,\vec{\vec{\mathrm{v}}}'_y(\star)\right]^{\perp},$$

which after simplification and transposition gives (5.46), concluding the proof. □

Obviously, the same technique as demonstrated in Remark 3.3 [p. 177] might be considered, in order to recover better directional information on $\vec{\sigma}_t(t)$ or $\vec{\vec{a}}_{2,t}$ for instance. However, the formal study of such a more complete and/or higher-order recovery is left for further research.

5.3.2 Interpretation and Comments

First let us contrast the arbitrage constraints imposed on the SImpV model in the Term Structure framework, as per Proposition 5.2 [p. 297], to those established in the simpler multidimensional case with Proposition 3.6 [p. 176].

We note immediately that the scalar products pertaining to the endogenous coefficient $\vec{\mathrm{v}}$ exhibit almost similar expressions: compare (5.35) and (5.36) respectively with (3.60) and (3.61). Naturally, the instantaneous volatility invoked in those expressions now belongs to the *Immediate* underlying, but this represents a relatively minor evolution, compared to the changes in the drift (more on this point shortly).

Following the argument of Remark 3.3 [p. 177] (which applied to a single asset) it becomes clear that, in a recovery logic, these two equations provide additional information on that Immediate volatility $\vec{\sigma}_t(t)$. Indeed, let us assume sufficient individual modulus and relative angle for the endogenous coefficient $\vec{\mathrm{v}}\,(\star)$ and its strike-differential $\vec{\mathrm{v}}\,'_y(\star)$, both taken IATM. Then the input of the two products does improve our knowledge of the direction of $\vec{\sigma}_t(t)$, and therefore mitigates the structural caveat of the new IATM identity (5.34). Whether this is sufficient to complete the picture is (again) dependent on modelling, in particular on the dimension of the map $\vec{\sigma}_t(t+\theta)$. Again, we refer to Remark 3.3 [p. 177] for a more detailed example of when the full immediate information can be recovered, which entails inductive y-differentiations of the ZDC.

Turning to the IATM drift expression (5.37) we observe again that its structure starts identically to (3.62), but that three additional terms have appeared: namely I_0, J_0 and K_0. This distinction comes as a consequence of the Term-Structure framework, and is carried over to the dynamic recovery formula: as (5.39) replaces (3.63), the first two TS terms are also present and the third is replaced by K_0'. It makes sense, therefore, to study all four terms together; it is initially apparent that terms I_0 and J_0 are both generated by the maturity-dependency of the underlying family $X_t(T)$, whereas term K_0' is induced by the very definition of $\vec{\sigma}_t(T)$ and its structural association to the numeraire family.

The first term I_0 is simply the slope compensation term $I(t, y, \theta)$ defined in (5.22) [p. 286], taken in the immediate ATM point. The nature and effects of this term have already been discussed within Sect. 5.2.1, which dealt with the ZDC. As mentioned in Remark 5.4, it can been seen as reporting the combination of a non-flat underlying curve and a skewed smile. Also, it stems entirely from the adoption of sliding coordinates for the SImpV model.

The second term J_0 is similar in nature to I_0, inasmuch as it is linked to the maturity-slope of the *instantaneous volatility* map $\vec{\sigma}\,'_T\,(t, t+\theta)$ taken at the origin (Immediate) point $\theta = 0$. Let us illustrate this point: the slope in question can be proxied by the difference, observed at current time t, between the instantaneous volatility of the *spot* 3Month-Libor rate, and the instantaneous volatility of a *forward* 3Month-Libor rate starting shortly, typically one day or one week after t.

Of course this variation is multi-dimensional and enables the points of the underlying curve $T \to X_t(T)$ to respond differently to various driver components, or "shocks". Think, for instance, of a yield curve within an HJM framework, where only two factors would define the curve movements: one for the short term and the other for the long term. In this simple framework, as T increases the respective weights allocated to each component of $\vec{\sigma}_t(T)$ would slowly shift from one coordinate to the other. This individual shift would be precisely the component-wise maturity-slope symbolised by $\vec{\sigma}'_T(t, t+\theta)$.

Looking back at the whole of J_0, we can therefore interpret this term as the part in the variation of the instantaneous volatility modulus (for the immediate underlying) which is due purely to the underlying riding down the maturity curve.

Also, we observe that this term is normalised, and thus exhibits the same physical "unit" as the modulus itself. This is better seen by rewriting term J_0 as

$$J_0 = -\frac{\partial_\theta \|\vec{\sigma}_t(t+\theta)\|}{\|\vec{\sigma}(t, t+\theta)\|}\,|_{\theta=0} = -\partial_\theta\, ln\left(\|\vec{\sigma}_t(t+\theta)\|\right)|_{\theta=0}.$$

Finally, let us comment simultaneously on terms K_0 and K'_0. We observe that these are the only terms in both expressions (5.37) and (5.39) invoking the numeraire volatility $\vec{\lambda}^N$ or risk premia $\vec{\lambda}^{\widetilde{\Sigma}}$ and $\vec{\lambda}^{\sigma}$. Let us then recall that these parametric processes are simply adjusters defining the measures chosen at the modelling stage, respectively for the dynamics of the underlying, implied volatility and instantaneous volatility families. Having opted for distinct measures, it comes as no surprise that some compensation term should appear.

Looking at term K_0, we realise that its origin can be traced back to Proposition 5.1 and its proof: refer in particular to Eq. (5.28) [p. 290]. We expressed the dynamics of the deflated option price[7] by employing Itô on the Black functional. This required to write simultaneously the dynamics of the underlying and of its implied volatility, which created the corrective term.

As for term K'_0 the same principle applies, but what is more surprising is the fact that $\vec{\lambda}^{\widetilde{\Sigma}}$ does not appear eventually. After all, we are expressing the dynamics of the instantaneous volatility from those of the implied volatility. Therefore we could expect any drift adjustment to invoke the difference $\vec{\lambda}^{\sigma} - \vec{\lambda}^{\widetilde{\Sigma}}$. It is not so, and it suggests that as far as the inverse problem and recovery is concerned, the choice of measure for the dynamics of $\widetilde{\Sigma}$ does not matter.

This last remark provides the ideal transition to commenting on the Recovery Theorem 5.1 proper, comparing expression (5.39) to its single underlying counterparts: (2.42) [p. 55] from Theorem 2.1 in the scalar case, and (3.63) [p. 178] from Theorem 3.3 in the multi-dimensional setup. We observe that, without surprise, the multi-dimensionality appears again to lessen the extent of the dynamic result.Indeed,

---

7 $\frac{V_t(X_t(T),T,K)}{N_t(T)} = C^{BS}\left(X_t(T), K, \Sigma_t(X_t(T), T, K)\cdot\sqrt{T-t}\right)$ (see p. 289).

as in (3.63) and contrary to (2.42), the dynamics of the modulus still rely on the endogenous coefficient $\vec{\nu}$. Let us therefore focus on the more specific influence of the new TS terms.

Despite their similarities, we note, however, a major difference between terms I_0 and J_0 when it comes to calibration and therefore solving the inverse problem. Indeed, in a market model such as HJM or LMM the initial yield curve and therefore the slope compensation I_0 represent model *inputs*, simply because they stem from the map $T \to X_t(T)$ *observed* at time t. Whereas term J_0 is determined by the term structure of instantaneous volatility $T \to \vec{\sigma}_t(T)$, which is precisely the target of the calibration procedure.

In terms of calibration again, it becomes necessary to decide up front whether the numeraire volatility $\vec{\lambda}^N$ and the premium $\vec{\lambda}^\sigma$ also belong to the target quantities. If they do, then it seems to add to the complexity of the dynamics recovery (5.39). In fact, barring some drastic simplification in the modelling, this choice would render the inverse problem totally ill-defined. However, we will see later that in some cases (caplets in an LMM framework for instance) the numeraire volatility $\vec{\lambda}^N$ can be expressed as a function of the underlyings $X_t(T)$ and of their instantaneous volatility $\vec{\sigma}_t(T)$. This feature then links term K_0 to the product $\vec{\nu}(\star)^\perp \vec{\sigma}_t(t)$ which is given through (5.35), and therefore does not add to the number of unknown variables of the inverse problem.

To summarise, in full generality the reconstruction of $d \parallel \vec{\sigma}^L(t) \parallel$ from purely the shape and dynamics of the sliding implied volatility $\widetilde{\Sigma}$ is structurally incomplete.

Even if the drift issue presented by term K_0' was removed, the dimensionality issue, along with term J_0, prevent the inverse problem from being well-defined.

Remark 5.5 The incompleteness of the recovery is true is the general case, but within a well-chosen parametric framework it is possible to solve the inverse problem. To illustrate this point, we show how one can reduce the above-mentioned "incompleteness" by selecting dedicated but realistic parametric families for $\vec{\sigma}_t(T)$.

Let us consider an HJM framework with two factors ($n_w = 2$). We have already noticed (see Remark 3.3 [p. 177]) that in this case, assuming the matrix

$$\left[\vec{\nu}(\star) \mid \vec{\nu}'_y(\star) \right]$$

is non-degenerate, then the static inverse problem is well-posed. For illustration purposes only, we now briefly present a "natural" stochastic implied volatility model (our input in the inverse problem) which satisfies this full-rank condition, before turning to the dynamic recovery.

We consider a *smile* market model driven by only three (independent) Wiener processes and defined by three dedicated scalar parametric processes: $a_t(\theta)$ for the smile (ATM) level, $b_t(\theta)$ for its skew and $c_t(\theta)$ for the curvature. Specifying these three defining processes is irrelevant for our purposes, except that we impose $a_t > 0$ and $b_t < 0$ *a.s.* Then the sliding smile is defined via

$$\widetilde{\Sigma}(t, y, \theta) = \widetilde{\Sigma}(0, y, \theta) + \int_0^t a_s(\theta)\, dB_s^1 + y \int_0^t b_s(\theta)\, dB_s^2 + \frac{1}{2} y^2 \int_0^t c_s(\theta)\, dB_s^3.$$

This specification entails that the only permitted smile dynamics consist of "heave", "roll" and "flapping".[8] Note that in practice this model can be calibrated[9] to the *dynamics* of an observed market smile, by analysis of four time series: the price processes of the underlying, and an ATM, OTM and ITM option.[10]

In order to cast this model within our framework, we now make an important choice: we allocate the "level" driver B_t^1 to the first component of our endogenous driver, W_t^1, while we distribute (using a correlation $\rho \neq 0$) the "skew" and "curvature" drivers B_t^2 and B_t^3 between the two remaining components, W_t^2 and Z_t. In financial terms, this is still a fairly realistic and flexible model. Indeed, the allocation does not, for instance, constrain the "backbone" (the ATM level as a function of the underlying value) be deterministic, thanks to the fact that a_t is still unspecified and could very well be driven by more factors. The allocation just states that both the underlying value and the ATM level respond to the same market shocks.

In this model, the endogenous coefficient and its y-differential are respectively

$$\vec{\nu}\,(t, y, \theta) = \begin{bmatrix} a_t(\theta) \\ y\,\rho\, b_t(\theta) + \frac{1}{2} y^2 \sqrt{1-\rho^2}\, c_t(\theta) \end{bmatrix}$$

$$\vec{\nu}\,'_y(t, y, \theta) = \begin{bmatrix} 0 \\ \rho\, b_t(\theta) + y \sqrt{1-\rho^2}\, c_t \end{bmatrix}.$$

It comes from our assumptions that $\vec{\nu}\,(t, 0, 0)$ and $\vec{\nu}\,'_y(t, 0, 0)$ are non-null and orthogonal, which satisfies the non-degeneracy condition, and therefore allows full recovery of the immediate ATM volatility $\vec{\sigma}_t(T)$.

Let us now turn to the dynamic inverse problem. First we choose to model the instantaneous volatility dynamics using the same driver as the underlying collection $T \to B_t(T)$. This is practically the norm with current published models, and allows us to dispense with term K_0'. As we mentioned above, term I_0 is an input, so that leaves us only with term J_0. We adopt the following parametric form for the instantaneous volatility vector:

---

[8] These are dynamic features, not to be confused with maturity-wise shapes such as "twist" or "flattening". The heave is a uniform vertical movement of the smile, analogous to a parallel movement of the yield curve in an HJM model. The roll is a uniform motion around the ATM axis, and flapping refers to the dynamics of the curvature.

[9] Assuming enough liquidity.

[10] This inference obviously has to be conducted maturity-by-maturity. The respective moneyness of the options should be roughly 90–95, 100 and 105–110 % to expect a decent precision from the finite difference approximations.

$$\vec{\sigma}_t(T) = \begin{bmatrix} \sigma_1 & e^{-\mu_1(T-t)^2} & f_1(t) \\ \sigma_2 & \left(1 - e^{-\mu_2(T-t)}\right) & f_2(t) \end{bmatrix}.$$

Again, this is a very classical setup, and corresponds to the short-term/long-term example mentioned above to illustrate the significance of $\vec{\sigma}'_T(t, t+\theta)$. Within this framework, σ_1 and σ_2 are respectively the short and long term volatility levels. Parameters μ_1 and μ_2 control the weight shift between the two coordinates and allow us to generate various maturity profiles for the Black volatility. As for the latter, it is possible in particular to opt for a "humped" function which seems to be both a market invariant and (consequently) a popular parametrisation. Also, the rationale for a parabolic decay on the short term component (as opposed to linear for the long term) is that shocks on the short term usually stem from sudden moves in central banks base rates, and die very quickly along the yield curve,[11] whereas moves on the long end are usually instigated by economic forecast, the influence of which tends to be more progressive and gradual in maturity. As for the functions $f_1(\cdot)$ and $f_2(\cdot)$, they establish that we are using separate variables (time and time-to-maturity) which in turn ensures a Markovian property.

With the model defined and justified, the maturity-slope of the volatility is

$$\vec{\sigma}'_T(t, \theta) = \begin{bmatrix} \sigma_1 & -2\,\mu_1\,\theta\, e^{-\mu_1\theta^2} & f_1(t) \\ \sigma_2 & \mu_2 e^{-\mu_2\theta} & f_2(t) \end{bmatrix}.$$

Observe that in $\theta = 0$ the two vectors are orthogonal, and therefore conclude that $J_0 = 0$. In consequence the dynamic recovery specified by (5.39) is now complete, at least at the origin point $(t, 0, 0)$.

5.4 Generating the SIV Surface: The First Layer

In this section we now deal with the *direct* problem, which is to infer (in practice, to approximate) the shape and dynamics of the sliding implied volatility $\widetilde{\Sigma}(t, y, \theta)$, given the input of the (absolute) instantaneous volatility $\vec{\sigma}_t(T)$ defining the dynamics of the maturity-dependent underlying $X_t(T)$.

As can be expected from the discussions on the inverse problem, we will see that the input of the full map (the "volatility profile") $T \to \vec{\sigma}_t(T)$ is not needed as such. Indeed, it will have to be approximated from its local shape (value and differentials w.r.t. the maturity T) taken at the origin ($T = t$, i.e. $\theta = 0$). Which is to say that in practice we will rely on a functional approximation, typically polynomial in a well-chosen variable.

[11] The ECB for instance uses standing facilities (marginal lending and deposit) which refer to the overnight rate, as well as open market operations (main refinancing, longer-term refinancing, fine-tuning and structural) which are mainly associated to maturities ranging from 1 week to 6 months.

5.4.1 Computing the Differentials

The previous identity result (5.34) provides the value of $\widetilde{\Sigma}$ in $(t, 0, 0)$, in other words the "level" of the functional approximation. Let us now focus on the local (immediate ATM) shape and dynamics. Beforehand, we introduce some normalised quantities and new notations:

Definition 5.1 We normalise the instantaneous volatility and its maturity-differentials, leading in particular to the following representations:

$$\vec{u}_t(T) \triangleq \frac{\vec{\sigma}_t(T)}{\|\vec{\sigma}_t(T)\|} \qquad \text{and} \qquad \vec{u}\,'_T(t,T) \triangleq \frac{\vec{\sigma}\,'_T(t,T)}{\|\vec{\sigma}_t(T)\|}.$$

Similarly, we normalise the relevant tensorial coefficients invoked in the dynamics of $\vec{\sigma}_t(T)$, which is consistent since these are written using a *Normal* convention:

$$\vec{c}_{1,t}(T) \triangleq \frac{\vec{a}_{1,t}(T)}{\|\vec{\sigma}_t(T)\|} \qquad \overset{\Rightarrow}{c}_{2,t}(T) \triangleq \frac{\overset{\Rightarrow}{a}_{2,t}(T)}{\|\vec{\sigma}_t(T)\|}$$

$$\overset{\Rightarrow}{c}_{3,t}(T) \triangleq \frac{\overset{\Rightarrow}{a}_{3,t}(T)}{\|\vec{\sigma}_t(T)\|} \qquad \overset{\Rrightarrow}{c}_{22,t}(T) \triangleq \frac{\overset{\Rrightarrow}{a}_{22,t}(T)}{\|\vec{\sigma}_t(T)\|}$$

Since the asymptotic expansions will call only on the initial value of these maps, i.e. only $T = t$, we introduce dedicated simplifying notations: omitting the arguments will imply that the tensor is evaluated at the immediate point $(t, T = t)$:

$$\begin{array}{llll} \vec{\sigma}_t(t) \leftrightarrow \vec{\sigma} & \vec{u}_t(t) \leftrightarrow \vec{u} & \|\vec{\sigma}_t(t)\| \leftrightarrow \sigma & \vec{\sigma}\,'_T(t,t) \leftrightarrow \vec{\sigma}\,'_T \\ \vec{a}_{1,t}(t) \leftrightarrow \vec{a}_1 & \overset{\Rightarrow}{a}_{2,t}(t) \leftrightarrow \overset{\Rightarrow}{a}_2 & \overset{\Rightarrow}{a}_{3,t}(t) \leftrightarrow \overset{\Rightarrow}{a}_3 & \overset{\Rrightarrow}{a}_{22,t}(t) \leftrightarrow \overset{\Rrightarrow}{a}_{22} \end{array}$$

As in physics, there are several reasons for such a normalisation. The first obvious objective is to *simplify* the expressions: as we have observed so far, because of the Lognormal nature of our baseline model (Black's model for the underlyings), most quantities indeed come as fractions with a power of $\|\vec{\sigma}_t(t)\|$ in the denominator.

The second reason is to exhibit *units* and check homogeneity, which is always a healthy concern, and often a powerful tool when the time comes for interpretation and generalisation. We note that this normalisation is not fully equivalent to running the method with the instantaneous volatility dynamics written in a Lognormal fashion. However, this convention emphasises the *directional information* of the instantaneous volatility, over its norm, and we happen to think that this feature is actually a pivotal point of the current term-structure framework.

Still on the matter of notations, and due again to the dimensionality of our setup, it will appear that one of our strongest practical constraints is simply to manage the heavy linear algebra involved. Indeed, the coming Theorem 5.2 invokes, for instance, $\overset{\Rightarrow}{a}_{22}$, which is a tensor of order 3, but also some fairly involved products.[12] The conventions used so far have proved efficient for clarity purposes, but seem less adapted to the sequel. Indeed, one of their shortfalls is to unnecessarily complicate the elementary products present in most coming expressions. As a consequence, some simplifications do not appear naturally and therefore, in a vicious circle, each Wiener chaos expression tend to conserve many redundant terms. Finally, although vector or matrix representations are intuitive, they tend to mask the true nature of the tensors, for which only indices matter.

Bearing in mind that all these observations will ring even more true for higher-order expansions, we believe that an effective and economical solution is to use "Modified Einstein" notation. These are inspired by conventions used intensively in mechanics, with a few added features, and should appear familiar and intuitive to most readers. Apart from simplicity and compactness, a significant advantage of these notations is that they can be programmed easily, which fits the announced need for a programmable induction method. In a nutshell, the summed indices are denoted in bold, while the remaining active coordinates are not, and are ordered alphabetically. Furthermore, the overall order of the tensor is recalled for convenience with the previous "arrow" conventions. For more detailed explanations and specific identities, we refer to Appendix D .

Theorem 5.2 (Generating the first layer in the TS framework) *Let us consider the SInsV model framework, defined by dynamics* (5.1)–(5.3) *for the underlying and* (5.9), (5.10) *for the volatility. Then the first layer of IATM differentials for the sliding IV* $\widetilde{\Sigma}(t, y, \theta)$ *(the* $\widetilde{\Sigma}$*-(2,0) group) can be expressed as follows. The notations and conventions are those of Definition* 5.1 *and Appendix D.*

With $(\star) \triangleq (t, 0, 0)$, *the IATM differentials of the shape process are*

$$\widetilde{\Sigma}'_y(\star) = \frac{1}{2}\left[u_{\mathbf{i}}u_{\mathbf{j}}\; c_{2\,\mathbf{ij}}\right] \tag{5.51}$$

$$\begin{aligned}\widetilde{\Sigma}''_{yy}(\star) = \sigma^{-1}\Bigg[&\frac{1}{3}\left[u_{\mathbf{i}}u_{\mathbf{k}}\left(c_{2\,\mathbf{ij}}\; c_{2\,\mathbf{jk}} + c_{2\,\mathbf{ji}}\; c_{2\,\mathbf{jk}} + c_{2\,\mathbf{ij}}\; c_{2\,\mathbf{kj}}\right)\right] - \frac{3}{2}\left[u_{\mathbf{i}}u_{\mathbf{j}}\; c_{2\,\mathbf{ij}}\right]^2\\ &+\frac{1}{3}\left[u_{\mathbf{i}}u_{\mathbf{k}}\; c_{3\,\mathbf{ij}}\; c_{3\,\mathbf{kj}}\right] + \frac{1}{3}\left[u_{\mathbf{i}}u_{\mathbf{j}}u_{\mathbf{k}}\; c_{22\,\mathbf{ijk}}\right]\Bigg]\end{aligned} \tag{5.52}$$

$$\begin{aligned}\widetilde{\Sigma}'_\theta(\star) = \frac{1}{2}\sigma\Bigg[&\frac{1}{2}\left[c_{2\,\mathbf{ij}}^{\,2}\right] - \frac{1}{3}\left[u_{\mathbf{i}}u_{\mathbf{k}}\left(c_{2\,\mathbf{ij}}\; c_{2\,\mathbf{jk}} + c_{2\,\mathbf{ji}}\; c_{2\,\mathbf{jk}} + c_{2\,\mathbf{ij}}\; c_{2\,\mathbf{kj}}\right)\right]\\ &+\frac{1}{2}\left[c_{3\,\mathbf{ij}}^{\,2}\right] + \frac{3}{4}\left[u_{\mathbf{i}}u_{\mathbf{j}}\; c_{2\,\mathbf{ij}}\right]^2 - \frac{1}{3}\left[u_{\mathbf{i}}u_{\mathbf{k}}\; c_{3\,\mathbf{ij}}\; c_{3\,\mathbf{kj}}\right] - \frac{1}{3}\left[u_{\mathbf{i}}u_{\mathbf{j}}u_{\mathbf{k}}\; c_{22\,\mathbf{ijk}}\right]\Bigg]\end{aligned}$$

---

[12] Such as $\|\vec{\sigma}_t(t)\|^{-3}\vec{\sigma}_t(t)^{\perp}\left[\overset{\Rightarrow}{a}_{2,t}(t) + \overset{\Rightarrow}{a}_{2,t}(t)^{\perp}\right]\overset{\Rightarrow}{a}_{2,t}(t)\vec{\sigma}_t(t)$, for instance.

$$+\frac{1}{2}\sigma\,\vec{u}^{\perp}\left[\vec{u_\theta}'+\vec{c}_1\right]-\frac{1}{4}\left[u_{\mathbf{i}}u_{\mathbf{j}}\,c_{2\,\mathbf{ij}}\right]\left[\frac{\widetilde{X}'_\theta}{\widetilde{X}}(t,0)-\sigma^2\right]$$

$$+\frac{1}{2}\sigma\overrightarrow{\left[c_{2\,\mathbf{i}j}\,u_{\mathbf{i}}\right]}^{\perp}\left[\vec{\lambda}^{N}_{t}(t)-\vec{\lambda}^{\sigma}_{t}(t)\right]. \tag{5.53}$$

The local differentials of the dynamics processes are:

$$\vec{\vec{\mathrm{v}}}(\star)=\vec{\vec{a}}_2^{\perp}\,\vec{u} \tag{5.54}$$

$$\vec{\vec{\mathrm{v}}}'_y(\star)=\frac{1}{2}\overrightarrow{\left[u_{\mathbf{i}}\left(c_{2\,\mathbf{ij}}+c_{2\,\mathbf{ji}}\right)c_{2\,\mathbf{j}k}\right]} \tag{5.55}$$

$$+\frac{1}{2}\overrightarrow{\left[u_{\mathbf{i}}u_{\mathbf{j}}\,c_{22\,\mathbf{ij}k}\right]}-\frac{3}{2}\overrightarrow{\left[u_{\mathbf{i}}u_{\mathbf{j}}u_{\mathbf{k}}\,c_{2\,\mathbf{ij}}\,c_{2\,\mathbf{k}l}\right]}$$

$$\vec{\vec{n}}(\star)=\vec{\vec{a}}_3^{\perp}\,\vec{u}\,. \tag{5.56}$$

Note again that we have $\widetilde{X}^{-1}\widetilde{X}'_\theta(t,0)=X^{-1}X'_T(t,t)$.

Proof

Step 1/3: Computing $\vec{\vec{\mathrm{v}}}(t,0,0)$, $\vec{\vec{n}}(t,0,0)$ **and** $\widetilde{\Sigma}'_y(t,0,0)$

We start with the static IATM identity (5.34) which we square. The resulting equation being valid *a.s.* and for all t, we have the following dynamics

$$d\,\|\,\vec{\sigma}_t(t)\,\|^2=d\widetilde{\Sigma}^2(t,0,0)$$

which we re-write, according to notation (D.2) and result (D.14), as:

$$\underbrace{2\vec{\sigma}_t(t)^{\perp}d\vec{\sigma}_t(t)+\langle\, d\vec{\sigma}_t(t)\,\rangle}_{\mathscr{L}}=\underbrace{2\widetilde{\Sigma}(\star)\,d\widetilde{\Sigma}(\star)+\langle\, d\widetilde{\Sigma}(\star)\,\rangle}_{\mathscr{R}}. \tag{5.57}$$

Let us expand each side separately.

- For the right-hand term, we get

$$\begin{aligned}\mathscr{R}&=2\widetilde{\Sigma}(\star)\left[\widetilde{b}(\star)dt+\vec{\vec{\mathrm{v}}}(\star)^{\perp}\,d\vec{W}_t^{\widetilde{\Sigma}(0)}+\vec{\vec{n}}(\star)^{\perp}\,d\vec{Z}_t\right]\\&\quad+\left[\|\,\vec{\vec{\mathrm{v}}}\,\|^2+\|\,\vec{\vec{n}}\,\|^2\right](\star)\,dt\\&=\left[2\widetilde{\Sigma}\widetilde{b}(\star)+\|\,\vec{\vec{\mathrm{v}}}(\star)\|^2+\|\,\vec{\vec{n}}(\star)\|^2-2\widetilde{\Sigma}\,\vec{\vec{\mathrm{v}}}(\star)^{\perp}\,\vec{\vec{\lambda}}_t^{\widetilde{\Sigma}}(0)\right]dt\\&\quad+2\widetilde{\Sigma}\,\vec{\vec{\mathrm{v}}}(\star)^{\perp}\,d\vec{W}_t+2\widetilde{\Sigma}\,\vec{\vec{n}}(\star)^{\perp}\,d\vec{Z}_t.\end{aligned} \tag{5.58}$$

- As for the left-hand term, we invoke the input dynamics of $\vec{\sigma}_t(T)$ as specified by (5.9), so that $\mathscr{L}$ becomes

$$\begin{aligned}\mathscr{L} &= 2\vec{\sigma}_t(t)^\perp\Big[\vec{a}_{1,t}(t)\,dt + \vec{\vec{a}}_{2,t}(t)\,d\vec{W}_t^{\sigma(t)} + \vec{\vec{a}}_{3,t}(t)\,d\vec{Z}_t\Big] \\ &\quad + \Big[\|\vec{\vec{a}}_{2,t}\|^2 + \|\vec{\vec{a}}_{3,t}\|^2\Big]\,dt \\ &= \Big[2\vec{\sigma}_t(t)^\perp\vec{a}_{1,t}(t) + \|\vec{\vec{a}}_{2,t}\|^2 + \|\vec{\vec{a}}_{3,t}\|^2 - 2\vec{\sigma}_t(t)^\perp\vec{\vec{a}}_{2,t}(t)\,\vec{\lambda}_t^{\sigma}(t)\Big]\,dt \\ &\quad + 2\vec{\sigma}_t(t)^\perp\vec{\vec{a}}_{2,t}(t)\,d\vec{W}_t + 2\vec{\sigma}_t(t)^\perp\vec{\vec{a}}_{3,t}(t)\,d\vec{Z}_t. \end{aligned} \tag{5.59}$$

Note that the risk-neutral driver $\vec{W}$ is used intentionally. Although this will not affect the computation of the first quantities (they only identify the driving coefficients, and are therefore insensitive to the choice of driver), later on the computation of the slope $\widetilde{\Sigma}'_\theta(t,0,0)$ will require us to match the drifts of different SDEs. In short, $\vec{W}$ will serve again as a common driver, in order to harmonise these dynamics.

We can now identify the diffusive terms in both sides of (5.57), which gives:

- with respect to $\vec{W}_t$:

$$\|\vec{\sigma}_t(t)\|\,\vec{\text{v}}\,(t,0,0)^\perp = \vec{\sigma}_t(t)^\perp\vec{\vec{a}}_{2,t}(t) \quad \text{thus} \quad \vec{\text{v}}\,(t,0,0) = \frac{\vec{\vec{a}}_{2,t}(t)^\perp\vec{\sigma}_t(t)}{\|\vec{\sigma}_t(t)\|},$$

  which after re-notation proves (5.54).
- with respect to $\vec{Z}_t$:

$$\|\vec{\sigma}_t(t)\|\,\vec{n}\,(t,0,0)^\perp = \vec{\sigma}_t(t)^\perp\vec{\vec{a}}_{3,t}(t) \quad \text{hence} \quad \vec{n}\,(t,0,0) = \frac{\vec{\vec{a}}_{3,t}(t)^\perp\vec{\sigma}_t(t)}{\|\vec{\sigma}_t(t)\|},$$

  which after re-notation proves (5.56).

We have now expressed both the endogenous and exogenous coefficients. In order to produce the IATM skew $\widetilde{\Sigma}'_y(\star)$, we exploit the SImpV model's arbitrage constraints. We borrow the static expression (5.35) [p. 297] for the product $\vec{\text{v}}\,(\star)^\perp\vec{\sigma}_t(t)$, and substitute the expression for $\vec{\text{v}}\,(\star)$ (5.54) obtained above to get:

$$\Big[\|\vec{\sigma}_t(t)\|^{-1}\vec{\vec{a}}_{2,t}(t)^\perp\vec{\sigma}_t(t)\Big]^\perp\vec{\sigma}_t(t) = 2\,\widetilde{\Sigma}^2(\star)\widetilde{\Sigma}'_y(\star)$$

from which we isolate the Immediate ATM sliding skew as

$$\widetilde{\Sigma}'_y(t,0,0) = \frac{1}{2}\,\|\vec{\sigma}_t(t)\|^{-3}\,\vec{\sigma}_t(t)^\perp\,\vec{\vec{a}}_{2,t}(t)\,\vec{\sigma}_t(t).$$

Simplifying and converting to "Modified Einstein" notation proves (5.51).

Step 2/3: Computation of the curvature $\widetilde{\Sigma}''_{yy}(\star)$

The strategy is to express the quantity $\vec{\vec{v}}'_y(t,0,0)$ as a function of the inputs, and then to invoke again the arbitrage constraints imposed on the SImpV model. More specifically, we shall use expression (5.36) which provides the scalar product $\vec{\sigma}_t(t)^\perp \vec{\vec{v}}'_y(t,0,0)$, in order to isolate the curvature.
First we differentiate once w.r.t. y the input SDE (5.13) driving the Sliding IV $\widetilde{\Sigma}$:

$$d\widetilde{\Sigma}'_y(t,y,\theta) = \widetilde{b}'_y(t,y,\theta)dt + \vec{\vec{v}}'_y(t,y,\theta)^\perp d\vec{W}_t^{\widetilde{\Sigma}(\theta)} + \vec{n}'_y(t,y,\theta)^\perp d\vec{Z}_t.$$

On one hand, if we take this expression in the immediate ATM point $(t,0,0)$, the endogenous coefficient will be $\vec{\vec{v}}'_y(t,0,0)$. On the other hand, if we now compute the same dynamics of $\widetilde{\Sigma}'_y(t,0,0)$ using the expression (5.51) just obtained for $\widetilde{\Sigma}'_y(\star)$, we can then identify the endogenous coefficient in both expressions, and therefore obtain $\vec{\vec{v}}'_y(t,0,0)$ as a function of our inputs.
So let us embark on computing the following bracket:

$$\left\langle d\left[\frac{1}{2}\,\|\vec{\sigma}_t(t)\|^{-3}\,\vec{\sigma}_t(t)^\perp\,\vec{\vec{a}}_{2,t}(t)\,\vec{\sigma}_t(t)\right],\; d\vec{W}_t\right\rangle.$$

Using Itô and simplified notations, we have the dynamics of the skew as

$$d\left[\frac{\vec{\sigma}^\perp\vec{\vec{a}}_2\vec{\sigma}}{2\sigma^3}\right] = \frac{1}{2}\sigma^{-3}\underbrace{d\left[\vec{\sigma}^\perp\vec{\vec{a}}_2\vec{\sigma}\right]} - \frac{1}{2}\left[\vec{\sigma}^\perp\vec{\vec{a}}_2\vec{\sigma}\right]\sigma^{-6}\underbrace{d\sigma^3} + [\cdots]\,dt. \tag{5.60}$$

Let us compute the two elementary dynamics under-braced above.

- Applying the generic result (D.15) with $p=3$, we obtain at once that

$$d\sigma^3 = 3\sigma\,\vec{\sigma}^\perp\vec{\vec{a}}_2\,d\vec{W}_t^{\sigma(t)} + [\cdots]\,dt + [\cdots]\,d\vec{Z}_t.$$

- While applying (D.17) provides at once

$$\begin{aligned}
d&\left[\vec{\sigma}^\perp\vec{\vec{a}}_2\vec{\sigma}\right] + [\cdots]\,dt + [\cdots]\,d\vec{Z}_t \\
&= \vec{\sigma}^\perp\vec{\vec{a}}_2^{\perp}\vec{\vec{a}}_2\,d\vec{W}_t^{\sigma(t)} + \vec{\sigma}^\perp\left[\vec{\vec{a}}_{22}\,d\vec{W}_t^{\sigma(t)}\right]\vec{\sigma} + \vec{\sigma}^\perp\vec{\vec{a}}_2\vec{\vec{a}}_2\,d\vec{W}_t^{\sigma(t)} \\
&= \vec{\sigma}^\perp\left[\vec{\vec{a}}_2 + \vec{\vec{a}}_2^{\perp}\right]\vec{\vec{a}}_2\,d\vec{W}_t^{\sigma(t)} + \vec{\sigma}^\perp\left[\vec{\vec{a}}_{22}\,d\vec{W}_t^{\sigma(t)}\right]\vec{\sigma}.
\end{aligned}$$

We can now re-inject both elementary dynamics into (5.60) and get:

$$d\left[\frac{\vec{\sigma}^{\perp}\vec{\vec{a}}_2\vec{\sigma}}{2\sigma^3}\right] = [\cdots]\,dt + [\cdots]\,d\vec{Z}_t + \frac{1}{2}\sigma^{-3}\vec{\sigma}^{\perp}\left(\vec{\vec{a}}_2 + \vec{\vec{a}}_2^{\perp}\right)\vec{\vec{a}}_2\,d\vec{W}_t^{\sigma(t)}$$
$$+\frac{1}{2}\sigma^{-3}\underbrace{\vec{\sigma}^{\perp}\left[\vec{\vec{a}}_{22}\,d\vec{W}_t^{\sigma(t)}\right]\vec{\sigma}}$$
$$-\frac{3}{2}\sigma^{-5}\left[\vec{\sigma}^{\perp}\vec{\vec{a}}_2\vec{\sigma}\right]\vec{\sigma}^{\perp}\vec{\vec{a}}_2\,d\vec{W}_t^{\sigma(t)}.$$

This expression looks fine, except for the under-braced term, which we can identify as a quadratic form but whose current form does not suit our purposes. We can, however, re-express it using modified Einstein notation as

$$\vec{\sigma}^{\perp}\left[\vec{\vec{a}}_{22}\,d\vec{W}_t^{\sigma(t)}\right]\vec{\sigma} = \sum_{i,j,k=1}^{N_w}\sigma_i\sigma_j\,a_{22\,ijk}\,d\vec{W}_k^{\sigma(t)} = \overrightarrow{\left[\sum_{i,j}\sigma_i\sigma_j\,a_{22\,ijk}\right]}^{\perp} d\vec{W}_t^{\sigma(t)}$$
$$= \overrightarrow{\left[\sigma_{\mathbf{i}}\sigma_{\mathbf{j}}\,a_{22\,\mathbf{ij}k}\right]}^{\perp} d\vec{W}_t^{\sigma(t)}.$$

Note that these conventions allow a more compact expression for other terms:

$$\left[\vec{\sigma}^{\perp}\vec{\vec{a}}_2\vec{\sigma}\right]\vec{\sigma}^{\perp}\vec{\vec{a}}_2 = \left[\sigma_{\mathbf{i}}\sigma_{\mathbf{j}}a_{2_{\mathbf{ij}}}\right]\overrightarrow{\left[\sigma_{\mathbf{k}}a_{2_{\mathbf{k}l}}\right]}^{\perp} = \overrightarrow{\left[\sigma_{\mathbf{i}}\sigma_{\mathbf{j}}\sigma_{\mathbf{k}}a_{2_{\mathbf{ij}}}a_{2_{\mathbf{k}l}}\right]}^{\perp}$$

and

$$\vec{\sigma}^{\perp}\left(\vec{\vec{a}}_2 + \vec{\vec{a}}_2^{\perp}\right)\vec{\vec{a}}_2 = \overrightarrow{\left[\sigma_{\mathbf{i}}\left(a_{2_{\mathbf{i}j}} + a_{2_{j\mathbf{i}}}\right)\right]}^{\perp}\vec{\vec{a}}_2 = \overrightarrow{\left[\sigma_{\mathbf{i}}\left(a_{2_{\mathbf{ij}}} + a_{2_{\mathbf{ji}}}\right)a_{2_{\mathbf{j}k}}\right]}^{\perp}.$$

Finally we are able to identify the endogenous coefficients and get:

$$\vec{\vec{\mathrm{v}}}'_y(t,0,0)^{\perp}$$
$$= \frac{1}{2}\sigma^{-3}\left[\overrightarrow{\left[\sigma_{\mathbf{i}}\left(a_{2_{\mathbf{ij}}} + a_{2_{\mathbf{ji}}}\right)a_{2_{\mathbf{j}k}}\right]}^{\perp} + \overrightarrow{\left[\sigma_{\mathbf{i}}\sigma_{\mathbf{j}}\,a_{22\,\mathbf{ij}k}\right]}^{\perp}\right] - \frac{3}{2}\sigma^{-5}\,\overrightarrow{\left[\sigma_{\mathbf{i}}\sigma_{\mathbf{j}}\sigma_{\mathbf{k}}a_{2_{\mathbf{ij}}}a_{2_{\mathbf{k}l}}\right]}^{\perp},$$

which after conjugation, normalisation and reverting to full notations proves (5.55). Following the announced strategy, we now invoke the arbitrage constraint (5.36) [p. 297] providing the scalar product

$$\vec{\vec{\mathrm{v}}}'_y(\star)^{\perp}\vec{\sigma}_t(t).$$

In this expression, all terms are now expressed as function of inputs, except for our target quantity which is the curvature $\widetilde{\Sigma}''_{yy}(\star)$. Let us recall from (5.36) [p. 297] that

$$\vec{\mathrm{v}}'_y(\star)^\perp \vec{\sigma}_t(t) = \frac{3}{2}\,\widetilde{\Sigma}^2\,\widetilde{\Sigma}''_{yy}(\star) + 3\widetilde{\Sigma}\,\widetilde{\Sigma}'^{\,2}_y(\star) - \frac{1}{2}\,\widetilde{\Sigma}^{-1}\left[\|\vec{\mathrm{v}}\|^2 + \|\vec{n}\|^2\right](\star)$$

which leads to

$$\widetilde{\Sigma}''_{yy}(\star) = \frac{2}{3}\sigma^{-2}\,\vec{\mathrm{v}}'_y(\star)^\perp\vec{\sigma}_t(t) \;-\; 2\sigma^{-1}\,\widetilde{\Sigma}'^{\,2}_y(\star) + \frac{1}{3}\sigma^{-3}\left[\|\vec{\mathrm{v}}\|^2 + \|\vec{n}\|^2\right](\star).$$

We can now develop this expression as a function of the inputs:

$$\widetilde{\Sigma}''_{yy}(\star) = \frac{2}{3}\sigma^{-2}\left[\frac{1}{2}\overrightarrow{\left[u_{\mathrm{i}}\left(c_{2\,\mathrm{ij}} + c_{2\,\mathrm{ji}}\right)c_{2\,\mathrm{j}k}\right]} + \frac{1}{2}\overrightarrow{\left[u_{\mathrm{i}}u_{\mathrm{j}}\;c_{22\,\mathrm{ij}k}\right]} - \frac{3}{2}\overrightarrow{\left[u_{\mathrm{i}}u_{\mathrm{j}}u_{\mathrm{k}}\;c_{2\,\mathrm{ij}}\;c_{2\,\mathrm{k}l}\right]}\right]^\perp \vec{\sigma}_t(t)$$
$$-\,2\sigma^{-1}\left[\frac{1}{2}\left[u_{\mathrm{i}}u_{\mathrm{j}}c_{2_{\mathrm{ij}}}\right]\right]^2 + \frac{1}{3}\sigma^{-3}\left[\|\vec{a}_2^{\perp}\vec{u}\|^2 + \|\vec{a}_3^{\perp}\vec{u}\|^2\right].$$

Noticing identities

$$\left[u_{\mathrm{i}}u_{\mathrm{j}}c_{2_{\mathrm{ij}}}\right]^2 = \left[u_{\mathrm{i}}u_{\mathrm{j}}c_{2_{\mathrm{ij}}}\right]\;\left[u_{\mathrm{k}}u_{\mathrm{l}}c_{2_{\mathrm{kl}}}\right] = \left[u_{\mathrm{i}}u_{\mathrm{j}}u_{\mathrm{k}}u_{\mathrm{l}}c_{2_{\mathrm{ij}}}c_{2_{\mathrm{kl}}}\right]$$

and

$$\|\vec{a}_2^{\perp}\vec{u}\|^2 = \|\overrightarrow{\left[u_{\mathrm{i}}a_{2_{\mathrm{i}j}}\right]}\|^2 = \sum\nolimits_j\;\left[u_{\mathrm{i}}a_{2_{\mathrm{i}j}}\right]^2 = \sum\nolimits_j\;\left[u_{\mathrm{i}}u_{\mathrm{k}}a_{2_{\mathrm{i}j}}a_{2_{\mathrm{k}j}}\right] = \left[u_{\mathrm{i}}u_{\mathrm{k}}a_{2_{\mathrm{ij}}}a_{2_{\mathrm{kj}}}\right]$$

we can rewrite

$$\widetilde{\Sigma}''_{yy}(\star) = \sigma^{-1}\left[\frac{1}{3}\left[u_{\mathrm{i}}u_{\mathrm{k}}\left(c_{2_{\mathrm{ij}}} + c_{2_{\mathrm{ji}}}\right)c_{2_{\mathrm{jk}}}\right] + \frac{1}{3}\left[u_{\mathrm{i}}u_{\mathrm{j}}u_{\mathrm{k}}\;c_{22\,\mathrm{ijk}}\right] - \left[u_{\mathrm{i}}u_{\mathrm{j}}c_{2_{\mathrm{ij}}}\right]^2\right]$$
$$-\frac{1}{2}\sigma^{-1}\left[u_{\mathrm{i}}u_{\mathrm{j}}c_{2_{\mathrm{ij}}}\right]^2 + \frac{1}{3}\sigma^{-3}\left[\left[u_{\mathrm{i}}u_{\mathrm{k}}a_{2_{\mathrm{ij}}}a_{2_{\mathrm{kj}}}\right] + \left[u_{\mathrm{i}}u_{\mathrm{k}}a_{3_{\mathrm{ij}}}a_{3_{\mathrm{kj}}}\right]\right].$$

Simplifying this expression, we eventually get

$$3\sigma\,\widetilde{\Sigma}''_{yy}(\star) = \left[u_{\mathrm{i}}u_{\mathrm{j}}u_{\mathrm{k}}\;c_{22\,\mathrm{ijk}}\right] + \left[u_{\mathrm{i}}u_{\mathrm{k}}c_{3_{\mathrm{ij}}}c_{3_{\mathrm{kj}}}\right]$$
$$+\left[u_{\mathrm{i}}u_{\mathrm{k}}\left(c_{2_{\mathrm{ij}}}c_{2_{\mathrm{jk}}} + c_{2_{\mathrm{ji}}}c_{2_{\mathrm{jk}}} + c_{2_{\mathrm{ij}}}c_{2_{\mathrm{kj}}}\right)\right] - \frac{9}{2}\left[u_{\mathrm{i}}u_{\mathrm{j}}c_{2_{\mathrm{ij}}}\right]^2,$$

which proves (5.52).

Step 3/3: Computation of the slope $\widetilde{\Sigma}'_\theta(t, 0, 0)$

Now identifying the drift terms in both sides of (5.57) brings:

$$2\vec{\sigma}_t(t)^\perp\vec{a}_{1,t}(t) + \|\vec{a}_{2,t}\|^2 + \|\vec{a}_{3,t}\|^2 - 2\vec{\sigma}_t(t)^\perp\vec{a}_{2,t}(t)\,\vec{\lambda}^{\sigma}_t(t)$$
$$= 2\widetilde{\Sigma}(\star)\widetilde{b}(\star) + \|\vec{\mathrm{v}}(\star)\|^2 + \|\vec{n}(\star)\|^2 - 2\widetilde{\Sigma}(\star)\,\vec{\mathrm{v}}(\star)^\perp\,\vec{\lambda}^{\widetilde{\Sigma}}_t(0). \qquad (5.61)$$

We now combine all three arbitrage constraints by substituting (5.35) and (5.36) into (5.37) in order to obtain the following expression for the IATM drift:

$$\begin{aligned}\widetilde{b}(\star) &= 2\widetilde{\Sigma}'_\theta(\star) + \widetilde{\Sigma}^2\left[\widetilde{\Sigma}''_{yy} - \widetilde{\Sigma}'_y\right](\star) + 3\widetilde{\Sigma}\ \widetilde{\Sigma}'^{\,2}_y(\star) \\ &\quad - \frac{1}{2}\widetilde{\Sigma}^{-1}\left[\|\vec{\widetilde{\mathrm{v}}}\|^2 + \|\vec{\widetilde{n}}\|^2\right](\star) + \widetilde{\Sigma}'_y(\star)\widetilde{X}^{-1}\widetilde{X}'_\theta(t,0) \\ &\quad - \widetilde{\Sigma}^{-1}(\star)\,\vec{\sigma}_t(t)^\perp\,\vec{\sigma}'_T(t,t) + \vec{\widetilde{\mathrm{v}}}(\star)^\perp\left[\vec{\widetilde{\lambda}}^{\widetilde{\Sigma}}_t(0) - \vec{\widetilde{\lambda}}^N_t(t)\right].\end{aligned}$$

Substituting the above expression into the right-hand side of (5.61) we obtain

$$\begin{aligned}&2\widetilde{\Sigma}\widetilde{b}(\star) + \|\vec{\widetilde{\mathrm{v}}}(\star)\|^2 + \|\vec{\widetilde{n}}(\star)\|^2 - 2\widetilde{\Sigma}\,\vec{\widetilde{\mathrm{v}}}(\star)^\perp\vec{\widetilde{\lambda}}^{\widetilde{\Sigma}}_t(0) \\ &= 2\widetilde{\Sigma}(\star)\Big[2\widetilde{\Sigma}'_\theta(\star) + \widetilde{\Sigma}^2\left[\widetilde{\Sigma}''_{yy} - \widetilde{\Sigma}'_y\right](\star) + 3\widetilde{\Sigma}\ \widetilde{\Sigma}'^{\,2}_y(\star) \\ &\qquad - \frac{1}{2}\widetilde{\Sigma}^{-1}\left[\|\vec{\widetilde{\mathrm{v}}}\|^2 + \|\vec{\widetilde{n}}\|^2\right](\star) + \widetilde{\Sigma}'_y(\star)\widetilde{X}^{-1}\widetilde{X}'_\theta(t,0) \\ &\qquad - \widetilde{\Sigma}^{-1}(\star)\vec{\sigma}_t(t)^\perp\,\vec{\sigma}'_T(t,t) + \vec{\widetilde{\mathrm{v}}}(\star)^\perp\left[\vec{\widetilde{\lambda}}^{\widetilde{\Sigma}}_t(0) - \vec{\widetilde{\lambda}}^N_t(t)\right]\Big] \\ &\quad + \|\vec{\widetilde{\mathrm{v}}}(\star)\|^2 + \|\vec{\widetilde{n}}(\star)\|^2 - 2\widetilde{\Sigma}\,\vec{\widetilde{\mathrm{v}}}(\star)^\perp\vec{\widetilde{\lambda}}^{\widetilde{\Sigma}}_t(0) \\ &= 2\sigma\left[2\widetilde{\Sigma}'_\theta(\star) + \sigma^2\left[\widetilde{\Sigma}''_{yy} - \widetilde{\Sigma}'_y\right](\star) + 3\sigma\ \widetilde{\Sigma}'^{\,2}_y(\star)\right. \\ &\qquad \left. + \widetilde{\Sigma}'_y(\star)\widetilde{X}^{-1}\widetilde{X}'_\theta(t,0) - \sigma^{-1}\vec{\sigma}_t(t)^\perp\,\vec{\sigma}'_T(t,t) - \vec{\widetilde{\mathrm{v}}}(\star)^\perp\vec{\widetilde{\lambda}}^N_t(t)\right].\end{aligned}$$

Note that this expression is now devoid of any direct reference to the IV drift adjustment $\vec{\widetilde{\lambda}}^{\widetilde{\Sigma}}$. Injecting (5.51)—which establishes the IATM sliding skew $\widetilde{\Sigma}'_y(\star)$— the expression evolves into

$$\begin{aligned}&2\widetilde{\Sigma}\widetilde{b}(\star) + \|\vec{\widetilde{\mathrm{v}}}(\star)\|^2 + \|\vec{\widetilde{n}}(\star)\|^2 - 2\widetilde{\Sigma}\,\vec{\widetilde{\mathrm{v}}}(\star)^\perp\vec{\widetilde{\lambda}}^{\widetilde{\Sigma}}_t(0) \\ &= 2\sigma\left[2\widetilde{\Sigma}'_\theta(\star) + \sigma^2\widetilde{\Sigma}''_{yy}(\star) + \frac{1}{2}\left[u_{\mathbf{i}}u_{\mathbf{j}}c_{2_{\mathbf{ij}}}\right]\left[\widetilde{X}^{-1}\widetilde{X}'_\theta(t,0) - \sigma^2\right]\right. \\ &\qquad \left. + 3\sigma\frac{1}{4}\left[u_{\mathbf{i}}u_{\mathbf{j}}c_{2_{\mathbf{ij}}}\right]^2 - \sigma^{-1}\vec{\sigma}_t(t)^\perp\,\vec{\sigma}'_T(t,t) - \vec{\widetilde{\mathrm{v}}}(\star)^\perp\vec{\widetilde{\lambda}}^N_t(t)\right]\end{aligned}$$

so that the full drift equalisation (5.61) becomes:

$$
\begin{aligned}
0 = {} & 4\sigma\,\widetilde{\Sigma}'_\theta(\star) + 2\sigma^3\,\widetilde{\Sigma}''_{yy}(\star) + \sigma\left[u_{\mathbf{i}}u_{\mathbf{j}}c_{2_{\mathbf{ij}}}\right]\left[\widetilde{X}^{-1}\widetilde{X}'_\theta(t,0) - \sigma^2\right] \\
& + \frac{3}{2}\sigma^2\left[u_{\mathbf{i}}u_{\mathbf{j}}c_{2_{\mathbf{ij}}}\right]^2 - 2\,\vec{\sigma}_t(t)^\perp\,\vec{\sigma}'_T(t,t) - 2\sigma\,\vec{\vec{\mathrm{v}}}(\star)^\perp\,\vec{\lambda}^N_t(t) \\
& - 2\,\vec{\sigma}_t(t)^\perp\,\vec{a}_{1,t}(t) - \|\vec{\vec{a}}_{2,t}\|^2 - \|\vec{\vec{a}}_{3,t}\|^2 + 2\,\vec{\sigma}_t(t)^\perp\,\vec{\vec{a}}_{2,t}(t)\,\vec{\lambda}^\sigma_t(t).
\end{aligned}
$$

At this stage, only the slope $\widetilde{\Sigma}'_\theta(t,0,0)$ remains undetermined, since all other terms can be expressed from the input SInsV model: the local[13] shape of the underlying[14] map $T \to X_t(T)$, the local shape and dynamics of its instantaneous volatility $\vec{\sigma}_t(T)$, and the immediate values of the model-dependent risk premia $T \to \vec{\lambda}^\sigma(T)$ and $T \to \vec{\lambda}^N(T)$. Isolating the slope $\widetilde{\Sigma}'_\theta(\star)$ and using (5.54), we simplify the above expression as

$$
\begin{aligned}
\widetilde{\Sigma}'_\theta(\star) = {} & -\frac{1}{2}\sigma^2\,\widetilde{\Sigma}''_{yy}(\star) - \frac{1}{2}\left[u_{\mathbf{i}}u_{\mathbf{j}}c_{2_{\mathbf{ij}}}\right]\left[\widetilde{X}^{-1}\widetilde{X}'_\theta(t,0) - \sigma^2\right] - \frac{3}{8}\sigma\left[u_{\mathbf{i}}u_{\mathbf{j}}c_{2_{\mathbf{ij}}}\right]^2 \\
& + \frac{1}{2}\,\vec{u}^\perp\left[\vec{\sigma}'_T(t,t) + \vec{a}_{1,t}(t)\right] + \frac{1}{2}\sigma\,\overrightarrow{\left[c_{2\,\mathbf{i}j}\,u_{\mathbf{i}}\right]}^\perp\left[\vec{\lambda}^N_t(t) - \vec{\lambda}^\sigma_t(t)\right] \\
& + \frac{1}{4}\sigma\,\|\vec{\vec{c}}_{2,t}\|^2 + \frac{1}{4}\sigma\,\|\vec{\vec{c}}_{3,t}\|^2.
\end{aligned}
$$

We can now replace $\widetilde{\Sigma}''_{yy}(t,0,0)$ with its expression (5.52) to obtain the slope with

$$
\begin{aligned}
2\sigma^{-1}\,\widetilde{\Sigma}'_\theta(\star) = {} & \frac{1}{2}\left[c^2_{2_{\mathbf{ij}}}\right] - \frac{1}{3}\left[u_{\mathbf{i}}u_{\mathbf{k}}\left(c_{2_{\mathbf{ij}}}c_{2_{\mathbf{jk}}} + c_{2_{\mathbf{ji}}}c_{2_{\mathbf{jk}}} + c_{2_{\mathbf{ij}}}c_{2_{\mathbf{kj}}}\right)\right] + \frac{3}{4}\left[u_{\mathbf{i}}u_{\mathbf{j}}c_{2_{\mathbf{ij}}}\right]^2 \\
& + \frac{1}{2}\left[c^2_{3_{\mathbf{ij}}}\right] - \frac{1}{3}\left[u_{\mathbf{i}}u_{\mathbf{k}}c_{3_{\mathbf{ij}}}c_{3_{\mathbf{kj}}}\right] - \frac{1}{3}\left[u_{\mathbf{i}}u_{\mathbf{j}}u_{\mathbf{k}}\;c_{22\,\mathbf{ijk}}\right] \\
& + \vec{u}^\perp\left[\vec{u}'_\theta + \vec{c}_1\right] - \sigma^{-1}\left[u_{\mathbf{i}}u_{\mathbf{j}}c_{2_{\mathbf{ij}}}\right]\left[\widetilde{X}^{-1}\widetilde{X}'_\theta(t,0) - \sigma^2\right] \\
& + \overrightarrow{\left[c_{2\,\mathbf{i}j}\,u_{\mathbf{i}}\right]}^\perp\left[\vec{\lambda}^N_t(t) - \vec{\lambda}^\sigma_t(t)\right],
\end{aligned}
$$

which proves (5.53) and concludes the proof. □

5.4.2 Interpretation and Comments

5.4.2.1 Comparison with the Single-Underlying Case

It is interesting to note that the normalisation effort we made did bear fruit. Indeed, one can observe that each expression in Theorem 5.2—excepting a single term in (5.53)—is homogeneous in the modulus $\|\vec{\sigma}_t(t)\|$.

---

[13] In $(t,0,0)$.

[14] Equivalent to the sliding underlying map $\theta \to \widetilde{X}_t(\theta)$.

Naturally our first task is to ensure that these results in the term structure framework do match their simpler equivalent in the single asset setup. Downgrading the Term-Structure equations is done in two steps. The first step consists in ignoring maturity-dependency (in particular by killing any maturity-differential) and aligning all endogenous drivers: we can then compare with the results of Theorem 3.4 [p. 181] in the multi-dimensional but single-underlying case. The second step corresponds to reducing all tensors to order and dimension 1, hence checking against the simples case of Theorem 2.2 [p. 62].

Among the six equations involved, some are simple and can therefore be compared at once. Indeed, it is immediately obvious that:

- The skew result (5.51) matches its single-asset equivalents, both multi-dimensional (3.73) and scalar (2.51).
- The equation (5.54) providing the endogenous coefficient is close to and clearly compatible with (3.76) and (2.55).
- The same is true of (5.56) giving the exogenous coefficient, which is consistent with (3.77) and (2.57).

As for the three remaining equations, it is preferable to downgrade the expressions in order to be convinced of the match:

- The term-structure curvature expression (5.52) becomes exactly (3.74) by removing the T-dependency on all tensors. Then reducing to the scalar case yields

$$\widetilde{\Sigma}''_{yy}(\star) = \sigma^{-1}\left[\frac{1}{3}\left[3\,c_2^2\right] - \frac{3}{2}\,c_2^2\right] + \frac{1}{3}\sigma^{-1}\,c_3^2 + \frac{1}{3}\sigma^{-1}\,c_{22}$$
$$= \frac{1}{\sigma_t^2}\left[\frac{a_{22}}{3}\right] + \frac{1}{\sigma_t^3}\left[\frac{a_3^2}{3} - \frac{a_2^2}{2}\right],$$

  which matches (2.52).
- Coming to the slope expression (5.53), we first exclude terms I_0, J_0 and K'_0. In the same manner, we then ignore the T-dependency which brings us exactly to (3.75). Then by reducing to a scalar framework we get

$$\widetilde{\Sigma}'_\theta(t,0,0)$$
$$= \frac{1}{2}\sigma\left[\frac{1}{2}c_2^2 - \frac{1}{3}\left[3\,c_2^2\right] + \frac{3}{4}c_2^2 + \frac{1}{2}c_3^2 - \frac{1}{3}c_3^2 - \frac{1}{3}c_{22}\right] + \frac{1}{2}\sigma\,c_1 + \frac{1}{4}\sigma^2\,c_2$$
$$= \sigma\left[\frac{1}{8}c_2^2 + \frac{1}{12}c_3^2 - \frac{1}{6}c_{22} + \frac{1}{2}c_1\right] + \frac{1}{4}\sigma^2\,c_2$$
$$= \sigma_t\left[\frac{1}{4}a_2\right] + \left[\frac{1}{2}a_1 - \frac{1}{6}a_{22}\right] + \frac{1}{\sigma_t}\left[\frac{1}{8}a_2^2 + \frac{1}{12}a_3^2\right],$$

  which is identical to (2.53).

- Finally (5.55), describing the skew endogenous coefficient, can be downgraded into the non-TS expression (3.78). Then reducing to the simplest case we get

$$\widetilde{\mathrm{v}}_y'(t,0,0) = \tfrac{1}{2}\left[(c_2 + c_2)\,c_2\right] + \tfrac{1}{2}c_{22} - \tfrac{3}{2}c_2^2 = \tfrac{1}{\sigma_t}\left[\tfrac{1}{2}a_{22}\right] + \tfrac{1}{\sigma_t^2}\left[-\tfrac{1}{2}a_2^2\right],$$

which matches (2.56).

We conclude, therefore, that this Theorem is compatible with previous results.

Among all the new terms, in our view the most interesting are those invoked in the slope expression (5.53):

- The term $\vec{u}^{\perp}\left[\vec{u_\theta}' + \vec{c}_1\right]$ comes from J_0, itself introduced in Sect. 5.3.1 and studied in Sect. 5.3.2. The group in brackets underlines that any drift written on the underlying's dynamics is equivalent to the volatility slide effect.
- Term I_0 is present again with $\left[u_\mathbf{i} u_\mathbf{j}\, c_{2\,\mathbf{ij}}\right]\left[\frac{\widetilde{X}_\theta'}{X}(t,0) - \sigma^2\right]$ and we gauge that the interpretation we gave in Sect. 5.3.2 is still valid. However, the term σ^2 seems out-of-place, mainly because it breaks the homogeneity of $\widetilde{\Sigma}_\theta'(t,0,0)$ (in σ).
- Finally the drift correction term $\left[\vec{\lambda}_t^N(t) - \vec{\lambda}_t^\sigma(t)\right]$ comes from K_0' but can be surprising, because it ignores the alignment of the sliding implied volatility process, namely $\vec{\lambda}^{\widetilde{\Sigma}}$. It is also apparent that this drift correction only appears in the slope term $\widetilde{\Sigma}_\theta'(t,0,0)$, which reminds us of the single-asset study, where the drift expansion terms (coefficients indexed with at least a 1) would only appear in θ-differentials.

5.4.2.2 Inconsistency of the Term-by-Term *Single Underlying* method

An important question that remains is the degree of compatibility between smile approximations (both in shape and dynamics) developed on one hand in this generic term-structure framework, and on the other hand by a term-by-term (fixing the maturity and therefore the numeraire) approach, using the results described in Part I for a single underlying.

The first aspect to take into account is that the term-structure approximation is *globally coherent*, whereas the individual term approximations cannot be. Let us illustrate this point: by fixing a maturity T, we also fix the underlying and the numeraire. Hence a single-asset approach on this underlying $X_t(T)$ will generate an implied volatility map, for all maturities until T. But this "smile" has only financial significance in maturity T,[15] simply because there is no liquid option on $X_t(T)$ with maturity $T^* \neq T$, whereas the maturity-dependent approach provides in all its (K,T)-domain an approximation for the actual target smile.

Apart from this fundamental caveat, what can we anticipate about the relative speed *vs* precision of the two methods? Let us start by considering the static smile.

---

15 At least in our framework, see the discussion on mid-curves.

Intuitively, in terms of precision the advantage is clearly on the side of the single underlying method. The reason is that for a given maturity T, the term-structure method will "estimate" the associated instantaneous volatility $\vec{\sigma}_t(T)$ from the immediate data $\vec{\sigma}_t(t)$ and $\vec{\sigma}'_T(t,t)$, as well as the following differentials if we push the chaos expansion higher. The same stands true for the underlying $X_t(T)$ itself. The TS method is therefore at a loss compared to the single underlying approach, which will take in the exact values through $\sigma_t \equiv \vec{\sigma}_t(T)$ and $S_t \equiv X_t(T)$. Evidently, should we pursue the computation to higher orders in θ with the TS method, more maturity-differentials would appear, providing a better prediction of $\vec{\sigma}_t(T)$, and consequently the gap between the two methods should close. The actual difference in precision will depend mainly on the regularity of the underlying and volatility profiles $T \to X_t(T)$ and $T \to \vec{\sigma}_t(T)$. In particular, a very rough profile, such as a piecewise constant function, will significantly deteriorate the performance of the term-structure method, whereas it will not affect the term-by-term method what so ever.

In terms of speed, and at a given expansion order, the term-by-term method will require the computation of as many sets of differentials as the number of maturities to be considered, whereas the TS method only requires one set. But, as we have seen, the single-asset case will be scalar by construction, while the calculations in the multi-dimensional case can be significantly more involved and compensate somewhat. We can therefore anticipate the same order of magnitude for the speed of the two methods.

In conclusion, for a given computing time, the precision advantage is probably with the single-asset method, and increasing the level of differentiation will asymptotically make the two approximations tally, but *only in T*.

This superior performance, however, comes with a price. Because the term-by-term method treats each maturity independently, we have to take extra care in order to establish some coherence throughout the smile. A doubtful choice, for instance, would be to use approximations of a different order for adjacent maturities: this would be guaranteed to generate inconsistencies, and possibly arbitrage opportunities. Such a mistake is clearly impossible within the term structure framework, which brings us back to the aforementioned global consistency.

So far we have only discussed the approximation of the shape, but what about dynamics? It is by now clear that considering smile deformation modes along the maturity axis will put the term-by-term method in a difficult position. Indeed, this method structurally ignores that the different maturities are related, and in particular it cannot incorporate the fact that the underlying will "slide" with time.

5.5 Extensions, Further Questions and Conclusion

We have extended the single-asset framework into a term-structure one. Maintaining this extension spirit, could we consider families in higher dimension? For instance, could we incorporate another maturity to the option field and consider $X_t(T_1, T_2)$?

Apart from a pure mathematical interest, the financial motivation is real, as it would enable us to represent more products.

The first type of underlyings/options that spring to mind are swap rates and swaptions. Since these are defined by both expiry and tenor, the current methodology is a priori limited to fixed-expiry or fixed-tenor data sets.

The second kind of product are so-called "mid-curve" options, which are usually defined by two points on a curve representation. A typical example would be CMS[16] spread options. This extension would get us closer to non-conform cases where the maturity of the underlying and that of the option do not match, and which are usually dealt with using convexity adjustments.

Having compared the term-structure and term-by-term approaches, it becomes clear that they are opposed in almost every aspect. The temptation is great, therefore, to combine them into a more powerful method. The idea would be to use the single-asset approach to correct the maturity-dependency of the full term-structure method. However, "interpolating" both methods between maturities is likely to give us "the worst of both worlds", and combine the pitfalls of each approach.

In the maturity-dependent framework, we have mentioned that the layer-1 approximation is determined by $\vec{\sigma}'_T(t,t)$ and $X'_T(t,t)$, ignoring the higher order differentials of these maps. A direct corollary is that if these maps cannot be reasonably approximated (with an affine function for instance), then the quality of our functional approximations is bound to suffer.

This leads to two trivial conclusions. The first is that in the term-structure framework, where both the underlying and its instantaneous volatility are mapped, obtaining higher order approximations in time-to-maturity θ is more important than in the single underlying case. The second conclusion is that for ACE to work well, we should promote the smoothest possible parametrisation in T for the model.

In the same spirit, and unless the method is automated, it makes sense to choose a model framework that simplifies the dimensionality issue. In particular, choosing a volatility structure that generates a simple form for the tensor $\vec{\vec{a}}_{22,t}$ would be sensible. In that respect, using separate variates (t and $T-t$) is usually a good technical start, as well as a realistic modelling approach. Indeed, the healthy desire to ensure *time-shift homogeneity* of the model by relying purely on Time-To-Maturity $T-t$ (see [4] in the LMM case, for instance) has to be balanced against the necessity to provide a good fit to the current smile (in particular, ATM options) which is made much easier by using functions of time t. An interesting example of this approach can be found in [5] (Sect. 7), also for the LMM, where a compromise is found by minimising the amplitude of the t-perturbation over the $T-t$ specification, using an intuitive penalisation function.

In general terms, relying heavily on time functions for the fit tends to create trader-induced hedging noise, and this is one of the reasons why we lean towards *stationary* and time-homogeneous models. In a sense, this discussion illustrates the notion of calibration to both statics and dynamics of the market, which we support since in principle it provides a reasonable *and* stable fit.

---

[16] Constant Maturity Swap.

References

1. El Karoui, N., Geman, H., Rochet, J.-C.: Changes of numéraire, changes of probability measure and option pricing. J. Appl. Probab. **32**(2), 443–458 (1995)
2. Brigo, D., Mercurio, F.: Interest Rate Models: Theory and Practice, 2nd edn. Springer Finance, Heidelberg (2006)
3. Hagan, P.S., Kumar, D., Lesniewski, A.S., Woodward, D.E.: Managing smile risk. Wilmott Mag. **1**, 84–108 (2002)
4. Schoenmakers, J.: Robust libor modelling and pricing of derivative products. In: Financial Mathematics. Chapman & Hall, London (2005)
5. Piterbarg, V.V.: Stochastic volatility model with time-dependent skew. Appl. Math. Financ. **12**, 147–185 (2005)

Chapter 6
Implied Dynamics in the SV-HJM Framework

Abstract In this chapter we apply the ACE methodology developed for the generic term structure (TS) framework in Chap. 5. We focus on very liquid interest rates derivatives products, valued within a universal Stochastic-Volatility (SV) Heath-Jarrow-Morton (HJM) modelling setup. Our aim is still to link the underlying's instantaneous (stochastic volatility) dynamics to the shape and dynamics of the implied volatility surface, with a natural emphasis on the direct problem (from model to smile) and on the first layer (second order in strike and first order in maturity). Thanks to the results provided in the generic framework, this application can be performed in only two steps. The first is rather conceptual and involves casting each product type (bond options, caplets, swaptions) into the generic framework, by allocating several TS (the underlying, the numeraire, the measure and the payoff). The second step is more computational and consists in computing the chaos dynamics for the underlying TS defined above, within the chosen SV-HJM parametrisation.

In Sect. 6.1 we set the *input* SV-HJM modelling framework in a generic, chaos diffusion style. In preparation for the intensive usage of tenor structures, we also introduce some dedicated notations. We define and justify our objectives in terms of the specific derivative products targeted, as well as the required precision of the *output* smile shapes and dynamics. We then impose some realistic simplifications on the interest rates environment, which will help in the modelling phase. Finally, we briefly discuss the close relationship and common potential of the SV-HJM and SV-LMM classes.

In Sect. 6.2 we expose some useful intermediate results, pertaining to the dynamics of rebased bonds, first single Zero-Coupons and then full-blown fixed structures. These results allow us to describe the chaos diffusion of a *driftless* process, which is a necessary element of our approach, and will simplify our computations later on.

In Sect. 6.3 we cover the first type of derivative product, the European Call options on fixed-coupon bonds. As mentioned above, we first immerse the problem into the generic setup, and then compute accordingly the chaos dynamics for the chosen underlying. The presence of the tenor structure generates complex results, which unfortunately lend themselves to only a basic interpretation.

In Sect. 6.4 we proceed along the same path for Caplets, expressing all the σ-(2,0) coefficients necessary to formulate the $\widetilde{\Sigma}$-(2,0) group of IATM differentials (both static dynamics). We exploit the relative simplicity of the Libor rate dynamics to

D. Nicolay, *Asymptotic Chaos Expansions in Finance*,
Springer Finance Lecture Notes, DOI 10.1007/978-1-4471-6506-4_6

provide an interpretation for them. Specifically we proceed with a semi-numerical analysis of the various terms, focusing on their sign and relative magnitudes, by introducing sequentially some realistic simplifications and assumptions.

Last but not least, in Sect. 6.5 we similarly cover the European payer Swaptions, which without surprise turns out to be the most involved case with regard to computation. In an even more pronounced manner that with bond options, we exploit as much as possible our earlier results pertaining to basket dynamics.

Finally, in Sect. 6.6 we discuss the (tempting) possibility of obtaining the same price surface information through an indirect approach, by considering options on *assets* rather than on the corresponding *rates*. Naturally this approach only concerns Caplets and Swaptions, which we treat separately.

6.1 Definitions, Notations and Objectives

6.1.1 The HJM Framework in a Chaos Context

An HJM framework can be defined equivalently by specifying the dynamics of either the forward instantaneous rate $f_t(T)$ or the Zero-Coupon (ZC) Bond $B_t(T)$. Due to the *market model* approach that we have chosen to take throughout this study, we shall use the latter.

Consequently, let us denote the no-arbitrage dynamics of the ZC family $B_t(T)$ as

$$dB_t(T) \;/\; B_t(T) = r_t\, dt + \overrightarrow{\Gamma}_t(T)^{\perp}\, d\overrightarrow{W}_t, \tag{6.1}$$

where r_t is the short rate and $\overrightarrow{W}_t$ is a Wiener process under the *risk-neutral* measure $\mathbb{Q}$. We then specify the dynamics of the family of instantaneous volatilities with

$$d\overrightarrow{\Gamma}_t(T) = \overrightarrow{\alpha}_{1,t}(T)\, dt + \overrightarrow{\overrightarrow{\alpha}}_{2,t}(T)\, d\overrightarrow{W}_t + \overrightarrow{\overrightarrow{\alpha}}_{3,t}(T)\, d\overrightarrow{Z}_t, \tag{6.2}$$

$$d\overrightarrow{\overrightarrow{\alpha}}_{2,t}(T) = \overrightarrow{\overrightarrow{[\cdots]}}\, dt + \overrightarrow{\overrightarrow{\overrightarrow{\alpha}}}_{22,t}(T)\, d\overrightarrow{W}_t + \overrightarrow{\overrightarrow{\overrightarrow{[\cdots]}}}\, d\overrightarrow{Z}_t. \tag{6.3}$$

Again, $\overrightarrow{W}_t$ and $\overrightarrow{Z}_t$ are independent Wiener processes with each a unit covariance matrix, while the instantaneous coefficients defining the volatility structure, $\overrightarrow{\Gamma}_t(T)$, $\overrightarrow{\alpha}_{1,t}(T)$, $\overrightarrow{\overrightarrow{\alpha}}_{2,t}(T)$, etc. are simply imposed to be adapted Itô processes. In particular, they can be driven by the exogenous Wiener process $\overrightarrow{Z}_t$, which justifies that this framework covers the SV-HJM model class in a very generic fashion, since the model state variables have not been defined.

Observe that the reference measure has been arbitrarily chosen as the risk-neutral one, but that this choice relates mainly to convention and the desire to exhibit the short rate r_t. Indeed, in many cases a forward or terminal measure $\mathbb{Q}^{T_*}$ is actually more pertinent and/or usual. Note that in this instance the associated numeraire becomes $B_t(T_*)$, so that *all* ZCs are martingale. Nevertheless, it is noticeable that whichever measure is selected, it has the potential to complexify the ZC dynamics further down the line, so that the chaos volatility structure ($\vec{\Gamma}_t(T)$, $\vec{\vec{\alpha}}_{2,t}(T)$, etc.) might come with complex drifts.

6.1.2 Tenor Structures and Simplified Notations

The HJM framework (6.1)–(6.3) relates to the yield curve and is therefore structurally continuous w.r.t. the maturity T. In the sequel, however, we will be often be dealing not with a full and continuous term structure, but with discrete and finite *tenor structures* or *schedules* $\{T_i\}_{0\le i\le N}$ which can be trivially re-parametrised using *accruals* $\{\delta_i\}_{1\le i\le N}$

$$[T_0,\ T_1,\dots,T_N] \iff [T_0,\ \delta_1,\dots,\delta_N] \qquad \text{with} \qquad \delta_i \triangleq T_i - T_{i-1}.$$

These tenor structures are associated essentially to bond and swap trades, which we will identify through their start date T_0, leaving the accruals to complete the specification of the schedule. This choice is obviously motivated by our generic term-structure approach, and justifies the following definitions ($\triangleq$) and compact notations ($\doteq$):

$$B_i \doteq B_{i,t} \doteq B_{i,t}(T_0) \triangleq B_t(T_i) \qquad\qquad \vec{\Gamma}_i \doteq \vec{\Gamma}_{i,t} \doteq \vec{\Gamma}_{i,t}(T_0) \triangleq \vec{\Gamma}_t(T_i)$$

$$\vec{\vec{\alpha}}_{i,2} \doteq \vec{\vec{\alpha}}_{i,2,t} \doteq \vec{\vec{\alpha}}_{i,2,t}(T_0) \triangleq \vec{\vec{\alpha}}_{2,t}(T_i) \qquad \vec{\vec{\alpha}}_{i,3} \doteq \vec{\vec{\alpha}}_{i,3,t} \doteq \vec{\vec{\alpha}}_{i,3,t}(T_0) \triangleq \vec{\vec{\alpha}}_{3,t}(T_i)$$

$$\vec{\alpha}_{i,1} \doteq \vec{\alpha}_{i,1,t} \doteq \vec{\alpha}_{i,1,t}(T_0) \triangleq \vec{\alpha}_{1,t}(T_i) \qquad \vec{\vec{\vec{\alpha}}}_{i,22} \doteq \vec{\vec{\vec{\alpha}}}_{i,22,t} \doteq \vec{\vec{\vec{\alpha}}}_{i,22,t}(T_0) \triangleq \vec{\vec{\vec{\alpha}}}_{22,t}(T_i)$$

Let us now move on and formally define the aims of this chapter.

6.1.3 Objectives and Assumptions

Our objectives are to establish the links between the SInsV and SImpV model classes, in an interest rates environment and within an SV-HJM modelling framework. More precisely, the dynamics of the Zero Coupons will be given by (6.1)–(6.3) and we will consider the option price surfaces corresponding to the following three product types: bond options, caplets and physical swaptions. We justify this choice by the fact that, at the time of writing, these are the most liquid vanilla[1] smiles.

The connection between the two stochastic volatility model classes will be demonstrated respectively for the σ-(2,0) group of instantaneous coefficients and the $\widetilde{\Sigma}$-(2,0) group of IATM differentials (refer to Definition 2.1 [p. 57]).

In order to apply the results of Chap. 5, we will have to immerse the product (underlying, numeraire, measure and option payoff), along with the HJM chaos dynamics (again, only up to the σ-(2,0) group of coefficients), into the former generic maturity-dependent framework.

In order to simplify our results (without altering their structure) or to give them more range, we either adopt or ignore certain approximations and assumptions that are commonly found in the interest rates literature:

Assumption 6.1 (*Simplification of the IR underlying market*) Unless specified otherwise:

- In all tenor structures considered, the rate frequencies are not necessarily perfectly regular. In other words we do NOT have $\delta_i \equiv \delta \ \forall i$, which would imply $T_i = T_{i-1} + \delta \ \forall 1 \le i \le N$. This accounts, in particular, for market-specific *bad day conventions*.
- For all coupon cashflows (either fixed- or floating-rate based) the accrual end date and the payment date will coincide. For all floating coupons, and in particular for those Libor-based, the fixing date and accrual start date will also be identical. Finally, the coupons of a given leg will be "spanning", in the sense that the end/payment date of one is the start/fixing date of the next. Overall, this will significantly simplify the legs/tenor structures considered: for N coupons there will only be $N + 1$ dates involved (instead of potentially $4N$).
- Any basis (e.g. tenor basis, currency basis, credit spreads) will be ignored, so that only one yield curve will be sufficient. In particular, funding and projection curves will be identical, which implies that collateral, credit or counterparty risk issues will be ignored. Note that some of these restrictions could be lifted without much technical difficulty, but at the expense of readability.
- In all swaps considered, the fixed and floating legs are not necessarily synchronous. In full generality we could have two distinct schedules, which would just increase the size of the date population to be considered. However, according to the previous assumptions, we can value the floating at par with

$$V_t\,(\text{floating leg}) = B_t(T_0) - B_t(T_N).$$

Therefore, for simplification purposes we will consider that the first and last dates (T_0 and T_N) do coincide on both legs.

[1] In fact, on Euro markets for instance, cash-settled swaptions are more liquid than physical ones. Similarly, CMS options can be considered as very liquid. But these are deemed *quasi*-vanilla products, since their valuation does not rely on a pure martingale argument. Indeed they require a further *curve assumption* to be made in order to infer the annuity *vs* swap rate relationship.

- The swaptions under consideration have an expiry date that coincides with the start date of the underlying swap.
- The swaptions considered are physically settled, as opposed to cash-settled.

6.1.4 Relative Pertinence of the SV-HJM and SV-LMM Classes

Note also that a further interest of this chapter is that a large proportion of the results it establishes will later be recycled for the Stochastic Volatility-*LMM* framework (see Chap. 7). This can be seen as a convenient computational trick, but in fact the link between the two model classes is strong and fruitful.

It is true that very few instances of an SV-HJM model have been published, and even fewer are used in practice. The popular IR model which is closest to that class is the low-dimensional Markovian family developed in [1, 2]. Compared to the rather large number of SV-LMM discussed in the literature (see [3] or [4] for instance), this seems a bit odd.

However, one of the main reasons for this situation is actually historical. Indeed, the HJM framework has been used since the early 90s in a downgraded, low-dimensional Markovian form, focusing on the availability of closed-form solutions for vanillas, and the pricing on lattices of callable products. The LMM on the other hand, became fashionable with practitioners ten years afterwards. It was focusing on the relevant state variables for the option world, and it was understood from the start that the model's potential lay in high dimension, so that pricing of exotics would have to go through Monte-Carlo.

The fact remains that not one framework is richer than the other, as these are just two distinct parametrisations of the same class. With local and early stochastic volatility extensions, there were indeed differences in performance, as for instance the Caplet smile was trading very lognormally (which is very difficult to represent in an HJM setup).

But with newer, full-blown stochastic volatility models, the overlap is much stronger. Indeed, an LMM framework can be re-expressed with dynamics of the rebased bonds, for instance. Clearly in our generic SInsV setup the two frameworks are identical, since we can have a bijection between the underlying maps, and since the state variables for volatility are ignored. Furthermore, we will see in the course of Chaps. 6 and 7 that the HJM framework makes computations easier for the group of three liquid options considered as a whole.

In our view, these reasons alone justify that we should deal with the SV-HJM model class on par with the SV-LMM, and advocates that both classes present an equally impressive potential in terms of calibrating and hedging for complex structured products.

Having now set the background and framework of our study, let us establish some intermediate results pertaining to the dynamics of rebased bonds, which will provide the driftless dynamics required by our asymptotic methodology.

6.2 Dynamics of Rebased Bonds

In the perspective of our generic framework, an obvious issue of the HJM parametrisation (6.1)–(6.3) is the presence of a drift in the asset dynamics (6.1), in the form of the short rate r_t. The classical answer is to choose a new numeraire, and an associated measure, making the Zero-Coupons martingale. We present such a re-basing in all its generality, and express the new chaos dynamics, as required, up to the σ-(2,0) level. Bearing in mind the target products (i.e. swaps and bonds), we then compute those same dynamics but for a fixed-weights basket of such re-based Zero Coupons.

6.2.1 Dynamics of the Rebased Zero Coupons

Let us consider a given but generic traded asset, whose *a.s.* non-null price process is denoted by $\{\pi_t\}_{t\geq 0}$. We specify its chaos dynamics under the *risk-neutral* measure and up to σ-(2,0) level as

$$\frac{d\pi_t}{\pi_t} = r_t\, dt + \vec{\gamma_t}^{\perp}\, d\vec{W}_t, \qquad \text{where } r_t \text{ is still the short rate,}$$

$$d\vec{\gamma_t} = \vec{c}_{1,t}\, dt + \vec{\vec{c}}_{2,t}\, d\vec{W}_t + \vec{\vec{c}}_{3,t}\, d\vec{Z}_t,$$

$$d\vec{\vec{c}}_{2,t} = \vec{[\cdot]}\, dt + \vec{\vec{\vec{c}}}_{22,t}\, d\vec{W}_t + \vec{\vec{[\cdot]}}\, d\vec{Z}_t.$$

Note that in full generality we could make the numeraire π_t maturity-dependent, i.e. we could have $\pi_t(T)$, but this additional complexity would bring no added value to the matter at hand. More specifically, although later in this chapter π_t will be chosen as the price of a bond, it happens that in the current generic section its financial nature is rather irrelevant. Indeed π_t and its dynamics unambiguously define, as usual, an associated measure $\mathbb{Q}^{\pi}$ and a Wiener driver $\vec{W}_t^{\pi}$. We then use this new numeraire to deflate[2] the whole HJM Zero-Coupon map:

$$d\vec{W}_t^{\pi} = d\vec{W}_t - \vec{\gamma_t}\, dt \qquad \text{and} \qquad \forall\, T \geq t \quad B_t^{\pi}(T) \triangleq \frac{B_t(T)}{\pi_t}.$$

We then have under $\mathbb{Q}^{\pi}$ some drift-less dynamics for the new deflated Zero Coupon map $B_t^{\pi}(T)$. Due to the additive nature of the new volatility map $T \to \vec{\Gamma}_t^{\pi}(T)$ it is also easy to present its associated chaos structure under the new measure, at least at the σ-(2,0) level. Formally we present the following utilitarian result:

[2] This technique is also commonly called re-scaling or rebasing in the literature.

Lemma 6.1 (Dynamics of the rebased Zeros) *The rebased Zero Coupons prices satisfy the following SDE system*

$$\forall\, T \geq t \qquad \frac{dB_t^\pi(T)}{B_t^\pi(T)} = \overrightarrow{\Gamma}_t^\pi \, d\overrightarrow{W}_t^\pi \qquad \textit{with} \qquad \overrightarrow{\Gamma}_t^\pi(T) \triangleq \overrightarrow{\Gamma}_t(T) - \overrightarrow{\gamma}_t \tag{6.4}$$

with the chaos dynamics continuing as

$$\begin{cases} d\overrightarrow{\Gamma}_t^\pi(T) = \overrightarrow{\alpha}_{1,t}^\pi(T)\, dt + \overrightarrow{\overrightarrow{\alpha}}_{2,t}^\pi(T)\, d\overrightarrow{W}_t^\pi + \overrightarrow{\overrightarrow{\alpha}}_{3,t}^\pi(T)\, d\overrightarrow{Z}_t \\ d\overrightarrow{\overrightarrow{\alpha}}_{2,t}^\pi(T) = \overrightarrow{\overrightarrow{[\cdots]}}\, dt + \overrightarrow{\overrightarrow{\overrightarrow{\alpha}}}_{22,t}^\pi(T)\, d\overrightarrow{W}_t^\pi + \overrightarrow{\overrightarrow{\overrightarrow{[\cdots]}}}\, d\overrightarrow{Z}_t \end{cases}$$

and where the coefficient correspondence is given by

$$\overrightarrow{\overrightarrow{\alpha}}_{2,t}^\pi(T) = \overrightarrow{\overrightarrow{\alpha}}_{2,t}(T) - \overrightarrow{\overrightarrow{c}}_{2,t}, \tag{6.5}$$

$$\overrightarrow{\alpha}_{1,t}^\pi(T) = \overrightarrow{\alpha}_{1,t}(T) - \overrightarrow{c}_{1,t} + \overrightarrow{\overrightarrow{\alpha}}_{2,t}^\pi(T)\overrightarrow{\gamma}_t, \tag{6.6}$$

$$\overrightarrow{\overrightarrow{\alpha}}_{3,t}^\pi(T) = \overrightarrow{\overrightarrow{\alpha}}_{3,t}(T) - \overrightarrow{\overrightarrow{c}}_{3,t}, \tag{6.7}$$

$$\overrightarrow{\overrightarrow{\overrightarrow{\alpha}}}_{22,t}^\pi(T) = \overrightarrow{\overrightarrow{\overrightarrow{\alpha}}}_{22,t}(T) - \overrightarrow{\overrightarrow{\overrightarrow{c}}}_{22,t}. \tag{6.8}$$

In the sequel we shall of course adapt the simplified notations of Sect. 6.1.2 to this new rebased environment, for instance

$$\overrightarrow{\Gamma}_i^\pi \doteq \overrightarrow{\Gamma}_{i,t}^\pi \doteq \overrightarrow{\Gamma}_{i,t}^\pi(T_0) \triangleq \overrightarrow{\Gamma}_t^\pi(T_i).$$

6.2.2 Dynamics of a Fixed-Weights Rebased Zero Coupon Basket

Let P_t^π be a *fixed-weights* basket of rebased ZCs, with current deflated price

$$P_t^\pi = \frac{P_t}{\pi_t} = \sum_{i=1}^{N} \beta_i \; B_t^\pi(T_i). \tag{6.9}$$

This basket can, of course, be seen as a rebased bond with fixed (possibly unequal albeit deterministic) coupons, but its use will be further-reaching. For the moment we wish to compute the σ-(2,0) coefficients of its chaos dynamics. We naturally apply straight the results of Corollary 3.4 [p. 202] relating to fixed-weights baskets, to obtain:

Corollary 6.1 (Bond basket's chaos dynamics under a generic mart. measure) *Following the definition presented in* (6.9), *we get the basket dynamics as*

$$\begin{cases} \dfrac{dP_t^\pi}{P_t^\pi} = \vec{\sigma}_t^{p,\pi\,\perp} \, d\vec{W}_t^\pi \\ d\,\vec{\sigma}_t{}^{p,\pi} = \vec{a}_{1,t}^{p,\pi}\, dt + \vec{\vec{a}}_{2,t}^{p,\pi}\, d\vec{W}_t^\pi + \vec{\vec{a}}_{3,t}^{p,\pi}\, d\vec{Z}_t \\ d\vec{\vec{a}}_{2,t}^{p,\pi} = \vec{\vec{[\cdot]}}\, dt + \vec{\vec{\vec{a}}}_{22,t}^{p,\pi}\, d\vec{W}_t^\pi + \vec{\vec{\vec{[\cdot]}}}\, d\vec{Z}_t. \end{cases}$$

These σ *-(2,0) coefficients are expressed as follows, both in the re-based and native forms. First the instantaneous volatility comes as*

$$\vec{\sigma}_t^{p,\pi} = \sum_{i=1}^{N} \omega_{i,t}\, \vec{\Gamma}_{i,t}^\pi = -\vec{\gamma}_t + \sum_{i=1}^{N} \omega_{i,t}\, \vec{\Gamma}_{i,t} \tag{6.10}$$

where

$$\omega_{i,t} \triangleq \frac{\beta_i\; B_t^\pi(T_i)}{P_t^\pi} = \frac{\beta_i\; B_t(T_i)}{P_t}. \tag{6.11}$$

Using simplified notations, the first-depth coefficients are expressed as

$$\vec{a}_{1,t}^{p,\pi} \triangleq -\vec{\vec{a}}_{2,t}^{p,\pi}\, \vec{\sigma}_t^{p,\pi} + \sum_{i=1}^{N} \omega_{i,t} \left[\vec{\alpha}_{i,1,t}^\pi + \vec{\vec{\alpha}}_{i,2,t}^\pi\, \vec{\Gamma}_{i,t}^\pi \right] \tag{6.12}$$

$$\begin{aligned} &= -\vec{\vec{a}}_{2,t}^{p,\pi}\, \vec{\sigma}_t^{p,\pi} - \vec{c}_{1,t} \\ &\quad - \vec{\vec{c}}_{2,t} \sum_{i=1}^{N} \omega_{i,t}\, \vec{\Gamma}_{i,t} + \sum_{i=1}^{N} \omega_{i,t} \left[\vec{\alpha}_{i,1,t} + \vec{\vec{\alpha}}_{i,2,t}\, \vec{\Gamma}_{i,t} \right] \end{aligned} \tag{6.13}$$

for the drift, then the endogenous term comes as

$$\vec{\vec{a}}_{2,t}^{p,\pi} \triangleq -\vec{\sigma}_t^{p,\pi\, 2\otimes} + \sum_{i=1}^{N} \omega_{i,t} \left[\vec{\vec{\alpha}}_{i,2,t}^\pi + \vec{\Gamma}_{i,t}^{\pi\; 2\otimes} \right] \tag{6.14}$$

$$= -\vec{\vec{c}}_{2,t} + \sum_{i=1}^{N} \omega_{i,t} \left(\vec{\vec{\alpha}}_{i,2,t} + \vec{\Gamma}_{i,t}^{\,2\otimes} \right) - \left[\sum_{i=1}^{N} \omega_{i,t} \vec{\Gamma}_{i,t} \right]^{2\otimes} \tag{6.15}$$

while the exogenous component is more simply expressed as

$$\vec{\vec{a}}_{3,t}^{p,\pi} \triangleq \sum_{i=1}^{N} \omega_{i,t}\, \vec{\vec{\alpha}}_{i,3,t}^\pi = -\vec{\vec{c}}_{3,t} + \sum_{i=1}^{N} \omega_{i,t}\, \vec{\vec{\alpha}}_{i,3,t}. \tag{6.16}$$

Finally the single, second-depth coefficient is

$$\overset{\Rrightarrow}{a}{}^{p,\pi}_{22,t} = -\overset{\Rightarrow}{a}{}^{p,\pi}_{2} \overset{\leftrightarrow}{\otimes} \vec{\sigma}^{\,p,\pi} + \sum_{i=1}^{N} \omega_i \, \overset{\Rightarrow}{\alpha}{}^{\pi}_{i,2} \overset{\leftrightarrow}{\otimes} \vec{\Gamma}^{\pi}_{i}$$
$$+ \sum_{i=1}^{N} w_i \left[\overset{\Rrightarrow}{\alpha}{}^{\pi}_{i,22} + \left[\left(\Gamma^{\pi}_{i} - \sigma^{p,\pi} \right)_k \alpha^{\pi}_{i,2\; jl} \right] \right.$$
$$\left. + \vec{\Gamma}^{\pi\; 3\otimes}_{i} - \vec{\Gamma}^{\pi}_{i} \otimes \vec{\sigma}^{\,p,\pi} \otimes \vec{\Gamma}^{\pi}_{i} \right] \tag{6.17}$$

which, with the native HIM parametrisation, gives

$$\overset{\Rrightarrow}{a}{}^{p,\pi}_{22,t} = -\overset{\Rrightarrow}{c}_{22} + \left[w_{\mathbf{i}} \, \overset{\Rrightarrow}{\alpha}_{\mathbf{i},22} \right] + \left[\omega_{\mathbf{i}} \, \overset{\Rightarrow}{\alpha}_{\mathbf{i},2} \overset{\leftrightarrow}{\otimes} \vec{\Gamma}_{\mathbf{i}} \right]$$
$$- \left[\omega_{\mathbf{i}} \, \overset{\Rightarrow}{\alpha}_{i,2} \right] \overset{\leftrightarrow}{\otimes} \left[\omega_{\mathbf{i}} \vec{\Gamma}_{\mathbf{i}} \right] + \left[w_{\mathbf{i}} \, \vec{\Gamma}^{3\otimes}_{\mathbf{i}} \right]$$
$$- \left[\omega_{\mathbf{i}} \, \vec{\Gamma}_{\mathbf{i}}^{\;2\otimes} \right] \overset{\leftrightarrow}{\otimes} \left[\omega_{\mathbf{i}} \vec{\Gamma}_{\mathbf{i}} \right] + 2 \left[\omega_{\mathbf{i}} \, \vec{\Gamma}_{\mathbf{i}} \right]^{3\otimes}$$
$$- \left[w_{\mathbf{i}} w_{\mathbf{j}} \, \vec{\Gamma}_{\mathbf{i}} \otimes \vec{\Gamma}_{\mathbf{j}} \otimes \vec{\Gamma}_{\mathbf{i}} \right] + \left[w_{\mathbf{i}} \left(\Gamma_{\mathbf{i}} - \left[w_{\mathbf{j}} \Gamma_{\mathbf{j}} \right] \right)_n \alpha_{\mathbf{i},2\; mp} \right]. \tag{6.18}$$

Proof These results come as a direct application of Corollary 3.4 [p. 202] and Lemma 6.1 [p. 329], where the population of N underlyings corresponds to the basket of Zero Coupons. First the volatility expression (6.10) follows simply as a combination of (3.94) and (6.4). Similarly, the exogenous expressions (3.97) and (6.5) yields (6.16). Then the first endogenous expression (6.14) is a direct application of (3.96), which we can re-formulate as a function of the original inputs using (6.4), (6.5) and (6.10):

$$\overset{\Rightarrow}{a}{}^{p,\pi}_{2,t} = -\left[\sum_{i=1}^{N} \omega_{i,t} \, \vec{\Gamma}_{i,t} - \vec{\gamma}_t \right]^{2\otimes} + \sum_{i=1}^{N} \omega_{i,t} \left[\overset{\Rightarrow}{\alpha}_{i,2,t} - \overset{\Rightarrow}{c}_{2,t} + \left[\vec{\Gamma}_{i,t} - \vec{\gamma}_t \right]^{2\otimes} \right]$$
$$= -\left[\sum_{i=1}^{N} \omega_{i,t} \vec{\Gamma}_{i,t} \right]^{2\otimes} + \vec{\gamma}_t \overset{\leftrightarrow}{\otimes} \left[\sum_{i=1}^{N} \omega_{i,t} \vec{\Gamma}_{i,t} \right] - \vec{\gamma}_t^{\,2\otimes} - \overset{\Rightarrow}{c}_{2,t}$$
$$+ \sum_{i=1}^{N} \omega_{i,t} \overset{\Rightarrow}{\alpha}_{i,2,t} + \sum_{i=1}^{N} \omega_{i,t} \vec{\Gamma}^{2\otimes}_{i,t} - \vec{\gamma}_t \overset{\leftrightarrow}{\otimes} \left[\sum_{i=1}^{N} \omega_{i,t} \vec{\Gamma}_{i,t} \right] + \vec{\gamma}_t^{\,2\otimes},$$

which after simplification proves (6.15). Turning to the drift coefficient, we apply the Corollary result 3.95 which provides (6.14). Then recalling the native volatility map, we get

$$\vec{a}^{\,p,\pi}_{1,t} \triangleq -\vec{\vec{a}}^{\,p,\pi}_{2,t}\,\vec{\sigma}^{\,p,\pi}_{t} + \sum_{i=1}^{N} \omega_{i,t}\left[\vec{\alpha}_{i,1,t} - \vec{c}_{1,t} + \left(\vec{\vec{\alpha}}_{i,2,t} - \vec{\vec{c}}_{2,t}\right)\vec{\gamma}_t\right]$$
$$+\sum_{i=1}^{N}\omega_{i,t}\left[\vec{\vec{\alpha}}_{i,2,t} - \vec{\vec{c}}_{2,t}\right]\left[\vec{\Gamma}_{i,t} - \vec{\gamma}_t\right],$$

which by using (6.4)–(6.6) we simplify into (6.13). Finally, a direct application of (3.98) provides the second-depth coefficient as per (6.17). Noting that

$$\vec{\Gamma}^{\pi\,3\otimes}_{i} - \vec{\Gamma}^{\pi}_{i} \otimes \vec{\sigma}^{\,p,\pi} \otimes \vec{\Gamma}^{\pi}_{i} = \vec{\Gamma}^{\pi}_{i} \otimes \left[\vec{\Gamma}^{\pi}_{i} - \vec{\sigma}^{\,p,\pi}\right] \otimes \vec{\Gamma}^{\pi}_{i}$$

we get by substituting the rebased coefficients with the native ones

$$\vec{\vec{a}}^{\,p,\pi}_{22,t} = \left[\vec{\vec{c}}_2 - \sum_{i=1}^{N}\omega_i\left(\vec{\vec{\alpha}}_{i,2} + \vec{\Gamma}_i^{\,2\otimes}\right) + \left[\sum_{i=1}^{N}\omega_i\vec{\Gamma}_i\right]^{2\otimes}\right]$$
$$\overset{\leftrightarrow}{\otimes}\left[\sum_{i=1}^{N}\omega_i\vec{\Gamma}_i - \vec{\gamma}\right] + \sum_{i=1}^{N}\omega_i\left[\vec{\vec{\alpha}}_{i,2} - \vec{\vec{c}}_2\right]\overset{\leftrightarrow}{\otimes}\left[\vec{\Gamma}_i - \vec{\gamma}\right]$$
$$+\sum_{i=1}^{N}w_i\left[\vec{\vec{\vec{\alpha}}}_{i,22} - \vec{\vec{\vec{c}}}_{22}\right] + \sum_{i=1}^{N}w_i\left[\left(\Gamma_i - \sum_{j=1}^{N}w_j\Gamma_j\right)_n\left(\alpha_{i,2} - c_2\right)_{mp}\right]$$
$$+\sum_{i=1}^{N}w_i\left[\vec{\Gamma}_i - \vec{\gamma}\right]\otimes\left[\vec{\Gamma}_i - \sum_{j=1}^{N}w_j\vec{\Gamma}_j\right]\otimes\left[\vec{\Gamma}_i - \vec{\gamma}\right].$$

We identify three terms on the r.h.s., which we expand and simplify individually.

Term 1

$$A \triangleq \left[\vec{\vec{c}}_2 - \sum_{i=1}^{N}\omega_i\left[\vec{\vec{\alpha}}_{i,2} + \vec{\Gamma}_i^{\,2\otimes}\right] + \left[\sum_{i=1}^{N}\omega_i\vec{\Gamma}_i\right]^{2\otimes}\right]\overset{\leftrightarrow}{\otimes}\left[\sum_{i=1}^{N}\omega_i\vec{\Gamma}_i - \vec{\gamma}\right]$$
$$+\sum_{i=1}^{N}\omega_i\left[\vec{\vec{\alpha}}_{i,2} - \vec{\vec{c}}_2\right]\overset{\leftrightarrow}{\otimes}\left[\vec{\Gamma}_i - \vec{\gamma}\right],$$

which simplifies as

$$A = \left[\left[\sum_{i=1}^{N} \omega_i \vec{\Gamma}_i\right]^{2\otimes} - \sum_{i=1}^{N} \omega_i \vec{\Gamma}_i^{2\otimes} - \sum_{i=1}^{N} \omega_i \vec{\vec{\alpha}}_{i,2}\right] \overset{\leftrightarrow}{\otimes} \left[\sum_{i=1}^{N} \omega_i \vec{\Gamma}_i - \vec{\gamma}\right]$$
$$+ \sum_{i=1}^{N} \omega_i \vec{\vec{\alpha}}_{i,2} \overset{\leftrightarrow}{\otimes} \left[\vec{\Gamma}_i - \vec{\gamma}\right]$$

and then expands into

$$A = 2\left[\sum_{i=1}^{N} \omega_i \vec{\Gamma}_i\right]^{3\otimes} - \left[\sum_{i=1}^{N} \omega_i \vec{\Gamma}_i^{2\otimes}\right] \overset{\leftrightarrow}{\otimes} \left[\sum_{i=1}^{N} \omega_i \vec{\Gamma}_i\right] + \sum_{i=1}^{N} \omega_i \left[\vec{\vec{\alpha}}_{i,2} \overset{\leftrightarrow}{\otimes} \vec{\Gamma}_i\right]$$
$$- \left[\sum_{i=1}^{N} \omega_i \vec{\vec{\alpha}}_{i,2}\right] \overset{\leftrightarrow}{\otimes} \left[\sum_{i=1}^{N} \omega_i \vec{\Gamma}_i\right] - \vec{\gamma} \overset{\leftrightarrow}{\otimes} \left[\sum_{i=1}^{N} \omega_i \vec{\Gamma}_i\right]^{2\otimes}$$
$$+ \vec{\gamma} \overset{\leftrightarrow}{\otimes} \left[\sum_{i=1}^{N} \omega_i \vec{\Gamma}_i^{2\otimes}\right].$$

Term 2
Since we have

$$\sum_{i=1}^{N} w_i \left[\vec{\Gamma}_i - \sum_{j=1}^{N} w_j \vec{\Gamma}_j\right] = \vec{0}$$

we can simplify as follows

$$\sum_{i=1}^{N} w_i \left[\vec{\vec{\alpha}}_{i,22} - \vec{\vec{c}}_{22}\right] + \sum_{i=1}^{N} w_i \left[\left(\Gamma_i - \sum_{j=1}^{N} w_j \Gamma_j\right)_n (\alpha_{i,2} - c_2)_{mp}\right]$$
$$= -\vec{\vec{c}}_{22} + \sum_{i=1}^{N} w_i \vec{\vec{\alpha}}_{i,22} + \sum_{i=1}^{N} w_i \left[\left(\Gamma_i - \sum_{j=1}^{N} w_j \Gamma_j\right)_n \alpha_{i,2\,mp}\right].$$

Term 3

$$B \triangleq \sum_{i=1}^{N} w_i \left[\vec{\Gamma}_i - \vec{\gamma}\right] \otimes \left[\vec{\Gamma}_i - \sum_{j=1}^{N} w_j \vec{\Gamma}_j\right] \otimes \left[\vec{\Gamma}_i - \vec{\gamma}\right]$$

can be re-organised as

$$B = \left[\sum_{i=1}^{N} w_i \vec{\Gamma}_i^{3\otimes}\right] + \vec{\gamma} \overset{\leftrightarrow}{\otimes} \left[\sum_{i=1}^{N} w_i \vec{\Gamma}_i\right]^{2\otimes}$$
$$- \vec{\gamma} \overset{\leftrightarrow}{\otimes} \left[\sum_{i=1}^{N} w_i \vec{\Gamma}_i^{2\otimes}\right] - \sum_{i=1}^{N}\sum_{j=1}^{N} w_i w_j \left[\vec{\Gamma}_i \otimes \vec{\Gamma}_j \otimes \vec{\Gamma}_i\right].$$

Gathering all three terms we get

$$\overset{\Rrightarrow}{\alpha}{}^{p,\pi}_{22,t} = 2\left[\sum_{i=1}^{N} \omega_i \vec{\Gamma}_i\right]^{3\otimes} - \left[\sum_{i=1}^{N} \omega_i \vec{\Gamma}_i^{2\otimes}\right] \overset{\leftrightarrow}{\otimes} \left[\sum_{i=1}^{N} \omega_i \vec{\Gamma}_i\right]$$
$$+ \sum_{i=1}^{N} \omega_i \left[\overset{\Rrightarrow}{\alpha}_{i,2} \overset{\leftrightarrow}{\otimes} \vec{\Gamma}_i\right] - \left[\sum_{i=1}^{N} \omega_i \overset{\Rrightarrow}{\alpha}_{i,2}\right] \overset{\leftrightarrow}{\otimes} \left[\sum_{i=1}^{N} \omega_i \vec{\Gamma}_i\right]$$
$$- \vec{\gamma} \overset{\leftrightarrow}{\otimes} \left[\sum_{i=1}^{N} \omega_i \vec{\Gamma}_i\right]^{2\otimes} + \vec{\gamma} \overset{\leftrightarrow}{\otimes} \left[\sum_{i=1}^{N} \omega_i \vec{\Gamma}_i^{2\otimes}\right] - \overset{\Rrightarrow}{c}_{22}$$
$$+ \sum_{i=1}^{N} w_i \overset{\Rrightarrow}{\alpha}_{i,22} + \sum_{i=1}^{N} w_i \left[\left(\Gamma_i - \sum_{j=1}^{N} w_j \Gamma_j\right)_n \alpha_{i,2\,mp}\right]$$
$$+ \left[\sum_{i=1}^{N} w_i \vec{\Gamma}_i^{3\otimes}\right] + \vec{\gamma} \overset{\leftrightarrow}{\otimes} \left[\sum_{i=1}^{N} w_i \vec{\Gamma}_i\right]^{2\otimes} - \vec{\gamma} \overset{\leftrightarrow}{\otimes} \left[\sum_{i=1}^{N} w_i \vec{\Gamma}_i^{2\otimes}\right]$$
$$- \sum_{i=1}^{N}\sum_{j=1}^{N} w_i w_j \left[\vec{\Gamma}_i \otimes \vec{\Gamma}_j \otimes \vec{\Gamma}_i\right].$$

This equation then simplifies as

$$\overset{\Rrightarrow}{\alpha}{}^{p,\pi}_{22,t} = -\overset{\Rrightarrow}{c}_{22} + \sum_{i=1}^{N} w_i \overset{\Rrightarrow}{\alpha}_{i,22} + \sum_{i=1}^{N} \omega_i \left[\overset{\Rrightarrow}{\alpha}_{i,2} \overset{\leftrightarrow}{\otimes} \vec{\Gamma}_i\right]$$
$$- \left[\sum_{i=1}^{N} \omega_i \overset{\Rrightarrow}{\alpha}_{i,2}\right] \overset{\leftrightarrow}{\otimes} \left[\sum_{i=1}^{N} \omega_i \vec{\Gamma}_i\right] + \left[\sum_{i=1}^{N} w_i \vec{\Gamma}_i^{3\otimes}\right]$$
$$- \left[\sum_{i=1}^{N} \omega_i \vec{\Gamma}_i^{2\otimes}\right] \overset{\leftrightarrow}{\otimes} \left[\sum_{i=1}^{N} \omega_i \vec{\Gamma}_i\right] + 2\left[\sum_{i=1}^{N} \omega_i \vec{\Gamma}_i\right]^{3\otimes}$$

$$
\begin{aligned}
&- \sum_{i=1}^{N} \sum_{j=1}^{N} w_i w_j \left[\vec{\Gamma}_i \otimes \vec{\Gamma}_j \otimes \vec{\Gamma}_i \right] \\
&+ \sum_{i=1}^{N} w_i \left[\left[\Gamma_i - \sum_{j=1}^{N} w_j \Gamma_j \right]_n \alpha_{i,2\,mp} \right].
\end{aligned}
$$

Introducing the modified Einstein notation provides (6.18) and concludes the proof. □

With the rebasing of bonds sorted in terms of dynamics, we are now ready to proceed with our asymptotic approach. Out of the three product families we choose to start with the oldest, the bond options, which is also the one demonstrating with the most clarity the benefits of that rebasing.

6.3 Bond Options

In this section we provide the input required by the asymptotic chaos expansion methodology, up to the first layer, for a term structure of bond options and within an HJM modelling framework. It seems that a European Call written on a vanilla bond, when using an HJM parametrisation, is a very natural instrument to consider. Indeed, the underlying consists in a fixed-weight basket of Zero-Coupons, which are themselves the native *state variables* for the yield curve.

6.3.1 Casting the Bond Options Into the Generic Framework

For simplification purposes, we will avoid *accrued coupon* issues by imposing that the *option maturity* T must coincide with an anniversary date T_0 of the underlying bond. In terms of maturities, tenor structures are bi-dimensional in nature. Therefore, in order to build the necessary option/underlying term structure we *fix the tenor* and allocate the framework maturity to the start date T_0, effectively making the bond *sliding* w.r.t. that date. In summary

$$
\begin{aligned}
& & T = T_0 & \qquad \text{is the option expiry,} \\
&\forall i \;/\; 1 \le i \le N & T_i = T_{i-1} + \delta_i & \qquad \text{are the coupon dates.}
\end{aligned}
$$

The underlying sliding bond is then defined by three elements: by its maturity T, by the sliding tenor structure $\delta \triangleq \{\delta_i\}_{1 \le i \le N}$ and by the fixed coupon c. In consequence, we shall denote its price process by $P_t^{\delta}(T; c)$ and we will employ the schedule notations of Sect. 6.1.2. As for the option payoff, it will assume a cash exchange[3] at time T for a value of

$$\Phi(T) \triangleq \left[P_T^\delta(T;c) - K\right]^+ .$$

Our choice of mapping w.r.t. an anniversary date and with a fixed tenor is not the sole alternative, but in our view both the simplest and the most useful choice. However, it cannot be used to represent all European options (all strikes and expiries) written on an existing, *running* bond. Although (conceptually) feasible, designing a framework to support the latter case is significantly more complex. Indeed, it must cope with the embedded *discontinuity* of the underlying and of the option settlements, which does not fare well in our *asymptotic* perspective.

Having defined the underlying and option continuum, let us move on to the SDE system. In order to make the bond's dynamics martingale, we shall use the usual method of re-basing *via* a given asset. In light of the simple payment sequence, that numeraire should naturally be the Zero Coupon $B_t(T)$ for the same maturity.[4] Adapting the notations of Sect. 6.2, we write the rebased Zeros and define the underlying Term Structure as per

$$B_t^T(U) \triangleq \frac{B_t(U)}{B_t(T)}$$

$$\text{and} \quad X_t(T) \longleftarrow P_t^{\delta,\mathbf{T}}(T;c) \triangleq \frac{P_t^\delta(T;c)}{B_t(T)} = c \sum_{i=1}^{N} \delta_i B_t^T(T_i).$$

The numeraire being selected, we have the martingale measure as the T-forward measure. In summary, the immersion continues with the following allocations:

$$N_t(T) \leftarrow B_t(T) \qquad \mathbb{Q}^{N(T)} \leftarrow \mathbb{Q}^T$$

$$\vec{W}_t^{N(T)} \leftarrow \vec{W}_t^T \quad \text{with} \quad d\vec{W}_t^T = d\vec{W}_t - \vec{\Gamma}_t(T)\,dt.$$

Finally, we verify that the actual payoff is coherent with the framework, which is true since

$$B_T(T) \equiv 1 \quad \Longrightarrow \quad \left[P_T^\delta(T;c) - K\right]^+ = B_T(T)\left[P_T^{\delta,\mathbf{T}}(T;c) - K\right]^+ .$$

At this point, the Black formula and the corresponding re-parametrisation *via* the Lognormal implied volatility presents no difficulty, and completes the formal cast.

---

[3] As opposed to a physical settlement, but unlike the swaption case here the valuations are identical.
[4] Note that, since for all considered indices i we have $T \le T_i$, the re-based ZC defined by $B_{i,t}^T(T) \triangleq B_t^T(T_i) \triangleq B_t(T_i)/B_t(T)$ is nothing else than the *Forward Zero Coupon* $B_t^f(T, T_i)$.

6.3.2 Dynamics of the Underlying Rebased Bond

It remains, however, to compute the first layer, i.e. the chaos dynamics under $\mathbb{Q}^T$ of the underlying term structure of rebased bonds. Since the scaling by the coupon c does not affect the Lognormal dynamics, we shall dispense with that parameter.

Lemma 6.2 (Bond chaos dynamics in an HJM framework) *The underlying bond, rebased by the maturity ZC, shows the following dynamics:*

$$\begin{aligned} \frac{dP_t^{\delta,T}(T)}{P_t^{\delta,T}(T)} &= \overrightarrow{\sigma}_t^{\,p}(T)^{\perp} d\overrightarrow{W}_t^T \\ d\,\overrightarrow{\sigma}_t^{\,p}(T) &= \overrightarrow{a}_{1,t}^{\,p}(T)\,dt + \overset{\Rightarrow}{a}_{2,t}^{\,p}(T)\,d\overrightarrow{W}_t^T + \overset{\Rightarrow}{a}_{3,t}^{\,p}(T)\,d\overrightarrow{Z}_t \\ d\overset{\Rightarrow}{a}_{2,t}^{\,p}(T) &= \overset{(2)}{\overrightarrow{[\cdots]}}\,dt + \overset{\Rrightarrow}{a}_{22,t}^{\,p}(T)\,d\overrightarrow{W}_t^T + \overset{(3)}{\overrightarrow{[\cdots]}}\,d\overrightarrow{Z}_t \end{aligned}$$

where $\overrightarrow{W}_t^T$ is the T-forward driver. The instantaneous volatility comes as

$$\overrightarrow{\sigma}_t^{\,p}(T) = -\overrightarrow{\Gamma_{0,t}}(T) + \sum_{i=1}^{N} \omega_{i,t}(T)\,\overrightarrow{\Gamma_{i,t}}(T) \quad \textit{where} \quad \omega_{i,t} \triangleq \frac{\delta_i\; B_{i,t}}{\sum_{k=1}^{N} \delta_k B_{k,t}}. \tag{6.19}$$

We thereafter use simplified notations, understating that each quoted quantity is a process parametrised by the maturity T, and that modified Einstein notation sums from indices 1 to N. We get

$$\begin{aligned} \overrightarrow{a}_{1,t}^{\,p}(T) &\triangleq -\overset{\Rightarrow}{a}_{2,t}^{\,p}\;\overrightarrow{\sigma}_t^{\,p} - \overrightarrow{\alpha}_{0,1} - \overset{\Rightarrow}{\alpha}_{0,2}\left[\omega_{\mathbf{i}}\;\overrightarrow{\Gamma_{\mathbf{i}}}\right] \\ &\quad + \left[\omega_{\mathbf{i}}\;\left(\overrightarrow{\alpha}_{\mathbf{i},1} + \overset{\Rightarrow}{\alpha}_{\mathbf{i},2}\;\overrightarrow{\Gamma_{\mathbf{i}}}\right)\right], \end{aligned} \tag{6.20}$$

$$\overset{\Rightarrow}{a}_{2,t}^{\,p}(T) \triangleq -\overset{\Rightarrow}{\alpha}_{0,2} + \left[\omega_{\mathbf{i}}\left(\overset{\Rightarrow}{\alpha}_{\mathbf{i},2} + \overrightarrow{\Gamma}_{\mathbf{i}}^{\,2\otimes}\right)\right] - \left[\omega_{\mathbf{i}}\overrightarrow{\Gamma}_{\mathbf{i}}\right]^{2\otimes}, \tag{6.21}$$

$$\overset{\Rightarrow}{a}_{3,t}^{\,p}(T) \triangleq -\overset{\Rightarrow}{\alpha}_{0,3} + \left[\omega_{\mathbf{i}}\;\overset{\Rightarrow}{\alpha}_{\mathbf{i},3}\right]. \tag{6.22}$$

Finally, the single second-depth coefficient is

$$\begin{aligned} \overset{\Rrightarrow}{a}_{22,t}^{\,p}(T) =& -\overset{\Rrightarrow}{\alpha}_{0,22} + \left[w_{\mathbf{i}}\;\overset{\Rrightarrow}{\alpha}_{\mathbf{i},22}\right] + \left[\omega_{\mathbf{i}}\;\overset{\Rightarrow}{\alpha}_{\mathbf{i},2} \overset{\leftrightarrow}{\otimes} \overrightarrow{\Gamma}_{\mathbf{i}}\right] - \left[\omega_{\mathbf{i}}\;\overset{\Rightarrow}{\alpha}_{i,2}\right] \overset{\leftrightarrow}{\otimes} \left[\omega_{\mathbf{i}}\overrightarrow{\Gamma}_{\mathbf{i}}\right] \\ & + \left[w_{\mathbf{i}}\;\overrightarrow{\Gamma}_{\mathbf{i}}^{3\otimes}\right] - \left[\omega_{\mathbf{i}}\;\overrightarrow{\Gamma}_{\mathbf{i}}^{\,2\otimes}\right] \overset{\leftrightarrow}{\otimes} \left[\omega_{\mathbf{i}}\overrightarrow{\Gamma}_{\mathbf{i}}\right] + 2\;\left[\omega_{\mathbf{i}}\;\overrightarrow{\Gamma}_{\mathbf{i}}\right]^{3\otimes} \\ & - \left[w_{\mathbf{i}}w_{\mathbf{j}}\;\overrightarrow{\Gamma}_{\mathbf{i}} \otimes \overrightarrow{\Gamma}_{\mathbf{j}} \otimes \overrightarrow{\Gamma}_{\mathbf{i}}\right] + \left[\;w_{\mathbf{i}}\left(\Gamma_{\mathbf{i}} - \left[w_{\mathbf{j}}\Gamma_{\mathbf{j}}\right]\right)_n\;\alpha_{\mathbf{i},2\,mp}\right]. \end{aligned} \tag{6.23}$$

Proof The case is obviously ripe to apply Corollary 6.1 [p. 330], dealing with fixed-weights baskets of Zero Coupons. Indeed, the basket is here defined without ambiguity, and the numeraire π_t will naturally be the Zero-Coupon $B_t(T)$, which gives us the result immediately. □

Armed with these results, we can now, for instance, invoke Theorem 5.2 [p. 310] in order to compute the IATM differentials of the smile.

6.3.3 Interpretation

Due to their complexity, interpreting the dynamics of the rebased (T-forward) bond, as exposed by Lemma 6.2, is not really straightforward. Logically, the IATM differentials that they imply—through the direct results of Chap. 5—are also quite involved. Nevertheless, it is possible to make a few initial observations.

A very noticeable feature is the important role played in the dynamics by the omnipresent numeraire, here $B_t(T)$. Indeed, in terms of overall magnitude impact, the numeraire's volatility (along with its further chaos-generated coefficients) seems to have roughly the same importance as the N Zero Coupons constituting the bond, *all put together*. The impact of the numeraire on the IATM differentials itself is more difficult to gauge. But as far as the IATM level is concerned, we can build some intuition. Indeed, we have

$$
\begin{aligned}
\widetilde{\Sigma}^p(t,0,0)^2 &= \|\vec{\sigma}_t^{\,p}(t)\|^2 \\
&= \|\overrightarrow{\Gamma_{0,t}}(t)\|^2 + \left\| \sum_{i=1}^N \omega_{i,t}(t)\, \overrightarrow{\Gamma_{i,t}}(t) \right\|^2 \\
&\quad - 2\, \overrightarrow{\Gamma_{0,t}}(t)^{\perp} \left[\sum_{i=1}^N \omega_{i,t}(t)\, \overrightarrow{\Gamma_{i,t}}(t) \right].
\end{aligned}
$$

However, since $\vec{\Gamma}_t(t)$ is the instantaneous Lognormal volatility of the immediate Zero $B_t(t)$, it must stay null almost surely. In other words, the IATM level of the smile is unaffected by the numeraire, which is reassuring since the immediate forward bond is nothing else than the spot bond.

An easy mistake to make would be to assume that all further IATM differentials are also immune to the numeraire specification. Looking at Theorem 5.2 [p. 310], and in particular at the slope specification (5.53), it is clear that the immediate shape of both the yield curve and its volatility map (i.e. $\partial_T B_t(T)$ and $\partial_T \vec{\Gamma}_t(T)$ taken in $T = t$) will have an impact on the maturity differentials of the smile, at the IATM point.

Another tempting mistake would be to consider that $B_t(T)$ has no impact on the IATM *price*: indeed the pricing argument (5.5) requires the multiplicative discounting by $B_t(T)$ in order to obtain the dollar price. From this operation, we deduce naturally that any positive correlation between the zero $B_t(T)$ and the bond $P_t(T)$ will have a *positive* effect on the price.

Of some interest to us is the repeated presence of the mixed volatility term

$$\sum_{i=1}^{N} \omega_{i,t}(T)\ \vec{\Gamma}_{i,t}(T)$$

which suggests that we should again exploit the basket analogy further.

It is clear that in traditional market circumstances, discount factors should decrease with expiry, so that the normalised weights $\omega_{i,t}$ will decrease with the index i. However, in practice, and in particular if the start date T_0 is far enough in the future, their magnitude will be similar. In consequence, the norm of that mixed volatility term will essentially report the *level of instantaneous correlation* within the yield curve segment $[T_1, T_N]$. Indeed, a high de-correlation translates by a large directional dispersion of the $\vec{\Gamma}_t(T_i)$ vectors.

Looking now at the skew as expressed by (5.51), we do note that the numeraire volatility coefficients are starting to interfere more deeply, in particular we observe the apparition of cross-terms which are difficult to gauge. But since those terms will also affect the IATM level, it makes sense to ignore them at the first order, as part of the normalisation issue. We then see that the *IATM skew* exhibits the typical sensitivity to dispersion that was observed with baskets, only this time the dispersion in question relates to the instantaneous correlation of the yield curve.

Let us now turn to the second of the three derivative families, the Caplets, which by its relative simplicity will allow us to be more interpretative than with the bond options.

6.4 Caplets

In this section we demonstrate the most straightforward approach allowing us to apply the asymptotic chaos expansion methodology to a Caplet smile, up to the first layer and in a stochastic instantaneous volatility HJM (SInsV HJM *a.k.a.* SV-HJM) framework.

This approach consists in formally defining a continuum in maturity (and strike) for the Caplet smile, and then in immersing the problem within the generic framework of Chap. 5. In turn, Proposition 5.2 [p. 297], Theorem 5.1 [p. 300] or Theorem 5.2 [p. 310], for instance, can then be invoked in order to link the SInsV and SImpV model classes through either the direct or the inverse problem.

In particular, these results allow us to compute the at-the-money short-expiry smile level, skew, curvature and slope, which are the main static descriptors of that surface. But the benefits go far beyond the pure calibration aspects, as in this interest rates context we can now approximate the joint dynamics of a term structure of marginal distributions.

6.4.1 Casting the Caplets into the Generic Framework

We consider a fixed accrual period δ (by convention less than a year, typically 1M, 3M, 6M, 1Y) and for a given fixing date T we denote by $L_t^\delta(T)$ the Forward Libor rate process corresponding to the forward Zero Coupon $B(t, T, T+\delta)$. Market conventions prescribe that we should express this rate in a linear fashion, in other words that it comes as a function of the HJM native state variables with

$$1 + \delta L_t^\delta(T) = B(t, T, T+\delta)^{-1} \quad \text{or equivalently} \quad L_t^\delta(T) = \frac{B_t(T) - B_t(T+\delta)}{\delta B_t(T+\delta)}.$$

For modelling purposes, we then assume a continuum of such rates w.r.t. their fixing T, and most naturally we cast the underlying map process $X_t(T)$ to these Libors. Let us stress a couple of practical points regarding these assumptions. First, we point out that in real markets the accrual period δ is not necessarily constant, as it can vary by a few days depending on day count conventions, etc. In order to simplify matters and in particular to maintain the continuity of the framework (a *day jump* is discontinuous) we nevertheless impose a common accrual $\delta(T) \equiv \delta$, but underline that this is not a blocking choice. Second, we note that building such a continuum within a coherent pricing system will require subjective choices, in particular w.r.t. interpolation benchmarks and methodology. Indeed, the Forward Libors are only available (traditionally by bootstrapping) for a discrete (rolling) set of dates, and their liquidity dwindles as their fixing increases.

Accordingly, the term structure of numeraires $N_t(T)$ becomes the Zero Coupon map $B_t(T+\delta)$ so that the associated term structure of martingale measures is simply the collection of $(T+\delta)$-forward measures, under which the parametric process $\vec{W}_t^{T+\delta}$ defined by

$$d\vec{W}_t^{T+\delta} \triangleq d\vec{W}_t - \vec{\Gamma}_t(T+\delta)\, dt$$

is a Brownian motion. In summary, we have simply instantiated the underlyings' part of the generic framework, specified by (5.1)–(5.3), with

$$X_t(T) \longleftarrow L_t^\delta(T) \qquad\qquad N_t(T) \longleftarrow B_t(T+\delta)$$

$$\vec{W}_t^{N(T)} \longleftarrow \vec{W}_t^{T+\delta} \qquad\qquad \vec{\lambda}_t^N(T) \longleftarrow \vec{\Gamma}_t(T+\delta)$$

Now let us move on to the option continuum, and again fix the (same) accrual period δ. The most liquid Caplets being those paid *in advance*, we will assume their cash payoff of $\delta\left[L_t^\delta(T) - K\right]^+$ to be transferred at time $T+\delta$.

Accordingly we consider a continuum of Caplet prices $C_t^\delta(L_t^\delta(T), K, T)$ both in strike K and in T. Note that, in conformity with the generic term structure framework defined in Sect. 5.1, each option is written on its own, private underlying. This case is a particularly simple example of the shared maturity indexation introduced in

Sect. 5.1.1 [p. 274]. In terms of vocabulary, the date T being now both the underlying's fixing and the option expiry, we may use both terms.

Now we must check that providing the underlying's instantiation that we have chosen, the corresponding generic payoff definition (5.4) [p. 275] does match the actual, real-life Caplet. Thankfully, the two contexts can indeed be made consistent, and furthermore in two equivalent ways. Indeed, the Caplet payoff can be discounted back to time T so that

$$\eta(T) \triangleq T \quad \Longrightarrow \quad \delta^{-1}\,\Phi(T) = B_T(T+\delta)\,\left[\,L_T^\delta(T) - K\,\right]^+$$
$$\eta(T) \triangleq T+\delta \quad \Longrightarrow \quad \delta^{-1}\,\Phi(T+\delta) = \underbrace{B_{T+\delta}(T+\delta)}_{1}\,\left[\,L_T^\delta(T) - K\,\right]^+$$

where we will opt subjectively for the first one, to get rid of the $\eta(\cdot)$ function.

Following Sect. 5.1.2, this surface of option *prices* is then associated to the implied volatility mapping via the usual normalised Black functional, with (5.5) instantiated as

$$\delta^{-1}\,C_t^\delta(L_t^\delta(T), K, T) = B_t(T+\delta)\,C^{BS}\,\left(L_t^\delta(T), K, \Sigma_t^\delta(L_t^\delta(T), T, K).\sqrt{T-t}\right). \tag{6.24}$$

Note that scaling the payoff by the accrual does not affect the nature of the problem.

We can then associate to these "absolute" variates their "sliding" counterparts:

$$\theta \triangleq T - t \qquad \widetilde{L}^\delta(t,\theta) \triangleq L^\delta(t,T) \qquad y \triangleq \ln\left(K/\widetilde{L}^\delta(t,\theta)\right)$$

$$\widetilde{C}^\delta\,(t,y,\theta) \triangleq C^\delta\left(t, L^\delta(t,T), K, T\right) \qquad \widetilde{\Sigma}^\delta\,(t,y,\theta) \triangleq \Sigma^\delta(t, L^\delta(t,T), K, T)$$

We can then define the dynamics of the Caplet stochastic *implied* volatility model in a formal manner, through (5.13) [p. 279]. But in this chapter we are starting from a generic HJM stochastic *instantaneous* volatility model, so that we will specify the underlying (i.e. the Forward Libor) chaos dynamics with the following SDE system

$$\begin{cases} \dfrac{dL_t^\delta(T)}{L_t^\delta(T)} = \vec{\sigma}_t^L(T)\,d\vec{W}_t^{T+\delta}, & (6.25)\\[2ex] d\vec{\sigma}_t^L(T) = \vec{a}_{1,t}^L(T)\,dt + \vec{\vec{a}}_{2,t}^L(T)\,d\vec{W}_t^{T+\delta} + \vec{\vec{a}}_{3,t}^L(T)\,d\vec{Z}_t, & (6.26)\\[2ex] d\vec{\vec{a}}_{2,t}^L(T) = \overset{(2)}{\overrightarrow{[\cdots]}}\,dt + \vec{\vec{\vec{a}}}_{22,t}^L(T)\,d\vec{W}_t^{T+\delta} + \overset{(3)}{\overrightarrow{[\cdots]}}\,d\vec{Z}_t. & (6.27) \end{cases}$$

In conclusion, all we need now in order to apply the generic results of Chap. 5 is to transfer the dynamics from the original HJM parametrisation to the generic maturity-dependent framework. In other words, we must compute the chaos dynamics of the Libor rates, knowing that all the modelling input comes from the Zero

Coupon map and its associated volatility structure, i.e. invoking only $B_t(T)$, $\vec{\Gamma}_t(T)$, $\vec{\vec{\alpha}}_{2,t}(T)$, etc.

In a sense (and as advertised within our objectives in Sect. 6.1) this method can be seen as the brute force approach. Indeed, it seems that reasoning on the assets (the bonds) directly, rather than on the corresponding rates, should be easier. Unfortunately, as will be discussed in Sect. 6.6, this straight approach is still the simplest if one wishes to capitalise on the results of Chap. 5.

Let us therefore compute the five coefficient maps constituting the σ-(2,0) group:

$$\vec{\sigma}_t^L(T), \quad \vec{a}_{1,t}^L(T), \quad \vec{\vec{a}}_{2,t}^L(T), \quad \vec{\vec{a}}_{3,t}^L(T) \quad \text{and} \quad \vec{\vec{a}}_{22,t}^L(T).$$

Note that in the sequel, for clarity we will omit the superscript δ wherever possible.

6.4.2 Dynamics of the Underlying Libor Rate

Let us start by making explicit the relevant coefficients of the Wiener chaos.

Lemma 6.3 (Libor rate chaos dynamics in an HJM framework) *In a generic SV-HJM framework defined in chaos by the SDE system* (6.1)–(6.3) *the σ-(2,0) group of maturity-dependent coefficients describing the Libor rate dynamics are given by* (6.25) *where*

$$\vec{\sigma}_t^L(T) = \frac{B_t(T)}{B_t(T) - B_t(T+\delta)} \left[\vec{\Gamma}_t(T) - \vec{\Gamma}_t(T+\delta) \right] \tag{6.28}$$

$$\begin{aligned} \vec{a}_{1,t}^L(T) = & \left[\frac{B_t(T+\delta)}{B_t(T)} \right] \|\vec{\sigma}_t^L(T)\|^2 \, \vec{\sigma}_t^L(T) \\ & + \left[\frac{B_t(T)}{B_t(T) - B_t(T+\delta)} \right] \left[\vec{\alpha}_{1,t}(T) - \vec{\alpha}_{1,t}(T+\delta) \right] \\ & - \left[\frac{B_t(T+\delta)}{B_t(T) - B_t(T+\delta)} \right] \left[\vec{\vec{\alpha}}_{2,t}(T) - \vec{\vec{\alpha}}_{2,t}(T+\delta) \right] \vec{\sigma}_t^L(T) \end{aligned} \tag{6.29}$$

$$\begin{aligned} \vec{\vec{a}}_{2,t}^L(T) = & \left[\frac{B_t(T)}{B_t(T) - B_t(T+\delta)} \right] \left[\vec{\vec{\alpha}}_{2,t}(T) - \vec{\vec{\alpha}}_{2,t}(T+\delta) \right] \\ & - \left[\frac{B_t(T+\delta)}{B_t(T)} \right] \vec{\sigma}_t^L(T)^{2\otimes} \end{aligned} \tag{6.30}$$

$$\vec{\vec{a}}_{3,t}^L(T) = \left[\frac{B_t(T)}{B_t(T) - B_t(T+\delta)} \right] \left[\vec{\vec{\alpha}}_{3,t}(T) - \vec{\vec{\alpha}}_{3,t}(T+\delta) \right] \tag{6.31}$$

$$\overset{\Rrightarrow}{a}{}^{L}_{22,t}(T) = \left[\frac{B_t(T)}{B_t(T)-B_t(T+\delta)}\right]\left[\overset{\Rrightarrow}{\alpha}_{22,t}(T) - \overset{\Rrightarrow}{\alpha}_{22,t}(T+\delta)\right]$$
$$+\left[\frac{B_t(T+\delta)}{B_t(T)}\left(1-\frac{B_t(T+\delta)}{B_t(T)}\right)\right]\vec{\sigma}{}^{L}_{t}(T)^{3\otimes}$$
$$-\left[\frac{B_t(T+\delta)}{B_t(T)}\right]\left[\overrightarrow{[a_2(T)_{ik}\,\sigma_j(T)]}^{(3)} + \overset{\Rightarrow}{a}_{2,t}(T)\overset{\leftrightarrow}{\otimes}\vec{\sigma}{}^{L}_{t}(T)\right] \quad (6.32)$$

where we use the linear algebra notations of Appendix D and omit the δ suffix, as no change in accrual period is required, and therefore no ambiguity exists.

Although simpler than for bond options, those dynamics do not really lend themselves to a straightforward, fully analytical interpretation, thus the same is true of the associated IATM differentials. The main reason for this apparent complexity is that the LMM setup is in essence a *maturity differential* of the HJM framework. This can also be seen asymptotically by linking the map of Zero Coupons to the term structure of the instantaneous forward rate:

$$B_t(T) = \exp\left(-\int_t^T f_t(s)\,ds\right) \quad\Longleftrightarrow\quad f_t(T) = -\partial_T \ln\left(B_t(T)\right).$$

At a macro rate/asset level, this feature explains the ever-present spreads and ratios between T- and $(T+\delta)$- indexed quantities. Nevertheless, instead of some *analytical* interpretation, it is interesting to build our intuition about these dynamics by gauging the sign and relative magnitudes of the terms that they involve. This is the object of Sect. 6.4.3, where a step-by-step analysis is conducted for this simplest of the three products.

Proof Let us first provide some useful conversion formulae between the HJM and LMM state variables, i.e. the Zero Coupons and the Libor rates. We have

$$L_t(T) = \delta^{-1}\left[\frac{B_t(T)}{B_t(T+\delta)} - 1\right] \text{ and } \frac{1+\delta L_t(T)}{\delta L_t(T)} = \frac{B_t(T)}{B_t(T)-B_t(T+\delta)}.$$

First up, the expression for the Libor volatility (6.28) is very well known and also quick to establish. Rewriting the rate with the ZCs as above and employing the usual change of numeraire technique, we have its dynamics as

$$\frac{dL_t(T)}{L_t(T)} = \frac{d\left[B_t(T)/B_t(T+\delta)\right]}{B_t(T)/B_t(T+\delta)-1} = \frac{B_t(T)/B_t(T+\delta)}{B_t(T)/B_t(T+\delta)-1}\,\frac{d\left[B_t(T)/B_t(T+\delta)\right]}{B_t(T)/B_t(T+\delta)}$$
$$= \frac{1+\delta L_t(T)}{\delta L_t(T)}\left[\vec{\Gamma}_t(T) - \vec{\Gamma}_t(T+\delta)\right]\left[d\vec{W}_t - \vec{\Gamma}_t(T+\delta)\right]. \quad (6.33)$$

On the r.h.s. we recognise the dynamics of the forward-neutral driver $\vec{W}_t^{T+\delta}$, and by changing back to the HJM state variables (6.33) provides (6.28). In order to compute the remaining coefficients, we could take the long route and apply Itô blindly on the above expression for the instantaneous volatility. However, noticing that $\vec{\sigma}_t^L(T)$ is itself expressed as a very simple two-element *tensorial basket*, we prefer to invoke the dedicated Lemma 3.6 [p. 192]. Our first step is therefore to obtain the dynamics of the two weights, which are

$$\omega_{1,t} \triangleq 1 + \delta^{-1} L_t^{-1} \qquad \text{and} \qquad \omega_{2,t} \triangleq -\omega_{1,t}.$$

Using Itô we have simply that

$$\frac{d\omega_{1,t}}{\omega_{1,t}} = \frac{d\omega_{2,t}}{\omega_{2,t}} = \frac{1}{1+\delta L_t} \left[\| \vec{\sigma}_t^L(T) \|^2 dt \; - \vec{\sigma}_t^L(T)^\perp d\vec{W}_t^{T+\delta} \right].$$

Since these dynamics incorporate no exogenous component, we can apply a simplified version of the Lemma's basket result (3.81) to get the first-depth coefficients.

First the **exogenous vol of vol** comes as

$$\vec{\vec{a}}_{3,t}(T) = \left[\frac{1+\delta L_t}{\delta L_t} \right] \left[\vec{\vec{\alpha}}_{3,t}(T) - \vec{\vec{\alpha}}_{3,t}(T+\delta) \right]$$

which after changing variables from forward Libor rates to ZCs gives (6.31).

As for the **endogenous vol of vol**, it is translated into

$$\begin{aligned}
\vec{\vec{a}}_{2,t}(T) &= \left[\frac{1+\delta L_t}{\delta L_t} \right] \left[\vec{\vec{\alpha}}_{2,t}(T) + \vec{\Gamma}_t(T) \otimes \left[\frac{-1}{1+\delta L_t} \right] \vec{\sigma}_t^L(T) \right. \\
&\qquad \left. - \vec{\vec{\alpha}}_{2,t}(T+\delta) \; - \vec{\Gamma}_t(T+\delta) \otimes \left[\frac{-1}{1+\delta L_t} \right] \vec{\sigma}_t^L(T) \right] \\
&= \frac{1+\delta L_t}{\delta L_t} \left[\vec{\vec{\alpha}}_{2,t}(T) - \vec{\vec{\alpha}}_{2,t}(T+\delta) \right] - \frac{1}{\delta L_t} \left[\vec{\Gamma}_t(T) - \vec{\Gamma}_t(T+\delta) \right] \otimes \vec{\sigma}_t^L(T)
\end{aligned}$$

and finally

$$\begin{aligned}
\vec{\vec{a}}_{2,t}(T) &= \left[\frac{1+\delta L_t}{\delta L_t} \right] \left[\vec{\vec{\alpha}}_{2,t}(T) - \vec{\vec{\alpha}}_{2,t}(T+\delta) \right] \\
&\quad - \frac{1}{1+\delta L_t} \vec{\sigma}_t^L(T)^{2\otimes}
\end{aligned} \tag{6.34}$$

which after substitution gives (6.30). Finally the **drift** comes with

$$\left[\frac{\delta L_t}{1+\delta L_t}\right] \vec{a}_{1,t}(T) = \vec{\alpha}_{1,t}(T) + \left[\frac{1}{1+\delta L_t}\right] \|\vec{\sigma}_t^L(T)\|^2\, \vec{\Gamma}_t(T)$$
$$- \vec{\alpha}_{1,t}(T+\delta) - \left[\frac{1}{1+\delta L_t}\right] \|\vec{\sigma}_t^L(T)\|^2\, \vec{\Gamma}_t(T+\delta)$$
$$+ \vec{\vec{\alpha}}_{2,t}(T)\left[\frac{-1}{1+\delta L_t}\right]\vec{\sigma}_t^L(T) - \vec{\vec{\alpha}}_{2,t}(T+\delta)\left[\frac{-1}{1+\delta L_t}\right]\vec{\sigma}_t^L(T)$$

hence

$$\begin{aligned}\vec{a}_{1,t}(T) &= \left[\frac{1+\delta L_t}{\delta L_t}\right]\left[\vec{\alpha}_{1,t}(T) - \vec{\alpha}_{1,t}(T+\delta)\right] \\ &\quad + \frac{1}{\delta L_t}\|\vec{\sigma}_t^L(T)\|^2\left[\vec{\Gamma}_t(T) - \vec{\Gamma}_t(T+\delta)\right] \\ &\quad - \frac{1}{\delta L_t}\left[\vec{\vec{\alpha}}_{2,t}(T) - \vec{\vec{\alpha}}_{2,t}(T+\delta)\right]\vec{\sigma}_t^L(T)\end{aligned}$$

which after substitution by the volatility and the native HJM state variables provides (6.29). The last step is slightly more involved as we now need to compute the (endogenous) dynamics of $\vec{\vec{a}}_{2,t}(T)$ in order to obtain **the second-depth coefficient** $\vec{\vec{a}}_{22,t}(T)$. To that end we shall exploit the fact that the former coefficient is naturally broken down into two distinct terms, the first pertaining to the vol of vol map, and the second to the volatility itself:

- The first term $\left[\frac{1+\delta L_t}{\delta L_t}\right]\left[\vec{\vec{\alpha}}_{2,t}(T) - \vec{\vec{\alpha}}_{2,t}(T+\delta)\right]$

is very similar in structure to the volatility $\vec{\sigma}_t^L(T)$, as a basket using the same weighting scheme. We can therefore apply Lemma 3.6 again, still with no exogenous component, to obtain its endogenous dynamics as

$$d\left[\left[\frac{1+\delta L_t}{\delta L_t}\right]\left[\vec{\vec{\alpha}}_{2,t}(T) - \vec{\vec{\alpha}}_{2,t}(T+\delta)\right]\right] = \vec{[\cdot]}\, dt + \vec{\vec{A}}_t\, d\vec{W}_t^{T+\delta} + \vec{\vec{[\cdot]}}\, d\vec{Z}_t$$

with

$$\begin{aligned}\vec{\vec{A}}_t &= \frac{1+\delta L_t}{\delta L_t}\left[\vec{\vec{\alpha}}_{22,t}(T) - \vec{\vec{\alpha}}_{2,t}(T)\otimes\frac{\vec{\sigma}_t^L(T)}{1+\delta L_t} - \vec{\vec{\alpha}}_{22,t}(T+\delta) + \vec{\vec{\alpha}}_{2,t}(T+\delta)\otimes\frac{\vec{\sigma}_t^L(T)}{1+\delta L_t}\right] \\ &= \frac{1+\delta L_t}{\delta L_t}\left[\vec{\vec{\alpha}}_{22,t}(T) - \vec{\vec{\alpha}}_{22,t}(T+\delta)\right] - \frac{1}{\delta L_t}\left[\vec{\vec{\alpha}}_{2,t}(T) - \vec{\vec{\alpha}}_{2,t}(T+\delta)\right]\otimes\vec{\sigma}_t^L(T)\end{aligned}$$

hence

$$\vec{A}_t = \frac{1+\delta L_t}{\delta L_t}\left[\overset{\Rightarrow}{\alpha}_{22,t}(T) - \overset{\Rightarrow}{\alpha}_{22,t}(T+\delta)\right] - \frac{1}{1+\delta L_t}\,\overset{\Rightarrow}{a}_{2,t}(T)\otimes\vec{\sigma}^{L}_{t}(T) - \frac{1}{(1+\delta L_t)^2}\,\vec{\sigma}^{L}_{t}(T)^{3\otimes}.$$

- The second term $\left[\frac{1}{1+\delta L}\right]\vec{\sigma}^{L}(T)^{2\otimes}$

is more involved due to the external product, hence its dynamics are more complex. Fortunately, we are only interested in the endogenous coefficient, and the dynamics of $\vec{\sigma}^{L}_{t}(T)^{2\otimes}$ come from (D.10) [p. 440] as

$$d\left[\vec{\sigma}^{L}(T)^{2\otimes}\right] = \left[\overrightarrow{\left[a_2(T)_{ik}\,\sigma_j(T)\right]}^{(3)} + \vec{\sigma}^{L}(T)\otimes\overset{\Rightarrow}{a}_2(T)\right] \times d\vec{W}^{T+\delta}_{t} + \overset{\Rightarrow}{[\cdot]}dt + \overset{\Rightarrow}{[\cdot]}d\vec{Z}_t$$

while the endogenous dynamics of $(1+\delta L)^{-1}$ are simply expressed as

$$d\,(1+\delta L)^{-1} + \overset{\Rightarrow}{[\cdot]}\,dt + \overset{\Rightarrow}{[\cdot]}\,d\vec{Z}_t = -\frac{\delta L}{(1+\delta L)^2}\,\vec{\sigma}^{L}_{t}(T)^{\perp}d\vec{W}^{T+\delta}_{t}.$$

Therefore the combined dynamics of the product come as

$$\begin{aligned} & d\left[(1+\delta L)^{-1}\,\vec{\sigma}^{L}(T)^{2\otimes}\right] + \overset{\Rightarrow}{[\cdot]}\,dt + \overset{\Rightarrow}{[\cdot]}\,d\vec{Z}_t \\ & = \frac{1}{1+\delta L}\left[\frac{-\delta L}{1+\delta L}\,\vec{\sigma}^{L}(T)^{3\otimes} + \overrightarrow{\left[a_2(T)_{ik}\,\sigma_j(T)\right]}^{(3)} + \vec{\sigma}^{L}(T)\otimes\overset{\Rightarrow}{a}_2(T)\right] d\vec{W}^{T+\delta}_{t}, \end{aligned}$$

so that eventually gathering both terms' dynamics, the $\overset{\Rightarrow}{a}_{22,t}(T)$ coefficient is

$$\begin{aligned} \overset{\Rightarrow}{a}_{22,t}(T) = {} & \frac{1+\delta L_t}{\delta L_t}\left[\overset{\Rightarrow}{\alpha}_{22,t}(T) - \overset{\Rightarrow}{\alpha}_{22,t}(T+\delta)\right] - \frac{1}{1+\delta L}\,\overrightarrow{\left[a_2(T)_{ik}\,\sigma_j(T)\right]}^{(3)} \\ & - \frac{1}{1+\delta L_t}\left[\overset{\Rightarrow}{a}_{2,t}(T)\otimes\vec{\sigma}^{L}_{t}(T) + \vec{\sigma}^{L}(T)\otimes\overset{\Rightarrow}{a}_2(T)\right] \\ & - \frac{1-\delta L_t}{(1+\delta L_t)^2}\,\vec{\sigma}^{L}_{t}(T)^{3\otimes}. \end{aligned}$$

Finally, substituting in that expression with the HJM native state variables, we obtain (6.32), which concludes the proof. □

6.4.3 Interpretation of the Libor Rate HJM Dynamics

Our intention is to proceed with a semi-numerical analysis of the Libor rate dynamics. We wish to gauge the sign and relative magnitude of all terms involved in its chaos coefficients, as described by Lemma 6.3. We choose to focus on the available endogenous coefficients:

$$\vec{\sigma}_t^L(T), \qquad \vec{\vec{a}}_{2,t}^L(T) \quad \text{and} \quad \vec{\vec{\vec{a}}}_{22,t}^L(T),$$

since they provide a good comparison base, by spanning three consecutive depths in the chaos expansion with increasingly rich expressions. Obviously this exercise requires a few simplifying assumptions in order to be demonstrative, but we will introduce them one by one, in order to observe their respective contributions.

6.4.3.1 Introducing a Low-Dimensionality Assumption

We start by breaking down the dimensionality issue by considering that the various tensors are uniform, which we formalise with

$$\vec{\Gamma}_t(T) \triangleq \Gamma_t(T)\ \vec{\mathbf{1}} \qquad \vec{\vec{\alpha}}_{2,t}(T) \triangleq \alpha_{2,t}(T)\ \vec{\vec{\mathbf{1}}} \qquad \vec{\vec{\vec{\alpha}}}_{22,t}(T) \triangleq \alpha_{22,t}(T)\ \vec{\vec{\vec{\mathbf{1}}}}$$

Agreed, this is indeed a very poor modelling environment, as it corresponds more or less to a one factor curve model. To put this statement in perspective, however, let us recall that historical PCA[5] usually indicates that around 80 % of the yield curve variance comes from parallel moves, i.e. from this first factor. Irrespective of its realism, this simplification definitely helps in giving us some idea of the respective signs and magnitudes. We then exploit the ever-present spread aspect by dedicating the following definitions and notations:

$$B \doteq B_t(T) \qquad \Gamma \doteq \vec{\Gamma}_t(T) \qquad \alpha_2 \doteq \alpha_{2,t}(T) \qquad \alpha_{22} \triangleq \alpha_{22,t}(T)$$

$$\Delta B \triangleq B_t(T) - B_t(T+\delta) \qquad \Delta\Gamma \triangleq \Gamma_t(T) - \Gamma(T+\delta)$$

$$\Delta\alpha_2 \triangleq \alpha_{2,t}(T) - \alpha_{2,t}(T+\delta) \qquad \Delta\alpha_{22} \triangleq \alpha_{22,t}(T) - \alpha_{22,t}(T+\delta)$$

Introducing in addition the ratio

$$\Lambda \triangleq B/\Delta B$$

we can then rewrite all the endogenous coefficients as

$$\vec{\sigma}_t^L(T) = \Lambda\ \Delta\Gamma\ \vec{\mathbf{1}} \tag{6.35}$$

[5] Principal Component Analysis.

$$\overset{\Rightarrow}{a}{}^{L}_{2,t}(T) = \frac{B}{\Delta B}\,\Delta\alpha_2\,\overset{\Rightarrow}{\mathbf{1}} - \left[1 - \frac{\Delta B}{B}\right]\left[\frac{B}{\Delta B}\Delta\Gamma\right]^2 \overset{\Rightarrow}{\mathbf{1}}$$
$$= \Lambda\,\left[\Delta\alpha_2 - [\Lambda - 1]\,(\Delta\Gamma)^2\right]\,\overset{\Rightarrow}{\mathbf{1}} \tag{6.36}$$

for the volatility and the endogenous vol of vol. The 2nd depth coefficient comes as

$$\overset{\Rrightarrow}{a}{}^{L}_{22,t}(T) = \frac{B}{\Delta B}\,\Delta\alpha_{22}\,\overset{\Rrightarrow}{\mathbf{1}} + \left[1 - \frac{\Delta B}{B}\right]\frac{\Delta B}{B}\left[\frac{B}{\Delta B}\Delta\Gamma\right]^3 \overset{\Rrightarrow}{\mathbf{1}}$$
$$- \left[1 - \frac{\Delta B}{B}\right]3\,\frac{B}{\Delta B}\left[\Delta\alpha_2 - \left[\frac{B}{\Delta B} - 1\right](\Delta\Gamma)^2\right]\frac{B}{\Delta B}\,\Delta\Gamma\,\overset{\Rightarrow}{\mathbf{1}}$$

and so, after simplification,

$$\overset{\Rrightarrow}{a}{}^{L}_{22,t}(T) = \Lambda\left[\Delta\alpha_{22} - 3\,(\Lambda - 1)\,\Delta\alpha_2\,\Delta\Gamma + (\Lambda - 1)\,(3\Lambda - 2)\,(\Delta\Gamma)^3\right]\overset{\Rrightarrow}{\mathbf{1}}. \tag{6.37}$$

It appears that the respective magnitudes of all the quoted terms depend on two types of specifications:

- The first model impact lies with the term structure of volatilities/coefficients, and in particular the relative sizes of their spread (between T and $T + \delta$).
- The second influential specification is the structure of the underlying zero-coupon map itself, but interestingly it only appears through the ratio Λ.

Note that in terms of modelling, in an HJM context the yield curve is a given input. As for the coefficient maps, the calibration process will see the modeller enjoy more and more freedom as their depth increases: $\vec{\Gamma}_t(T)$ is very constrained by the ATM level, $\overset{\Rightarrow}{a}_{2,t}(T)$ presents a little more flexibility as it influences mainly the skew, while $\overset{\Rrightarrow}{a}_{22,t}(T)$ should be the least restricted.

We stress that so far no assumption has been made, apart from the dimensionality simplification. Therefore, let us examine the respective influences of these two items by imposing on them some realistic conditions.

6.4.3.2 Introducing Some Yield Curve Assumptions

Let us examine first the impact of the yield curve on these three coefficient expressions (6.35), (6.36) and (6.37).Our priority is to gauge the magnitude of the ratio Λ in different realistic market situations. Obviously this will depend on the shape of the curve, on the fixing date T and on the accrual δ. Since the market imposes $\delta \leq 1Y$,

Table 6.1 Ratio Λ for combinations of δ and R

$\delta\backslash R$	1 %	2 %	5 %	10 %
1M	1201	601	241	121
3M	401	201	81	41
6M	201	101	41	21
1Y	101	51	21	11

we can reasonably assume a flat curve between T and $T+\delta$. Taking a continuous rate convention,[6] the ratio comes as

$$\Lambda = \frac{B_t(T)}{B_t(T) - B_t(T+\delta)} = \frac{e^{-R(T-t)}}{e^{-R(T-t)} - e^{-R(T+\delta-t)}} = \frac{1}{1-e^{-R\delta}}$$

and is therefore not only independent from the maturity T, but also only affected by the product $R\,\delta$: at the first order, Λ is inversely proportional to $R\,\delta$. This also means that, for these matters, our focus on the short term (which is associated to the asymptotic concern) will have no impact. Plotting the ratio Λ in realistic ranges for the rate R and for the accrual δ[7] we get Table 6.1.

It is clear that in all configurations we can safely make the simplifying assumption of $\Lambda \gg 1$. The endogenous coefficients can then be approximated by

$$\begin{aligned}
\vec{\sigma}_t^{\,L}(T) &= \Lambda\,\Delta\Gamma\;\vec{\mathbf{1}},\\
\vec{\vec{a}}^{\,L}_{2,t}(T) &\approx \Lambda\left[\Delta\alpha_2 - \Lambda\;(\Delta\Gamma)^2\right]\vec{\vec{\mathbf{1}}},\\
\vec{\vec{\vec{a}}}^{\,L}_{22,t}(T) &\approx \Lambda\left[\Delta\alpha_{22} - 3\,\Lambda\,\Delta\alpha_2\,\Delta\Gamma + 3\,\Lambda^2\;(\Delta\Gamma)^3\right]\vec{\vec{\vec{\mathbf{1}}}}.
\end{aligned}$$

At this point we need to introduce a further assumption in order to continue the comparison, to relate the respective magnitudes of the three endogenous volatility spreads: $\Delta\Gamma$, $\Delta\alpha_2$ and $\Delta\alpha_{22}$.

6.4.3.3 Introducing Some Term Structure Assumptions for the Coefficients

We propose two examples of such a hypothesis, based on the two most common market conventions when it comes to dynamics. Recall that our instantaneous chaos dynamics, as defined in the generic framework by the SDE system (5.1)–(5.10) and in our current HJM context by (6.1)–(6.3), are of a hybrid nature from a convention point of view. They describe the underlying's volatility in Lognormal terms, while the two endogenous coefficients further down are expressed in Normal fashion. Our take is therefore to use the same convention throughout in order to make numerical values comparable, which is fairly intuitive and usually preferred by practitioners.

---

[6] A continuous rate convention is defined by $B_t(T) = \exp(-(T-t)R_t(T))$.

[7] We use simplified business days conventions, i.e. 250 Days per Year.

6.4.3.4 First Case: Homogeneous Lognormal Dynamics

A first possibility is to consider that the chaos dynamics are uniformly written in *Lognormal* fashion, with all Lognormal volatilities quoted by such a system exhibiting roughly the same level. In that case we get

$$\Gamma \approx \frac{\alpha_2}{\Gamma} \approx \frac{\alpha_{22}}{\alpha_2} \qquad \text{hence} \qquad \alpha_2 \approx \Gamma^2 \qquad \text{and} \qquad \alpha_{22} \approx \Gamma^3. \tag{6.38}$$

We are interested in the spreads, hence we approximate them at first order in δ:

$$\Gamma_t(T+\delta) \approx \Gamma_t(T)\ (1+\lambda\ \delta), \qquad \alpha_{2,t}(T+\delta) \approx \alpha_{2,t}(T)\ (1+\lambda_2\ \delta),$$

$$\alpha_{22,t}(T+\delta) \approx \alpha_{22,t}(T)\ (1+\lambda_{22}\ \delta).$$

For these approximations to hold in conjunction with (6.38) we need

$$\lambda_2 = 2\ \lambda \qquad \text{and} \qquad \lambda_{22} = 3\ \lambda$$

so that the endogenous coefficients become

$$\begin{aligned}
\vec{\sigma}_t^L(T) &= \Lambda\ \lambda\ \delta\ \Gamma\ \vec{\mathbf{1}},\\
\vec{\vec{a}}_{2,t}^L(T) &\approx \Lambda\ \left[2\ \lambda\ \delta\ \Gamma^2 - \Lambda\ \lambda^2\ \delta^2\ \Gamma^2\right] \vec{\vec{\mathbf{1}}} = \Lambda\ \Gamma^2\ \lambda\ \delta\ [\,2 - \Lambda\ \lambda\ \delta\,]\ \vec{\vec{\mathbf{1}}},\\
\vec{\vec{\vec{a}}}_{22,t}^L(T) &\approx \Lambda\ \left[3\ \lambda\ \delta\ \Gamma^3 - 3\ \Lambda\ 2\ \lambda^2\ \delta^2\ \Gamma^3 + 3\ \Lambda^2\ \lambda^3\ \delta^3\ \Gamma^3\right] \vec{\vec{\vec{\mathbf{1}}}}\\
&\approx 3\ \Lambda\ \lambda\ \delta\ \Gamma^3\ [1 - \Lambda\ \lambda\ \delta]^2\ \vec{\vec{\vec{\mathbf{1}}}}.
\end{aligned}$$

Hence the leading term is identified by the magnitude of $\Lambda\ \lambda\ \delta$.

Recall that λ is essentially the proportional rate at which the ZC volatility structure evolves with maturity, taken locally in T. Assuming that this value is reasonably stable and uniform across all maturities, it seems realistic to impose that the volatility coefficients should decrease from their maximum at the short end to almost a standstill at the long end. Given a horizon of 50 years, that gives us $\lambda \approx \frac{1}{50} = 2 \cdot 10^{-2}$.

Recall also that we have $\Lambda \approx (\delta R)^{-1}$, therefore

$$\Lambda\ \lambda\ \delta \approx \frac{2 \cdot 10^{-2}}{R}.$$

In summary, with such a set of reasonable assumptions we have shown that all terms are of similar magnitude and therefore cannot be discarded at leading order. Note that we have made no assumption on the level of the volatility itself.

Going back to the original expression (6.32) for $\overset{\Rrightarrow}{a}{}^{L}_{22,t}(T)$, a careful examination would show that only the second term on the r.h.s., the only one *explicitly* in $\vec{\sigma}{}^{L}_{t}(T)^{3\otimes}$, has disappeared.

6.4.3.5 Second Case: Homogeneous Normal Dynamics

A second, alternative possibility is to consider that all maps exhibit approximately the same level of instantaneous *Normal* volatility. Then we obtain

$$B_t(T)\,\Gamma \;\approx\; \alpha_2 \;\approx\; \alpha_{22}.$$

We shall take the magnitude of the Zero Coupon as unit, as a concession to short maturities. Using again the first-order expansion in δ, we show that we must have $\lambda = \lambda_2 = \lambda_{22}$. The endogenous coefficients become

$$\begin{aligned}
\vec{\sigma}{}^{L}_{t}(T) &= \Lambda\,\lambda\,\delta\,\Gamma\;\vec{\mathbf{1}},\\
\overset{\Rightarrow}{a}{}^{L}_{2,t}(T) &\approx \Lambda\,\lambda\,\delta\,\Gamma\;[1-\Lambda\,\lambda\,\delta\,\Gamma]\;\overset{\Rightarrow}{\mathbf{1}},\\
\overset{\Rrightarrow}{a}{}^{L}_{22,t}(T) &\approx \Lambda\,\lambda\,\delta\,\Gamma\;\Big[1-3\,\Lambda\,\lambda\,\delta\,\Gamma\;+3\;(\Lambda\,\lambda\,\delta\,\Gamma\,)^2\;\Big]\;\overset{\Rrightarrow}{\mathbf{1}}.
\end{aligned}$$

This time the term whose magnitude will drive the dominance is $\Lambda\,\lambda\,\delta\,\Gamma$. The assumptions made earlier w.r.t. λ remain valid so that with rates in the region of 2 % the relevant magnitude will be that of $\Gamma_t(T)$. To infer that value, let us assume a typical Lognormal volatility of 20 % for the Zeros. Then with same short expiry assumption, we end up with a Normal volatility of the same order. It then become clear that the lowest powers of $\Lambda\,\lambda\,\delta\,\Gamma$ will dominate.

6.4.3.6 Conclusion

We observe that when using common conventions and magnitudes it seems difficult to identify any clearly leading term within the Libor rate dynamics. In other words, all terms must be taken into account, which does not help with building our modelling intuition. This is not to say that this kind of rough qualitative analysis is pointless, since we *can* gauge with precision the impact of the HJM specification on each coefficient of the Libor rate chaos diffusion. Also, it is worth remembering that we have presented a theoretical, generic framework, where all terms are *a priori* important since we hold no information on them. In practical models, however, (typically a combined local and stochastic volatility), the structure of endogenous coefficients is often simple and shows some redundancy, making the above analysis easier.

Leaving the Caplets for now, we turn to the most complex but also the most liquid vanilla interest rate products, i.e. the European payer swaptions.

6.5 Swaptions

The goal of this section is to provide all the information necessary to link the SImpV specification of the *swaption smile* (in practice its $\widetilde{\Sigma}$-(2,0) IATM differentials) to the SInsV model class defined by the Stochastic Volatility HJM dynamics (again, up to σ-(2,0) order).

6.5.1 Casting the Swaptions into the Generic Framework

Following the market simplifications of Assumption 6.1 we will consider a plain vanilla swap, whose floating and fixed legs share the first and last dates, and also whose start date coincides[8] with the *option expiry* T. As for frequencies, and as discussed in Assumption 6.1, they *can* be different on both legs, but in our limited context this point will be irrelevant. With these simplifications, the tenor structure now has two main degrees of freedom: the expiry/maturity date and the tenor. In order to build the option- and underlying-term structure defining the generic framework of Sect. 5.1.1, we choose to fix the tenor and therefore we allocate the generic maturity T to the start date T_0. Subsequently, both legs and therefore the swap itself have been made sliding with regard to that single date. In summary we have

$$\underset{\text{option expiry/swap start date}}{T = T_0} \qquad \text{and} \qquad \underset{\text{relevant coupon dates}}{T_i = T_{i-1} + \delta_i \quad \forall i \;/\; 1 \le i \le N.}$$

The underlying swap is then defined by its maturity T and the tenor structure

$$\delta \triangleq \{\delta_i\}_{1 \le i \le N}$$

hence we can employ the notations defined specifically in Sect. 6.1.2 for sliding schedules. Moving from the asset to the corresponding rate, we recall that the absence of arbitrage opportunity between the fixed and floating legs prescribes, within the simplified context of Assumption 6.1, that the *forward par swap rate* comes as

$$S_t^\delta(T) \triangleq \frac{B_{0,t}(T) - B_{N,t}(T)}{\sum_{i=1}^{N} \delta_i B_{i,t}(T)}.$$

---

[8] Note that, as discussed in Sect. 5.1 we could deal with mid-curve options, but this is beyond the scope of this book.

Our modelling approach then assumes a *continuum* of such rates w.r.t. their fixing T, which we cast into the underlying map process $X_t(T)$. We note that the denominator, the embedded *annuity* or *level*, has also been made sliding:

$$A_t^{\delta}(T) \triangleq \sum_{i=1}^{N} \delta_i B_{i,t}(T)$$

and becomes the maturity-dependent numeraire $N_t(T)$. The three SDEs (5.1)–(5.10) defining the first layer dynamics are respectively instantiated as

$$\frac{dS_t^{\delta}(T)}{S_t^{\delta}(T)} = \vec{\sigma}_t^{s}(T)^{\perp} d\vec{W}_t^{A(T)}, \tag{6.39}$$

$$d\vec{\sigma}_t^{s}(T) = \vec{a}_{1,t}^{s}(T)\, dt + \vec{\vec{a}}_{2,t}^{s}(T)\, d\vec{W}_t^{A(T)} + \vec{\vec{a}}_{3,t}^{s}(T)\, d\vec{Z}_t, \tag{6.40}$$

$$d\vec{\vec{a}}_{2,t}^{s}(T) = \overrightarrow{[\cdots]}^{(2)}\, dt + \vec{\vec{\vec{a}}}_{22,t}^{s}(T)\, d\vec{W}_t^{A(T)} + \overrightarrow{[\cdots]}^{(3)}\, d\vec{Z}_t. \tag{6.41}$$

The Annuity driver is selected to express those dynamics, as (5.2) becomes

$$d\vec{W}_t^{A(T)} \triangleq d\vec{W}_t - \vec{\sigma}_t^{a}(T)\, dt \tag{6.42}$$

with $\vec{W}_t$ the R-N driver and $\vec{\sigma}_t^{a}(T)$ defining the numeraire dynamics akin to (5.3)

$$\frac{dA_t(T)}{A_t(T)} = r_t\, dt + \vec{\sigma}_t^{a}(T)^{\perp} d\vec{W}_t. \tag{6.43}$$

Having defined the term structure (TS) of underlyings, numeraires and martingale measures, let us now move on to the option field. We must check that the TS of payoffs associated to our choices for the generic framework does correspond to a continuum of actual *physically settled* European payer swaptions. Which is the case, since discounting all the physical swaption cashflows down to date T yields

$$\begin{aligned}\Phi(T) &\triangleq \sum_{i=1}^{N} \delta_i\ B_{i,T}(T)\ \left[S_T^{\delta}(T) - K\right]^+ = A_T(T)\ \left[S_T^{\delta}(T) - K\right]^+ \\ &= N_{\eta(T)}(T)\ [X_T(T) - K]^+ \qquad \text{with} \quad \eta(x) \equiv x.\end{aligned}$$

Note that in terms of modelling scope we face the same limitations as with bond options, in the sense that we chose not to cover a continuum of swaptions written on an existing, running swap. It is also worth remembering the remarks made in Sect. 6.4.1 with regard to the issues involved in practice when building such a continuum (Day Count Conventions, liquidity, benchmark definition, interpolation, etc.).

Assuming a continuum in strike K for these options as per Sect. 5.1.2, we can then associate this surface of prices to the implied volatility mapping via the normalised Black functional. Then the generic Eq. (5.5) becomes

$$C_t^\delta(S_t^\delta(T), K, T) = A_t^\delta(T)\, C^{BS}\left(S_t^\delta(T), K, \Sigma_t^\delta(S_t^\delta(T), T, K) \cdot \sqrt{T-t}\right). \tag{6.44}$$

These absolute quantities are then re-parametrised into sliding counterparts *via*

$$\theta \triangleq T - t, \qquad \widetilde{S}^\delta(t,\theta) \triangleq S^\delta(t,T), \qquad y \triangleq \ln\left(K/\widetilde{S}^\delta(t,\theta)\right),$$

$$\widetilde{C}^\delta\,(t,y,\theta) \triangleq C^\delta\left(t, S^\delta(t,T), K, T\right), \qquad \widetilde{\Sigma}^\delta\,(t,y,\theta) \triangleq \Sigma^\delta(t, S^\delta(t,T), K, T).$$

In the sequel, since the tenor structure is fixed throughout, we will often omit the δ superscript. The dynamics of the swaption (sliding) stochastic implied volatility (SImpV) model are then envisaged formally as per (5.13) [p. 279], which completes the formal cast.

From that point, in order to apply the generic results of Chap. 5 we need "only" express the SInsV dynamics in the original SV-HJM parametrisation. More precisely, we must compute:

- the σ-(2,0) chaos dynamics of the term structure of par swap rates:

$$\vec{\sigma}_t^{\,s}(T) \qquad \vec{a}_{1,t}^{\,s}(T) \qquad \vec{\vec{a}}_{2,t}^{\,s}(T) \qquad \vec{\vec{a}}_{3,t}^{\,s}(T) \qquad \text{and} \qquad \vec{\vec{\vec{a}}}_{22,t}^{\,s}(T);$$

- under the relevant, maturity-dependent annuity measure;
- taking as input the $B_t(T)$ map and its associated volatility structure:

$$\vec{\Gamma}_t(T) \qquad \vec{\alpha}_{1,t}(T) \qquad \vec{\vec{\alpha}}_{2,t}(T) \qquad \vec{\vec{\alpha}}_{3,t}(T) \qquad \text{and} \qquad \vec{\vec{\vec{\alpha}}}_{22,t}(T).$$

These relatively heavy computations are the object of the next Sect. 6.5.2.

6.5.2 Dynamics of the Underlying Swap Rate

Let us provide the only missing information that enables us to exploit the generic framework results, in other words the underlying dynamics within the *native* HJM SInsV model class.

Lemma 6.4 (Swap rate chaos dynamics in an HJM framework) *In an SV-HJM framework defined in chaos by the system* (6.1)–(6.3) *the* σ-(2,0) *group of coefficients describing the par swap rate dynamics are given by*

$$\vec{\sigma}_t^{\,s}(T) \triangleq \left[\left(\omega_{\mathbf{i}}^f - \omega_{\mathbf{i}}^a\right)\vec{\Gamma}_{\mathbf{i}}\right] \tag{6.45}$$

$$
\begin{aligned}
\vec{a}^{\,s}_{1,t}(T) \triangleq\; & \vec{\vec{a}}^{\,s}_{2}\left[\omega^a_{\mathbf{i}}\,\vec{\Gamma}_{\mathbf{i}}\right] + \left[\left(\omega^f_{\mathbf{i}} - \omega^a_{\mathbf{i}}\right)\vec{\alpha}_{\mathbf{i},1}\right] \\
& + \left[\left(\omega^f_{\mathbf{i}} - \omega^a_{\mathbf{i}}\right)\vec{\vec{\alpha}}_{\mathbf{i},2}\,\vec{\Gamma}_{\mathbf{i}}\right] - \left[\omega^f_{\mathbf{i}}\,\vec{\vec{\alpha}}_{\mathbf{i},2}\right]\left[\omega^f_{\mathbf{i}}\,\vec{\Gamma}_{\mathbf{i}}\right] + \left[\omega^a_{\mathbf{i}}\,\vec{\vec{\alpha}}_{\mathbf{i},2}\right]\left[\omega^a_{\mathbf{i}}\,\vec{\Gamma}_{\mathbf{i}}\right] \\
& + \left[\omega^f_{\mathbf{i}}\,\vec{\Gamma}_{\mathbf{i}}\right]^{3\otimes} - \left[\omega^a_{\mathbf{i}}\,\vec{\Gamma}_{\mathbf{i}}\right]^{3\otimes} - \left[\omega^f_{\mathbf{i}}\,\vec{\Gamma}_{\mathbf{i}}^{2\otimes}\right]\left[\omega^f_{\mathbf{i}}\,\vec{\Gamma}_{\mathbf{i}}\right] \\
& + \left[\omega^a_{\mathbf{i}}\,\vec{\Gamma}_{\mathbf{i}}^{2\otimes}\right]\left[\omega^a_{\mathbf{i}}\,\vec{\Gamma}_{\mathbf{i}}\right]
\end{aligned}
\tag{6.46}
$$

$$
\vec{\vec{a}}^{\,s}_{2,t}(T) \triangleq \left[\omega^a_{\mathbf{i}}\,\vec{\Gamma}_{\mathbf{i}}\right]^{2\otimes} - \left[\omega^f_{\mathbf{i}}\,\vec{\Gamma}_{\mathbf{i}}\right]^{2\otimes} + \left[\left(\omega^f_{\mathbf{i}} - \omega^a_{\mathbf{i}}\right)\ \left(\vec{\vec{\alpha}}_{\mathbf{i},2} + \vec{\Gamma}_{\mathbf{i}}^{2\otimes}\right)\right] \tag{6.47}
$$

$$
\vec{\vec{a}}^{\,s}_{3,t}(T) \triangleq \left[\left(\omega^f_{\mathbf{i}} - \omega^a_{\mathbf{i}}\right)\ \vec{\vec{\alpha}}_{\mathbf{i},3}\right] \tag{6.48}
$$

for the volatility and the first-depth coefficients, and by

$$
\begin{aligned}
\vec{\vec{\vec{a}}}^{\,s}_{22,t}(T) \triangleq\; & \left[\left(\omega^f_{\mathbf{i}} - \omega^a_{\mathbf{i}}\right)\vec{\vec{\vec{\alpha}}}_{\mathbf{i},22}\right] + \left[\left(\omega^f_{\mathbf{i}} - \omega^a_{\mathbf{i}}\right)\vec{\vec{\alpha}}_{\mathbf{i},2}\overset{\leftrightarrow}{\otimes}\vec{\Gamma}_{\mathbf{i}}\right] \\
& - \left[\omega^f_{\mathbf{i}}\vec{\vec{\alpha}}_{\mathbf{i},2}\right]\overset{\leftrightarrow}{\otimes}\left[\omega^f_{\mathbf{i}}\vec{\Gamma}_{\mathbf{i}}\right] + \left[\omega^a_{\mathbf{i}}\vec{\vec{\alpha}}_{\mathbf{i},2}\right]\overset{\leftrightarrow}{\otimes}\left[\omega^a_{\mathbf{i}}\vec{\Gamma}_{\mathbf{i}}\right] \\
& - \left[\omega^f_{\mathbf{i}}\,\vec{\Gamma}_{\mathbf{i}}^{2\otimes}\right]\overset{\leftrightarrow}{\otimes}\left[\omega^f_{\mathbf{i}}\vec{\Gamma}_{\mathbf{i}}\right] + \left[\omega^a_{\mathbf{i}}\,\vec{\Gamma}_{\mathbf{i}}^{2\otimes}\right]\overset{\leftrightarrow}{\otimes}\left[\omega^a_{\mathbf{i}}\vec{\Gamma}_{\mathbf{i}}\right] \\
& + \left[(\omega^f_{\mathbf{i}} - \omega^a_{\mathbf{i}})\,\vec{\Gamma}_{\mathbf{i}}^{3\otimes}\right] + 2\left[\omega^f_{\mathbf{i}}\vec{\Gamma}_{\mathbf{i}}\right]^{3\otimes} - 2\left[\omega^a_{\mathbf{i}}\vec{\Gamma}_{\mathbf{i}}\right]^{3\otimes} \\
& - \left[\omega^f_{\mathbf{i}}\,\vec{\Gamma}_{\mathbf{i}}\otimes\left[\omega^f_{\mathbf{j}}\,\vec{\Gamma}_{\mathbf{j}}\right]\otimes\vec{\Gamma}_{\mathbf{i}}\right] + \left[\omega^a_{\mathbf{i}}\vec{\Gamma}_{\mathbf{i}}\otimes\left[\omega^a_{\mathbf{j}}\,\vec{\Gamma}_{\mathbf{j}}\right]\otimes\vec{\Gamma}_{\mathbf{i}}\right] \\
& + \left[\omega^f_{\mathbf{i}}\ \left[\left(\Gamma_{\mathbf{i}} - \left[\omega^f_{\mathbf{j}}\ \Gamma_{\mathbf{j}}\right]\right)_l\ \alpha_{\mathbf{i},2\,km}\right]\right] \\
& - \left[\omega^a_{\mathbf{i}}\ \left[\left(\Gamma_{\mathbf{i}} - \left[\omega^a_{\mathbf{j}}\ \Gamma_{\mathbf{j}}\right]\right)_l\ \alpha_{\mathbf{i},2\,km}\right]\right]
\end{aligned}
\tag{6.49}
$$

for the second-depth coefficient, with the weighting scheme defined as

$$
\omega^f_{0,t} \triangleq \frac{B_{0,t}}{B_{0,t} - B_{N,t}} \qquad \omega^f_{i,t} \triangleq 0 \quad \forall i \notin \{0, N\} \qquad \omega^f_{N,t} \triangleq \frac{-B_{N,t}}{B_{0,t} - B_{N,t}}
$$

$$
\omega^a_{0,t} \triangleq 0 \qquad \omega^a_{i,t} \triangleq \frac{\delta_i\ B_{i,t}}{\sum_{k=1}^{N} \delta_k B_{k,t}} \qquad \forall i \in [1, N]
$$

and where we use simplified notations. Indeed, since all quantities involved are processes parametrised by the maturity, we can omit arguments t and T. Also, the modified Einstein notation refers to summations from indices 0 to N.

Proof Our approach is to make use of Corollary 6.1 [p. 330] that gives the dynamics of a single, rebased, fixed-coupon bond. Indeed, for a generic traded asset π_t we can re-express the par swap rate as

$$S_t^\delta(T_0) = \frac{F_t^{\delta,\pi}(T_0)}{A_t^{\delta,\pi}(T_0)} \tag{6.50}$$

$$\text{where} \quad \begin{cases} F_t^{\delta,\pi}(T_0) \triangleq B_t^\pi(T_0) - B_t^\pi(T_N), \\ \\ A_t^{\delta,\pi}(T_0) \triangleq \sum_{i=1}^{N} \delta_i \; B_t^\pi(T_i). \end{cases}$$

The numerator F_t^M represents the rebased floating leg, while the denominator A_t^M corresponds to the rebased annuity associated to the fixed leg. In this instance we will take as numeraire π_t the *Money Market Account (MMA)* process κ_t:

$$\pi_t \;\leftarrow\; \kappa_t \triangleq \exp\left(\int_0^t r_s ds \right).$$

In order to fall back into our generic framework, we need only the par swap rate $S_t^\delta(T_0)$ to be martingale under *some* measure: there is no such explicit requirement w.r.t. either the numerator (floating leg) or denominator (annuity). So why re-base F_t and A_t, why by the same numeraire, and why specifically use the money-market account?

The rationale is purely technical and aims at simplifying our computations. Rebasing both numerator and denominator kills their drift when using the risk-neutral driver, and we take the same numeraire on both sides to avoid rebasing the swap rate itself. The choice of the MMA maintains the original HJM volatility structure $\vec{\Gamma}_t(T)$, a property that extends to its chaos dynamics and will prove useful to express those of the rebased annuity. In summary we have

$$\forall T \quad \begin{array}{lll} \vec{\Gamma}_t^\kappa(T) \equiv \vec{\Gamma}_t(T) & d\vec{W}_t^\kappa \equiv d\vec{W}_t & \vec{\alpha}_{1,t}^\kappa(T) \equiv \vec{\alpha}_{1,t}(T) \\ \vec{\vec{\alpha}}_{2,t}^\kappa(T) \equiv \vec{\vec{\alpha}}_{2,t}(T) & \vec{\vec{\alpha}}_{3,t}^\kappa(T) \equiv \vec{\vec{\alpha}}_{2,t}(T) & \vec{\vec{\vec{\alpha}}}_{22,t}^\kappa(T) \equiv \vec{\vec{\vec{\alpha}}}_{22,t}(T) \end{array}$$

The first consequence of these choices is that each leg $F_t^{\delta,\pi}$ and $A_t^{\delta,\pi}$ has driftless dynamics when using the associated MMA driver, which is the risk-neutral driver itself. Therefore both numeraire and martingale measure are here unrelated to the maturity T considered.

The second consequence is that the par swap rate is now martingale under the measure associated to the *rebased* annuity, which is nothing else than the annuity measure proper.

We now need to compute the dynamics of both numerator F_t^M and denominator A_t^M, before combining them to obtain the ratios. To that end, we capitalise on previous results pertaining to fixed-weights baskets of Zero-Coupons.

Dynamics of the MMA-rebased floating leg F_t^κ

We shall denote the chaos dynamics of that basket as

$$\begin{cases} \dfrac{dF_t^\kappa(T)}{F_t^\kappa(T)} = \vec{\sigma}_t^{f,\kappa}(T)^\perp d\vec{W}_t, \\ d\vec{\sigma}_t^{f,\kappa}(T) = \vec{a}_{1,t}^{f,\kappa}(T)\,dt + \vec{\vec{a}}_{2,t}^{f,\kappa}(T)\,d\vec{W}_t + \vec{\vec{a}}_{3,t}^{f,\kappa}(T)\,d\vec{Z}_t, \\ d\vec{\vec{a}}_{2,t}^{f,\kappa}(T) = \overrightarrow{[\cdots]}^{(2)}\,dt + \vec{\vec{\vec{a}}}_{22,t}^{f,\kappa}(T)\,d\vec{W}_t + \overrightarrow{[\cdots]}^{(3)}\,d\vec{Z}_t. \end{cases}$$

Let us express the σ-(2,0) coefficients invoked above, as given by Corollary 6.1 [p. 330]. First, the instantaneous volatility comes as

$$\vec{\sigma}_t^{f,\kappa}(T) = \omega_{0,t}^f\,\overrightarrow{\Gamma_{0,t}}(T) + \omega_{N,t}^f\,\overrightarrow{\Gamma_{N,t}}(T) \qquad \text{i.e.} \qquad \omega_{i,t}^f(T) \equiv 0 \;\; \forall i \notin \{0, N\} \tag{6.51}$$

where

$$\omega_{0,t}^f(T) \triangleq \frac{B_{0,t}(T)}{B_{0,t}(T) - B_{N,t}(T)} \qquad \text{and} \qquad \omega_{N,t}^f(T) \triangleq \frac{-B_{N,t}(T)}{B_{0,t}(T) - B_{N,t}(T)}.$$

Then the first-depth coefficients are expressed as

$$\vec{a}_{1,t}^{f,\kappa}(T) \triangleq -\vec{\vec{a}}_{2,t}^{f,\kappa}\,\vec{\sigma}_t^{f,\kappa}(T) + \sum_{i\in\{0,N\}} \omega_{i,t}^f(T)\left[\vec{\alpha}_{i,1,t} + \vec{\vec{\alpha}}_{i,2,t}\,\overrightarrow{\Gamma_{i,t}}\right](T)$$

$$\vec{\vec{a}}_{2,t}^{f,\kappa}(T) \triangleq -\vec{\sigma}_t^{f,\kappa}(T)^{2\otimes} + \sum_{i\in\{0,N\}} \omega_{i,t}^f(T)\left[\vec{\vec{\alpha}}_{i,2,t} + \overrightarrow{\Gamma_{i,t}}^{2\otimes}\right](T)$$

$$\vec{\vec{a}}_{3,t}^{f,\kappa}(T) \triangleq \sum_{i\in\{0,N\}} \omega_{i,t}^f\,\vec{\vec{\alpha}}_{i,3,t}(T).$$

Finally, the last, second-depth coefficient comes as

$$\vec{\vec{\vec{a}}}_{22,t}^{f,\kappa}(T) = \left[-\vec{\vec{a}}_{2,t}^{f,\kappa} \overset{\leftrightarrow}{\otimes} \vec{\sigma}_t^{f,\kappa} + \sum_{i\in\{0,N\}} \omega_{i,t}^f\,\vec{\vec{\alpha}}_{i,2,t} \overset{\leftrightarrow}{\otimes} \overrightarrow{\Gamma_{i,t}} + \sum_{i\in\{0,N\}} \omega_{i,t}^f\,x_{i,t}\right](T)$$

with the temporary notation

$$x^f_{i,t}(T) \triangleq \left[\overrightarrow{\overrightarrow{\alpha}}_{i,22,t} + (\Gamma_{i,t} - \sigma^{f,\kappa}_t)_k \, \alpha_{i,2,t\;jl} + \overrightarrow{\Gamma_{i,t}}^{3\otimes} - \overrightarrow{\Gamma_{i,t}} \otimes \overrightarrow{\sigma}^{f,\kappa}_t(T) \otimes \overrightarrow{\Gamma_{i,t}} \right](T).$$

Dynamics of the MMA-rebased annuity A^κ_t

We denote the dynamics of this rebased basket with the following SDE system

$$\begin{cases} \dfrac{dA^\kappa_t(T)}{A^\kappa_t(T)} = \overrightarrow{\sigma}^{a,\kappa}_t(T)^\perp d\overrightarrow{W}_t \\ d\overrightarrow{\sigma}^{a,\kappa}_t(T) = \overrightarrow{a}^{a,\kappa}_{1,t}(T)\,dt + \overrightarrow{\overrightarrow{a}}^{a,\kappa}_{2,t}(T)\,d\overrightarrow{W}_t + \overrightarrow{\overrightarrow{a}}^{a,\kappa}_{3,t}(T)\,d\overrightarrow{Z}_t \\ d\overrightarrow{\overrightarrow{a}}^{a,\kappa}_{2,t}(T) = \overrightarrow{[\cdots]}^{(2)}\,dt + \overrightarrow{\overrightarrow{\overrightarrow{a}}}^{a,\kappa}_{22,t}(T)\,d\overrightarrow{W}_t + \overrightarrow{[\cdots]}^{(3)}\,d\overrightarrow{Z}_t \end{cases}$$

where, again from Corollary 6.1 [p. 330], we have the instantaneous volatility as

$$\overrightarrow{\sigma}^{a,\kappa}_t(T) = \sum_{i=1}^N \omega^a_{i,t}(T)\,\overrightarrow{\Gamma}_{i,t}(T) \quad \text{where} \quad \omega^a_{i,t}(T) \triangleq \frac{\delta_i\;B_{i,t}(T)}{\sum_{k=1}^N \delta_k B_{k,t}(T)}. \tag{6.52}$$

The first-depth coefficients come as

$$\overrightarrow{a}^{a,\kappa}_{1,t}(T) \triangleq -\overrightarrow{\overrightarrow{a}}^{a,\kappa}_{2,t}\;\overrightarrow{\sigma}^{a,\kappa}_t(T) + \sum_{i=1}^N \omega^a_{i,t}\left[\overrightarrow{\alpha}_{i,1,t} + \overrightarrow{\overrightarrow{\alpha}}_{i,2,t}\,\overrightarrow{\Gamma_{i,t}}\right](T),$$

$$\overrightarrow{\overrightarrow{a}}^{a,\kappa}_{2,t}(T) \triangleq -\overrightarrow{\sigma}^{a,\kappa}_t(T)^{2\otimes} + \sum_{i=1}^N \omega^a_{i,t}\left[\overrightarrow{\overrightarrow{\alpha}}_{i,2,t} + \overrightarrow{\Gamma_{i,t}}^{2\otimes}\right](T),$$

$$\overrightarrow{\overrightarrow{a}}^{a,\kappa}_{3,t}(T) \triangleq \sum_{i=1}^N \omega^a_{i,t}\;\overrightarrow{\overrightarrow{\alpha}}_{i,3,t}(T).$$

Finally, the second-depth coefficient is expressed as

$$\overrightarrow{\overrightarrow{\overrightarrow{\alpha}}}^{a,\kappa}_{22,t}(T) = -\overrightarrow{\overrightarrow{a}}^{a,\kappa}_{2,t} \overset{\leftrightarrow}{\otimes} \overrightarrow{\sigma}^{a,\kappa}_t(T) + \sum_{i=1}^N \left[\omega^a_{i,t}\;\overrightarrow{\overrightarrow{\alpha}}_{i,2,t} \overset{\leftrightarrow}{\otimes} \overrightarrow{\Gamma_{i,t}}\right](T) + \sum_{i=1}^N \omega^a_{i,t}\;x^a_{i,t}(T)$$

with

$$x^a_{i,t}(T) = \left[\overrightarrow{\overrightarrow{\overrightarrow{\alpha}}}_{i,22,t} + \left[(\Gamma_{i,t} - \sigma^{a,\kappa}_t)_k\;\alpha_{i,2,t\;jl}\right] + \overrightarrow{\Gamma_{i,t}}^{3\otimes} - \overrightarrow{\Gamma_{i,t}} \otimes \overrightarrow{\sigma}^{a,\kappa}_t \otimes \overrightarrow{\Gamma_{i,t}}\right](T).$$

Dynamics of the par swap rate $S_t^\delta(T)$

Considering the expression for the par swap rate (6.50), Because both numerator and denominator have been simultaneously scaled by the same numeraire, it is clear that the choice of the latter will have no impact on the dynamics of the ratio itself.

Having chosen the Money Market Account guarantees to maintain the ZC volatility structure. In particular, the martingale measure is still the risk-neutral one and therefore by construction

$$\vec{\sigma}_t^{a}(T) \equiv \vec{\sigma}_t^{a,\kappa}(T).$$

However, the dynamics of the par swap rate are martingale under the measure associated to its denominator A_t^δ, so that using the usual change of numeraire technique under the Lognormal convention, we get (6.39)–(6.41) with

$$\vec{\sigma}_t^{s}(T) \triangleq \vec{\sigma}_t^{f,\kappa}(T) - \vec{\sigma}_t^{a,\kappa}(T). \tag{6.53}$$

Since the volatility presents an additive structure, the dynamic chaos structure can be split into two distinct parts, each pertaining either to the numerator or to the denominator. Gathering the previous results for both the floating leg and the annuity, and taking into account the change of driver, we therefore have the following expressions for the dynamic coefficients:

$$\vec{\vec{a}}_{2,t}^{s}(T) = \left[\vec{\vec{a}}_{2,t}^{f,\kappa} - \vec{\vec{a}}_{2,t}^{a,\kappa}\right](T), \qquad \vec{a}_{1,t}^{s}(T) = \left[\vec{a}_{1,t}^{f,\kappa} - \vec{a}_{1,t}^{a,\kappa} + \vec{\vec{a}}_{2,t}^{s}\,\vec{\sigma}_t^{a,\kappa}\right](T),$$

$$\vec{\vec{a}}_{3,t}^{s}(T) = \left[\vec{\vec{a}}_{3,t}^{f,\kappa} - \vec{\vec{a}}_{3,t}^{a,\kappa}\right](T), \qquad \vec{\vec{\vec{a}}}_{22,t}^{s}(T) = \left[\vec{\vec{\vec{a}}}_{22,t}^{f,\kappa} - \vec{\vec{\vec{a}}}_{22,t}^{a,\kappa}\right](T).$$

In order to develop these expressions, we use the aforementioned notations with regard to the arguments (t and T) and to the Einstein summation indices. We immediately get the simplest expressions (6.45), (6.47) and (6.48). As for the two remaining coefficients, by replacement only the drift comes as

$$\begin{aligned}
\vec{a}_{1,t}^{s}(T) = & \left[\left[\omega_{\mathbf{i}}^{f}\,\vec{\Gamma}_{\mathbf{i}}\right]^{2\otimes} - \left[\omega_{\mathbf{i}}^{f}\,\left(\vec{\vec{\alpha}}_{\mathbf{i},2} + \vec{\Gamma}_{\mathbf{i}}^{2\otimes}\right)\right]\right]\left[\omega_{\mathbf{i}}^{f}\,\vec{\Gamma}_{\mathbf{i}}\right] \\
& + \left[\omega_{\mathbf{i}}^{f}\,\left(\vec{\alpha}_{\mathbf{i},1} + \vec{\vec{\alpha}}_{\mathbf{i},2}\,\vec{\Gamma}_{\mathbf{i}}\right)\right] - \left[\omega_{\mathbf{i}}^{a}\,\left(\vec{\alpha}_{\mathbf{i},1} + \vec{\vec{\alpha}}_{\mathbf{i},2}\,\vec{\Gamma}_{\mathbf{i}}\right)\right] \\
& + \left[\left[\omega_{\mathbf{i}}^{a}\,\left(\vec{\vec{\alpha}}_{\mathbf{i},2} + \vec{\Gamma}_{\mathbf{i}}^{2\otimes}\right)\right] - \left[\omega_{\mathbf{i}}^{a}\,\vec{\Gamma}_{\mathbf{i}}\right]^{2\otimes}\right]\left[\omega_{\mathbf{i}}^{a}\,\vec{\Gamma}_{\mathbf{i}}\right] \\
& + \vec{\vec{a}}_{2}^{s}\,\left[\omega_{\mathbf{i}}^{a}\,\vec{\Gamma}_{\mathbf{i}}\right]
\end{aligned}$$

which provides (6.46) after some re-ordering of the terms, whose aim is to underline the *discrete covariance* structures. Finally, the second-depth coefficient comes as

$$\begin{aligned}
\overset{\Rightarrow}{a}{}^{s}_{22,t} \triangleq\ & \left[\left[\omega^f_{\mathbf{i}}\ \vec{\Gamma}_{\mathbf{i}}\right]^{2\otimes} - \left[\omega^f_{\mathbf{i}}\ (\overset{\Rightarrow}{\alpha}_{\mathbf{i},2} + \vec{\Gamma}_{\mathbf{i}}^{2\otimes})\right]\right] \overset{\leftrightarrow}{\otimes} \left[\omega^f_{\mathbf{i}}\ \vec{\Gamma}_{\mathbf{i}}\right] + \left[\omega^f_{\mathbf{i}}\ \overset{\Rightarrow}{\alpha}_{\mathbf{i},2} \overset{\leftrightarrow}{\otimes} \vec{\Gamma}_{\mathbf{i}}\right] \\
& + \left[\omega^f_{\mathbf{i}} \left[\overset{\Rightarrow}{\alpha}_{\mathbf{i},22} + \left[\left(\Gamma_{\mathbf{i}} - \left[\omega^f_{\mathbf{j}}\ \Gamma_{\mathbf{j}}\right]\right)_l \alpha_{\mathbf{i},2\,km}\right] + \vec{\Gamma}_{\mathbf{i}}^{3\otimes} - \vec{\Gamma}_{\mathbf{i}} \otimes \left[\omega^f_{\mathbf{j}}\ \vec{\Gamma}_{\mathbf{j}}\right] \otimes \vec{\Gamma}_{\mathbf{i}}\right]\right] \\
& - \left[\left[\omega^a_{\mathbf{i}}\ \vec{\Gamma}_{\mathbf{i}}\right]^{2\otimes} - \left[\omega^a_{\mathbf{i}}\ (\overset{\Rightarrow}{\alpha}_{\mathbf{i},2} + \vec{\Gamma}_{\mathbf{i}}^{2\otimes})\right]\right] \overset{\leftrightarrow}{\otimes} \left[\omega^a_{\mathbf{i}}\ \vec{\Gamma}_{\mathbf{i}}\right] - \left[\omega^a_{\mathbf{i}}\ \overset{\Rightarrow}{\alpha}_{\mathbf{i},2} \overset{\leftrightarrow}{\otimes} \vec{\Gamma}_{\mathbf{i}}\right] \\
& - \left[\omega^a_{\mathbf{i}} \left[\overset{\Rightarrow}{\alpha}_{\mathbf{i},22} + \left[\left(\Gamma_{\mathbf{i}} - \left[\omega^a_{\mathbf{j}}\ \Gamma_{\mathbf{j}}\right]\right)_l \alpha_{\mathbf{i},2\,km}\right] + \vec{\Gamma}_{\mathbf{i}}^{3\otimes} - \vec{\Gamma}_{\mathbf{i}} \otimes \left[\omega^a_{\mathbf{j}}\ \vec{\Gamma}_{\mathbf{j}}\right] \otimes \vec{\Gamma}_{\mathbf{i}}\right]\right]
\end{aligned}$$

and by re-ordering the terms we get (6.49), which concludes the proof. □

Given their complexity, the swap rate dynamics are not ideally suited for an intuitive interpretation. There are, however, a few remarks to be made regarding the proof itself and the associated techniques.

First, note that moving to the Annuity driver can be done at alternative steps during the computation. For instance, one could start from (6.53) and treat $\vec{\sigma}_t^{\,D}(A)$ separately by computing its dynamics under $\mathbb{Q}^{\delta,T}$. Indeed, this is a basket with stochastic weights $\omega_i = \delta_i / A_t$, but those are martingale under the annuity measure, which significantly simplifies the computations.

The choice of driver will influence (only) the drift term, which is measure-specific. Statically, that drift value can be recovered easily: ignoring the tensorial aspect, it is simple to get $a_{\ldots 1}$ from $a_{\ldots 2}$ (and possibly $a_{\ldots 3}$) when the driver is modified. However, as soon as the drift dynamics are involved we need to start from the correct measure. Granted, within the σ-(2,0) group we do not have to compute the dynamics of $\vec{a}_{1,t}(T)$, so the moment at which we adopt the annuity measure makes no difference. However, we have to assume that one might want, in the future, to continue these derivations to higher orders.

Still on the matter of the proof, note that we could obtain the swap rate chaos dynamics by alternative means. Still using our Corollaries, we could, for instance, write the swap rate as a whole in the form of a bi-dimensional, stochastic weights basket:

$$\begin{aligned}
S_t^{\delta}(T) &= \omega_{0,t}\ B_t^{\kappa}(T_0) + \omega_{N,t}\ B_t^{\kappa}(T_N) \\
\text{with} \quad \omega_{0,t} &\triangleq -\omega_{N,t} \triangleq \left[A_t^{\delta,\kappa}\right]^{-1}.
\end{aligned} \tag{6.54}$$

It is also tempting to try and interpret these complex formulas by exploiting further the basket technique/analogy and invoking the discrete probability scheme. For a more detailed description of this subject, we refer to Sects. 3.5.6.1 [p. 205], 3.5.6.2 [p. 205] and 3.5.6.3 [p. 206]. Indeed, the discrete insight is often able to formulate complex tensorial blocks in terms of variance-covariance, or if required as higher "moments" of the population of Zero-Coupons. The simplest application would involve rewriting terms that are recurring in (6.45)–(6.49), such as

$$\sum_i \omega^*_{i,t}(T)\, \overrightarrow{\Gamma_{i,t}}^{2\otimes}(T) - \left[\sum_i \omega^{\cdot}_{i,t}(T)\, \overrightarrow{\Gamma_{i,t}}(T)\right]^{2\otimes} = \mathbb{V}^{\omega}_{\otimes}\left[\overrightarrow{\Gamma_{\cdot,t}}(T)\right].$$

The issue, however, is that we would need *two* discrete probability measures, one for the floating leg and one for the fixed leg. Otherwise we would end up with a non-unit mass, in a similar manner as (6.54) would present a null mass.

To summarise, the discrete probabilistic interpretation would certainly be possible but a lot less appealing, which is why we choose not to present it. Instead, we postpone this kind of analysis to the LMM parametrisation in Chap. 7, which offers a more appropriate decomposition of the swap rate as a basket of Libors. It is indeed *that* basket analogy which is commonly used by practitioners, usually with a freezing of the weights.

Having covered the three families of vanilla options, to conclude this SV-HJM chapter we discuss a less obvious, parallel approach. Indeed, among the three products two are officially written on a rate, which is a re-parametrisation of a given asset. In terms of hedging and of underlying it is therefore possible, at least in principle, to use the latter.

6.6 Indirect Approaches: Assets *vs* Rates

In this section we discuss the general principle and the technical feasibility (or rather, lack thereof) of applying our asymptotic chaos expansions to caplets and swaptions, but in an *indirect* fashion, by matching the underlyings not to the *rates* (as was previously done) but instead to the associated *assets* themselves.

6.6.1 Applying the Asymptotic Approach to Caplets

This indirect approach stems directly from a classical property of the interest rates world (see [5] for instance) that allows us to define the same Caplet as one of two equivalent products:

Remark 6.1 (Equivalence between rate and asset options for a Caplet) An in-advance Caplet written on the Forward Libor rate $L_t(T, T+\delta)$, with notional M and strike K, presents the same payoff and therefore the same price as a Put written on the Zero-Coupon bond $B_t(T+\delta)$, with notional $(1+\delta K)$, strike $(1+\delta K)^{-1}$ and maturity T. This equivalence is summarised in Table 6.2.

Table 6.2 Caplet payoff equivalence

Underlying S	Option	Notional	Strike X	Expiry	Payment
$L_t(T, T+\delta)$	Call : $[S - X]^+$	M	K	T	$T+\delta$
$B_t(T+\delta)$	Put : $[X - S]^+$	$M\ (1+\delta K)$	$(1+\delta K)^{-1}$	T	T

This standard identity is easily proven by using a no-arbitrage argument. Indeed, discounting the Caplet payoff from date $T+\delta$ down to T, we have

$$\begin{aligned}\Phi(T) &= B_T\ (T+\delta)\ \delta\ [\ L_T\ (T, T+\delta) - K\ ]^+ \\ &= (1+\delta K)\left[(1+\delta K)^{-1} - B_T\ (T+\delta)\right]^+ .\end{aligned}$$

Here we observe three distinctive features deviating from the generic framework:

- The first difference is that we are now facing a European Put, instead of a Call.
- The second novelty is that the strike is now defined as $K_B \triangleq (1+\delta K)^{-1}$. Note that this value is indeed homogeneous to a Zero Coupon and does correspond to the forward $B_t(T, T+\delta)$ if $L_t^{\delta}(T) = K$.
- The third difference is that we now have a strike-dependent notional.

The pivotal item of the asymptotic methodology is to establish *some* Zero Drift Condition, therefore let us see how we could proceed in order to reach a similar situation. Thereafter we follow through the steps of Sects. 5.1 and 5.2, discussing if and how the generic methodology can be either instantiated or adapted to produce a workable ZDC.

6.6.1.1 Defining the Underlyings, Numeraires, Measures and Payoffs

First of all, we need to define the underlying $X_t(T)$ and the numeraire $N_t(T)$, the latter's volatility $\vec{\lambda}_t^N(T)$ defining a risk premium, which itself defines the driver $\vec{W}_t^{N(T)}$ and the martingale measure. To obtain driftless dynamics for an option's underlying which happens to be a traded asset, the usual approach is to rebase this underlying by a another asset. Given the payment sequence, here we choose to rebase with the Zero-Coupon $B_t(T)$ as numeraire.[9] We denote by $B_t^T(T+\delta)$ the new underlying, which is nothing else than the forward zero-coupon. Schematically, for Caplets we have

$$\begin{aligned} X_t(T) &\longleftarrow B_t^T(T+\delta) \triangleq \frac{B_t(T+\delta)}{B_t(T)} = B_t(T, T+\delta) \\ N_t(T) &\longleftarrow B_t(T) \qquad \vec{\lambda}_t^N(T) \longleftarrow \vec{\Gamma}_t(T) \\ d\vec{W}_t^{N(T)} &\longleftarrow d\vec{W}_t^T \triangleq d\vec{W}_t - \vec{\Gamma}_t(T)dt \end{aligned}$$

[9] Note that at maturity we have $X_T(U) = B_T^T(U) = B_T(U)$ and also $N_T(T) = 1$.

In summary, at this stage we have managed to instantiate the generic dynamics (5.1)–(5.3). But in order to match the actual product, we must adapt the previous generic payoff definition (5.4) into

$$h\ \left(X_T(T),\ N_T(T),\ K_B\right) \triangleq \frac{1}{K_B}\ \underbrace{N_T(T)}_{1}\ \left[K_B - X_T(T)\right]^+ \qquad \text{paid at time } T \tag{6.55}$$

where the novelty consists in the strike factor $1/K_B$. In itself, since the function $K \longrightarrow (1+\delta K)^{-1}$ is monotonous, this modification is not an issue. It can be interpreted as a change of strike unit and/or as a different definition of moneyness.

6.6.1.2 Defining the Implied Volatility Surface

First of all we need to decide on using either an absolute or a sliding representation for the implied volatility surface. As mentioned before, this is more of a cosmetic decision but it will significantly affect the resulting equations, in particular the ZDC. With a concern for consistency and therefore comparability, we choose to maintain a *sliding* SImpV parametrisation.

In order to define the sliding quantities we keep the usual time-to-maturity θ, but we logically define the new moneyness as

$$y \triangleq \ln\left(\frac{K_B}{X_t(T)}\right).$$

The next step is to re-parametrise *via* the IV, but we elect to replace (5.5) by

$$\begin{aligned} &V_t(X_t(T), T, K_B) \\ &= N_t(T)\ \frac{1}{K_B}\ P^{BS}\left(X_t(T),\ K_B,\ \Sigma_t(X_t(T), T, K_B)\cdot\sqrt{T-t}\right) \end{aligned} \tag{6.56}$$

where $P^{BS}(\cdot,\cdot,\cdot)$ is the normalised Black Put function:

$$P^{BS}\ (f,k,v) = k\ \left[1-\mathcal{N}(d_2)\right]\ - f\ \left[1-\mathcal{N}(d_1)\right] \tag{6.57}$$

$$\text{with} \qquad d_{1/2}(f,k,v) = -y/v \pm \frac{1}{2}v \qquad \text{and} \qquad y\ (f,k) = \ln\left(k/f\right).$$

We then take the same formal dynamics for the SImpV model as per (5.13). The question is therefore if and how this modification will affect the ZDC. In particular, we want to be dealing with sliding quantities only, so that the new (absolute) strike factor in (6.55) and in (6.56) seems inconvenient.

6.6.1.3 Deriving the Zero Drift Condition

Following Sect. 5.2.1, there is nothing stopping us from defining the sliding underlying, and then deriving its dynamics along with those of the absolute implied volatility surface. It is important to note that these computations ignore the nature of these two parametric processes, instead they represent purely a change of variable and rely on Itô-Kunita. Therefore, as long as we take K_B as our reference strike, we obtain precisely the same SDEs as in Lemma 5.1 [p. 283]. Consequently, we have the dynamics of the absolute implied volatility surface expressed from those of the sliding one.

We can now continue through the proof establishing the ZDC (see p. 289). From (6.56) we get that the fixed-strike, rebased Put

$$\frac{V_t(X_t(T), T, K_B)}{N_t(T)} = \frac{1}{K_B}\ P^{BS}\left(X_t(T),\ K_B,\ \Sigma_t(X_t(T), T, K_B)\cdot\sqrt{T-t}\right)$$

is a martingale under $\mathbb{Q}^{N(T)}$. We can write its dynamics, isolate the drift and write that the latter is null *a.s.* and for all times t: clearly the presence of the strike factor $1/K_B$ has no bearing on the corresponding PDE. Besides, and as discussed in Sect. 3.3.1.2 [p. 149], since the Greeks involved are the same, the Zero Drift Condition will be identical for a Call and for a Put.

In conclusion, we end up with exactly the same ZDC as in the generic case, only the relationship between the implied volatility and the price surface has changed. Indeed, the strike unit has changed, and the Black formula must be used with a scaling strike factor as per (6.56).

Note that beyond the fact that its price interpretation is different, the ZDC will be *numerically* different from the one obtained through the classical Libor rate approach (exposed in Sect. 6.4). This is because both the risk premium and the instantaneous volatility correspond to distinct processes.

6.6.2 Applying the Asymptotic Approach to Swaptions

In the same spirit, we can re-express a European payer physical Swaption as a Put on an *in fine* fixed-coupon bond. Indeed, that structure bundles together the fixed and the floating legs:

Remark 6.2 (Equivalence between rate and asset option for a Swaption) For a given tenor structure $[T_0, T_1, \ldots, T_N]$, the payoff and price of a payer physical swaption with notional M and strike K are identical to those of a Put written on an in fine bond with the same tenor structure and a fixed coupon of K, with notional δM, unit strike and expiry T_0. This equivalence is represented in Table 6.3.

Table 6.3 Payer physical swaption payoff equivalence

Underlying S	Option	Notional	Strike X	Expiry	Payment
$S_t^\delta(T_0)$	Call : $[S - X]^+$	M	K	T_0	Physical
$P_t^{\delta,\text{infine}}(T_0; K)$	Put : $[X - S]^+$	$M\ \delta$	1	T_0	T_0

Again, this identity is proven by comparing the value of both products at time T_0:

$$\Phi = A_{T_0}^\delta\ \delta\ \left[S_{T_0}^\delta - K\right]^+ = A_{T_0}^\delta\ \delta\ \left[\frac{1 - B_{T_0}(T_N)}{A_{T_0}^\delta} - K\right]^+$$
$$= \delta\left[1 - \left(\sum_{i=1}^{N-1} K\ B_{T_0}(T_i) + (1+K)\ B_{T_0}(T_N)\right)\right]^+ .$$

Let us now discuss how this asset *vs* rate equivalence would translate in terms of smile asymptotics. Again, we observe three main differences with the generic framework:

- The first novelty is that we are again facing a Put, as opposed to a Call.
- The second difference is that the strike is now fixed as unit: $K = 1$.
- The third distinctive feature is that the underlying is now strike-dependent.

Note that the scaling of the principal M by the accrual δ is not an issue at all and will not be discussed. We are facing a completely new setup, where the strike specification has been transferred from the option down to the underlying. Re-basing again by the maturity Zero $B_t(T)$, we can formally define our underlying map as

$$X_t(T; K) \longleftarrow P_t^{\delta,fine,T}(T) = \sum_{i=1}^{N-1} K\ B_t(T, T_i) + (1+K)\ B_t(T, T_N)$$

and consequently take the martingale measure to be the T-forward. We would have dynamics for both the numeraire and the underlying all right, but then the payoff definition would deviate completely from the generic framework.

In a sense, one could be forgiven for considering this context to be a logical extension of our modelling momentum. Indeed, in the single-underlying case of Chap. 2, both strike K and maturity T belonged solely to the option. Then in the generic term-structure framework of Chap. 5 each maturity had its own underlying. Now every point (K, T) on the smile corresponds to a different underlying. But in fact we have not really made the framework any richer, since the option has lost one degree of freedom.

Another consequence of these changes is that the notion of moneyness has been altered. Indeed, it is now the underlying bond whose strike makes it *at par* which becomes our reference. With all these modifications, it may be possible to apply an asymptotic approach to the problem. It would, however, be so removed from the mainstream methodology exposed so far that little recycling would be possible. In short, we would have to re-develop the computation process from scratch, and in terms of concept such an *ad hoc* approach sits precisely at the opposite corner of our methodology, which is focused on a generic methodology. For that reason, we leave that avenue to further research and will not explore it any further in the current study.

References

1. Cheyette, O.: Term structure dynamics and mortgage valuation. J. Fixed Income **1**, 28–41 (1992)
2. Ritchken, P., Sankarasubramanian, L.: Volatility structures of forward rates and the dynamics of the term structure. Math. Financ. **5**, 55–72 (1995)
3. Andersen, L., Andreasen, J.: Volatile volatilities. Risk Mag. (2002)
4. Piterbarg, V.V.: Stochastic volatility model with time-dependent skew. Appl Math. Financ. **12**, 147–185 (2005)
5. El Karoui, N.: Processus stochastiques et produits dérivés. Lecture Notes. University of Paris VI, Paris (2009)

Chapter 7
Implied Dynamics in the SV-LMM Framework

Abstract In this chapter we apply the generic term structure framework defined in Chap. 5 to the particular case of interest rates options, within a universal Stochastic Volatility Libor Market Model (SV-LMM). As in Chap. 6, our main focus is to solve the direct problem (generating the smile's shape and dynamics from the model specification) up to the first layer (which includes the smile's curvature and slope). We target some of the most liquid option types, namely caplets, swaptions and bond options. For technical reasons we exploit a model re-parametrisation via the rebased Zero Coupons, which allows us to recycle some of the SV-HJM results of Chap. 6. Likewise, in order to manage swaptions we use the basket results of Sect. 3.5. This enables us, in particular, to compute the systematic error of the usual *frozen weights* approximation.

The academic and practical interest for the LMM model class is currently intense. Indeed, this model—in its various flavours—has become the market reference for high-dimensional IR modelling. In its theoretical and continuous version, the LMM is a re-parametrisation of the HJM class, and hence offers a very rich framework. In its practical and discretised form, it describes the dynamics of a finite number of Libor rates (usually on a given tenor structure). Thus using observable, liquid market instruments as state variables it gains in tractability while retaining most of its capacity to deliver a complex correlation structure between maturities.

The class has evolved significantly, from the first Lognormal version of [1], up to the stochastic volatility instances of [2–5], and the jump-diffusion version of [6], via the local volatility extension of [7]. More recent SV versions exhibit even better compatibility with the vanilla smiles (both statically and dynamically) by incorporating dynamics close to SABR, the standard for European options (see Sect. 4.2). On that topic one can refer to [8–10] (Chap. 8), [11–15]. The class will certainly continue to improve, since research threads such as [16] announce ever more precise SV extensions.

The difficulty, however, lies not so much in *specifying* complex dynamics, but in providing fast and accurate approximations for European option pricing, required by the calibration and possibly the hedging algorithms. Deriving such approximations for a single Caplet in a Libor rate model relates to the single-underlying, multi-dimensional problem. However, deriving swaption price proxies involves the dynamics of a collection, in a term-structure context where measure issues dominate.

D. Nicolay, *Asymptotic Chaos Expansions in Finance*,
Springer Finance Lecture Notes, DOI 10.1007/978-1-4471-6506-4_7

One approach is to model the swap rates themselves: this is the swap market model (SMM) (see [17–19]). This framework offers several advantages, for instance it makes it easier to match the prices of co-initial or co-terminal swaptions, which itself is interesting for product-specific calibration. Conversely, the SMM is not necessarily adapted for calibrating in a more agnostic fashion (which is arguably a strength of the LMM), to match a regular set of expiries and tenors, or when targeting caplet volatilities. The LMM can be seen as a special (co-sliding) case of the SMM (see [19]) but the arbitrage constraints ensure that neither framework supersedes the other. Apart from specific applications, adopting one or the other tends to simply displace the issues.

In consequence, adapting our ACE results to provide good whole-smile extrapolations, for the most liquid options, does present a strong practical interest. It should help in using these—theoretically interesting—SV-LMM families for day-today trading and risk management.

In Sect. 7.1 we present the generic LMM modelling framework, including its specific chaos dynamics. We also provide specialised notations for the dominant context of sliding tenor structures, and recall some simplifying assumptions with regard to the I.R. market. Finally, we state our objectives, which are still to link the SInsV model specification to three SImpV frameworks, each corresponding to one of the most liquid interest rates derivatives.

In Sect. 7.2 we proceed with an intermediate step, by expressing the dynamics of the rebased Zero Coupons with the LMM parametrisation. We explain the rationale for this transform, which is both theoretical and practical. In the latter respect, these new chaos dynamics will be used extensively to simplify the computations in Sects. 7.3 and 7.5.

In Sect. 7.3 we cover the first liquid product defining an implied volatility surface, the bond option. Capitalising on the work of Chap. 5 we need only define a continuum, immerse the problem into the generic term structure framework, and then derive the chaos dynamics.

In Sect. 7.4 we deal with the simplest case, which is the Caplet smile. Indeed, this is the natural instrument in an LMM parametrisation, since the state variable matches the underlying. Since the chaos dynamics are both given and simple, we exploit the situation to provide some interpretation w.r.t. the IATM level, skew and maturity slope.

In Sect. 7.5 we treat the physical swaptions, which entices the most complex computation of the three products. This is where the previous work done both on baskets (in Sect. 3.5 [p. 185]) and on rebased bonds (Sect. 7.2) can be called upon, providing faster and in a relatively compact way the σ-(2,0) coefficients.

Finally in Sect. 7.6 we discuss the usual ("Jamshidian") approximation for swaptions' volatility, which employs a fixed-weight basket of Libor rates as a proxy underlying. We express the exact and multi-dimensional error generated on the weighting scheme, then specialise this result to an idealised *affine Libor curve* setting, and provide a numerical example.

7.1 Definitions, Notations and Objectives

The presentation of the LMM context is quite a regular—and therefore frustrating—feature of the interest rates literature. It remains, however, that our notations must be established, but more importantly we must describe the chaos inherent to the SInsV framework.

7.1.1 *The LMM Framework in a Chaos Context*

In the context of stochastic interest rates, we assume a full yield curve, in other words a continuum of Zero Coupon bond processes (or discount factors) $B_t(T)$ for all[1] maturities T. For a fixed accrual, this spot curve equivalently[2] defines a continuum of forward Zero-Coupon processes $B_t(T, T+\delta)$ which can be re-parametrised through the linear Libor rate process $L_t(T, T+\delta)$ *via* the definition

$$L_t^\delta(T) = \frac{B(t,T) - B(t,T+\delta)}{\delta B(t,T+\delta)}. \tag{7.1}$$

The role of this rather long introduction is to stress the fact that the LMM framework structurally carries the same information as the HJM framework. Accordingly, many of our approaches will re-use some of the results and recycle some of the discussions conducted in Chap. 6. Again, we will opt for the market model approach and represent the dynamics of the Libor rate itself, rather than those of the instantaneous forward rate $f_t(T)$. This requires us to fix the accrual period δ so that we can set

$$dL_t^\delta(T)/L_t^\delta(T) = \vec{\sigma}_t^{L,\delta}(T)^\perp \, d\vec{W}_t^{T+\delta} \tag{7.2}$$

where the endogenous driver $\vec{W}_t^{T+\delta}$ is a Wiener process under the $T+\delta$-forward measure $\mathbb{Q}^{T+\delta}$ associated to the numeraire $B_t(T+\delta)$. We can then continue to specify the chaos dynamics of the term structure of instantaneous volatility, up to the σ-(2, 0) order:

$$d\vec{\sigma}_t^{L,\delta}(T) = \vec{a}_{1,t}^{L,\delta}(T)dt + \overset{\Rightarrow}{a}{}_{2,t}^{L,\delta}(T)d\vec{W}_t^{T+\delta} + \overset{\Rightarrow}{a}{}_{3,t}^{L,\delta}(T)d\vec{Z}_t, \tag{7.3}$$

$$d\overset{\Rightarrow}{a}{}_{2,t}^{L,\delta}(T) = \overset{(2)}{\overrightarrow{[\cdots]}}dt + \overset{\Rrightarrow}{a}_{22,t}(T)d\vec{W}_t^{T+\delta} + \overset{(3)}{\overrightarrow{[\cdots]}}d\vec{Z}_t. \tag{7.4}$$

The exogenous driver $\vec{Z}_t$ is assumed to be a Brownian motion orthogonal to $\vec{W}_t^{T+\delta}$ and also with unit covariance matrix. In accordance with the generic

[1] In practice we could restrict to a finite horizon, but this is a moot point.

[2] With the proviso that the instantaneous forward rates $f_t(s)$ for $t \le s < \delta$ are also given.

framework of Chap. 5, the instantaneous coefficient maps constituting the σ-(2,0) group, i.e.

$$\vec{\sigma}_t^{L,\delta}(T), \quad \vec{a}_{1,t}^{L,\delta}(T), \quad \vec{\vec{a}}_{2,t}^{L,\delta}(T), \quad \vec{\vec{a}}_{3,t}^{L,\delta}(T) \quad \text{and} \quad \vec{\vec{\vec{a}}}_{22,t}^{L,\delta}(T)$$

are simply assumed to be adapted (parametric and tensorial) processes. In particular there is no mention of any state variable, which justifies for this framework the status of generic Stochastic Volatility Libor Market Model (SV-LMM), as an instance of the term structure SInsV model class (see Sect. 5.1.3.1 [p. 277]). Naturally, most published SV-LMM models can be cast into our generic framework, with the noticeable exception of those specifications exploiting an infinite-dimensional driver: see [20] (Chap. 4), [21] or [22] for instance.

7.1.2 Tenor Structures and Simplified Notations

The LMM framework defined by (7.2)–(7.4) is structurally continuous with regard to the maturity parameter T. However, the most liquid derivatives products, in particular bond options and swaptions, exhibit payoffs that depend exclusively on a given, finite collection of Libor rates. In other words, they involve discrete *tenor structures* or *schedules* $\{T_i\}_{0\le i\le N}$ which are usually regular. Since our generic term structure is defined for a *single* expiry continuum, we will re-parametrise these tenor structures, using a combination of their start date T_0 and of the *accrual sequence* $\{\delta_i\}_{1\le i\le N}$. Formally, we define the equivalent representation as per

$$[T_0, T_1, \ldots, T_N] \iff [T_0, \delta_1, \ldots, \delta_N] \quad \text{with} \quad \delta_i \triangleq T_i - T_{i-1}.$$

Transferring this sliding parametrisation to the underlyings, we introduce the following definitions ($\triangleq$) and compact notations ($\doteq$) for the Libor rates and their dynamics:

$$L_i \doteq L_{i,t} \doteq L_{i,t}(T_0) \triangleq L_t(T_i) \qquad \vec{\sigma}_i^L \doteq \vec{\sigma}_{i,t}^L \doteq \vec{\sigma}_{i,t}^L(T_0) \triangleq \vec{\sigma}_t^L(T_i)$$

$$\vec{\vec{a}}_{i,2}^L \doteq \vec{\vec{a}}_{i,2,t}^L \doteq \vec{\vec{a}}_{i,2,t}^L(T_0) \triangleq \vec{\vec{a}}_{2,t}^L(T_i) \qquad \vec{\vec{a}}_{i,3}^L \doteq \vec{\vec{a}}_{i,3,t}^L \doteq \vec{\vec{a}}_{i,3,t}^L(T_0) \triangleq \vec{\vec{a}}_{3,t}^L(T_i)$$

$$\vec{a}_{i,1}^L \doteq \vec{a}_{i,1,t}^L \doteq \vec{a}_{i,1,t}^L(T_0) \triangleq \vec{a}_{1,t}^L(T_i) \qquad \vec{\vec{\vec{a}}}_{i,22}^L \doteq \vec{\vec{\vec{a}}}_{i,22,t}^L \doteq \vec{\vec{\vec{a}}}_{i,22,t}^L(T_0) \triangleq \vec{\vec{\vec{a}}}_{22,t}^L(T_i)$$

where the reference to the individual Libor rate's accrual period L^δ has been omitted. Indeed, we will only consider *spanning* rates, in the sense that

$$L_t(T_i) \doteq L_t(T_i, T_i + \delta_{i+1}).$$

when considering tenor structures, we will denote by δ the entire accrual sequence defining the swap rate or the bond: $\delta \triangleq \{\delta_i\}_{1 \le i \le N}$. Let us now define the assumptions and goals of the present chapter.

7.1.3 Objectives and Assumptions

First, we must simplify the modelling context, hence for simplicity and consistency reasons we adopt the same conventions as laid out in Chap. 6 with Assumption 6.1. These can be briefly summarised with the following principles:

- The tenor structures considered are not necessarily regular, neither are the fixed and floating legs synchronous, but their first and last dates (T_0 and T_N) do coincide. Also, all Libor fixing and value dates are identical, while in all schedules we have matching calendar, fixing, accrual and payment dates.
- Any basis is ignored so that a single yield curve is sufficient.
- The swaptions under consideration are physically settled, and their expiry date matches the underlying swap's start date.

Our aims are to link the SInsV model class provided by the SV-LMM framework to the SImpV class associated to the three most liquid smiles. More precisely, the instantaneous dynamics will be given by (7.2)–(7.4) and the products will be bond options, *in-advance* caplets and physical swaptions.

Our obvious strategy is to immerse the SInsV and the different smile dynamics in the generic framework of Sect. 5.1, respectively up to the σ-(2,0) and the $\widetilde{\Sigma}$-(2,0) levels. In so doing we can then inherit all the generic results of Chap. 5, in particular those of Sect. 5.3 dealing with the recovery, and those of Sect. 5.4 providing the solution to the direct problem.

7.2 Chaos Dynamics of the Zeros in an LMM Framework

In this Section we provide some intermediate results that will be used in Sect. 7.3 (dedicated to bond options) and Sect. 7.5 (dealing with swaptions). They express the (rebased) Zero Coupons, both statically and dynamically, using an LMM parametrisation. This transfer of dynamics will allow us to recycle the HJM results of Chap. 6.

7.2.1 State Variables and Rationale for Rebasing

Transferring the LMM dynamics onto rebased bonds is an efficient re-parametrisation. One of its most noticeable advantages is that it enforces martingale dynamics for the new state variables. This is especially beneficial for numerical simulation, as the LMM drift is one of the main issues of the Monte-Carlo discretisation: see [23, 24].

Expressing these ZC dynamics will be done within the context of a tenor structure $[T_0, T_1, \ldots, T_N]$ where the Libor rates are defined by (7.1) and the notations are those of Sect. 7.1.2. The Zeros considered will be both the standard map $B_t(T)$ and the T_0-rebased term structure $B_t^0(T)$ defined by

$$B_t^0(T) \triangleq \frac{B_t(T)}{B_t(T_0)}. \tag{7.5}$$

For all three state variables maps, the dynamics will be described up to the first layer, a.k.a. the σ-(2,0) group of coefficients in the chaos diffusion. The LMM is naturally described by the SDEs (7.2)–(7.4) introduced in the current chapter. As for the zeros, the standard version had their dynamics introduced in Sect. 6.1 [p. 324] with (6.1)–(6.3). From there, the SDEs driving the rebased version comes easily with a change of measure and is given by specialising Lemma 6.1 [p. 329]. Formally, we have

$$\frac{dB_t(T)}{B_t(T)} = r_t dt + \vec{\Gamma}_t(T)^{\perp} d\vec{W}_t, \qquad \frac{dB_t^0(T)}{B_t^0(T)} = \vec{\Gamma}_t^0(T)^{\perp} d\vec{W}_t^0,$$

with the chaos continuing as

$$d\vec{\Gamma}_t(T) = \vec{\alpha}_{1,t}(T)dt + \vec{\vec{\alpha}}_{2,t}(T)d\vec{W}_t + \vec{\vec{\alpha}}_{3,t}(T)d\vec{Z}_t$$
$$d\vec{\vec{\alpha}}_{2,t}(T) = \overrightarrow{\overrightarrow{[\cdot]}}dt + \vec{\vec{\vec{\alpha}}}_{22,t}(T)d\vec{W}_t + \vec{\vec{\vec{[\cdot]}}}d\vec{Z}_t$$

and

$$d\vec{\Gamma}_t^0(T) = \vec{\alpha}_{1,t}^0(T)dt + \vec{\vec{\alpha}}_{2,t}^0(T)d\vec{W}_t^0 + \vec{\vec{\alpha}}_{3,t}^0(T)d\vec{Z}_t$$
$$d\vec{\vec{\alpha}}_{2,t}^0(T) = \overrightarrow{\overrightarrow{[\cdot]}}dt + \vec{\vec{\vec{\alpha}}}_{22,t}^0(T)d\vec{W}_t^0 + \vec{\vec{\vec{[\cdot]}}}d\vec{Z}_t$$

where $\vec{W}_t$ and $\vec{W}_t^0$ are respectively the risk-neutral and T_0-forward drivers, and where the standard and rebased processes are linked by the following simple relationships:

$$d\vec{W}_t^0 = d\vec{W}_t - \vec{\Gamma}_t(T_0)dt \qquad \vec{\Gamma}_t^0(T) = \vec{\Gamma}_t(T) - \vec{\Gamma}_t(T_0)$$
$$\vec{\alpha}_{1,t}^0(T) = \vec{\alpha}_{1,t}(T) - \vec{\alpha}_{1,t}(T_0) + \vec{\vec{\alpha}}_{2,t}^0(T)\vec{\Gamma}_t(T_0)$$
$$\vec{\vec{\alpha}}_{2,t}^0(T) = \vec{\vec{\alpha}}_{2,t}(T) - \vec{\vec{\alpha}}_{2,t}(T_0) \qquad \vec{\vec{\alpha}}_{3,t}^0(T) = \vec{\vec{\alpha}}_{3,t}(T) - \vec{\vec{\alpha}}_{3,t}(T_0)$$
$$\vec{\vec{\vec{\alpha}}}_{22,t}^0(T) = \vec{\vec{\vec{\alpha}}}_{22,t}(T) - \vec{\vec{\vec{\alpha}}}_{22,t}(T_0).$$

First we relate the state variables between both environments. Inverting the forward Libor rate definition (7.1) gives the ZCs in an LMM parametrisation: for $1 \le i \le N$

$$B_{i,t}(T) = \frac{B_{0,t}(T)}{\prod_{k=0}^{i-1}\left[1+\delta_k L_{k,t}(T)\right]}, \qquad \text{hence } B^0_{i,t}(T) = \prod_{k=0}^{i-1}\frac{1}{\left[1+\delta_k L_{k,t}(T)\right]}. \tag{7.6}$$

Moving on to the dynamics of these Zero Coupons, we recall that the SDE driving the Libor rates is expressed in the HJM specification by (6.33) [p. 343] as:

$$\frac{dL_t^\delta(T)}{L_t^\delta(T)} = \frac{1+\delta L_t^\delta(T)}{\delta L_t^\delta(T)}\left[\vec{\Gamma}_t(T) - \vec{\Gamma}_t(T+\delta)\right] d\vec{W}_t^{T+\delta}.$$

Note that this relationship alone justifies that in a given model, only one of the two specifications (LMM and HJM) can exhibit a *deterministic* volatility: at best, the other one will be local. In our generic stochastic instantaneous volatility (SInsV) context, however, the coefficient maps are simply taken as adapted processes. No information is provided w.r.t. other state variables, such as a volatility process, so that both specifications are theoretically equivalent.[3]

Our first objective is to invert that relationship, in a tenor structure context defined in Sect. 7.1.2. In other words, we wish to express the Zero Coupon volatilities $\vec{\Gamma}_t(T_i)$ from the Libor rates $Ł_t(T_i)$ and from their volatilities $\vec{\sigma}_t^{\delta,L}(T_i)$. Inverting (6.33) within the tenor structure we get

$$\forall k \in [0, N-1]\; \vec{\Gamma}_t(T_k) - \vec{\Gamma}_t(T_{k+1}) = \frac{\delta_k L_t(T_k)}{1+\delta_k L_t(T_k)}\vec{\sigma}_t^L(T_k)$$

hence

$$\forall i \in [1, N] \sum_{k=0}^{i-1} \vec{\Gamma}_t(T_k) - \vec{\Gamma}_t(T_{k+1}) = \sum_{k=0}^{i-1}\frac{\delta_k L_t(T_k)}{1+\delta_k L_t(T_k)}\vec{\sigma}_t^L(T_k).$$

By referencing the schedule w.r.t. T_0 and δ and by introducing the following notation

$$M_{i,t}(T) \doteq M_t(T_i) \triangleq \frac{\delta_i L_t(T_i)}{1+\delta_i L_t(T_i)} = 1 - B_t(T_i, T_{i+1}) \tag{7.7}$$

the equation for the Zeros volatility becomes

$$\vec{\Gamma}_{i,t} = \vec{\Gamma}_{0,t} - \sum_{k=0}^{i-1} M_{k,t}\vec{\sigma}_{k,t}^L \quad \vec{\Gamma^0_{i,t}} = -\sum_{k=0}^{i-1} M_{k,t}\vec{\sigma}_{k,t}^L \quad \forall i \in [1, N]. \tag{7.8}$$

[3] However, some specific parametrisations might perform better in one framework than in the other.

We observe in (7.8) a classical feature of this change of yield curve representation. Indeed, after changing to an LMM representation and without rebasing, some of the native HJM specification is still present through a single Zero volatility $\vec{\Gamma}_t(T_0)$. This is similar to the change in state variable (7.6), where a single Zero remained. This restriction is sometimes referred to as the Libor representation of the yield curve presenting a *blind spot* for maturities between t and $t+\delta$. It appears then that, unless the current date t belongs to the schedule (i.e. $\exists k/t = T_k$), the simplest way to remove that dependency to the initial Zero $B_t(T_0)$ and to its volatility $\vec{\Gamma}_t(T_0)$ is to rebase the HJM structure with respect to the same Zero. It is this assessment that justifies most of our interest in the rebased map.

Note that in (7.8) the minus sign in front of the sum makes perfect sense: on the left-hand side we are dealing with the volatility of an *asset*, while on the right-hand side the dynamics are those of *rates*, which by construction move in opposite directions.

7.2.2 Computing the Chaos Dynamics

We are now intent on computing the chaos dynamics of $\overrightarrow{\Gamma_{i,t}(T)}$ and $\vec{\Gamma}^0_{i,t}(T)$. Because of the additive structure of the former volatility, we choose to work on the latter, which presents itself as a tensor basket with stochastic weights.

Note that, in principle, (and we will follow this route) the derivation of these dynamics must be achieved by using consistently throughout the $\vec{W}^0_t$ driver invoked in their definition. Indeed, changing the driver modifies the drift coefficients, which is fine in a static perspective, but in a chaos perspective it makes it difficult to derive the dynamics for these coefficients. However, given the specific depth of the dynamics computed here, i.e. the σ-(2,0) group, we could get away with selecting any driver for the computation on the basis of convenience, and then switching at the last moment. We present these dynamics with the following lemma:

Lemma 7.1 (First layer chaos dynamics of rebased ZCs under SV-LMM) *In a generic SV-LMM framework whose chaos dynamics are defined up to the first layer by* (7.2)–(7.4) *the dynamics of the T_0-rebased Zero Coupons defined in* Sect. 7.2.1 *are given by* (7.8) *for the volatility, and by*

$$\begin{aligned}\vec{\alpha}^0_{i,1} \triangleq & -\sum_{k=0}^{i-1} M_{k,t}\left[\vec{a}^L_{k,1,t} + \vec{\vec{a}}^L_{k,2,t}\,\vec{\sigma}^L_{k,t}\right] \\ & -\sum_{k=0}^{i-1} M_{k,t}\left[\vec{\vec{a}}^L_{k,2,t} + \frac{1}{1+\delta_k L_{k,t}}\,\vec{\sigma}^{L\ 2\otimes}_{k,t}\right]\left[\sum_{j=0}^{k-1} M_{j,t}\,\vec{\sigma}^L_{j,t}\right]\end{aligned} \tag{7.9}$$

$$\overset{\Rightarrow 0}{\alpha}_{i,2,t} \triangleq - \sum_{k=0}^{i-1} M_{k,t} \left[\overset{\Rightarrow L}{a}_{k,2,t} + \frac{1}{1+\delta_k L_{k,t}} \vec{\sigma}_{k,t}^{L\ 2\otimes} \right] \tag{7.10}$$

$$\overset{\Rightarrow 0}{\alpha}_{i,3,t} \triangleq - \sum_{k=0}^{i-1} M_{k,t} \overset{\Rightarrow L}{a}_{k,3,t} \tag{7.11}$$

$$\overset{\Rrightarrow 0}{\alpha}_{i,22,t} \triangleq - \sum_{k=0}^{i-1} \frac{M_k}{1+\delta_k L_k} \left[\frac{1-\delta_k L_k}{1+\delta_k L_k} \vec{\sigma}_k^{L\ 3\otimes} + \overset{\Rightarrow L}{a}_{k,2} \overset{\leftrightarrow}{\otimes} \vec{\sigma}_k^{L} + \left[\sigma_{k\ n}^{L}\ a_{k,2\ mp}^{L} \right] \right]. \tag{7.12}$$

Note that all processes listed above are parametrised by the same date T_0 anchoring the tenor structure. Also, by convention a sum with incompatible bounds (e.g. $\sum_{k=0}^{-1}$) is taken as null.

Proof The dynamics of those weights $M_{i,t}$ against the T_0-driver come as

$$dM_{i,t} = d\left[\frac{\delta_i L_{i,t}}{1+\delta_i L_{i,t}} \right] = \delta_i \left(1+\delta_i L_{i,t}\right)^{-2} dL_{i,t} - \delta_i^2 \left(1+\delta_i L_{i,t}\right)^{-3} \langle dL_{i,t} \rangle$$

hence the Lognormal dynamics

$$\begin{aligned} \frac{dM_{i,t}}{M_{i,t}} &= -\frac{\delta_i L_{i,t}}{\left(1+\delta_i L_{i,t}\right)^2} \| \vec{\sigma}_{i,t}^{L} \|^2 dt + \frac{1}{1+\delta_i L_{i,t}} \vec{\sigma}_{i,t}^{L\ \perp} d\vec{W}_t^{i+1} \\ &= -\frac{\delta_i L_{i,t}}{\left(1+\delta_i L_{i,t}\right)^2} \| \vec{\sigma}_{i,t}^{L} \|^2 dt + \frac{1}{1+\delta_i L_{i,t}} \vec{\sigma}_{i,t}^{L\ \perp} \left[d\vec{W}_t^0 - \overrightarrow{\Gamma_{i+1,t}^0} dt \right] \end{aligned}$$

and finally

$$\frac{dM_{i,t}}{M_{i,t}} = \frac{-1}{1+\delta_i L_{i,t}} \left[\frac{\delta_i L_{i,t}}{1+\delta_i L_{i,t}} \| \vec{\sigma}_{i,t}^{L} \|^2 + \vec{\sigma}_{i,t}^{L\ \perp} \vec{\Gamma}_{i+1,t}^{0} \right] dt + \frac{\vec{\sigma}_{i,t}^{L\ \perp}}{1+\delta_i L_{i,t}} d\vec{W}_t^0.$$

Similarly, the dynamics of the individual Libor volatilities come against the same T_0-driver as

$$d\vec{\sigma}_{i,t}^{L} = \left[\vec{a}_{i,1,t}^{L} - \overset{\Rightarrow L}{a}_{i,2,t} \vec{\Gamma}_{i+1,t}^{0} \right] dt + \overset{\Rightarrow L}{a}_{i,2,t} d\vec{W}_t^0 + \overset{\Rightarrow L}{a}_{i,3,t} d\vec{Z}_t.$$

Noting that the weights dynamics involve no exogenous component, we can now use Lemma 3.6 [p. 192] to obtain the basket dynamics as

$$d\overrightarrow{\Gamma_{i,t}^{0}} = \vec{\alpha}_{i,1,t}^{0} dt + \overset{\Rightarrow 0}{\alpha}_{i,2,t} d\vec{W}_t^0 + \overset{\Rightarrow 0}{\alpha}_{i,3,t} d\vec{Z}_t$$

where, omitting the time argument, the coefficients come as

$$\vec{\alpha}^{\,0}_{i,1} \triangleq -\sum_{k=0}^{i-1} M_k \Bigg[\vec{a}^{\,L}_{k,1} - \overset{\Rightarrow}{a}^{L}_{k,2} \vec{\Gamma}^{\,0}_{k+1} + \frac{1}{1+\delta_k L_k} \overset{\Rightarrow}{a}^{L}_{k,2} \vec{\sigma}^{\,L}_{k}$$

$$- \frac{1}{1+\delta_k L_k} \left[\frac{\delta_k L_k}{1+\delta_k L_k} \vec{\sigma}^{\,L\,\perp}_{k} \vec{\sigma}^{\,L}_{k} + \vec{\sigma}^{\,L\,\perp}_{k} \vec{\Gamma}^{\,0}_{k+1} \right] \vec{\sigma}^{\,L}_{k} \Bigg]$$

$$\overset{\Rightarrow}{\alpha}^{0}_{i,2} \triangleq -\sum_{k=0}^{i-1} M_k \left[\overset{\Rightarrow}{a}^{L}_{k,2} + \frac{1}{1+\delta_k L_{k,t}} \vec{\sigma}^{\,L}_{k} \otimes \vec{\sigma}^{\,L}_{k} \right]$$

$$\overset{\Rightarrow}{\alpha}^{0}_{i,3} \triangleq -\sum_{k=0}^{i-1} M_k \overset{\Rightarrow}{a}^{L}_{k,3},$$

which immediately gives (7.10) and (7.11). Then rewriting the drift equation as

$$\vec{\alpha}^{\,0}_{i,1} \triangleq -\sum_{k=0}^{i-1} M_k \left[\vec{a}^{\,L}_{k,1} + \overset{\Rightarrow}{a}^{L}_{k,2} \left[\frac{1}{1+\delta_k L_k} \vec{\sigma}^{\,L}_{k} + \sum_{j=0}^{k} M_{j,t} \vec{\sigma}^{\,L}_{j,t} \right] \right]$$

$$+ \sum_{k=0}^{i-1} M_k \frac{1}{1+\delta_k L_k} \vec{\sigma}^{\,L\,2\otimes}_{k} \left[\frac{\delta_k L_k}{1+\delta_k L_k} \vec{\sigma}^{\,L}_{k} - \sum_{j=0}^{k} M_{j,t} \vec{\sigma}^{\,L}_{j,t} \right]$$

we obtain (7.9) after simplification. The dynamics of $\overset{\Rrightarrow}{\alpha}^{0}_{i,2,t}$ come by applying to (7.10) the same process, moreover with the same weights. The individuals' endogenous dynamics are

$$d\left[\overset{\Rightarrow}{a}^{L}_{k,2} + \frac{1}{1+\delta_i L_{k,t}} \vec{\sigma}^{\,L\,2\otimes}_{k} \right] = \overset{(2)}{\overrightarrow{[\cdots]}} dt + \overset{(3)}{\overrightarrow{[\cdots]}} d\vec{Z}_t + \overset{\Rrightarrow}{a}^{L}_{k,2}$$

$$- \frac{\delta_k L_k}{(1+\delta_k L_k)^2} \vec{\sigma}^{\,L\,3\otimes}_{k} + \frac{1}{1+\delta_k L_k} \vec{\sigma}^{\,L}_{k} \otimes \overset{\Rightarrow}{a}^{L}_{k,2}$$

$$+ \frac{1}{1+\delta_k L_k} \left[\sigma^{L}_{k\;n}\, a^{L}_{k,2\;mp} \right].$$

We can therefore apply Lemma 3.6 again to obtain

$$d\overset{\Rightarrow}{\alpha}^{0}_{i,2,t} = \overset{(2)}{\overrightarrow{[\cdots]}} dt + \overset{\Rrightarrow}{\alpha}^{0}_{i,22,t} \vec{W}^{0}_{t} + \overset{(3)}{\overrightarrow{[\cdots]}} d\vec{Z}_t$$

where the relevant coefficient comes as

$$\overset{\Rightarrow}{\alpha}{}^{0}_{i,22,t} = -\sum_{k=0}^{i-1} M_k \left[-\frac{\delta_k L_k}{(1+\delta_k L_k)^2} \vec{\sigma}_k^{L\,3\otimes} + \frac{1}{1+\delta_k L_k} \vec{\sigma}_k^L \otimes \overset{\Rightarrow}{a}{}^L_{k,2} \right.$$

$$\left. + \frac{1}{1+\delta_k L_k} \left[\sigma^L_{k\ n}\, a^L_{k,2\ mp} \right] + \frac{1}{1+\delta_k L_{k,t}} \left[\overset{\Rightarrow}{a}{}^L_{k,2} + \frac{1}{1+\delta_i L_{k,t}} \vec{\sigma}_k^{L\,2\otimes} \right] \otimes \vec{\sigma}_k^L \right]$$

which after simplification provides (7.12) and concludes the proof. □

7.3 Bond Options

In this section we apply our generic term-structure asymptotic methodology to the case of bond options and within a general Stochastic Volatility Libor Market Model parametrisation.

In order to exploit the results of Chap. 5 we start by immersing this specific option field into the generic framework, by selecting appropriate maps for the underlying, the numeraire, the martingale measure and the option payoff. We then express the dynamic coefficients for the underlying, in a chaos diffusion at the σ-(2,0) group level. To do so, several approaches are now possible given the available intermediate results (in particular the basket results of Sect. 3.5). We choose the easiest method by transferring the equivalent results of Sect. 6.3 established in the HJM context, therefore invoking the technical Lemma 6.2 of Sect. 6.3.

With these coefficients we will then have sufficient information to link the σ-(2,0) SInsV group to its $\widetilde{\Sigma}$-(2,0) SImpV equivalent, in other words to relate the LMM stochastic instantaneous volatility model class to the statics and dynamics of its associated implied volatility surface.

7.3.1 Casting the Bond Options into the Generic Framework

This short section essentially represents a summary of its HJM equivalent 6.3, to which the reader should refer for more details, with the exception of the fixed-coupons structure. We assume that the option maturity T matches an anniversary date T_0 of the underlying bond, thus avoiding accrued coupons. We fix the tenor component of the bond's tenor structure and allocate the option/underlying maturity of the generic framework to the start date T_0:

$T = T_0$ is the option expiry,

$T_i = T_{i-1} + \delta_i$ is the payment date for coupon c_i with $1 \leq i \leq N$.

We end up with a sliding bond [4] defined by its maturity T, by its sliding tenor structure $\delta \triangleq \{\delta_i\}$ for $1 \le i \le N$ and by its fixed coupon structure $c \triangleq \{c_i\}$. We denote its price process by $P_t^\delta(T;c)$ and employ the schedule notations of Sect. 6.1.2. As for the actual option payoff, it consists in a cash exchange at time T for a value of

$$\Phi(T) \triangleq \left[P_T^\delta(T;c) - K\right]^+.$$

We move from the underlying and option continuum to their driving SDE system, re-basing all bonds by the ZC $B_t(T)$ and defining the underlying term structure by

$$B_t^T(U) \triangleq \frac{B_t(U)}{B_t(T)}, \quad X_t(T) \longleftarrow P_t^{\delta,\mathbf{T}}(T;c) \triangleq \frac{P_t^\delta(T;c)}{B_t(T)} = \sum_{i=1}^N c_i\delta_i B_t^T(T_i).$$

Hence the martingale measure is the T-forward $\mathbb{Q}^T$ and the immersion comes as

$$N_t(T) \leftarrow B_t(T) \qquad \mathbb{Q}^{N(T)} \leftarrow \mathbb{Q}^T$$

$$\vec{W}_t^{N(T)} \leftarrow \vec{W}_t^T \;\text{ with }\; d\vec{W}_t^T = d\vec{W}_t - \vec{\Gamma}_t(T)dt.$$

It follows that the associated payoff does match the actual bond option:

$$B_T(T) \equiv 1 \quad \Longrightarrow \quad \left[P_T^\delta(T;c) - K\right]^+ = B_T(T)\left[P_T^{\delta,\mathbf{T}}(T;c) - K\right]^+.$$

Following Sect. 5.1.2 exactly, we then re-parametrise the absolute price mapping into the absolute Lognormal implied volatility surface *via* the normalised Black formula. We then introduce the sliding quantities and thus define the term-structure, sliding stochastic implied volatility (SImpV) model whose dynamics are only formally given through (5.13).

We have now completed the formal cast and it remains to express the first layer coefficients in the chaos dynamics of the rebased bond, within the LMM parametrisation.

7.3.2 Dynamics of the Underlying Rebased Bond

Our first task is to express the static underlying as a function of the Libor rates. We get from the immersion Sect. 7.3.1 and from the intermediate result 7.6 that

$$P_t^{\delta,\mathbf{T}}(T;c) = \sum_{i=1}^N c_i\delta_i B_{i,t}^0(T) = \sum_{i=1}^N c_i\delta_i \prod_{k=0}^{i-1} \frac{1}{\left[1+\delta_k L_{k,t}(T)\right]}. \tag{7.13}$$

[4] Recall that these choices imply that we cannot represent all European options written on a running bond.

Our strategy to express the underlying's chaos dynamics under $\mathbb{Q}^T$ is to capitalise on the results established for a fixed-coupon bond in a rebased HJM framework in Sect. 6.2.2. Thus exploiting Corollary 6.1 [p. 330] and incorporating the coupon structure c we get:

Corollary 7.1 (Bond chaos dynamics in an LMM framework) *In an LMM framework defined by* (7.2)–(7.4) *the dynamics of the T-rebased bond defined by* (7.13) *are as follows:*

$$\frac{dP_t^{\delta,T}(T)}{P_t^{\delta,T}(T)} = \vec{\sigma}_t^P(T)^{\perp} d\vec{W}_t^T$$

$$d\vec{\sigma}_t^P(T) = \vec{a}_{1,t}^P(T)dt + \vec{\vec{a}}_{2,t}^P(T)d\vec{W}_t^T + \vec{\vec{a}}_{3,t}^P(T)d\vec{Z}_t$$

$$d\vec{\vec{a}}_{2,t}^P(T) = \overrightarrow{[\cdots]}^{(2)}dt + \vec{\vec{\vec{a}}}_{22,t}^P(T)d\vec{W}_t^T + \overrightarrow{[\cdots]}^{(3)}d\vec{Z}_t$$

*where $\vec{W}_t^T$ is the T-forward driver. Apart from the coupons c_i and accruals δ_i, all quantities invoked are maturity-indexed processes, hence we omit arguments t and T whenever possible. Using modified Einstein notation to sum indices **from** 0 **to** $N-1$, we get the instantaneous volatility as*

$$\vec{\sigma}_t^P(T) = -\left[M_{\mathbf{i}}\Omega_{\mathbf{i}}\vec{\sigma}_{\mathbf{i}}^L\right] \tag{7.14}$$

with $M_i(T)$ defined from (7.7) *and where the weighting scheme is defined by*

$$\underset{1\le i\le N}{\omega_{i,t}(T)} \triangleq \frac{c_i\delta_i\prod_{k=0}^{i-1}\left[1+\delta_k L_{k,t}(T)\right]^{-1}}{\sum_{j=1}^{N}c_j\delta_j\prod_{k=0}^{j-1}\left[1+\delta_k L_{k,t}(T)\right]^{-1}}$$

$$\text{and}\quad \underset{0\le i\le N-1}{\Omega_{i,t}(T)} \triangleq \sum_{j=i+1}^{N}\omega_{j,t}(T). \tag{7.15}$$

At first depth we get the drift as

$$\vec{a}_{1,t}^P(T) = -\vec{\vec{a}}_{2,t}^P\vec{\sigma}_t^P - \left[M_{\mathbf{i}}\Omega_{\mathbf{i}}\vec{a}_{\mathbf{i},1}^L\right] - \left[M_{\mathbf{i}}\Omega_{\mathbf{i}}\vec{\vec{a}}_{\mathbf{i},2}^L\vec{\sigma}_{\mathbf{i}}^L\right]$$
$$+\sum_{i=0}^{N-1}\sum_{j=\mathbf{i}}^{N-1}\Omega_{\mathbf{j}}M_iM_j\left[\vec{\vec{a}}_{i,2}^L + \frac{1}{1+\delta_i L_i}\vec{\sigma}_i^{L\,2\otimes}\right]\vec{\sigma}_j^L \tag{7.16}$$

while the endogenous and exogenous coefficients read as

$$\vec{\vec{a}}^{\,p}_{2,t}(T) \triangleq -\vec{\sigma}^{\,p\,2\otimes}_{t} - \left[M_{\mathbf{i}}\Omega_{\mathbf{i}}\vec{\vec{a}}^{\,L}_{\mathbf{i},2}\right] + \left[\Omega_{max(\mathbf{i},\mathbf{j})}M_{\mathbf{i}}M_{\mathbf{j}}\vec{\sigma}^{\,L}_{\mathbf{i}} \otimes \vec{\sigma}^{\,L}_{\mathbf{j}}\right] - \left[\frac{M_{\mathbf{i}}\Omega_{\mathbf{i}}}{1+\delta_{\mathbf{i}}L_{\mathbf{i}}}\vec{\sigma}^{\,L\,2\otimes}_{\mathbf{i}}\right] \tag{7.17}$$

$$\vec{\vec{a}}^{\,p}_{3,t}(T) \triangleq -\left[M_{\mathbf{i}}\Omega_{\mathbf{i}}\vec{\vec{a}}^{\,L}_{\mathbf{i},3}\right]. \tag{7.18}$$

Finally we get the single second-depth coefficient as

$$\begin{aligned}
\vec{\vec{a}}^{\,p}_{22,t} = & -\vec{\vec{a}}^{\,p}_{2} \overset{\leftrightarrow}{\otimes} \vec{\sigma}^{\,p} + \left[\Omega_{max(\mathbf{i},\mathbf{j})}M_{\mathbf{i}}M_{\mathbf{j}}\vec{\vec{a}}^{\,L}_{\mathbf{i},2} \overset{\leftrightarrow}{\otimes} \vec{\sigma}^{\,L}_{\mathbf{j}}\right] - \left[\frac{\Omega_{\mathbf{i}}M_{\mathbf{i}}}{1+\delta_{\mathbf{i}}L_{\mathbf{i}}}\vec{\vec{a}}^{\,L}_{\mathbf{i},2} \overset{\leftrightarrow}{\otimes} \vec{\sigma}^{\,L}_{\mathbf{i}}\right] \\
& + \left[\Omega_{max(\mathbf{i},\mathbf{j})}\frac{M_{\mathbf{i}}M_{\mathbf{j}}}{1+\delta_{\mathbf{i}}L_{\mathbf{i}}}\vec{\sigma}^{\,L\,2\otimes}_{\mathbf{i}} \overset{\leftrightarrow}{\otimes} \vec{\sigma}^{\,L}_{\mathbf{j}}\right] - \left[\Omega_{\mathbf{i}}\frac{\delta_{\mathbf{i}}L_{\mathbf{i}}(1-\delta_{\mathbf{i}}L_{\mathbf{i}})}{(1+\delta_{\mathbf{i}}L_{\mathbf{i}})^3}\vec{\sigma}^{\,L\,3\otimes}_{\mathbf{i}}\right] \\
& - \left[\Omega_{max(\mathbf{i},\mathbf{j})}M_{\mathbf{i}}M_{\mathbf{j}}\vec{\sigma}^{\,L}_{\mathbf{i}} \otimes \vec{\sigma}^{\,p} \otimes \vec{\sigma}^{\,L}_{\mathbf{j}}\right] \\
& - \left[\Omega_{max(\mathbf{i},\mathbf{j},\mathbf{k})}M_{\mathbf{i}}M_{\mathbf{j}}M_{\mathbf{k}}\vec{\sigma}^{\,L}_{\mathbf{i}} \otimes \vec{\sigma}^{\,L}_{\mathbf{j}} \otimes \vec{\sigma}^{\,L}_{\mathbf{k}}\right] \\
& - \left[\frac{\Omega_{\mathbf{i}}M_{\mathbf{i}}}{1+\delta_{\mathbf{i}}L_{\mathbf{i}}}\left[\sigma^{L}_{\mathbf{i}\,n}\, a^{L}_{\mathbf{i},2\,mp}\right]\right] + \left[\Omega_{max(\mathbf{i},\mathbf{j})}M_{\mathbf{i}}M_{\mathbf{j}}\left[\sigma^{L}_{\mathbf{i}\,n}\, \vec{\vec{a}}^{\,L}_{\mathbf{j},2\,mp}\right]\right] \\
& + \left[\Omega_{max(\mathbf{i},\mathbf{j})}\frac{M_{\mathbf{i}}M_{\mathbf{j}}}{1+\delta_{\mathbf{j}}L_{\mathbf{j}}}\left[\sigma^{L}_{\mathbf{j}\,m}\,\sigma^{L}_{\mathbf{i}\,n}\,\sigma^{L}_{\mathbf{j}\,p}\right]\right] \\
& + \left[\Omega_{\mathbf{i}}M_{\mathbf{i}}\left[\sigma^{p}_{\ n}\, a^{L}_{\mathbf{i},2\,mp}\right]\right] + \left[\Omega_{\mathbf{i}}\frac{M_{\mathbf{i}}}{1+\delta_{\mathbf{i}}L_{\mathbf{i}}}\left[\sigma^{L}_{\mathbf{i}\,m}\,\sigma^{p}_{\ n}\,\sigma^{L}_{\mathbf{i}\,p}\right]\right].
\end{aligned} \tag{7.19}$$

We note that the volatility of the re-based bond does come as a basket of Libor rates volatilities, but that contrary to the HJM specification the weights are *not* normalised. Consequently, the first Libors have more impact on the dynamics than those located at the long end of the tenor structure. Note also that the *discrete* covariance structure of the endogenous vol of vol renders it insensitive to the addition of any vector (such as $\vec{\Gamma}_{0,t}(T)$) to the volatility structure.

Proof Note that we can use either of two sets of results, based on convenience. The first set is Corollary 6.1 [p. 330], which pertains to the dynamics of a fixed-coupon bond, but in a generically rebased context with the associated volatility structure. The second set of results is Lemma 6.2 [p. 337], which expresses the dynamics of a specifically T_0-rebased fixed-coupon bond, and this time in the native HJM parametrisation.

We start with either (6.10) [p. 330] or (6.19) [p. 337], which gives

$$\vec{\sigma}_t^p(T) = \sum_{i=1}^{N} \omega_{i,t} \vec{\Gamma}_{i,t}^0(T)$$

where the normalised weights are re-expressed from the LMM state variables as

$$\omega_{i,t}(T) \triangleq \frac{c_i\delta_i B_{i,t}(T)}{\sum_{j=1}^N \delta_j B_{j,t}(T)} = \frac{c_i\delta_i \prod_{k=0}^{i-1} \frac{1}{[1+\delta_k L_{k,t}(T)]}}{\sum_{j=1}^N \delta_j \prod_{k=0}^{j-1} \frac{1}{[1+\delta_k L_{k,t}(T)]}}$$

and where we can substitute the individual rebased ZC volatilities with (7.8), giving

$$\vec{\sigma}_t^p(T) = -\sum_{i=1}^{N} \omega_{i,t} \sum_{j=0}^{i-1} M_{j,t} \vec{\sigma}_{j,t}^L = -\sum_{i=0}^{N-1} M_{i,t} \left[\sum_{j=i+1}^{N} \omega_{j,t}\right] \vec{\sigma}_{i,t}^L,$$

which proves (7.14). The exogenous coefficient is also easy to translate, since substituting (7.11) [p. 375] into either (6.16) [p. 330] or (6.22) [p. 337] gives us

$$\begin{aligned}
\vec{\vec{a}}_{3,t}^{p}(T) &= \sum_{i=1}^{N} \omega_{i,t} \vec{\vec{\alpha}}_{i,3,t}^{0} = -\sum_{i=1}^{N} \omega_{i,t} \sum_{j=0}^{i-1} M_{j,t} \vec{\vec{a}}_{j,3,t}^{L} \\
&= -\sum_{i=0}^{N-1} M_{i,t} \left[\sum_{j=i+1}^{N} \omega_{j,t}\right] \vec{\vec{a}}_{i,3,t}^{L},
\end{aligned}$$

which proves (7.18). The computation for the endogenous vol of vol is only marginally more involved, using (6.14) [p. 330] instead of (6.21) [p. 337]. Injecting the LMM parametrisation (7.8) and (7.10) and omitting arguments t and T gives

$$\begin{aligned}
\vec{\vec{a}}_{2,t}^{p} \triangleq & -\vec{\sigma}^{p\,2\otimes} + \sum_{i=1}^{N} \omega_i \left[\sum_{k=0}^{i-1} M_k \vec{\sigma}_k^L\right]^{2\otimes} \\
& \underbrace{-\sum_{i=1}^{N} \omega_i \left[\sum_{k=0}^{i-1} M_k \left[\vec{\vec{a}}_{k,2}^{L} + \frac{1}{1+\delta_k L_k} \vec{\sigma}_k^{L\,2\otimes}\right]\right]}
\end{aligned}$$

where the under-braced sum is

$$\sum_{i=1}^{N} \omega_i \left[\sum_{k=0}^{i-1} \frac{M_k}{1+\delta_k L_k} \vec{\sigma}_k^{L\,2\otimes}\right] - \sum_{i=1}^{N} \omega_i \left[\sum_{j=0}^{i-1}\sum_{k=0}^{i-1} M_j M_k \vec{\sigma}_j^L \otimes \vec{\sigma}_k^L\right].$$

Using Fubini to re-organise the sums we get finally

$$\begin{aligned}\vec{\vec{a}}^{\,p}_{2,t} = &-\vec{\sigma}^{\,p\,2\otimes}_t - \sum_{i=0}^{N-1} \Omega_i M_i \vec{\vec{a}}^{\,L}_{i,2}\\ &+\sum_{i=0}^{N-1}\sum_{j=0}^{N-1} \Omega_{max(i,j)}\, M_i M_j \vec{\sigma}^{\,L}_i \otimes \vec{\sigma}^{\,L}_j - \sum_{i=0}^{N-1} \frac{\Omega_i M_i}{1+\delta_i L_i}\, \vec{\sigma}^{\,L\,2\otimes}_i,\end{aligned}$$

which provides (7.17). Coming to the drift, we adapt (6.12) [p. 330] into (again, omitting the arguments)

$$\vec{a}^{\,p}_{1,t}(T) = -\vec{\vec{a}}^{\,p}_2 \vec{\sigma}^{\,p} + \sum_{i=1}^{N} \omega_i \vec{\vec{\alpha}}^{\,0}_{i,2} \vec{\Gamma}^0_i + \sum_{i=1}^{N} \omega_i\, \vec{\alpha}^{\,0}_{i,1}$$

which after injecting the LMM parametrisation (7.8)–(7.10) becomes

$$\begin{aligned}\vec{a}^{\,p}_{1,t}(T) = &-\vec{\vec{a}}^{\,p}_2 \vec{\sigma}^{\,p}\\ &+\sum_{i=1}^{N} \omega_i \left[\sum_{k=0}^{i-1} M_k \left[\vec{\vec{a}}^{\,L}_{k,2} + \frac{1}{1+\delta_k L_k}\, \vec{\sigma}^{\,L\,2\otimes}_k\right]\right]\left[\sum_{j=0}^{i-1} M_j \vec{\sigma}^{\,L}_j\right]\\ &-\sum_{i=1}^{N} \omega_i \sum_{k=0}^{i-1} M_k \left[\vec{a}^{\,L}_{k,1} + \vec{\vec{a}}^{\,L}_{k,2} \vec{\sigma}^{\,L}_k + \left[\vec{\vec{a}}^{\,L}_{k,2} + \frac{1}{1+\delta_k L_k}\, \vec{\sigma}^{\,L\,2\otimes}_k\right]\left[\sum_{j=0}^{k-1} M_j \vec{\sigma}^{\,L}_j\right]\right].\end{aligned}$$

We split and factorise this expression into

$$\begin{aligned}\vec{a}^{\,p}_{1,t}(T) = &-\vec{\vec{a}}^{\,p}_{2,t} \vec{\sigma}^{\,p}_t - \sum_{i=1}^{N} \omega_i \sum_{k=0}^{i-1} M_k \vec{a}^{\,L}_{k,1} - \sum_{i=1}^{N} \omega_i \sum_{k=0}^{i-1} M_k \vec{\vec{a}}^{\,L}_{k,2} \vec{\sigma}^{\,L}_k\\ &+\sum_{i=1}^{N} \omega_i \sum_{k=0}^{i-1} M_k \left[\vec{\vec{a}}^{\,L}_{k,2} + \frac{1}{1+\delta_k L_k}\, \vec{\sigma}^{\,L\,2\otimes}_k\right]\left[\sum_{j=k}^{i-1} M_j \vec{\sigma}^{\,L}_j\right]\end{aligned}$$

where re-organising the indices gives us

$$\begin{aligned}\vec{a}^{\,p}_{1,t}(T) = &-\vec{\vec{a}}^{\,p}_{2,t} \vec{\sigma}^{\,p}_t - \sum_{i=0}^{N-1} M_i \Omega_i \vec{a}^{\,L}_{i,1} - \sum_{i=0}^{N-1} M_i \Omega_i \vec{\vec{a}}^{\,L}_{i,2} \vec{\sigma}^{\,L}_i\\ &+\sum_{i=0}^{N-1}\sum_{\mathbf{j}=\mathbf{i}}^{N-1} \Omega_{\mathbf{j}} M_i M_j \left[\vec{\vec{a}}^{\,L}_{i,2} + \frac{1}{1+\delta_i L_i}\, \vec{\sigma}^{\,L\,2\otimes}_i\right] \vec{\sigma}^{\,L}_j,\end{aligned}$$

which proves (7.16). Finally, to obtain the 2nd level coefficient we adapt the rebased bond basket result (6.17) (as opposed to the native HJM formula (6.23)) into

$$
\begin{aligned}
\overset{\Rightarrow}{a}{}^p_{22,t} = & -\overset{\Rightarrow}{a}{}^p_2 \overset{\leftrightarrow}{\otimes} \vec{\sigma}^p + \sum_{i=1}^{N} \omega_i \overset{\Rightarrow}{\alpha}{}^0_{i,2} \overset{\leftrightarrow}{\otimes} \vec{\Gamma}^0_i + \sum_{i=1}^{N} \omega_i \overset{\Rightarrow}{\alpha}{}^0_{i,22} \\
& + \sum_{i=1}^{N} \omega_i \left[\left(\Gamma^0_i - \sigma^p \right)_n \alpha^0_{i,2\,mp} \right] + \sum_{i=1}^{N} \omega_i \, \vec{\Gamma}{}^{0\;3\otimes}_i - \sum_{i=1}^{N} \omega_i \vec{\Gamma}^0_i \otimes \vec{\sigma}^p \otimes \vec{\Gamma}^0_i .
\end{aligned}
$$

Injecting the LMM parametrisation (7.8), (7.10) and (7.12) we get

$$
\begin{aligned}
\overset{\Rightarrow}{a}{}^p_{22,t} = & -\overset{\Rightarrow}{a}{}^p_2 \overset{\leftrightarrow}{\otimes} \vec{\sigma}^p \\
& + \sum_{i=1}^{N} \omega_i \left[-\sum_{j=0}^{i-1} M_j \left[\overset{\Rightarrow}{a}{}^L_{j,2} + \frac{1}{1+\delta_j L_j} \vec{\sigma}{}^{L\;2\otimes}_j \right] \right] \overset{\leftrightarrow}{\otimes} \left[-\sum_{k=0}^{i-1} M_k \vec{\sigma}^L_k \right] \\
& + \sum_{i=1}^{N} \omega_i \left[-\sum_{k=0}^{i-1} \frac{M_k}{1+\delta_k L_k} \left[\frac{1-\delta_k L_k}{1+\delta_k L_k} \vec{\sigma}{}^{L\;3\otimes}_k + \overset{\Rightarrow}{a}{}^L_{k,2} \overset{\leftrightarrow}{\otimes} \vec{\sigma}^L_k + \left[\sigma^L_{k\;n} \, a^L_{k,2\,mp} \right] \right] \right] \\
& + \sum_{i=1}^{N} \omega_i \left[\left(-\sum_{k=0}^{i-1} M_k \vec{\sigma}^L_k - \vec{\sigma}^p \right)_n \left(-\sum_{k=0}^{i-1} M_k \left[\overset{\Rightarrow}{a}{}^L_{k,2} + \frac{1}{1+\delta_k L_{k,t}} \vec{\sigma}{}^{L\;2\otimes}_k \right] \right)_{mp} \right] \\
& + \sum_{i=1}^{N} \omega_i \left[-\sum_{k=0}^{i-1} M_k \vec{\sigma}^L_k \right]^{3\otimes} - \sum_{i=1}^{N} \omega_i \left[-\sum_{j=0}^{i-1} M_j \vec{\sigma}^L_j \right] \otimes \vec{\sigma}^p \otimes \left[-\sum_{k=0}^{i-1} M_k \vec{\sigma}^L_k \right]
\end{aligned}
$$

which we develop using (D.11) into

$$
\begin{aligned}
\overset{\Rightarrow}{a}{}^p_{22,t} = & -\overset{\Rightarrow}{a}{}^p_2 \overset{\leftrightarrow}{\otimes} \vec{\sigma}^p + \sum_{i=1}^{N} \omega_i \left[\sum_{j=0}^{i-1} M_j \overset{\Rightarrow}{a}{}^L_{j,2} \right] \overset{\leftrightarrow}{\otimes} \left[\sum_{k=0}^{i-1} M_k \vec{\sigma}^L_k \right] \\
& - \sum_{i=1}^{N} \omega_i \sum_{k=0}^{i-1} \frac{M_k}{1+\delta_k L_k} \overset{\Rightarrow}{a}{}^L_{k,2} \overset{\leftrightarrow}{\otimes} \vec{\sigma}^L_k - \sum_{i=1}^{N} \omega_i \sum_{k=0}^{i-1} \frac{\delta_k L_k (1-\delta_k L_k)}{(1+\delta_k L_k)^3} \vec{\sigma}{}^{L\;3\otimes}_k \\
& + \sum_{i=1}^{N} \omega_i \left[\sum_{j=0}^{i-1} \frac{M_j}{1+\delta_j L_j} \vec{\sigma}{}^{L\;2\otimes}_j \right] \overset{\leftrightarrow}{\otimes} \left[\sum_{k=0}^{i-1} M_k \vec{\sigma}^L_k \right] \\
& - \sum_{i=1}^{N} \omega_i \sum_{j=0}^{i-1} \sum_{k=0}^{i-1} \sum_{l=0}^{i-1} M_j M_l \vec{\sigma}^L_j \otimes \left[\vec{\sigma}^p + M_k \vec{\sigma}^L_k \right] \otimes \vec{\sigma}^L_l \\
& - \sum_{i=1}^{N} \omega_i \sum_{k=0}^{i-1} \frac{M_k}{1+\delta_k L_k} \left[\sigma^L_{k\;n} \, a^L_{k,2\,mp} \right] + \sum_{i=1}^{N} \omega_i \sum_{j=0}^{i-1} \sum_{k=0}^{i-1} M_j M_k
\end{aligned}
$$

$$\times \left[\vec{\sigma}^L_{j\ n} \ \vec{\vec{a}}^L_{k,2\ mp} \right] + \sum_{i=1}^{N} \omega_i \sum_{j=0}^{i-1} \sum_{k=0}^{i-1} \frac{M_j M_k}{1+\delta_k L_k} \left[\sigma^L_{k\ m}\ \sigma^L_{j\ n}\ \sigma^L_{k\ p} \right]$$

$$+ \sum_{i=1}^{N} \omega_i \sum_{j=0}^{i-1} M_j \left[\sigma^P{}_n \ \vec{\vec{a}}^L_{j,2\ mp} \right] + \sum_{i=1}^{N} \omega_i \sum_{j=0}^{i-1} \frac{M_j}{1+\delta_j L_j} \left[\sigma^L_{j\ m}\ \sigma^P{}_n\ \sigma^L_{j\ p} \right].$$

Re-organising the sums we get

$$\vec{\vec{a}}^P_{22,t} = -\vec{\vec{a}}^P_2 \overset{\leftrightarrow}{\otimes} \vec{\sigma}^P - \sum_{i=0}^{N-1} \Omega_i \frac{\delta_i L_i (1-\delta_i L_i)}{(1+\delta_i L_i)^3} \vec{\sigma}^{L\ 3\otimes}_i$$

$$- \sum_{i=0}^{N-1} \frac{\Omega_i M_i}{1+\delta_i L_i} \vec{\vec{a}}^L_{i,2} \overset{\leftrightarrow}{\otimes} \vec{\sigma}^L_i + \sum_{i=0}^{N-1} \sum_{j=0}^{N-1} \Omega_{max(i,j)} \frac{M_i M_j}{1+\delta_i L_i} \vec{\sigma}^{L\ 2\otimes}_i \overset{\leftrightarrow}{\otimes} \vec{\sigma}^L_j$$

$$+ \sum_{i=0}^{N-1} \sum_{j=0}^{N-1} \Omega_{max(i,j)} M_i M_j \vec{\vec{a}}^L_{i,2} \overset{\leftrightarrow}{\otimes} \vec{\sigma}^L_j - \sum_{i=0}^{N-1} \frac{\Omega_i M_i}{1+\delta_i L_i} \left[\sigma^L_{i\ n}\ a^L_{i,2\ mp} \right]$$

$$- \sum_{i=0}^{N-1} \sum_{j=0}^{N-1} \sum_{k=0}^{N-1} \Omega_{max(i,j,k)} M_i M_j M_k \vec{\sigma}^L_i \otimes \vec{\sigma}^L_j \otimes \vec{\sigma}^L_k$$

$$- \sum_{i=0}^{N-1} \sum_{j=0}^{N-1} \Omega_{max(i,j)} M_i M_j \vec{\sigma}^L_i \otimes \vec{\sigma}^P \otimes \vec{\sigma}^L_j$$

$$+ \sum_{i=0}^{N-1} \Omega_i \frac{M_i}{1+\delta_i L_i} \left[\sigma^L_{i\ m}\ \sigma^P{}_n\ \sigma^L_{i\ p} \right] + \sum_{i=0}^{N-1} \Omega_i M_i \left[\sigma^P{}_n \ \vec{\vec{a}}^L_{i,2\ mp} \right]$$

$$+ \sum_{i=0}^{N-1} \sum_{j=0}^{N-1} \Omega_{max(i,j)} M_i M_j \left[\sigma^L_{i\ n}\ a^L_{j,2\ mp} \right]$$

$$+ \sum_{i=0}^{N-1} \sum_{j=0}^{N-1} \Omega_{max(i,j)} \frac{M_i M_j}{1+\delta_j L_j} \left[\sigma^L_{j\ m}\ \sigma^L_{i\ n}\ \sigma^L_{j\ p} \right],$$

which proves (7.18) and concludes the proof. □

7.4 Caplets

In this section we apply the asymptotic chaos expansion methodology, up to the first layer, to a term structure of Caplets and within an LMM framework. Obviously a Caplet, by nature a European Call written on a Libor rate, is the simplest option to consider when modelling with the LMM parametrisation. In effect, the (term structure of) option underlyings are precisely the native *state variables* describing the yield curve.

This asymptotic approach involves the modelling of the Caplet smile as a continuum in maturity and strike, and then the immersion of that context into the generic framework of Chap. 5. The results of that chapter[5] can then be invoked in order to connect the SInsV and SImpV model classes. This work therefore provides useful information, for the static calibration problem as well as for the analysis of joint marginal dynamics within an LMM term structure model.

7.4.1 Casting the Caplets into the Generic Framework

We will move quickly on this immersion, as it is similar to the HJM Caplet treatment of Sect. 6.4.1 [p. 340]. Our first task is to define an underlying continuum, which involves the numeraire map and the term structure of martingale measures (see Sect. 5.1). We consider a fixed accrual period δ. Recall that by definition of this linear rate, it is expressed from the Zeros with (7.1) so that its dynamics will be driftless under the $T+\delta$ forward measure, associated to its numeraire $B_t(T+\delta)$. Assuming a continuum in T of such rates, we cast the generic equations (5.1)–(5.3) with the following:

$$X_t(T) \longleftarrow L_t^\delta(T) \qquad N_t(T) \longleftarrow B_t(T+\delta)$$

$$\vec{W}_t^{N(T)} \longleftarrow \vec{W}_t^{T+\delta} \qquad \vec{\lambda}_t^N(T) \longleftarrow \vec{\Gamma}_t(T+\delta)$$

To define an option surface, we assume *in-advance* Caplet prices $C^\delta(t, L_t^\delta(T), T, K)$ to provide a market continuum, both in K and in T. We consider δ still fixed and their individual payoff of $\delta[L_t^\delta(T) - K]^+$ to be transferred at time $T+\delta$. We check that the payoff now associated to the generic framework corresponds to a real Caplet, by choosing $\eta(T) \triangleq T$ and discounting down to time T:

$$\delta^{-1}\Phi(T) = B_T(T+\delta)\left[L_T^\delta(T) - K\right]^+ .$$

This surface of option prices is then re-parametrised via the LNIV map:

$$\delta^{-1}C_t^\delta(L_t^\delta(T), K, T) = B_t(T+\delta)C^{BS}\left(L_t^\delta(T), K, \Sigma_t^\delta(L_t^\delta(T), T, K)\cdot\sqrt{T-t}\right) \tag{7.20}$$

with $C^{BK}(f, k, v)$ the time-normalised Black functional, as per Sect. 5.1.2.

[5] Proposition 5.2 [p. 297], Theorem 5.1 [p. 300] or Theorem 5.2 [p. 310], for instance.

We then derive these absolute quantities into their sliding counterparts, with:

$$\theta \triangleq T - t \qquad \widetilde{L}^{\delta}(t,\theta) \triangleq L^{\delta}(t,T) \qquad y \triangleq \ln\left(K/\widetilde{L}^{\delta}(t,\theta)\right)$$

$$\widetilde{C}^{\delta}(t,y,\theta) \triangleq C^{\delta}\left(t, L_t^{\delta}(T), K, T\right) \qquad \widetilde{\Sigma}^{\delta}(t,y,\theta) \triangleq \Sigma^{\delta}(t, L^{\delta}(t,T), K, T)$$

To complete the generic framework, we consider the same generic SImpV model as in (5.13). Contrary to the HJM case, the native LMM model provides the SInsV dynamics in the right format, hence we can immediately apply the results of Chap. 5 linking both model classes.

Note that the dynamics chosen as input are respectively absolute for the forward Libor $L_t^{\delta}(T)$ and relative for the implied volatility map $\widetilde{\Sigma}_t^{\delta}(y,\theta)$. This configuration has been selected for the reader's convenience, since it is rare that a Libor Market Model be defined from the sliding forward Libor $\widetilde{L}_t^{\delta}(\theta)$. In any case, the dynamics of both counterparts quantities are made explicit in the course of the proofs within Chap. 5.

7.4.2 IATM Differentials of the Caplet Smile

As discussed above, the fact that the option underlying framework is so close to the native term structure of state variables makes for straightforward post-immersion SInsV dynamics. This is therefore the ideal situation in which to instantiate the generic results for the *direct* problem, in other words to express some of the interesting IATM differentials for the Caplet smile.

First, the IATM level is given by Corollary 5.2 [p. 295] where adapting (5.34) gives:

Corollary 7.2 (IATM level of the SV-LMM Caplet smile) *In a generic Stochastic Volatility—Libor Market Model framework defined by chaos dynamics* (7.2)–(7.4), *the IATM level of the sliding Caplet smile is*

$$\widetilde{\Sigma}\,(t,0,0) = \|\vec{\sigma}_t^{\,\delta,L}(t)\|. \tag{7.21}$$

On the right-hand side of (7.21), the modulus belongs to the *spot* Libor rate $L_t^{\delta}(0)$ for the considered accrual period δ, which has significant implications in terms of predictive power of the methodology. On that subject, it is worth running through the various levels of approximation affecting the generation of the implied volatility surface.

The first level is inherent to the asymptotic approach, and in (7.21) it imposes the current time t as the argument of the instantaneous volatility $\vec{\sigma}_t^{\,\delta,L}(T)$. As we know from Part I, the level of precision in that respect is usually good and can be controlled at the cost of higher-order differentiation.

The second level of approximation is due to the dimensionality, i.e. modulus versus full vectorial information, but we have covered this subject (in particular for the baskets) and we know that this is only an issue for the inverse problem.

The last and newest level comes with the presence of the immediate maturity ($T = t$) on the r.h.s. of (7.21). This implies that if we aim at approximating the ATM level of implied volatility for a 10Y-maturity 3M Caplet, our first order will be the instantaneous/implied volatility of the spot 3M Libor rate. In comparison, a term-by-term single-underlying approach for the same Caplet, as was developed in Part I, would also aggregate the current instantaneous volatility but for the exact maturity, in other words:

$$\text{With a single-underlying approach} \quad \widetilde{\Sigma}\,(t,0,0) = \|\vec{\sigma}_t^{\,\delta,L}(\mathbf{T})\|. \tag{7.22}$$

Balancing these two approaches, we see pros and cons on either side. First it must be stressed that in (7.22) this IATM level corresponds to an option that is out-of-context and certainly not liquid. Indeed, we are looking at a sort of mid-curve option, whose expiry is asymptotically short but whose underlying is the *forward* Libor rate fixing in 10 years. By contrast, the term-structure result (7.21) is perfectly coherent, as is the whole smile that it relates to. As has been discussed before, the two smiles considered (irrespective of the approximation) are *a priori* different everywhere, except at the single maturity T which is the unique point where they correspond to the same underlying, numeraire and product.

In our view, an important point to take home is that—with regard to precision now—the term-structure approach is likely to require more higher-order differentials in the maturity direction, starting with the slope $\widetilde{\Sigma}'_\theta(t,0,0)$. It does present the advantage, however, of representing all maturities at once, so that all individual marginals are proxied *jointly*, in a consistent manner, statically and dynamically. In contrast, the term-by-term approach cannot guarantee such homogeneity.

Let us therefore move on to applying the generic IATM differential results of Theorem 5.2 [p. 310]. Still on the static side, we have the IATM skew as:

Corollary 7.3 (IATM skew of the LMM Caplet smile) *In an LMM framework described by the chaos dynamics* (7.2)–(7.4)*, the IATM skew of the sliding Caplet smile is*

$$\widetilde{\Sigma}'_y\,(t,0,0) = \widetilde{\Sigma}^{-3}(t,0,0)\left[\vec{\sigma}_t^{\,\delta,L}(t)^{\perp}\,\vec{\vec{a}}_{2,t}^{\,\delta,L}(t)\,\vec{\sigma}_t^{\,\delta,L}(t)\right] \tag{7.23}$$

which leads us to anticipate the same type of structural differences between the single-underlying and the term-structure approaches. Note, however, that to improve the precision of the skew w.r.t. maturity we would have to fetch the *twist* $\widetilde{\Sigma}''_{y\theta}(t,0,0)$. However, that differential comes as an output of the more involved second layer (see Sect. 3.2).

Anticipating the same kind of conclusions from the IATM curvature $\widetilde{\Sigma}''_{yy}(t,0,0)$, we prefer to move on directly to the more interesting slope $\widetilde{\Sigma}'_\theta(\star)$. We express it in a symbolic, simplified version in order to stress the importance of the maturity specification.

Corollary 7.4 (IATM slope of the LMM Caplet smile) *In an LMM framework* (7.2)–(7.4), *where the SInsV and SImpV drivers are assumed aligned, the IATM slope of the sliding Caplet smile comes symbolically as*

$$\widetilde{\Sigma}'_\theta(\star) = [\cdots]_t\,(t)$$
$$\underbrace{+\frac{1}{2}\widetilde{\Sigma}^2\,\widetilde{\Sigma}'_y(\star)}_{\textcircled{1}}\;\underbrace{-\frac{1}{2}\widetilde{\Sigma}'_y(\star)\frac{\partial_T L_t^\delta(t)}{L_t^\delta(t)}}_{\textcircled{2}}\;\underbrace{+\frac{1}{2}\widetilde{\Sigma}^{-1}(\star)\vec{\sigma}_t^{\delta,L}(t)^\perp\left[\partial_T\vec{\sigma}_t^{\delta,L}(t)\right]}_{\textcircled{1}} \qquad (7.24)$$

where the symbolic term $[\cdots]_t\,(t)$ *is homogeneous to the IATM level* $\widetilde{\Sigma}(\star)$ *and is a scalar function purely of the endogenous and exogenous coefficients,*[6] *all taken in* $(T = t)$.

On the r.h.s. we observe several features which are specific to this static IATM differential:

- Term ① is familiar, as it matches the single-underlying, multi-dimensional expression (3.75) [p. 181]. It confirms an idiosyncracy of the maturity slope, among the IATM differentials of the first layer: its inhomogeneity w.r.t. the modulus $\widetilde{\Sigma}(\star)$. Indeed, the skew is dimensionless, so that term ① comes in $\widetilde{\Sigma}^2(\star)$. Drawing a parallel with the underlying dynamics, what we observe is a typical feature of drift terms, simply as a consequence of the quadratic variation coming into play.
- Term ② is a consequence of the individual Libor rate switching as θ increases. This has nothing to do with the subjective choice of a sliding implied volatility, but rather comes from the very existence of a term structure for the underlying.
- Term ① is a consequence of the volatility switching as θ increases. This is due again to the underlying term structure, but also to the TS of numeraires. Indeed, the volatility is implicitly defined by (5.1) or (7.2) in regard of a specific numeraire.

Overall, the maturity slope seems to be the IATM differential which concentrates most of the term-structure novelty.

7.5 Swaptions

The goal of this section is to link the SInsV specification of the generic Stochastic Volatility Libor Market Model class to the statics and dynamics of the *swaption* implied volatility surface (the SImpV specification). In particular, we are interested in the Immediate ATM differentials of that smile, which corresponds to the most liquid options in the interest rates market.

---

[6] These coefficients are $\vec{\sigma}_t^{\delta,L}(T)$, $\overset{\Rightarrow}{a}{}^{\delta,L}_{2,t}(T)$, $\overset{\Rightarrow}{a}{}^{\delta,L}_{3,t}(T)$ and $\overset{\Rrightarrow}{a}{}^{\delta,L}_{22,t}(T)$.

Note that given the aforementioned higher liquidity of the swaption market, several versions of the associated market model have been published. Those flavours of the Swap Market Model include the co-terminal, co-initial and co-sliding families (see [25]). Indeed, since a curve model offers only one dimension in terms of maturity, the modelled swap rates must share some common feature: either and respectively their end date, start date, tenor, etc. In particular, the finite collection of state variables allows the bootstrapping of Libor rates or discount factors. So in essence the SMM is not a richer framework that the Libor Market Model, simply a re-parametrised version of a discretised LMM or HJM model.

However, parametrisation *does* matter, so that it is generally easier to match the swaption smile with an SMM than with any of the two other frameworks. Indeed, to get a perfect *static* fit it would theoretically suffice to use a local volatility diffusion for the swap rates. However, the issue would then be to price Caplets, and to a lesser extent bond options. There is no perfect alternative and, if only for cultural reasons, the Libor market model remains the reference framework to price complex IR exotics, and hence must be calibrated to all (pertinent) liquid products. This explains why a large part of the LMM literature concerns itself with the calibration to the swaption smile, hence the motivation for the current section.

In that respect our asymptotic approach distinguishes itself from the other available approximations by a new feature. Not only is it generic (it can handle most SV-LMM models) and does it provide control on the proxy error (the attainable differentiation order is in principle infinite) but it exploits the *real* dynamics of the swap rate. Indeed, most available methods start by approximating the SDE driving the par swap rate (see [5] on that matter). Basket as well as functional approximations are common, so that often the methods are specific to certain LMM dynamics. It remains that this step inevitably degrades the quality of the approximation, and in a manner which is not always measurable and even less controllable. By contrast, using the exact dynamics in a chaos context leads us to quite complex expressions (see (7.33), for instance) and this is just a dynamic coefficient, not an IATM differential. There is, of course, the argument that, once programmed, the complexity of the method loses its hindrance. However, for analysis and intuition-building, simplicity is necessary. This is why we will later examine (in Sect. 7.6) the most common simplification method for the swap rate dynamics, which is the basket approach.

7.5.1 Casting the Swaptions into the Generic Framework

As per Assumption 6.1, we shall consider a plain vanilla swap. The first fixing occurs at the start date T_0, which matches the option expiry. The first fixed coupon payment is in T_1 and the last netted payment in T_N. Furthermore, the floating and fixed legs share T_0 and T_N. With the accrual sequence $\delta = \{\delta_i\}_{1 \leq i \leq N}$ defined by

$$\forall i/1 \leq i \leq N T_i = T_{i-1} + \delta_i$$

the tenor structure is completely specified by T_0 and by δ. The generic framework (see Sect. 5.1.1) requiring a single term structure, we allocate the generic maturity T to the start date T_0. The underlying swap is now sliding and we can employ the specific notations of Sect. 6.1.2. As for the forward par swap rate, recall that by arbitrage and by Assumption 6.1 it comes as

$$S_t^\delta(T) \triangleq \frac{B_{0,t}(T) - B_{N,t}(T)}{A_t^\delta(T)}$$

where

$$A_t^\delta(T) \triangleq \sum_{i=1}^{N} \delta_i B_{i,t}(T) \quad \text{is the sliding annuity.}$$

We assume continua w.r.t. T for the par swap rate and for the annuity, which become respectively the term structure processes for the underlying $X_t(T)$ and for the numeraire $N_t(T)$. The dynamics (5.1), (5.9) and (5.10) are then instantiated as

$$\frac{dS_t^\delta(T)}{S_t^\delta(T)} = \vec{\sigma}_t^s(T)^\perp d\vec{W}_t^{A(T)} \tag{7.25}$$

$$d\vec{\sigma}_t^s(T) = \vec{a}_{1,t}^s(T)dt + \vec{\vec{a}}_{2,t}^s(T)d\vec{W}_t^{A(T)} + \vec{\vec{a}}_{3,t}^s(T)d\vec{Z}_t \tag{7.26}$$

$$d\vec{\vec{a}}_{2,t}^s(T) = \overrightarrow{[\cdots]}^{(2)}dt + \vec{\vec{\vec{a}}}_{22,t}^s(T)d\vec{W}_t^{A(T)} + \overrightarrow{[\cdots]}^{(3)}d\vec{Z}_t \tag{7.27}$$

where the maturity-dependent annuity's driver $\vec{W}_t^{A(T)}$ and volatility $\vec{\sigma}_t^a(T)$ are defined from the risk-neutral $\vec{W}_t$ by

$$d\vec{W}_t^{A(T)} = d\vec{W}_t - \vec{\sigma}_t^a(T)dt \quad \text{and} \quad \frac{dA_t(T)}{A_t(T)} = r_t dt + \vec{\sigma}_t^a(T)^\perp d\vec{W}_t.$$

Having thereby instantiated, respectively, (5.2) and (5.3), the specification of term structure of underlyings, numeraires and martingale measures is complete. As for the option field, we check that the payoff associated to our choices indeed corresponds to a real physically settled European payer swaption:

$$\Phi(T) \triangleq \sum_{i=1}^{N} \delta_i B_{i,T}(T)\left[S_T^\delta(T) - K\right]^+ = A_T(T)\left[S_T^\delta(T) - K\right]^+.$$

We define the price surface by assuming a continuum in strike K for these options. Note that on actual markets and with a free tenor, this product class is specified in three dimensions (expiry/tenor/strike), which justifies the vocable of *swaption grid* or *swaption cube*. In fact, due to further refinements in the collateral agreements and in the funding conventions for instance, further dimensions can be added to that field (at least internally to an institution). Nevertheless, in our simplified setup we can re-parametrise the bi-dimensional price surface into the implied volatility map, using the normalised Black functional:

$$C_t^\delta(S_t^\delta(T), K, T) = A_t^\delta(T) C^{BS}\left(S_t^\delta(T), K, \Sigma_t^\delta(S_t^\delta(T), T, K) \cdot \sqrt{T-t}\right). \quad (7.28)$$

Moving on to sliding quantities we define as usual

$$\theta \triangleq T - t, \qquad \widetilde{S}^\delta(t,\theta) \triangleq S^\delta(t,T), \qquad y \triangleq \ln\left(K/\widetilde{S}^\delta(t,\theta)\right),$$

$$\widetilde{C}^\delta(t,y,\theta) \triangleq C^\delta\left(t, S^\delta(t,T), K, T\right), \qquad \widetilde{\Sigma}^\delta(t,y,\theta) \triangleq \Sigma^\delta(t, S^\delta(t,T), K, T).$$

The instantiation is then completed by the formal SImpV dynamics of (5.13) [p. 279].

Now we must compute the SInsV dynamics of the term structure of par swap rates, under the annuity measures, using the Stochastic Volatility—LMM specification. Limiting ourselves to the first layer, this means that we need to express

$$S_t^\delta(T) \qquad \vec{\sigma}_t^{\,S}(T) \qquad \vec{a}_{1,t}^{\,S}(T) \qquad \vec{\vec{a}}_{2,t}^{\,S}(T) \qquad \vec{\vec{a}}_{3,t}^{\,S}(T) \qquad \vec{\vec{\vec{a}}}_{22,t}^{\,S}(T)$$

from

$$L_t(T) \qquad \vec{\sigma}_t^{\,L}(T) \qquad \vec{a}_{1,t}^{\,L}(T) \qquad \vec{\vec{a}}_{2,t}^{\,L}(T) \qquad \vec{\vec{a}}_{3,t}^{\,L}(T) \qquad \vec{\vec{\vec{a}}}_{22,t}^{\,L}(T)$$

which is the topic of Sect. 7.5.2.

7.5.2 Dynamics of the Underlying Par Swap Rate

Let us now provide the remaining information necessary to exploit the generic framework results, which is the underlying chaos dynamics within the *native* SV-LMM SInsV model class.

Lemma 7.2 (Swap rate chaos dynamics in an LMM framework) *In an SV-LMM framework defined in chaos by the system* (7.2)–(7.4) *the* σ-(2,0) *group of coefficients describing the par swap rate dynamics are given by*

$$\vec{\sigma}_t^{\,S}(T) \triangleq -\left[M_{\mathbf{i}} \Omega_{\mathbf{i}}^S \vec{\sigma}_{\mathbf{i}}^{\,L}\right] \quad (7.29)$$

$$\vec{a}^{\,s}_{1,t}(T) \triangleq -\vec{\vec{a}}^{\,F,0}_{2,t}\,\vec{\sigma}^{\,S}_{t} - \left[\Omega^S_{\mathbf{i}} M_{\mathbf{i}} \vec{a}^{\,L}_{\mathbf{i},1}\right] - \left[\Omega^S_{\mathbf{i}} M_{\mathbf{i}} \vec{\vec{a}}^{\,L}_{\mathbf{i},2}\,\vec{\sigma}^{\,L}_{\mathbf{i}}\right] + \sum_{i=0}^{N-1}\sum_{j=i}^{N-1} \Omega^S_{\mathbf{j}} M_i M_j \left[\vec{\vec{a}}^{\,L}_{i,2} + \frac{1}{1+\delta_i L_i}\,\vec{\sigma}^{\,L\,2\otimes}_{i}\right]\vec{\sigma}^{\,L}_{j} \tag{7.30}$$

$$\vec{\vec{a}}^{\,s}_{2,t}(T) \triangleq \vec{\sigma}^{\,A,0\,2\otimes}_{t} - \vec{\sigma}^{\,F,0\,2\otimes}_{t} - \left[M_i \Omega^S_i \vec{\vec{a}}^{\,L}_{i,2}\right] + \left[\Omega^S_{max(i,j)} M_i M_j \vec{\sigma}^{\,L}_{i} \otimes \vec{\sigma}^{\,L}_{j}\right] - \left[\frac{M_i \Omega^S_i}{1+\delta_i L_i}\,\vec{\sigma}^{\,L\,2\otimes}_{i}\right] \tag{7.31}$$

$$\vec{\vec{a}}^{\,s}_{3,t}(T) \triangleq -\left[M_{i,t}\Omega^S_{i,t}\vec{\vec{a}}^{\,L}_{i,3,t}\right] \tag{7.32}$$

$$\begin{aligned}
\vec{\vec{\vec{a}}}^{\,S}_{22,t}(T) \triangleq\; & \vec{\vec{a}}^{\,A,0}_{2} \overset{\leftrightarrow}{\otimes} \vec{\sigma}^{\,A,0} - \vec{\vec{a}}^{\,F,0}_{2} \overset{\leftrightarrow}{\otimes} \vec{\sigma}^{\,F,0} + \left[\Omega^S_{max(\mathbf{i},\mathbf{j})} M_{\mathbf{i}} M_{\mathbf{j}} \vec{\vec{a}}^{\,L}_{\mathbf{i},2} \overset{\leftrightarrow}{\otimes} \vec{\sigma}^{\,L}_{\mathbf{j}}\right] \\
& - \left[\frac{\Omega^S_{\mathbf{i}} M_{\mathbf{i}}}{1+\delta_{\mathbf{i}} L_{\mathbf{i}}}\,\vec{\vec{a}}^{\,L}_{\mathbf{i},2} \overset{\leftrightarrow}{\otimes} \vec{\sigma}^{\,L}_{\mathbf{i}}\right] + \left[\Omega^S_{max(\mathbf{i},\mathbf{j})} \frac{M_{\mathbf{i}} M_{\mathbf{j}}}{1+\delta_{\mathbf{i}} L_{\mathbf{i}}}\,\vec{\sigma}^{\,L\,2\otimes}_{\mathbf{i}} \overset{\leftrightarrow}{\otimes} \vec{\sigma}^{\,L}_{\mathbf{j}}\right] \\
& - \left[\Omega^S_{\mathbf{i}} \frac{\delta_{\mathbf{i}} L_{\mathbf{i}}(1-\delta_{\mathbf{i}} L_{\mathbf{i}})}{(1+\delta_{\mathbf{i}} L_{\mathbf{i}})^3}\,\vec{\sigma}^{\,L\,3\otimes}_{\mathbf{i}}\right] - \left[\Omega^S_{max(\mathbf{i},\mathbf{j})} M_{\mathbf{i}} M_{\mathbf{j}} \vec{\sigma}^{\,L}_{\mathbf{i}} \otimes \vec{\sigma}^{\,p} \otimes \vec{\sigma}^{\,L}_{\mathbf{j}}\right] \\
& - \left[\Omega^S_{max(\mathbf{i},\mathbf{j},\mathbf{k})} M_{\mathbf{i}} M_{\mathbf{j}} M_{\mathbf{k}} \vec{\sigma}^{\,L}_{\mathbf{i}} \otimes \vec{\sigma}^{\,L}_{\mathbf{j}} \otimes \vec{\sigma}^{\,L}_{\mathbf{k}}\right] - \left[\frac{\Omega^S_{\mathbf{i}} M_{\mathbf{i}}}{1+\delta_{\mathbf{i}} L_{\mathbf{i}}}\left[\sigma^L_{\mathbf{i}\,n}\, a^L_{\mathbf{i},2\,mp}\right]\right] \\
& + \left[\Omega^S_{max(\mathbf{i},\mathbf{j})} M_{\mathbf{i}} M_{\mathbf{j}}\left[\sigma^L_{\mathbf{i}\,n}\, a^L_{\mathbf{j},2\,mp}\right]\right] + \left[\Omega^S_{max(\mathbf{i},\mathbf{j})} \frac{M_{\mathbf{i}} M_{\mathbf{j}}}{1+\delta_{\mathbf{j}} L_{\mathbf{j}}}\left[\sigma^L_{\mathbf{j}\,m}\, \sigma^L_{\mathbf{i}\,n}\, \sigma^L_{\mathbf{j}\,p}\right]\right] \\
& + \left[\Omega^S_{\mathbf{i}} M_{\mathbf{i}}\left[\sigma^p{}_n\, a^L_{\mathbf{i},2\,mp}\right]\right] + \left[\Omega^S_{\mathbf{i}} \frac{M_{\mathbf{i}}}{1+\delta_{\mathbf{i}} L_{\mathbf{i}}}\left[\sigma^L_{\mathbf{i}\,m}\, \sigma^p{}_n\, \sigma^L_{\mathbf{i}\,p}\right]\right]
\end{aligned} \tag{7.33}$$

where the invoked sub-terms are, with $\bullet \in \{A, F\}$*:*

$$\vec{\sigma}^{\,\bullet,0}_{t} \triangleq -\left[M_{\mathbf{i}} \Omega^{\bullet}_{\mathbf{i}}\,\vec{\sigma}^{\,L}_{\mathbf{i}}\right]$$

$$\vec{\vec{a}}^{\,\bullet,0}_{2,t}(T) \triangleq -\vec{\sigma}^{\,\bullet,0\,2\otimes}_{t} - \left[M_i \Omega^{\bullet}_i \vec{\vec{a}}^{\,L}_{i,2}\right] + \left[\Omega^{\bullet}_{max(i,j)} M_i M_j \vec{\sigma}^{\,L}_{i} \otimes \vec{\sigma}^{\,L}_{j}\right] - \left[\frac{M_i \Omega^{\bullet}_i}{1+\delta_i L_i}\,\vec{\sigma}^{\,L\,2\otimes}_{i}\right].$$

The $M_i(T)$ term structure is defined by (see (7.7))

$$M_{i,t}(T) \triangleq \frac{\delta_i L_t(T_i)}{1+\delta_i L_t(T_i)}$$

and the weighting scheme is specified by

$$\underset{1\leq i\leq N-1}{\omega^F_{i,t}} \triangleq 0 \quad \omega^F_N \triangleq \left[1-\prod_{k=0}^{N-1}[1+\delta_k L_k]\right]^{-1} \qquad \underset{0\leq i\leq N-1}{\Omega^F_i} \triangleq \omega^F_N \tag{7.34}$$

$$\underset{1\leq i\leq N}{\omega^A_i} \triangleq \frac{\delta_i \prod_{k=0}^{i-1}\left[1+\delta_k L_{k,t}(T)\right]^{-1}}{\sum_{j=1}^{N}\delta_j\prod_{k=0}^{j-1}\left[1+\delta_k L_{k,t}(T)\right]^{-1}} \qquad \underset{0\leq i\leq N-1}{\Omega^A_i} = \sum_{j=i+1}^{N}\omega^A_j \tag{7.35}$$

$$\underset{1\leq i\leq N}{\omega^S_i} = \omega^F_i - \omega^A_i \qquad \underset{0\leq i\leq N-1}{\Omega^S_i} = \sum_{j=i+1}^{N}\omega^S_j = \Omega^F_i - \Omega^A_i. \tag{7.36}$$

In these expressions we use simplified notations, in the sense that, apart from the accrual δ_i, all quantities involved are processes, parametrised by the maturity. Hence we can omit arguments t and T without ambiguity. Also, the modified Einstein notation $[\cdots]$ refers to summations ***from indices*** 0 ***to*** $N-1$*, which means that it involves all Libor rates in the tenor structure.*

Note that in (7.33) the first couple of terms can be rewritten as

$$\begin{aligned}
\overset{\Rightarrow}{a}{}_2^{A,0} \overset{\leftrightarrow}{\otimes} \vec{\sigma}^{A,0} - \overset{\Rightarrow}{a}{}_2^{F,0} \overset{\leftrightarrow}{\otimes} \vec{\sigma}^{F,0} &= \left[\overset{\Rightarrow}{a}{}_2^{F,0} - \overset{\Rightarrow}{a}{}_2^{S,0}\right] \overset{\leftrightarrow}{\otimes} \vec{\sigma}^{A,0} - \overset{\Rightarrow}{a}{}_2^{F,0} \overset{\leftrightarrow}{\otimes} \vec{\sigma}^{F,0} \\
&= \overset{\Rightarrow}{a}{}_2^{F,0} \overset{\leftrightarrow}{\otimes} \left[\vec{\sigma}^{A,0} - \vec{\sigma}^{F,0}\right] - \overset{\Rightarrow}{a}{}_2^{S,0} \overset{\leftrightarrow}{\otimes} \vec{\sigma}^{A,0} \\
&= -\overset{\Rightarrow}{a}{}_2^{F,0} \overset{\leftrightarrow}{\otimes} \vec{\sigma}^{S,0} - \overset{\Rightarrow}{a}{}_2^{S,0} \overset{\leftrightarrow}{\otimes} \vec{\sigma}^{A,0}
\end{aligned}$$

but apart from reducing the number of sub-terms involved in the expressions of the chaos coefficients, this type of simplification presents no tangible interest. Instead we have elected to stress the symmetry of the roles played by the floating and fixed legs. Overall, there are many ways to present these expressions, and we have settled both for compactness and for an inductive basket structure on the Libors' volatilities and dynamic coefficients.

In that respect, it is interesting to note that the weighting scheme does relay the dual way of considering the trade: either as an exchange of legs or as an option on a bond. Indeed, recall from Remark 6.2 [p. 364] that the payoff of our swaption (with notional M and strike K) matches the payoff for a Put written on an in fine bond with the same tenor structure. The latter must have a fixed coupon of K, a notional of δM, a strike of 1 and T_0 for expiry. It is therefore conceivable to interpret the Ω^A structure as accounting for the coupons of that fictional bond, while the Ω^F scheme corresponds to the principal repayment.

Proof As in the HJM context, our strategy is to decompose the swap rate as a fraction of floating leg and annuity, seen as fixed-coupon bonds. We deflate both by a common ZC and thus use Corollary 7.1 [p. 379] to obtain their dynamics. Formally, recall from (6.50) that for a generic traded asset π_t we can re-express the par swap rate as

$$S_t^\delta(T_0) = \frac{F_t^{\delta,\pi}(T_0)}{A_t^{\delta,\pi}(T_0)}$$

$$\text{where} \quad \begin{cases} F_t^{\delta,\pi}(T_0) \triangleq B_t^\pi(T_0) - B_t^\pi(T_N) \\ A_t^{\delta,\pi}(T_0) \triangleq \sum_{i=1}^N \delta_i B_t^\pi(T_i). \end{cases}$$

In the sequel we omit the δ superscript whenever there is no ambiguity on the tenor structure. In this instance we take as numeraire π_t the initial Zero Coupon $B_{0,t}(T)$:

$$\frac{dF_t^0(T)}{F_t^0(T)} = \vec{\sigma}_t^{F,0}(T)^\perp d\vec{W}_t^0 \quad \frac{dA_t^0(T)}{A_t^0(T)} = \vec{\sigma}_t^{A,0}(T)^\perp d\vec{W}_t^0$$

so that, using the classical change of numeraire tool, the swap rate dynamics read as

$$\frac{dS_t^\delta(T)}{S_t^\delta(T)} = \left[\vec{\sigma}_t^{F,0} - \vec{\sigma}_t^{A,0}\right](T)\left[d\vec{W}_t^0 - \vec{\sigma}_t^{A,0}dt\right].$$

But since by definition the right bracket is

$$\begin{aligned} d\vec{W}_t^0 - \vec{\sigma}_t^{A,0}dt &= d\vec{W}_t - \vec{\Gamma}_{0,t}(T)dt - \left[\vec{\sigma}_t^A - \vec{\Gamma}_{0,t}\right](T)dt = d\vec{W}_t - \vec{\sigma}_t^A(T)dt \\ &= d\vec{W}_t^A \end{aligned}$$

which is the annuity driver, the swap rate volatility is formally

$$\vec{\sigma}_t^s(T) = \vec{\sigma}_t^{F,0} - \vec{\sigma}_t^{A,0}. \tag{7.37}$$

We note—as expected—that the choice of the numeraire does not affect the result. The rationale for selecting $B_t(T_0)$ is clearly the simplification that it brings, when transferring the rebased bond dynamics into the LMM parametrisation. Recalling that in (7.37) the dynamics of the r.h.s. are defined w.r.t. the T_0-driver, we have

$$\begin{aligned} d\vec{\sigma}_t^s(T) &= d\vec{\sigma}_t^{F,0} - d\vec{\sigma}_t^{A,0} \\ &= \left[\vec{a}_{1,t}^{F,0} - \vec{a}_{1,t}^{A,0}\right]dt + \left[\vec{\vec{a}}_{2,t}^{F,0} - \vec{\vec{a}}_{2,t}^{A,0}\right]d\vec{W}_t^0 + \left[\vec{\vec{a}}_{3,t}^{F,0} - \vec{\vec{a}}_{3,t}^{A,0}\right]d\vec{Z}_t \end{aligned}$$

hence

$$\vec{\sigma}_t^s(T) = \left[\vec{a}_{1,t}^{F,0} - \vec{a}_{1,t}^{A,0}\right] dt$$
$$+ \left[\vec{\vec{a}}_{2,t}^{F,0} - \vec{\vec{a}}_{2,t}^{A,0}\right] \left[d\vec{W}_t^A + \vec{\sigma}_t^{A,0} dt\right] + \left[\vec{\vec{a}}_{3,t}^{F,0} - \vec{\vec{a}}_{3,t}^{A,0}\right] d\vec{Z}_t,$$

so that the first-depth coefficients in the swap rate dynamics come formally as

$$\vec{a}_{1,t}^{S}(T) = \vec{a}_{1,t}^{F,0}(T) - \vec{a}_{1,t}^{A,0}(T) + \vec{\vec{a}}_{2,t}^{S}(T)\vec{\sigma}_t^{A,0}(T) \tag{7.38}$$

$$\vec{\vec{a}}_{2,t}^{S}(T) = \vec{\vec{a}}_{2,t}^{F,0}(T) - \vec{\vec{a}}_{2,t}^{A,0}(T) \tag{7.39}$$

$$\vec{\vec{a}}_{3,t}^{S}(T) = \vec{\vec{a}}_{3,t}^{F,0}(T) - \vec{\vec{a}}_{3,t}^{A,0}(T) \tag{7.40}$$

Then from the endogenous coefficient (7.39) we get

$$d\vec{\vec{a}}_{2,t}^{S}(T) = \overset{(2)}{\overrightarrow{[\cdots]}}dt + \left[\vec{\vec{\vec{a}}}_{22,t}^{F,0}(T) - \vec{\vec{\vec{a}}}_{22,t}^{A,0}(T)\right] d\vec{W}_t^0 + \overset{(3)}{\overrightarrow{[\cdots]}}d\vec{Z}_t.$$

Changing driver from $\vec{W}_t^0$ to $\vec{W}_t^A$ naturally does not modify the bracket in front of it, hence the second-depth coefficient is formally

$$\vec{\vec{\vec{a}}}_{22,t}^{S}(T) = \vec{\vec{\vec{a}}}_{22,t}^{F,0}(T) - \vec{\vec{\vec{a}}}_{22,t}^{A,0}(T). \tag{7.41}$$

It therefore remains *only* to express, in the LMM parametrisation, the first-layer coefficients of the rebased, fixed-coupons bonds F_t^0 and A_t^0.

Dynamics of the T_0-rebased floating leg F_t^0

In the fashion of Corollary 7.1 [p. 379], dedicated to Bond chaos dynamics in an LMM framework, following (7.13) we rewrite the floating leg somewhat artificially as a fixed-weights basket, but on the *whole* rebased ZC population, i.e. for maturities spanning T_0 to T_N:

$$F_t^0 = B_{0,t}^0(T) - B_{N,t}^0(T) = 1 - B_{N,t}^0(T) = \sum_{i=\mathbf{0}}^{N} c_i^F B_{i,t}^0(T)$$

where we formally consider a common, unit accrual ($\delta_i = 1 \forall i$) and where the fixed weights are non-zero only for the first and last Zeros:

$$c_0^F = 1 \qquad c_i^F = 0 \;\text{ if }\; 1 \le i \le N-1 \qquad c_N^F = -1. \tag{7.42}$$

However, the inclusion in the basket population of the first Zero $B_{0,t}(T)$, which is precisely our rebasing numeraire, places us outside of Corollary 7.1 framework. But since the additional element is actually a constant ($B^0_{0,t}(T) \equiv 1$ *a.s.* $\forall t \, \forall T$), we have several simple ways to fall back in line. We can start by considering our current basket as a limit case of the classical framework. Indeed, we can make the first accrual vanish, while maintaining the first *effective coupon*:

$$\delta_1 \searrow 0 \qquad \text{with} \quad c_1 \delta_1 \equiv 1.$$

This shows that we can integrate $B^0_{0,t}(T)$ as the first zero within the exact framework of Corollary 7.1 (with $N+1$ elements) and under the same fixed weighting scheme (7.42). But since the volatility structure of $B^0_{0,t}(T)$ is immaterial, when deriving the chaos we can simplify that weighting convention by taking $\omega^F_{0,t} \longleftarrow 0$ without affecting the results.

Eventually, we assume, therefore, that the basket spans only the same N elements as the annuity, from $B^0_{1,t}$ to $B^0_{N,t}$, so that only the terminal weight remains:

$$\underset{1 \le i \le N-1}{\omega^F_{i,t}} \triangleq 0 \quad \omega^F_N \triangleq \left[1 - \prod_{k=0}^{N-1} [1 + \delta_k L_k] \right]^{-1}. \tag{7.43}$$

Note that the total mass is not unit any more but this is not problematic here. It follows that the aggregated weights are all equal to the last weight:

$$\text{for} \quad 0 \le i \le N-1 \quad \Omega^F_{i,t} \triangleq \sum_{j=i+1}^{N} \omega^F_{j,t} = \omega^F_{N,t}. \tag{7.44}$$

Hence combining (7.43) and (7.44) establishes the whole normalised weighting scheme (7.34). Taking notation (7.7) for $M_{i,t}(T)$ and ignoring arguments t and T whenever possible, we use modified Einstein notation to sum indices 0 to $N-1$. Then applying Corollary 7.1, we get first the instantaneous volatility from (7.14) as

$$\vec{\sigma}^{F,0}_t(T) = -\left[M_{\mathbf{i}} \Omega^F_{\mathbf{i}} \vec{\sigma}^L_{\mathbf{i}} \right]. \tag{7.45}$$

From (7.16) we get the drift as:

$$\begin{aligned} \vec{a}^{F,0}_{1,t}(T) = & -\vec{\vec{a}}^{F,0}_{2,t} \vec{\sigma}^{F,0}_t - \left[M_{\mathbf{i}} \Omega^F_{\mathbf{i}} \vec{a}^L_{\mathbf{i},1} \right] - \left[M_{\mathbf{i}} \Omega^F_{\mathbf{i}} \vec{\vec{a}}^L_{\mathbf{i},2} \vec{\sigma}^L_{\mathbf{i}} \right] \\ & + \sum_{i=0}^{N-1} \sum_{j=i}^{N-1} \Omega^F_{\mathbf{j}} M_i M_j \left[\vec{\vec{a}}^L_{i,2} + \frac{1}{1 + \delta_i L_i} \vec{\sigma}^{L\, 2\otimes}_i \right] \vec{\sigma}^L_j. \end{aligned} \tag{7.46}$$

Then from (7.17) and (7.18) the endogenous and exogenous coefficients are respectively

$$\begin{aligned}\overset{\Rightarrow}{a}{}^{F,0}_{2,t}(T) \triangleq{}& -\vec{\sigma}{}^{F,0\,2\otimes}_{t} - \left[M_{\mathbf{i}}\Omega_{\mathbf{i}}\overset{\Rightarrow}{a}{}^{L}_{\mathbf{i},2}\right]\\ &+ \left[\Omega_{max(\mathbf{i},\mathbf{j})}M_{\mathbf{i}}M_{\mathbf{j}}\vec{\sigma}{}^{L}_{\mathbf{i}}\otimes\vec{\sigma}{}^{L}_{\mathbf{j}}\right] - \left[\frac{M_{\mathbf{i}}\Omega_{\mathbf{i}}}{1+\delta_{\mathbf{i}}L_{\mathbf{i}}}\vec{\sigma}{}^{L\,2\otimes}_{\mathbf{i}}\right]\end{aligned} \tag{7.47}$$

$$\overset{\Rightarrow}{a}{}^{p}_{3,t}(T) \triangleq -\left[M_{\mathbf{i}}\Omega_{\mathbf{i}}\overset{\Rightarrow}{a}{}^{L}_{\mathbf{i},3}\right]. \tag{7.48}$$

Finally, from (7.19) we get the second-depth coefficient as

$$\begin{aligned}\overset{\Rightarrow}{a}{}^{F,0}_{22,t} ={}& -\overset{\Rightarrow}{a}{}^{p}_{2}\overset{\leftrightarrow}{\otimes}\vec{\sigma}{}^{p} + \left[\Omega_{max(\mathbf{i},\mathbf{j})}M_{\mathbf{i}}M_{\mathbf{j}}\overset{\Rightarrow}{a}{}^{L}_{\mathbf{i},2}\overset{\leftrightarrow}{\otimes}\vec{\sigma}{}^{L}_{\mathbf{j}}\right] - \left[\frac{\Omega_{\mathbf{i}}M_{\mathbf{i}}}{1+\delta_{\mathbf{i}}L_{\mathbf{i}}}\overset{\Rightarrow}{a}{}^{L}_{\mathbf{i},2}\overset{\leftrightarrow}{\otimes}\vec{\sigma}{}^{L}_{\mathbf{i}}\right]\\ &+ \left[\Omega_{max(\mathbf{i},\mathbf{j})}\frac{M_{\mathbf{i}}M_{\mathbf{j}}}{1+\delta_{\mathbf{i}}L_{\mathbf{i}}}\vec{\sigma}{}^{L\,2\otimes}_{\mathbf{i}}\overset{\leftrightarrow}{\otimes}\vec{\sigma}{}^{L}_{\mathbf{j}}\right] - \left[\Omega_{\mathbf{i}}\frac{\delta_{\mathbf{i}}L_{\mathbf{i}}(1-\delta_{\mathbf{i}}L_{\mathbf{i}})}{(1+\delta_{\mathbf{i}}L_{\mathbf{i}})^3}\vec{\sigma}{}^{L\,3\otimes}_{\mathbf{i}}\right]\\ &- \left[\Omega_{max(\mathbf{i},\mathbf{j})}M_{\mathbf{i}}M_{\mathbf{j}}\vec{\sigma}{}^{L}_{\mathbf{i}}\otimes\vec{\sigma}{}^{p}\otimes\vec{\sigma}{}^{L}_{\mathbf{j}}\right] - \left[\Omega_{max(\mathbf{i},\mathbf{j},\mathbf{k})}M_{\mathbf{i}}M_{\mathbf{j}}M_{\mathbf{k}}\vec{\sigma}{}^{L}_{\mathbf{i}}\otimes\vec{\sigma}{}^{L}_{\mathbf{j}}\otimes\vec{\sigma}{}^{L}_{\mathbf{k}}\right]\\ &- \left[\frac{\Omega_{\mathbf{i}}M_{\mathbf{i}}}{1+\delta_{\mathbf{i}}L_{\mathbf{i}}}\left[\sigma^{L}_{\mathbf{i}\,n}\,a^{L}_{\mathbf{i},2\,mp}\right]\right] + \left[\Omega_{max(\mathbf{i},\mathbf{j})}M_{\mathbf{i}}M_{\mathbf{j}}\left[\sigma^{L}_{\mathbf{i}\,n}\,\overset{\Rightarrow}{a}{}^{L}_{\mathbf{j},2\,mp}\right]\right]\\ &+ \left[\Omega_{max(\mathbf{i},\mathbf{j})}\frac{M_{\mathbf{i}}M_{\mathbf{j}}}{1+\delta_{\mathbf{j}}L_{\mathbf{j}}}\left[\sigma^{L}_{\mathbf{j}\,m}\,\sigma^{L}_{\mathbf{i}\,n}\,\sigma^{L}_{\mathbf{j}\,p}\right]\right]\\ &+ \left[\Omega_{\mathbf{i}}M_{\mathbf{i}}\left[\sigma^{p}_{n}\,a^{L}_{\mathbf{i},2\,mp}\right]\right] + \left[\Omega_{\mathbf{i}}\frac{M_{\mathbf{i}}}{1+\delta_{\mathbf{i}}L_{\mathbf{i}}}\left[\sigma^{L}_{\mathbf{i}\,m}\,\sigma^{p}_{n}\,\sigma^{L}_{\mathbf{i}\,p}\right]\right].\end{aligned} \tag{7.49}$$

Dynamics of the T_0-rebased annuity leg A^0_t

We exploit the natural basket structure of the rebased annuity by writing

$$A^{\delta,0}_t(T) = \sum_{i=1}^{N} c_i\delta_i B^0_{i,t}(T) \quad \text{which matches (7.13) [p. 378] with } c_i = 1 \;\; \forall i.$$

With modified Einstein notation summing indices 0 **to** $N-1$, the instantaneous volatility comes from (7.14) as

$$\vec{\sigma}{}^{A,0}_{t}(T) = -\left[M_{\mathbf{i}}\Omega^{A}_{\mathbf{i}}\vec{\sigma}{}^{L}_{\mathbf{i}}\right] \tag{7.50}$$

where $M_i(T)$ is still defined by (7.7) and the weights are specified by (7.15) as

$$\underset{1\le i\le N}{\omega^A_{i,t}(T)} \triangleq \frac{\delta_i \prod_{k=0}^{i-1}\left[1+\delta_k L_{k,t}(T)\right]^{-1}}{\sum_{j=1}^{N}\delta_j \prod_{k=0}^{j-1}\left[1+\delta_k L_{k,t}(T)\right]^{-1}} \quad \text{and} \quad \underset{0\le i\le N-1}{\Omega^A_{i,t}(T)} = \sum_{j=i+1}^{N} \omega^A_{j,t}(T),$$

which establishes (7.35). The drift, endogenous and exogenous coefficients come respectively from (7.16)–(7.18) as

$$\begin{aligned}
\vec{a}^{A,0}_{1,t}(T) = &-\overset{\Rightarrow}{a}^{A,0}_{2,t}\vec{\sigma}^{A,0}_{t} - \left[M_{\mathbf{i}}\Omega^A_{\mathbf{i}}\vec{a}^{L}_{\mathbf{i},1}\right] - \left[M_{\mathbf{i}}\Omega^A_{\mathbf{i}}\overset{\Rightarrow}{a}^{L}_{\mathbf{i},2}\vec{\sigma}^{L}_{\mathbf{i}}\right] \\
&+ \sum_{i=0}^{N-1}\sum_{j=i}^{N-1}\Omega^A_{\mathbf{j}} M_i M_j \left[\overset{\Rightarrow}{a}^{L}_{i,2} + \frac{1}{1+\delta_i L_i}\vec{\sigma}^{L\,2\otimes}_{i}\right]\vec{\sigma}^{L}_{j}
\end{aligned} \tag{7.51}$$

$$\begin{aligned}
\overset{\Rightarrow}{a}^{A,0}_{2,t}(T) \triangleq &-\vec{\sigma}^{A,0\,2\otimes}_{t} - \left[M_{\mathbf{i}}\Omega^A_{\mathbf{i}}\overset{\Rightarrow}{a}^{L}_{\mathbf{i},2}\right] \\
&+ \left[\Omega^A_{max(\mathbf{i},\mathbf{j})} M_{\mathbf{i}} M_{\mathbf{j}} \vec{\sigma}^{L}_{\mathbf{i}} \otimes \vec{\sigma}^{L}_{\mathbf{j}}\right] - \left[\frac{M_{\mathbf{i}}\Omega^A_{\mathbf{i}}}{1+\delta_{\mathbf{i}}L_{\mathbf{i}}}\vec{\sigma}^{L\,2\otimes}_{\mathbf{i}}\right]
\end{aligned} \tag{7.52}$$

$$\overset{\Rightarrow}{a}^{A,0}_{3,t}(T) \triangleq -\left[M_{\mathbf{i}}\Omega^A_{\mathbf{i}}\overset{\Rightarrow}{a}^{L}_{\mathbf{i},3}\right]. \tag{7.53}$$

We get the single second-depth coefficient from (7.19) as

$$\begin{aligned}
\overset{\Rrightarrow}{a}^{A,0}_{22,t} = &-\overset{\Rightarrow}{a}^{A,0}_{2} \overset{\leftrightarrow}{\otimes} \vec{\sigma}^{A,0} + \left[\Omega^A_{max(\mathbf{i},\mathbf{j})} M_{\mathbf{i}} M_{\mathbf{j}} \overset{\Rightarrow}{a}^{L}_{\mathbf{i},2} \overset{\leftrightarrow}{\otimes} \vec{\sigma}^{L}_{\mathbf{j}}\right] \\
&- \left[\frac{\Omega^A_{\mathbf{i}} M_{\mathbf{i}}}{1+\delta_{\mathbf{i}}L_{\mathbf{i}}}\overset{\Rightarrow}{a}^{L}_{\mathbf{i},2} \overset{\leftrightarrow}{\otimes} \vec{\sigma}^{L}_{\mathbf{i}}\right] - \left[\Omega^A_{max(\mathbf{i},\mathbf{j})} M_{\mathbf{i}} M_{\mathbf{j}} \vec{\sigma}^{L}_{\mathbf{i}} \otimes \vec{\sigma}^{p} \otimes \vec{\sigma}^{L}_{\mathbf{j}}\right] \\
&- \left[\Omega^A_{max(\mathbf{i},\mathbf{j},\mathbf{k})} M_{\mathbf{i}} M_{\mathbf{j}} M_{\mathbf{k}} \vec{\sigma}^{L}_{\mathbf{i}} \otimes \vec{\sigma}^{L}_{\mathbf{j}} \otimes \vec{\sigma}^{L}_{\mathbf{k}}\right] \\
&+ \left[\Omega^A_{max(\mathbf{i},\mathbf{j})} \frac{M_{\mathbf{i}} M_{\mathbf{j}}}{1+\delta_{\mathbf{i}}L_{\mathbf{i}}} \vec{\sigma}^{L\,2\otimes}_{\mathbf{i}} \overset{\leftrightarrow}{\otimes} \vec{\sigma}^{L}_{\mathbf{j}}\right] + \left[\Omega^A_{\mathbf{i}} \frac{M_{\mathbf{i}}}{1+\delta_{\mathbf{i}}L_{\mathbf{i}}} \left[\sigma^{L}_{\mathbf{i}\,m}\, \sigma^{p}_{\ n}\, \sigma^{L}_{\mathbf{i}\,p}\right]\right] \\
&+ \left[\Omega^A_{max(\mathbf{i},\mathbf{j})} M_{\mathbf{i}} M_{\mathbf{j}} \left[\sigma^{L}_{\mathbf{i}\,n}\, \overset{\Rightarrow}{a}^{L}_{\mathbf{j},2\,mp}\right]\right] + \left[\Omega^A_{max(\mathbf{i},\mathbf{j})} \frac{M_{\mathbf{i}} M_{\mathbf{j}}}{1+\delta_{\mathbf{j}}L_{\mathbf{j}}} \left[\sigma^{L}_{\mathbf{j}\,m}\, \sigma^{L}_{\mathbf{i}\,n}\, \sigma^{L}_{\mathbf{j}\,p}\right]\right] \\
&+ \left[\Omega^A_{\mathbf{i}} M_{\mathbf{i}} \left[\sigma^{p}_{\ n}\, \overset{\Rightarrow}{a}^{L}_{\mathbf{i},2\,mp}\right]\right] - \left[\Omega^A_{\mathbf{i}} \frac{\delta_{\mathbf{i}}L_{\mathbf{i}}(1-\delta_{\mathbf{i}}L_{\mathbf{i}})}{(1+\delta_{\mathbf{i}}L_{\mathbf{i}})^3} \vec{\sigma}^{L\,3\otimes}_{\mathbf{i}}\right] \\
&- \left[\frac{\Omega^A_{\mathbf{i}} M_{\mathbf{i}}}{1+\delta_{\mathbf{i}}L_{\mathbf{i}}} \left[\sigma^{L}_{\mathbf{i}\,n}\, a^{L}_{\mathbf{i},2\,mp}\right]\right].
\end{aligned} \tag{7.54}$$

Aggregation into the dynamics of the par swap rate

Let us note first that the expressions for $\vec{\sigma}^{F,0}_t(T)$, $\vec{\sigma}^{A,0}_t(T)$, $\vec{\vec{a}}^{F,0}_{22,t}(T)$ and $\vec{\vec{a}}^{A,0}_{22,t}(T)$ quoted in the lemma have already been established, respectively, by (7.45), (7.50), (7.47) and (7.52). Now we combine the two weighting schemes as per (7.36) by defining

$$\underset{1\le i\le N}{\omega^S_{i,t}(T)} \triangleq \omega^F_{i,t}(T) - \omega^A_{i,t}(T)$$

and

$$\underset{0\le i\le N-1}{\Omega^S_{i,t}(T)} \triangleq \sum_{j=i+1}^{N} \omega^S_{j,t}(T) = \Omega^F_{i,t}(T) - \Omega^A_{i,t}(T).$$

Substituting (7.45) and (7.50) into (7.37) we then get the swap rate volatility as

$$\vec{\sigma}^S_t(T) = -\left[M_{\mathbf{i}}\Omega^S_{\mathbf{i}}\vec{\sigma}^L_{\mathbf{i}}\right]$$

which proves (7.29). The drift comes by substituting (7.46)–(7.51) into (7.38):

$$\begin{aligned}\vec{a}^S_{1,t}(T) &= \vec{\vec{a}}^{A,0}_{2,t}\vec{\sigma}^{A,0}_t - \vec{\vec{a}}^{F,0}_{2,t}\vec{\sigma}^{F,0}_t + \vec{\vec{a}}^S_{2,t}\vec{\sigma}^{A,0}_t - \left[\Omega^S_{\mathbf{i}}M_{\mathbf{i}}\vec{\vec{a}}^L_{\mathbf{i},2}\vec{\sigma}^L_{\mathbf{i}}\right] \\ &\quad - \left[\Omega^S_{\mathbf{i}}M_{\mathbf{i}}\vec{a}^L_{\mathbf{i},1}\right] + \sum_{i=0}^{N-1}\sum_{j=i}^{N-1}\Omega^S_{j,t}M_{i,t}M_{j,t}\left[\vec{\vec{a}}^L_{i,2,t} + \frac{\vec{\sigma}^{L\ 2\otimes}_{i,t}}{1+\delta_i L_{i,t}}\right]\vec{\sigma}^L_{j,t}\end{aligned}$$

where the first three terms simplify into

$$\vec{\vec{a}}^{A,0}_{2,t}\vec{\sigma}^{A,0}_t - \vec{\vec{a}}^{F,0}_{2,t}\vec{\sigma}^{F,0}_t + \vec{\vec{a}}^S_{2,t}\vec{\sigma}^{A,0}_t = \vec{\vec{a}}^{F,0}_{2,t}\vec{\sigma}^{A,0}_t - \vec{\vec{a}}^{F,0}_{2,t}\vec{\sigma}^{F,0}_t = -\vec{\vec{a}}^{F,0}_{2,t}\vec{\sigma}^S_t$$

so that eventually the coefficient rewrites as

$$\begin{aligned}\vec{a}^S_{1,t}(T) &= -\vec{\vec{a}}^{F,0}_{2,t}\vec{\sigma}^S_t - \left[\Omega^S_{\mathbf{i}}M_{\mathbf{i}}\vec{a}^L_{\mathbf{i},1}\right] - \left[\Omega^S_{\mathbf{i}}M_{\mathbf{i}}\vec{\vec{a}}^L_{\mathbf{i},2}\vec{\sigma}^L_{\mathbf{i}}\right] \\ &\quad + \sum_{i=0}^{N-1}\sum_{j=i}^{N-1}\Omega^S_{j,t}M_{i,t}M_{j,t}\left[\vec{\vec{a}}^L_{i,2,t} + \frac{1}{1+\delta_i L_{i,t}}\vec{\sigma}^{L\ 2\otimes}_{i,t}\right]\vec{\sigma}^L_{j,t},\end{aligned}$$

proving (7.30). To obtain $\vec{\vec{a}}^S_{2,t}(T)$ we inject (7.47)–(7.52) into (7.39), which gives

$$\begin{aligned}\vec{\vec{a}}^S_{2,t}(T) &= \vec{\sigma}^{A,0\ 2\otimes}_t - \vec{\sigma}^{F,0\ 2\otimes}_t - \left[M_{\mathbf{i}}\Omega^S_{\mathbf{i}}\vec{\vec{a}}^L_{\mathbf{i},2}\right] \\ &\quad + \left[\Omega^S_{max(\mathbf{i},\mathbf{j})}M_{\mathbf{i}}M_{\mathbf{j}}\vec{\sigma}^L_{\mathbf{i}}\otimes\vec{\sigma}^L_{\mathbf{j}}\right] - \left[\frac{M_{\mathbf{i}}\Omega^S_{\mathbf{i}}}{1+\delta_{\mathbf{i}}L_{\mathbf{i}}}\vec{\sigma}^{L\ 2\otimes}_{\mathbf{i}}\right]\end{aligned}$$

and proves (7.31). Then substituting (7.48) and (7.53) into (7.40) we obtain

$$\overset{\Rightarrow S}{a}_{3,t}(T) = -\left[M_{\mathbf{i}} \Omega_{\mathbf{i}}^{S} \overset{\Rightarrow L}{a}_{\mathbf{i},3}\right],$$

which proves (7.32). Finally, substituting (7.49) and (7.54) into (7.41) we get

$$\begin{aligned}
\overset{\Rightarrow S}{a}_{22,t}(T) &= \overset{\Rightarrow A,0}{a}_2 \overset{\leftrightarrow}{\otimes} \vec{\sigma}^{A,0} - \overset{\Rightarrow F,0}{a}_2 \overset{\leftrightarrow}{\otimes} \vec{\sigma}^{F,0} \\
&+ \left[\Omega^S_{max(\mathbf{i},\mathbf{j})} M_{\mathbf{i}} M_{\mathbf{j}} \overset{\Rightarrow L}{a}_{\mathbf{i},2} \overset{\leftrightarrow}{\otimes} \vec{\sigma}^L_{\mathbf{j}}\right] - \left[\frac{\Omega_{\mathbf{i}}^S M_{\mathbf{i}}}{1+\delta_{\mathbf{i}} L_{\mathbf{i}}} \overset{\Rightarrow L}{a}_{\mathbf{i},2} \overset{\leftrightarrow}{\otimes} \vec{\sigma}^L_{\mathbf{i}}\right] \\
&+ \left[\Omega^S_{max(\mathbf{i},\mathbf{j})} \frac{M_{\mathbf{i}} M_{\mathbf{j}}}{1+\delta_{\mathbf{i}} L_{\mathbf{i}}} \vec{\sigma}^{L\,2\otimes}_{\mathbf{i}} \overset{\leftrightarrow}{\otimes} \vec{\sigma}^L_{\mathbf{j}}\right] \\
&- \left[\Omega_{\mathbf{i}}^S \frac{\delta_{\mathbf{i}} L_{\mathbf{i}}(1-\delta_{\mathbf{i}} L_{\mathbf{i}})}{(1+\delta_{\mathbf{i}} L_{\mathbf{i}})^3} \vec{\sigma}^{L\,3\otimes}_{\mathbf{i}}\right] \\
&- \left[\Omega^S_{max(\mathbf{i},\mathbf{j})} M_{\mathbf{i}} M_{\mathbf{j}} \vec{\sigma}^L_{\mathbf{i}} \otimes \vec{\sigma}^p \otimes \vec{\sigma}^L_{\mathbf{j}}\right] \\
&- \left[\Omega^S_{max(\mathbf{i},\mathbf{j},\mathbf{k})} M_{\mathbf{i}} M_{\mathbf{j}} M_{\mathbf{k}} \vec{\sigma}^L_{\mathbf{i}} \otimes \vec{\sigma}^L_{\mathbf{j}} \otimes \vec{\sigma}^L_{\mathbf{k}}\right] \\
&- \left[\frac{\Omega_{\mathbf{i}}^S M_{\mathbf{i}}}{1+\delta_{\mathbf{i}} L_{\mathbf{i}}} \left[\sigma^L_{\mathbf{i}\,n}\, a^L_{\mathbf{i},2\,mp}\right]\right] + \left[\Omega^S_{max(\mathbf{i},\mathbf{j})} M_{\mathbf{i}} M_{\mathbf{j}} \left[\sigma^L_{\mathbf{i}\,n}\, \overset{\Rightarrow L}{a}_{\mathbf{j},2\,mp}\right]\right] \\
&+ \left[\Omega^S_{max(\mathbf{i},\mathbf{j})} \frac{M_{\mathbf{i}} M_{\mathbf{j}}}{1+\delta_{\mathbf{j}} L_{\mathbf{j}}} \left[\sigma^L_{\mathbf{j}\,m}\, \sigma^L_{\mathbf{i}\,n}\, \sigma^L_{\mathbf{j}\,p}\right]\right] + \left[\Omega_{\mathbf{i}}^S M_{\mathbf{i}} \left[\sigma^p{}_n\, \overset{\Rightarrow L}{a}_{\mathbf{i},2\,mp}\right]\right] \\
&+ \left[\Omega_{\mathbf{i}}^S \frac{M_{\mathbf{i}}}{1+\delta_{\mathbf{i}} L_{\mathbf{i}}} \left[\sigma^L_{\mathbf{i}\,m}\, \sigma^p{}_n\, \sigma^L_{\mathbf{i}\,p}\right]\right],
\end{aligned}$$

which proves (7.33) and concludes the proof. □

7.6 Approximating the Swap Rate Volatility

It seems clear from the swaption Sect. 7.5 that, within an LMM framework, the swap rate dynamics become quite complex. This feature suggests the use of some kind of *proxy*, usually achieved by a fixed-weight basket of Libor rates, and the most common of these is called the *freezing* approximation. Our aim in this section is to measure the impact of that proxy, in the general case and then in a simplified situation.

7.6.1 The Basket Approximation for Swap Rates

First of all, let us rewrite the swap rate as a basket of Libor rates:

$$\begin{aligned} S^{\delta}(T_0) &= \frac{B_t(T_0) - B_t(T_N)}{\sum_{k=1}^{N} \delta_k B_t(T_k)} = \frac{\sum_{i=0}^{N-1} \left[B_t(T_i) - B_t(T_{i+1})\right]}{\sum_{k=1}^{N} \delta_k B_t(T_k)} \\ &= \sum_{i=0}^{N-1} \frac{\delta_{i+1} B_t(T_{i+1}) L_t(T_i)}{\sum_{k=1}^{N} \delta_k B_t(T_k)}. \end{aligned}$$

Thus we face an *asset basket* with *stochastic weights*, as defined by (3.82) [p. 194]:

$$S_t^{\delta}(T) = \sum_{i=0}^{N-1} \lambda_{i,t} L_{i,t}(T) \quad \text{with} \quad \lambda_{i,t} \triangleq \frac{\delta_{i+1} B_t(T_{i+1})}{\sum_{k=1}^{N} \delta_k B_t(T_k)}.$$

For future use, we note that the absolute weights $\lambda_{i,t}$ are actually normalised:

$$\sum_{i=0}^{N-1} \lambda_{i,t} = \frac{\sum_{i=0}^{N-1} \delta_{i+1} B_t(T_{i+1})}{\sum_{k=1}^{N} \delta_k B_t(T_k)} = 1. \tag{7.55}$$

Then the *weight freezing* approximation consists in defining a swap rate proxy with

$$S_t^{\star}(T) \triangleq \sum_{i=0}^{N-1} \lambda_{i,\mathbf{0}} L_{i,t}(T) \quad \text{for } t \geq 0. \tag{7.56}$$

If by construction the initial value $S_0^{\star}(T)$ of the proxy does match the actual swap rate $S_0(T)$, the same cannot be said of its dynamics, and therefore of future values. Especially, the frozen-weights proxy swap rate is *not* a martingale under the natural annuity measure.[7]

Lemma 7.3 (Dynamics of the frozen swap rate) *Under the annuity measure* $\mathbb{Q}^A$ *the dynamics of the proxy swap rate* (7.56) *are*

$$\frac{dS_t^{\star}(T)}{S_t^{\star}(T)} = \mu_t^{S^{\star},A} dt + \vec{\sigma}_t^{S^{\star}}(T)^{\perp} d\vec{W}_t^{A(T)} \tag{7.57}$$

[7] To make the proxy dynamics (7.58) martingale, we would have to consider the following numeraire: $N_t \triangleq \prod_{i=1}^{N} \delta_i B_t(T_i)$. However, this process is *not* a recognised asset.

with

$$\mu_t^{S^\star,A}(T) \triangleq \sum_{i=0}^{N-1} \left[\omega_{i,t}^* \mu_{i,t}^{L,A}\right](T) \;\; and \;\; \vec{\sigma}_t^{S^\star}(T) \triangleq \sum_{i=0}^{N-1} \left[\omega_{i,t}^* \vec{\sigma}_{i,t}^{L}\right](T),$$

where $\mu_{i,t}^{L,A}$ is the individual Libor's $\mathbb{Q}^A$-drift. The weighting scheme is defined by

$$\omega_{i,t}^*(T) \triangleq \frac{\lambda_{i,0} L_{i,t}^\delta(T)}{\sum_{k=0}^{N-1} \lambda_{k,0} L_{k,t}(T)}. \tag{7.58}$$

Furthermore, the overall drift resultant can be rewritten as (omitting argument T)

$$\mu_t^{S^\star,A} \triangleq \left[\sum_{i=0}^{N-1} \omega_{i,t}^* \vec{\sigma}_{i,t}^{L}\right]^\perp \left[\sum_{i=0}^{N-1} \lambda_{i,t} \vec{\Gamma}_{i+1,t}\right] - \sum_{i=0}^{N-1} \omega_{i,t}^* \left[\vec{\sigma}_{i,t}^{L\,\perp} \vec{\Gamma}_{i+1,t}\right], \tag{7.59}$$

which features a pseudo-covariance structure and is a priori a.s. non-null.

Again for future use, we define a collection of maturity-dependent processes:

$$\vec{\gamma}_{i,t}(T) \triangleq \vec{\Gamma}_{i+1,t}(T) - \vec{\sigma}_t^A(T). \tag{7.60}$$

Proof Let us recall the instantaneous volatility of the annuity process, characterising $\mathbb{Q}^A$:

$$\frac{dA_t^\delta(T)}{A_t^\delta(T)} = r_t dt + \vec{\sigma}_t^A(T)^\perp d\vec{W}_t \;\; \text{and} \;\; d\vec{W}_t^{A(T)} \triangleq d\vec{W}_t - \vec{\sigma}_t^A(T)dt.$$

The dynamics of the individual Libor rates are

$$\begin{aligned}
\frac{dL_{i,t}^\delta(T)}{L_{i,t}^\delta(T)} &= \vec{\sigma}_{i,t}^L(T)^\perp d\vec{W}_t^{T_{i+1}} \\
&= \vec{\sigma}_{i,t}^L(T)^\perp \left[d\vec{W}_t^{A(T)} + \vec{\sigma}_t^A(T)dt - \vec{\Gamma}_t(T_{i+1})dt\right] \\
&= \mu_{i,t}^{L,A} dt + \vec{\sigma}_{i,t}^L(T)^\perp d\vec{W}_t^{A(T)} \;\; \text{with} \;\; \mu_{i,t}^L \triangleq -\vec{\sigma}_t^L(T)^\perp \vec{\gamma}_{i,t}(T).
\end{aligned} \tag{7.61}$$

We can now express the proxy swap rate's dynamics under the Annuity measure:

$$\begin{aligned}
dS_t^\star(T) &= \sum_{i=0}^{N-1} \lambda_{i,0}\, dL_{i,t}(T) \\
&= \sum_{i=0}^{N-1} \lambda_{i,0} \left[L_{i,t}^\delta(T)\mu_{i,t}^L dt + L_{i,t}^\delta(T)\vec{\sigma}_{i,t}^L(T)^\perp d\vec{W}_t^{A(T)}\right].
\end{aligned}$$

Introducing the stochastic weights (7.58) we then rewrite in Lognormal fashion, proving (7.57). Now focusing on the drift and omitting the maturity argument, we inject (7.60) and expand as

$$
\begin{aligned}
\mu_t^{S^\star,A} &= \sum_{i=0}^{N-1} \omega_{i,t}^* \, \vec{\sigma}_{i,t}^{L\ \perp} \left[\vec{\sigma}_t^A - \vec{\Gamma}_{i+1,t} \right] \\
&= \left[\sum_{i=0}^{N-1} \omega_{i,t}^* \vec{\sigma}_{i,t}^L \right]^{\perp} \vec{\sigma}_t^A - \sum_{i=0}^{N-1} \omega_{i,t}^* \left[\vec{\sigma}_{i,t}^{L\ \perp} \vec{\Gamma}_{i+1,t} \right].
\end{aligned}
$$

However, in the above equation the instantaneous volatility of the annuity comes as

$$
\vec{\sigma}_t^A(T) = \frac{\sum_{i=1}^N \delta_i B_{i,t}(T) \vec{\Gamma}_{i,t}(T)}{\sum_{i=1}^N \delta_i B_{i,t}(T)} = \sum_{i=0}^{N-1} \lambda_{i,t} \vec{\Gamma}_{i+1,t}(T),
$$

which after replacement yields (7.59) and concludes the proof. □

It seems clear that, as it exhibits a drift, this $S^\star$ *frozen* proxy is not really adapted for *simulating* the swap rate. This is not to say that the principle of approximating the latter with a Libor *basket* is fundamentally flawed, but just that the specific choice of these *frozen* weights would not be optimal for dynamic purposes. Instead, it is actually possible to design different fixed weighting schemes, specifically to provide a better dynamic fit. Such weights can be chosen for instance by matching *some* of the coefficients in the swap rate's *theoretical* Wiener chaos. In that respect, our IATM results can prove useful as they identify the influence of each such coefficient on the smile, allowing to optimise the weighting scheme w.r.t. a clear market-related objective function. Alternatively, *empirical* fitting of the weights can be achieved, for example, by moment matching on a given set of Monte-Carlo paths.

In practice, however, the proxy is not used for such a close dynamic match. Instead, practitioners working in the LMM framework use only the volatility expression in (7.57), i.e.

$$
\vec{\sigma}_t^{S^\star}(T) = \sum_{i=0}^{N-1} \omega_{i,t}^*(T) \vec{\sigma}_{i,t}^L(T), \tag{7.62}
$$

in order to estimate the liquid part of the smile for European swaptions (some of the most traded IR derivatives). Hence the question that we address thereafter is the accuracy of this *volatility* proxy. The priority is obviously to assess the precision of the ATM level, but we could also examine other smile descriptors, such as the ATM skew. In summary, it is the discrepancy generated by (7.62) as a vectorial basket of Libor *volatilities* that we wish to compute.

7.6.2 Exact Swap Rate Dynamics in the Basket Representation

We have already computed in Sect. 7.5.2 the exact swap rate dynamics within an LMM representation. However, besides being quite complex and therefore difficult to interpret, the form of the resulting expression is not really adapted to a comparison with the freezing proxy (7.62). Let us therefore provide an alternative formulation.

Lemma 7.4 (Exact swap rate dynamics in an LMM basket representation) *In an LMM parametrisation, the exact swap rate dynamics can be written as*

$$\frac{dS_t^\delta(T)}{S_t^\delta(T)} = \vec{\sigma}_t^S(T)^\perp d\vec{W}_t^A \quad \text{with} \quad \vec{\sigma}_t^S = \sum_{i=0}^{N-1} \omega_{i,t} \vec{\sigma}_{i,t}^L + \sum_{i=0}^{N-1} \omega_{i,t} \vec{\gamma_{i,t}}, \tag{7.63}$$

where the normalised weights are specified by

$$\omega_{i,t}(T) \triangleq \frac{\lambda_{i,t}(T) L_{i,t}(T)}{S_t^\delta(T)}$$

and where $\vec{\gamma_{i,t}}$ *is defined by* (7.60)*. Furthermore, the second volatility component rewrites as*

$$\sum_{i=0}^{N-1} \omega_{i,t} \vec{\gamma_{i,t}} = \sum_{i=0}^{N-1} \left[\Omega_i^A - \Omega_i^\gamma \right] M_i \vec{\sigma}_i^L, \tag{7.64}$$

where the Annuity aggregated weights $\left\{ \Omega_i^A \right\}$ *are defined by* (7.35) [p. 393]*, while*

$$\Omega_i^\gamma(T) \triangleq \sum_{j=i}^{N-1} \omega_{i,t}(T) \tag{7.65}$$

defines a new aggregated scheme, associated to the stochastic weights.

Proof In order to compute the swap rate's exact chaos dynamics, and in particular its instantaneous volatility, we apply Proposition 3.7 [p. 194], dedicated to generic asset baskets. Since the swap rate is martingale under $\mathbb{Q}^{A(T)}$, we need the chaos dynamics of both the Libor population $\left\{ L_{i,t}(T) \right\}$ and the associated weights $\left\{ \lambda_{i,t} \right\}$ under that same Annuity measure. However, the *absolute* weights $\lambda_{i,t}$ are by construction martingale under $\mathbb{Q}^{A(T)}$. Indeed, exploiting the usual change of numeraire argument and using the notation (7.60) provides their dynamics as

$$\frac{d\lambda_{i,t}}{\lambda_{i,t}} = \left[\vec{\Gamma}_{i+1,t}(T) - \vec{\sigma}_t^A(T) \right]^\perp d\vec{W}_t^{A(T)} = \vec{\gamma_{i,t}}(T)^\perp d\vec{W}_t^{A(T)}.$$

Following Definition 3.8 [p. 194], the *normalised* weight processes are introduced as per (3.83):

$$\omega_{i,t} \triangleq \frac{\lambda_{i,t} L_{i,t}}{S_t^\delta} = \frac{\left[\delta_{i+1} B_{i+1,t}/A_t^\delta\right] L_{i,t}}{\sum_{k=0}^{N-1}\left[\delta_{k+1} B_{k+1,t}/A_t^\delta\right] L_{k,t}} = \frac{B_{i,t} - B_{i+1,t}}{\sum_{k=0}^{N-1}\left[B_{k,t} - B_{k+1,t}\right]}$$
$$= \frac{B_{i,t} - B_{i+1,t}}{B_{0,t} - B_{N,t}},$$

hence we verify—as we should—that these normalised weights *do* sum to unity:

$$\sum_{i=1}^{N} \omega_{i,t} = 1.$$

Applying (3.84) and (3.85) [p. 195], the basket's dynamics rewrite formally as

$$\frac{dS_t^\delta(T)}{S_t^\delta(T)} = \mu_t^S(T)dt + \vec{\sigma}_t^{\,S}(T)^\perp d\vec{W}_t^A$$

where, by invoking (7.61), the drift reassuringly vanishes:

$$\mu_t^S = \sum_{i=0}^{N-1} \omega_{i,t} \underbrace{\left[\mu_{i,t}^L + \vec{\sigma}_{i,t}^{\,L}(T)^\perp \vec{\gamma}_{i,t}\right]}_{=0} = 0.$$

As for the volatility of the exact swap rate, it comes from the Proposition as

$$\vec{\sigma}_t^{\,S}(T) = \sum_{i=0}^{N-1} \omega_{i,t} \left[\vec{\sigma}_{i,t}^{\,L}(T) + \vec{\gamma}_{i,t}(T)\right],$$

which proves (7.63). Now let us rewrite the volatility of the absolute weights as a function of the HJM volatility structure, adopting simplified notations:

$$\vec{\gamma}_{i,t} \triangleq \vec{\Gamma}_{i+1,t} - \vec{\sigma}_t^{\,A} = \vec{\Gamma}_{i+1,t}^{\,0} - \vec{\sigma}_t^{\,A,0} = \vec{\Gamma}_{i+1,t}^{\,0} - \sum_{j=0}^{N-1} \lambda_{j,t} \vec{\Gamma}_{j+1,t}^{\,0}.$$

Injecting (7.8) to recover the LMM parametrisation, then applying Fubini, gives

$$\vec{\gamma}_i = -\sum_{k=0}^{i} M_k \vec{\sigma}_k^{\,L} + \sum_{j=0}^{N-1}\sum_{k=0}^{j} \lambda_j M_k \vec{\sigma}_k^{\,L}$$
$$= \sum_{k=0}^{N-1}\left[\sum_{j=k}^{N-1} \lambda_j\right] M_k \vec{\sigma}_k^{\,L} - \sum_{k=0}^{i} M_k \vec{\sigma}_k^{\,L}.$$

From (7.35) [p. 393] we identify the absolute scheme as an offset of the Annuity's:

$$\lambda_j \equiv \omega_{j+1}^A.$$

Then we can re-express the aggregated absolute weights as

$$\sum_{j=k}^{N-1} \lambda_j = \sum_{j=k+1}^{N} \omega_j^A = \Omega_k^A \quad \text{for } 0 \le k \le N-1.$$

Eventually each individual $\vec{\gamma_i}$ rewrites as

$$\vec{\gamma_i} = \sum_{j=0}^{N-1} \Omega_j^A M_k \vec{\sigma}_k^L - \sum_{k=0}^{i} M_k \vec{\sigma}_k^L.$$

It follows that the weight-induced part of the swap rate volatility is expressed as

$$\sum_{i=0}^{N-1} \omega_{i,t} \overrightarrow{\gamma_{i,t}}(T) = \sum_{i=0}^{N-1} \Omega_i^A M_i \vec{\sigma}_i^L - \sum_{i=1}^{N} \omega_{i,t} \sum_{k=0}^{i} M_k \vec{\sigma}_k^L.$$

Invoking Fubini again, as well as (7.65), we get

$$\begin{aligned}
\sum_{i=0}^{N-1} \omega_{i,t} \overrightarrow{\gamma_{i,t}}(T) &= \sum_{i=0}^{N-1} \Omega_i^A M_i \vec{\sigma}_i^L - \sum_{i=0}^{N-1} \left[\sum_{j=i}^{N-1} \omega_{j,t} \right] M_i \vec{\sigma}_i^L \\
&= \sum_{i=0}^{N-1} \Omega_i^A M_i \vec{\sigma}_i^L - \sum_{i=0}^{N-1} \Omega_i^\gamma M_i \vec{\sigma}_i^L \\
&= \sum_{i=0}^{N-1} \left[\Omega_i^A - \Omega_i^\gamma \right] M_i \vec{\sigma}_i^L,
\end{aligned}$$

which proves (7.64) and concludes the proof. □

We find noteworthy that if the absolute weights $\lambda_{i,t}$ can be interpreted as a relative *level* of the yield curve—the ratio of the local zero $B_t(T_{i+1})$ over some *average* zero—the normalised weights $\omega_{i,t}$ can be seen as relative *slope* of that same curve. Hence the swap rate itself can be understood as the *ratio of the average slope over the average level*. As we shall see shortly, it is the dispersion of the individual levels (among the Libor rate population) that deteriorates the quality of the frozen weights proxy.

7.6.3 Impact of the Freezing Approximation in the General Case

Considering Lemma 7.4, it seems only natural to split the exact swap rate volatility (7.63) as the sum of the Libors' and of their stochastic weights' contributions:

$$\vec{\sigma}_t^{S}(T) = \vec{\sigma}_t^{S,L}(T) + \vec{\sigma}_t^{S,\omega}(T),$$

where we define

$$\vec{\sigma}_t^{S,L}(T) \triangleq \sum_{i=0}^{N-1} \omega_{i,t}\vec{\sigma}_{i,t}^{L}(T) \quad \text{and} \quad \vec{\sigma}_t^{S,\omega}(T) \triangleq \sum_{i=0}^{N-1} \omega_{i,t}\vec{\gamma}_{i,t}(T).$$

We investigate first the precision of the proxy volatility (7.62) taken at the initial time $t = 0$. Indeed, this is the *freezing time* of the weights, and should correspond to the maximum accuracy. But since the definition (7.58) ensures that

$$\omega_{i,0}^{*} = \omega_{i,0} \quad \forall i \;\; 0 \le i \le N-1$$

we can identify the initial proxy volatility to the initial pure Libors' contribution:

$$\vec{\sigma}_0^{S\star}(T) = \vec{\sigma}_0^{S,L}(T). \tag{7.66}$$

This result is well-known and quite intuitive, but all it allows us to state at this point is that the proxy exhibits *some a priori* discrepancy at the initial time. As for predicting the ATM level of a T-expiry European swaption (which is the usual goal), we observe that the multi-dimensional LMM term-structure of volatility significantly complicates the picture. Indeed, expressing instead the corresponding ATM implied variance we have:

$$\underbrace{\widetilde{\Sigma}_0^2(\text{ATM},T) \approx}_{\text{IATM identity}} \|\vec{\sigma}_0^{S}(T)\|^2 = \underbrace{\|\vec{\sigma}_0^{S\star}(T)\|^2}_{\substack{\text{proxy}\\ \text{contribution}}} + \underbrace{\|\vec{\sigma}_0^{S,\omega}(T)\|^2}_{\substack{\text{weights}\\ \text{contribution}}} + \underbrace{2\left[\vec{\sigma}_0^{\star\perp}\,\vec{\sigma}_0^{S,\omega}\right](T)}_{\text{interference term}}.$$

The usual rationale for freezing the weights is that their volatilities are an order of magnitude smaller than the Libor's. This heuristic is empirically verified, as it has been repeatedly tested with market-calibrated models (see [26] for instance). But it relies on considering only the ATM implied volatility, for short to medium expiries, which as we know represents a rough and scalar aggregate of the full volatility information. In particular, our knowledge of the quantitative impact of the interference term is limited at this stage. We can only bound its modulus, and conjecture its sign by noting that the weights are related to the Zeros, so that they should be negatively correlated to the Libors. Our position, therefore, is that the freezing approximation does not accurately reflect the complexity of the problem:

- The initial error should be considered in regard of the multi-dimensional volatility $\vec{\sigma_0}^{S^\star}(T)$, rather than simply the modulus of that vector.
- This vectorial discrepancy will deteriorate at future times, hence we need to envisage it as a process and gauge its dynamics.

Although the formal impact of the freezing approximation has (surprisingly) not motivated many studies [26] represents one of the best references. In particular, that thesis proposes a *corrective* term for the proxy's Call price, when the underlying is a generic basket of Lognormal assets. To that end, it uses a classical asymptotic expansion technique on the price itself, an approach presenting similarities with [27], for instance. Also, restricting the LMM framework to some manageable LN dynamics, and using the L^2 norm (i.e. the modulus) as a measure of precision, it provides an upper bound for the discrepancy of the swap rate volatility.

Instead, and in compliance with our above-stated position, we would rather *adopt a finer resolution* for this problem, by *focusing on the individual weighting error for each Libor's volatility*. Also, we wish to *measure* the influence of any dynamic assumption *directly on the smile*, which tends to make interpretation easier. Now, the question that we would like to address is how our previous ACE results can complement those of [26]. For that purpose, let us first review some general considerations, as well as the several avenues that seem open to us.

First, we note that as our approach exploits the chaos dynamics of the swap rate process, and since we know that the basket structure is propagated along that chaos (see Lemma 3.6 [p. 192]) we can implement the freezing approximation *at different levels of that chaos*. In other words, we *do* have the tools to examine the impact of locking the weights in the asset definition, which is the usual and above-described freezing approximation. But, potentially, we *can also* freeze the weights in the volatility expression (7.62), or in the endogenous coefficient expression that would follow in the chaos, etc. Furthermore, these different levels of coefficient freeze can be combined, which gives us a high number of possible combinations, depending on the depth of the chaos expansion. Overall, this flexibility allows us an additional degree of freedom, and the latter offers new configurations in the necessary tradeoff of simplicity versus precision, associated to any sort of approximation. However, for reasons of interpretation and therefore of simplicity, we will limit ourselves to freezing at the first level only.

Also, the chaos coefficients affected by the above freezing will generate the IATM smile differentials, in a way that is known to us. Hence we can now *measure* the freezing error in more intuitive terms: in the form of an uncertainty on the smile level, and/or skew, and/or curvature, etc. But again, as will shortly be obvious, even in a simplified case the associated formulae are quite involved, so that we shall limit ourselves to examining the ATM level.

In order to justify and introduce the simplified case of Sect. 7.6.4, let us now state a few elementary facts about the problem at hand. First, it is clear that the absolute weights λ_j are decreasing, but as mentioned before they sum to unity. Therefore, if all Libors rates were to be (initially) identical, then we would have $\lambda_i = \omega_i \forall i$. This observation that a *flat Libor curve* maximises the precision of the freezing formula is a well-known feature. But even in that case, the curve cannot *stay* flat for ever, so that a degradation will occur for longer maturities.

Turning to the individual error on each Libor's volatility, as provided by (7.64), we observe that the aggregated weights $\Omega^\star_\cdot$ also live in [0, 1]. Indeed, they start at their maximum value of 1 and then keep decreasing. In turn, this gives us a blunt upper bound, or worst case scenario, in terms of absolute and individual volatility error:

$$\left|\left[\Omega_i^A - \Omega_i^\gamma\right] M_i\right| \leq M_i = \frac{\delta_i L_i}{1+\delta_i L_i} \leq \delta_i L_i$$

which seems small since the accrual δ_i lies by market convention in [0, 1], while L_i is typically of the order of a few percent. However, this error has no real interpretation until we can compare it to the actual or frozen weight of that individual Libor volatility. In other words, the *relative* error defined by

$$\frac{\left[\Omega_i^A - \Omega_i^\gamma\right] M_i}{\omega_i}$$

seems more pertinent to evaluate the impact of using the freezing approximation.

Also, we remark that the simplification effect brought by a flat Libor curve assumption is not limited to the swap rate's freezing proxy. Indeed, the maturity-differential $\partial_T L_t^\delta(T)$ intervenes in the term-Structure ZDC (5.21) [p. 286] as the *slope compensation term* $I(t, y, \theta)$: see (5.22). As a consequence, it significantly influences the term structure of implied volatility for both Caplets and Swaptions.

7.6.4 Impact of the Freezing Approximation in a Simplified Case

We have seen that a flat Libor curve provides the highest precision for the swap rate's proxy volatility (7.62) [p. 403]. Hence it makes sense to compute the sensitivity of that discrepancy when the curve deviates from the flat shape. Since the first deformation mode of the curve is usually horizontal, while the second corresponds to the slope, we will consider a first-order *affine* perturbation. Let us formalise this simplified case:

Assumption 7.1 (*Idealised affine Libor curve*) We consider a simplistic tenor structure and yield curve where

- All accruals are identical:

$$\forall i / 0 \leq i \leq N-1 \qquad \delta_i = \delta \qquad \text{or equivalently} \qquad T_{i+1} = T_i + \delta.$$

- The Libor rate structure is currently affine in maturity:

$$\forall i / 0 \leq i \leq N-1 \qquad L_t(T_i, T_{i+1}) = L_t(T_0, T_1)\left[1 + i\varepsilon\right].$$

The ε parameter should be considered small

$$\varepsilon N \ll 1$$

as it denotes a deviation from a flat term structure of Libor rates, for which we know that the (value of the) swap rate's instantaneous volatility is exactly matched by the fixed-weights basket.

Then we can compute the error on the swap rate volatility as a first-order expansion on ε.

Lemma 7.5 (Weighting discrepancies under the affine Libor assumption) *Under Assumption* 7.1 *the weight discrepancy for each individual Libor rate volatility* $\vec{\sigma}_i^L$ *as a component of the swap rate instantaneous volatility is, as a function of the slope* ε*:*

In absolute terms:

$$\left[\left(\Omega_i^A - \Omega_i^\gamma\right) M_i\right](\varepsilon) = \frac{1-B}{1-B^N}\left[\frac{NB^N}{1-B^N} - iB^i - \frac{NB^N}{1-B^N}B^i\right]\varepsilon + o(\varepsilon^2). \tag{7.67}$$

In relative terms:

$$\left[\frac{\left(\Omega_i^A - \Omega_i^\gamma\right) M_i}{\omega_i}\right](\varepsilon) = \frac{1}{B^i}\left[\frac{NB^N}{1-B^N} - iB^i - \frac{NB^N}{1-B^N}B^i\right]\varepsilon + o(\varepsilon^2), \tag{7.68}$$

where the initial Zero is defined by

$$B \triangleq \frac{1}{1+\delta L_t(T_0,T_1)}.$$

By rewriting the bracket shared by (7.67) and (7.68) as a ratio and observing its numerator, it is clear that (in non-pathological cases) both the absolute and relative weighting errors will be negative. This does not mean, however, that the overall modulus of the swap rate's volatility will be systematically underestimated. To gauge the contribution of each Libor's volatility to the overall error (and, in particular, to the directional information) we choose to use an example. We select a tenor structure which is typical of the centre grid: $10Y$ expiry, $20Y$ tenor, semi-annual Libor rate. As for market data, we used a stylised configuration where the $10Y$ Zero rate is set at 5 %, the first ($10Y$) forward Libor rate also at 5 % and the last ($29Y6M$) forward Libor rate at 10 %. The result of this test is shown in Fig. 7.1 and leads us to two main observations, pertaining respectively to *shape* and *scale*.

First of all, the inverted bell-curve shape, combined with the almost-linear decreasing weight structure, means that the centre of the tenor structure will bear the bulk of the error. Then the magnitude of the individual error, for these central Libor rates, is

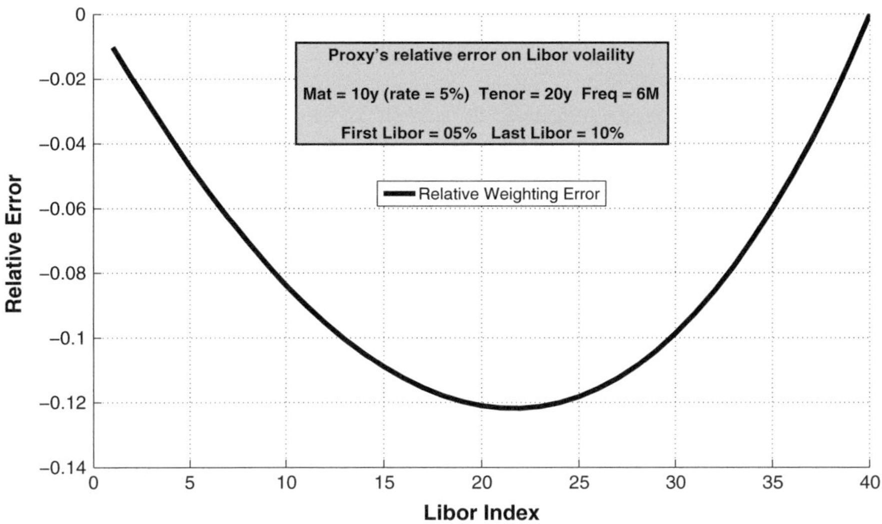

Fig. 7.1 Freezing proxy for the swap rate: relative weighting Libor error

not negligible: a 12 % error on a 20 % Lognormal volatility represents a discrepancy of more than 200 bpvol. Before going into the simple but lengthy proof of Lemma 7.5, let us recall some elementary but useful identities.

Lemma 7.6 (Elementary notations for derivatives of geometric series) *For argument* $q \in]0, 1[$ *and with bounds* $0 \leq j < k$ *we have*

$$G_j^k(q) \triangleq \sum_{i=j}^{k-1} q^i = \frac{q^k - q^j}{q-1}, \tag{7.69}$$

$$G_j^{k'}(q) = \sum_{i=j}^{k-1} i q^{i-1} = \frac{(kq^{k-1} - jq^{j-1})(q-1) - (q^k - q^j)}{(q-1)^2}, \tag{7.70}$$

$$G_j^{k''}(q) = \sum_{i=j}^{k-1} i(i-1) q^{i-2}. \tag{7.71}$$

Proof The basic result on geometric series provides (7.69), which we reformulate as:

$$\sum_{i=j}^{k-1} q^i = \frac{q^k - q^j}{q-1} \triangleq \frac{A(q)}{B(q)} \quad \text{where } B'(q) \equiv 1.$$

Differentiating once w.r.t. q we get

$$\sum_{i=j}^{k-1} iq^{i-1} = \partial_q \sum_{i=j}^{k-1} q^i = \partial_q \frac{A(q)}{B(q)} = \frac{(kq^{k-1} - jq^{j-1})(q-1) - (q^k - q^j)}{(q-1)^2},$$

proving (7.70), which with the obvious (7.71) concludes this very simple proof. □

We can now move on to the main proof.

Proof of Lemma 7.5

Let us work with a slightly more general formulation than Assumption 7.1 by assuming some constant c and setting

$$L_i = L_0 + ic\varepsilon.$$

Our strategy is obviously to compute the first-order ε-expansions of all relevant quantities, at the initial (flat curve) position corresponding to $\varepsilon = 0$. To simplify notations, and unless otherwise specified, we will consider all G functions and their derivatives to be taken in $q = B$:

$$G_j^{k\,(p)} \equiv G_j^{k\,p}(B).$$

Step 1: expanding the rebased zeros and the annuity

First the T_0-rebased Zero Coupons are, without surprise,

$$B_i^0(\varepsilon) = \left[\prod_{j=0}^{i-1} [1 + \delta L_j(\varepsilon)]\right]^{-1} = \left[\prod_{j=0}^{i-1} [1 + \delta L_0 + \delta jc\varepsilon]\right]^{-1}.$$

From now on we will consider all assets to be rebased by $B_t(T_0)$.

Hence, with

$$B \doteq B(\delta, L_0) \triangleq \frac{1}{1 + \delta L_0} \quad \text{or equivalently} \quad L_0 = \frac{1 - B}{\delta B},$$

we then have the Zeros' initial value as

$$B_i^0(\varepsilon = 0) = B^i. \tag{7.72}$$

As for the first-order ε-derivative we get

$$\partial_\varepsilon B_i(\varepsilon) = -\delta c B_i(\varepsilon) \sum_{j=0}^{i-1} \frac{j}{[1 + \delta L_0 + \delta j\varepsilon]}.$$

Evaluating this expression at $\varepsilon = 0$ we obtain

$$\partial_\varepsilon B_i(0) = -\frac{1}{2}\delta c i(i-1)B^{i+1}. \tag{7.73}$$

Turning to the annuity

$$A(\varepsilon) = \delta \sum_{i=1}^{N} B_i(\varepsilon)$$

we get its initial value for $\varepsilon = 0$ as

$$A(0) = \delta \sum_{i=1}^{N} B^i = \delta B G_0^N \tag{7.74}$$

and from (7.73) we get its generic ε-differential as

$$\partial_\varepsilon A(\varepsilon = 0) = \delta \sum_{i=1}^{N} \partial_\varepsilon B_i(0) = -\frac{1}{2}\delta^2 c B^3 \sum_{i=1}^{N} i(i-1)B^{i-2}.$$

Applying (7.71) with summation bounds $j = 1$ and $k = N+1$ we get

$$\partial_\varepsilon A(\varepsilon = 0) = -\frac{1}{2}\delta^2 c B^3 G_1^{(N+1)''}. \tag{7.75}$$

Step 2: expanding the individual Libor weights and the swap rate

We can now turn to *absolute* weights of the Libor basket, which in our context are

$$\lambda_i(\varepsilon) = \frac{B_{i+1}}{\sum_{j=1}^{N} B_j} = \frac{\delta B_{i+1}(\varepsilon)}{A(\varepsilon)}.$$

In $\varepsilon = 0$ we can substitute (7.72) and (7.74) so that these weights become

$$\lambda_i(0) = \frac{B^{i+1}}{B G_0^N} = \frac{B^i}{G_0^N}. \tag{7.76}$$

Hence their ε-differential is

$$\partial_\varepsilon \lambda_i(\varepsilon) = \frac{A(\varepsilon)\partial_\varepsilon B_{i+1}(\varepsilon) - B_{i+1}(\varepsilon)\partial_\varepsilon A(\varepsilon)}{A^2(\varepsilon)}.$$

Rather than expanding this expression, we instead evaluate it directly with $\varepsilon = 0$ and substitute the previous expressions (7.72)–(7.75) to obtain the ε-differential of the absolute weights as:

$$\partial_\varepsilon \lambda_i(0) = -\frac{c}{2}\frac{B^{i+1}}{G_0^N}\left[i(i+1) - B\frac{G_1^{(N+1)''}}{G_0^N}\right]. \tag{7.77}$$

We can now express the ε-differential of the individual product in the basket as

$$\partial_\varepsilon\left[\lambda_i L_i\right](\varepsilon) = L_i \partial_\varepsilon \lambda_i(\varepsilon) + \lambda_i \partial_\varepsilon L_i(\varepsilon)$$

which, by evaluating at $\varepsilon = 0$ and substituting (7.77), gives

$$\partial_\varepsilon\left[\lambda_i L_i\right](0) = -\frac{c}{2}\frac{B}{G_0^N} L_0 B^i\left[i(i+1) - B\frac{G_1^{(N+1)''}}{G_0^N}\right] + \frac{B^i}{G_0^N} ic \tag{7.78}$$

$$= c\frac{B^i}{G_0^N}\left[\frac{1}{2}BL_0\left[B\frac{G_1^{(N+1)''}}{G_0^N} - i(i+1)\right] + i\right]. \tag{7.79}$$

Moving on to the basket, we have the swap rate as

$$S(\varepsilon) = \sum_{i=0}^{N-1} \lambda_i(\varepsilon)(L_0 + i\delta\varepsilon).$$

Hence, using the absolute weights property (7.55), its value for $\varepsilon = 0$ is

$$S(0) = \sum_{i=0}^{N-1} \lambda_i(0) L_0 = L_0 \sum_{i=0}^{N-1} \lambda_i(0) = L_0 = \frac{1-B}{\delta B}.$$

The ε-differential of the swap rate in $\varepsilon = 0$ follows from (7.78) as

$$\partial_\varepsilon S(0) = \sum_{i=0}^{N-1} \partial_\varepsilon\left[\lambda_i L_i\right](0)$$

$$= -\frac{c}{2}\frac{B}{G_0^N} L_0 \sum_{i=0}^{N-1} B^i\left[i(i+1) - B\frac{G_1^{(N+1)''}}{G_0^N}\right] + \frac{c}{G_0^N}\sum_{i=0}^{N-1} iB^i.$$

Then identities (7.69) and (7.71) make the first sum simply vanish:

$$\sum_{i=0}^{N-1} B^i \left[i(i+1) - B\frac{G_1^{(N+1)''}}{G_0^N} \right] = BG_1^{(N+1)''} - BG_1^{(N+1)''} = 0$$

For the second term, we use (7.70) so that the swap rate initial differential comes as

$$\partial_\varepsilon S(0) = c\frac{B}{G_0^N} \sum_{i=0}^{N-1} iB^{i-1} = cB\frac{G_0^{N'}}{G_0^N}. \tag{7.80}$$

We turn now to the normalised weights, which are expressed as

$$\omega_i(\varepsilon) = \frac{\lambda_i L_i(\varepsilon)}{\sum_{k=0}^{N-1} \lambda_k(\varepsilon) L_k(\varepsilon)} = \frac{\lambda_i L_i(\varepsilon)}{S(\varepsilon)}$$

and for $\varepsilon = 0$ we see that the absolute and normalised weights are identical:

$$\omega_i(0) = \frac{\lambda_i(0) L_0}{L_0} = \lambda_i(0) = \frac{B^i}{G_0^N}. \tag{7.81}$$

Their generic differential is

$$\partial_\varepsilon \omega_i(\varepsilon) = \frac{1}{S^2(\varepsilon)} \left[S(\varepsilon)\partial_\varepsilon[\lambda_i L_i](\varepsilon) - \lambda_i L_i(\varepsilon)\partial_\varepsilon S(\varepsilon) \right].$$

Evaluating this expression at $\varepsilon = 0$ and substituting (7.76), (7.79) and (7.80) gives

$$\begin{aligned}
\partial_\varepsilon \omega_i(0) &= \frac{1}{S^2(0)} \left[S(0)\partial_\varepsilon[\lambda_i L_i](0) - \lambda_i(0) L_0 \partial_\varepsilon S(0) \right] \\
&= \frac{c}{L_0} \frac{B^i}{G_0^N} \left[\frac{1}{2} B L_0 \left[B\frac{G_1^{(N+1)''}}{G_0^N} - i(i+1) \right] + i - B\frac{G_0^{N'}}{G_0^N} \right].
\end{aligned} \tag{7.82}$$

Step 3: expanding the absolute and normalised aggregate weights

Let us turn to the cumulated weights, starting with the fixed-weight basket:

$$\Omega_i^A(\varepsilon) = \sum_{j=i}^{N-1} \lambda_j(\varepsilon)$$

which, taken with $\varepsilon = 0$ and invoking (7.69) gives

$$\Omega_i^A(0) = \sum_{j=i}^{N-1} \frac{B^i}{g_0^N} = \frac{G_i^N}{G_0^N}. \tag{7.83}$$

Using (7.77), the ε-differential of $\Omega_i^A(\varepsilon)$ with $\varepsilon = 0$ becomes

$$\partial_\varepsilon \Omega_i^A(0) = \sum_{j=i}^{N-1} \partial_\varepsilon \lambda_j(0) = -\frac{c}{2}\frac{B}{G_0^N} \sum_{j=i}^{N-1} B^j \left[j(j+1) - B\frac{G_1^{(N+1)''}}{G_0^N} \right]. \tag{7.84}$$

Then for the weights contribution we get

$$\Omega_i^\gamma(\varepsilon) = \sum_{j=i}^{N-1} \omega_j(\varepsilon).$$

Hence, invoking (7.81) we have the initial value of the sub-sums for free:

$$\Omega_i^\gamma(0) = \Omega_i^A(0). \tag{7.85}$$

Invoking (7.82), the ε-differential of Ω_i^γ with $\varepsilon = 0$ is

$$\begin{aligned} \partial_\varepsilon \Omega_i^\gamma(0) &= \frac{c}{L_0}\frac{1}{G_0^N} \sum_{j=i}^{N-1} B^j \left[\frac{1}{2} B L_0 \left[B\frac{G_1^{(N+1)''}}{G_0^N} - j(j+1) \right] + j - B\frac{G_0^{N'}}{G_0^N} \right] \\ &= \frac{cB}{L_0} \left[\frac{G_i^{N'}}{G_0^N} - \frac{G_i^N\, G_0^{N'}}{G_0^{N\,2}} \right] + \frac{cB}{2G_0^N} \sum_{j=i}^{N-1} B^j \left[B\frac{G_1^{(N+1)''}}{G_0^N} - j(j+1) \right]. \end{aligned} \tag{7.86}$$

Step 4: expanding the individual discrepancy

The discrepancy in weight for *each individual* Libor volatility is therefore

$$\left[\Omega_i^A - \Omega_i^\gamma\right](\varepsilon) = \left[\Omega_i^A - \Omega_i^\gamma\right](0) + \partial_\varepsilon \left[\Omega_i^A - \Omega_i^\gamma\right](0) + o(\varepsilon^2).$$

Invoking (7.83) and (7.85) we indeed have the initial discrepancy value as

$$\left[\Omega_i^A - \Omega_i^\gamma\right](0) = 0,$$

which was the starting point of the present first-order expansion. Looking now at the actual individual discrepancy of each Libor volatility, we write

$$\begin{aligned}\partial_\varepsilon \left[\left[\Omega_i^A - \Omega_i^\gamma\right] M_i\right](0) &= M_i(0)\partial_\varepsilon\left[\Omega_i^A - \Omega_i^\gamma\right](0) + \underbrace{\left[\Omega_i^A - \Omega_i^\gamma\right](0)\partial_\varepsilon M_i(0)}_{=0} \\ &= (1-B)\partial_\varepsilon\left[\Omega_i^A - \Omega_i^\gamma\right](0).\end{aligned}$$

Furthermore, the *relative* discrepancy in weight for each Libor is by expansion

$$\frac{\left[\Omega_i^A - \Omega_i^\gamma\right] M_i}{\omega_i}(\varepsilon) = \frac{\Omega_i^A - \Omega_i^\gamma}{\omega_i}(0) + \partial_\varepsilon\left[\frac{\Omega_i^A - \Omega_i^\gamma}{\omega_i}\right](0) + o(\varepsilon^2)$$

and the ε-differential of the ratio comes as

$$\partial_\varepsilon\left[\frac{\Omega_i^A - \Omega_i^\gamma}{\omega_i}\right](\varepsilon) = \frac{\left[\partial_\varepsilon\Omega_i^A - \partial_\varepsilon\Omega_i^\gamma\right]\omega_i - \left[\Omega_i^A - \Omega_i^\gamma\right]\partial_\varepsilon\omega_i}{{\omega_i}^2}(\varepsilon).$$

Evaluating this expression at $\varepsilon = 0$ we get more simply

$$\partial_\varepsilon\left[\frac{\Omega_i^A - \Omega_i^\gamma}{\omega_i}\right](0) = \frac{\partial_\varepsilon\left[\Omega_i^A - \Omega_i^\gamma\right]}{\omega_i}(0). \tag{7.87}$$

Going back to the initial *absolute* discrepancy differential, we use (7.84) and (7.86) to get

$$\begin{aligned}\partial_\varepsilon\left[\Omega_i^A - \Omega_i^\gamma\right](0) = &-\frac{c}{2}\frac{B}{G_0^N}\sum_{j=i}^{N-1} B^j\left[j(j+1) - B\frac{G_1^{(N+1)''}}{G_0^N}\right] \\ &-\frac{c}{2}\frac{B}{G_0^N}\sum_{j=i}^{N-1} B^j\left[B\frac{G_1^{(N+1)''}}{G_0^N} - j(j+1)\right] \\ &+\frac{c}{L_0}\frac{B}{{G_0^N}^2}\left[G_i^N\, G_0^{N'} - G_0^N\, G_i^{N'}\right].\end{aligned}$$

The first two terms cancel out, and in the last term the bracket simplifies into, using (7.70):

$$\left[G_i^N\, G_0^{N'} - G_0^N\, G_i^{N'}\right] = \frac{1}{(B-1)^2}\left[NB^{N-1} + i(B^N - 1)B^{i-1} - NB^{N-1}B^i\right].$$

Going back to the discrepancy term and injecting (7.69) we get

$$\partial_\varepsilon \left[\Omega_i^A - \Omega_i^\gamma\right](0) = \frac{c}{L_0}\frac{1}{(B^N-1)^2}\left[NB^N + i(B^N-1)B^i - NB^N B^i\right].$$

Hence, finally, we present the absolute error as

$$\partial_\varepsilon \left[\left[\Omega_i^A - \Omega_i^\gamma\right]M_i\right](0) = \frac{c}{L_0}\frac{1-B}{1-B^N}\left[\frac{NB^N}{1-B^N} - iB^i - \frac{NB^N}{1-B^N}B^i\right],$$

which by taking $c \leftarrow L_0$ proves (7.67). As for the relative error, it follows from (7.81) and (7.87) that

$$\partial_\varepsilon \left[\frac{\Omega_i^A - \Omega_i^\gamma}{\omega_i}\right](0) = \frac{c}{L_0}\frac{1}{B^i}\left[\frac{NB^N}{1-B^N} - iB^i - \frac{NB^N}{1-B^N}B^i\right],$$

which, again with $c \leftarrow L_0$, proves (7.68) and concludes the proof. □

References

1. Brace, A., Gatarek, D., Musiela, M.: The market model of interest rates dynamics. Math. Financ. **7**(2), 127–155 (1997)
2. Andersen, L., Brotherton-Ratcliffe, R.: Extended libor market models with stochastic volatility. Technical report, Bank of America (2001)
3. Joshi, M., Rebonato, R.: A stochastic-volatility, displaced-diffusion extension of the libor market model. Working paper, Royal Bank of Scotland (2001)
4. Andersen, L., Andreasen, J.: Volatile volatilities. RISK Magazine (2002)
5. Piterbarg, V.V.: Stochastic volatility model with time-dependent skew. Appl. Math. Finance **12**, 147–185 (2005)
6. Glasserman, P., Kou, S.: The term structure of simple forward rates with jump risk. Working paper, Columbia Unversity, Columbia (1999)
7. Andersen, L., Andreasen, J.: Volatility skews and extension of the Libor market model. Appl. Math. Finance **7**, 1–32 (2000)
8. Hagan, P., Lesniewski, A.: LIBOR market model with SABR style stochastic volatility. Technical report, JP Morgan (2008)
9. Henry-Labordere, P.: Unifying the BGM and SABR Models: a short ride in hyperbolic geometry. Technical report, Societe Generale (2006)
10. Henry-Labordere, P.: Analysis, Geometry and Modeling in Finance—Advanced Methods in Option Pricing. CRC Financial Mathematics Series. Chapman & Hall, London (2008)
11. Rebonato, R., White, R.: Linking Caplets and Swaptions Prices in the LMM-SABR Model. Technical report, Imperial College London, Tanaka Business School (2007)
12. Rebonato, R.: A time-homogeneous, SABR-consistent extension of the LMM: calibration and numerical results. Technical report, Imperial College London, Tanaka Business School (2007)
13. Rebonato, R., McKay, K., White, R.: The SABR/LIBOR Market Model: Pricing, Calibration and Hedging for Complex Interest Rate Derivatives. Wiley, New York (2009)
14. Rebonato, R.: No-arbitrage dynamics for a tractable SABR term structure libor model. Technical report, Bloomberg (2007)

15. Nawalkha, S.K.: The LIBOR/SABR market models: a critical review. Technical report, University of Massachusets Amherst (2009)
16. Shiraya, K., Takahashi, A., Yamazaki, A.: Pricing swaptions under the libor market model of interest rates with local-stochastic volatility models. Technical report, Graduate School of Economics, the University of Tokyo (2010)
17. Jamshidian, F.: Libor and swap market models and measures. Finance Stochast. **1**, 293–330 (1997)
18. Galluccio, S., Hunter, C.: The co-initial swap market model. Econ. Notes **33**, 209–232 (2004)
19. Galluccio, S., Ly, J.-M., Scaillet, O.: Theory and calibration of swap market models. Math. Financ. **17**, 111–141 (2007)
20. Filipovic, D.: Consistency Problems for Heath-Jarrow-Morton Interest Rate Models. Lecture Notes in Mathematics. Springer, Berlin (2001)
21. Carmona, R., Tehranchi, M.: Interest rate models: an infinite-dimensional stochastic analysis perspective. Springer Finance (2006)
22. Cont, R.: Modeling term structure dynamics: an infinite dimensional approach. Technical report, Ecole Polytechnique (2004)
23. Glasserman, P., Zhao, X.: Arbitrage-free discretization of lognormal forward Libor and swap rate models. Finance stochast. **4**, 35–68 (2000)
24. Glasserman, P., Wang, H.: Discretization of deflated bond prices. Adv. Appl. Probab. **32**, 540–563 (2001)
25. Musiela, M., Rutkowski, M.: Martingale Methods in Financial Modelling, 2nd edn. Stochastic Modelling and Applied Probability. Springer, Berlin (2004)
26. Aspremont, A.D.: Interest rate model calibration and risk-management using semidefinite programming. Ph.D. thesis, Ecole Polytechnique (2003)
27. Fournie, E., Lebuchoux, J., Touzi, N.: Small noise expansion and importance sampling. Asymptot. Anal. **14**(4), 361–376 (1997)

Chapter 8
Conclusion

As the time now comes to summarise and assess this presentation of ACE, let us first recall our original mandate. Our intention was to establish an explicit and non-arbitrable connection between some of the SV model classes, which are capable of describing the joint dynamics of an underlying and of its associated European options. That connection could be approximate, provided that its precision was known and if possible controllable. We also demanded a generic treatment in terms of covered models, and were aiming for some practical, efficient algorithm. We now offer our views on which of these objectives have been attained, and on which still remain open subjects.

Starting with the down side, the correspondence presented herein concerns only a pair of SV classes: on one hand, the stochastic instantaneous volatility (SInsV) models and, on the other hand, the stochastic implied volatility (SImpV) framework. Also, this connection is more developed in one direction (from SInsV to SImpV, the direct problem) than in the other (the inverse problem). In fact, the SImpV framework has throughout been considered as a formal object, rather than as an actual modelling tool. Besides, this connection is also a *partial* link, in the sense that it connects subsets of SInsV chaos coefficients to groups of IATM differentials of the SImpV surfaces, both static and dynamic: the link is performed layer-wise.

Then on the upside, connecting the SInsV class *to* static and dynamic differentials of the smile is the most *productive* link between all *market models* (MM). First, because it suits well current applications, i.e. the sensitivity-based calibration and hedging algorithms within popular instances of the SInsV class. Also, because the latter plays a central role among the MM family, seemingly linking (through their asymptotics) the SImpV, SVarS and SLocV frameworks. Therefore the method employed for the SImpV class should be transferrable to the two remaining frameworks. Another positive outcome is the actual generic nature of the connection, both in model coverage and in differential order, meaning that—assuming holomorphic smile functionals—the direct link is complete. Furthermore, the (unforeseen) extensions, first to a multi-dimensional and then to a term structure setup, illustrate the versatility of the approach.

D. Nicolay, *Asymptotic Chaos Expansions in Finance*,
Springer Finance Lecture Notes, DOI 10.1007/978-1-4471-6506-4_8

Overall, we have extended the work of [1] and of its sequels in several directions, which we now cover in more detail, by categorising them artificially as either theoretical or applicative.

8.1 Summary of Achievements

Let us first review the most significant *theoretical derivations* performed as part of this study. First of all, we have *formalised* the approach of [1] (mainly in Chap. 2 and Sect. 3.1) by organising the ACE methodology around the Zero-Drift Condition (2.18) in a sliding context. That choice has enabled us to expose in a new light the structural links—both static and dynamic—existing between the SInsV and sliding SImpV classes. More specifically, we have established the connection between, on one hand, the chaos coefficients of the SInsV representation, and, on the other hand, the Immediate ATM differentials of the SImpV framework. We have shown that these elements can be organised in coherent groups called layers, along which the inductive derivation sequence of ACE is built, according to the ladder constraint. By making explicit the correspondence between these subsets of the SInsV and SImpV models, we have also shown the natural asymmetry between instantaneous and implied volatility (especially in the multi-dimensional case) which is due to the latter *aggregating* the dynamic information.

The same effort towards formalising our approach has contributed to decomposing the ACE procedure into a programmable algorithm, which does not require a genuine formal calculus engine. In particular, we have shown how the direct problem can be solved up to any order, which proves especially simple in the local volatility case, and illustrated this algorithm by manually computing two additional layers within a generic bi-dimensional model. We have thereby provided universal expressions for the main IATM differentials (or *descriptors*) of the smile, which so far were only available for a handful of SV models. With these expressions we have demonstrated the analytic power of ACE by measuring the impact of ubiquitous model features, such as vol of vol or mean-reversion, on market observables (typically the flattening of the IV surface).

Still in the single-underlying context, we have either discussed or investigated multiple avenues for extension and generalisation. In particular, we have described several methods for transferring ACE results to different baseline models, with special emphasis on Normal dynamics.

Then we have extended ACE to the multi-dimensional framework. Observing the inverse problem becoming ill-posed from the first layer up, we have identified some sufficient conditions for a partial recovery. By contrast, we have shown the limited additional complexity of the direct problem, for which the dimensionality proves cumbersome but not materially detrimental.

We have then taken a significant structural step, by extending the setup to a term-structure context, detailing in particular the associated numeraire and measure aspects. We have identified the origin (sliding representation, multi-dimensionality or term structure) and discussed the interpretation of each new, additive term. Indeed, the latter start appearing as early as the Zero Drift Condition, filtering through the differentiation and the asymptotics into the IATM expressions, and provide the main contrast between a term-by-term and a global approach.

Let us now mention some of the main *applicative subjects* covered within this book. We have first employed the fundamental and the first layer results on the concept of local volatility, seen both as a dynamic interpretation of the marginals (or of the smile, see [2]) and as a class of instantaneous volatility models. Incidentally, we feel that there is still much to learn from that notion, which is deceptively simple but very demonstrative. We have then compared the exact differentials provided by the ACE approach to the main published results, thereby identifying structural discrepancies in Gatheral's *most probable path* heuristic formula.

Afterwards we have moved on to combined local-stochastic volatility models, specifically the Extended Skew Market Model (ESMM), a container class covering most bi-dimensional SInsV models in production (such as SABR or Heston). We have provided and analysed the first layer results for this ESMM, demonstrating the structural, clean decoupling of local and stochastic volatility effects. In particular, and as the name suggests, we have generalised the *LV slope vs correlation* effect on the skew that was clearly exposed in [3] for the SABR model.

Turning to specific model classes to perform the computations at higher orders, we have then illustrated the generic second and third layers for the FL-SV and SABR classes. As part of these computations, we have demonstrated the use of induction, of symmetry and of other simplification techniques to expedite the derivation of chaos dynamics within most model classes. Then, for the CEV-SABR model, we have investigated the numerical aspects of whole-smile extrapolations, concluding that a stand-alone ACE approach does outperform Hagan's formula, but that it is also well suited to complement other methods.

We have applied the multi-dimensional framework to tensorial and stochastic weights baskets, hence providing the IATM differentials of the first layer for a wide range of more complex, multi-asset financial products. This instance has confirmed that the multi-dimensionality is in practice no issue for the direct problem, and that the basket structure allows us to define artificial and discrete probability structures, which help with interpretation.

Finally, in the interest rates environment we have provided the first layer formulae for three liquid smiles, corresponding to bond options, to caplets and to swaptions. We have done so under both a Stochastic Volatility (SV) Heath-Jarrow-Morton and a SV Libor Market Model framework, illustrating the interests and difficulties of the full-smile (*vs* term-by-term) approach.

8.2 Advantages of the Methodology

In this section our aim is to assess the approach itself, independently of our own efforts or merit in developing it. We consider ACE in its current form only, and in comparison with published alternatives.

Regarding approximate vanilla pricing within local and stochastic volatility models, the state of the art is mostly asymptotic (see the Introduction or Chap. 4). But from singular perturbations to Wiener chaos expansions, and as far as practical algorithms are concerned, these approaches are restricted to specific model classes and/or to a given expansion order.

By contrast, Asymptotic Chaos Expansions are as generic as can be expected: they view stochastic instantaneous volatility only as a formal, adapted process defined in a chaos (either scalar, tensorial or maturity-dependent) and can provide the smile expansions at any required order. This means that within a trading system, the static calibration of (regular) SInsV models onto European options can be integrated in an abstract fashion. Likewise, within a Monte-Carlo pricing architecture the same information can be used to design systematic control variates or to employ importance sampling (see [4]). As for hedging, fast Vega (through the internal model parameters and the Jacobian) or even pathwise Vega is made more practical and precise. Besides, although the methodology can be programmed in an imperative and functional way, it can *also* be implemented in a rule-based logic. In other words, combining ACE with symbolic calculus is not necessary, but brings its generic features to the next level.

In more practical terms, the choice of the order providing the optimal precision/complexity tradeoff depends on the model, the option specifics and the user's preferences. However, in that choice ACE offers another advantage: the alignment of the method's precision with the market's bid-ask spread. Indeed, the latter is usually an increasing function of liquidity, itself a decreasing function of time-to-maturity θ and (absolute) log-moneyness $|y|$.

This property is in contrast with the more traditional asymptotic methods, where the deviation from the base case usually concerns the *model* itself. Using thereafter an expansion on the SDE, on the backward/forward PDE or on the infinitesimal generator, their solutions tend to be *global* with regard to the smile. But in mathematical terms, all these asymptotic methods provide exact information *only* for $\varepsilon = 0$. Hence they differ in their embedded *extrapolation* method from ACE, which itself focuses exclusively on the IATM point. This distinction suggests a strong potential in combining various *global* approaches with ACE, hence exploiting their respective qualities. For instance, pre-processing a non-stationary model with the Markovian projection approach of [5] provides efficient parameters that can then be fed to ACE. Conversely, one can correct the IATM differentials of a whole-smile extrapolation using high-order, exact results.

Because of this versatility, ACE is well suited for the design and the performance analysis of new models. It can also be included in abstract fashion within generic trading systems, to help with calibration, with pricing and with hedging. Typically, it allows us to modify the functional form of a diffusion coefficient without requiring the derivation of a complete new proxy.

We consider its dynamic focus to be another main attraction of the ACE methodology, as it provides a proxy of the (underlying and smile) joint diffusion at no additional cost. Indeed, the structure of the ZDC allocates the same *status* to the dynamic coefficients as to the static ones. This property transpires through all the layers of IATM arbitrage constraints for the SImpV class (see Proposition 2.2 [p. 54], for instance) so that eventually the dynamic differentials come as part of the overall IATM information. In other words, the generic-order ACE methodology (see Sect. 3.1) offers information relative to the smile's drift and volatility, as a by-product of a computation which is traditionally focused on the shape. Also, this additional knowledge enables us to link more efficiently the various kind of SV models allowing autonomous movements of the smile (the SInsV, SImpV and SLocV classes).

By comparison, the other available approximations do not offer similar information within SInsV models (see p. 6). This is unfortunate, since backbone and skew behaviour (for instance) represent a major interest of that class. Although some of these methods could possibly be extended to provide dynamics' proxies, we are not aware of any publication in that spirit. Hence, by default we would have to introduce the SDEs driving their static approximations. In practical terms, this means that ACE yields a significant competitive advantage when the hedge strategy uses such vanilla options in a dynamic fashion. This feature is interesting for callable options and for some volatility derivatives, but also for *dynamic calibration* purposes or to assess relative value.

Another attractive feature of Asymptotic Chaos Expansions is that, as a methodology, they offer many degrees of freedom which are either non-existent or unpractical with other approaches. Let us review the most important of these controls.

As we have mentioned early on (see Sect. 2.1.2.3 [p. 29]), from a mathematical perspective the role of the moneyness (sliding) re-parametrisation is more cosmetic than fundamental. It can, however, contribute to a simplification of the ZDC (and therefore of the asymptotic expressions) and—from a modelling perspective—to a more stationary representation of the smile. It is therefore beneficial that switching the log-moneyness for another functional presents a priori no fundamental difficulty within ACE, as discussed in Sect. 3.3.1.1 [p. 148].[1]

Similarly, we have shown in Sects. 3.3.1.3 [p. 150] and 3.3.2 [p. 156] that transferring the existing framework and derivations to another baseline was not only possible, but in some cases surprisingly effortless (as with the Normal instance). Such a capability is especially important for applicative concerns, in particular for the quality of whole-smile extrapolations, which is dependent on the similarity between the baseline and target models.

Another control is offered by the type of output for the direct problem, which comes as a set of pure IATM differentials, as opposed to some global or whole-smile solution. Indeed, such localisation of the asymptotics leaves the user free to specify better variables for the expansion itself, on an ad hoc basis. For instance, as we have discussed in Sect. 4.1 [p. 212] and observed in Sect. 4.5 [p. 245], the choice of the space variable significantly conditions the precision and validity of the extreme strike regions.

By contrast, most other asymptotic approaches will be *hooked* on a given representation for the price functional (i.e. the baseline and its implied parameter) as well as for the time and space coordinates (including the option parameters). For

[1] See also [6] on this subject.

example, most methods based on the backward PDE will be heavily dependent on the representation of the state variables. Typically, changing from $\log(S_t)$ back to S_t for space modifies first the structure of the mathematical problem, then inevitably the properties of the solution and of a potential induction feature.

This type of flexibility issue seems to be more pregnant with the analytical than with the probabilistic approaches, following the typology of the Introduction. However, in that respect ACE presents the interesting feature of being as much SDE- as PDE-based. Indeed, the generic algorithm—as exposed in Sect. 3.1 [p. 118]—alternates constantly between differentiating the ZDC (i.e a stochastic PDE constraining the SImpV model) and applying Itô calculus to formal expressions (involving the chaos coefficients of the SInsV model). Such a combination is quite unique among asymptotic approaches, and seems to confer to ACE a large part of its versatility. Overall, this methodology does not only extend and diversify the practitioner's toolbox, it also introduces brand new capabilities for the modeller.

8.3 Limitations of the Methodology

Some of the shortfalls of the ACE approach stem from its very foundations. For instance, the quoted, sufficient regularity assumptions are not *constructive*: in practice it is difficult to prove a priori the existence of short-expiry limits for the smile's shape and dynamics. This also means that the model cannot be modified *a posteriori* in order to fulfil these criteria, although we have not yet met a case where such manipulation would prove necessary.

This is not to say that designing a model *specifically* to fail the ACE methodology is impossible. For instance, employing dynamics which are strongly non-stationary, explosive (see [7]) or more generally non-integrable, i.e. features that the approach *cannot* capture, would probably succeed in that enterprise.

Other limitations can be established in comparison to the state of the art, even according to our criteria. In our view, some of the best benchmarks would be [4, 8–10] or [11], against which ACE does not necessarily represent a *systematic* improvement. It is especially noticeable (but not surprising) that by limiting the modelling framework, especially by using a finite number of explicit state variables, it becomes possible to derive powerful methods. Some of these algorithms are not only extendable to any differential order, but they can also be (at least partially) *integral* in nature. This last property in particular appears interesting, in order to handle time- and space-inhomogeneous parametrisations. By contrast and due to its differential nature, the ACE approach can only manage some limited non-stationarity, such as time-dependent volatility, for instance. In particular, its performance is degraded when the SInsV diffusion incorporate non-parametric, irregular of time-dependent coefficients.

However, it is important to note that ACE does not *have* to be pitched against other approaches in the perspective of replacing them. Indeed, it can easily be used for corrective measures instead, as discussed in Sect. 4.5.3 [p. 269], for instance.

8.4 Extensions and Further Work

A natural follow-up would be to mature some of the topics edited from the final cut (see p. 16). Among these, the extreme strikes focus (see [12]) is certainly interesting from a practical perspective. We also anticipate powerful theoretical and applied results from the SLocV context (see [13, 14]), since most of the ACE methodology—developed for the SImpV framework—can be transferred to its sister class. It has already allowed us to derive the foundations of an asymptotic description for the stochastic (or *dynamic*) local volatility. But we think that the potential benefits of developing these results further are significant, for instance, for the analysis of local-stochastic SInsV models, or by exploiting the simplicity of the validity constraints.

Following on the regularity issue evoked above, there is clearly ample academic interest in determining some necessary and/or tighter sufficient regularity conditions. As mentioned in the Introduction, this study was geared at the onset towards exploring a wide envelope of attainable results, including their interpretation and their applicability, rather than towards determining a minimum assumption set. This new vein would clearly constitute a technical and difficult subject. However, in the process of exploring the boundaries of ACE, even finding some demonstrative counter-examples would be beneficial.

Another ambitious endeavour would be to try and derive an *integral* version of our (differential) method. In that respect, a good starting point would be the asymptotic methods of a probabilistic nature, such as those presented in the Introduction. However, this new tack would realistically involve a significant deviation from the existing ACE structure. Hence it would probably represent another difficult project, albeit with a significant payoff.

At a lesser complexity level, it would be worth experimenting with alternative parametrisations of the SInsV dynamics. One possible benefit would be to provide more compact results, especially in the multi-dimensional case. The same rationale is of course valid for the SImpV framework, since the free specification of the moneyness—for instance—has not been fully exploited.

References

1. Durrleman, V.: From limplied to spot volatilities. Ph.D. thesis, Princeton University (2003)
2. Dupire, B.: Pricing with a smile. Risk **7**(1), 18–20 (1994)
3. Hagan, P.S., Kumar, D., Lesniewski, A.S., Woodward. D.E.: Managing smile risk. Wilmott, pp. 84–108 (2002)
4. Fournie, E., Lebuchoux, J., Touzi, N.: Small noise expansion and importance sampling. Asymptot. Anal. **14**(4), 361–376 (1997)
5. Piterbarg, V.V.: Markovian projection for volatility calibration. Risk Magazine, vol. 20, 84–89 (2007)
6. Hafner, R.: Stochastic implied volatility. Lecture Notes in Economics and Mathematical Systems, vol. 545. Springer, Berlin (2004)
7. Andersen, L.B.G., Piterbarg, V.V.: Moment explosions in stochastic volatility models. Finance Stochast. **11**(9), 29–50 (2006)

8. Kunitomo, N., Takahashi, A.: Applications of the asymptotic expansion approach based on Malliavin-Watanabe calculus in financial problems. University of Tokyo, Graduate School of Mathematical Sciences, Report (2003)
9. Medvedev, A.: Asymptotic methods for computing implied volatilities under stochastic volatility. Technical report, National Center of Competence in Research (2004)
10. Osajima, Y.: General asymptotics of Wiener functionals and application to mathematical finance. University of Tokyo, Graduate school of mathematical sciences, Report (2007)
11. Benhamou, E., Gobet, E., Miri, M.: Expansion formulas for European options in a local volatility model. Forthcoming Int. J. Theor. Appl. Finance **13**, 603 (2010)
12. Lee, R.W.: The moment formula for implied volatility at extreme strikes. Math. Financ. **14**(3), 469–480 (2004)
13. Carmona, R., Nadtochiy, S.: An infinite dimensional stochastic analysis approach to local volatility models. Commun. Stoch. Anal **2**(1), 109–123 (2008)
14. Carmona, R., Nadtochiy, S.: Local volatility dynamic models. Finance Stochast. **13**(1), 1–48 (2009)

Appendix A
Itô-Kunita Formula

Theorem A.1 (Itô-Kunita formula) *Let $\overrightarrow{W}_t$ be an n-dimensional Wiener process.*

Let $X(t, \overrightarrow{\beta})$ be a parametric Itô process with values in $\mathbb{R}$, satisfying the SDE

$$dX(t, \overrightarrow{\beta}) = a(t, \overrightarrow{\beta})dt + \overrightarrow{\sigma}(t, \overrightarrow{\beta})^{\perp} d\overrightarrow{W}_t$$

and where X is twice differentiable w.r.t. its parameter $\overrightarrow{\beta}$, which lives in $\mathbb{R}^m$.

Let $\overrightarrow{\alpha}_t$ be an m-dimensional real-valued process, satisfying the SDE:

$$d\overrightarrow{\alpha_t} = \overrightarrow{b}_t\, dt + \overrightarrow{\overrightarrow{\nu}}_t \cdot d\overrightarrow{W}_t$$

with $\overrightarrow{\overrightarrow{\nu}}_t$ a volatility process with values in $\mathbb{R}^{m\times n}$.

Now let us provide a dynamic for the fixed parameter $\overrightarrow{\beta}$ by replacing it with the process $\overrightarrow{\alpha}_t$. Then we have the resulting new dynamics for the process X:

$$\begin{aligned} dX(t, \overrightarrow{\alpha}_t) =& a(t, \overrightarrow{\alpha}_t)\, dt + \overrightarrow{\sigma}(t, \overrightarrow{\alpha}_t)^{\perp} d\overrightarrow{W}_t \\ &+ \sum_{i=1}^{m} \frac{\partial X}{\partial \beta^i}(t, \overrightarrow{\alpha}_t)\, d\alpha_t^i + \frac{1}{2} \sum_{i,j=1}^{m} \frac{\partial^2 X}{\partial \beta^i\, \partial \beta^j}(t, \overrightarrow{\alpha}_t)\langle d\alpha_t^i \mid d\alpha_t^j\rangle \\ &+ \sum_{i=1}^{m} \langle \frac{\partial \overrightarrow{\sigma}}{\partial \beta^i} \cdot d\overrightarrow{W}_t \mid d\alpha_t^i\rangle \end{aligned}$$

with

$$\langle d\alpha_t^i \mid d\alpha_t^j\rangle = \sum_{k=1}^{n} \nu_t^{i,k}\ \nu_t^{j,k}\ dt$$

and

$$\langle \frac{\partial \overrightarrow{\sigma}}{\partial \beta^i} \cdot d\overrightarrow{W}_t \mid d\alpha_t^i = \left[\frac{\partial \overrightarrow{\sigma}}{\partial \beta^i}(t, \overrightarrow{\alpha}_t)^{\perp}\ \overrightarrow{\nu}_t^i\right] dt = \sum_{k=1}^{n} \frac{\partial \sigma^k}{\partial \beta^i}(t, \overrightarrow{\alpha}_t)\ \nu_t^{i,k}\ dt.$$

D. Nicolay, *Asymptotic Chaos Expansions in Finance,*
Springer Finance Lecture Notes, DOI 10.1007/978-1-4471-6506-4

Appendix B
Transition Formulae

In this section we make explicit some transfers between the sliding and absolute coordinates, for a series of relevant differentials. The functional under consideration is taken as generic and is denoted $X(t, S_t, K, T)$ in its absolute form, while $\widetilde{X}(t, y, \theta)$ indicates its sliding counterpart. As usual, we use compact argument notations with

$$(\circ) \triangleq (t, y, \theta) \qquad \text{and} \qquad (\propto) \triangleq (t, S_t, K, T)\,.$$

Then we have easily the following differentials w.r.t. the underlying S_t:

$$\frac{\partial X}{\partial S_t}(\propto) = -\frac{1}{S_t}\,\widetilde{X}'_y(\circ) \qquad \text{and} \qquad \frac{\partial^2 X}{\partial S_t^2}(\propto) = \frac{1}{S_t^2}\left[\widetilde{X}''_{yy} + \widetilde{X}'_y\right](\circ). \tag{B.1}$$

The maturity differential is trivial, since it comes simply as

$$\frac{\partial X}{\partial T}(\propto) = \widetilde{X}'_\theta(\circ).$$

Finally the first four differentials w.r.t. strike K are

$$\frac{\partial X}{\partial K}(\propto) = \frac{1}{K}\,\widetilde{X}'_y(\circ), \tag{B.2}$$
$$\frac{\partial^2 X}{\partial K^2}(\propto) = \frac{1}{K^2}\left[\widetilde{X}''_{yy} - \widetilde{X}'_y\right](\circ), \tag{B.3}$$
$$\frac{\partial^3 X}{\partial K^3}(\propto) = \frac{1}{K^3}\left[\widetilde{X}'''_{yyy} - 3\widetilde{X}''_{yy} + 2\widetilde{X}'_y\right](\circ),$$
$$\frac{\partial^4 X}{\partial K^4}(\propto) = \frac{1}{K^4}\left[\widetilde{X}^{4'}_{y^4} - 6\,\widetilde{X}'''_{yyy} + 11\,\widetilde{X}''_{yy} - 6\,\widetilde{X}'_y\right](\circ).$$

D. Nicolay, *Asymptotic Chaos Expansions in Finance*,
Springer Finance Lecture Notes, DOI 10.1007/978-1-4471-6506-4

Appendix C
Black and Bachelier Differentials

In this appendix we compute and gather a number of differential expressions for the normalised Black-Scholes and Bachelier (a.k.a. *Normal Black*) formulae. Most of these expressions are invoked in the derivation of the main Zero Drift Conditions.

Normalised Black-Scholes

Let us first recall the definition of the normalised BS functional:

$$C^{BS}(x,k,v) \triangleq x\,\mathcal{N}(d_1) - k\,\mathcal{N}(d_2) \qquad \text{with} \qquad d_{1/2}(x,k,v) = \frac{-y}{v} \pm \frac{1}{2}v,$$

$$\text{where} \qquad \mathcal{N}(z) = \int_{-\infty}^{z} \frac{1}{\sqrt{2\pi}}\, e^{-\frac{1}{2}s^2}\, ds \qquad \text{and} \qquad y(x,k) = \ln\left(\frac{k}{x}\right).$$

We note some useful properties of the Normal cumulative:

$$\mathcal{N}''(x) = -x\,\mathcal{N}'(x) \qquad x\,\mathcal{N}'(d_1) = k\,\mathcal{N}'(d_2) \qquad e^{-y}\mathcal{N}'(d_1) = \mathcal{N}'(d_2).$$

First we compute the first-order differentials for the elementary d_1 and d_2 terms:

$$\frac{\partial d_{1/2}}{\partial x} = \frac{1}{xv} \qquad \frac{\partial d_{1/2}}{\partial v} = \frac{y}{v^2} \pm \frac{1}{2} = \frac{-d_{\mathbf{2/1}}}{v} \qquad \frac{\partial d_{1/2}}{\partial k} = -\frac{1}{kv}.$$

We can then derive the three first-order differentials, including Delta and Vega:

$$C_v^{BS'} = x\,\mathcal{N}'(d1) = k\,\mathcal{N}'(d_2), \tag{C.1}$$

$$C_x^{BS'} = \mathcal{N}(d_1), \qquad C_k^{BS'} = -\mathcal{N}(d_2).$$

D. Nicolay, *Asymptotic Chaos Expansions in Finance*,
Springer Finance Lecture Notes, DOI 10.1007/978-1-4471-6506-4

We then get the Gamma, Volga, Vanna and other second-order differentials:

$$C_{vv}^{BS''} = k\; v\; \mathcal{N}'(d_2)\left[\frac{y^2}{v^4} - \frac{1}{4}\right], \tag{C.2}$$

$$B_{xx}^{BS''} = \frac{1}{xv}\;\mathcal{N}'(d_1), \qquad C_{xv}^{BS''} = \mathcal{N}'(d_1), \left[\frac{y}{v^2} + \frac{1}{2}\right],$$

$$C_{kk}^{BS''} = \frac{1}{kv}\;\mathcal{N}'(d_2), \qquad C_{kv}^{BS''} = \frac{d_1}{v}\;\mathcal{N}'(d_2).$$

Normalised Bachelier

Call prices in a Normal (i.e. Gaussian) model come with the *normalised Bachelier* functional as

$$B^n(z, v) = v\;\;\mathcal{G}\left(\frac{z}{v}\right) - z\left[1 - \mathcal{N}\left(\frac{z}{v}\right)\right] \tag{C.3}$$

where $\mathcal{G}(\cdot)$ and $\mathcal{N}(\cdot)$ denote, respectively, the standard Gaussian density and cumulative density functions. We provide the simple proof for consistency.

Proof The model dynamics and their solution are simply

$$dS_t = \sigma\; dW_t \qquad \text{and} \qquad S_T = S_t + \sigma\;\;(W_T - W_t) \overset{law}{\sim} N\left(S_t, \sigma\sqrt{T-t}\right).$$

To compute Call prices, we remark first that if $X \sim N(\mu, \sigma)$ is a Normal random variate, then its first two conditional moments are

$$\mathbb{P}(X \geq K) = \int_K^{+\infty} \mathcal{G}_{\mu,\sigma}(x)\; dx = \int_{\frac{K-\mu}{\sigma}}^{+\infty} \mathcal{G}_{0,1}(y)\; dy\; = 1 - \mathcal{N}\left(\frac{K-\mu}{\sigma}\right),$$

$$\begin{aligned}\mathbb{E}[X \mid X \geq K] &= \int_K^{+\infty} x\; \mathcal{G}_{\mu,\sigma}(x)\; dx = \int_{\frac{K-\mu}{\sigma}}^{+\infty} (\mu + \sigma y)\; \mathcal{G}_{0,1}(y)\; dy \\ &= \mu\left[1 - \mathcal{N}\left(\frac{K-\mu}{\sigma}\right)\right] + \sigma\; \mathcal{G}\left(\frac{K-\mu}{\sigma}\right).\end{aligned}$$

In consequence the Call price comes as

$$\begin{aligned} C(t, S_t, K, T) &= \mathbb{E}[X \mid X \geq K] - K\,\mathbb{P}(X \geq K) \\ &= \sigma\sqrt{T-t}\ \mathscr{G}\left(\frac{K - S_t}{\sigma\sqrt{T-t}}\right) - (K - S_t)\left[1 - \mathscr{N}\left(\frac{K - S_t}{\sigma\sqrt{T-t}}\right)\right], \end{aligned}$$

hence

$$C(t, S_t, K, T) = B^n\left(K - S_t,\ \sigma\sqrt{T-t}\right),$$

which concludes the proof. □

Therefore the volatility differentials are

$$B^{n\,\prime}_v = \mathscr{G}\left(\frac{z}{v}\right) \qquad \text{and} \qquad B^{n\,\prime\prime}_{vv} = \frac{z^2}{v^3}\,\mathscr{G}\left(\frac{z}{v}\right). \tag{C.4}$$

Similarly, we get the first-order space differential as

$$B^{n\,\prime}_z = \mathscr{N}\left(\frac{z}{v}\right) - 1, \tag{C.5}$$

so that the second order differentials are given by

$$B^{n\,\prime\prime}_{zz} = \frac{1}{v}\,\mathscr{G}\left(\frac{z}{v}\right) \qquad \text{and} \qquad B^{n\,\prime\prime}_{zv} = -\frac{z}{v^2}\,\mathscr{G}\left(\frac{z}{v}\right). \tag{C.6}$$

Appendix D
Linear Algebra Toolbox

In this appendix we define some notations and we gather several useful results, all pertaining to Itô calculus with tensor processes. We focus on some specific types of products (e.g. outer, entwined) and on their dynamics. These results, for instance those related to *modified Einstein* notation, contribute to simplifying our computations in the multi-dimensional framework.

Token Processes and Dynamics

The overhead array system (for instance $\overrightarrow{[\cdot]}$, $\overset{\Rightarrow}{[\cdot]}$, $\overset{\Rrightarrow}{[\cdot]}$ and, in the general case, $\overset{(n)}{\overrightarrow{[\cdot]}}$) denotes tensor processes of order 1, 2, 3 and higher. Their time argument is generally omitted, and a quantity in **bold** denotes a generic-order tensor. Whenever left unspecified, the dimensions are assumed to ensure compatibility for all operations. Throughout this appendix we will consider the generic (token) vectorial processes $\overrightarrow{X}_t$ and $\overrightarrow{Y}_t$. Their dynamics, for instance for the former, read as

$$d\overrightarrow{X}_t = \overrightarrow{D}_t^X \, dt + \overset{\Rightarrow}{\Sigma}_t^X \, d\overrightarrow{W}_t + [\cdots] \, d\overrightarrow{Z}_t, \tag{D.1}$$

where $\overrightarrow{W}_t$ and $\overrightarrow{Z}_t$ are, as usual, standard and orthogonal Wiener processes with unit covariance. When possible, we omit suffixes, including time t (e.g. $\overset{\Rightarrow}{\Sigma}_t^X$ becoming $\overset{\Rightarrow}{\Sigma}$) and limit ourselves to endogenous dynamics. Also, for any tensor (e.g. a matrix) we denote the Frobenius norm by

$$\|\overset{\Rightarrow}{B}\|^2 \triangleq \sum_{i=1}^{n} \sum_{j=1}^{m} b_{ij}^2.$$

D. Nicolay, *Asymptotic Chaos Expansions in Finance*,
Springer Finance Lecture Notes, DOI 10.1007/978-1-4471-6506-4

Brackets of Tensor Dynamics

We employ the following notations for these finite variation terms

$$\langle d\overrightarrow{X}^{\perp} |\, d\overrightarrow{Y}\rangle \triangleq \sum_i \langle dx_i, dy_i\rangle, \qquad \langle d\overrightarrow{X}\rangle \triangleq \langle d\overrightarrow{X}^{\perp} | d\overrightarrow{X}\rangle = \sum_i \langle dx_i\rangle \tag{D.2}$$

and

$$\langle d\overrightarrow{\overrightarrow{A}} |\, d\overrightarrow{X}\rangle \triangleq \overrightarrow{Y} \qquad \text{with} \qquad Y_i = \sum_j \langle da_{ij}, dx_i\rangle.$$

Setting endogenous dynamics for vectorial processes $\overrightarrow{X}$ and $\overrightarrow{Y}$ according to (D.1) we define a specific covariation with

$$\langle d\overrightarrow{X} | d\overrightarrow{Y}^{\perp}\rangle = \overrightarrow{\overrightarrow{A}}\, dt \qquad \text{with} \qquad a_{i,j} \triangleq \langle dx_i | dy_j\rangle = \overrightarrow{\Sigma}^{X}_{i,\cdot}{}^{\perp}\ \overrightarrow{\Sigma}^{Y}_{j,\cdot}\ dt,$$

so that this bracket is determined by

$$\overrightarrow{\overrightarrow{A}} = \overrightarrow{\overrightarrow{\Sigma}}^{X}\ \overrightarrow{\overrightarrow{\Sigma}}^{Y^{\perp}}.$$

The Outer Product $\otimes$ and Commutative Outer Product $\overleftrightarrow{\otimes}$

If some tensor **A** is of order N, while $\overrightarrow{X}$ and $\overrightarrow{Y}$ are two vectors of compatible dimensions, we define the tensor **B** of order $N+1$ classically as:

$$\mathbf{B} = \mathbf{A} \otimes \overrightarrow{X} \qquad \Longleftrightarrow \qquad B_{ijk} = A_{ij} X_k.$$

This *outer product* notably exhibits the following static property:

$$\mathbf{A}\overrightarrow{X}^{\perp}\overrightarrow{Y} = \left[\mathbf{A} \otimes \overrightarrow{X}\right]\overrightarrow{Y} = \left[\mathbf{A} \otimes \overrightarrow{Y}\right]\overrightarrow{X} = \overrightarrow{X}^{\perp}\overrightarrow{Y}\mathbf{A}. \tag{D.3}$$

In particular, if **A** is itself a vector and/or if $\overrightarrow{X}$ is identical to either $\overrightarrow{Y}$ or **A** then

$$\left[\overrightarrow{X}^{\perp}\overrightarrow{Y}\right]\overrightarrow{Z} = \left[\overrightarrow{Z} \otimes \overrightarrow{Y}\right]\overrightarrow{X} = \left[\overrightarrow{Z} \otimes \overrightarrow{X}\right]\overrightarrow{Y},$$

$$\left[\mathbf{A} \otimes \overrightarrow{X}\right]\overrightarrow{X} = \left\|\overrightarrow{X}\right\|^2 \mathbf{A}, \tag{D.4}$$

$$\left[\overrightarrow{X}^{\perp}\overrightarrow{Y}\right]\overrightarrow{X} = \overrightarrow{X}^{2\otimes}\,\overrightarrow{Y}. \tag{D.5}$$

We then introduce the notation for the *commutative outer product* (COP):

$$\mathbf{A} \overset{\leftrightarrow}{\otimes} \mathbf{B} \triangleq \mathbf{A} \otimes \mathbf{B} + \mathbf{B} \otimes \mathbf{A}. \tag{D.6}$$

This symmetrical operator will appear naturally within the dynamics of some outer products, themselves involved in the computation of multi-dimensional chaos dynamics for ACE. In other words, the COP helps keep our most complex expressions (reasonably) compact.

Entwined Outer Product and Modified Einstein Notation

To introduce both concepts at once, we start by assuming generic dynamics for two vectorial processes $\overrightarrow{X}$ and $\overrightarrow{Y}$ as per (D.1), and then write the dynamics of their outer product as

$$d[\overrightarrow{X} \otimes \overrightarrow{Y}] = d[\overrightarrow{X}\,\overrightarrow{Y}^{\perp}] = [d\overrightarrow{X}]\,\overrightarrow{Y}^{\perp} + \overrightarrow{X}\,[d\overrightarrow{Y}^{\perp}] + \langle d\overrightarrow{X} \mid d\overrightarrow{Y}^{\perp}\rangle,$$

so that

$$\begin{aligned} d[\overrightarrow{X} \otimes \overrightarrow{Y}] = {} & [\overrightarrow{D}^X dt + \overrightarrow{\overrightarrow{\Sigma}}^X d\overrightarrow{W}_t]\overrightarrow{Y}^{\perp} + \overrightarrow{X}[\overrightarrow{D}^{Y^{\perp}} dt + d\overrightarrow{W}_t^{\perp}\ \overrightarrow{\overrightarrow{\Sigma}}^{Y^{\perp}}] \\ & + \overrightarrow{\overrightarrow{\Sigma}}^X\,\overrightarrow{\overrightarrow{\Sigma}}^{Y^{\perp}} dt + \overrightarrow{\overrightarrow{[\cdot]}}\, d\overrightarrow{Z}_t. \end{aligned} \tag{D.7}$$

Simplifying the notation, the matrices $\overrightarrow{\overrightarrow{\Sigma}}\,\overrightarrow{W}\,\overrightarrow{Y}^{\perp}$ and $\overrightarrow{X}\,\overrightarrow{W}^{\perp}\overrightarrow{\overrightarrow{\Sigma}}^{\perp}$ are given by

$$\begin{aligned} [\overrightarrow{\overrightarrow{\Sigma}}\overrightarrow{W}\,\overrightarrow{Y}^{\perp}]_{ij} &= \sum_k \sigma_{ik}[\overrightarrow{W}\,\overrightarrow{Y}^{\perp}]_{kj} = \sum_k \sigma_{ik}\ y_j\ w_k \\ \text{i.e. } \overrightarrow{\overrightarrow{\Sigma}}\,\overrightarrow{W}\,\overrightarrow{Y}^{\perp} &= \overrightarrow{\overrightarrow{\overrightarrow{[\sigma_{ik}\ y_j]}}}\,\overrightarrow{W}, \\ [\overrightarrow{X}\,\overrightarrow{W}^{\perp}\overrightarrow{\overrightarrow{\Sigma}}^{\perp}]_{ij} &= \sum_k [\overrightarrow{X}\,\overrightarrow{W}^{\perp}]_{ik}[\overrightarrow{\overrightarrow{\Sigma}}^{\perp}]_{kj} = \sum_k \sigma_{jk}\ x_i\ w_k \\ \text{i.e. } \overrightarrow{X}\,\overrightarrow{W}^{\perp}\overrightarrow{\overrightarrow{\Sigma}}^{\perp} &= \left[\overrightarrow{X} \otimes \overrightarrow{\overrightarrow{\Sigma}}\right]\overrightarrow{W}. \end{aligned}$$

Finally, substituting these two expressions into (D.7) we obtain

$$\begin{aligned} d[\overrightarrow{X} \otimes \overrightarrow{Y}] = {} & \left[\overrightarrow{D}^X\overrightarrow{Y}^{\perp} + \overrightarrow{X}\ \overrightarrow{D}^{Y^{\perp}} + \overrightarrow{\overrightarrow{\Sigma}}^X\ \overrightarrow{\overrightarrow{\Sigma}}^{Y^{\perp}}\right] dt \\ & + \left[[\sigma_{ik}^X\ y_j] + \overrightarrow{X} \otimes \overrightarrow{\overrightarrow{\Sigma}}^Y\right] d\overrightarrow{W}_t + \overrightarrow{\overrightarrow{[\cdot]}} d\overrightarrow{Z}_t. \end{aligned} \tag{D.8}$$

The *modified* Einstein (ME) notation helps to make such computations easier and more compact. Indeed, these simple expressions allow us to define a product not only involving *several* tensors, but also defined along one or several *chosen* coordinates. This feature is most useful when such a product cannot be simply represented in a classical form (such as *via* inner or outer products, or even using transposition), and is then referred to as *entwined*.

To set writing conventions, let us consider a typical instance of ME notation:

$$\overrightarrow{\overrightarrow{[x_m \; a_{i\mathbf{k}} \; z_{\mathbf{k}}]}}.$$

In that expression, the bracket $[\cdots]$ serves only to delimit the tensor itself, and can be capped by an overhead array to indicate its order. The bold index(es), such as **k** in this example, are those which are summed within a specified range or set. Contrary to the *classical* Einstein notation, these indices are not necessarily repeated. Thus we also use the ME notation to compactify standard sums.

The allocation of the remaining coordinates then follows the relative alphabetical order: in this instance, index i goes to the first and m to the second order.

More involved but useful examples include the following static relationships:

$$\begin{aligned} \overset{\Rightarrow}{C}\,\vec{X}\,\vec{Y} &= \overset{(2)}{\overrightarrow{\left[c_{ij\mathbf{k}}\, x_{\mathbf{k}}\right]}}\,\vec{Y} = \overrightarrow{\left[c_{i\mathbf{jk}}\, x_{\mathbf{k}}\, y_{\mathbf{j}}\right]} = \overset{(2)}{\overrightarrow{\left[c_{i\mathbf{j}k}\, y_{\mathbf{j}}\right]}}\,\vec{X}, \\ \vec{X}^{\perp}\,\overset{\Rightarrow}{C}\,\vec{X} &= \left[x_{\mathbf{i}}\, x_{\mathbf{j}}\, \overrightarrow{c_{\mathbf{ij}k}}\right]. \end{aligned} \tag{D.9}$$

Using the ME notation we can easily compute the outer product's dynamics as

$$\begin{aligned} d\left[\vec{X}\otimes\vec{Y}\right] &= d\left[x_i\; y_j\right] = \left[y_j\; dx_i + x_i\; dy_j\right] + [\cdots]\,dt \\ &= \left[y_j\; \sigma^X_{i\mathbf{k}}\; dW_{\mathbf{k}} + x_i\; \sigma^Y_{j\mathbf{k}}\; dW_{\mathbf{k}}\right] + [\cdots]\,dt + [\cdots]\,d\vec{Z}_t, \end{aligned}$$

hence

$$d\left[\vec{X}\otimes\vec{Y}\right] = \left[\left[y_j\; \sigma^X_{ik}\right] + \vec{X}\otimes\overset{\Rightarrow}{\Sigma}{}^Y\right] d\vec{W}_t + [\cdots]\,dt + [\cdots]\,d\vec{Z}_t. \tag{D.10}$$

Also, let us mention a basic property of the entwined product. By definition, we have

$$\overset{(3)}{\vec{Z}} = \left[[\vec{X}]_n[\overset{\Rightarrow}{Y}]_{mp}\right] \overset{\Delta}{\Longleftrightarrow} z_{nmp} = x_n\; y_{mp},$$

hence $\overset{(3)}{\vec{Z}} = \left[[\vec{X}_1 + \vec{X}_2]_n[!\overset{\Rightarrow}{Y}_1 + \overset{\Rightarrow}{Y}_2]_{mp}\right] \Longleftrightarrow z_{nmp} = \left(x_{1,n} + x_{2,n}\right)\left(y_{1,mp} + y_{2.mp}\right),$

which shows that the entwined product is conveniently *distributive*:

$$\left[[\overrightarrow{X_1}+\overrightarrow{X_2}]_n[\overrightarrow{\overrightarrow{Y_1}}+\overrightarrow{\overrightarrow{Y_2}}]_{mp}\right]=\left[[\overrightarrow{X_1}]_n[\overrightarrow{\overrightarrow{Y_1}}]_{mp}\right]+\left[[\overrightarrow{X_1}]_n[\overrightarrow{\overrightarrow{Y_2}}]_{mp}\right]$$
$$+\left[[\overrightarrow{X_2}]_n[\overrightarrow{\overrightarrow{Y_1}}]_{mp}\right]+\left[[\overrightarrow{X_2}]_n[\overrightarrow{\overrightarrow{Y_2}}]_{mp}\right]. \quad \text{(D.11)}$$

Furthermore, let us offer a more compact but more limited alternative to the entwined product, with the dual-argument and component-wise operator $\otimes$, which allows us to define a matrix with

$$\overrightarrow{\overrightarrow{C}} \underset{j}{\otimes} \overrightarrow{X} \triangleq \overrightarrow{[\, x_{\mathbf{j}}\; c_{i\mathbf{j}k}]}.$$

Dynamics of a Scalar-Vectorial Product

Specifying the dynamics of a vectorial process $\overrightarrow{X}$ with (D.1), and of a scalar process S_t with

$$dS_t = \mu\, dt + \overrightarrow{\gamma}^{\perp} d\overrightarrow{W}_t$$

we then get for their product

$$\begin{aligned}
d\left[S_t\,\overrightarrow{X_t}\right] &= d\left[S_t X_{1,t},\; S_t X_{2,t}, \ldots, S_t X_{N,t}\right]^{\perp} \\
&= \overrightarrow{\begin{bmatrix} \vdots \\ X_{i,t}\, dS_t + S_t\, dX_{i,t} + \langle dS_t, dX_{i,t}\rangle \\ \vdots \end{bmatrix}} \\
&= \mu\,\overrightarrow{X_t}\,dt + [\overrightarrow{X_t}\otimes\overrightarrow{\gamma}\,]\,d\overrightarrow{W}_t + S_t\,d\overrightarrow{X_t} + [\overrightarrow{\overrightarrow{\Sigma}}^X\,\overrightarrow{\gamma}\,]\,dt. \quad \text{(D.12)}
\end{aligned}$$

Dynamics of Inner Products and Quadratic Forms

Let us first recall that dynamics and transposition are trivially exchangeable:

$$d[\overrightarrow{X}^{\perp}] = [d\overrightarrow{X}]^{\perp}.$$

Then combining (D.2) with endogenous dynamics of type (D.1),we express a vector's bracket dynamics as

$$\begin{aligned}
\langle d\overrightarrow{X}\rangle &= \sum_i \langle \sum_j \sigma_{ij}\, dW_t^j \mid \sum_k \sigma_{ik}\, dW_t^j\rangle = \sum_i\sum_j \sigma_{ij}^2\, dt \\
&= \|\overrightarrow{\overrightarrow{\Sigma}}\|^2\, dt. \quad \text{(D.13)}
\end{aligned}$$

Turning to the dynamics of the squared modulus, we then have

$$d\left[\|\overrightarrow{X}\|^2\right] = d\left[\sum_i x_i^2\right] = \sum_i [2\ x_i\ dx_i + \langle dx_i\rangle]$$
$$= 2\ \overrightarrow{X}^{\perp} d\overrightarrow{X} + \langle d\overrightarrow{X}\rangle, \tag{D.14}$$

which is generalised to any higher power by

$$d\left[\|\overrightarrow{X}\|^p\right] = d\left[\left(\|\overrightarrow{X}\|^2\right)^{\frac{p}{2}}\right] = \frac{p}{2}\ \|\overrightarrow{X}\|^{p-2}\ d\left[\|\overrightarrow{X}\|^2\right] + [\cdot]\, dt$$
$$= p\ \|\overrightarrow{X}\|^{p-2}\ \left[d\overrightarrow{X}^{\perp}\overrightarrow{X}\right] + [\cdot]\, dt. \tag{D.15}$$

We can now move on to a *quadratic* form, which we define by

$$Q_{\overrightarrow{\overrightarrow{A}}}(\overrightarrow{X}) \triangleq \overrightarrow{X}^{\perp}\overrightarrow{\overrightarrow{A}}\ \overrightarrow{X} = \overrightarrow{X}^{\perp}\ \overrightarrow{\overrightarrow{A}}^{\perp}\ \overrightarrow{X}. \tag{D.16}$$

Establishing first that

$$d[\overrightarrow{\overrightarrow{A}}\overrightarrow{X}] = \overrightarrow{Y} \qquad \text{with} \qquad Y_i = d\left[a_{i\mathbf{j}}x_{\mathbf{j}}\right] = \left[x_{\mathbf{j}}\ d\ a_{i\mathbf{j}} + a_{i\mathbf{j}}\ d\ x_{\mathbf{j}} + \langle d\ a_{i\mathbf{j}},\ d\ x_{\mathbf{j}}\rangle\right],$$

we can rewrite

$$d[\overrightarrow{\overrightarrow{A}}\overrightarrow{X}] = [d\ \overrightarrow{\overrightarrow{A}}]\overrightarrow{X} + \overrightarrow{\overrightarrow{A}}[d\ \overrightarrow{X}] + \langle d\overrightarrow{\overrightarrow{A}}|d\overrightarrow{X}\rangle.$$

We then get the dynamics of the quadratic process as

$$d\ Q_{\overrightarrow{\overrightarrow{A}}}(\overrightarrow{X}) = d[\overrightarrow{X}^{\perp}\overrightarrow{\overrightarrow{A}}\overrightarrow{X}] = [d\overrightarrow{X}]^{\perp}\overrightarrow{\overrightarrow{A}}\overrightarrow{X} + \overrightarrow{X}^{\perp}d[\overrightarrow{\overrightarrow{A}}\overrightarrow{X}] + \langle dX^{\perp}|\ d[\overrightarrow{\overrightarrow{A}}\overrightarrow{X}]\rangle$$

and finally

$$d\ Q_{\overrightarrow{\overrightarrow{A}}}(\overrightarrow{X}) = \overrightarrow{X}^{\perp}\overrightarrow{\overrightarrow{A}}^{\perp}\ d\overrightarrow{X} + \overrightarrow{X}^{\perp}\left[[d\overrightarrow{\overrightarrow{A}}]\overrightarrow{X} + \overrightarrow{\overrightarrow{A}}[d\overrightarrow{X}] + \langle d\overrightarrow{\overrightarrow{A}}, d\overrightarrow{X}\rangle\right]$$
$$+ \langle dX^{\perp}\ |\ [d\overrightarrow{\overrightarrow{A}}]\overrightarrow{X} + \overrightarrow{\overrightarrow{A}}[d\ \overrightarrow{X}]\rangle. \tag{D.17}$$

Appendix E
Computation of the 2nd and 3rd Layers

This appendix gathers the computations of the direct problem, for the further layers discussed in Sect. 3.2 [p. 135]. They follow the generic methodology established in Sect. 3.1 [p. 118] and, more specifically, the roadmap laid out in Fig. 3.2 [p. 138].

Preliminary Results

Dynamics of a Recurring Ratio

As detailed in Sect. 3.1, the procedure is very inductive and hence will benefit from several generic, utilitarian results. With our choice of parametrisation, we are constantly dealing with the dynamics of a specific type of ratio, at the denominator of which is found a power of the instantaneous volatility σ_t. The following lemma aims at providing these generic dynamics.

Lemma E.1 (Dynamics of a generic volatility-power scaling) *Let x_t be a scalar Itô process whose dynamics are formally described by:*

$$d\,x_t = c_1\,dt + c_2\,dW_t + c_3\,dZ_t.$$

Then we have, for $n \in \mathbb{N}^*$

$$d\left[\frac{x_t}{\sigma_t^n}\right] = \left[\frac{c_1}{\sigma_t^n} - \frac{n}{\sigma_t^{n+1}}\left(a_2c_2 + a_3c_3 + x_ta_1\right) + \frac{n(n+1)}{2\sigma_t^{n+2}}x_t\left(a_2^2 + a_3^2\right)\right]dt$$
$$+\left[\frac{c_2}{\sigma_t^n} - \frac{nx_ta_2}{\sigma_t^{n+1}}\right]dW_t + \left[\frac{c_3}{\sigma_t^n} - \frac{nx_ta_3}{\sigma_t^{n+1}}\right]dZ_t.$$

D. Nicolay, *Asymptotic Chaos Expansions in Finance*,
Springer Finance Lecture Notes, DOI 10.1007/978-1-4471-6506-4

Proof It suffices to apply Itô's Lemma to the function $f(x_t, \sigma_t)$ with $f(x, y) = \frac{x}{y^n}$, which gives

$$\partial_x f = \frac{1}{y^n}, \qquad \partial^2_{x^2} f = 0, \qquad \partial^2_{xy} f = -\frac{n}{y^{n+1}},$$
$$\partial^2_{y^2} f = n(n+1)\frac{x}{y^{n+2}}, \qquad \partial_y f = -n\frac{x}{y^{n+1}}. \qquad \square$$

Term-By-Term Differentiation of the ZDC

Other intermediate steps can be pre-computed, in particular we have seen in Sect. 3.1 that:

- For the first column in Fig. 3.2, all pure-strike IATM differentials $\widetilde{\Sigma}^{(k,0)}(t,0,0)$ are obtained by successive y-differentiations of the Immediate ZDC (2.28), before evaluating the resulting equation at the IATM point in $(t, 0, 0)$. In other words, we need only compute $\nabla^{(m,0)}F(t,0,0)$.
- For the other columns, we need to cross-differentiate the main ZDC, before evaluating it at $(t, 0, 0)$. The resulting IATM (m, p)-ZDC expression will involve $\nabla^{(m,p-1)}D(t,0,0)$, $\nabla^{(m,p)}E(t,0,0)$ and $\nabla^{(m,p+1)}F(t,0,0)$ on its right-hand side. For more details, refer to the ZDC (2.18) [p. 43] and to Sect. 3.1.

In order to speed up the coming computations we propose, therefore, to provide the corresponding differential expressions pertaining to the $E(t, y, \theta)$ and $F(t, y, \theta)$ terms.

Note that the *macro* terms constituting the ZDC, *i.e.* $D(t, y, \theta)$, $E(t, y, \theta)$ and $F(t, y, \theta)$ can lend themselves to a mathematical interpretation. In particular, $F(t, y, 0)$ gives us the Immediate ZDC (3.31), and we could probably extract more information from the differentials of all three terms. However, we see the internal computations involved in deriving these differentials (see terms A, B and C later on) as essentially low-level, with no apparent interpretation potential. Furthermore, they are naturally easier to carry out sequentially and at once, rather than on demand throughout the various proofs. In our view, these reasons justify their outsourcing into the current section.

Lemma E.2 (Relevant differentials of the $F(t, y, \theta)$ term) *We have in (t, y, θ) the pure strike differentials:*

$$\begin{aligned}\nabla^{(3,0)}F(t,y,\theta) &= 12\,\widetilde{\Sigma}\widetilde{\Sigma}'^{\,3}_y + 18\,\widetilde{\Sigma}^2\widetilde{\Sigma}'_y\widetilde{\Sigma}''_{yy} + 2\left[\widetilde{\Sigma}^3 + \sigma_t^2\widetilde{\Sigma}\right]\widetilde{\Sigma}'''_{yyy}\\ &\quad + 3\,\sigma_t\widetilde{\Sigma}''_{yy}\widetilde{\mathrm{v}} - 3\,\sigma_t\widetilde{\Sigma}\widetilde{\mathrm{v}}''_{yy} - 6\,\widetilde{\mathrm{v}}\widetilde{\mathrm{v}}'_y - 6\,\widetilde{n}\widetilde{n}'_y + y\,[\cdot]\,, \qquad \text{(E.1)}\end{aligned}$$

$$\nabla^{(4,0)} F(t, y, \theta) = 12\, \widetilde{\Sigma}'^{\,4}_{y} + 72\, \widetilde{\Sigma}\widetilde{\Sigma}'^{\,2}_{y}\, \widetilde{\Sigma}''_{yy} + 24\, \widetilde{\Sigma}^2 \widetilde{\Sigma}'_y \widetilde{\Sigma}'''_{yyy}$$
$$+ 3\left[6\widetilde{\Sigma}^2 - \sigma_t^2\right] \widetilde{\Sigma}''^{\,2}_{yy} + \left[2\widetilde{\Sigma}^3 + 3\, \sigma_t^2 \widetilde{\Sigma}\right] \widetilde{\Sigma}^{(4)}_{y^4} + 8\, \sigma_t \widetilde{\mathrm{v}} \widetilde{\Sigma}'''_{yyy}$$
$$+ 12\, \sigma_t \widetilde{\mathrm{v}}'_y \widetilde{\Sigma}''_{yy} - 4\, \sigma_t \widetilde{\mathrm{v}}'''_{yyy} \widetilde{\Sigma} - 12\, \widetilde{\mathrm{v}}'^{\,2}_{y} - 12\, \widetilde{\mathrm{v}}\widetilde{\mathrm{v}}''_{yy}$$
$$- 12\, \widetilde{n}'^{\,2}_{y} - 12 \widetilde{n}\widetilde{n}''_{yy} + y\,[\cdot]\,. \tag{E.2}$$

The only relevant maturity differential reads as

$$F^{(0,2)}(t, y, \theta) = \left[6\widetilde{\Sigma}^2 - \sigma_t^2\right] \widetilde{\Sigma}'^{\,2}_{\theta} + \left[2\widetilde{\Sigma}^2 - \sigma_t^2\right] \widetilde{\Sigma}\, \widetilde{\Sigma}''_{\theta\theta} + y\,[\cdot]\,. \tag{E.3}$$

Finally, in terms of cross-differentials, we have

$$\nabla^{(1,1)} F(t, y, \theta) = 6\widetilde{\Sigma}^2 \widetilde{\Sigma}'_{\theta} \widetilde{\Sigma}'_{y} + 2\widetilde{\Sigma}^3 \widetilde{\Sigma}''_{y\theta} - \sigma_t \left[\widetilde{\Sigma}'_{\theta}\widetilde{\mathrm{v}} + \widetilde{\Sigma}\widetilde{\mathrm{v}}'_{\theta}\right] + y\,[\cdot]\,, \tag{E.4}$$

$$\nabla^{(2,1)} F(t, y, \theta) = 12\, \widetilde{\Sigma}\widetilde{\Sigma}'_{\theta}\, \widetilde{\Sigma}'^{\,2}_{y} + 12\, \widetilde{\Sigma}^2 \widetilde{\Sigma}'_y \widetilde{\Sigma}''_{y\theta} + \left[6\, \widetilde{\Sigma}^2 + \sigma_t^2\right] \widetilde{\Sigma}'_{\theta} \widetilde{\Sigma}''_{yy}$$
$$- 2\, \sigma_t \widetilde{\Sigma}'_{\theta}\widetilde{\mathrm{v}}'_y + \left[2\, \widetilde{\Sigma}^3 + \sigma_t^2 \widetilde{\Sigma}\right] \widetilde{\Sigma}'''_{yy\theta} - 2\, \sigma_t \widetilde{\Sigma}\widetilde{\mathrm{v}}''_{y\theta}$$
$$- 2\widetilde{\mathrm{v}}\widetilde{\mathrm{v}}'_{\theta} - 2\, \widetilde{n}\widetilde{n}'_{\theta} + y\,[\cdot]\,. \tag{E.5}$$

Note that our ultimate concern is asymptotic, and in particular that we shall focus on $y = 0$. This explains why, for simplicity's sake, we have chosen not to present the blocks that have a factor of y in all the expressions of Lemma E.2.

Proof We decompose the F term defined by (2.21) into three components as per

$$F(t, y, \theta) = A(t, y, \theta) - y\; B(t, y, \theta) - \frac{1}{2}\, y^2\; C(t, y, \theta), \tag{E.6}$$

where

$$A(t, y, \theta) = \frac{1}{2}\widetilde{\Sigma}^4 - \frac{1}{2}\sigma_t^2 \widetilde{\Sigma}^2, \tag{E.7}$$
$$B(t, y, \theta) = \sigma_t \widetilde{\Sigma}\left[\widetilde{\mathrm{v}} - \sigma_t \widetilde{\Sigma}'_y\right] = \sigma_t \widetilde{\Sigma}\widetilde{\mathrm{v}} - \sigma_t^2 \widetilde{\Sigma}\widetilde{\Sigma}'_y, \tag{E.8}$$
$$C(t, y, \theta) = \left[\widetilde{\mathrm{v}} - \sigma_t \widetilde{\Sigma}'_y\right]^2 + \widetilde{n}^2 = \widetilde{\mathrm{v}}^2 + \sigma_t^2\, \widetilde{\Sigma}'^{\,2}_{y} - 2\sigma_t \widetilde{\mathrm{v}}\widetilde{\Sigma}'_y + \widetilde{n}^2. \tag{E.9}$$

Our roadmap is to compute the relevant differentials for the three components A, B and C (in that order), before aggregating them into F. We take and compute all expressions in the generic argument $(\circ) = (t, y, \theta)$, which will be omitted on the r.h.s. for simplicity.

Relevant Differentials of the $A(t, y, \theta)$ Component

Let us first concentrate on the pure-strike differentials, starting from Definition (E.7) and then differentiating sequentially w.r.t. y:

$$A'_y(t, y, \theta) = 2\widetilde{\Sigma}^3 \widetilde{\Sigma}'_y - \sigma_t^2 \widetilde{\Sigma} \widetilde{\Sigma}'_y, \tag{E.10}$$

$$A''_{yy}(t, y, \theta) = 6\widetilde{\Sigma}^2 \, \widetilde{\Sigma}'^{\,2}_y + 2\widetilde{\Sigma}^3 \widetilde{\Sigma}''_{yy} - \sigma_t^2 \left[\widetilde{\Sigma}'^{\,2}_y + \widetilde{\Sigma} \widetilde{\Sigma}''_{yy} \right], \tag{E.11}$$

$$A^{(3,0)}_{y^3}(t, y, \theta) = 12\widetilde{\Sigma} \widetilde{\Sigma}'^{\,3}_y + 18\widetilde{\Sigma}^2 \widetilde{\Sigma}'_y \widetilde{\Sigma}''_{yy} + 2\widetilde{\Sigma}^3 \widetilde{\Sigma}'''_{yyy} - \sigma_t^2 \left[3\widetilde{\Sigma}'_y \widetilde{\Sigma}''_{yy} + \widetilde{\Sigma} \widetilde{\Sigma}'''_{yyy} \right], \tag{E.12}$$

$$\begin{aligned} A^{(4,0)}(t, y, \theta) = {} & 12 \, \widetilde{\Sigma}'^{\,4}_y + 72 \, \widetilde{\Sigma} \, \widetilde{\Sigma}'^{\,2}_y \, \widetilde{\Sigma}''_{yy} + 18 \, \widetilde{\Sigma}^2 \, \widetilde{\Sigma}''^{\,2}_{yy} + 24\widetilde{\Sigma}^2 \widetilde{\Sigma}'_y \widetilde{\Sigma}'''_{yyy} \\ & + 2\widetilde{\Sigma}^3 \widetilde{\Sigma}^{(4)}_{y^4} - \sigma_t^2 \left[3 \, \widetilde{\Sigma}''^{\,2}_{yy} + 4 \, \widetilde{\Sigma}'_y \widetilde{\Sigma}'''_{yyy} + \widetilde{\Sigma} \widetilde{\Sigma}^{(4)}_{y^4} \right]. \end{aligned} \tag{E.13}$$

As for the only relevant maturity differential, we get

$$A^{(0,2)}(t, y, \theta) = \left[6\widetilde{\Sigma}^2 - \sigma_t^2 \right] \, \widetilde{\Sigma}'^{\,2}_\theta + \left[2\widetilde{\Sigma}^2 - \sigma_t^2 \right] \widetilde{\Sigma} \widetilde{\Sigma}''_{\theta\theta}. \tag{E.14}$$

Turning to cross-differentials, we obtain $A^{(1,1)}(t, y, \theta)$ and $A^{(2,1)}(\circ)$ by differentiating, respectively, (E.10) and (E.11) once w.r.t. θ:

$$A''_{y\theta}(t, y, \theta) = 6\widetilde{\Sigma}^2 \widetilde{\Sigma}'_\theta \widetilde{\Sigma}'_y + 2\widetilde{\Sigma}^3 \widetilde{\Sigma}''_{y\theta} - \sigma_t^2 \left[\widetilde{\Sigma}'_\theta \widetilde{\Sigma}'_y + \widetilde{\Sigma} \widetilde{\Sigma}''_{y\theta} \right], \tag{E.15}$$

$$\begin{aligned} A^{(2,1)}(t, y, \theta) = {} & 12\widetilde{\Sigma} \widetilde{\Sigma}'_\theta \, \widetilde{\Sigma}'^{\,2}_y + 12\widetilde{\Sigma}^2 \widetilde{\Sigma}'_y \widetilde{\Sigma}''_{y\theta} + 6\widetilde{\Sigma}^2 \widetilde{\Sigma}'_\theta \widetilde{\Sigma}''_{yy} + 2\widetilde{\Sigma}^3 \widetilde{\Sigma}'''_{yy\theta} \\ & - \sigma_t^2 \left[2\widetilde{\Sigma}'_y \widetilde{\Sigma}''_{y\theta} + \widetilde{\Sigma}'_\theta \widetilde{\Sigma}''_{yy} + \widetilde{\Sigma} \widetilde{\Sigma}'''_{yy\theta} \right]. \end{aligned} \tag{E.16}$$

Relevant Differentials of the $B(t, y, \theta)$ Component

Again, we start with pure strike differentials, in a sequential fashion starting from (E.8)

$$B'_y(t, y, \theta) = \sigma_t \left[\widetilde{\Sigma}'_y \widetilde{\mathrm{v}} + \widetilde{\Sigma} \widetilde{\mathrm{v}}'_y \right] - \sigma_t^2 \left[\widetilde{\Sigma}'^{\,2}_y + \widetilde{\Sigma} \widetilde{\Sigma}''_{yy} \right], \tag{E.17}$$

$$B''_{yy}(t, y, \theta) = \sigma_t \left[\widetilde{\Sigma}''_{yy} \widetilde{\mathrm{v}} + 2 \, \widetilde{\Sigma}'_y \widetilde{\mathrm{v}}'_y + \widetilde{\Sigma} \widetilde{\mathrm{v}}''_{yy} \right] - \sigma_t^2 \left[3\widetilde{\Sigma}'_y \widetilde{\Sigma}''_{yy} + \widetilde{\Sigma} \widetilde{\Sigma}'''_{yyy} \right], \tag{E.18}$$

$$\begin{aligned} B^{(3,0)}(t, y, \theta) = {} & \sigma_t \left[\widetilde{\Sigma}'''_{yyy} \widetilde{\mathrm{v}} + 3\widetilde{\Sigma}''_{yy} \widetilde{\mathrm{v}}'_y + 3\widetilde{\Sigma}'_y \widetilde{\mathrm{v}}''_{yy} + \widetilde{\Sigma} \widetilde{\mathrm{v}}'''_{yyy} \right] \\ & - \sigma_t^2 \left[3 \, \widetilde{\Sigma}''^{\,2}_{yy} + 4\widetilde{\Sigma}'_y \widetilde{\Sigma}'''_{yyy} + \widetilde{\Sigma} \widetilde{\Sigma}^{(4)}_{y^4} \right]. \end{aligned} \tag{E.19}$$

Turning now to maturity and cross-differentials, we obtain $B^{(0,1)}$ and $B^{(1,1)}$ respectively from (E.8) and (E.17) by differentiating w.r.t. θ:

$$B^{(0,1)}(t, y, \theta) = \sigma_t \left[\widetilde{\Sigma}'_\theta \widetilde{\mathrm{v}} + \widetilde{\Sigma}\widetilde{\mathrm{v}}'_\theta\right] - \sigma_t^2 \left[\widetilde{\Sigma}'_\theta \widetilde{\Sigma}'_y + \widetilde{\Sigma}\,\widetilde{\Sigma}''_{y\theta}\right], \tag{E.20}$$

$$\begin{aligned} B^{(1,1)}(t, y, \theta) = {} & \sigma_t \left[\widetilde{\Sigma}''_{y\theta} \widetilde{\mathrm{v}} + \widetilde{\Sigma}'_y \widetilde{\mathrm{v}}'_\theta + \widetilde{\Sigma}'_\theta \widetilde{\mathrm{v}}'_y + \widetilde{\Sigma}\widetilde{\mathrm{v}}''_{y\theta}\right] \\ & - \sigma_t^2 \left[2\, \widetilde{\Sigma}'_y \widetilde{\Sigma}''_{y\theta} + \widetilde{\Sigma}'_\theta \widetilde{\Sigma}''_{yy} + \widetilde{\Sigma}\,\widetilde{\Sigma}'''_{yy\theta}\right]. \end{aligned} \tag{E.21}$$

Relevant Differentials of the $C(t, y, \theta)$ Component

Again, starting from Definition (E.9) we differentiate sequentially w.r.t. y:

$$C'_y(t, y, \theta) = 2\, \widetilde{\mathrm{v}}\widetilde{\mathrm{v}}'_y + 2\, \sigma_t^2 \widetilde{\Sigma}'_y \widetilde{\Sigma}''_{yy} - 2\, \sigma_t \left[\widetilde{\mathrm{v}}'_y \widetilde{\Sigma}'_y + \widetilde{\mathrm{v}}\widetilde{\Sigma}''_{yy}\right] + 2\, \widetilde{n}\widetilde{n}'_y, \tag{E.22}$$

$$\begin{aligned} C^{(2,0)}(t, y, \theta) = {} & 2\left[\widetilde{\mathrm{v}}'^{\,2}_y + \widetilde{\mathrm{v}}\widetilde{\mathrm{v}}''_{yy}\right] + 2\, \sigma_t^2 \left[\widetilde{\Sigma}''^{\,2}_{yy} + \widetilde{\Sigma}'_y \widetilde{\Sigma}'''_{yyy}\right] \\ & - 2\, \sigma_t \left[\widetilde{\mathrm{v}}''_{yy} \widetilde{\Sigma}'_y + 2\widetilde{\mathrm{v}}'_y \widetilde{\Sigma}''_{yy} + \widetilde{\mathrm{v}}\widetilde{\Sigma}'''_{yyy}\right] + 2\left[\widetilde{n}'^{\,2}_y + \widetilde{n}\widetilde{n}''_{yy}\right]. \end{aligned} \tag{E.23}$$

Turning to cross-differentials, we get $C^{(0,1)}(\circ)$ by differentiating (E.9) into

$$C^{(0,1)}(t, y, \theta) = 2\widetilde{\mathrm{v}}\widetilde{\mathrm{v}}'_\theta + 2\, \sigma_t^2 \widetilde{\Sigma}'_y \widetilde{\Sigma}''_{y\theta} - 2\, \sigma_t \widetilde{\mathrm{v}}'_\theta \widetilde{\Sigma}'_y - 2\, \sigma_t \widetilde{\mathrm{v}}\widetilde{\Sigma}''_{y\theta} + 2\, \widetilde{n}\widetilde{n}'_\theta. \tag{E.24}$$

Aggregation into the Differentials of $F(t, y, \theta)$

Let us first establish a generic formula for the pure strike-differentials of F, taken at the generic point (t, y, θ). Recalling the elementary results (3.3) and (3.4) [p. 125] we obtain

$$\begin{aligned} \nabla^{(m,0)} F(t, y, \theta) = {} & \left[A^{(m,0)} - m\, B^{(m-1,0)} - \frac{1}{2}\, \mathbf{1}_{\{m \geq 2\}}\, m(m-1)\, C^{(m-2,0)}\right] \\ & - y\, \left[B^{(m,0)} + m\, C^{(m-1,0)}\right] - \frac{1}{2}\, y^2\, C^{(m,0)}. \end{aligned} \tag{E.25}$$

Let us first focus on pure-strike differentials. Applying (E.25) with $m = 3$ and $m = 4$ gives

$$F^{(3,0)}(t, y, \theta) = A^{(3,0)} - 3\, B^{(2,0)} - 3\, C^{(1,0)} + y\,[\cdot]\,,$$

$$F^{(4,0)}(t, y, \theta) = A^{(4,0)} - 4\, B^{(3,0)} - 6\, B^{(3,0)} + y\,[\cdots]\,.$$

Then substituting (E.12), (E.18), (E.22) into the first equation, and (E.13), (E.19), (E.23) into the second, after simplification we get, respectively, (E.1) and (E.2). As for the single, pure-maturity differential, from expression (E.6) we easily get

$$F^{(0,2)}(t, y, \theta) = A^{(0,2)}(t, y, \theta) - y\, B^{(0,2)}(t, y, \theta) - \frac{1}{2} y^2\, C^{(0,2)}(t, y, \theta), \quad \text{(E.26)}$$

where injecting (E.14) provides (E.3). Turning to cross-differentials, we apply (E.25) with $m = 1$ and $m = 2$ to get $F^{(1,1)}(t, y, \theta)$ and $F^{(2,1)}(t, y, \theta)$, then θ-differentiate once to obtain

$$\nabla^{(1,1)} F(t, y, \theta) = \left[A^{(1,1)}(\circ) - B^{(0,1)}(\circ)\right] + y\,[\cdot],$$

$$\nabla^{(2,1)} F(t, y, \theta) = \left[A^{(2,1)}(\circ) - 2B^{(1,1)}(\circ) - C^{(0,1)}(\circ)\right] + y\,[\cdot].$$

Substituting, respectively, (E.15)–(E.20) and (E.16), (E.21), (E.24), then simplifying gives us, respectively, (E.4) and (E.5), which concludes the proof. □

Let us now examine the $E(t, y, \theta)$ term, whose relevant differentials are given by:

Lemma E.3 (Relevant differentials of the $E(t, y, \theta)$ term) *At the generic point* $(\circ) \triangleq (t, y, \theta)$ *we have the following expressions for the relevant pure strike-differentials and the first maturity differential:*

$$E^{(1,0)}(t, y, \theta) = 3\widetilde{\Sigma}^2 \widetilde{\Sigma}'_y \left[\widetilde{\Sigma}'_\theta - \frac{1}{2}\sigma_t^2 \widetilde{\Sigma}''_{yy} + \sigma_t \widetilde{\mathrm{V}}'_y - \frac{1}{2}\sigma_t \widetilde{\mathrm{V}}\right]$$
$$+ \widetilde{\Sigma}^3 \left[\widetilde{\Sigma}''_{y\theta} - \frac{1}{2}\sigma_t^2 \widetilde{\Sigma}'''_{yyy} + \sigma_t \widetilde{\mathrm{V}}''_{yy} - \frac{1}{2}\sigma_t \widetilde{\mathrm{V}}'_y\right], \quad \text{(E.27)}$$

$$E^{(2,0)}(t, y, \theta) = \left[6\widetilde{\Sigma}\, \widetilde{\Sigma}'^{\,2}_y + 3\widetilde{\Sigma}^2 \widetilde{\Sigma}''_{yy}\right]\left[\widetilde{\Sigma}'_\theta - \frac{1}{2}\sigma_t^2 \widetilde{\Sigma}''_{yy} - \frac{1}{2}\sigma_t \widetilde{\mathrm{V}} + \sigma_t \widetilde{\mathrm{V}}'_y\right]$$
$$+ 6\widetilde{\Sigma}^2 \widetilde{\Sigma}'_y \left[\widetilde{\Sigma}''_{y\theta} - \frac{1}{2}\sigma_t^2 \widetilde{\Sigma}'''_{yyy} - \frac{1}{2}\sigma_t \widetilde{\mathrm{V}}'_y + \sigma_t \widetilde{\mathrm{V}}''_{yy}\right]$$
$$+ \widetilde{\Sigma}^3 \left[\widetilde{\Sigma}'''_{yy\theta} - \frac{1}{2}\sigma_t^2 \widetilde{\Sigma}^{(4')}_{y^4} - \frac{1}{2}\sigma_t \widetilde{\mathrm{V}}''_{yy} + \sigma_t \widetilde{\mathrm{V}}'''_{yyy}\right], \quad \text{(E.28)}$$

$$E^{(0,1)}(t, y, \theta) = 3\widetilde{\Sigma}^2 \widetilde{\Sigma}'_\theta \left[\widetilde{\Sigma}'_\theta - \frac{1}{2}\sigma_t^2 \widetilde{\Sigma}''_{yy} + \sigma_t \widetilde{\mathrm{V}}'_y - \frac{1}{2}\sigma_t \widetilde{\mathrm{V}}\right]$$
$$+ \widetilde{\Sigma}^3 \left[\widetilde{\Sigma}''_{\theta\theta} - \frac{1}{2}\sigma_t^2 \widetilde{\Sigma}'''_{yy\theta} + \sigma_t \widetilde{\mathrm{V}}''_{y\theta} - \frac{1}{2}\sigma_t \widetilde{\mathrm{V}}'_\theta\right]. \quad \text{(E.29)}$$

Proof From the definition (2.20) [p. 44] of $E(t, y, \theta)$ we get (E.27) by a first y-differentiation. Repeating the process gives us, still at the generic point $(\circ) \triangleq (t, y, \theta)$

$$E^{(2,0)}(t,y,\theta) = \left[6\widetilde{\Sigma}\,\widetilde{\Sigma}_y'^{\,2} + 3\widetilde{\Sigma}^2\widetilde{\Sigma}_{yy}''\right]\left[\widetilde{\Sigma}_\theta' - \frac{1}{2}\sigma_t^2\widetilde{\Sigma}_{yy}'' - \frac{1}{2}\sigma_t\widetilde{\nu} + \sigma_t\widetilde{\nu}_y'\right]$$
$$+ 3\widetilde{\Sigma}^2\widetilde{\Sigma}_y'\left[\widetilde{\Sigma}_{y\theta}'' - \frac{1}{2}\sigma_t^2\widetilde{\Sigma}_{yyy}''' - \frac{1}{2}\sigma_t\widetilde{\nu}_y' + \sigma_t\widetilde{\nu}_{yy}''\right]$$
$$+ 3\widetilde{\Sigma}^2\widetilde{\Sigma}_y'\left[\widetilde{\Sigma}_{y\theta}'' - \frac{1}{2}\sigma_t^2\widetilde{\Sigma}_{yyy}''' - \frac{1}{2}\sigma_t\widetilde{\nu}_y' + \sigma_t\widetilde{\nu}_{yy}''\right]$$
$$+ \widetilde{\Sigma}^3\left[\widetilde{\Sigma}_{yy\theta}''' - \frac{1}{2}\sigma_t^2\widetilde{\Sigma}_{y^4}^{(4')} - \frac{1}{2}\sigma_t\widetilde{\nu}_{yy}'' + \sigma_t\widetilde{\nu}_{yyy}''\right],$$

which after simplification provides (E.28). As for the maturity differential, a single derivation of (2.20) w.r.t. θ gives us (E.29). □

Computation of the Hyperskew

Here we are eventually interested in the pure strike differential $\widetilde{\Sigma}^{(3,0)}(t, 0, 0)$, so let us start with the associated dynamic coefficients.

Expressing $\widetilde{b}_y'(\star)$ *and* $\widetilde{n}_y'(\star)$

Since we have only expressed $\widetilde{\nu}_y'(\star)$ within Theorem 2.2, which focused on the $\widetilde{\Sigma}$-(2,0) group,let us derive the missing drift and exogenous coefficient. We have:

Proposition E.1 (Expressing $\widetilde{b}_y'(\star)$ and $\widetilde{n}_y'(\star)$ in the generic bi-dimensional case) *For a generic bi-dimensional model, the IATM dynamics involve*

$$\widetilde{b}_y'(t,0,0) = \frac{1}{\sigma_t}\left[\frac{1}{2}a_{21}\right] + \frac{1}{\sigma_t^2}\left[-\frac{1}{2}a_1a_2 - \frac{1}{2}a_2a_{22} - \frac{1}{2}a_3a_{23}\right]$$
$$+ \frac{1}{\sigma_t^3}\left[\frac{1}{2}a_2^3 + \frac{1}{2}a_2a_3^2\right], \tag{E.30}$$

$$\widetilde{n}_y'(t,0,0) = \frac{1}{\sigma_t}\left[\frac{1}{2}a_{23}\right] + \frac{1}{\sigma_t^2}\left[-\frac{1}{2}a_2a_3\right]. \tag{E.31}$$

Proof We have from (2.51):

$$\widetilde{\Sigma}'_y(t,0,0) = \frac{1}{2}\,\frac{a_2}{\sigma_t}.$$

Lemma E.1 with $x_t = a_{2,t}$ and $n = 1$ gives the dynamics of the IATM skew $d\,\widetilde{\Sigma}'_y(\star)$. The uniqueness of Itô's decomposition allows us to identify term by term to get

- for the finite variation term:

$$\widetilde{b}'_y(t,0,0) = \frac{1}{2}\left[\frac{a_{21}}{\sigma_t} - \frac{1}{\sigma_t^2}\left(a_2 a_{22} + a_3 a_{23} + a_2 a_1\right) + \frac{2}{2\sigma_t^3} a_2 \left(a_2^2 + a_3^2\right)\right],$$

- for the exogenous coefficient:

$$\widetilde{n}'_y(t,0,0) = \frac{a_{23}}{2\sigma_t} - \frac{a_2 a_3}{2\sigma_t^2},$$

which prove, respectively, (E.30) and (E.31). □

Expressing $\widetilde{b}''_{yy}(\star)$, $\widetilde{\nu}''_{yy}(\star)$ and $\widetilde{n}''_{yy}(\star)$

We can now initiate the induction proper, starting with the full $\nabla^{(2,0)}$ dynamics:

Proposition E.2 (Expressing $\widetilde{b}''_{yy}(\star)$, $\widetilde{\nu}''_{yy}(\star)$ and $\widetilde{n}''_{yy}(\star)$ in the bi-dimensional case) *For a generic bi-dimensional model we have*

$$\widetilde{b}''_{yy}(\star) = \frac{1}{\sigma_t^2}\left[\frac{1}{3}a_{221}\right] + \frac{1}{\sigma_t^3}\left[x_{-3,t}\right] + \frac{1}{\sigma_t^4}\left[x_{-4,t}\right] + \frac{1}{\sigma_t^5}\left[x_{-5,t}\right] \tag{E.32}$$

with

$$\begin{aligned} 6\,x_{-3,t} &\triangleq -4a_1 a_{22} - a_2\left[6a_{21} + 4a_{222}\right] \\ &\quad + a_3\left[4a_{31} - 4a_{223}\right] - 3a_{22}^2 - 3a_{23}^2 + 2a_{32}^2 + 2a_{33}^2 \\ 2\,x_{-4,t} &\triangleq a_1\left[3a_2^2 - 2a_3^2\right] + 8a_2^2 a_{22} + a_2 a_3\left[6a_{23} - 4a_{32}\right] \\ &\quad + a_3^2\left[2a_{22} - 4a_{33}\right] \\ x_{-5,t} &\triangleq -3a_2^4 - a_2^2 a_3^2 + 2a_3^4 \end{aligned}$$

for the drift and

$$\widetilde{\nu}''_{yy}(\star) = \frac{1}{\sigma_t^2}\left[\frac{1}{3}a_{222}\right] + \frac{1}{\sigma_t^3}\left[\frac{2}{3}a_3a_{32} - \frac{5}{3}a_2a_{22}\right] + \frac{1}{\sigma_t^4}\left[\frac{3}{2}a_2^3 - a_2a_3^2\right] \quad \text{(E.33)}$$

$$\widetilde{n}''_{yy}(\star) = \frac{a_{223}}{3\sigma_t^2} + \frac{1}{\sigma_t^3}\left[\frac{2}{3}a_3\left(a_{33} - a_{22}\right) - a_2a_{23}\right] + \frac{a_3}{\sigma_t^4}\left[\frac{3}{2}a_2^2 - a_3^2\right] \quad \text{(E.34)}$$

for the endogenous and exogenous coefficients.

Proof From (2.52) we have

$$\widetilde{\Sigma}''_{yy}(t,0,0) = \frac{1}{\sigma_t^2}\left[\frac{1}{3}a_{22}\right] + \frac{1}{\sigma_t^3}\left[-\frac{1}{2}a_2^2 + \frac{1}{3}a_3^2\right].$$

Let us compute the dynamics of this expression, by applying Lemma E.1 to each term. For the first term, with $x_t = a_{22}$ and $n = 2$ we get

$$d\left[\frac{a_{22}}{\sigma_t^2}\right] = \left[\frac{a_{221}}{\sigma_t^2} - \frac{2}{\sigma_t^3}\left(a_2a_{222} + a_3a_{223} + a_1a_{22}\right) + \frac{6}{2\sigma_t^4}a_{22}\left(a_2^2 + a_3^2\right)\right]dt$$
$$+\left[\frac{a_{222}}{\sigma_t^2} - \frac{2a_2a_{22}}{\sigma_t^3}\right]dW_t + \left[\frac{a_{223}}{\sigma_t^2} - \frac{2a_3a_{22}}{\sigma_t^3}\right]dZ_t. \quad \text{(E.35)}$$

For the second term, with $x_t = a_2^2$ and $n = 3$ we start by computing

$$d\,a_2^2 = \left[2a_2a_{21} + a_{22}^2 + a_{23}^2\right]dt + 2a_2a_{22}dW_t + 2a_2a_{23}dZ_t.$$

Then applying Lemma E.1 we get

$$d\left[\frac{a_2^2}{\sigma_t^3}\right] = \left[\frac{2a_2a_{22}}{\sigma_t^3} - \frac{3a_2^3}{\sigma_t^4}\right]dW_t + \left[\frac{2a_2a_{23}}{\sigma_t^3} - \frac{3a_2^2a_3}{\sigma_t^4}\right]dZ_t$$
$$+\left[\frac{2a_2a_{21} + a_{22}^2 + a_{23}^2}{\sigma_t^3} - \frac{3}{\sigma_t^4}\left[2a_2^2a_{22} + 2a_2a_3a_{23} + a_1a_2^2\right]\right.$$
$$\left. + \frac{6}{\sigma_t^5}a_2^2\left(a_2^2 + a_3^2\right)\right]dt. \quad \text{(E.36)}$$

Finally for the third term, with $x_t = a_3^2$ and $n = 3$ we first express

$$d\,a_3^2 = \left[2a_3a_{31} + a_{32}^2 + a_{33}^2\right]dt + 2a_3a_{32}dW_t + 2a_3a_{33}dZ_t,$$

so that Lemma E.1 gives us

$$d\left[\frac{a_3^2}{\sigma_t^3}\right] = \left[\frac{2a_3a_{32}}{\sigma_t^3} - \frac{3a_2a_3^2}{\sigma_t^4}\right]dW_t + \left[\frac{2a_3a_{33}}{\sigma_t^3} - \frac{3a_3^3}{\sigma_t^4}\right]dZ_t$$
$$+ \left[\frac{2a_3a_{31} + a_{32}^2 + a_{33}^2}{\sigma_t^3} - \frac{3}{\sigma_t^4}\left[2a_2a_3a_{32} + 2a_3^2a_{33} + a_1a_3^2\right]\right.$$
$$\left. + \frac{6}{\sigma_t^5}a_3^2\left(a_2^2 + a_3^2\right)\right]dt. \tag{E.37}$$

Gathering (E.35)–(E.37), by identification we get the three coefficients (drift, endogenous, exogenous) as (E.32)–(E.34), which concludes the proof. □

Expressing the $(3,0)$*-IZDC and the Hyperskew* $\widetilde{\Sigma}'''_{yyy}(\star)$

We are now equipped for our main result, the third pure-strike static differential:

Proposition E.3 (Expressing the hyperskew $\widetilde{\Sigma}'''_{yyy}(\star)$ in the bi-dim. case)
For a generic bi-dimensional model we have for $(\star) \triangleq (t,0,0)$

$$\widetilde{\Sigma}'''_{yyy}(\star) = \frac{1}{\sigma_t^3}\left[\frac{a_{222}}{4}\right] + \frac{1}{\sigma_t^4}\left[-\frac{3}{2}a_2a_{22} + \frac{3}{4}a_3a_{23} + \frac{1}{2}a_3a_{32}\right]$$
$$+ \frac{1}{\sigma_t^5}\left[\frac{3}{2}a_2^3 - \frac{5}{2}a_2a_3^2\right]. \tag{E.38}$$

Proof Applying (E.1) at the IATM point $(t,0,0)$ we get the $(3,0)$-IZDC:

$$0 = F^{(3,0)}(t,0,0) = 12\,\sigma_t\,\widetilde{\Sigma}'^{\,3}_y + 18\,\sigma_t^2\,\widetilde{\Sigma}'_y\widetilde{\Sigma}''_{yy} + 4\,\sigma_t^3\,\widetilde{\Sigma}'''_{yyy}$$
$$- 3\,\sigma_t^2\,\widetilde{\nu}''_{yy} - 6\,\widetilde{\nu}\widetilde{\nu}'_y + 3\,\sigma_t\,\widetilde{\nu}\widetilde{\Sigma}''_{yy} - 6\,\widetilde{n}\widetilde{n}'_y. \tag{E.39}$$

Isolating $\widetilde{\Sigma}'''_{yyy}(\star)$ and replacing all other terms by their input-expressions gives

$$\widetilde{\Sigma}'''_{yyy}(\star) = -\frac{3}{\sigma_t^2}\left(\frac{a_2}{2\sigma_t}\right)^3 - \frac{9}{2\sigma_t}\left(\frac{a_2}{2\sigma_t}\right)\left[\frac{1}{\sigma_t^2}\left[\frac{a_{22}}{3}\right] + \frac{1}{\sigma_t^3}\left[\frac{a_3^2}{3} - \frac{a_2^2}{2}\right]\right]$$
$$+ \frac{3}{4\sigma_t}\left[\frac{a_{222}}{3\sigma_t^2} + \frac{1}{\sigma_t^3}\left[\frac{2}{3}a_3a_{32} - \frac{5}{3}a_2a_{22}\right] + \frac{a_2}{\sigma_t^4}\left[\frac{3}{2}a_2^2 - a_3^2\right]\right]$$

$$+\frac{3}{2\sigma_t^3}a_2\left[\frac{a_{22}}{2\sigma_t}-\frac{a_2^2}{2\sigma_t^2}\right]-\frac{3}{4\sigma_t^2}a_2\left[\frac{1}{\sigma_t^2}\left[\frac{a_{22}}{3}\right]+\frac{1}{\sigma_t^3}\left[\frac{a_3^2}{3}-\frac{a_2^2}{2}\right]\right]$$
$$+\frac{3}{2\sigma_t^3}a_3\left[\frac{a_{23}}{2\sigma_t}-\frac{a_2a_3}{2\sigma_t^2}\right].$$

Rearranging the terms by power of σ_t, then simplifying, proves (E.38) and concludes the proof. □

Computation of the Hypercurve

We are now interested in the next pure-strike differential $\widetilde{\Sigma}^{(4,0)}(t,0,0)$, and again we start by computing the differentials of the coefficients.

Expressing $\widetilde{b}'''_{yyy}(\star)$, $\widetilde{\nu}''_{yyy}(\star)$ *and* $\widetilde{n}'''_{yyy}(\star)$

Proposition E.4 (Expressing $\widetilde{b}'''_{yyy}(\star)$, $\widetilde{\nu}''_{yyy}(\star)$, $\widetilde{n}'''_{yyy}(\star)$ in the bi-dimensional case) *For a generic bi-dimensional model we have, for* $(\star)\triangleq(t,0,0)$

$$\widetilde{b}'''_{yyy}(\star)=\frac{1}{\sigma_t^3}\left[\frac{1}{4}a_{2221}\right]+\frac{x_{4,t}}{\sigma_t^4}+\frac{x_{5,t}}{\sigma_t^5}+\frac{x_{6,t}}{\sigma_t^6}+\frac{x_{7,t}}{\sigma_t^7} \tag{E.40}$$

with

$$\begin{aligned}
4\,x_{4,t}&\triangleq -3a_1a_{222}-a_2\left[6a_{221}+3a_{2222}\right]+a_3\left[2a_{321}+3a_{231}-3a_{2223}\right]\\
&\quad-6a_{22}\left[a_{21}+a_{222}\right]+a_{23}\left[3a_{31}-6a_{223}\right]\\
&\quad+a_{32}\left[2a_{31}+3a_{232}+2a_{322}\right]+a_{33}\left[3a_{233}+2a_{323}\right]\\
2\,x_{5,t}&\triangleq a_1a_2\left[12a_{22}\right]-a_1a_3\left[6a_{23}+4a_{32}\right]\\
&\quad+2a_2a_3\left[-5a_{31}+6a_{223}-3a_{232}-2a_{322}\right]\\
&\quad+a_2\left[21a_{22}^2+9a_{23}^2-6a_{23}a_{32}-9a_{32}^2-5a_{33}^2\right]+a_2^2\left[9a_{21}+15a_{222}\right]\\
&\quad+a_3^2\left[-5a_{21}+3a_{222}-6a_{233}-4a_{323}\right]\\
&\quad+2a_3\left[a_{22}\left(6a_{23}-5a_{32}\right)-a_{33}\left(8a_{23}+2a_{32}\right)\right]\\
\frac{2}{5}\,x_{6,t}&\triangleq a_1a_2\left[5a_3^2-3a_2^2\right]-15\,a_2^3a_{22}+6\,a_2^2a_3\left[2a_{32}-a_{23}\right]\\
&\quad+a_2a_3^2\left[10a_{33}-a_{22}\right]+2a_3^3\left[4a_{23}+a_{32}\right]\\
2\,x_{7,t}&\triangleq 15\,a_2\left[3\,a_2^4-2\,a_2^2a_3^2-5\,a_3^4\right]
\end{aligned}$$

for the drift coefficient,

$$
\begin{aligned}
\widetilde{\mathrm{v}}''_{yyy}(\star) &= \frac{1}{\sigma_t^3}\left[\frac{a_{2222}}{4}\right] \\
&+ \frac{1}{\sigma_t^4}\left[-\frac{9}{4}a_2a_{222} + \frac{1}{2}a_3a_{322} + \frac{3}{4}a_3a_{232} - \frac{3}{2}a_{22}^2 + \frac{3}{4}a_{23}a_{32} + \frac{1}{2}a_{32}^2\right] \\
&+ \frac{1}{\sigma_t^5}\left[-7a_2a_3a_{32} - 3a_2a_3a_{23} + \frac{21}{2}a_2^2a_{22} - \frac{5}{2}a_3^2a_{22}\right] \\
&+ \frac{1}{\sigma_t^6}\left[-\frac{15}{2}a_2^4 + \frac{25}{2}a_2^2a_3^2\right] \qquad \text{(E.41)}
\end{aligned}
$$

for the endogenous term, while the exogenous coefficient comes as

$$
\begin{aligned}
\widetilde{n}'''_{yyy}(\star) = \frac{1}{\sigma_t^3}\left[\frac{1}{4}a_{2223}\right] &+ \frac{1}{\sigma_t^6}\left[-\frac{15}{2}a_2^3a_3 + \frac{25}{2}a_2a_3^3\right] \\
+ \frac{1}{\sigma_t^4}\Bigg[&-\frac{3}{2}a_2a_{223} - \frac{3}{4}a_3a_{222} + \frac{3}{4}a_3a_{233} + \frac{1}{2}a_3a_{323} \\
&- \frac{3}{2}a_{22}a_{23} + \frac{3}{4}a_{23}a_{33} + \frac{1}{2}a_{32}a_{33}\Bigg] \\
+ \frac{1}{\sigma_t^5}\Bigg[&+\frac{9}{2}a_2^2a_{23} + 6a_2a_3a_{22} - 5a_2a_3a_{33} - \frac{11}{2}a_3^2a_{23} - 2a_3^2a_{32}\Bigg].
\end{aligned}
\qquad \text{(E.42)}
$$

Proof Let us compute the dynamics of input-expression (E.38) [p. 452] for $\widetilde{\Sigma}'''_{yyy}(\star)$.

Computing the Dynamics of the First Term of $\widetilde{\Sigma}'''_{yyy}(\star)$

Applying Lemma E.1 we get:

$$
\begin{aligned}
d\left[\frac{a_{222}}{4\sigma_t^3}\right] &= \frac{1}{4}\left[\frac{a_{2221}}{\sigma_t^3} - \frac{3}{\sigma_t^4}\left(a_2a_{2222} + a_3a_{2223} + a_1a_{222}\right) + \frac{12}{2\sigma_t^5}a_{222}\left(a_2^2 + a_3^2\right)\right]dt \\
&+ \frac{1}{4}\left[\frac{a_{2222}}{\sigma_t^3} - \frac{3a_2a_{222}}{\sigma_t^4}\right]dW_t + \frac{1}{4}\left[\frac{a_{2223}}{\sigma_t^3} - \frac{3a_3a_{222}}{\sigma_t^4}\right]dZ_t. \qquad \text{(E.43)}
\end{aligned}
$$

Computing the Dynamics of the Second Term of $\widetilde{\Sigma}'''_{yyy}(\star)$

We first express the dynamics of the bracket:

$$
d\left[a_3\left(\frac{1}{2}a_{32} + \frac{3}{4}a_{23}\right) - \frac{3}{2}a_2a_{22}\right] = x_{1,t}\,dt + x_{2,t}\,dW_t + x_{3,t}\,dZ_t
$$

with

$$x_{1,t} \triangleq \frac{1}{2}\left(a_{32}a_{31} + a_3 a_{321} + a_{32}a_{322} + a_{33}a_{323}\right)$$
$$+ \frac{3}{4}\left(a_{23}a_{31} + a_3 a_{231} + a_{32}a_{232} + a_{33}a_{233}\right)$$
$$- \frac{3}{2}\left(a_{22}a_{21} + a_2 a_{221} + a_{22}a_{222} + a_{23}a_{223}\right)$$
$$x_{2,t} \triangleq \frac{1}{2}\left(a_{32}^2 + a_3 a_{322}\right) + \frac{3}{4}\left(a_{23}a_{32} + a_3 a_{232}\right) - \frac{3}{2}\left(a_{22}^2 + a_2 a_{222}\right)$$
$$x_{3,t} \triangleq \frac{1}{2}\left(a_{32}a_{33} + a_3 a_{323}\right) + \frac{3}{4}\left(a_{23}a_{33} + a_3 a_{233}\right) - \frac{3}{2}\left(a_{22}a_{23} + a_2 a_{223}\right).$$

Then applying Lemma E.1 gives

$$d\left[\frac{1}{\sigma_t^4}\left[a_3\left(\frac{1}{2}a_{32} + \frac{3}{4}a_{23}\right) - \frac{3}{2}a_2 a_{22}\right]\right] = X_{1,t}\, dt + X_{2,t}\, dW_t + X_{3,t}\, dZ_t,$$

where

$$X_{1,t} \triangleq \frac{1}{\sigma_t^4}\, x_{1,t} - \frac{4}{\sigma_t^5}\, X_{1,5,t} + \frac{10}{\sigma_t^6}\, X_{1,6,t} \tag{E.44}$$

$$X_{1,5,t} \triangleq a_2\left[\frac{1}{2}\left(a_{32}^2 + a_3 a_{322}\right) + \frac{3}{4}\left(a_{23}a_{32} + a_3 a_{232}\right) - \frac{3}{2}\left(a_{22}^2 + a_2 a_{222}\right)\right]$$
$$+ a_3\left[\frac{1}{2}\left(a_{32}a_{33} + a_3 a_{323}\right) + \frac{3}{4}\left(a_{23}a_{33} + a_3 a_{233}\right) - \frac{3}{2}\left(a_{22}a_{23} + a_2 a_{223}\right)\right]$$
$$+ a_1\left[a_3\left(\frac{1}{2}a_{32} + \frac{3}{4}a_{23}\right) - \frac{3}{2}a_2 a_{22}\right]$$
$$X_{1,6,t} \triangleq \left[a_3\left(\frac{1}{2}a_{32} + \frac{3}{4}a_{23}\right) - \frac{3}{2}a_2 a_{22}\right]\left(a_2^2 + a_3^2\right)$$
$$X_{2,t} \triangleq \frac{1}{\sigma_t^4}\left[\frac{1}{2}\left(a_{32}^2 + a_3 a_{322}\right) + \frac{3}{4}\left(a_{23}a_{32} + a_3 a_{232}\right) - \frac{3}{2}\left(a_{22}^2 + a_2 a_{222}\right)\right]$$
$$- \frac{4a_2}{\sigma_t^5}\left[a_3\left(\frac{1}{2}a_{32} + \frac{3}{4}a_{23}\right) - \frac{3}{2}a_2 a_{22}\right]$$
$$X_{3,t} \triangleq \frac{1}{\sigma_t^4}\left[\frac{1}{2}\left(a_{32}a_{33} + a_3 a_{323}\right) + \frac{3}{4}\left(a_{23}a_{33} + a_3 a_{233}\right) - \frac{3}{2}\left(a_{22}a_{23} + a_2 a_{223}\right)\right]$$
$$- \frac{4a_3}{\sigma_t^5}\left[a_3\left(\frac{1}{2}a_{32} + \frac{3}{4}a_{23}\right) - \frac{3}{2}a_2 a_{22}\right].$$

Computing the Dynamics of the Third Term of $\widetilde{\Sigma}'''_{yyy}(\star)$

Similarly, the dynamics of the bracket come as

$$d\left[a_2\left(\frac{3}{2}a_2^2-\frac{5}{2}a_3^2\right)\right]=x_{1,t}dt+\left[\frac{3}{2}\left[3a_2^2a_{22}\right]-\frac{5}{2}\left[a_3^2a_{22}+2a_2a_3a_{32}\right]\right]dW_t$$
$$+\left[\frac{3}{2}\left[3a_2^2a_{23}\right]-\frac{5}{2}\left[a_3^2a_{23}+2a_2a_3a_{33}\right]\right]dZ_t,$$

with

$$x_{1,t}\triangleq\frac{3}{2}\left[3a_2^2a_{21}+3a_2\left(a_{22}^2+a_{23}^2\right)\right]$$
$$-\frac{5}{2}\left[a_3^2a_{21}+a_2\left(2a_3a_{31}+a_{32}^2+a_{33}^2\right)+2a_3a_{22}a_{32}+2a_3a_{23}a_{33}\right].$$

Applying Lemma E.1 we get the ratio's dynamics as

$$d\left[\frac{1}{\sigma_t^5}\left[a_2\left(\frac{3}{2}a_2^2-\frac{5}{2}a_3^2\right)\right]\right]=X_{1,t}\,dt$$
$$+\left[\frac{1}{\sigma_t^5}\left[\frac{3}{2}\left[3a_2^2a_{22}\right]-\frac{5}{2}\left[a_3^2a_{22}+2a_2a_3a_{32}\right]\right]-\frac{5a_2^2}{\sigma_t^6}\left[\frac{3}{2}a_2^2-\frac{5}{2}a_3^2\right]\right]dW_t$$
$$+\left[\frac{1}{\sigma_t^5}\left[\frac{3}{2}\left[3a_2^2a_{23}\right]-\frac{5}{2}\left[a_3^2a_{23}+2a_2a_3a_{33}\right]\right]-\frac{5a_2a_3}{\sigma_t^6}\left[\frac{3}{2}a_2^2-\frac{5}{2}a_3^2\right]\right]dZ_t, \tag{E.45}$$

with

$$X_{1,t}\triangleq\frac{1}{\sigma_t^5}\,X_{1,5,t}-\frac{5}{\sigma_t^6}\,X_{1,6,t}+\frac{15}{\sigma_t^7}\left[a_2\left(\frac{3}{2}a_2^2-\frac{5}{2}a_3^2\right)\right]\left(a_2^2+a_3^2\right)$$

$$X_{1,5,t}\triangleq\frac{3}{2}\left[3a_2^2a_{21}+3a_2\left(a_{22}^2+a_{23}^2\right)\right]$$
$$-\frac{5}{2}\left[a_3^2a_{21}+a_2\left(2a_3a_{31}+a_{32}^2+a_{33}^2\right)+2a_3a_{22}a_{32}+2a_3a_{23}a_{33}\right]$$

$$X_{1,6,t}\triangleq a_1\left[a_2\left(\frac{3}{2}a_2^2-\frac{5}{2}a_3^2\right)\right]+a_2\left[\frac{3}{2}\left[3a_2^2a_{22}\right]-\frac{5}{2}\left[a_3^2a_{22}+2a_2a_3a_{32}\right]\right]$$
$$+a_3\left[\frac{3}{2}\left[3a_2^2a_{23}\right]-\frac{5}{2}\left[a_3^2a_{23}+2a_2a_3a_{33}\right]\right].$$

Aggregation of the Three Terms

Combining (E.43)–(E.45), we can write the hyperskew's dynamics as

$$d\,\widetilde{\Sigma}'''_{yyy}(t,0,0) = \widetilde{b}'''_{yyy}(t,0,0)\,dt + \widetilde{\nu}''_{yyy}(t,0,0)\,dW_t + \widetilde{n}'''_{yyy}(t,0,0)\,dZ_t.$$

Identifying and simplifying provides (E.40)–(E.42), concluding the proof. □

Note that the expressions for $\widetilde{b}'''_{yyy}(\star)$ and $\widetilde{n}'''_{yyy}(\star)$ have been computed for the sake of completeness and will neither be used nor interpreted.

Expressing the (4, 0)*-IZDC and the Hypercurve* $\widetilde{\boldsymbol{\Sigma}}^{(4')}_{\mathbf{y}^4}(\star)$

We are now ready for our main static result.

Proposition E.5 (Expressing the hypercurve $\widetilde{\Sigma}^{(4')}_{y^4}(\star)$ in the bi-dim. case) *For a generic bi-dimensional stochastic instantaneous volatility model we have*

$$\begin{aligned}
&\widetilde{\Sigma}^{(4)}_{y^4}(\star)\\
&= \frac{1}{\sigma_t^4}\left[\frac{1}{5}a_{2222}\right] + \frac{1}{\sigma_t^5}\left[-2a_2a_{222} + \frac{4}{5}a_3a_{223} + \frac{3}{5}a_3a_{232} + \frac{2}{5}a_3a_{322}\right.\\
&\qquad\left. - \frac{4}{3}a_{22}^2 + \frac{3}{5}a_{23}^2 + \frac{3}{5}a_{23}a_{32} + \frac{2}{5}a_{32}^2\right]\\
&\quad + \frac{1}{\sigma_t^6}\left[+10a_2^2a_{22} - 9a_2a_3a_{23} - 6a_2a_3a_{32} - \frac{14}{3}a_3^2a_{22} + \frac{8}{5}a_3^2a_{33}\right]\\
&\quad + \frac{1}{\sigma_t^7}\left[-\frac{15}{2}a_2^4 + 22a_2^2a_3^2 - \frac{41}{15}a_3^4\right].
\end{aligned} \tag{E.46}$$

Proof Applying the preliminary result (E.2) at $(t,0,0)$ we get

$$\begin{aligned}
\nabla^{(4,0)}F(t,0,0) &= 12\,\widetilde{\Sigma}'^{\,4}_{y}(\star) + 72\,\sigma_t\,\widetilde{\Sigma}'^{\,2}_{y}\,\widetilde{\Sigma}''_{yy}(\star) + 15\,\sigma_t^2\,\widetilde{\Sigma}''^{\,2}_{yy}(\star)\\
&\quad + 24\,\sigma_t^2\,\widetilde{\Sigma}'_y\,\widetilde{\Sigma}'''_{yyy}(\star) + 5\,\sigma_t^3\,\widetilde{\Sigma}^{(4)}_{y^4}(\star) + 8\,\sigma_t\widetilde{\nu}\,\widetilde{\Sigma}'''_{yyy}(\star)\\
&\quad + 12\,\sigma_t\widetilde{\nu}'_y\,\widetilde{\Sigma}''_{yy}(\star) - 4\,\sigma_t^2\widetilde{\nu}''_{yyy}(\star) - 12\,\widetilde{\nu}'^{\,2}_{y}(\star)\\
&\quad - 12\,\widetilde{\nu\nu}''_{yy}(\star) - 12\,\widetilde{n}'^{\,2}_{y}(\star) - 12\,\widetilde{nn}''_{yy}(\star).
\end{aligned} \tag{E.47}$$

Furthermore, we have the (4, 0)-IZDC, which taken at $(t,0,0)$ reads as

$$\nabla^{(4,0)}F(t,0,0) = 0.$$

Combining both results, isolating $\widetilde{\Sigma}^{(4)}_{y^4}(t,0,0)$ and replacing all other quantities by their input-specification, we get

$$
\begin{aligned}
&\widetilde{\Sigma}^{(4)}_{y^4}(t,0,0)\\
&= -\frac{12}{5\sigma_t^3}\left[\frac{a_2}{2\sigma_t}\right]^4 - \frac{12}{5\sigma_t^2}\left[6\left[\frac{a_2}{2\sigma_t}\right]^2 + \frac{a_{22}}{2\sigma_t} - \frac{a_2^2}{2\sigma_t^2}\right]\left[\frac{1}{\sigma_t^2}\left[\frac{a_{22}}{3}\right] + \frac{1}{\sigma_t^3}\left[\frac{a_3^2}{3} - \frac{a_2^2}{2}\right]\right]\\
&\quad - \frac{3}{\sigma_t}\left[\frac{1}{\sigma_t^2}\left[\frac{a_{22}}{3}\right] + \frac{1}{\sigma_t^3}\left[\frac{a_3^2}{3} - \frac{a_2^2}{2}\right]\right]^2 - \left[\frac{24}{5\sigma_t}\left[\frac{a_2}{2\sigma_t}\right] + \frac{8}{5\sigma_t^2}a_2\right]\\
&\quad \times \left[\frac{1}{\sigma_t^3}\left[\frac{a_{222}}{4}\right] + \frac{1}{\sigma_t^4}\left[\frac{1}{2}a_3a_{32} + \frac{3}{4}a_3a_{23} - \frac{3}{2}a_2a_{22}\right]\right.\\
&\quad \left. + \frac{1}{\sigma_t^5}\left[\frac{3}{2}a_2^3 - \frac{5}{2}a_2a_3^2\right]\right] + \frac{4}{5\sigma_t}\left[\frac{1}{\sigma_t^3}\left[\frac{a_{2222}}{4}\right]\right.\\
&\quad + \frac{1}{\sigma_t^4}\left[-\frac{9}{4}a_2a_{222} + \frac{1}{2}a_3a_{322} + \frac{3}{4}a_3a_{232} - \frac{3}{2}a_{22}^2 + \frac{3}{4}a_{23}a_{32} + \frac{1}{2}a_{32}^2\right]\\
&\quad + \frac{1}{\sigma_t^5}\left[-7a_2a_3a_{32} - 3a_2a_3a_{23} + \frac{21}{2}a_2^2a_{22} - \frac{5}{2}a_3^2a_{22}\right]\\
&\quad \left. + \frac{1}{\sigma_t^6}\left[\frac{-15}{2}a_2^4 + \frac{25}{2}a_2^2a_3^2\right]\right] + \frac{12}{5\sigma_t^3}\left[\frac{a_{22}}{2\sigma_t} - \frac{a_2^2}{2\sigma_t^2}\right]^2\\
&\quad + \frac{12}{5\sigma_t^3}a_2\left[\frac{a_{222}}{3\sigma_t^2} + \frac{1}{\sigma_t^3}\left[\frac{2}{3}a_3a_{32} - \frac{5}{3}a_2a_{22}\right] + \frac{a_2}{\sigma_t^4}\left[\frac{3}{2}a_2^2 - a_3^2\right]\right] + \frac{12}{5\sigma_t^3}\left[\frac{a_{23}}{2\sigma_t} - \frac{a_2a_3}{2\sigma_t^2}\right]^2\\
&\quad + \frac{12}{5\sigma_t^3}a_3\left[\frac{1}{\sigma_t^2}\frac{a_{223}}{3} + \frac{1}{\sigma_t^3}\left[\frac{2}{3}a_3a_{33} - \frac{2}{3}a_3a_{22} - a_2a_{23}\right] + \frac{1}{\sigma_t^4}\left[\frac{3}{2}a_2^2a_3 - a_3^3\right]\right].
\end{aligned}
$$

Expanding this expression, grouping the terms according to the power of σ_t, and finally simplifying gives (E.46) and concludes the proof. □

Computation of the Twist

Here the target quantity is our first cross-differential $\widetilde{\Sigma}''_{y\theta}(t,0,0)$, and as always we must start with some differentials of the dynamic coefficients, only this time with respect to time-to-maturity θ.

Expressing $\widetilde{b}'_\theta(\star)$, $\widetilde{\nu}'_\theta(\star)$ and $\widetilde{n}'_\theta(\star)$

In this section we derive the three coefficients for further reference , but it is $\widetilde{\nu}'_\theta(t,0,0)$ that we require in the short term, in order to compute the target static differential (the twist).

Proposition E.6 (Expressing $\widetilde{b}'_\theta(\star)$, $\widetilde{\nu}'_\theta(\star)$ and $\widetilde{n}'_\theta(\star)$ in the bi-dimensional case) *For a generic bi-dimensional stochastic instantaneous volatility model we have*

$$
\begin{aligned}
&\widetilde{b}'_\theta(t,0,0)\\
&\quad= \sigma_t\left[\frac{a_{21}}{4}\right] + \left[\frac{1}{2}a_{11} - \frac{1}{6}a_{221} + \frac{1}{4}a_1a_2 + \frac{1}{4}a_2a_{22} + \frac{1}{4}a_3a_{23}\right]\\
&\quad + \frac{1}{\sigma_t}\left[\frac{1}{4}a_2a_{21} + \frac{1}{8}a_{22}^2 + \frac{1}{8}a_{23}^2 + \frac{1}{6}a_3a_{31} + \frac{1}{12}a_{32}^2 + \frac{1}{12}a_{33}^2\right]\\
&\quad + \frac{1}{\sigma_t^2}\left[-\frac{1}{4}a_2^2a_{22} - \frac{1}{4}a_2a_3a_{23} - \frac{1}{8}a_1a_2^2 - \frac{1}{6}a_2a_3a_{32} - \frac{1}{6}a_3^2a_{33} - \frac{1}{12}a_1a_3^2\right]\\
&\quad + \frac{1}{\sigma_t^3}\left[\frac{1}{8}a_2^4 + \frac{5}{24}a_2^2a_3^2 + \frac{1}{12}a_3^4\right]
\end{aligned}
\tag{E.48}
$$

for the drift,

$$
\begin{aligned}
\widetilde{\nu}'_\theta(t,0,0) &= \sigma_t\left[\frac{a_{22}}{4}\right] + \left[\frac{1}{2}a_{12} - \frac{1}{6}a_{222} + \frac{1}{4}a_2^2\right]\\
&\quad + \frac{1}{\sigma_t}\left[\frac{1}{4}a_2a_{22} + \frac{1}{6}a_3a_{32}\right] + \frac{1}{\sigma_t^2}\left[-\frac{1}{8}a_2^3 - \frac{1}{12}a_2a_3^2\right]
\end{aligned}
\tag{E.49}
$$

for the endogenous coefficient, and

$$
\begin{aligned}
\widetilde{n}'_\theta(t,0,0) &= \sigma_t\left[\frac{a_{23}}{4}\right] + \left[\frac{1}{2}a_{13} - \frac{1}{6}a_{223} + \frac{1}{4}a_2a_3\right]\\
&\quad + \frac{1}{\sigma_t}\left[\frac{1}{4}a_2a_{23} + \frac{1}{6}a_3a_{33}\right] + \frac{1}{\sigma_t^2}\left[-\frac{1}{8}a_2^2a_3 - \frac{1}{12}a_3^3\right]
\end{aligned}
$$

for the exogenous coefficient.

Proof Let us compute the dynamics $d\widetilde{\Sigma}'_\theta(\star)$ of the slope expression (2.53) [p. 62]: We split the dynamics into four sub-terms, where the two simplest come as

$$
d\left[\frac{a_1}{2} - \frac{a_{22}}{6}\right] = \left[\frac{a_{11}}{2} - \frac{a_{221}}{6}\right]dt + \left[\frac{a_{12}}{2} - \frac{a_{222}}{6}\right]dW_t + \left[\frac{a_{13}}{2} - \frac{a_{223}}{6}\right]dZ_t,
$$

$$
\begin{aligned}
d\left[\frac{1}{4}\sigma_t a_2\right] &= \frac{1}{4}\left[a_1a_2 + \sigma_t a_{21} + a_2a_{22} + a_3a_{23}\right]dt\\
&\quad + \frac{1}{4}\left[a_2^2 + \sigma_t a_{22}\right]dW_t + \frac{1}{4}\left[a_2a_3 + \sigma_t a_{23}\right]dZ_t.
\end{aligned}
$$

For the remaining two terms we apply Lemma E.1 to obtain

$$
\begin{aligned}
d\left[\frac{1}{8}\frac{a_2^2}{\sigma_t}\right] &= \frac{1}{8}\left[\frac{2a_2a_{22}}{\sigma_t} - \frac{a_2^3}{\sigma_t^2}\right]dW_t + \frac{1}{8}\left[\frac{2a_2a_{23}}{\sigma_t} - \frac{a_2^2a_3}{\sigma_t^2}\right]dZ_t \\
&\quad + \frac{1}{8}\left[\frac{1}{\sigma_t}\left[2a_2a_{21} + a_{22}^2 + a_{23}^2\right]\right. \\
&\qquad \left. - \frac{1}{\sigma_t^2}\left[2a_2^2a_{22} + 2a_2a_3a_{23} + a_1a_2^2\right] + \frac{a_2^2}{\sigma_t^3}\left[a_2^2 + a_3^2\right]\right]dt, \\
d\left[\frac{1}{12}\frac{a_3^2}{\sigma_t}\right] &= \frac{1}{12}\left[\frac{2a_3a_{32}}{\sigma_t} - \frac{a_2a_3^2}{\sigma_t^2}\right]dW_t + \frac{1}{12}\left[\frac{2a_3a_{33}}{\sigma_t} - \frac{a_3^3}{\sigma_t^2}\right]dZ_t \\
&\quad + \frac{1}{12}\left[\frac{1}{\sigma_t}\left[2a_3a_{31} + a_{32}^2 + a_{33}^2\right]\right. \\
&\qquad \left. - \frac{1}{\sigma_t^2}\left[2a_2a_3a_{32} + 2a_3^2a_{33} + a_1a_3^2\right] + \frac{a_3^2}{\sigma_t^3}\left[a_2^2 + a_3^2\right]\right]dt.
\end{aligned}
$$

Identifying the three coefficients ($\widetilde{b}'_\theta(\star)$, $\widetilde{\nu}'_\theta(\star)$ and $\widetilde{n}'_\theta(\star)$) then reorganising the terms according to the power of σ_t, gives us the desired results. □

Expressing the Twist $\widetilde{\boldsymbol{\Sigma}}''_{y\theta}(t, 0, 0)$

We can now move on to our main static result:

Proposition E.7 (Expression of the twist $\widetilde{\Sigma}''_{y\theta}(\star)$ in the bi-dimensional case) *For a generic bi-dimensional SInsV model the twist is given by*

$$
\begin{aligned}
\widetilde{\Sigma}''_{y\theta}(\star) &= \left[\frac{1}{6}a_{22}\right] + \frac{1}{\sigma_t}\left[-\frac{1}{24}a_2^2 + \frac{1}{6}a_{12} + \frac{1}{6}a_{21} - \frac{1}{8}a_{222}\right] \\
&\quad + \frac{1}{\sigma_t^2}\left[-\frac{1}{4}a_1a_2 + \frac{1}{4}a_2a_{22} - \frac{1}{24}a_3a_{23} - \frac{1}{12}a_3a_{32}\right] \\
&\quad + \frac{1}{\sigma_t^3}\left[-\frac{7}{48}a_2^3 + \frac{1}{24}a_2a_3^2\right].
\end{aligned}
\tag{E.50}
$$

Reassuringly, this result is coherent with [1] (refer to Appendix E of [2]).

Proof

Step 1: Differentiation of the ZDC

From (3.7) [p. 130] we get the (1, 0)-ZDC, taken at the IATM point $(\star) = (t, 0, 0)$:

$$3\widetilde{\Sigma}^2 \widetilde{\Sigma}'_y \widetilde{b}(\star) + \widetilde{\Sigma}^3 \widetilde{b}'_y(\star) = E^{(1,0)}(\star) + F^{(1,1)}(\star). \tag{E.51}$$

Step 2: Developing the right-hand side

Evaluating the pre-computed results (E.4) and (E.27) at $(t, 0, 0)$ we obtain

$$\begin{aligned} E^{(1,0)}(\star) + F^{(1,1)}(\star) = &-\frac{1}{2}\sigma_t^5 \widetilde{\Sigma}'''_{yyy}(\star) + \left[9\sigma_t^2 \widetilde{\Sigma}'_y(\star) - \sigma_t \widetilde{\mathrm{v}}(\star)\right] \widetilde{\Sigma}'_\theta + \sigma_t^4 \widetilde{\mathrm{v}}''_{yy}(\star) \\ &- \frac{3}{2}\sigma_t^4 \widetilde{\Sigma}'_y \widetilde{\Sigma}''_{yy}(\star) + \left[3\sigma_t^3 \widetilde{\Sigma}'_y(\star) - \frac{1}{2}\sigma_t^4\right] \widetilde{\mathrm{v}}'_y \\ &- \frac{3}{2}\sigma_t^3 \widetilde{\Sigma}'_y \widetilde{\mathrm{v}}(\star) - \sigma_t^2 \widetilde{\mathrm{v}}'_\theta(\star) + 3\sigma_t^3 \widetilde{\Sigma}''_{y\theta}(\star). \end{aligned}$$

Step 3: Input expression of the right-hand side

First we pre-compute the brackets on the r.h.s.

$$9\sigma_t^2 \widetilde{\Sigma}'_y(\star) - \sigma_t \widetilde{\mathrm{v}}(\star) = \sigma_t \left[\frac{7}{2}a_2\right] \qquad -\frac{3}{2}\sigma_t^4 \widetilde{\Sigma}'_y(\star) = \sigma_t^3 \left[-\frac{3}{4}a_2\right].$$

Then we replace the right-hand side by its input expressions, expand the resulting expression, group the terms according to the power of σ_t and finally simplify. Then the r.h.s. comes as

$$\begin{aligned} E^{(1,0)}(\star) + F^{(1,1)}(\star) = &\sigma_t^3 \left[-\frac{1}{2}a_{22}\right] + \sigma_t^2 \left[\frac{1}{8}a_2^2 - \frac{1}{2}a_{12} + \frac{3}{8}a_{222}\right] \\ &+ \sigma_t \left[\frac{7}{4}a_1 a_2 - \frac{5}{4}a_2 a_{22} - \frac{3}{8}a_3 a_{23} + \frac{3}{12}a_3 a_{32}\right] \\ &+ \left[\frac{15}{16}a_2^3 + \frac{3}{8}a_2 a_3^2\right] + 3\sigma_t^3 \widetilde{\Sigma}''_{y\theta}. \end{aligned}$$

Step 4: Input expression of the left-hand side

We replace the IATM differentials by their input expressions to get

$$\begin{aligned} 3\, \widetilde{\Sigma}^2 \widetilde{\Sigma}'_y \widetilde{b}(\star) + \widetilde{\Sigma}^3 \widetilde{b}'_y(\star) = &\sigma_t^2 \left[\frac{1}{2}a_{21}\right] + \sigma_t \left[a_1 a_2 - \frac{1}{2}a_2 a_{22} - \frac{1}{2}a_3 a_{23}\right] \\ &+ \left[\frac{1}{2}a_2^3 + \frac{1}{2}a_2 a_3^2\right]. \end{aligned}$$

Step 5: Solving for $\widetilde{\Sigma}''_{y\theta}(t,0,0)$
Putting left- and right-hand sides together, we get after simplification

$$\begin{aligned}
&3\sigma_t^3\,\widetilde{\Sigma}''_{y\theta}(\star)\\
&\quad=\sigma_t^3\left[\frac{1}{2}a_{22}\right]+\sigma_t^2\left[-\frac{1}{8}a_2^2+\frac{1}{2}a_{12}+\frac{1}{2}a_{21}-\frac{3}{8}a_{222}\right]\\
&\quad\quad+\sigma_t\left[-\frac{3}{4}a_1a_2+\frac{3}{4}a_2a_{22}-\frac{1}{8}a_3a_{23}-\frac{3}{12}a_3a_{32}\right]+\left[-\frac{7}{16}a_2^3+\frac{1}{8}a_2a_3^2\right],
\end{aligned}$$

which leads to (E.50) and concludes the proof. □

Computation of the Flattening

We are now focusing on the next cross-differential $\widetilde{\Sigma}'''_{yy\theta}(t,0,0)$.

Expressing $\widetilde{b}''_{y\theta}(\star)$, $\widetilde{\mathrm{v}}''_{y\theta}(\star)$ and $\widetilde{n}''_{y\theta}(\star)$

Proposition E.8 (Expressing $\widetilde{b}''_{y\theta}(\star)$, $\widetilde{\mathrm{v}}''_{y\theta}(\star)$ and $\widetilde{n}''_{y\theta}(\star)$ in the bi-dimensional case) *For a generic bi-dimensional SInsV model we have*

$$\widetilde{b}''_{y\theta}(\star)=\frac{1}{6}a_{221}+\frac{x_{1,t}}{\sigma_t}+\frac{x_{2,t}}{\sigma_t^2}+\frac{x_{3,t}}{\sigma_t^3}+\frac{x_{4,t}}{\sigma_t^4}+\frac{x_{5,t}}{\sigma_t^5} \tag{E.52}$$

with

$$x_{1,t}\triangleq-\frac{1}{12}a_2a_{21}-\frac{1}{24}a_{22}^2-\frac{1}{24}a_{23}^2+\frac{1}{6}a_{121}+\frac{1}{6}a_{211}-\frac{1}{8}a_{2221}$$

$$\begin{aligned}
x_{2,t}\triangleq\;&\frac{1}{24}a_1a_2^2-\frac{1}{6}a_1a_{12}-\frac{5}{12}a_1a_{21}+\frac{1}{8}a_1a_{222}+\frac{1}{12}a_2^2a_{22}-\frac{1}{4}a_2a_{11}-\frac{1}{6}a_2a_{122}\\
&-\frac{1}{6}a_2a_{212}+\frac{1}{4}a_2a_{221}+\frac{1}{8}a_2a_{2222}+\frac{1}{12}a_2a_3a_{23}-\frac{1}{6}a_3a_{123}-\frac{1}{6}a_3a_{213}\\
&-\frac{1}{24}a_3a_{231}-\frac{1}{12}a_3a_{321}+\frac{1}{8}a_3a_{2223}-\frac{1}{4}a_{12}a_{22}-\frac{1}{4}a_{13}a_{23}+\frac{1}{4}a_{22}a_{21}\\
&+\frac{1}{4}a_{22}a_{222}+\frac{1}{4}a_{23}a_{223}-\frac{1}{24}a_{23}a_{31}-\frac{1}{12}a_{32}a_{31}-\frac{1}{24}a_{32}a_{232}\\
&-\frac{1}{12}a_{32}a_{322}-\frac{1}{24}a_{33}a_{233}-\frac{1}{12}a_{33}a_{323}
\end{aligned}$$

$$
\begin{aligned}
x_{3,t} \triangleq\ & \frac{1}{2}a_1^2 a_2 + \frac{7}{12}a_1 a_3 a_{23} + \frac{1}{6}a_1 a_3 a_{32} - \frac{1}{24}a_2^4 - \frac{1}{24}a_2^2 a_3^2 + \frac{2}{3}a_2^2 a_{12} \\
& - \frac{13}{48}a_2^2 a_{21} - \frac{5}{8}a_2^2 a_{222} + \frac{1}{2}a_2 a_3 a_{13} + \frac{1}{12}a_2 a_3 a_{31} \\
& - \frac{1}{2}a_2 a_3 a_{223} + \frac{1}{12}a_2 a_3 a_{232} + \frac{1}{6}a_2 a_3 a_{322} - \frac{15}{16}a_2 a_{22}^2 \\
& - \frac{7}{16}a_2 a_{23}^2 + \frac{1}{12}a_2 a_{23} a_{32} + \frac{5}{24}a_2 a_{32}^2 + \frac{1}{24}a_2 a_{33}^2 + \frac{1}{6}a_3^2 a_{12} \\
& + \frac{5}{24}a_3^2 a_{21} - \frac{1}{8}a_3^2 a_{222} + \frac{1}{12}a_3^2 a_{233} + \frac{1}{6}a_3^2 a_{323} - \frac{1}{2}a_3 a_{22} a_{23} \\
& + \frac{1}{12}a_3 a_{22} a_{32} + \frac{1}{6}a_3 a_{23} a_{33} + \frac{1}{6}a_3 a_{32} a_{33}
\end{aligned}
$$

$$
\begin{aligned}
x_{4,t} \triangleq\ & \frac{7}{16}a_1 a_2^3 - \frac{3}{4}a_1 a_2 a_2^2 - \frac{7}{8}a_1 a_2 a_3^2 + \frac{33}{16}a_2^3 a_{22} \\
& + \frac{19}{16}a_2^2 a_3 a_{23} - \frac{1}{2}a_2^2 a_3 a_{32} + \frac{5}{8}a_2 a_3^2 a_{22} \\
& - \frac{1}{4}a_2 a_3^2 a_{33} - \frac{1}{4}a_3^3 a_{23} - \frac{1}{4}a_3^3 a_{32}
\end{aligned}
$$

$$
x_{5,t} \triangleq -\frac{7}{8}a_2^5 - \frac{5}{8}a_2^3 a_3^2 + \frac{1}{4}a_2 a_3^4
$$

for the drift coefficient, while

$$
\begin{aligned}
\widetilde{\nu}''_{y\theta}(\star) =\ & \frac{1}{6}a_{222} + \frac{1}{\sigma_t}\left[-\frac{1}{12}a_2 a_{22} + \frac{1}{6}a_{122} + \frac{1}{6}a_{212} - \frac{1}{8}a_{2222}\right] \\
& + \frac{1}{\sigma_t^2}\left[-\frac{1}{4}a_1 a_{22} + \frac{1}{24}a_2^3 - \frac{5}{12}a_2 a_{12} - \frac{1}{6}a_2 a_{21} + \frac{3}{8}a_2 a_{222} \right. \\
& \qquad \left. - \frac{1}{24}a_3 a_{232} - \frac{1}{12}a_3 a_{322} + \frac{1}{4}a_{22}^2 - \frac{1}{24}a_{23} a_{32} - \frac{1}{12}a_{32}^2\right] \\
& + \frac{1}{\sigma_t^3}\left[\frac{1}{2}a_1 a_2^2 - \frac{15}{16}a_2^2 a_{22} + \frac{1}{12}a_2 a_3 a_{23} + \frac{1}{4}a_2 a_3 a_{32} + \frac{1}{24}a_3^2 a_{22}\right] \\
& + \frac{1}{\sigma_t^4}\left[\frac{7}{16}a_2^4 - \frac{1}{8}a_2^2 a_3^2\right] \qquad \text{(E.53)}
\end{aligned}
$$

expresses the endogenous coefficient, and finally the exogenous coefficient reads as

$$\widetilde{n}''_{y\theta}(\star) = \frac{1}{6}a_{223} + \frac{1}{\sigma_t}\left[-\frac{1}{12}a_2a_{23} + \frac{1}{6}a_{123} + \frac{1}{6}a_{213} - \frac{1}{8}a_{2223}\right]$$
$$+\frac{1}{\sigma_t^2}\left[-\frac{1}{4}a_1a_{23} + \frac{1}{24}a_2^2a_3 - \frac{1}{4}a_2a_{13} + \frac{1}{4}a_2a_{223} - \frac{1}{6}a_3a_{12} - \frac{1}{6}a_3a_{21}\right.$$
$$\left.+\frac{1}{8}a_3a_{222} - \frac{1}{24}a_3a_{233} - \frac{1}{12}a_3a_{323} + \frac{1}{4}a_{22}a_{23} - \frac{1}{24}a_{23}a_{33} - \frac{1}{12}a_{32}a_{33}\right]$$
$$+\frac{1}{\sigma_t^3}\left[+\frac{1}{2}a_1a_2a_3 - \frac{7}{16}a_2^2a_{23} - \frac{1}{2}a_2a_3a_{22} + \frac{1}{12}a_2a_3a_{33} + \frac{1}{8}a_3^2a_{23} + \frac{1}{6}a_3^2a_{32}\right]$$
$$+\frac{1}{\sigma_t^4}\left[+\frac{7}{16}a_2^3a_3 - \frac{1}{8}a_2a_3^3\right]. \tag{E.54}$$

Proof Let us consider each of the four terms of input expression (E.50) for the twist $\widetilde{\Sigma}''_{y\theta}(\star)$. The dynamics of the first term are trivial, so we move to the

Dynamics of the Second Term

For the bracket we have

$$d\left[\frac{a_{12}}{6} - \frac{a_2^2}{24} + \frac{a_{21}}{6} - \frac{a_{222}}{8}\right] = \left[\frac{a_{121}}{6} - \frac{a_2a_{21}}{12} - \frac{a_{22}^2}{24} - \frac{a_{23}^2}{24} + \frac{a_{211}}{6} - \frac{a_{2221}}{8}\right]dt$$
$$+\left[\frac{a_{122}}{6} - \frac{a_2a_{22}}{12} + \frac{a_{212}}{6} - \frac{a_{2222}}{8}\right]dW_t$$
$$+\left[\frac{a_{123}}{6} - \frac{a_2a_{23}}{12} + \frac{a_{213}}{6} - \frac{a_{2223}}{8}\right]dZ_t.$$

Therefore, using Lemma E.1, we get the ratio's dynamics as:

$$d\left[\frac{1}{\sigma_t}\left[-\frac{1}{24}a_2^2 + \frac{1}{6}a_{12} + \frac{1}{6}a_{21} - \frac{1}{8}a_{222}\right]\right] = x_{1,t}\,dt + x_{2,t}\,dW_t + x_{3,t}\,dZ_t \tag{E.55}$$

with

$$x_{1,t} = \frac{x_{1,1,t}}{\sigma_t} - \frac{x_{1,2,t}}{\sigma_t^2} + \frac{x_{1,3,t}}{\sigma_t^3}$$

$$x_{1,1,t} = -\frac{a_2a_{21}}{12} - \frac{a_{22}^2}{24} - \frac{a_{23}^2}{24} + \frac{a_{121}}{6} + \frac{a_{211}}{6} - \frac{a_{2221}}{8}$$

$$x_{1,2,t} = a_1\left[-\frac{a_2^2}{24} + \frac{a_{12}}{6} + \frac{a_{21}}{6} - \frac{a_{222}}{8}\right] + a_2\left[-\frac{a_2a_{22}}{12} + \frac{a_{122}}{6} + \frac{a_{212}}{6} - \frac{a_{2222}}{8}\right]$$
$$+a_3\left[-\frac{a_2a_{23}}{12} + \frac{a_{123}}{6} + \frac{a_{213}}{6} - \frac{a_{2223}}{8}\right]$$

$$x_{1,3,t} = \left[-\frac{a_2^2}{24} + \frac{a_{12}}{6} + \frac{a_{21}}{6} - \frac{a_{222}}{8}\right]\left[a_2^2 + a_3^2\right]$$

$$x_{2,t} = \frac{1}{\sigma_t}\left[\frac{a_{122}}{6} - \frac{a_2 a_{22}}{12} + \frac{a_{212}}{6} - \frac{a_{2222}}{8}\right] - \frac{1}{\sigma_t^2}\left[a_2\left[\frac{a_{12}}{6} - \frac{a_2^2}{24} + \frac{a_{21}}{6} - \frac{a_{222}}{8}\right]\right]$$

$$x_{3,t} = \frac{1}{\sigma_t}\left[\frac{a_{123}}{6} - \frac{a_2 a_{23}}{12} + \frac{a_{213}}{6} - \frac{a_{2223}}{8}\right] - \frac{1}{\sigma_t^2}\left[a_3\left[\frac{a_{12}}{6} - \frac{a_2^2}{24} + \frac{a_{21}}{6} - \frac{a_{222}}{8}\right]\right].$$

Dynamics of the Third Term

Again, the bracket's dynamics come as

$$d\left[-\frac{1}{4}a_1 a_2 + \frac{1}{4}a_2 a_{22} - \frac{1}{24}a_3 a_{23} - \frac{1}{12}a_3 a_{32}\right] = x_{1,t}\,dt + x_{2,t}\,dW_t + x_{3,t}\,dZ_t$$

with

$$\begin{aligned}
x_{1,t} &= -\frac{1}{4}\left[a_2 a_{11} + a_1 a_{21} + a_{12} a_{22} + a_{13} a_{23}\right] \\
&\quad + \frac{1}{4}\left[a_{22} a_{21} + a_2 a_{221} + a_{22} a_{222} + a_{23} a_{223}\right] \\
&\quad - \frac{1}{24}\left[a_{23} a_{31} + a_3 a_{231} + a_{32} a_{232} + a_{33} a_{233}\right] \\
&\quad - \frac{1}{12}\left[a_{32} a_{31} + a_3 a_{321} + a_{32} a_{322} + a_{33} a_{323}\right] \\
x_{2,t} &\triangleq -\frac{1}{4}\left[a_2 a_{12} + a_1 a_{22}\right] + \frac{1}{4}\left[a_{22}^2 + a_2 a_{222}\right] \\
&\quad - \frac{1}{24}\left[a_{23} a_{32} + a_3 a_{232}\right] - \frac{1}{12}\left[a_{32}^2 + a_3 a_{322}\right] \\
x_{3,t} &\triangleq -\frac{1}{4}\left[a_2 a_{13} + a_1 a_{23}\right] + \frac{1}{4}\left[a_{22} a_{23} + a_2 a_{223}\right] \\
&\quad - \frac{1}{24}\left[a_{23} a_{33} + a_3 a_{233}\right] - \frac{1}{12}\left[a_{32} a_{33} + a_3 a_{323}\right].
\end{aligned}$$

Therefore, using Lemma E.1 again, we have the dynamics of the ratio as

$$d\left[\frac{1}{\sigma_t^2}\left[\frac{a_2 a_{22}}{4} - \frac{a_1 a_2}{4} - \frac{a_3 a_{23}}{24} - \frac{a_3 a_{32}}{12}\right]\right] = X_{1,t}\,dt + X_{2,t}\,dW_t + X_{3,t} dZ_t \tag{E.56}$$

with

$$X_{1,t} \triangleq \frac{x_{1,t}}{\sigma_t^2} - \frac{2\, X_{1,3,t}}{\sigma_t^3} + \frac{3\, X_{1,4,t}}{\sigma_t^4}$$

$$\begin{aligned}
X_{1,3,t} \triangleq\ & a_2 \left[-\frac{1}{4} [a_2 a_{12} + a_1 a_{22}] + \frac{1}{4} \left[a_{22}^2 + a_2 a_{222} \right] \right. \\
& \left. -\frac{1}{24} [a_{23} a_{32} + a_3 a_{232}] - \frac{1}{12} \left[a_{32}^2 + a_3 a_{322} \right] \right] \\
& + a_3 \left[-\frac{1}{4} [a_2 a_{13} + a_1 a_{23}] + \frac{1}{4} [a_{22} a_{23} + a_2 a_{223}] \right. \\
& \left. -\frac{1}{24} [a_{23} a_{33} + a_3 a_{233}] - \frac{1}{12} [a_{32} a_{33} + a_3 a_{323}] \right] \\
& + a_1 \left[-\frac{1}{4} a_1 a_2 + \frac{1}{4} a_2 a_{22} - \frac{1}{24} a_3 a_{23} - \frac{1}{12} a_3 a_{32} \right]
\end{aligned}$$

$$X_{1,4,t} \triangleq \left[-\frac{1}{4} a_1 a_2 + \frac{1}{4} a_2 a_{22} - \frac{1}{24} a_3 a_{23} - \frac{1}{12} a_3 a_{32} \right] \left[a_2^2 + a_3^2 \right]$$

$$\begin{aligned}
X_{2,t} \triangleq\ & \frac{1}{\sigma_t^2} \left[-\frac{1}{4} (a_2 a_{12} + a_1 a_{22}) + \frac{1}{4} \left(a_{22}^2 + a_2 a_{222} \right) \right. \\
& \left. -\frac{1}{24} [a_{23} a_{32} + a_3 a_{232}] - \frac{1}{12} \left[a_{32}^2 + a_3 a_{322} \right] \right] \\
& - \frac{1}{\sigma_t^3} \left[2 a_2 \left[-\frac{1}{4} a_1 a_2 + \frac{1}{4} a_2 a_{22} - \frac{1}{24} a_3 a_{23} - \frac{1}{12} a_3 a_{32} \right] \right]
\end{aligned}$$

$$\begin{aligned}
X_{3,t} \triangleq\ & \frac{1}{\sigma_t^2} \left[-\frac{1}{4} [a_2 a_{13} + a_1 a_{23}] + \frac{1}{4} [a_{22} a_{23} + a_2 a_{223}] \right. \\
& \left. -\frac{1}{24} [a_{23} a_{33} + a_3 a_{233}] - \frac{1}{12} [a_{32} a_{33} + a_3 a_{323}] \right] \\
& - \frac{1}{\sigma_t^3} \left[2 a_3 \left[-\frac{1}{4} a_1 a_2 + \frac{1}{4} a_2 a_{22} - \frac{1}{24} a_3 a_{23} - \frac{1}{12} a_3 a_{32} \right] \right].
\end{aligned}$$

Dynamics of the Fourth Term

The bracket exhibits the following dynamics:

$$d \left[-\frac{7}{48} a_2^3 + \frac{1}{24} a_2 a_3^2 \right] = x_{1,t}\, dt + x_{2,t}\, dW_t + x_{3,t}\, dZ_t$$

with

$$
\begin{aligned}
x_{1,t} &\triangleq -\frac{7}{48}\left[3a_2^2a_{21} + 3a_2a_{22}^2 + 3a_2a_{23}^2\right] \\
&\quad + \frac{1}{24}\left[a_3^2a_{21} + 2a_2a_3a_{31} + a_2a_{32}^2 + a_2a_{33}^2 + 2a_3a_{22}a_{32} + 2a_3a_{23}a_{33}\right] \\
x_{2,t} &\triangleq -\frac{7}{48}\left[3a_2^2a_{22}\right] + \frac{1}{24}\left[a_3^2a_{22} + 2a_2a_3a_{32}\right] \\
x_{3,t} &\triangleq -\frac{7}{48}\left[3a_2^2a_{23}\right] + \frac{1}{24}\left[a_3^2a_{23} + 2a_2a_3a_{33}\right].
\end{aligned}
$$

Therefore, using Lemma E.1, we have the ratio's dynamics as

$$
d\left[\frac{1}{\sigma_t^3}\left[-\frac{7}{48}a_2^3 + \frac{1}{24}a_2a_3^2\right]\right] = X_{1,t}\,dt + X_{2,t}\,dW_t + X_{3,t}\,dW_t \tag{E.57}
$$

with

$$
\begin{aligned}
X_{1,t} &\triangleq \frac{x_{1,t}}{\sigma_t^3} - \frac{1}{\sigma_t^4}X_{1,4,t} + \frac{6}{\sigma_t^5}\left[-\frac{7}{48}a_2^3 + \frac{1}{24}a_2a_3^2\right]\left[a_2^2 + a_3^2\right] \\
X_{1,4,t} &\triangleq 3a_2\left[-\frac{7}{48}\left[3a_2^2a_{22}\right] + \frac{1}{24}\left[a_3^2a_{22} + 2a_2a_3a_{32}\right]\right] \\
&\quad + 3a_3\left[-\frac{7}{48}\left[3a_2^2a_{23}\right] + \frac{1}{24}\left[a_3^2a_{23} + 2a_2a_3a_{33}\right]\right] \\
&\quad + 3a_1\left[-\frac{7}{48}a_2^3 + \frac{1}{24}a_2a_3^2\right] \\
X_{2,t} &\triangleq \frac{1}{\sigma_t^3}\left[-\frac{7}{48}\left[3a_2^2a_{22}\right] + \frac{1}{24}\left[a_3^2a_{22} + 2a_2a_3a_{32}\right]\right] \\
&\quad - \frac{1}{\sigma_t^4}\left[3a_2\left[-\frac{7}{48}a_2^3 + \frac{1}{24}a_2a_3^2\right]\right] \\
X_{3,t} &\triangleq \frac{1}{\sigma_t^3}\left[-\frac{7}{48}\left[3a_2^2a_{23}\right] + \frac{1}{24}\left[a_3^2a_{23} + 2a_2a_3a_{33}\right]\right] \\
&\quad - \frac{1}{\sigma_t^4}\left[3a_3\left[-\frac{7}{48}a_2^3 + \frac{1}{24}a_2a_3^2\right]\right].
\end{aligned}
$$

Gathering dynamics (E.55)–(E.57) formally provides those of the twist:

$$
d\widetilde{\Sigma}''_{y\theta}(t,0,0) = \widetilde{b}''_{y\theta}(t,0,0)\,dt + \widetilde{\nu}''_{y\theta}(t,0,0)\,dW_t + \widetilde{n}''_{y\theta}(t,0,0)\,dZ_t,
$$

where, after identification and simplification, $\widetilde{b}''_{y\theta}(t,0,0)$ comes as (E.52), $\widetilde{\nu}''_{y\theta}(t,0,0)$ as (E.53), and $\widetilde{n}''_{y\theta}(t,0,0)$ as (E.54), concluding the proof. □

Note that neither (E.52) nor (E.54) will be used in the sequel, and have been computed here simply for the sake of completeness. Apart from being available for further expansions, these expressions confirm the inductive nature of the computation, including for the $\widetilde{b}$, $\widetilde{\nu}$ and $\widetilde{n}$ coefficients.

Expressing the Flattening $\widetilde{\Sigma}'''_{yy\theta}(\star)$

We now have the necessary information to derive the targeted static differential.

Proposition E.9 (Expressing the flattening $\widetilde{\Sigma}'''_{yy\theta}(\star)$ in the bi-dim. case) *For a generic bi-dimensional model we have the flattening as:*

$$\begin{aligned}\widetilde{\Sigma}'''_{yy\theta}(\star) = &\frac{1}{\sigma_t}\left[\frac{1}{8}a_{222}\right] + \frac{x_{2,t}}{\sigma_t^2} + \frac{x_{3,t}}{\sigma_t^3} + \frac{x_{4,t}}{\sigma_t^4}\\ &+ \frac{1}{\sigma_t^5}\left[\frac{15}{32}a_2^4 - \frac{3}{4}a_2^2a_3^2 + \frac{4}{45}a_3^4\right]\end{aligned} \tag{E.58}$$

with

$$\begin{aligned}x_{2,t} &= -\frac{1}{3}a_2a_{22} + \frac{1}{8}a_3a_{23} + \frac{1}{12}a_3a_{32} + \frac{1}{12}a_{122}\\ &\quad + \frac{1}{12}a_{212} + \frac{1}{12}a_{221} - \frac{1}{10}a_{2222},\\ x_{3,t} &= -\frac{1}{3}a_1a_{22} + \frac{5}{24}a_2^3 - \frac{1}{12}a_2a_3^2 - \frac{1}{3}a_2a_{12} - \frac{1}{3}a_2a_{21} + \frac{3}{8}a_2a_{222}\\ &\quad + \frac{1}{4}a_3a_{13} + \frac{1}{6}a_3a_{31} - \frac{3}{20}a_3a_{223} - \frac{2}{15}a_3a_{232} - \frac{7}{60}a_3a_{322}\\ &\quad + \frac{2}{9}a_{22}^2 - \frac{1}{20}a_{23}^2 - \frac{2}{15}a_{23}a_{32} - \frac{1}{30}a_{32}^2 + \frac{1}{12}a_{33}^2,\\ x_{4,t} &= \frac{3}{4}a_1a_2^2 - \frac{5}{12}a_1a_3^2 - \frac{23}{24}a_2^2a_{22} + \frac{13}{24}a_2a_3a_{23}\\ &\quad + \frac{7}{12}a_2a_3a_{32} + \frac{13}{36}a_3^2a_{22} - \frac{13}{60}a_3^2a_{33}.\end{aligned}$$

Proof

Step 1: Differentiation of the ZDC

Differentiating (E.51) [p. 461] gives the (2, 0)-ZDC, which is then taken at $(\star) \triangleq (t, 0, 0)$:

$$\begin{aligned}E^{(2,0)}(\star) + F^{(2,1)}(\star) = &\, 6\,\widetilde{\Sigma}\widetilde{\Sigma}_y'^{\,2}\,\widetilde{b}(\star) + 3\,\widetilde{\Sigma}^2\widetilde{\Sigma}''_{yy}\widetilde{b}(\star)\\ &+ 6\,\widetilde{\Sigma}^2\widetilde{\Sigma}'_y\widetilde{b}'_y(\star) + \widetilde{\Sigma}^3\widetilde{b}''_{yy}(\star).\end{aligned} \tag{E.59}$$

Step 2: Developing the right-hand side

Applying the pre-computed expression (E.5) for $F^{(2,1)}(t, y, \theta)$ at $(t, 0, 0)$ gives

$$\begin{aligned} F^{(2,1)}(t,0,0) = {} & 12\sigma_t \widetilde{\Sigma}'_\theta \, {\widetilde{\Sigma}'_y}^2(\star) - 2\sigma_t \widetilde{\Sigma}'_\theta \widetilde{\nu}'_y(\star) \\ & + 12\sigma_t^2 \widetilde{\Sigma}'_y \widetilde{\Sigma}''_{y\theta}(\star) + 7\sigma_t^2 \widetilde{\Sigma}'_\theta \widetilde{\Sigma}''_{yy}(\star) \\ & - 2\sigma_t^2 \widetilde{\nu}''_{y\theta}(\star) + 3\sigma_t^3 \widetilde{\Sigma}'''_{yy\theta}(\star) - 2\widetilde{\nu\nu}'_\theta(\star) - 2\widetilde{nn}'_\theta(\star). \end{aligned} \tag{E.60}$$

Similarly, we can apply (E.28) in $(t, 0, 0)$ to obtain

$$\begin{aligned} E^{(2,0)}(\star) = {} & 6\sigma_t \, {\widetilde{\Sigma}'_y}^2 \, \widetilde{\Sigma}'_\theta(\star) - 3\sigma_t^3 \, {\widetilde{\Sigma}'_y}^2 \, \widetilde{\Sigma}''_{yy}(\star) - 3\sigma_t^2 \, {\widetilde{\Sigma}'_y}^2 \, \widetilde{\nu}(\star) \\ & + 6\sigma_t^2 \, {\widetilde{\Sigma}'_y}^2 \, \widetilde{\nu}'_y(\star) + 3\sigma_t^2 \widetilde{\Sigma}''_{yy} \widetilde{\Sigma}'_\theta(\star) - \frac{3}{2}\sigma_t^4 \, {\widetilde{\Sigma}''_{yy}}^2(\star) \\ & - \frac{3}{2}\sigma_t^3 \widetilde{\Sigma}''_{yy} \widetilde{\nu}(\star) + 3\sigma_t^3 \widetilde{\Sigma}''_{yy} \widetilde{\nu}'_y(\star) + 6\sigma_t^2 \widetilde{\Sigma}'_y \widetilde{\Sigma}''_{y\theta}(\star) \\ & - 3\sigma_t^4 \widetilde{\Sigma}'_y \widetilde{\Sigma}'''_{yyy}(\star) - 3\sigma_t^3 \widetilde{\Sigma}'_y \widetilde{\nu}'_y(\star) + 6\sigma_t^3 \widetilde{\Sigma}'_y \widetilde{\nu}''_{yy}(\star) \\ & + \sigma_t^3 \widetilde{\Sigma}'''_{yy\theta}(\star) - \frac{1}{2}\sigma_t^5 \widetilde{\Sigma}^{(4')}_{y^4}(\star) - \frac{1}{2}\sigma_t^4 \widetilde{\nu}''_{yy}(\star) + \sigma_t^4 \widetilde{\nu}''_{yyy}(\star). \end{aligned} \tag{E.61}$$

Then combining (E.60) with (E.61) we have the r.h.s. of (E.59) as

$$\begin{aligned} E^{(2,0)}(\star) + F^{(2,1)}(\star) = {} & \underline{4\sigma_t^3 \widetilde{\Sigma}'''_{yy\theta}} \; \overbrace{-\frac{1}{2}\sigma_t^5 \widetilde{\Sigma}^{(4')}_{y^4}}^{R_1} + \overbrace{18\sigma_t^2 \widetilde{\Sigma}'_y \widetilde{\Sigma}''_{y\theta}}^{R_2} \\ & \overbrace{-3\sigma_t^4 \widetilde{\Sigma}'_y \widetilde{\Sigma}'''_{yyy} + \sigma_t^4 \widetilde{\nu}''_{yyy}}^{R_3} + \overbrace{\widetilde{\Sigma}'_\theta \left[18\sigma_t \, {\widetilde{\Sigma}'_y}^2 + 10\sigma_t^2 \widetilde{\Sigma}''_{yy} - 2\sigma_t \widetilde{\nu}'_y\right]}^{R_4} \\ & \underbrace{-\frac{1}{2}\sigma_t^4 \widetilde{\nu}''_{yy} - 2\widetilde{\nu\nu}'_\theta - 2\widetilde{nn}'_\theta}_{R_5} \; \underbrace{+\widetilde{\Sigma}''_{yy}\left[-\frac{3}{2}\sigma_t^4 \widetilde{\Sigma}''_{yy} - 3\sigma_t^3 \, {\widetilde{\Sigma}'_y}^2 - \frac{3}{2}\sigma_t^3 \widetilde{\nu} + 3\sigma_t^3 \widetilde{\nu}'_y\right]}_{R_6} \\ & \underbrace{-3\sigma_t^2 \, {\widetilde{\Sigma}'_y}^2 \, \widetilde{\nu} + 6\sigma_t^2 \, {\widetilde{\Sigma}'_y}^2 \, \widetilde{\nu}'_y - 3\sigma_t^3 \widetilde{\Sigma}'_y \widetilde{\nu}'_y + 6\sigma_t^3 \widetilde{\Sigma}'_y \widetilde{\nu}''_{yy}}_{R_7} \; \underbrace{-2\sigma_t^2 \widetilde{\nu}''_{y\theta}}_{R_8}, \end{aligned}$$

where the last term is our target quantity.

Step 3: Input expression of the right-hand side

Let us take each group of the right-hand term, starting with

$$R_1 \triangleq -\frac{1}{2}\sigma_t^5 \, \widetilde{\Sigma}^{(4')}_{y^4}(\star).$$

This term converts into

$$R_1 = \sigma_t \left[-\frac{1}{10} a_{2222} \right] + \left[\frac{2}{3} a_{22}^2 + a_2 a_{222} - \frac{1}{5} a_3 a_{322} - \frac{3}{10} a_3 a_{232} - \frac{3}{10} a_{23} a_{32} - \frac{1}{5} a_{32}^2 - \frac{3}{10} a_{23}^2 - \frac{2}{5} a_3 a_{223} \right]$$
$$+ \frac{1}{\sigma_t} \left[\frac{7}{3} a_3^2 a_{22} - 5 a_2^2 a_{22} + 3 a_2 a_3 a_{32} + \frac{9}{2} a_2 a_3 a_{23} - \frac{4}{5} a_3^2 a_{33} \right]$$
$$+ \frac{1}{\sigma_t^2} \left[\frac{15}{4} a_2^4 + \frac{41}{30} a_3^4 - 11 a_2^2 a_3^2 \right].$$

Then let us turn to

$$R_2 \triangleq 18\, \sigma_t^2\ \widetilde{\Sigma}_y' \widetilde{\Sigma}_{y\theta}''(\star), \qquad \text{where} \qquad 18\, \sigma_t^2 \widetilde{\Sigma}_y'(\star) = \sigma_t \left[9 a_2 \right],$$

and therefore

$$R_2 = \sigma_t \left[\frac{3}{2} a_2 a_{22} \right] + \left[-\frac{3}{8} a_2^3 + \frac{3}{2} a_2 a_{12} + \frac{3}{2} a_2 a_{21} - \frac{9}{8} a_2 a_{222} \right]$$
$$+ \frac{1}{\sigma_t} \left[-\frac{9}{4} a_1 a_2^2 + \frac{9}{4} a_2^2 a_{22} - \frac{3}{8} a_2 a_3 a_{23} - \frac{3}{4} a_2 a_3 a_{32} \right]$$
$$+ \frac{1}{\sigma_t^2} \left[-\frac{21}{16} a_2^4 + \frac{3}{8} a_2^2 a_3^2 \right].$$

Now let us tackle

$$R_3 \triangleq -3\sigma_t^4 \widetilde{\Sigma}_y' \widetilde{\Sigma}_{yyy}'''(\star) + \sigma_t^4 \widetilde{\mathrm{v}}_{yyy}''(\star), \qquad \text{where} \qquad -3\sigma_t^4 \widetilde{\Sigma}_y'(\star) = \sigma_t^3 \left[-\frac{3}{2} a_2 \right].$$

Using (E.38) [p. 452] and (E.41) [p. 454] we can write

$$R_3 = \sigma_t \left[\frac{1}{4} a_{2222} \right] + \left[-\frac{21}{8} a_2 a_{222} + \frac{3}{4} a_3 a_{232} + \frac{1}{2} a_3 a_{322} - \frac{3}{2} a_{22}^2 + \frac{3}{4} a_{23} a_{32} + \frac{1}{2} a_{32}^2 \right]$$
$$+ \frac{1}{\sigma_t} \left[+\frac{51}{4} a_2^2 a_{22} - \frac{33}{8} a_2 a_3 a_{23} - \frac{3}{4} a_2 a_3 a_{32} - 7 a_2 a_3 a_{32} - \frac{5}{2} a_3^2 a_{22} \right]$$
$$+ \frac{1}{\sigma_t^2} \left[-\frac{39}{4} a_2^4 + \frac{65}{4} a_2^2 a_3^2 \right].$$

We now address

$$R_4 \triangleq \widetilde{\Sigma}_\theta'(\star) \left[18\sigma_t\ \widetilde{\Sigma}_y'^{\,2}(\star) + 10\sigma_t^2 \widetilde{\Sigma}_{yy}''(\star) - 2\sigma_t \widetilde{\mathrm{v}}_y'(\star) \right]$$

where the bracket reads as

$$18\sigma_t\, \widetilde{\Sigma}_y^{\prime\,2}(\star) + 10\sigma_t^2\widetilde{\Sigma}_{yy}''(\star) - 2\sigma_t\widetilde{\mathrm{v}}_y'(\star)$$
$$= \left[\frac{7}{3}a_{22}\right] + \frac{1}{\sigma_t}\left[\frac{1}{2}a_2^2 + \frac{10}{3}a_3^2\right]$$

so that

$$R_4 = \left[\sigma_t\left[\frac{a_2}{4}\right] + \left[\frac{a_1}{2} - \frac{a_{22}}{6}\right] + \frac{1}{\sigma_t}\left[\frac{a_2^2}{8} + \frac{a_3^2}{12}\right]\right]$$
$$\times\left[\left[\frac{7a_{22}}{3}\right] + \frac{1}{\sigma_t}\left[\frac{a_2^2}{2} + \frac{10a_3^2}{3}\right]\right].$$

After developing and simplifying this expression we get

$$R_4 = \sigma_t\left[\frac{7}{12}a_2a_{22}\right] + \left[\frac{7}{6}a_1a_{22} + \frac{1}{8}a_2^3 + \frac{5}{6}a_2a_3^2 - \frac{7}{18}a_{22}^2\right]$$
$$+ \frac{1}{\sigma_t}\left[\frac{1}{4}a_1a_2^2 + \frac{5}{3}a_1a_3^2 + \frac{5}{24}a_2^2a_{22} - \frac{13}{36}a_3^2a_{22}\right]$$
$$+ \frac{1}{\sigma_t^2}\left[\frac{1}{16}a_2^4 + \frac{11}{24}a_2^2a_3^2 + \frac{5}{18}a_3^4\right].$$

Let us now compute

$$R_5 \triangleq -\frac{1}{2}\sigma_t^4\widetilde{\mathrm{v}}_{yy}''(\star) - 2\widetilde{\mathrm{v}\mathrm{v}}_\theta'(\star) - 2\widetilde{nn}_\theta'(\star),$$

which translates into

$$R_5 = -\frac{1}{2}\sigma_t^4\left[\frac{a_{222}}{3\sigma_t^2} + \frac{1}{\sigma_t^3}\left[\frac{2a_3a_{32}}{3} - \frac{5a_2a_{22}}{3}\right] + \frac{a_2}{\sigma_t^4}\left[\frac{3a_2^2}{2} - a_3^2\right]\right]$$
$$-2a_2\left[\sigma_t\left[\frac{a_{22}}{4}\right] + \left[\frac{a_{12}}{2} - \frac{a_{222}}{6} + \frac{a_2^2}{4}\right]\right.$$
$$\left. + \frac{1}{\sigma_t}\left[\frac{a_2a_{22}}{4} + \frac{a_3a_{32}}{6}\right] + \frac{1}{\sigma_t^2}\left[-\frac{a_2^3}{8} - \frac{a_2a_3^2}{12}\right]\right]$$
$$-2a_3\left[\sigma_t\left[\frac{a_{23}}{4}\right] + \left[\frac{a_{13}}{2} - \frac{a_{223}}{6} + \frac{a_2a_3}{4}\right]\right.$$
$$\left. + \frac{1}{\sigma_t}\left[\frac{a_2a_{23}}{4} + \frac{a_3a_{33}}{6}\right] + \frac{1}{\sigma_t^2}\left[-\frac{a_2^2a_3}{8} - \frac{a_3^3}{12}\right]\right].$$

Expanding and simplifying the expression, we get

$$\begin{aligned}
R_5 = \sigma_t^2 &\left[-\frac{1}{6}a_{222}\right] + \sigma_t\left[\frac{1}{3}a_2a_{22} - \frac{1}{2}a_3a_{23} - \frac{1}{3}a_3a_{32}\right] \\
&+ \left[-\frac{5}{4}a_2^3 + \frac{1}{3}a_2a_{222} - a_2a_{12} - a_3a_{13} + \frac{1}{3}a_3a_{223}\right] \\
&+ \frac{1}{\sigma_t}\left[-\frac{1}{2}a_2^2a_{22} - \frac{1}{2}a_2a_3a_{23} - \frac{1}{3}a_2a_3a_{32} - \frac{1}{3}a_3^2a_{33}\right] \\
&+ \frac{1}{\sigma_t^2}\left[\frac{1}{4}a_2^4 + \frac{5}{12}a_2^2a_3^2 + \frac{1}{6}a_3^4\right].
\end{aligned}$$

Let us now turn to

$$R_6 \triangleq \widetilde{\Sigma}''_{yy}(\star)\left[-\frac{3}{2}\sigma_t^4\,\widetilde{\Sigma}''_{yy}(\star) - 3\sigma_t^3\ {\widetilde{\Sigma}'_y}^{\,2}(\star) - \frac{3}{2}\sigma_t^3\widetilde{\mathrm{v}}(\star) + 3\sigma_t^3\widetilde{\mathrm{v}}'_y(\star)\right],$$

where the bracket converts into

$$\begin{aligned}
-\frac{3}{2}\sigma_t^4\,\widetilde{\Sigma}''_{yy}(\star) &- 3\sigma_t^3\ {\widetilde{\Sigma}'_y}^{\,2}(\star) - \frac{3}{2}\sigma_t^3\widetilde{\mathrm{v}}(\star) + 3\sigma_t^3\widetilde{\mathrm{v}}'_y(\star) \\
&= \sigma_t^3\left[-\frac{3}{2}a_2\right] + \sigma_t^2\,[a_{22}] + \sigma_t\left[-\frac{3}{2}a_2^2 - \frac{1}{2}a_3^2\right]
\end{aligned}$$

and therefore the whole term comes as

$$\begin{aligned}
R_6 = &\left[\frac{1}{\sigma_t^2}\left[\frac{a_{22}}{3}\right] + \frac{1}{\sigma_t^3}\left[\frac{a_3^2}{3} - \frac{a_2^2}{2}\right]\right] \\
&\times\left[\sigma_t^3\left[-\frac{3a_2}{2}\right] + \sigma_t^2\,[a_{22}] + \sigma_t\left[-\frac{3a_2^2}{2} - \frac{a_3^2}{2}\right]\right] \\
= \sigma_t &\left[-\frac{a_2a_{22}}{2}\right] + \left[\frac{3a_2^3}{4} - \frac{a_2a_3^2}{2} + \frac{a_{22}^2}{3}\right] \\
&+ \frac{1}{\sigma_t}\left[-a_2^2a_{22} + \frac{a_3^2a_{22}}{6}\right] + \frac{1}{\sigma_t^2}\left[\frac{3a_2^4}{4} - \frac{a_2^2a_3^2}{4} - \frac{a_3^4}{6}\right].
\end{aligned}$$

Now let us compute

$$R_7 \triangleq -3\sigma_t^2\ {\widetilde{\Sigma}'_y}^{\,2}\,\widetilde{\mathrm{v}}(\star) + \left[6\sigma_t^2\ {\widetilde{\Sigma}'_y}^{\,2} - 3\sigma_t^3\,\widetilde{\Sigma}'_y\right]\widetilde{\mathrm{v}}'_y(\star) + 6\sigma_t^3\,\widetilde{\Sigma}'_y\widetilde{\mathrm{v}}''_{yy}(\star)$$

which comes as

$$\begin{aligned}
R_7 &= -3\sigma_t^2\left(\frac{a_2}{2\sigma_t}\right)^2 a_2 + \left[6\sigma_t^2\left(\frac{a_2}{2\sigma_t}\right)^2 - 3\sigma_t^3\frac{a_2}{2\sigma_t}\right]\left[\frac{1}{\sigma_t}\left[\frac{1}{2}a_{22}\right] + \frac{1}{\sigma_t^2}\left[-\frac{1}{2}a_2^2\right]\right] \\
&\quad + 6\sigma_t^3\frac{a_2}{2\sigma_t}\left[\frac{1}{\sigma_t^2}\left[\frac{1}{3}a_{222}\right] + \frac{1}{\sigma_t^3}\left[\frac{2}{3}a_3a_{32} - \frac{5}{3}a_2a_{22}\right] + \frac{1}{\sigma_t^4}\left[\frac{3}{2}a_2^3 - a_2a_3^2\right]\right] \\
&= \sigma_t\left[-\frac{3a_2a_{22}}{4}\right] + [a_2a_{222}] + \frac{1}{\sigma_t}\left[-\frac{17a_2^2a_{22}}{4} + 2a_2a_3a_{32}\right] + \frac{1}{\sigma_t^2}\left[\frac{15a_2^4}{4} - 3a_2^2a_3^2\right].
\end{aligned}$$

As for the last term

$$R_8 \triangleq -2\sigma_t^2\widetilde{\nu}''_{y\theta}(\star)$$

we use (E.53) [p. 463] to write at once

$$\begin{aligned}
R_8 &= \sigma_t^2\left[-\frac{a_{222}}{3}\right] + \sigma_t\left[+\frac{a_2a_{22}}{6} - \frac{a_{122}}{3} - \frac{a_{212}}{3} + \frac{a_{2222}}{4}\right] \\
&\quad + \left[\frac{a_1a_{22}}{2} - \frac{a_2^3}{12} + \frac{5a_2a_{12}}{6} + \frac{a_2a_{21}}{3} - \frac{3a_2a_{222}}{4}\right. \\
&\quad\quad \left. + \frac{a_3a_{232}}{12} + \frac{a_3a_{322}}{6} - \frac{a_{22}^2}{2} + \frac{a_{23}a_{32}}{12} + \frac{a_{32}^2}{6}\right] \\
&\quad + \frac{1}{\sigma_t}\left[-a_1a_2^2 + \frac{15a_2^2a_{22}}{8} - \frac{a_2a_3a_{23}}{6} - \frac{a_2a_3a_{32}}{2} - \frac{a_3^2a_{22}}{12}\right] \\
&\quad + \frac{1}{\sigma_t^2}\left[-\frac{7a_2^4}{8} + \frac{a_2^2a_3^2}{4}\right].
\end{aligned}$$

Finally, we aggregate R_1 to R_8 to obtain the complete right-hand side of (E.59):

$$\begin{aligned}
&E^{(2,0)}(\star) + F^{(2,1)}(\star) \\
&= \sigma_t^2\left[-\frac{a_{222}}{2}\right] + \sigma_t\left[+\frac{4a_2a_{22}}{3} - \frac{a_3a_{23}}{2} - \frac{a_3a_{32}}{3} - \frac{a_{122}}{3} - \frac{a_{212}}{3} + \frac{2a_{2222}}{5}\right] \\
&\quad + \left[\frac{5a_1a_{22}}{3} - \frac{5a_2^3}{6} + \frac{a_2a_3^2}{3} + \frac{4a_2a_{12}}{3} + \frac{11a_2a_{21}}{6} - \frac{13a_2a_{222}}{6} - a_3a_{13}\right. \\
&\quad\quad \left. - \frac{a_3a_{223}}{15} + \frac{8a_3a_{232}}{15} + \frac{7a_3a_{322}}{15} - \frac{25a_{22}^2}{18} - \frac{3a_{23}^2}{10} + \frac{8a_{23}a_{32}}{15} + \frac{7a_{32}^2}{15}\right] \\
&\quad + \frac{1}{\sigma_t}\left[\frac{5a_1a_3^2}{3} - 3a_1a_2^2 + \frac{19a_2^2a_{22}}{3} - \frac{2a_2a_3a_{23}}{3} - \frac{13a_2a_3a_{32}}{3}\right. \\
&\quad\quad \left. - \frac{4a_3^2a_{22}}{9} - \frac{17a_3^2a_{33}}{15}\right] + \frac{1}{\sigma_t^2}\left[-\frac{27}{8}a_2^4 + \frac{7}{2}a_2^2a_3^2 + \frac{74}{45}a_3^4\right] + \underline{4\sigma_t^3\widetilde{\Sigma}'''_{yy\theta}}.
\end{aligned}$$

Input Expression of the Left-Hand Side

Substituting (E.30) and (E.32), expanding and simplifying, we get the l.h.s. as

$$
\begin{aligned}
6\widetilde{\Sigma}\widetilde{\Sigma}_y'^{\,2}\,\widetilde{b}(\star) + 3\widetilde{\Sigma}^2\widetilde{\Sigma}_{yy}''\widetilde{b}(\star) + 6\widetilde{\Sigma}^2\widetilde{\Sigma}_y'\widetilde{b}_y'(\star) + \widetilde{\Sigma}^3\widetilde{b}_{yy}''(\star) &= \sigma_t\left[\frac{a_{221}}{3}\right] \\
&+ \left[\frac{a_1a_{22}}{3} + \frac{a_2a_{21}}{2} - \frac{2a_2a_{222}}{3} + \frac{2a_3a_{31}}{3} - \frac{2a_3a_{223}}{3} - \frac{a_{22}^2}{2} - \frac{a_{23}^2}{2} + \frac{a_{32}^2}{3} + \frac{a_{33}^2}{3}\right] \\
&+ \frac{1}{\sigma_t}\left[+\frac{5a_2^2a_{22}}{2} + \frac{3a_2a_3a_{23}}{2} - 2a_2a_3a_{32} + a_3^2a_{22} - 2a_3^2a_{33}\right] \\
&+ \frac{1}{\sigma_t^2}\left[-\frac{3a_2^4}{2} + \frac{a_2^2a_3^2}{2} + 2a_3^4\right].
\end{aligned}
$$

Solving for $\widetilde{\Sigma}_{yy\theta}'''(\star)$

Reuniting the left- and right-hand sides, and isolating our target quantity, we get

$$
\begin{aligned}
&4\sigma_t^3\,\widetilde{\Sigma}_{yy\theta}'''(\star) \\
&= \sigma_t^2\left[\frac{a_{222}}{2}\right] + \sigma_t\left[-\frac{4a_2a_{22}}{3} + \frac{a_3a_{23}}{2} + \frac{a_3a_{32}}{3} + \frac{a_{122}}{3} + \frac{a_{212}}{3} + \frac{a_{221}}{3} - \frac{2a_{2222}}{5}\right] \\
&+ \left[-\frac{4a_1a_{22}}{3} + \frac{5a_2^3}{6} - \frac{a_2a_3^2}{3} - \frac{4a_2a_{12}}{3} - \frac{4a_2a_{21}}{3} + \frac{3a_2a_{222}}{2} + a_3a_{13} + \frac{2a_3a_{31}}{3}\right. \\
&\quad\left. - \frac{3a_3a_{223}}{5} - \frac{8a_3a_{232}}{15} - \frac{7a_3a_{322}}{15} + \frac{8a_{22}^2}{9} - \frac{a_{23}^2}{5} - \frac{8a_{23}a_{32}}{15} - \frac{2a_{32}^2}{15} + \frac{a_{33}^2}{3}\right] \\
&+ \frac{1}{\sigma_t}\left[+3a_1a_2^2 - \frac{5a_1a_3^2}{3} - \frac{23a_2^2a_{22}}{6} + \frac{13a_2a_3a_{23}}{6} + \frac{7a_2a_3a_{32}}{3} + \frac{13a_3^2a_{22}}{9}\right. \\
&\quad\left. - \frac{13a_3^2a_{33}}{15}\right] + \frac{1}{\sigma_t^2}\left[+\frac{15}{8}a_2^4 - 3a_2^2a_3^2 + \frac{16}{45}a_3^4\right].
\end{aligned}
$$

Finally multiplying both sides by $\frac{1}{4}\,\sigma_t^{-3}$ provides the flattening (E.58) and concludes the proof. □

Computation of the Arch

We are now interested in the last static differential, the arch $\widetilde{\Sigma}_{\theta\theta}''(t,0,0)$. We have already computed all the necessary dynamic coefficients, therefore we can state the final result:

Proposition E.10 (Expression of the arch $\widetilde{\Sigma}''_{\theta\theta}(\star)$ in the bi-dimensional case) *For a generic bi-dimensional model we have the arch as:*

$$
\begin{aligned}
\widetilde{\Sigma}''_{\theta\theta}(\star) = {} & \sigma_t^2\left[\frac{a_{22}}{12}\right] + \sigma_t\left[\frac{a_2^2}{12} - \frac{a_3^2}{12} + \frac{a_{12}}{6} + \frac{a_{21}}{6} - \frac{a_{222}}{8}\right] \\
& + \left[\frac{a_1 a_2}{4} + \frac{a_2 a_{22}}{6} + \frac{5 a_3 a_{23}}{24} + \frac{a_3 a_{32}}{12} + \frac{a_{11}}{3} - \frac{a_{122}}{12} - \frac{a_{212}}{12} - \frac{a_{221}}{12} + \frac{a_{2222}}{20}\right] \\
& + \frac{1}{\sigma_t}\left[\frac{a_1^2}{12} + \frac{a_2^3}{48} - \frac{a_2 a_3^2}{24} + \frac{a_2 a_{12}}{6} + \frac{a_2 a_{21}}{6} - \frac{a_2 a_{222}}{8} + \frac{a_3 a_{13}}{12} + \frac{a_3 a_{31}}{6}\right. \\
& \qquad \left. - \frac{a_3 a_{223}}{20} - \frac{a_3 a_{232}}{60} + \frac{a_3 a_{322}}{60} + \frac{a_{23}^2}{15} - \frac{a_{23} a_{32}}{60} + \frac{a_{32}^2}{10} + \frac{a_{33}^2}{12}\right] \\
& + \frac{1}{\sigma_t^2}\left[-\frac{a_1 a_2^2}{8} - \frac{a_1 a_3^2}{6} + \frac{a_2^2 a_{22}}{8} - \frac{a_2 a_3 a_{23}}{24} - \frac{a_2 a_3 a_{32}}{12}\right. \\
& \qquad \left. + \frac{a_3^2 a_{22}}{12} - \frac{11 a_3^2 a_{33}}{60}\right] + \frac{1}{\sigma_t^3}\left[-\frac{3 a_2^4}{64} - \frac{a_2^2 a_3^2}{48} + \frac{7 a_3^4}{80}\right].
\end{aligned}
\tag{E.62}
$$

Proof

Step 1: Differentiation of the ZDC

Let us consider the (0, 1)-ZDC, given by (3.7) and taken at $(t, 0, 0)$:

$$
3\widetilde{\Sigma}^2 \widetilde{\Sigma}'_\theta \widetilde{b}(\star) + \widetilde{\Sigma}^3 \widetilde{b}'_\theta(\star) = D(\star) + E^{(0,1)}(\star) + \frac{1}{2} F^{(0,2)}(\star). \tag{E.63}
$$

Step 2: Input-expression of the right-hand side

Evaluating, respectively, (2.19), (E.29) and (E.3) [pp. 43, 448, 445] at the IATM point gives

$$
D(t, 0, 0) = \frac{1}{8}\sigma_t^4\left[\left(\widetilde{\nu}(\star) - \sigma_t \widetilde{\Sigma}'_y(\star)\right)^2 + \widetilde{n}^2(\star)\right] \tag{E.64}
$$

$$
\begin{aligned}
E^{(0,1)}(t, 0, 0) = {} & 3\sigma_t^2 \widetilde{\Sigma}'_\theta(\star)\left[\widetilde{\Sigma}'_\theta(\star) - \frac{1}{2}\sigma_t^2 \widetilde{\Sigma}''_{yy}(\star) + \sigma_t \widetilde{\nu}'_y(\star) - \frac{1}{2}\sigma_t \widetilde{\nu}(\star)\right] \\
& + \sigma_t^3\left[\widetilde{\Sigma}''_{\theta\theta}(\star) - \frac{1}{2}\sigma_t^2 \widetilde{\Sigma}'''_{yy\theta}(\star) + \sigma_t \widetilde{\nu}''_{y\theta}(\star) - \frac{1}{2}\sigma_t \widetilde{\nu}'_\theta(\star)\right]
\end{aligned}
\tag{E.65}
$$

$$
F^{(0,2)}(t, 0, 0) = 5\sigma_t^2\, \widetilde{\Sigma}'^{\,2}_\theta(\star) + \sigma_t^3 \widetilde{\Sigma}''_{\theta\theta}(\star). \tag{E.66}
$$

Combining (E.64) with (E.65) and (E.66), the r.h.s. of (E.63) rewrites as

$$\begin{aligned}
&D(\star) + E^{(0,1)}(\star) + \frac{1}{2}F^{(0,2)}(\star) \\
&= \underbrace{\frac{1}{8}\sigma_t^4\left[\left(\widetilde{\mathrm{v}}(\star) - \sigma_t\widetilde{\Sigma}'_y(\star)\right)^2 + \widetilde{n}^2(\star)\right]}_{R_1} \\
&\quad + \underbrace{\sigma_t^2\widetilde{\Sigma}'_\theta(\star)\left[\frac{11}{2}\widetilde{\Sigma}'_\theta(\star) - \frac{3}{2}\sigma_t^2\widetilde{\Sigma}''_{yy}(\star) + 3\sigma_t\widetilde{\mathrm{v}}'_y(\star) - \frac{3}{2}\sigma_t\widetilde{\mathrm{v}}(\star)\right]}_{R_2} \\
&\quad \underbrace{-\frac{1}{2}\sigma_t^5\widetilde{\Sigma}'''_{yy\theta}(\star)}_{R_3} \underbrace{+\sigma_t^4\widetilde{\mathrm{v}}''_{y\theta}(\star) - \frac{1}{2}\sigma_t^4\widetilde{\mathrm{v}}'_\theta(\star)}_{R_4} + \underline{\frac{3}{2}\sigma_t^3\widetilde{\Sigma}''_{\theta\theta}(\star)}.
\end{aligned}$$

Let us express each of the terms R_1 to R_4 as a function of the inputs.

We start with

$$R_1 \triangleq \frac{1}{8}\sigma_t^4\left[\left(\widetilde{\mathrm{v}}(\star) - \sigma_t\widetilde{\Sigma}'_y(\star)\right)^2 + \widetilde{n}^2(\star)\right] = \sigma_t^4\left[\frac{1}{32}a_2^2 + \frac{1}{8}a_3^2\right].$$

We then turn to

$$R_2 \triangleq \sigma_t^2\widetilde{\Sigma}'_\theta(\star)\left[\frac{11}{2}\widetilde{\Sigma}'_\theta(\star) - \frac{3}{2}\sigma_t^2\widetilde{\Sigma}''_{yy}(\star) + 3\sigma_t\widetilde{\mathrm{v}}'_y(\star) - \frac{3}{2}\sigma_t\widetilde{\mathrm{v}}(\star)\right].$$

Invoking (2.53), (2.52), (2.55) and (2.56), we express the term in brackets as

$$\begin{aligned}
&\frac{11}{2}\widetilde{\Sigma}'_\theta(\star) - \frac{3}{2}\sigma_t^2\widetilde{\Sigma}''_{yy}(\star) + 3\sigma_t\widetilde{\mathrm{v}}'_y(\star) - \frac{3}{2}\sigma_t\widetilde{\mathrm{v}}(\star) \\
&\quad = \sigma_t\left[-\frac{1}{8}a_2\right] + \left[\frac{11}{4}a_1 + \frac{1}{12}a_{22}\right] + \frac{1}{\sigma_t}\left[-\frac{1}{16}a_2^2 - \frac{1}{24}a_3^2\right],
\end{aligned}$$

so that, using (2.53) again, expanding and simplifying, we get

$$\begin{aligned}
R_2 = {} & \sigma_t^4\left[-\frac{1}{32}a_2^2\right] + \sigma_t^3\left[\frac{5}{8}a_1a_2 + \frac{1}{24}a_2a_{22}\right] \\
& + \sigma_t^2\left[+\frac{11}{8}a_1^2 - \frac{5}{12}a_1a_{22} - \frac{1}{32}a_2^3 - \frac{1}{48}a_2a_3^2 - \frac{1}{72}a_{22}^2\right] \\
& + \sigma_t\left[\frac{5}{16}a_1a_2^2 + \frac{5}{24}a_1a_3^2 + \frac{1}{48}a_2^2a_{22} + \frac{1}{72}a_3^2a_{22}\right] \\
& + \left[-\frac{1}{128}a_2^4 - \frac{1}{96}a_2^2a_3^2 - \frac{1}{288}a_3^4\right].
\end{aligned}$$

Let us now examine

$$R_3 \triangleq -\frac{1}{2}\sigma_t^5 \widetilde{\Sigma}'''_{yy\theta}(\star)$$

where using (E.58) gives immediately

$$R_3 = \sigma_t^4\left[-\frac{a_{222}}{16}\right] + \sigma_t^3\left[\frac{a_2a_{22}}{6} - \frac{a_3a_{23}}{16} - \frac{a_3a_{32}}{24} - \frac{a_{122}}{24} - \frac{a_{212}}{24} - \frac{a_{221}}{24} + \frac{a_{2222}}{20}\right]$$
$$+ \sigma_t^2\left[\frac{a_1a_{22}}{6} - \frac{5a_2^3}{48} + \frac{a_2a_3^2}{24} + \frac{a_2a_{12}}{6} + \frac{a_2a_{21}}{6} - \frac{3a_2a_{222}}{16} - \frac{a_3a_{13}}{8} - \frac{a_3a_{31}}{12}\right.$$
$$\left. + \frac{3a_3a_{223}}{40} + \frac{a_3a_{232}}{15} + \frac{7a_3a_{322}}{120} - \frac{a_{22}^2}{9} + \frac{a_{23}^2}{40} + \frac{a_{23}a_{32}}{15} + \frac{a_{32}^2}{60} - \frac{a_{33}^2}{24}\right]$$
$$+ \sigma_t\left[\frac{5a_1a_3^2}{24} - \frac{3a_1a_2^2}{8} + \frac{23a_2^2a_{22}}{48} - \frac{13a_2a_3a_{23}}{48} - \frac{7a_2a_3a_{32}}{24}\right.$$
$$\left. - \frac{13a_3^2a_{22}}{72} + \frac{13a_3^2a_{33}}{120}\right] + \left[-\frac{15a_2^4}{64} + \frac{3a_2^2a_3^2}{8} - \frac{2a_3^4}{45}\right].$$

Finally we turn to

$$R_4 \triangleq \sigma_t^4 \widetilde{\mathrm{V}}''_{y\theta}(\star) - \frac{1}{2}\sigma_t^4 \widetilde{\mathrm{V}}'_{\theta}(\star)$$

where substituting (E.53) and (E.49) gives, after simplification

$$R_4 = \sigma_t^5\left[-\frac{1}{8}a_{22}\right] + \sigma_t^4\left[-\frac{1}{8}a_2^2 - \frac{1}{4}a_{12} + \frac{1}{4}a_{222}\right]$$
$$+ \sigma_t^3\left[-\frac{5}{24}a_2a_{22} - \frac{1}{12}a_3a_{32} + \frac{1}{6}a_{122} + \frac{1}{6}a_{212} - \frac{1}{8}a_{2222}\right]$$
$$+ \sigma_t^2\left[-\frac{1}{4}a_1a_{22} + \frac{5}{48}a_2^3 + \frac{1}{24}a_2a_3^2 - \frac{5}{12}a_2a_{12} - \frac{1}{6}a_2a_{21} + \frac{3}{8}a_2a_{222}\right.$$
$$\left. - \frac{1}{24}a_3a_{232} - \frac{1}{12}a_3a_{322} + \frac{1}{4}a_{22}^2 - \frac{1}{24}a_{23}a_{32} - \frac{1}{12}a_{32}^2\right]$$
$$+ \sigma_t\left[\frac{1}{2}a_1a_2^2 - \frac{15}{16}a_2^2a_{22} + \frac{1}{12}a_2a_3a_{23} + \frac{1}{4}a_2a_3a_{32} + \frac{1}{24}a_3^2a_{22}\right]$$
$$+ \left[\frac{7}{16}a_2^4 - \frac{1}{8}a_2^2a_3^2\right].$$

Finally, gathering R_1 to R_4 and simplifying, we express the r.h.s. of (E.63) as

$$\begin{aligned}
&D(\star) + E^{(0,1)}(\star) + \frac{1}{2}F^{(0,2)}(\star)\\
&= \sigma_t^5\left[-\frac{a_{22}}{8}\right] + \sigma_t^4\left[\frac{a_3^2}{8} - \frac{a_2^2}{8} - \frac{a_{12}}{4} + \frac{3a_{222}}{16}\right]\\
&\quad + \sigma_t^3\left[\frac{5a_1a_2}{8} - \frac{a_3a_{23}}{16} - \frac{a_3a_{32}}{8} + \frac{a_{122}}{8} + \frac{a_{212}}{8} - \frac{a_{221}}{24} - \frac{3a_{2222}}{40}\right]\\
&\quad + \sigma_t^2\left[+\frac{11a_1^2}{8} - \frac{a_1a_{22}}{2} - \frac{a_2^3}{32} + \frac{a_2a_3^2}{16} - \frac{a_2a_{12}}{4} + \frac{3a_2a_{222}}{16} - \frac{a_3a_{13}}{8} - \frac{a_3a_{31}}{12}\right.\\
&\qquad \left. + \frac{3a_3a_{223}}{40} + \frac{a_3a_{232}}{40} - \frac{a_3a_{322}}{40} + \frac{a_{22}^2}{8} + \frac{a_{23}^2}{40} + \frac{a_{23}a_{32}}{40} - \frac{a_{32}^2}{15} - \frac{a_{33}^2}{24}\right]\\
&\quad + \sigma_t\left[\frac{7a_1a_2^2}{16} + \frac{5a_1a_3^2}{12} - \frac{7a_2^2a_{22}}{16} - \frac{3a_2a_3a_{23}}{16} - \frac{a_2a_3a_{32}}{24} - \frac{a_3^2a_{22}}{8} + \frac{13a_3^2a_{33}}{120}\right]\\
&\quad + \left[\frac{25a_2^4}{128} + \frac{23a_2^2a_3^2}{96} - \frac{23a_3^4}{480}\right] + \frac{3}{2}\sigma_t^3\,\widetilde{\Sigma}''_{\theta\theta}(\star).
\end{aligned}$$

Step 3: Input expression of the left-hand side

Substituting (2.53) and (E.48) [pp. 62, 459] into the l.h.s., developing and simplifying, we get

$$\begin{aligned}
&3\widetilde{\Sigma}^2\,\widetilde{\Sigma}'_\theta\,\widetilde{b}(\star) + \widetilde{\Sigma}^3\,\widetilde{b}'_\theta(\star)\\
&= \sigma_t^4\left[\frac{a_{21}}{4}\right] + \sigma_t^3\left[a_1a_2 + \frac{a_2a_{22}}{4} + \frac{a_3a_{23}}{4} + \frac{a_{11}}{2} - \frac{a_{221}}{6}\right]\\
&\quad + \sigma_t^2\left[\frac{3}{2}a_1^2 - \frac{a_1a_{22}}{2} + \frac{a_2a_{21}}{4} + \frac{a_3a_{31}}{6} + \frac{a_{22}^2}{8} + \frac{a_{23}^2}{8} + \frac{a_{32}^2}{12} + \frac{a_{33}^2}{12}\right]\\
&\quad + \sigma_t\left[\frac{a_1a_2^2}{4} + \frac{a_1a_3^2}{6} - \frac{a_2^2a_{22}}{4} - \frac{a_2a_3a_{23}}{4} - \frac{a_2a_3a_{32}}{6} - \frac{a_3^2a_{33}}{6}\right]\\
&\quad + \left[\frac{a_2^4}{8} + \frac{5a_2^2a_3^2}{24} + \frac{a_3^4}{12}\right].
\end{aligned}$$

Step 4: Solving for $\widetilde{\Sigma}''_{\theta\theta}(t,0,0)$

Gathering both left- and right-hand sides, we get after simplification

$$
\begin{aligned}
&\frac{3}{2}\sigma_t^3\,\widetilde{\Sigma}''_{\theta\theta}(\star)\\
&\quad=\sigma_t^5\left[\frac{a_{22}}{8}\right]+\sigma_t^4\left[\frac{a_2^2}{8}-\frac{a_3^2}{8}+\frac{a_{12}}{4}+\frac{a_{21}}{4}-\frac{3a_{222}}{16}\right]\\
&\qquad+\sigma_t^3\left[\frac{3a_1a_2}{8}+\frac{a_2a_{22}}{4}+\frac{5a_3a_{23}}{16}+\frac{a_3a_{32}}{8}+\frac{a_{11}}{2}-\frac{a_{122}}{8}-\frac{a_{212}}{8}-\frac{a_{221}}{8}+\frac{3a_{2222}}{40}\right]\\
&\qquad+\sigma_t^2\left[\frac{a_1^2}{8}+\frac{a_2^3}{32}-\frac{a_2a_3^2}{16}+\frac{a_2a_{12}}{4}+\frac{a_2a_{21}}{4}-\frac{3a_2a_{222}}{16}+\frac{a_3a_{13}}{8}+\frac{a_3a_{31}}{4}\right.\\
&\qquad\qquad\left.-\frac{3a_3a_{223}}{40}-\frac{a_3a_{232}}{40}+\frac{a_3a_{322}}{40}+\frac{a_{23}^2}{10}-\frac{a_{23}a_{32}}{40}+\frac{3a_{32}^2}{20}+\frac{a_{33}^2}{8}\right]\\
&\qquad+\sigma_t\left[-\frac{3a_1a_2^2}{16}-\frac{a_1a_3^2}{4}+\frac{3a_2^2a_{22}}{16}-\frac{a_2a_3a_{23}}{16}-\frac{a_2a_3a_{32}}{8}+\frac{a_3^2a_{22}}{8}-\frac{11a_3^2a_{33}}{40}\right]\\
&\qquad+\left[-\frac{9a_2^4}{128}-\frac{a_2^2a_3^2}{32}+\frac{21a_3^4}{160}\right].
\end{aligned}
$$

Finally, multiplying both sides by $\frac{2}{3}\sigma_t^{-3}$ provides (E.62) and concludes the proof. □

References

1. Durrleman, V.: From implied to spot volatilities. Mathematical Finance (2006).
2. Nicolay, D.: Volatility Dynamics. Ph.D. thesis, Ecole Polytechnique (2011).

Solutions

Solution to Exercise 4.1

The purpose of this exercise [p. 244] was to express the basic induction block of the model, which will be used to derive the chaos coefficients. The solution is as follows:

Lemma E.4 (Elementary block dynamics for the FL-SV model) *Let $\alpha(\cdot)$, $k(\cdot)$ and $h(\cdot)$ be some respectively C^1, C^2 and C^2 abstract functions.*

Let us introduce the following notations for some of their derived functions:

$$
\begin{aligned}
k_1(x) &\triangleq k'(x)\ \psi(x) & \alpha_1(x) &\triangleq \alpha'(x) \\
k_2(x) &\triangleq k''(x)\ \psi^2(x) & \alpha_2(x) &\triangleq \alpha(x)\ \gamma(x) \\
k_3(x) &\triangleq k(x)\ b(x) & \alpha_3(x) &\triangleq \alpha(x)\ \gamma^2(x) \\
k_4(x) &\triangleq k(x)\ b^2(x) & h_1(x) &\triangleq h'(x)\ \varphi(x) \\
k_5(x) &\triangleq k'(x)\ c(x) & h_2(x) &\triangleq h''(x)\ \varphi^2(x)
\end{aligned}
$$

Then we have, for the processes S_t and V_t of the Extended FL-SV model (4.73)

$$d\left[\alpha(t)\ k(V_t)\ h(S_t)\right] = \mu_t\ dt\ \alpha_2(t)\ k_3\left(V_t\right)\ h_1\left(S_t\right)\ dW_t + \varepsilon\ \alpha(t)\ k_1\left(V_t\right)\ h(S_t)\ dZ_t$$

with

$$\mu_t = \alpha_1(t)\ k(V_t)\ h(S_t) + \alpha(t)\left[\kappa\ k_5 + \frac{1}{2}\ \varepsilon^2\ k_2\right](V_t)\ h(S_t) + \frac{1}{2}\ \alpha_3(t)\ k_4\left(V_t\right)\ h_2\left(S_t\right).$$

Omitting the arguments for clarity, and applying this relation to the functions defining the generic FL-SV setup, we have in particular that

$$
\begin{aligned}
d\left[\gamma_m\ b_n\ g_p\right] = &\left[\gamma_m\left[\kappa\ b_{n5} + \frac{1}{2}\ \varepsilon^2\ b_{n2}\right]\ g_p + \gamma_{m1}\ b_n\ g_p + \frac{1}{2}\ \gamma_{m3}\ b_{n4}\ g_{p2}\right] dt \\
&+ \gamma_{m2}\ b_{n3}\ g_{p1}\ dW_t + \varepsilon\ \gamma_m\ b_{n1}\ g_p\ dZ_t.
\end{aligned} \tag{E.67}
$$

D. Nicolay, *Asymptotic Chaos Expansions in Finance*,
Springer Finance Lecture Notes, DOI 10.1007/978-1-4471-6506-4

Proof of Lemma E.4.
Exploiting the independence condition, we get simply by Itô:

$$d\left[\alpha(t)\,k(V_t)\,h(S_t)\right] = \alpha'(t)\,k(V_t)\,h(S_t)\,dt + \alpha(t)\,k'(V_t)\,h(S_t)\,dV_t + \alpha(t)\,k(V_t)\,h'(S_t)\,dS_t + \frac{1}{2}\,\alpha(t)\,k''(V_t)\,h(S_t)\,\langle dV_t\rangle + \frac{1}{2}\,\alpha(t)\,k(V_t)\,h''(S_t)\,\langle dS_t\rangle$$

which gives

$$d\left[\alpha(t)\,k(V_t)\,h(S_t)\right] = \mu_t\,dt + \alpha(t)\gamma(t)\,k(V_t)b(V_t)\,h'(S_t)\varphi(S_t)\,dW_t + \varepsilon\,\alpha(t)\,k'(V_t)\psi(V_t)\,h(S_t)\,dZ_t$$

with

$$\mu_t = \alpha'(t)\,k(V_t)\,h(S_t) + \kappa\,\alpha(t)\,k'(V_t)\,c(V_t)\,h(S_t) + \frac{1}{2}\,\varepsilon^2\,\alpha(t)\,k''(V_t)\psi^2(V_t)\,h(S_t) + \frac{1}{2}\,\alpha(t)\gamma^2(t)\,k(V_t)b^2(V_t)\,h''(S_t)\varphi^2(S_t).$$

Using the given compact notations, we obtain the desired result. □

Lemma E.4 shows a recurring structure that will enable us to work efficiently with induction. This situation is similar to the SABR treatment of Sect. 4.2. Also, and without surprise, it appears that the drift terms are significantly more involved than the coefficients of the non-finite variation terms. Fortunately, these drift terms are the ones least invoked by the generic formulae for the IATM differentials, developed in Chap. 2 and in Sect. 3.2.

Solution to Exercise 4.2

The aim of this exercise [p. 244] was to express the generic chaos coefficients invoked in the expressions of the static IATM differentials, for the first three layers. The solution is as follows:

Lemma E.5 (Chaos dynamics up to *3rd* layer for the FL-SV model) *Let us consider the Generic FL-SV framework defined by* (4.73), *then the chaos coefficients invoked in layers one, two and three are as follows. For the sake of compactness we have omitted the functions' arguments, understating that they are taken in the current position, i.e.*

$$g_m \ \textit{denotes}\ g_m(t) \qquad b_n \ \textit{denotes}\ b_n(V_t) \qquad g_p \ \textit{denotes}\ g_p(S_t)$$

First-level coefficients:

$$\sigma_t = \gamma\ b\ g \tag{E.68}$$

$$a_{1,t} = \gamma\ [\frac{1}{2}\ \varepsilon^2\ b_2 + \kappa\ b_5]\ \ g + \gamma_1\ b\ g + \frac{1}{2}\ \gamma_3\ b_4\ g_2 \tag{E.69}$$

$$a_{2,t} = \gamma_2\ b_3\ g_1 \tag{E.70}$$

$$a_{3,t} = \varepsilon\ \gamma\ b_1\ g \tag{E.71}$$

Second-level coefficients:

$$\begin{aligned} a_{11,t} = \gamma &\left[\frac{1}{4}\ \varepsilon^4\ b_{22} + \frac{1}{2}\ \kappa\varepsilon^2\ \ (b_{25} + b_{52}) + \kappa^2\ b_{55}\right] g \\ &+ \gamma_1 \left[\varepsilon^2\ b_2 + 2\kappa\ b_5\right]\ g + \gamma_3 \left[\frac{1}{4}\ \varepsilon^2\ \ (\ b_{24} + b_{42}) + \frac{\kappa}{2}\ \ (b_{45} + b_{54})\right] g_2 \\ &+ \gamma_{11}\ b\ g + \frac{1}{2}\ \ (\gamma_{13} + \gamma_{31})\ \ b_4\ g_2 + \frac{1}{4}\ \gamma_{33}\ b_{44}\ g_{22} \end{aligned} \tag{E.72}$$

$$a_{12,t} = \gamma_2\ \left[\frac{1}{2}\varepsilon^2 b_{23} + \kappa b_{53}\right] g_1 + \gamma_{12}\ b_3\ g_1 + \frac{1}{2}\ \gamma_{32}\ b_{43}\ g_{21} \tag{E.73}$$

$$a_{13,t} = \gamma \left[\frac{1}{2}\varepsilon^3\ b_{21} + \kappa\varepsilon\ b_{51}\right] g + \varepsilon\ \gamma_1\ b_1\ g + + \frac{1}{2}\varepsilon\ \gamma_3\ b_{41}\ g_2 \tag{E.74}$$

$$a_{21,t} = \gamma_2 \left[\frac{1}{2}\ \varepsilon^2\ b_{32} + \kappa\ b_{35}\right] g_1 + \gamma_{21}\ b_3\ g_1 + \frac{1}{2}\ \gamma_{23}\ b_{34}\ g_{12} \tag{E.75}$$

$$a_{22,t} = \gamma_{22}\ b_{33}\ g_{11} \tag{E.76}$$

$$a_{23,t} = \varepsilon\ \gamma_2\ b_{31}\ g_1 \tag{E.77}$$

$$a_{31,t} = \gamma \left[\frac{1}{2}\ \varepsilon^3\ b_{12} + \kappa\varepsilon\ b_{15}\right] g + \varepsilon\ \gamma_1\ b_1\ g + \frac{1}{2}\ \varepsilon\ \gamma_3\ b_{14}\ g_2 \tag{E.78}$$

$$a_{32,t} = \varepsilon\ \gamma_2\ b_{13}\ g_1 \tag{E.79}$$

$$a_{33,t} = \varepsilon^2\ \gamma\ b_{11}\ g \tag{E.80}$$

Third-level coefficients:

$$a_{122,t} = \gamma_{22} \left[\frac{1}{2}\varepsilon^2 b_{233} + \kappa b_{533}\right]\ g_{11} + \gamma_{122}\ b_{33}\ g_{11} + \frac{1}{2}\ \gamma_{322}\ b_{433}\ g_{211} \tag{E.81}$$

$$a_{212,t} = \gamma_{22} \left[\frac{1}{2}\ \varepsilon^2\ b_{323} + \kappa\ b_{353}\right] g_{11} + \gamma_{212}\ b_{33}\ g_{11} + \frac{1}{2}\ \gamma_{232}\ b_{343}\ g_{121} \tag{E.82}$$

$$a_{221,t} = \gamma_{22} \left[\frac{1}{2}\ \varepsilon^2\ b_{332} + \kappa\ b_{335}\right] g_{11} + \gamma_{221}\ b_{33}\ g_{11} + \frac{1}{2}\ \gamma_{223}\ b_{334}\ g_{112} \tag{E.83}$$

$$a_{222,t} = \gamma_{222}\ b_{333}\ g_{111} \tag{E.84}$$

$$a_{223,t} = \varepsilon\ \gamma_{22}\ b_{331}\ g_{11} \tag{E.85}$$

$$a_{232,t} = \varepsilon\ \gamma_{22}\ b_{313}\ g_{11} \tag{E.86}$$

$$a_{322,t} = \varepsilon\ \gamma_{22}\ b_{133}\ g_{11} \tag{E.87}$$

And finally:

$$a_{2222,t} = \gamma_{2222}\ b_{3333}\ g_{1111} \tag{E.88}$$

Proof of Lemma E.5.
Using (2.36) and (4.74), we have the instantaneous volatility as

$$\sigma_t = \gamma(t)\ b(V_t)\ g(S_t),$$

which proves (E.68). Then applying Lemma E.4 come the dynamics

$$\begin{aligned} d\sigma_t = {} & \gamma_2(t)\ b_3\,(V_t)\ \ g_1\,(S_t)\ \ dW_t + \varepsilon\ \gamma(t)\ b_1\,(V_t)\ \ g(S_t)\ dZ_t \\ & + \Big[\gamma_1(t) b(V_t) g(S_t) + \gamma(t)\Big[\kappa b_5 + \frac{1}{2}\ \varepsilon^2 b_2\Big](V_t)\ g(S_t) \\ & + \frac{1}{2}\ \gamma_3(t) b_4\,(V_t)\ g_2\,(S_t)\Big]\ dt. \end{aligned}$$

Omitting the arguments t, V_t, S_t as advertised, we therefore have

$$\begin{cases} a_1 = \gamma\Big[\frac{1}{2}\ \varepsilon^2\ b_2 + \kappa\ b_5\Big]\ g + \gamma_1\ b\ g + \frac{1}{2}\ \gamma_3\ b_4\ g_2 \\ a_2 = \gamma_2\ b_3\ g_1 \\ a_3 = \varepsilon\ \gamma\ b_1\ g \end{cases}$$

which proves (E.69)–(E.71). Let us first compute $d\ a_1$: using (E.67) we get the four elementary dynamics

$$\begin{aligned} d\big[\gamma\ b_2\ g\big] = {} & \gamma_2\ b_{23}\ g_1\ dW_t + \varepsilon\ \gamma\ b_{21}\ g\ dZ_t \\ & + \Big[\gamma_1\ b_2\ g + \gamma\Big[\kappa\ b_{25} + \frac{1}{2}\ \varepsilon^2\ b_{22}\Big]\ g + \frac{1}{2}\ \gamma_3\ b_{24}\ g_2\Big]\ dt \\ d\big[\gamma\ b_5\ g\big] = {} & \gamma_2\ b_{53}\ g_1\ dW_t + \varepsilon\ \gamma\ b_{51}\ g\ dZ_t \\ & + \Big[\gamma_1\ b_5\ g + \gamma\ \Big[\kappa\ b_{55} + \frac{1}{2}\ \varepsilon^2\ b_{52}\Big]\ g + \frac{1}{2}\ \gamma_3\ b_{54}\ g_2\Big]\ dt \\ d\big[\gamma_1\ b\ g\big] = {} & \gamma_{12}\ b_3\ g_1\ dW_t + \varepsilon\ \gamma_1\ b_1\ g\ dZ_t \\ & + \Big[\gamma_{11}\ b\ g + \gamma_1\Big[\kappa\ b_5 + \frac{1}{2}\ \varepsilon^2\ b_2\Big]\ g + \frac{1}{2}\ \gamma_{13}\ b_4\ g_2\Big]\ dt \end{aligned}$$

$$d\left[\gamma_3\ b_4\ g_2\right] = \gamma_{32}\ b_{43}\ g_{21}\ dW_t + \varepsilon\ \gamma_3\ b_{41}\ g_2\ dZ_t$$
$$+\left[\gamma_{31}\ b_4\ g_2 + \gamma_3\Big[\kappa\ b_{45} + \frac{1}{2}\ \varepsilon^2\ b_{42}\Big]\ g_2 + \frac{1}{2}\ \gamma_{33}\ b_{44}\ g_{22}\right] dt.$$

Therefore the drift of $a_{1,t}$ is

$$a_{11} = \frac{1}{2}\ \varepsilon^2\left[\gamma_1\ b_2\ g + \gamma\Big[\kappa\ b_{25} + \frac{1}{2}\ \varepsilon^2\ b_{22}\Big]\ g + \frac{1}{2}\ \gamma_3\ b_{24}\ g_2\right]$$
$$+\kappa\left[\gamma_1\ b_5\ g + \gamma\Big[\kappa\ b_{55} + \frac{1}{2}\ \varepsilon^2\ b_{52}\Big]\ g + \frac{1}{2}\ \gamma_3\ b_{54}\ g_2\right]$$
$$+\left[\gamma_{11}\ b\ g + \gamma_1\Big[\kappa\ b_5 + \frac{1}{2}\ \varepsilon^2\ b_2\Big]\ g + \frac{1}{2}\ \gamma_{13}\ b_4\ g_2\right]$$
$$+\frac{1}{2}\left[\gamma_{31}\ b_4\ g_2 + \gamma_3\Big[\kappa\ b_{45} + \frac{1}{2}\ \varepsilon^2\ b_{42}\Big]g_2 + \frac{1}{2}\ \gamma_{33}\ b_{44}\ g_{22}\right]$$

which after simplification provides

$$a_{11} = \gamma\left[\frac{1}{4}\varepsilon^4\ b_{22} + \frac{1}{2}\kappa\varepsilon^2\ b_{25} + \frac{1}{2}\kappa\varepsilon^2\ b_{52} + \kappa^2\ b_{55}\right] g + \gamma_1\left[\varepsilon^2\ b_2 + 2\kappa\ b_5\right] g$$
$$+\gamma_3\left[\frac{1}{4}\ \varepsilon^2\ \ b_{24} + \frac{1}{4}\ \varepsilon^2\ b_{42} + \frac{1}{2}\kappa\ b_{45} + \frac{1}{2}\kappa\ b_{54}\right]\ g_2 + \gamma_{11}\ b\ g$$
$$+\frac{1}{2}\ (\gamma_{13} + \gamma_{31})\ \ b_4\ g_2 + \frac{1}{4}\ \gamma_{33}\ b_{44}\ g_{22}$$

and proves (E.72). Similarly, we have the two coefficients for the non-finite variation terms:

$$a_{12} = \gamma_2\ \left[\frac{1}{2}\varepsilon^2 b_{23} + \kappa b_{53}\right]\ g_1 + \gamma_{12}\ b_3\ g_1 + \frac{1}{2}\ \gamma_{32}\ b_{43}\ g_{21},$$
$$a_{13} = \gamma\left[\frac{1}{2}\varepsilon^3\ b_{21} + \kappa\varepsilon\ b_{51}\right] g + \varepsilon\ \gamma_1\ b_1\ g + +\frac{1}{2}\varepsilon\ \gamma_3\ b_{41}\ g_2,$$

which proves (E.73) and (E.74). To compute $d\ a_2$ we again invoke Lemma E.4, which leads to

$$d\left[\gamma_2\ b_3\ g_1\right] = \gamma_{22}\ b_{33}\ g_{11}\ dW_t + \varepsilon\ \gamma_2\ b_{31}\ g_1\ dZ_t$$
$$+\left[\gamma_{21}\ b_3\ g_1 + \gamma_2\Big[\kappa\ b_{35} + \frac{1}{2}\ \varepsilon^2\ b_{32}\Big]\ g_1 + \frac{1}{2}\ \gamma_{23}\ b_{34}\ g_{12}\right] dt,$$

so that we get all three coefficients at once:

$$\begin{cases} a_{21} = \gamma_2 \left[\frac{1}{2}\,\varepsilon^2\, b_{32} + \,\kappa\, b_{35}\right]\, g_1 + \gamma_{21}\, b_3\, g_1 + \frac{1}{2}\,\gamma_{23}\, b_{34}\, g_{12} \\ a_{22} = \gamma_{22}\, b_{33}\, g_{11} \\ a_{23} = \varepsilon\,\gamma_2\, b_{31}\, g_1, \end{cases}$$

which proves (E.75)–(E.77). As for $d\,a_3$, we have by similarity:

$$\begin{aligned} d\left[\gamma\, b_1\, g\right] &= \gamma_2\, b_{13}\, g_1\, dW_t + \varepsilon\,\gamma\, b_{11}\, g\, dZ_t \\ &\quad + \left[\gamma_1\, b_1\, g + \gamma\left[\frac{1}{2}\,\varepsilon^2\, b_{12} + \kappa\, b_{15}\right] g + \frac{1}{2}\,\gamma_3\, b_{14}\, g_2\right] dt, \end{aligned}$$

hence the three coefficients

$$\begin{cases} a_{31} = \gamma\left[\frac{1}{2}\,\varepsilon^3\, b_{12} + \kappa\varepsilon\, b_{15}\right] g + \varepsilon\,\gamma_1\, b_1\, g + \frac{1}{2}\,\varepsilon\,\gamma_3\, b_{14}\, g_2 \\ a_{32} = \varepsilon\,\gamma_2\, b_{13}\, g_1 \\ a_{33} = \varepsilon^2\,\gamma\, b_{11}\, g \end{cases}$$

which proves (E.78)–(E.80). Let us now move on to the third-level coefficients. The expressions (E.73) for $a_{12,t}$ and (E.75) for $a_{21,t}$, used in conjunction with Lemma E.4, give us straight away the next two coefficients:

$$\begin{aligned} a_{122} &= \gamma_{22}\left[\frac{1}{2}\varepsilon^2 b_{233} + \kappa b_{533}\right] g_{11} + \gamma_{122}\, b_{33}\, g_{11} + \frac{1}{2}\,\gamma_{322}\, b_{433}\, g_{211}, \\ a_{212} &= \gamma_{22}\left[\frac{1}{2}\,\varepsilon^2\, b_{323} + \kappa\, b_{353}\right] g_{11} + \gamma_{212}\, b_{33}\, g_{11} + \frac{1}{2}\,\gamma_{232}\, b_{343}\, g_{121}, \end{aligned}$$

which proves (E.81) and (E.82). To compute $d\,a_{22}$ we apply Lemma E.4 to (E.76):

$$\begin{aligned} d\left[\gamma_{22}\, b_{33}\, g_{11}\right] &= \gamma_{222}\, b_{333}\, g_{111}\, dW_t + \varepsilon\,\gamma_{22}\, b_{331}\, g_{11}\, dZ_t \\ &\quad + \left[\gamma_{22}\left[\frac{1}{2}\,\varepsilon^2\, b_{332} + \kappa\, b_{335}\right] g_{11} + \gamma_{221}\, b_{33}\, g_{11} \right. \\ &\qquad \left. + \frac{1}{2}\,\gamma_{223}\, b_{334}\, g_{112}\right] dt, \end{aligned}$$

so that

$$\begin{cases} a_{221} = \gamma_{22}\left[\frac{1}{2}\,\varepsilon^2\, b_{332} + \kappa\, b_{335}\right] g_{11} + \gamma_{221}\, b_{33}\, g_{11} + \frac{1}{2}\,\gamma_{223}\, b_{334}\, g_{112} \\ a_{222} = \gamma_{222}\, b_{333}\, g_{111} \\ a_{223} = \varepsilon\,\gamma_{22}\, b_{331}\, g_{11} \end{cases},$$

which proves (E.83)–(E.85). Again using Lemma E.4 but with (E.77) and (E.79) we get, respectively,

$$a_{232} = \varepsilon\ \gamma_{22}\ b_{313}\ g_{11} \qquad \text{and} \qquad a_{322} = \varepsilon\ \gamma_{22}\ b_{133}\ g_{11},$$

which proves (E.86) and (E.87). Finally, expression (E.84) for $a_{222,t}$ gives

$$a_{2222} = \gamma_{2222}\ b_{3333}\ g_{1111},$$

which proves (E.84) and concludes the proof. □

Solution to Exercise 4.3

The purpose of this exercise was simply to express the static IATM differentials of the first layer, as a function of the FL-SV parameters.

Proposition E.11 (Extended FL-SV model: 1st layer static IATM differentials) *The IATM smile shape differentials of the Extended FL-SV model class* (4.73) *are*

$$\widetilde{\Sigma}(\star) = \gamma(t)\ b(V_t)\ g(S_t) \tag{E.89}$$

$$\widetilde{\Sigma}'_y(\star) = \frac{1}{2}\ \gamma(t)\ b(V_t)\ S_t\ g'(S_t) \tag{E.90}$$

$$\widetilde{\Sigma}''_{yy}(\star) = \varepsilon^2\left[\frac{1}{3}\ \frac{b_1^2}{\gamma\ b^3\ g}\right] + \gamma\ b\left(\frac{1}{3}\frac{g_{11}}{g^2} - \frac{1}{2}\frac{g_1^2}{g^3}\right) \tag{E.91}$$

$$\widetilde{\Sigma}'_\theta(\star) = \varepsilon^2\left[\frac{1}{4}\ \gamma\left(b_2 + \frac{1}{3}\ \frac{b_1^2}{b}\right) g\right] + \kappa\left[\frac{1}{2}\ \gamma\ b_5\ g\right] \tag{E.92}$$

$$+ \gamma^3\ b^3\left[\frac{1}{4}\ g g_1 + \frac{1}{8}\ \frac{g_1^2}{g} + \frac{1}{4}\ g_2 - \frac{1}{6}\ g_{11}\right] + \frac{1}{2}\ \gamma_1\ b\ g.$$

Note the recurrence of the term $\frac{g'(x)}{g(x)} = \frac{\partial}{\partial x}\ \ln(g(x))$ due to the Lognormal baseline.

Proof The IATM level comes simply from combining (2.36) with (E.68), which proves (E.89). We have the IATM skew from (2.51) and (E.68)–(E.70) as

$$\begin{aligned}\widetilde{\Sigma}'_y(t,0,0) &= \frac{a_2}{2\sigma_t} = \frac{\gamma_2\ b_3\ g_1}{2\ \gamma\ b\ g} = \frac{1}{2}\ \gamma(t)\ b(V_t)\ \frac{g'(S_t)}{g(S_t)}\ \varphi(S_t)\\ &= \frac{1}{2}\ \gamma(t)\ b(V_t)\ S_t\ g'(S_t),\end{aligned}$$

which proves (E.90). Then rewriting (2.52) with (E.68)–(E.76) gives

$$
\begin{aligned}
\widetilde{\Sigma}''_{yy}(t,0,0) &= \frac{1}{3}\,\frac{a_{22}}{\sigma_t^2} + \frac{1}{\sigma_t^3}\left[\frac{1}{3}\,a_3^2 - \frac{1}{2}\,a_2^2\right] \\
&= \frac{1}{3}\,\frac{\gamma_{22}\,b_{33}\,g_{11}}{\gamma^2\,b^2\,g^2} + \frac{1}{\gamma^3 b^3 g^3}\left[\frac{1}{3}\,\varepsilon^2\gamma^2 b_1^2 g^2 - \frac{1}{2}\,\gamma_2^2 b_3^2 g_1^2\right]
\end{aligned}
$$

which, after replacing the compact notations, provides (E.91). We turn to the slope, combining (2.53) with (E.68)–(E.76):

$$
\begin{aligned}
\widetilde{\Sigma}'_{\theta}(\star) = {} & \sigma_t\left[\frac{1}{4}\,a_2\right] + \left[\frac{1}{2}\,a_1 - \frac{1}{6}\,a_{22}\right] + \frac{1}{\sigma_t}\left[\frac{1}{8}\,a_2^2 + \frac{1}{12}\,a_3^2\right] \\
= {} & \gamma b g\left[\frac{1}{4}\,\gamma_2 b_3 g_1\right] + \frac{1}{2}\left[\gamma\left[\frac{1}{2}\,\varepsilon^2\,b_2 + \kappa\,b_5\right]g + \gamma_1\,b\,g + \frac{1}{2}\,\gamma_3\,b_4\,g_2\right] \\
& - \frac{1}{6}\,\gamma_{22} b_{33} g_{11} + \frac{1}{\gamma b g}\left[\frac{1}{8}\,[\gamma_2 b_3 g_1]^2 + \frac{1}{12}\,[\varepsilon\gamma b_1 g]^2\right]
\end{aligned}
$$

which we develop into

$$
\begin{aligned}
\widetilde{\Sigma}'_{\theta}(\star) = {} & \frac{1}{4}\,\gamma\gamma_2\;bb_3\;gg_1 + \varepsilon^2\left[\frac{1}{4}\,\gamma\;b_2\;g\right] + \kappa\left[\frac{1}{2}\,\gamma\;b_5\;g\right] + \frac{1}{2}\,\gamma_1\;b\;g \\
& + \frac{1}{4}\,\gamma_3\;b_4\;g_2 - \frac{1}{6}\,\gamma_{22}\;b_{33}\;g_{11} + \frac{1}{8}\,\frac{\gamma_2^2\;b_3^2\;g_1^2}{\gamma b g} + \varepsilon^2\left[\frac{1}{12}\,\gamma\;\frac{b_1^2}{b}\;g\right]
\end{aligned}
$$

and which after simplification and replacement provides (E.92). □

Index

D. Nicolay, *Asymptotic Chaos Expansions in Finance*,
Springer Finance Lecture Notes, DOI 10.1007/978-1-4471-6506-4

MIX
Papier aus verantwortungsvollen Quellen
Paper from responsible sources
FSC® C105338

If you have any concerns about our products,
you can contact us on
ProductSafety@springernature.com

In case Publisher is established outside the EU,
the EU authorized representative is:
Springer Nature Customer Service Center GmbH
Europaplatz 3, 69115 Heidelberg, Germany

Printed by Libri Plureos GmbH
in Hamburg, Germany